Advances in Transport Phenomena in Porous Media

NATO ASI Series

Advanced Science Institutes Series

A Series presenting the results of activities sponsored by the NATO Science Committee, which aims at the dissemination of advanced scientific and technological knowledge, with a view to strengthening links between scientific communities.

The Series is published by an international board of publishers in conjunction with the NATO Scientific Affairs Division

A	Life Sciences	Plenum Publishing Corporation
B	Physics	London and New York
C	Mathematical and Physical Sciences	D. Reidel Publishing Company Dordrecht, Boston, Lancaster and Tokyo
D	Behavioural and Social Sciences	Martinus Nijhoff Publishers Boston, Dordrecht and Lancaster
E	Applied Sciences	
F	Computer and Systems Sciences	Springer-Verlag Berlin, Heidelberg, New York
G	Ecological Sciences	London, Paris, Tokyo
H	Cell Biology	

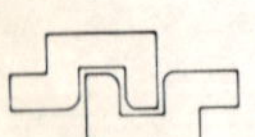

Series E: Applied Sciences – No. 128

Advances in Transport Phenomena in Porous Media

edited by:

Jacob Bear

Department of Civil Engineering
Technion-Israel Institute of Technology
Technion City
Haifa 32000
Israel

M. Yavuz Corapcioglu

Department of Civil Engineering
City College of New York
New York, NY 10031
USA

1987 **Martinus Nijhoff Publishers**
Dordrecht / Boston / Lancaster
Published in cooperation with NATO Scientific Affairs Division

Proceedings of the NATO Advanced Study Institute on "Fundamentals of Transport Phenomena in Porous Media", Newark, Delaware, USA, July 14-23, 1985

Library of Congress Cataloging in Publication Data

NATO Advanced Study Institute on Fundamentals of
Transport Phenomena in Porous Media (1985 : Newark,
Del.)
Advances in transport phenomena in porous media.

(NATO Advanced Study Institutes series. Series E,
Applied sciences ; v. 128)
"Proceedings of the NATO Advanced Study Institute on
Fundamentals of Transport Phenomena in Porous Media,
Newark, Delaware, USA, July 14-23, 1985"--Verso t.p.
"Published in cooperation with NATO Scientific Affairs
Division."
Includes index.
1. Groundwater flow--Congresses. 2. Porous materials--
Congresses. 3. Transport theory--congresses. I. Bear,
Jacob. II. Corapcioglu, M. Yavuz. III. North Atlantic
Treaty Organization. Scientific Affairs Division.
IV. Title. V. Series.
TC176.N388 1985 621.1'064 87-11202

ISBN 90-247-3533-3 (this volume)
ISBN 90-247-2689-1 (series)

Distributors for the United States and Canada: Kluwer Academic Publishers, P.O. Box 358, Accord-Station, Hingham, MA 02018-0358, USA

Distributors for the UK and Ireland: Kluwer Academic Publishers, MTP Press Ltd, Falcon House, Queen Square, Lancaster LA1 1RN, UK

Distributors for all other countries: Kluwer Academic Publishers Group, Distribution Center, P.O. Box 322, 3300 AH Dordrecht, The Netherlands

Printed in The Netherlands

PREFACE

This volume contains the lectures presented at the NATO ADVANCED STUDY INSTITUTE that took place at Newark, Delaware, U.S.A., July 14-23, 1985. The objective of this meeting was to present and discuss selected topics associated with transport phenomena in porous media. By their very nature, porous media and phenomena of transport of extensive quantities that take place in them, are very complex. The solid matrix may be rigid, or deformable (elastically, or following some other constitutive relation), the void space may be occupied by one or more fluid phases. Each fluid phase may be composed of more than one component, with the various components capable of interacting among themselves and/or with the solid matrix. The transport process may be isothermal or non-isothermal, with or without phase changes.

Porous medium domains in which extensive quantities, such as mass of a fluid phase, component of a fluid phase, or heat of the porous medium as a whole, are being transported occur in the practice in a variety of disciplines. For example, we encounter transport in porous media in Civil Engineering, in connection with the flow and pollution of groundwater, or the movement of moisture and heat through building materials, in Soil Mechanics, in dealing with soil compaction and land subsidence, in Reservoir Engineering, where we encounter multiphase flow, often under non-isothermal conditions, or in connection with enhanced oil recovery techniques, in Geothermal Reservoir Engineering, and in Chemical Engineering, where processes take place in packed beds. In all these cases, decisions related to the operation of the system have to be made. Rate and location of water injection in an oil reservoir, rate of pumping, or artificial recharge of an aquifer and rate of production from a geothermal reservoir, may serve as examples of such management decision. To make these decisions, the planner needs a tool that will represent the excitation-response behavior of the system. This tool is the model. The model is a simplified version of the complex real system, that simulates the behavior of those parts of the real system that are relevant to the management problem. It enables the planner to predict the outcome of implementing proposed management schemes. The type of model and its required accuracy are, therefore, also dictated by the model's use.

The first step in constructing a model for a given problem of transport, in a given porous medium domain, is to state the simplifying assumptions that transform the real (complex) world into the model. We often refer to this set of assumptions as the conceptual model. The model is then cast into a mathematical (or numerical) format. For most cases of practical interest, due to the complexity of the model, only a numerical solution is possible.

No model can be used, unless the numerical values of the various coefficients appearing in it are known. These can be estimated by solving the "inverse problem". In such problems, the input is data on field observations and the output includes the values of the model's coefficients, or parameters. Although some methods for solving the inverse problem are available, many problems still remain unresolved. Among them we may mention the question of uniqueness and criteria for obtaining the best set of coefficients.

Because of the heterogeneity inherent in the real porous medium domain, and in view of the relatively small number of samples, or observations, that we have in large heterogeneous domain, there is always uncertainty associated with the values of the domain's parameters and their spatial distribution. This means that uncertainty is also inherent in the model's predictions (that serve as input to the management problems). This feature of uncertainty should, therefore, be represented in the transport model. In recent years, much progress has been made toward this goal.

With this background in mind, a small number of subjects was selected for presentation and discussion in the 1985 NATO/ASI. These lectures are assembled in this volume.

Chapter 1 is devoted to the modeling of heat and mass transport in porous media. Both single and multiple fluid phases are considered. The treatment of porous medium deformability is also included. Chapter 2 introduces the topic of particle transport in porous media as encountered in reservoir engineering and in filtration. Chapter 3 deals with transport phenomena in fractured rock and fractured porous rock domains. The first section presents the theory and the practice of water flow and the transport of pollutants in fractured domains.

Chapter 4 introduces the important subject of uncertainty in models, arising from the complexity of the modeled system, and especially the spatial variability of the porous matrix properties. The question of the sensitivity of the model's predicted results (here water levels) to changes in model parameters is introduced in a number of sections. The problem of parameter estimation is also presented and discussed.

In recent years, there is a growing interest in two numerical methods, the Eulerian-Lagrangian models and point (or particle) tracking techniques. Although numerical methods was not a central topic in the 1985 NATO/ASI, it was felt that some advances in these two techniques should be included.

Chapter 5 includes lectures on these two techniques, in addition to lectures on numerical models of multiphase flow, a subject discussed in detail in Chapter 1.

In a way, this volume is a sequel to the book that contains the lectures of the 1982 NATO Advanced Study Institute (published in 1984 by Martinus Nijhoff Publishers). For convenience, the table of contents of the 1984 book is included at the end of this volume. A number of lectures in the present volume include references to the previous one.

A number of persons have contributed their invaluable time and effort to the organization of the ASI and/or the preparation of this book, and without their help such a task could not be undertaken. We take great pleasure in acknowledging the contributions of M. Özden Corapcioglu, Nancy Diffenderfer, Akhter Hossain, Sorab Panday, and Carol Wong at various stages of organization and manuscript preparation.

We greatly appreciate the financial support of NATO without which this Institute would not be possible. We are also grateful to the authors for accepting the invitation to lecture and to prepare written papers, and to all the participants for their contributions during the discussions.

We hope that this volume, like the 1984 one, will serve as a further step in the formulation of a unified approach to the modeling of transport in porous media.

July, 1985

J. Bear M. Y. Corapcioglu

TABLE OF CONTENTS

PART 1

HEAT AND MASS TRANSPORT IN SINGLE AND MULTIPHASE SYSTEMS

ON THE CONCEPT AND SIZE OF A REPRESENTATIVE ELEMENTARY VOLUME (REV)

Yehuda Bachmat and Jacob Bear

ON THE CONCEPT AND SIZE OF A REPRESENTATIVE ELEMENTARY VOLUME (REV)

Yehuda Bachmat

Director
Hydrological Service
Jerusalem, Israel

Jacob Bear

Albert and Anne Mansfield
Professor of Water Resources
Department of Civil Engineering
Technion, Haifa 32000, Israel

ABSTRACT

This chapter discusses the concept of the Representative Elementary Volume (REV) that serves as a cornerstone in the continuum modeling of transport phenomena in porous media (4). Following the presentation of the concept, quantitative criteria are presented for the selection of the size of the REV.

1. INTRODUCTION

The concept of a Representative Elementary Volume (REV) underlies the *continuum* approach to the modelling of transport of extensive quantities in porous media. At the *microscopic* level of description, the transport of an extensive quantity of a phase is modelled in terms of state variables at points *inside* the domain occupied by *that phase*. Interphase surfaces serve as boundaries. In the continuum approach, a passage is made from the microscopic level of description to a *macroscopic* one, in which to *every* point within an investigated porous medium domain, we assign values to state variables of all phases present in the domain. The *macroscopic* model is then expressed in terms of these *macroscopic state variables*. The value of a macroscopic state variable at a point is obtained by averaging the microscopic values of this variable over a certain volume of porous medium, centered at that point. This characteristic volume is called the *Representative Elementary Volume*.

It is worth noting that although the continuum model of a porous medium eliminates the need for specifying the microscopic configuration of interphase boundaries, its effects appear at the macroscopic level in the form of matrix coefficients. Similary, the effects of the microscopic variations of state variables within each phase also appear at the macroscopic level. To express these effects in terms of averaged quantities, statistical models of the microscopic variations will, in general, be required.

The continuum approach to modelling transport in porous media, using the REV concept, is well known and need not be repeated here (see, for example, 1, 2, 4, 5, 7, 8, 9). Also the concept of the REV has been around for some time. (In addition to the above references, see also: 3, 10, 11, 12). The objective of this paper, which may be considered as supplementing the authors' paper presented at the 1982 NATO Advanced Study Institute (4), is to re-examine this concept and its usefulness and to propose criteria for determining the size of an REV.

2. THE POROUS MEDIUM

For the purpose of this work we define a porous medium as a multiphase material body characterized by the following distinct features: a) *A Representative Elementary Volume (REV)* can be determined, such that no matter where we place it within the porous medium domain, it will always contain two persistent subdomains: a solid phase and a void space. b) The void space of any REV contains a multiply-connected subdomain referred to as the *interconnected void space*. c) The size of the REV is such that parameters that represent the distribution of void and solid within it are statistically meaningful. The quantification of this requirement is discussed in detail below.

In principle, *any Arbitrary Elementary Volume (AEV)* may be selected as an averaging volume for passing from the microscopic level of description to the macroscopic one. Obviously, different AEV's will yield different averaged values for each quantity of interest and there is no sense in asking which of them is more "correct". The selection of an averaging volume in any particular case depends only on the model's objectives. Also, the size of the "window" of the instrument that measures an averaged value should correspond to that of the selected AEV, so that, within the range of error introduced by the conceptual model of the process, the predicted and measured averaged values always be the same. The main drawback of this approach is that since every averaged value may strongly depend on the size of the selected AEV, it must be labelled by the size of the AEV (like a yardstick) over which it was taken. To circumvent this difficulty, rather than selecting the volume of averaging arbitrarily, we need a universal criterion which is based on measurable characteristics of any porous medium and that determines, for any given porous medium, a range of averaging volumes within which these characteristics remain, more or less, constant. As long as instrument's "window" is in that range, observed and computed values will be close, within a prescribed level of error.

An averaging volume which belongs to that range will be referred to as a *Representative Elementary Volume (REV)*.

3. SELECTION OF REV SIZE

Let the spatial distribution of the void space in a porous medium domain, (D), be represented by the *characteristic function*

$$\gamma(\underset{\sim}{x},t) = \begin{cases} 1 \text{ if } \underset{\sim}{x} \text{ is in the void space} \\ 0 \text{ if } \underset{\sim}{x} \text{ is in the solid matrix} \end{cases} \tag{3.1}$$

where $\underset{\sim}{x}$ denotes the position vector of a point and t is time. Also, let (U) be a domain centered at a point $\underset{\sim}{x}_o$ within (D). Consider the averages

$$\bar{\gamma}(\underset{\sim}{x}_o,U) \equiv \overline{\gamma(\underset{\sim}{x})}\Big|_{\underset{\sim}{x}_o,U} \equiv \frac{1}{U}\int_{(U)} \gamma(\underset{\sim}{x})\, dU = \frac{1}{U}\int_{(U)} dU_v = n(\underset{\sim}{x}_o,U) \tag{3.2}$$

where n is the porosity of the medium within (U), and dU_v represents a volume element of the void space,

$$\overline{\mathring{\gamma}(\underset{\sim}{x})\;\mathring{\gamma}(\underset{\sim}{x}+\underset{\sim}{h})}\Big|_{\underset{\sim}{x}_o,U,\underset{\sim}{h}} \equiv \frac{1}{U}\int_{(U)} [\gamma(\underset{\sim}{x}) - \bar{\gamma}(\underset{\sim}{x}_o)][\gamma(\underset{\sim}{x}+\underset{\sim}{h}) \tag{3.3}$$

$$- \bar{\gamma}(\underset{\sim}{x}_o + \underset{\sim}{h})]dU = \frac{1}{U}\int_{(U)} \gamma(\underset{\sim}{x})\gamma(\underset{\sim}{x}+\underset{\sim}{h})dU - n(\underset{\sim}{x}_o,U)\,n(\underset{\sim}{x}_o+\underset{\sim}{h},U)$$

where $\mathring{\gamma}(\underset{\sim}{x}) = \gamma(\underset{\sim}{x}) - \bar{\gamma}(\underset{\sim}{x}_o)$, $\underset{\sim}{x} \in (U)$. Eq. (3.3) is a measure of the distribution of the void space within (U).

A particular case of Eq. (3.3) is

$$\overline{(\mathring{\gamma})^2}\Big|_{\underset{\sim}{x}_o,U} = \frac{1}{U}\int_{(U)} \{\gamma^2(\underset{\sim}{x}) - \bar{\gamma}^2(\underset{\sim}{x}_o,U)\}dU = n(1-n)\Big|_{\underset{\sim}{x}_o,U} \tag{3.4}$$

The construction of a mathematical continuum model of a porous medium imposes certain restrictions on the size of the REV. Foremost is the requirement that the *value of any averaged geometrical characteristic of the microstructure of the porous material at any point in the porous medium domain be a single valued function of the location of that point and of time only, but independent of the size of the REV.*

Accordingly, we now define a volume $U=U_o$ as a *Representative Elementary Volume (REV)*, if

$$\left.\frac{\partial n(\underset{\sim}{x}_o,U)}{\partial U}\right|_{U=U_o} \equiv \left.\frac{\partial \bar{\gamma}(\underset{\sim}{x}_o,U)}{\partial U}\right|_{U=U_o} = 0 \tag{3.5}$$

and

$$\left.\frac{\partial\, \overline{\overset{o}{\gamma}(\underset{\sim}{x})\ \overset{o}{\gamma}(\underset{\sim}{x}+\underset{\sim}{h})}\Big|_{\underset{\sim}{x}_o,U}}{\partial U}\right|_{U=U_o} = 0 \tag{3.6}$$

In principle, one can visualize an experiment consisting of a succession of gradually increasing volumes $U_1<U_2<U_3$ ..., all *centered at* $\underset{\sim}{x}_o$, and a concurrent determination of $\bar{\gamma}$ and $\overset{o}{\gamma}(\underset{\sim}{x})\overset{o}{\gamma}(\underset{\sim}{x}+\underset{\sim}{h})$, hoping that a $U=U_o$, which satisfies both Eqs.(3.5) and (3.6), will be found. After repeating this experiment at all points $\underset{\sim}{x}_o \in (D)$, one can replace the actual porous medium within (D) by a model of a fictitious continuum, provided $U=U_o$ is uniform throughout (D). However, this is obviously an impossible task since it is impractical to observe *all* points within (D).

The question then arises as to the possibility of making inferences about the size of U_o from its relationships with macroscopic measurable parameters of the microscopic configuration of the void space.

An answer to this question can be obtained by regarding $\gamma(\underset{\sim}{x})$ as a random function of position, $\underset{\sim}{x}$. Thus, if $\gamma(\underset{\sim}{x})$ is a stationary *random function* of $\underset{\sim}{x}$, i.e., if the domain under consideration, (U_o), is statistically homogeneous with respect to the geometrical characteristics of the void space, as expressed by the moments of $\gamma(\underset{\sim}{x})$, and if $\gamma(\underset{\sim}{x})$ possesses the *ergodic property*, then, for a sufficiently large (U_o) (13)

$$\bar{\gamma}(\underset{\sim}{x}_o) = \frac{1}{U_o}\int_{(U_o)} \gamma(\underset{\sim}{x})\,dU = n(\underset{\sim}{x}_o) \cong E(\gamma\,\big|_{\underset{\sim}{x}_o})$$

$$\left.\overline{\overset{\circ}{\gamma}(\underset{\sim}{x})\ \overset{\circ}{\gamma}(\underset{\sim}{x}+\underset{\sim}{h})}\ \right|_{U_o,\underset{\sim}{h}} = \frac{1}{U_o}\int_{(U_o)} \overset{\circ}{\gamma}(\underset{\sim}{x})\overset{\circ}{\gamma}(\underset{\sim}{x}+\underset{\sim}{h})\,dU \cong \mathrm{Cov}_\gamma(\underset{\sim}{h}) \quad (3.7)$$

$$= \mathrm{Var}[\gamma(\underset{\sim}{x}_o)]\tau_\gamma(\underset{\sim}{h}) = n(1-n)\tau_\gamma(\underset{\sim}{h})$$

where Var $\gamma = n(1-n)$, $\tau_\gamma(\underset{\sim}{h})$ is the *correlation coefficient* between values of γ at points spaced an oriented distance $\underset{\sim}{h}$ apart and $n\ (=U_{ov}/U_o)$ is the *porosity*, where U_{ov} denotes the volume of voids within U_o. By definition

$$\tau_\gamma(0) = 1; \quad \tau_\gamma(\underset{\sim}{h}) \to 0 \text{ as } |\underset{\sim}{h}| \to \infty \quad (3.8)$$

In fact, the volume U_o of an REV should be sufficiently large, so that the volumetric averages can be considered as satisfactory estimates of *all* relevant population parameters of the void space configuration at $\underset{\sim}{x}_o$, i.e., estimates which are free of errors caused by the size of the sample and its random choice.

As was shown by Debye et al. (6), for an isotropic porous medium

$$\left.\frac{\partial \tau_\gamma}{\partial h}\right|_{h=0} = -\frac{1}{4\Delta(1-n)} \quad (3.9)$$

where $\Delta = U_{ov}/S_{vs}$ is the *hydraulic radius* of the void space (volume U_{ov} and area of contact with the solid, S_{vs}). An approximate expression for $\tau_\gamma(\underset{\sim}{h})$, for an isotropic medium, with a random distribution of void and solid spaces, is given by Debye et al. (6),

$$\tau_\gamma \cong \exp\{-h/4\Delta(1-n)\}\ ;\ h = |\underset{\sim}{h}| \quad (3.10)$$

It follows that a necessary condition for obtaining *nonrandom estimates* of the geometric characteristics of the void space at any point $\underset{\sim}{x}_o$ which serves as a centroid of a sphere of volume U_o and diameter ℓ, is

$$h_{max} = \ell \gg \Delta \quad (3.11)$$

The magnitude of ℓ_{min} is determined by the chosen accuracy and reliability levels of the parameter estimates. Thus, as a conceptual experiment for estimating the porosity, n, of a porous medium at a point $\underset{\sim}{x}_o$, let the volume U_o of a cubical REV centered at that point be split into N disjoint elementary subdomains, $\delta U = U_o/N$, such that in each of them one may encounter (more or less) *either* solid *or* void. The average of γ over the N samples is taken as an estimate, $\hat{n}$, of the porosity, n, at $\underset{\sim}{x}_o$, i.e.

$$\hat{n} = \sum_{i=1}^{N} \gamma_i/N \left(= \frac{1}{N\delta U} \sum_{i=1}^{N} \gamma_i \delta U\right) \tag{3.12}$$

By definition and Eq. (3.7), we have

$$\sigma^2_{\hat{n}} = \frac{1}{N^2} \sum_{p=1}^{N} \sum_{q=1}^{N} \mathrm{Cov}(\gamma_p, \gamma_q) = \frac{n(1-n)}{N^2} \sum_{p=1}^{N} \sum_{q=1}^{N} \tau_\gamma(h_{pq}) \tag{3.13}$$

where $\sigma^2_{\hat{n}}$ is variance of the estimate of $\hat{n}$, and $h_{pq} = |\underset{\sim}{x}_p - \underset{\sim}{x}_q|$ is the distance between points $\underset{\sim}{x}_p$ and $\underset{\sim}{x}_q$.

Employing Eq.(3.10), we obtain

$$\sigma^2_{\hat{n}} = \frac{n(1-n)}{N^2} \left(N + \sum_{\substack{p=1 \\ p\neq q}}^{N} \sum_{q=1}^{N} \exp\{-h_{pq}/4(1-n)\Delta\} \right) \tag{3.14}$$

From Eq. (3.14) it follows that, since h_{pq} is expressed in units of Δ, $N = N(n, \sigma^2_{\hat{n}})$.

Figure 1 shows the relationship $\sigma^2_{\hat{n}}(n,N)$. For example, for $n = 0.3$, $N = 8000$, $\sigma^2_{\hat{n}} = 0.00347$,

Thus, for a cubical REV, $U_{o,min}$ $(\equiv \ell^{(n)3}_{min}) = N\delta U$, with each elementary volume $\delta U = (C_\Delta \Delta)^3$, we have

$$\ell^{(n)}_{min} = \{N(n, \sigma^2_{\hat{n}})\}^{1/3} C_\Delta \Delta; \qquad C_\Delta \equiv \left. \frac{\ell^{(n)}_{min}}{\Delta} \right|_{N=1} \tag{3.15}$$

In the above example, this means $\ell^{(n)}_{min} = 20C_\Delta\Delta$, where we have added superscript (n) to emphasize that we have been considering the porosity, n, as the macroscopic geometrical characteristic.

According to Chebyshev's inequality, the probability that the magnitude of the estimation error exceeds a prescribed level, say ε, is bounded from above by

$$P(|\hat{n} - n| \geqq \varepsilon) < \sigma^2_{\hat{n}}/\varepsilon^2 \tag{3.16}$$

Let β denote a probability such that $\sigma^2_{\hat{n}}/\varepsilon^2 = \beta$. Then, N would represent the smallest number of subdomains of (U_o) which is sufficient to ensure, with a reliability $1 - \beta$, that the estimation error $|\hat{n} - n|$ will not exceed ε. In the above example, this means for example, that for $\varepsilon = 0.1$, $\beta = 0.35$. Obviously, any reduction of ε and β (in this example) will require a larger value of N.

It is of interest to note that in order to determine $\ell^{(n)}_{min}$ by Eq. (3.15), one has to make use of a preliminary estimate of n and $C_\Delta \Delta$.

The requirement of ergodicity also sets an upper bound on the size of the REV, namely $\ell < \ell_{max}$, where ℓ_{max} is the distance between points in the porous medium domain beyond which the domain of averaging ceases to be statistically homogeneous with respect to the moments of $\gamma(\underset{\sim}{x})$.

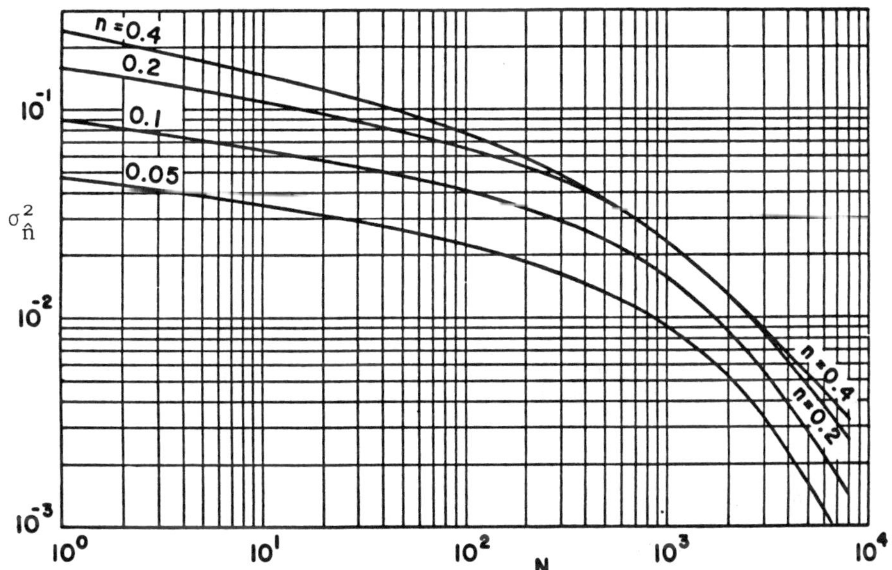

Figure 1. Variance of the estimate, $\hat{n}$, of porosity as a function of number, N, of elementary subdomains.

In reality, the requirement of homogeneity is seldom satisfied as the macroscopic parameters of the void geometry usually vary from point to point. However, even for a domain that is heterogeneous with respect to these parameters, one can define around every point a sufficiently small subdomain, within which these parameters may still be considered uniform, up to a prescribed error level. The size of such a subdomain at a point serves as the upper bound for the size of the REV at that point.

In order to determine this upper bound, let us define a domain (U) centered at a point $\underset{\sim}{x}_o$ as homogeneous, if for all $\underset{\sim}{x} \in (U)$

$$E\gamma(\underset{\sim}{x}) \equiv n(\underset{\sim}{x}) = \text{Constant} = n_o$$

and

$$\text{Cov}[\gamma(\underset{\sim}{x} + \underset{\sim}{h}), \gamma(\underset{\sim}{x})] = f(\underset{\sim}{h})$$

Then

$$\text{Var}\ \gamma(\underset{\sim}{x}) = f(0) = \text{Constant} = n_o(1 - n_o)$$

For a heterogeneous domain (U), $n = n(\underset{\sim}{x})$. However, for a sufficiently small (U), any continuous and differentiable function $n(\underset{\sim}{x})$ can be *estimated* by its linear part (Figure 2). Therefore, we may select an estimator $\hat{n}(\underset{\sim}{x})$ that has the linear form

$$\hat{n}(\underset{\sim}{x}) = n_o + \underset{\sim}{b}\cdot\underset{\sim}{h}\ ; \quad n_o \equiv n\Big|_{\underset{\sim}{x}_o}\ ; \quad \underset{\sim}{h} \equiv \underset{\sim}{x} - \underset{\sim}{x}_o \tag{3.17}$$

where $\underset{\sim}{b} = \text{grad}\ n\Big|_{\underset{\sim}{x}_o}$. By definition

$$\hat{n}(\underset{\sim}{x}) = n(\underset{\sim}{x}) + \varepsilon(\underset{\sim}{x}) \tag{3.18}$$

Assuming that $E\varepsilon(\underset{\sim}{x}) = 0$ for all $\underset{\sim}{x}$, $\hat{n}(\underset{\sim}{x})$ is an unbiased estimator of $n(\underset{\sim}{x})$.

We shall refer to the domain (U) as *approximately homogeneous* if for any point $\underset{\sim}{x}(=\underset{\sim}{x}_o + \underset{\sim}{h})$ within it, we have

$$\frac{|\Delta\hat{n}|}{n_o} \leq \delta = \frac{|\Delta\hat{n}|_{max}}{n_o}\ ; \quad \Delta\hat{n} = \hat{n}(\underset{\sim}{x}) - n_o \tag{3.19}$$

and $0 < \delta \leq 1$ is an arbitrarily selected small number, representing an acceptable average relative error introduced by replacing $n(\underset{\sim}{x})$ by n_o.

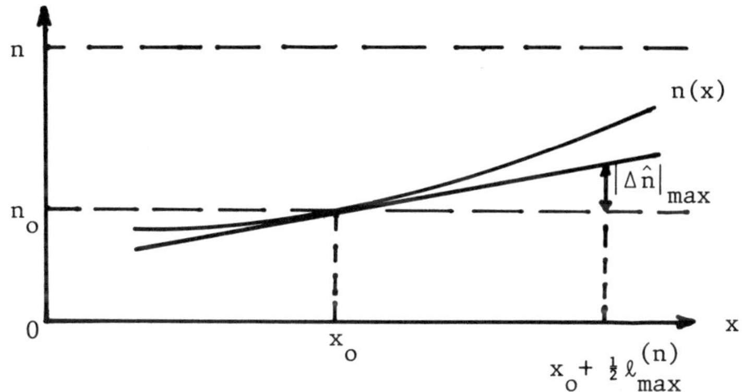

Figure 2. Conceptual determination of $\ell^{(n)}_{max}$ by Eq. (3.19).

From Eq. (3.17) and the definition of δ, it follows that the sought upper bound, denoted by $\ell^{(n)}_{max}$, is given by

$$\ell^{(n)}_{max} = \frac{2n_o}{|\text{grad } n|_{\underset{\sim}{x}_o}} \delta \tag{3.20}$$

Thus

$$\ell^{(n)}_{min} << \ell^{(n)} << \ell^{(n)}_{max} \tag{3.21}$$

The distance $\ell^{(n)}_{max}$ (with respect to porosity) is thus the upper limit for the size of the REV at a point $\underset{\sim}{x}_o$, at the selected error level. We have to scan all points $\underset{\sim}{x}_o$ within the given domain in order to determine the smallest value of $\ell^{(n)}_{max}$.

If $\ell^{(n)}_{max} \leq \ell^{(n)}_{min}$ at $\underset{\sim}{x}_o$, an REV cannot be defined there. On the other hand, if a non-zero range of $\ell^{(n)}$ can be found, which is common to all points within a given spatial domain of a porous medium, one can adopt the continuum model for the porous medium within that domain.

Finally, we have to relate $\ell^{(n)}$ to the dimensions of the considered domain. If L^* is a characteristic length of the domain, we require that

$$\ell^{(n)} << L^* \tag{3.22}$$

in order to ensure that the boundary region of the domain, which has a width $\ell^{(n)}$, and in which the continuum approach is not applicable, be small compared to the size of the domain itself.

So far, the concept and size of the REV have been related to porosity as a geometrical porous medium property. We have indicated this fact by using the superscript (n). Whenever, additional characteristics of the porous medium appear in the macroscopic model describing a transport problem, e.g., permeability, a range for REV has to be determined for each of them. If a common REV range can be found, a continuum model of the porous medium can be employed.

One of the requirements for the range of the REV is that $\partial n/\partial U = 0$ within it, as defined by Eq. (3.5). This does not necessarily imply that $n(\underset{\sim}{x})$ is uniform within U_o. To illustrate this point, consider the ratio $U_v(\underset{\sim}{x}_o)/U(\underset{\sim}{x}_o)$, where $U(\underset{\sim}{x}_o)$ is the volume of a sphere centered at an arbitrary point $\underset{\sim}{x}_o$ within (D) and $U_v(\underset{\sim}{x}_o)$ is the volume of the void space within $U(\underset{\sim}{x}_o)$.

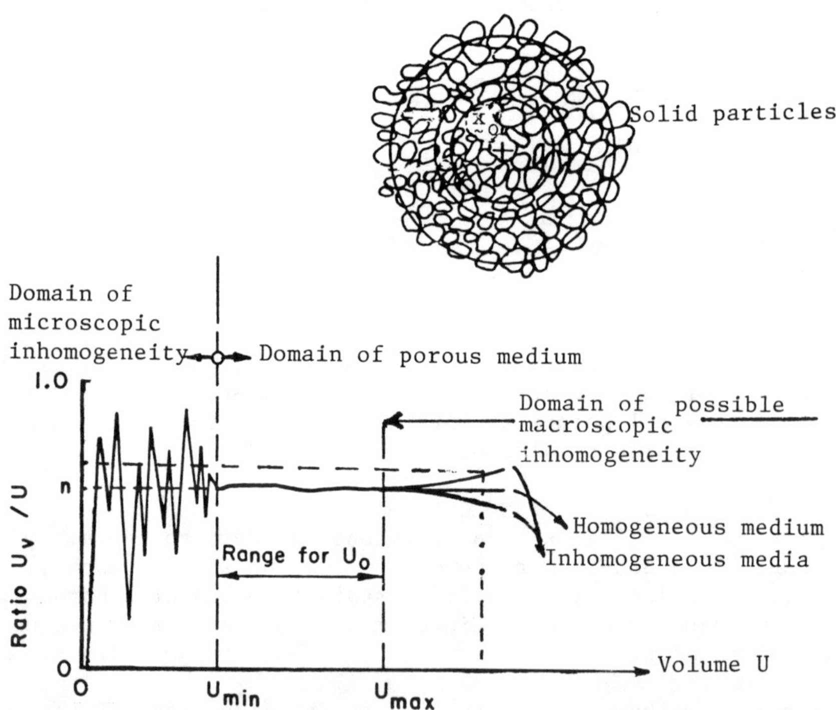

Figure 3. Definition of $\ell^{(n)}_{min}$, Representative Elementary Volume and porosity at $\underset{\sim}{x}_o$.

Figure 3 shows the variations of the ratio U_v/U as U increases. For very small values of $U(\underset{\sim}{x}_o)$, the above ratio is one or zero, depending on whether $\underset{\sim}{x}_o$ happens to fall in the void space or in the solid. As $U(\underset{\sim}{x}_o)$ increases, we note large fluctuations in U_v/U. However, as U continues to grow, these fluctuations are gradually attenuated, until, above some value $U=U_{min}$, they decay, leaving only small amplitude fluctuations around some constant value.

In order to examine the behavior of the function $n(\underset{\sim}{x}_o)$ in the domain in which $\partial n/\partial U|_{\underset{\sim}{x}_o} = 0$, consider a domain of averaging in the form of a rectangular prism contered at $\underset{\sim}{x}_o$, with edges parallel to the coordinate axes. By definition, its volume is $U|_{\underset{\sim}{x}_o} = \Delta x_1 \Delta x_2 \Delta x_3$ and

$$n(\underset{\sim}{x}_o) \equiv \overline{\gamma}(\underset{\sim}{x}_o) =$$

$$\frac{1}{\Delta x_1 \Delta x_2 \Delta x_3} \int_{x_{10}-\frac{1}{2}\Delta x_1}^{x_{10}+\frac{1}{2}\Delta x_1} \int_{x_{20}-\frac{1}{2}\Delta x_2}^{x_{20}+\frac{1}{2}\Delta x_2} \int_{x_{30}-\frac{1}{2}\Delta x_3}^{x_{30}+\frac{1}{2}\Delta x_3} \gamma(x_1,x_2,x_3)\,dx_1 dx_2 dx_3 =$$

$$\frac{1}{\Delta x_1} \int_{x_{10}-\frac{1}{2}\Delta x_1}^{x_{10}+\frac{1}{2}\Delta x_1} \tilde{\gamma}^1(x_1,x_{20},x_{30})\,dx_1 \qquad (3.23)$$

where

$$\tilde{\gamma}^1(x_1,x_{20},x_{30}) = \frac{1}{\Delta x_2 \Delta x_3} \int_{x_{20}-\frac{1}{2}\Delta x_2}^{x_{20}+\frac{1}{2}\Delta x_2} \int_{x_{30}-\frac{1}{2}\Delta x_3}^{x_{30}+\frac{1}{2}\Delta x_3} \gamma(x_1,x_2,x_3)\,dx_2 dx_3$$

is the areal average of γ over a surface normal to the x_1 - axis.

Now

$$\frac{\partial n}{\partial(\Delta x_1)} = -\frac{1}{(\Delta x_1)^2}\int_{x_{10}-\frac{1}{2}\Delta x_1}^{x_{10}+\frac{1}{2}\Delta x_1} \tilde{\gamma}^1(x_1)dx_1 + \frac{1}{2\Delta x_1}[\tilde{\gamma}^1(x_{10} - \frac{\Delta x_1}{2})$$

$$+ \tilde{\gamma}^1(x_{10} + \frac{\Delta x_1}{2})]$$

$$= \frac{1}{\Delta x_1}\{ - n(\underset{\sim}{x}_o) + \frac{1}{2}[\tilde{\gamma}^1(x_{10} + \frac{\Delta x_1}{2}, x_{20}, x_{30})$$

$$+ \tilde{\gamma}^1(x_{10} - \frac{\Delta x_1}{2}, x_{20}, x_{30})]\}$$

Hence, for $\partial n/\partial(\Delta x_1) = 0$, in the range $(\Delta x_1)_{min} < \Delta x_1 < (\Delta x_1)_{max}$, we obtain

$$n(\underset{\sim}{x}_o) \equiv \bar{\gamma}(\underset{\sim}{x}_o) = \frac{1}{2}[\tilde{\gamma}^1(x_{10} + \frac{\Delta x_1}{2}, x_{20}, x_{30}) +$$

$$+ \tilde{\gamma}^1(x_{10} - \frac{\Delta x_1}{2}, x_{20}, x_{30})] \qquad (3.24)$$

Assuming that $\tilde{\gamma}^1(x_1, x_{20}, x_{30})$ is differentiable with respect to x_1 in the domain $|x_1 - x_{10}| \leq \Delta x_1/2$, and expanding the terms on the R.H.S. into power series about x_{10}, we obtain

$$\tilde{\gamma}^1(x_{10} + \frac{\Delta x_1}{2}, x_{20}, x_{30}) = \tilde{\gamma}^1(x_{10}, x_{20}, x_{30}) +$$

$$+ \frac{\partial \tilde{\gamma}^1}{\partial x_1}\bigg|_{x_{10},x_{20},x_{30}} \frac{\Delta x_1}{2} + \frac{1}{2}\frac{\partial^2 \tilde{\gamma}^1}{\partial x^2}\bigg|_{\underset{\sim}{x}_o} \left(\frac{\Delta x_1}{2}\right)^2 + \ldots$$

$$\tilde{\gamma}(x_{10} - \frac{\Delta x_1}{2}, x_{20}, x_{30}) = \tilde{\gamma}^1(x_{10}, x_{20}, x_{30}) -$$

$$- \frac{\partial \tilde{\gamma}^1}{\partial x_1}\bigg|_{x_{10},x_{20},x_{30}} \frac{\Delta x_1}{2} + \frac{1}{2}\frac{\partial^2 \tilde{\gamma}^1}{\partial x_1^2}\bigg|_{\underset{\sim}{x}_o} \left(\frac{\Delta x_1}{2}\right)^2 + \ldots.$$

Substituting these expressions into Eq. (3.24), yields

$$n(\underset{\sim}{x}_o) \equiv \bar{\gamma}(\underset{\sim}{x}_o) = \tilde{\gamma}^1(\underset{\sim}{x}_o) + \frac{\partial^2 \tilde{\gamma}^1}{\partial x_1^2}\bigg|_{\underset{\sim}{x}_o} \left(\frac{\Delta x_1}{2}\right)^2 + \frac{2}{4!}\frac{\partial^4 \tilde{\gamma}^1}{\partial x_1^4}\bigg|_{\underset{\sim}{x}_o} \left(\frac{\Delta x_1}{2}\right)^4$$

$$+ \frac{2}{6!}\frac{\partial^6 \tilde{\gamma}^1}{\partial x_1^6}\bigg|_{\underset{\sim}{x}_o} \left(\frac{\Delta x_1}{2}\right)^6 + \ldots. \tag{3.25}$$

In order that $\bar{\gamma}(\underset{\sim}{x}_o)$ retains its value for any Δx_1 in the range where $\partial n/\partial(\Delta x_1) = 0$, all terms containing Δx_1 in the last equation must vanish, i.e., $\tilde{\gamma}^1$ must be a linear function of x_1 in that range, i.e.,

$$\tilde{\gamma}^1(x_1, x_{20}, x_{30}) = a_o + b_1 x_1 + f(x_{20}, x_{30})$$

However, since

$$\tilde{\gamma}^1(\underset{\sim}{x}_o) = \tilde{\gamma}^2(\underset{\sim}{x}_o) = \tilde{\gamma}^3(\underset{\sim}{x}_o)$$

it follows that

$$n(\underset{\sim}{x}) = a_o + \underset{\sim}{b} \cdot \underset{\sim}{x} \quad ; \quad \underset{\sim}{b} = \text{Constant} \tag{3.26}$$

It may thus be concluded that if the function $n(U|\underset{\sim}{x}_o)$ has a plateaux within a given range of U, then $n(\underset{\sim}{x}|U)$ is a linear function of $\underset{\sim}{x}$ in that range, and vice versa.

If $U(\underset{\sim}{x}_o)$ is further increased, say beyond some value $U = U_{max}$, we may observe a trend in the considered ratio, due to a systematic variation in the latter. The Representative Elementary Volume is that volume $U_o(\underset{\sim}{x}_o)$, within the range of $U_{min} < U < U_{max}$ that will make the ratio U_v/U independent of U, and hence a single valued function of $\underset{\sim}{x}_o$ only. For $U = U_o$, the ratio U_v/U represents the porous medium's *porosity*, n, at $\underset{\sim}{x}_o$. By definition, for the REV, the volumetric fraction of the solid, 1-n, is also a single valued function of $\underset{\sim}{x}_o$.

Following the discussion leading to the definition of $\ell^{(n)}_{max}$ by Eq. (3.20), the upper limit for $\ell^{(n)}$, for a given δ, is defined by $\ell^{(n)}_{max}$, which, in turn, depends on U_{max} that indicates the point of deviation from the plateaux, produced by the linearity of $n(\underset{\sim}{x}_o)$ in the vicinity of $\underset{\sim}{x}_o$.

We note that the determination of the size of the REV and the porosity (and other geometrical parameters') distribution for a given REV is an iterative process, due to their dependence on each other.

Once an REV has been selected, we use it to define the averaged values of all state variables within the context of a continuum approach.

4. ACKNOWLEDGEMENT

This work is part of a cooperative research program of the Department of Civil Engineering, Division of Water Resources, Technion - Israel Institute of Technology, and the Hydrological Service of Israel on Transport Phenomena in Porous Media. Partial support was provided by the Fund for the Promotion of Research at Technion.

5. REFERENCES

1. Bachmat, Y., "Spatial Macroscopization of Processes in Heterogeneous Systems," *Israel Journal of Technology*, Vol. 10, No. 5, 1972, pp. 391 - 403.
2. Bachmat, Y., and Bear, J., "Mathematical Formulation of Transport Phenomena in Porous Media," *Proceedings of the Symposium on Fundamentals of Transport Phenomena in Porous Media*, University of Guelph, Guelph, Ontario, Canada, Vol. 1, 1972, pp. 174 - 193.
3. Baveye, P., and Sposito, G., "The Operational Significance of the Continuum Hypothesis in the Theory of Water Movement Through Soils and Aquifers," *Water Resources Research*, Vol. 20, No. 5, 1984, pp. 521 - 530.
4. Bear, J., and Bachmat, Y., "Transport Phenomena in Porous Media-Basic Equations," *Fundamentals of Transport Phenomena in Porous Media*, J. Bear and M. Y. Corapcioglu, eds., Martinus Nijhoff Publishers BV, The Hague, The Netherlands, 1984, pp. 3 - 62.
5. Carbonell, R. G., and Whitaker, S., "Heat and Mass Transfer in Porous Media," *Fundamentals of Transport Phenomena in Porous Media*, J. Bear and M. Y. Corapcioglu, eds., Martinus Nijhoff Publishers BV, The Hague, The Netherlands, 1984, pp. 121 - 198.
6. Debye, P., Anderson, H. R., and Brumberger, H., "Scattering by an Inhomogeneous Solid II. The Correlation Function and its Application," *Journal of Applied Physics*, Vol. 28, No. 4, 1957, pp. 679 - 683.
7. Hassanizadeh, M., and Gray, W. G., "General Conservation Equations for Multiphase Systems. 1. Averaging Procedure," *Advances in Water Resources*, Vol. 2, No. 3, 1979, pp. 131 - 144.
8. Hassanizadeh, M., and Gray, W. G., "General Conservation Equations for Multiphase Systems, 2. Mass,Momenta, Energy and Entrophy Equations," *Advances in Water Resources*, Vol. 2, No. 4, 1979, p. 191.
9. Hassanizadeh, M. and Gray, W. G., "General Conservation Equations for Multiphase Systems, 3. Constitutive Theory for Porous Media Flow," *Advances in Water Resources*, Vol. 3, No. 1, 1980, pp. 25 - 40.
10. Hubbert, M. K., "Darcy's Law and the Field Equations for the Flow of Underground Fluids," *Transactions*, American Institute of Mining and Metallurgical Engineers, Vol. 207, 1956, pp. 222 - 239.
11. Nikolaevski, V. N., Basniev, K. S., Gorbunov, A. T., and Zotov, G. A. "Mechanics of Saturated Porous Media," (in Russian), *NEDRA*, Moscow, 1970, p. 334.
12. Prager, S., "Viscous Flow Through Porous Media," *The Physics of Fluids*, Vol. 4, no. 12, 1961, pp. 1477 - 1482.
13. Venzel, E. S., "Probability Theory" (in Russian), *FIZMATGIZ*, Moscow, 1962, p. 560.

6. LIST OF SYMBOLS

$\underset{\sim}{b}$	$\left\| \text{grad } n \right\|_{\underset{\sim}{x}_o}$
E( )	Expected value of ( )
$\underset{\sim}{h}$	An oriented distance
ℓ	Diameter of REV (also ℓ_{min}, ℓ_{max})
L^*	Characteristic length of domain
n	Porosity ($n(\underset{\sim}{x})$ = porosity at $\underset{\sim}{x}$)
N	Number of items
p	Probability
S_{vs}	Surface area between void space and solid phase within an REV
U	Volume (U_o = volume of REV; U_{ov} = volume of voids in REV)
Var( )	Variance of ( )
$\underset{\sim}{x}$	Position vector of point (also $\underset{\sim}{x}_o$)
β	Probability level ($=\sigma^2_{\hat{n}}/\varepsilon^2$)
$\gamma(\underset{\sim}{x})$	Characteristic function
Δ	Hydraulic radius (= U_{ov}/S_{vs})
δ	$\|\Delta\hat{n}\|_{max}/n_o$
ε	Prescribed error level
$\sigma^2_{\hat{n}}$	Variance of $\hat{n}$ ($\equiv$ Var $\hat{n}$)
τ	Correlation coefficient
$(\overset{o}{\ })$	Deviation of value ( ) at a microscopic point from its average over REV
$(\hat{\ })$	Estimated value of ( )

ADVECTIVE AND DIFFUSIVE FLUXES IN POROUS MEDIA

Jacob Bear and Yehuda Bachmat

ADVECTIVE AND DIFFUSIVE FLUXES IN POROUS MEDIA

Jacob Bear

Albert and Anne Mansfield
Professor of Water Resources
Department of Civil Engineering
Technion, Haifa 32000, Israel

Yehuda Bachmat

Director
Hydrological Service
Jerusalem, Israel

ABSTRACT

Advective, dispersive and diffusive fluxes appear in the macroscopic balance equation of an extensive quantity (e.g., mass, mass of a component, heat) transported in a porous medium domain (2). The objective of this chapter is to develop expressions for the macroscopic advective and diffusive fluxes of mass and heat in terms of macroscopic state variables.

1. INTRODUCTION

In the 1982 NATO Advanced Study Institute, the authors presented the continuum approach to modelling the transport of such extensive quantities as mass of a phase, mass of a component of a phase, momentum and energy, in porous medium domains (2). The passage from the *microscopic* level of description to the *macroscopic one* was achieved by averaging the former over a Representative Elementary Volume (REV) of the porous medium. The reader is referred to the above mentioned reference. Three kinds of macroscopic fluxes appear in the balance equation of any extensive quantity: an advective flux, a dispersive one and a diffusive one. Our objective in this chapter is to modify and expand the chapter on macroscopic fluxes appearing in that reference, focusing our attention on the advective and diffusive fluxes of fluid mass and of heat in a porous medium domain.

2. ADVECTIVE MASS FLUX OF A PHASE

The macroscopic advective mass flux of an α-phase, $\underset{\sim}{q}_\alpha$, is expressed by

$$\underset{\sim}{q}_\alpha = \theta_\alpha \overline{\rho}_\alpha^\alpha \overline{\underset{\sim}{V}}_\alpha^\alpha \tag{2.1}$$

where θ_α is the volumetric fraction of the α-phase, ρ_α is its density and $\underset{\sim}{V}_\alpha$ is its volume averaged velocity. The symbol $\overline{(\ )}^\alpha$ denotes the *intrinsic phase average* of ( ) taken over the volume $U_{o\alpha}$ occupied by the α-phase within the REV of volume U_o (2). Our objective in what follows is to express $\overline{\underset{\sim}{V}}_\alpha^\alpha$ for a fluid α-phase in terms of macroscopic (averaged) state variables (e.g., fluid pressure). We shall limit the discussion to θ_α = n = porosity, i.e., to the case of a single fluid that occupies the entire void space, with some comments on multiphase flow.

As a point of departure, we start from the microscopic momentum balance of an α-phase

$$\rho_\alpha \frac{D^{m_\alpha} \underset{\sim}{V}^{m_\alpha}}{Dt} = \nabla . \underset{\approx}{\tau}_\alpha - \nabla p_\alpha - \rho_\alpha g \nabla z \tag{2.2}$$

where; $\underset{\sim}{V}^{m_\alpha}$ is the mass weighted velocity of the α-phase, $\underset{\approx}{\tau}_\alpha$ is the viscous stress tensor, p_α is the pressure, z is the vertical coordinate (positive upward) and $D^{m_\alpha}(\ )/Dt$ denotes the material derivative of ( ) with respect to an observer moving at the velocity $\underset{\sim}{V}^{m_\alpha}$. By averaging Eq. (2.2) over the REV, we obtain

$$\theta_\alpha \overline{\rho}_\alpha^\alpha \frac{D^{m_\alpha} \overline{\underset{\sim}{V}^{m_\alpha}}^\alpha}{Dt} = \theta_\alpha \overline{\nabla . \underset{\approx}{\tau}_\alpha}^\alpha - \theta_\alpha \overline{\nabla p_\alpha}^\alpha - \theta_\alpha \overline{\rho_\alpha g \nabla z}^\alpha \tag{2.3}$$

where we have introduced the approximation

$$\overline{\rho_\alpha \frac{D^{m_\alpha} \underset{\sim}{V}^{m_\alpha}}{Dt}}^\alpha \cong \overline{\rho}_\alpha^\alpha \overline{\frac{D^{m_\alpha} \underset{\sim}{V}^{m_\alpha}}{Dt}}^\alpha \cong \overline{\rho}_\alpha^\alpha \frac{D^{m_\alpha} \overline{\underset{\sim}{V}^{m_\alpha}}^\alpha}{Dt}$$

assuming also that $\nabla . \underset{\sim}{V}^{m_\alpha} = 0$, i.e., isochoric mass motion prevails within $U_{o\alpha}$.

Let us assume that pressure varies monotonously, i.e., with no minimum or maximum, within $(U_{o\alpha})$. Such pressure distribution is characterized by

$$\nabla^2 p = 0 \text{ in } (U_{o\alpha}) \tag{2.4}$$

In Eq. (2.3) we have averages of spatial derivaties of state variables, e.g., in the form of $\overline{\nabla p_\alpha}^\alpha$. In order to express such averages in terms of derivatives of the averages of state variables, we have developed a modified form of the averaging rule for a spatial derivative (e.g.,(2)). In view of Eq. (2.4), we may now apply Eq. (5.8) of Appendix A to p_α, obtaining

$$\theta_\alpha \overline{\frac{\partial p_\alpha}{\partial x_j}}^\alpha = \theta_\alpha \frac{\partial \overline{p_\alpha}^\alpha}{\partial x_i} T^*_{\alpha ij} + \frac{1}{U_o} \int_{(S_{\alpha s})} \frac{\partial p_\alpha}{\partial x_i} \overset{o}{x}_j \nu_{\alpha i} dS \tag{2.5}$$

Our next objective is to study the boundary condition $\partial p_\alpha / \partial x_i$ on $(S_{\alpha s})$.

Assuming that in the vicinity of the fluid-solid interface, the inertial force of the fluid is small relative to the viscous resistance, and that the components of the resistance force normal to this interface, $(\nabla . \underset{\approx}{\tau}_\alpha) . \underset{\sim}{\nu}_\alpha$. are much smaller than the tangential ones, we obtain

$$(\nabla . \underset{\approx}{\tau}_\alpha) . \underset{\sim}{\nu}_\alpha = (\nabla p_\alpha + \rho_\alpha g \nabla z) . \underset{\sim}{\nu}_\alpha \cong 0 \qquad \text{on } (S_{\alpha s}) \tag{2.6}$$

or

$$\frac{\partial p_\alpha}{\partial x_i} \nu_{\alpha i} = -\rho_\alpha \, g \frac{\partial z}{\partial x_i} \nu_{\alpha i} = -\rho_\alpha \, g \delta_{3i} \nu_{\alpha i} \qquad \text{on } (S_{\alpha s}) \tag{2.7}$$

where $z = x_3$. By inserting this expression into Eq. (2.5), we obtain

$$\theta_\alpha \overline{\frac{\partial p_\alpha}{\partial x_j}}^\alpha = \theta_\alpha \frac{\partial \overline{p_\alpha}^\alpha}{\partial x_i} T^*_{\alpha ij} - \frac{1}{U_o} \int_{(S_{\alpha s})} \rho_\alpha \, g \delta_{3i} \overset{o}{x}_j \nu_{\alpha i} \, dS \tag{2.8}$$

Assuming that $\overset{o}{\rho}_\alpha << \overline{\rho}_\alpha^\alpha$, the last term on the R.H.S of Eq. (2.8) may be approximated by

$$\frac{1}{U_o}\int_{(S_{\alpha s})} \rho_\alpha g \delta_{3i} \overset{o}{x}_j \nu_{\alpha i} dS \cong \overline{\rho}_\alpha^\alpha \, g \, \frac{1}{U_o}\int_{(S_{\alpha s})} \overset{o}{x}_j \nu_{\alpha 3} dS$$

$$= \theta_\alpha \overline{\rho}_\alpha^\alpha \, g(\delta_{3j} - T^*_{\alpha 3j}) \tag{2.9}$$

In writing Eq. (2.9), we have made use of Eq. (5.9) and of the relationship

$$\int_{(S_{\alpha s})} \overset{o}{x}_j \nu_{\alpha i} dS + \int_{(S_{\alpha\alpha})} \overset{o}{x}_j \nu_{\alpha i} dS = U_{o\alpha} \delta_{ij} \tag{2.10}$$

The last term of the R.H.S of Eq. (2.3) can be expressed by

$$\overline{\rho_\alpha g \frac{\partial z}{\partial x_j}} = \theta_\alpha \overline{\rho}_\alpha^\alpha \, g \delta_{3j} \equiv \theta_\alpha \overline{\rho}_\alpha^\alpha \, g \frac{\partial z}{\partial x_j} \tag{2.11}$$

By inserting Eq. (2.8), Eq. (2.9) and Eq. (2.11) into Eq. (2.3), we obtain

$$\overline{\rho}_\alpha^\alpha \frac{D^{m_\alpha} \overline{V_j^{m_\alpha}}^\alpha}{Dt} = \frac{\overline{\partial \tau_{\alpha ij}}^\alpha}{\partial x_i} - \frac{\partial \overline{p}_\alpha^\alpha}{\partial x_i} T^*_{\alpha ij} + \overline{\rho}_\alpha^\alpha \, g(\delta_{3j} - T^*_{\alpha 3j}) - \overline{\rho}_\alpha^\alpha \, g\delta_{3j}$$

$$= \frac{\overline{\partial \tau_{\alpha ij}}^\alpha}{\partial x_i} - \left(\frac{\partial \overline{p}_\alpha^\alpha}{\partial x_i} + \overline{\rho}_\alpha^\alpha \, g \frac{\partial z}{\partial x_i}\right) T^*_{\alpha ij} \tag{2.12}$$

Making use of (5.1), the first term on the R.H.S of Eq. (2.12) can be rewritten in the form

$$\theta_\alpha \overline{\frac{\partial \tau_{\alpha ij}}{\partial x_i}}^\alpha = \frac{\partial \theta_\alpha \overline{\tau_{\alpha ij}}^\alpha}{\partial x_i} + \frac{1}{U_o} \int_{(S_{\alpha s})} \tau_{\alpha ij} \nu_{\alpha i} dS \tag{2.13}$$

Our next objective is to express the internal viscous resistance force, $\nabla . \theta_\alpha \overline{\underset{\sim}{\tau}_\alpha}^\alpha$, appearing in Eq. (2.13) in terms of the averaged fluid velocity. We shall first limit the discussion to an incompressible Newtonian fluid, for which the constitutive relationship is given by

$$\tau_{\alpha ij} = \mu_\alpha \left(\frac{\partial V_i^{m_\alpha}}{\partial x_j} + \frac{\partial V_j^{m_\alpha}}{\partial x_i}\right) \tag{2.14}$$

Other types of fluids may be considered by the same methodology. By using Eq. (5.1) to average Eq. (2.14), we obtain

$$\theta_\alpha \overline{\tau_{\alpha ij}}^\alpha = \mu_\alpha \left(\frac{\partial \theta_\alpha \overline{V_i^{m_\alpha}}^\alpha}{\partial x_j} + \frac{\partial \theta_\alpha \overline{V_j^{m_\alpha}}^\alpha}{\partial x_i}\right) + \frac{\mu_\alpha}{U_o} \int_{(S_{\alpha s})} (V_i^{m_\alpha} \nu_{\alpha j} + V_j^{m_\alpha} \nu_{\alpha i}) dS \tag{2.15}$$

where μ_α is assumed constant within the REV.

As a special case of interest, let us assume that (a) the fluid (that occupies the entire void space) adheres to the solid (= *no slip* condition, i.e., $\underset{\sim}{V}^{m_\alpha}\big|_{S_{\alpha s}} \equiv \underset{\sim}{V}_s\big|_{S_{\alpha s}}$), and (b) $\overset{\sim}{\underset{\sim}{V}}_s{}^\alpha \equiv \overline{\underset{\sim}{V}_s}^s$, i.e., the solid is approximately rigid, and, therefore, its averaged velocity on $(S_{\alpha s})$ is equal to its average over (U_{os}). Then Eq. (2.15) can be approximated by

$$n\, \overline{\tau_{\alpha ij}}^\alpha = \mu_\alpha \left\{ \left(\frac{\partial n \overline{V_i^{m_\alpha}}^\alpha}{\partial x_j} + \frac{\partial n \overline{V_j^{m_\alpha}}^\alpha}{\partial x_i}\right) - \left(\overline{V_{si}}^s \frac{\partial n}{\partial x_j} + \overline{V_{sj}}^s \frac{\partial n}{\partial x_i}\right) \right\}$$

$$= \mu_\alpha\{\frac{\partial n(\overline{V_i^{m_\alpha}}^\alpha - \overline{V}_{si}^s)}{\partial x_j} + \frac{\partial n(\overline{V_j^{m_\alpha}}^\alpha - \overline{V}_{sj}^s)}{\partial x_i} + n(\frac{\partial \overline{V}_{si}^s}{\partial x_j} + \frac{\partial \overline{V}_{sj}^s}{\partial x_i})\} \quad (2.16)$$

in which n replaces θ_α and we have made use of the relationship

$$\nabla\theta_\alpha = -\frac{1}{U_o}\int_{(S_{\alpha\beta})} \underset{\sim}{\nu}_\alpha dS = \frac{1}{U_o}\int_{(S_{\alpha\alpha})} \underset{\sim}{\nu}_\alpha dS \quad (2.17)$$

derived by applying Eq. (5.1) to $G_\alpha = \theta_\alpha$. Since we have assumed that the solid is approximately (macroscopically) rigid, and hence

$$\frac{\partial \overline{V}_{si}^s}{\partial x_j} + \frac{\partial \overline{V}_{sj}^s}{\partial x_i} = 0 \quad (2.18)$$

Eq. (2.16) reduces to

$$n\,\overline{\tau_{\alpha ij}}^\alpha = \mu_\alpha\{\frac{\partial n(\overline{V_i^{m_\alpha}}^\alpha - \overline{V}_{si}^s)}{\partial x_j} + \frac{\partial n(\overline{V_j^{m_\alpha}}^\alpha - \overline{V}_{sj}^s)}{\partial x_i}\} \quad (2.19)$$

Henceforth, we shall use Eq. (2.19) as an approximation of Eq. (2.16) also for a non-rigid solid phase.

From Eq. (2.19), we now obtain

$$\frac{\partial n\overline{\tau_{\alpha ij}}^\alpha}{\partial x_i} = \mu_\alpha\{\frac{\partial^2 n(\overline{V_i^{m_\alpha}}^\alpha - \overline{V}_{si}^s)}{\partial x_i\,\partial x_j} + \frac{\partial^2 n(\overline{V_j^{m_\alpha}}^\alpha - \overline{V}_{sj}^s)}{\partial x_i\,\partial x_i}\} \quad (2.20)$$

$$= \mu_\alpha(\frac{\partial^2 q_{ri}^{m_\alpha}}{\partial x_i \partial x_j} + \frac{\partial^2 q_{rj}^{m_\alpha}}{\partial x_i \partial x_i})$$

where $\underset{\sim}{q}_r^{m_\alpha}/n = (\overline{\underset{\sim}{V}^{m_\alpha}}^\alpha - \overline{\underset{\sim}{V}}_s^s)$ is the mass weighted fluid velocity relative to that of the solid, and $\underset{\sim}{q}_r^{m_\alpha}$ is the relative (mass weighted) specific discharge.

For a stationary solid (or approximately so, i.e., $\underset{\sim}{\overline{V}}_s^s = 0$), and neglecting the diffusive mass flux of the fluid due to molecular diffusion (i.e., assuming $\underset{\sim}{V}^{m_\alpha} \cong \underset{\sim}{V}_\alpha$ = volume weighted velocity), Eq. (2.20) reduces to

$$\frac{\partial n\overline{\tau_{\alpha ij}}^\alpha}{\partial x_i} = \mu_\alpha\left(\frac{\partial^2 q_{\alpha i}}{\partial x_i\,\partial x_j} + \frac{\partial^2 q_{\alpha j}}{\partial x_i\,\partial x_i}\right) \tag{2.21}$$

where $\underset{\sim}{q}_\alpha$ ($= \underset{\sim}{q}_{\alpha r} = n\underset{\sim}{V}^{m_\alpha} \cong n\underset{\sim}{V}_\alpha$) is the fluid's specific discharge relative to a fixed coordinate system.

For the special case of *macroscopically isochoric flow*, where $\nabla\cdot\underset{\sim}{q}_\alpha = 0$, Eq. (2.21) reduces to

$$\frac{\partial n\overline{\tau_{\alpha ij}}^\alpha}{\partial x_i} = \frac{\partial^2 q_{\alpha j}}{\partial x_i\,\partial x_i} \qquad \nabla.n\overline{\underset{\approx}{\tau}_\alpha}^\alpha = \nabla^2 \underset{\sim}{q}_\alpha \tag{2.22}$$

For a *macroscopically uniform flow*, where $\underset{\sim}{q}_\alpha$ = const., $\nabla.n\overline{\underset{\approx}{\tau}_\alpha}^\alpha = 0$.

We now turn to the evaluation of the surface integral over $(S_{\alpha s})$ in Eq. (2.13), that expresses the transfer of momentum from the solid to the fluid. Actually, this term expresses the force resisting the flow of the fluid per unit volume of porous medium. Let us assume that

$$\frac{\partial V_j^{m_\alpha}}{\partial x_i} = \frac{\partial V_j^{m_\alpha}}{\partial s_\nu}\nu_{\alpha i} + \frac{\partial V_j^{m_\alpha}}{\partial s_{t'}}t'_{\alpha i} + \frac{\partial V_j^{m_\alpha}}{\partial s_{t''}}t''_{\alpha i} \cong \frac{\partial V_j^{m_\alpha}}{\partial s_\nu}\nu_{\alpha i} \tag{2.23}$$

where $\underset{\sim}{\nu}_\alpha$, $\underset{\sim}{t}'_\alpha$, $\underset{\sim}{t}''_\alpha$ are the normal unit vector (=principal normal) and the two tangential ones on $(S_{\alpha s})$ in the "local coordinates" at a point on $(S_{\alpha s})$. By employing the approximation $\partial V_j^{m_\alpha}/\partial x_i \cong (\partial V_j^{m_\alpha}/\partial s_\nu)\nu_{\alpha i}$, we have neglected the components of the velocity gradient in the local tangent plane to $(S_{\alpha s})$.

With this approximation, and for the incompressible Newtonian fluid considered here, the surface integral in Eq. (2.13) becomes

$$\frac{1}{U_o}\int_{(S_{\alpha s})} \tau_{\alpha ij}\nu_{\alpha i} dS = \mu_\alpha \frac{1}{U_o}\int_{(S_{\alpha s})} \left(\frac{\partial V_i^{m_\alpha}}{\partial x_j} + \frac{\partial V_j^{m_\alpha}}{\partial x_i}\right)\nu_{\alpha i} dS$$

$$= \mu_\alpha \frac{1}{U_o}\int_{(S_{\alpha s})} \left(\frac{\partial V_i^{m_\alpha}}{\partial s_\nu}\nu_{\alpha j} + \frac{\partial V_j^{m_\alpha}}{\partial s_\nu}\nu_{\alpha i}\right)\nu_{\alpha i} dS$$

$$= \mu_\alpha \frac{1}{U_o}\int_{(S_{\alpha s})} \frac{\partial V_i^{m_\alpha}}{\partial s_\nu}(\nu_{\alpha j}\nu_{\alpha i} + \delta_{ij}) dS$$

$$= \mu_\alpha(\widetilde{\nu_{\alpha j}\nu_{\alpha i}}^{\alpha s} + \delta_{ij}) \frac{1}{U_o}\int_{(S_{\alpha s})} \frac{\partial V_i^{m_\alpha}}{\partial s_\nu} dS \qquad (2.24)$$

where s_ν is the length measured along $\underset{\sim}{\nu}_\alpha$ and $\widetilde{\nu_{\alpha i}\nu_{\alpha j}}^{\alpha s} = \frac{1}{S_{\alpha s}}\int_{(S_{\alpha s})} \nu_{\alpha i}\nu_{\alpha j} dS$ is a coefficient that is related to the microscopic configuration of the $S_{\alpha s}$ - surface.

We now introduce the approximation

$$\frac{1}{U_o}\int_{(S_{\alpha s})} \frac{\partial V_i^{m_\alpha}}{\partial s_\nu} dS \cong \frac{\widetilde{V_i^{m_\alpha}}^{\alpha s} - \overline{V_i^{m_\alpha}}}{\Delta}\frac{S_{\alpha s}}{U_o} \cong \frac{\overline{V}^s_{si} - \overline{V_i^{m_\alpha}}^{\alpha}}{\Delta}\frac{S_{\alpha s}}{U_o} \qquad (2.25)$$

where $\widetilde{V_i^{m_\alpha}}^{\alpha s}$ is the average velocity of the α-phase on the $S_{\alpha s}$ - surface. Since we have assumed that the fluid adheres to the solid wall, we have $\widetilde{V_i^{m_\alpha}}^{\alpha s} = \widetilde{V}_{si}^{\alpha s}$. We now add the approximation $\widetilde{V}_{si}^{\alpha s} \cong \overline{V}_{si}^{s}$

with the approximation sign changing into an equality one when the solid (not the solid matrix!) is rigid.

The distance Δ appearing in Eq. (2.25) is a characteristic distance from the solid walls to the interior of the phase. In a granular material, it is some measure of the size of pores. For example, it can be taken as proportional to the *hydraulic radius*, i.e., $\Delta = C_\alpha U_{o\alpha}/S_{\alpha s}$, where C_α is a shape factor. With this definition of Δ, Eq. (2.25) becomes

$$\mu_\alpha \frac{1}{U_o} \int_{(S_{\alpha s})} \frac{\partial V_i^{m_\alpha}}{\partial s_\nu} dS \cong - \mu_\alpha C_\alpha n(\overline{V_i^{m_\alpha}}^\alpha - \overline{V}_{si}^s)/\Delta^2 \qquad (2.26)$$

Thus, in terms of the fluid's mass weighted specific discharge $\overline{\underset{\sim}{q}_r^{m_\alpha}}^\alpha = n(\overline{\underset{\sim}{V}^{m_\alpha}} - \overline{\underset{\sim}{V}}_s^s)$, the averaged momentum balance equation for a single incompressible Newtonian fluid that completely fills the void space, is

$$n\overline{\rho}_\alpha^\alpha \left(\frac{\partial \overline{V_j^{m_\alpha}}^\alpha}{\partial t} + \overline{V_i^{m_\alpha}}^\alpha \frac{\partial \overline{V_j^{m_\alpha}}^\alpha}{\partial x_i}\right) - \mu_\alpha\left(\frac{\partial^2 q_{ri}^{m_\alpha}}{\partial x_i \partial x_j} + \frac{\partial^2 q_{rj}^{m_\alpha}}{\partial x_i \partial x_i}\right)$$

$$+ n\left(\frac{\partial \overline{p}_\alpha^\alpha}{\partial x_i} + \overline{\rho}_\alpha^\alpha g \frac{\partial z}{\partial x_i}\right) T^*_{\alpha ij} + \frac{C_\alpha \mu_\alpha}{\Delta^2} \left(\overset{\sim\sim\sim\sim\sim\sim\sim\alpha s}{\nu_{\alpha i} \nu_{\alpha j}} + \delta_{ij}\right) q_{ri}^{m_\alpha} = 0 \qquad (2.27)$$

Finally, for the sake of simplicity, let us assume that the molecular diffusion flux, due to $\nabla\rho_\alpha \neq 0$ within $(U_{o\alpha})$, can be neglected, so that $\underset{\sim}{V}^{m_\alpha} = \underset{\sim}{V}_\alpha$. Then Eq. (2.27) reduces to

$$n\overline{\rho}_\alpha^\alpha \left(\frac{\partial (q_{\alpha j}/n)}{\partial t} + \frac{q_{\alpha i}}{n} \frac{\partial (q_{\alpha j}/n)}{\partial x_i}\right) - \mu_\alpha\left(\frac{\partial^2 q_{r\alpha i}}{\partial x_i \partial x_j} + \frac{\partial^2 q_{r\alpha j}}{\partial x_i \partial x_i}\right)$$

$$+ n\left(\frac{\partial \overline{p}_{\alpha}^{\alpha}}{\partial x_i} + \overline{\rho}_{\alpha}^{\alpha} \; g \frac{\partial z}{\partial x_i}\right) T^*_{\alpha ij} + \frac{C_\alpha \mu_\alpha}{\Delta^2} (\overline{\tilde{\nu}_{\alpha j}\tilde{\nu}_{\alpha i}}^{\alpha s} + \delta_{ij}) q_{ri} = 0 \qquad (2.28)$$

Eq. (2.28) is commonly used as a good approximation also for a compressible fluid.

To recapitulate, the first term on the L.H.S. of Eq. (2.28) represents the inertial force, the second term represents the resistance force due to shear inside the fluid, the third term represents the force acting on the fluid as a result of pressure gradient and gravity and the last term expresses the force exerted by the solid phase on the flowing fluid.

Let us mention two special cases of Eq. (2.28):
a)When the inertial effects are negligible, and so are the effects of internal friction, Eq. (2.28) reduces to the well known *Darcy law.*

$$q_{rj}^{m_\alpha} = n(\overline{V_j^{m\alpha}}^{\alpha} - \overline{V}_{sj}^{s}) = - \frac{k_{\alpha jm}}{\mu_\alpha} \left(\frac{\partial \overline{p}_{\alpha}^{\alpha}}{\partial x_m} + \overline{\rho}_{\alpha}^{\alpha} g \frac{\partial z}{\partial x_m}\right) \qquad (2.29)$$

where

$$k_{\alpha jm} = \frac{n\Delta^2}{C_\alpha} (\overline{\tilde{\nu}_{\alpha j}\tilde{\nu}_{\alpha i}}^{\alpha s} + \delta_{ij})^{-1} \; T^*_{\alpha im} = \frac{\Delta n^2}{\Sigma'_{\alpha s}} (\overline{\tilde{\nu}_{\alpha j}\tilde{\nu}_{\alpha i}}^{\alpha s} + \delta_{ij})^{-1} \; T^*_{\alpha im}$$

$$= \frac{C_\alpha n^3}{(\Sigma'_{\alpha s})^2} (\overline{\tilde{\nu}_{\alpha i}\tilde{\nu}_{\alpha j}}^{\alpha s} + \delta_{ij})^{-1} \; T^*_{\alpha im} \quad ; \quad \Sigma'_{\alpha s} = \frac{S_{\alpha s}}{U_o} \qquad (2.30)$$

is a coefficient related only to macroscopic parameters that describe the microscopic configuration of the fluid-solid interface. The coefficient $k_{\alpha jm}$ - a second rank symmetric tensor - is called the *permeability* of the porous medium. We recall that $\underset{\approx}{T}^*_{\alpha}$ is defined by Eq. (5.9) and (5.15).

Eq. (2.29) is the basic form of the motion equation for saturated flow in an anisotropic porous medium at low Reynolds numbers(defined by $N_{Re} = q\overline{\rho}_{\alpha}^{\alpha}\Delta/\mu_\alpha$).

b) When the inertial effects are negligible, but we do wish to include the effects of internal friction, then Eq. (2.28) reduces to

$$- \frac{k_{\alpha jp}(T^*_{\alpha pm})^{-1}}{n}\left(\frac{\partial^2 q^{m_\alpha}_{\alpha ri}}{\partial x_i\ \partial x_j} + \frac{\partial^2 q^{m_\alpha}_{\alpha rj}}{\partial x_i\ \partial x_i}\right) + \frac{k_{\alpha mj}}{\mu_\alpha}\left(\frac{\partial \overline{p}^\alpha_\alpha}{\partial x_j} + \overline{\rho}^\alpha_\alpha\, g\, \frac{\partial z}{\partial x_j}\right)$$

$$+ q_{\alpha rm} = 0 \qquad (2.31)$$

For an isotropic porous medium, $T^*_\alpha = \frac{1}{3}\, T^*_{\alpha ii}$, and macroscopically isochoric flow, i.e., $\nabla . \underset{\sim}{q}^{m_\alpha}_r = 0$, Eq. (2.31) reduces to

$$\frac{\mu_\alpha}{nT^*_\alpha}\ \nabla . \nabla \underset{\sim}{q}^{m_\alpha}_{\alpha r} - (\nabla \overline{p}^\alpha_\alpha + \overline{\rho}^\alpha_\alpha g \nabla z) - \frac{\mu_\alpha}{k_\alpha}\ \underset{\sim}{q}^{m_\alpha}_{\alpha r} = 0 \qquad (2.32)$$

This equation, with $\underset{\sim}{q}^{m_\alpha}_{\alpha r}$ replaced by $\underset{\sim}{q}_\alpha$, without the coefficient $1/nT^*_\alpha$ and without the gravity term, was proposed by Brinkman (3) and is known as *Brinkman's equation*.

The entire discussion presented above can also be extended to multiphase flow, i.e., when two or more immiscible fluids occupy the void space. Without going into a detailed discussion, let us accept the conceptual model in which each fluid occupies a certain portion of the void space. In principle each fluid then has an interface with the solid as well as with each of the other fluids. Momentum can then be transferred across any of these internal surfaces. Accordingly, if we consider two fluids: a wetting (w) one and a nonwetting (nw) one, we should replace the integral over $S_{\alpha s}$ in Eq. (2.26) by a sum of integrals over the surfaces S_{ws}, S_{wnw}. If then we assume $S_{ws} >> S_{wnw}$, $S_{nws} >> S_{wnw}$, and that, therefore, the fluid-fluid momentum transfer is much smaller then the fluid-solid one, we rewrite Eq. (2.28) twice: Once for the wetting fluid, with α replaced by w and θ_α by θ_w, and once for the nonwetting fluid, with α replaced by nw and n replaced by θ_{nw}. In the modified equations, we shall then identify the effective permeabilities, $k_{wij}(\theta_w, \Delta_w(\theta_w))$ and $k_{nwij}(\theta_{nw}, \Delta_{nw}(\theta_{nw}))$ for the wetting and non-wetting fluids, respectively, with the possibility that the relations $\Delta_w(\theta_w)$ and $\Delta_{nw}(\theta_{nw})$ are non-unique.

3. DIFFUSIVE MASS FLUX

The diffusive mass flux, $\underset{\sim}{J}^m_{\alpha\gamma}$, of the mass of a γ - component of an α-phase continuum (i.e., at the microscopic level) is defined by

$$\underset{\sim}{J}^{m}_{\alpha\gamma} = \rho_{\alpha\gamma}(\underset{\sim}{V}^{m}_{\alpha\gamma} - \underset{\sim}{V}_{\alpha}) \tag{3.1}$$

where $\rho_{\alpha\gamma}$ is the density of the γ-component of the α-phase, $\underset{\sim}{V}^{m\alpha\gamma}$ is its mass weighted velocity and $\underset{\sim}{V}_{\alpha}$ is the volume weighted velocity of the α-phase. We also refer to this flux as *molecular diffusion*. Note that $\sum_{(\gamma)} \underset{\sim}{J}^{m}_{\alpha\gamma} \neq 0$, except when $\underset{\sim}{V}_{m_\alpha} \cong \underset{\sim}{V}_{\alpha}$ (single component).

At the microscopic level, disregarding coupling between transport phenomena, this flux, denoted by $\underset{\sim}{J}^{d}_{\alpha\gamma}$, is expressed by *Fick's law*. For a binary system this law takes the form

$$\underset{\sim}{J}^{d}_{\alpha\gamma} = - D^{d}_{\alpha\gamma} \nabla c_{\alpha\gamma} \tag{3.2}$$

where $D^{d}_{\alpha\gamma}$ is the coefficient of molecular diffusion of the γ-component in the α-phase and $c_{\alpha\gamma}$ is the concentration of the γ-component in the α-phase. We assume that $D^{d}_{\alpha\gamma}$ is independent of $c_{\alpha\gamma}$.

The macroscopic flux of molecular diffusion is obtained by applying the averaging rule of spatial derivatives (e.g., 2) to Eq. (3.2). We obtain

$$\overline{\underset{\sim}{J}^{d}_{\alpha\gamma}} = - D^{d}_{\alpha\gamma}\theta_{\alpha}\overline{\nabla c_{\alpha\gamma}}^{\alpha} = - \theta_{\alpha}D^{d}_{\alpha\gamma}(\nabla\overline{c}^{\alpha}_{\alpha\gamma} + \frac{1}{U_o}\int_{(S_{\alpha s})} \overset{o}{c}_{\alpha\gamma}\underset{\sim}{\nu}_{\alpha}dS) \tag{3.3}$$

in which $\overset{o}{c}_{\alpha\gamma}$ denotes the deviation of $c_{\alpha\gamma}$ from its average over the REV, i.e., $\overset{o}{c}_{\alpha\gamma}(\underset{\sim}{x}, t; \underset{\sim}{x}_o) = c_{\alpha\gamma}(\underset{\sim}{x}, t; \underset{\sim}{x}_o) - \overline{c}^{\alpha}_{\alpha\gamma}(\underset{\sim}{x}_o, t)$. The difficulty in employing Eq. (3.3) as an expression for the macroscopic diffusive flux is that the integral appearing in it involves information on the microscopic configuration of the $S_{\alpha s}$ - surface and on the distribution of $c_{\alpha\gamma}$ on it. We need a way to overcome albeit as an approximation, the lack of this information.

Let us assume that (a) the γ-component does not interact with the solid, e.g., in the form of adsorption (this interaction is considered in (1)), and (b) that no sources of γ are present within U_o. Under such conditions, at a given instant of time, the concentration, $c_{\alpha\gamma}$, varies monotonously within $(U_{o\alpha})$ and hence, satisfies the condition

$$\nabla^2 c_{\alpha\gamma} = 0 \quad \text{in } (U_{o\alpha}) \tag{3.4}$$

In view of assumption (a) above, the α-phase - solid interface $(S_{\alpha s})$, acts as a material surface for the γ-component, i.e.

$$\{c_{\alpha\gamma}(\underset{\sim}{V}_\alpha - \underset{\sim}{u}) + \overset{d}{\underset{\sim}{J}}_{\alpha\gamma}\}\cdot\underset{\sim}{\nu}_\alpha = 0 \qquad \text{on } (S_{\alpha s}) \tag{3.5}$$

where $\underset{\sim}{u}$ is the velocity of $(S_{\alpha s})$. Note that Eq. (3.5), written at the microscopic level, states that no γ-component crosses $S_{\alpha s}$, whether by advection or by diffusion.

At the same time, the solid acts also as a material surface with respect to the total fluid α-phase mass, and, therefore, $(\underset{\sim}{V}_\alpha-\underset{\sim}{u})\cdot\underset{\sim}{\nu}_\alpha=0$. Hence, in view of Eq. (3.5), we have

$$\overset{d}{\underset{\sim}{J}}_{\alpha\gamma}\cdot\underset{\sim}{\nu}_\alpha = 0, \qquad \text{or} \qquad (\nabla c_{\alpha\gamma})\cdot\underset{\sim}{\nu}_\alpha = 0 \qquad \text{on } (S_{\alpha s}) \tag{3.6}$$

With $c_{\alpha\gamma}$ satisfying Eqs. (3.4) and (3.6), the case under consideration corresponds to *Case A* defined by Eqs. (5.3) and (5.10), with $G_\alpha \equiv c_{\alpha\gamma}$. Hence, making use of Eq. (5.11), the macroscopic flux in this case takes the form

$$\overline{\overset{d}{\underset{\sim}{J}}_{\alpha\gamma}} = -\overline{\overset{d}{D}_{\alpha\gamma}\nabla c_{\alpha\gamma}} = -n\overset{d}{D}_{\alpha\gamma}\overline{\nabla c_{\alpha\gamma}}^{\alpha} = -n\overset{d}{D}_{\alpha\gamma}\underset{\approx}{T}^{*}_{\alpha}\cdot\nabla\overline{c}^{\alpha}_{\alpha\gamma}$$

or, in indicial notation

$$\overline{\overset{d}{J}_{\alpha i}} = -n\overset{d}{D}_{\alpha\gamma}T^{*}_{\alpha ij}\frac{\partial c_{\alpha\gamma}}{\partial x_j} = -n(\overset{d*}{D}_{\alpha\gamma})_{ij}\frac{\partial\overline{c}^{\alpha}_{\alpha\gamma}}{\partial x_j} \tag{3.7}$$

where n is porosity and $\overset{d*}{D}_{\alpha\gamma} = \overset{d}{D}_{\alpha\gamma}\underset{\approx}{T}^{*}_{\alpha}$, a second rank symmetric tensor, is the *coefficient of molecular diffusion in a porous medium*. The coefficient $\underset{\approx}{T}^{*}_{\alpha}$ is defined and discussed in Appendix B. We have thus achieved our objective of replacing the missing microscopic information by a macroscopic coefficient that represents it.

In a multiphase system, the fluid α-phase occupies only part of the void space. Then the interface between the α-phase and all other phases within the REV is a material surface with respect to both the total mass of the α-phase and the mass of the γ-component. Therefore, the discussion presented above remains valid. Replacing the porosity n by the volumetric fraction, θ_α, of the α-phase, we obtain

$$\overline{\underset{\sim}{J}^{d}_{\alpha\gamma}} = -\theta_\alpha \underset{\approx}{D}^{d*}_{\alpha\gamma} \cdot \nabla \overline{c}^{\alpha}_{\alpha\gamma} \tag{3.8}$$

except that in this case, $\underset{\approx}{D}^{d*}_{\alpha\gamma} = \underset{\approx}{D}^{d*}_{\alpha\gamma}(\theta_\alpha)$.

4. DIFFUSIVE HEAT FLUX

We now consider the diffusive heat flux (=heat conduction) within a fluid α-phase that completely fills the void space of a porous medium. We shall first assume that both the fluid α-phase (denoted here by f) and the solid phase (denoted here by s) may conduct heat, with thermal conductivities λ_f and λ_s, respectively, assumed constant within an REV.

The diffusive heat flux, $\underset{\sim}{J}^{h}_{f}$, within the fluid phase occupying U_{of}, is expressed by *Fourier's Law*

$$\underset{\sim}{J}^{h}_{f} = -\lambda_f \nabla T_f \tag{4.1}$$

where T_f is the temperature of the fluid phase. To obtain the corresponding macroscopic flux, we employ an averaging rule in the form of Eq. (3.3) in which $c_{\alpha\gamma}$ is replaced by T_α. As in the case of molecular diffusion, here also we have to overcome the lack of information on the (microscopic) configuration of (S_{fs}) and the distribution of T_α on it.

To achieve this goal, similar to the case of molecular diffusion, let us assume that no sources or sinks of heat are present within (U_{of}) and (U_{os}), and that within each phase, the temperature (T_f in the fluid phase and T_s in the solid one) varies monotonously, such that

$$\nabla^2 T_f = 0 \quad \text{in} \quad (U_{of})\ ; \qquad \nabla^2 T_s = 0 \quad \text{in} \quad (U_{os}) \tag{4.2}$$

However, in this case, the total heat flux leaving one phase, say the fluid, is absorbed, without any loss, by the other phase. This observation is expressed by the conditions

$$-\lambda_f \frac{\partial T_f}{\partial x_i} \nu_{fi} \bigg|_{\substack{\text{f-side} \\ \text{of } (S_{fs})}} = \lambda_s \frac{\partial T_s}{\partial x_i} \nu_{si} \bigg|_{\substack{\text{s-side} \\ \text{of } (S_{fs})}} \tag{4.3}$$

and

$$T_f \Big|_{\substack{f\ side \\ of\ (S_{fs})}} = T_s \Big|_{\substack{s\ side \\ of\ (S_{fs})}} \tag{4.4}$$

(i.e., the interphase surface is no more material with respect to heat). Because the solid is impervious to fluid, the (S_{fs}) - surface is material with respect to fluid mass, and hence no heat advection takes place through it.

By comparing Eq. (4.3) and Eq. (4.4) with Eqs. (5.12) and (5.13 we conclude that the case on hand is identical to *Case B* of Appendix, with $T_f \equiv G_\alpha$ and $T_s \equiv G_\beta$.

Hence, making use of Eq. (5.17), the expression for the macroscopic heat flux of a fluid phase that fills the entire void space ($\theta_f \equiv n$) takes the form

$$\overline{\underset{\sim}{J}_f^h} = - \lambda_f \overline{\nabla T_f} = - n\lambda_f \overline{\nabla T_f}^f$$

$$= - \frac{n\lambda_f}{\lambda_f - \lambda_s}[(\lambda_f \nabla \overline{T}_f^f - \lambda_s \nabla \overline{T}_s^s) \cdot \underset{\approx}{T}_f^* - \frac{\lambda_s}{n} \nabla\{n(\overline{T}_f^f - \overline{T}_s^s)\}] \tag{4.5}$$

An analogous expression, in terms of $\theta_s = 1 - n$ and $\theta_s \underset{\approx}{T}_s^* = \underset{\approx}{\delta} - \theta_f \underset{\approx}{T}_f^*$ (see Eq. (5.15)) can be written for the macroscopic heat flux, in the solid phase.

We have thus achieved our goal of expressing the macroscopic heat fluxes in terms of the macroscopic state variables.

It is interesting to note the basic difference between Eq. (3.8) and Eq. (4.5). Because the fluid-solid interface is "impervious to the diffusive mass flux", the relationship between the microscopic flux and the macroscopic one, expressed by Eq. (3.8), depends only on what happens in the fluid phase. The coefficient $\underset{\approx}{T}_f^*$ may be called a *tortuosity* (of the void space, or of the (S_{fs})-configuration). On the other hand, in the case of heat transport, the (S_{fs})-surface is "pervious to heat", with conditions (Eqs. (4.3) and (4.4)) on it. Hence, we note the coupling between the heat transport in the two domains, (U_{of}) and (U_{os}). Heat is exchanged continuously between the two phases. Under such conditions, $\underset{\approx}{T}_f^*$ has still the meaning of a tortuosity of the fluid phase.

Let us consider the special cases $\overline{T}_f^f = \overline{T}_s^s$. Then Eq. (4.5) reduces to

$$\overline{\underset{\sim}{J}_f^h} = - n\lambda_f \underset{\approx}{T}_f^* . \nabla \overline{T}_f^f = - n\underset{\approx}{\lambda}_f^* . \nabla \overline{T}_f^f \qquad (4.6)$$

which is similar to Eq. (3.7), with $\underset{\approx}{\lambda}_f^* = \lambda_f \underset{\approx}{T}_f^*$. This result is obvious since we have no heat exchange (on the average) between the two phases. Hence, Eq. (4.6) is also valid for $\lambda_s = 0$

When $\overline{T}_f^f = \overline{T}_s^s$, the total heat flux in both phases is given by

$$\underset{\sim}{J}_f^h + \underset{\sim}{J}_s^h = - (n\underset{\approx}{\lambda}_f^* + (1-n)\underset{\approx}{\lambda}_s^*) . \nabla \overline{T}_f^f = - \underset{\approx}{\Lambda} . \nabla \overline{T}_f^f \qquad (4.7)$$

where $\underset{\approx}{\Lambda} = n\lambda_f^* + (1-n)\underset{\approx}{\lambda}_s^*$ is the thermal conductivity of the saturated porous medium as a whole.

5. APPENDIX A

The objective of this Appendix is to develop, following Bachmat and Bear (1) a modified form of the commonly used averaging rule for a spatial derivative

$$\theta_\alpha \overline{\frac{\partial G_{jk\ell\ldots}}{\partial x_i}}^\alpha = \frac{\partial (\theta_\alpha \overline{G_{jk\ell\ldots}}^\alpha)}{\partial x_i} + \frac{1}{U_o} \int_{(S_{\alpha\beta})} G_{jk\ell\ldots} \nu_{\alpha i} dS \qquad (5.1)$$

where $G_{jk\ell\ldots}$ is a tensorial property of a phase, U_o is the volume of the REV, θ_α is the volumetric fraction of the α-phase, $S_{\alpha\beta}$ is the surface area of contact of the α-phase with all other phases within the REV, $\overline{(\ )}^\alpha$ denotes the intrinsic phase average and $\underset{\sim}{\nu}_\alpha$ is the normal outward unit vector on $(S_{\alpha\beta})$ (e.g. Bear and Bachmat (2)).

The integral on the R.H.S. of Eq. (5.1) requires information on both the geometry of the $S_{\alpha\beta}$ - boundary and on the values of G_α on it (first type boundary condition). Let us develop a modified form of Eq. (5.1) for a scalar G_α in which the required information is on the normal component of ∇G_α on $(S_{\alpha\beta})$ (i.e., second type boundary condition for which information is sometimes available). To this end, consider the quantity $\overset{o}{x}_j \partial G_\alpha / \partial x_i$.

By Gauss Theorem applied to the domain $(U_{o\alpha})$ of volume $U_{o\alpha}$ of an α-phase within the domain (U_o) of volume U_o, we obtain

$$\int_{(U_{o\alpha})} \frac{\partial}{\partial x_i} (\overset{o}{x}_j \frac{\partial G_\alpha}{\partial x_i}) dU = \int_{(S_{o\alpha})} \overset{o}{x}_j \frac{\partial G_\alpha}{\partial x_i} \nu_{\alpha i} dS \tag{5.2}$$

where $(S_{o\alpha})$ is the total surface of the closed area surrounding $(U_{o\alpha})$.

We shall limit the following discussion to G_α's that satisfy the following two conditions:
(a) G_α attains no maximum or minimum value within $(U_{o\alpha})$, i.e., G_α varies monotonously within $(U_{o\alpha})$. Under this condition we have (4)

$$\frac{\partial^2 G_\alpha}{\partial x_i \, \partial x_i} = 0 \qquad \text{within } (U_{o\alpha}) \tag{5.3}$$

(b)

$$\int_{(S_{\alpha\alpha})} \frac{\partial G_\alpha}{\partial x_i} \overset{o}{x}_j \nu_{\alpha i} dS \cong \frac{\widetilde{\partial G_\alpha}^{\alpha\alpha}}{\partial x_i} \int_{(S_{\alpha\alpha})} \overset{o}{x}_j \nu_{\alpha i} dS \cong \frac{\partial \overline{G}_\alpha^{\alpha}}{\partial x_i} \int_{(S_{\alpha\alpha})} \overset{o}{x}_j \nu_{\alpha i} dS \tag{5.4}$$

where

$$\frac{\widetilde{\partial G_\alpha}^{\alpha\alpha}}{\partial x_i} = \frac{1}{S_{\alpha\alpha}} \int_{(S_{\alpha\alpha})} \frac{\partial G_\alpha}{\partial x_i} dS \tag{5.5}$$

i.e., we assume that the average of the gradient of G_α on the α-α portion of the outer surface of the REV is equal to the gradient of the average of G_α over the volume $U_{o\alpha}$ of the α-phase within the REV.

By definition, and by Eq. (5.3) we have

$$\frac{\partial}{\partial x_i} (\overset{o}{x}_j \frac{\partial G_\alpha}{\partial x_i}) = \frac{\partial \overset{o}{x}_j}{\partial x_i} \frac{\partial G_\alpha}{\partial x_i} + \overset{o}{x}_j \frac{\partial^2 G_\alpha}{\partial x_i \partial x_i} = \delta_{ij} \frac{\partial G_\alpha}{\partial x_i} = \frac{\partial G_\alpha}{\partial x_j} \tag{5.6}$$

Hence, Eq. (5.2) reduces to

$$\int_{(U_{o\alpha})} \frac{\partial G_\alpha}{\partial x_j} dU_\alpha = \int_{(S_{\alpha\alpha})} \overset{o}{x}_j \frac{\partial G_\alpha}{\partial x_i} \nu_{\alpha i} dS + \int_{(S_{\alpha\beta})} \overset{o}{x}_j \frac{\partial G_\alpha}{\partial x_i} \nu_{\alpha i} dS \tag{5.7}$$

or, with Eq. (5.4)

$$\overline{\frac{\partial G_\alpha}{\partial x_i}}^\alpha = \frac{\partial \overline{G_\alpha}^\alpha}{\partial x_i} T^*_{\alpha ij} + \frac{1}{U_{o\alpha}} \int_{(S_{\alpha\beta})} \overset{\circ}{x}_j \frac{\partial G_\alpha}{\partial x_i} \nu_{\alpha i} \, dS \tag{5.8}$$

where

$$T^*_{\alpha ij} = \frac{1}{U_{o\alpha}} \int_{(S_{\alpha\alpha})} \overset{\circ}{x}_j \nu_{\alpha i} \, dS \tag{5.9}$$

Thus, Eq. (5.8) is another form of the averaging rule for ∇G_α, this time requiring information on $\nabla G_\alpha \cdot \underset{\sim}{\nu}_\alpha$ on $(S_{\alpha\beta})$.

Two cases, the physical interpretation of which is presented in this paper, may be considered.

Case A.

$$\frac{\partial G_\alpha}{\partial x_i} \nu_{\alpha i} = 0 \qquad \text{on} \quad (S_{\alpha\beta}) \tag{5.10}$$

In this case, Eq. (5.8) reduces to

$$\overline{\frac{\partial G_\alpha}{\partial x_j}}^\alpha = \frac{\partial \overline{G_\alpha}^\alpha}{\partial x_i} T^*_{\alpha ij} \tag{5.11}$$

Case B.

In this case, we have $\partial G_\alpha / \partial \nu_\alpha \neq 0$ on $(S_{\alpha\beta})$. Let the following two boundary conditions hold on $(S_{\alpha\beta})$:

$$- \lambda_\alpha \frac{\partial G_\alpha}{\partial \nu_\alpha} \Bigg|_{\substack{\alpha\text{-side} \\ \text{of } (S_{\alpha\beta})}} = \lambda_\beta \frac{\partial G_\beta}{\partial \nu_\beta} \Bigg|_{\substack{\beta\text{-side} \\ \text{of } (S_{\alpha\beta})}} \tag{5.12}$$

and

$$G_\alpha \Big|_{\substack{\alpha\text{-side} \\ \text{of } (S_{\alpha\beta})}} = G_\beta \Big|_{\substack{\beta\text{-side} \\ \text{of } (S_{\alpha\beta})}} \tag{5.13}$$

where λ_α and λ_β are two coefficients that depend on the physical nature of G and the α and β phases, respectively, and G_β denotes the value of G in $(U_{o\beta})$.

By applying Eq. (5.8) first to the α-phase and multiplying the result by λ_α, then to the β-phase and multiplying the result by λ_β and adding the two resulting equations, employing Eq. (5.12), we obtain

$$\lambda_\alpha \theta_\alpha \overline{\frac{\partial G_\alpha}{\partial x_j}}^\alpha + \lambda_\beta \theta_\beta \overline{\frac{\partial G_\beta}{\partial x_j}}^\beta = \lambda_\alpha \theta_\alpha \frac{\partial \overline{G_\alpha}^\alpha}{\partial x_i} T^*_{\alpha ij} + \lambda_\beta \theta_\beta \frac{\partial \overline{G_\beta}^\beta}{\partial x_i} T^*_{\beta ij} \tag{5.14}$$

Now, by Eq. (5.9)

$$\int_{(S_o)} \overset{o}{x}_j \nu_{\alpha i} \, dS = \int_{(S_{\alpha\alpha})} \overset{o}{x}_j \nu_{\alpha i} \, dS + \int_{(S_{\beta\beta})} \overset{o}{x}_j \nu_{\alpha i} \, dS$$

$$= U_{o\alpha} T^*_{\alpha ij} + U_{o\beta} T^*_{\beta ij} = U_o \, \delta_{ij}$$

whence

$$\theta_\alpha T^*_{\alpha ij} + \theta_\beta T^*_{\beta ij} = \delta_{ij} \tag{5.15}$$

Also, by writing Eq. (5.1) twice, once for the α-phase and once for the β-phase and adding the two equations, employing the condition Eq. (5.13), we obtain

$$\theta_\alpha \overline{\frac{\partial G_\alpha}{\partial x_j}}^\alpha + \theta_\beta \overline{\frac{\partial G_\beta}{\partial x_j}}^\beta = \frac{\partial (\theta_\alpha \overline{G_\alpha}^\alpha)}{\partial x_j} + \frac{\partial (\theta_\beta \overline{G_\beta}^\beta)}{\partial x_j} \tag{5.16}$$

Finally, multiplying Eq. (5.16) by λ_β and subtracting the result from Eq. (5.14), yields

$$\frac{\partial \overline{G}_\alpha^\alpha}{\partial x_j} = \frac{1}{\lambda_\alpha - \lambda_\beta} \{(\lambda_\alpha \frac{\partial \overline{G}_\alpha^\alpha}{\partial x_i} - \lambda_\beta \frac{\partial \overline{G}_\beta^\beta}{\partial x_i}) T^*_{\alpha ij} - \frac{\lambda_\beta}{\theta_\alpha} \frac{\partial(\theta_\alpha(\overline{G}_\alpha^\alpha - \overline{G}_\beta^\beta))}{\partial x_j}\} \tag{5.17}$$

where Eq. (5.13) and the relationship $\theta_\alpha + \theta_\beta = 1$ were employed.

In the particular case $\lambda_\beta = 0$, Eq. (5.17) reduces to Eq. (5.11), corresponding to *Case A*. The same holds when $\lambda_\alpha \neq \lambda_\beta$, but $\overline{G}_\alpha^\alpha = \overline{G}_\beta^\beta$ throughout the entire porous medium domain.

6. APPENDIX B: THE COEFFICIENT $\underset{\approx}{T}^*_\alpha$

The coefficient $\underset{\approx}{T}^*_\alpha$, defined in Eq. (5.9) represents the *static moment of oriented areal elements of* $S_{\alpha\alpha}$ *, with respect to planes passing through* $\underset{\sim}{x}_o$*, per unit volume of the* α *phase within* (U_o).

To obtain an estimate of the magnitude of the components $T^*_{\alpha ij}$, consider a spherical REV of radius R. Then, Eq. (5.9) can be written in the form

$$T^*_{\alpha ij} = \frac{1}{\theta_\alpha U_o} \int_{(S_{\alpha\alpha})} R \nu_{\alpha i} \nu_{\alpha j} \, dS$$

$$= \frac{\theta_\alpha^s S_o R}{\theta_\alpha U_o} \{\frac{1}{S_{\alpha\alpha}} \int_{(S_{\alpha\alpha})} \nu_{\alpha i} \nu_{\alpha j} \, dS\} = \frac{3\theta_\alpha^s}{\theta_\alpha} \widetilde{\nu_{\alpha i} \nu_{\alpha j}}^{\alpha\alpha} \tag{6.1}$$

where θ_α^s (= $S_{\alpha\alpha}/S_o$) denotes the α-α fraction of the surface S_o, and $\widetilde{\nu_{\alpha i} \nu_{\alpha j}}^{\alpha\alpha}$ represents the average of $\nu_{\alpha i}\nu_{\alpha j}$ on $(S_{\alpha\alpha})$.

The term $\nu_{\alpha i}\nu_{\alpha j}$ is a symmetric second rank tensor. Hence, $\widetilde{\nu_{\alpha i} \nu_{\alpha j}}^{\alpha\alpha}$ which is a linear combination of $\nu_{\alpha i}\nu_{\alpha j}$, is also symmetric. Therefore, there exists at least one set of three mutually orthogonal planes of symmetry for $\widetilde{\nu_{\alpha i} \nu_{\alpha j}}^{\alpha\alpha}$, and three principal axes normal to these planes. In the coordinate system of the principal axes, $\widetilde{\nu_{\alpha i} \nu_{\alpha j}}^{\alpha\alpha}$ can be expressed in the form

$$\widetilde{\nu}_{\alpha i}\widetilde{\nu}_{\alpha j}^{\alpha\alpha} = a_1\delta_{1i}\delta_{1j} + a_2\delta_{2i}\delta_{2j} + a_3\delta_{3i}\delta_{3j} \tag{6.2}$$

where a_1, a_2 and a_3 are the *principal values* of $\widetilde{\nu}_{\alpha i}\widetilde{\nu}_{\alpha j}^{\alpha\alpha}$, i.e., $\widetilde{\nu^2_{\alpha 1}}^{\alpha\alpha}$, $\widetilde{\nu^2_{\alpha 2}}^{\alpha\alpha}$ and $\widetilde{\nu^2_{\alpha 3}}^{\alpha\alpha}$, respectively. Hence

$$0 < a_i < 1 \qquad \text{for} \qquad i = 1, 2, 3 \tag{6.3}$$

For an isotropic porous medium, with respect to $\widetilde{\nu}_{\alpha i}\widetilde{\nu}_{\alpha j}^{\alpha\alpha}$, $a_1 = a_2 = a_3 = a$, and Eq. (6.2) reduces to

$$\widetilde{\nu}_{\alpha i}\widetilde{\nu}_{\alpha j}^{\alpha\alpha} = a\delta_{ij} \tag{6.4}$$

Now, $\Sigma_{(i)}\ \widetilde{\nu}_{\alpha i}\widetilde{\nu}_{\alpha i}^{\alpha\alpha} = a\Sigma_{(i)}\ \delta_{ii} = 3a$. On the other hand, by definition, $\nu_{\alpha i}\nu_{\alpha i} \equiv \Sigma_{(i)}\ \widetilde{\nu}_{\alpha i}\widetilde{\nu}_{\alpha i}^{\alpha\alpha} = 1$. Hence

$$a = \frac{1}{3}\ ; \qquad \widetilde{\nu}_{\alpha i}\widetilde{\nu}_{\alpha j}^{\alpha\alpha} = \frac{1}{3}\ \delta_{ij} \tag{6.5}$$

By inserting this result into Eq. (6.1), we obtain for an isotropic porous medium

$$T^*_{\alpha ij} = \frac{\theta^s_\alpha}{\theta_\alpha}\ \delta_{ij} \tag{6.6}$$

For porous media for which $\theta^s_\alpha < \theta_\alpha$ (5), Eq. (6.6) yields

$$0 < T^*_{\alpha ii} < 1 \qquad \text{for any } i \ \text{(no summation on } i) \tag{6.7}$$

However, in the general case

$$T^*_{\alpha ij} = 3\ \frac{\theta^s_\alpha}{\theta_\alpha}\ (\widetilde{\nu^2_{\alpha 1}}^{\alpha\alpha}\ \delta_{1i}\delta_{1j} + \widetilde{\nu^2_{\alpha 2}}^{\alpha\alpha}\ \delta_{2i}\delta_{2j} + \widetilde{\nu^2_{\alpha 3}}^{\alpha\alpha}\ \delta_{3i}\delta_{3j}) \tag{6.8}$$

Hence $T^{*}_{\alpha ii} < 1$, when $\frac{3\theta^{s}_{\alpha}}{\theta_{\alpha}} \mathrm{Max}(\tilde{\tilde{\nu}}^{2\,\alpha\alpha}_{\alpha i}) < 1$

The ratio $\theta^{s}_{\alpha}/\theta_{\alpha}$ is a measure of the tortuosity of the void space, while the term $\tilde{\tilde{\nu}}_{\alpha i}\tilde{\tilde{\nu}}^{\alpha\alpha}_{\alpha j}$ represents the effect of anisotropy on the tortuosity (1).

7. ACKNOWLEDGEMENT

This work is part of a cooperative research program of the Department of Civil Engineering, Division of Water Resources, Technion - Israel Institute of Technology, and the Hydrological Service of Israel on Transport Phenomena in Porous Media. Partial support was provided by the Fund for the Promotion of Research at the Technion.

8. REFERENCES

1. Bachmat, Y. and Bear, J., "Macroscopic Modeling of Transport Phenomena in Porous Media 1:The Continuum Approach," *Transport in Porous Media*, Vol. 1, No. 3, 1986, pp. 213.
2. Bear,J. and Bachmat, Y., "Transport Phenomena in Porous Media-Basic Equations," *Fundamentals of Transport Phenomena in Porous Media*, J. Bear and M.Y. Corapcioglu, eds., Martinus Nijhoff Publishers BV, The Hague, The Netherlands, 1984, pp. 3-61.
3. Brinkman, H.C., "Calculations of the Flow of Heterogeneous Mixtures Through Porous Media," *Applied Science Society*, Vol. 2, 1948, pp. 81 - 86.
4. Morse, P.M. and Feshbach, H., *Methods of Theoretical Physics*, McGraw Hill Book Co., 1953.
5. Santalo, L.A., *Integral Geometry and Geometric Probability*, Addision Wesley, 1976.

9. LIST OF SYMBOLS

a_i	Principal values of $\tilde{\tilde{\nu}}_{\alpha i}\tilde{\tilde{\nu}}^{\alpha\alpha}_{\alpha j}$, i = 1, 2, 3
C_{α}	Shape factor of $(U_{o\alpha})$
$c_{\alpha\gamma}$	Concentration of γ-component in α-phase
$D^{d}_{\alpha\gamma}$	Coefficient of molecular diffusion of a γ-component in a binary system α-phase
$\underset{\approx}{D}^{d*}_{\alpha\gamma}$	Coefficient of molecular diffusion in a porous medium (= $D^{d}_{\alpha\gamma}\ \underset{\approx}{T}^{*}_{\alpha}$)
f	Subscript indicating fluid phase

G_α	Tensorial property of a α-phase
g	Gravity of acceleration
$\underset{\sim}{J}^d_{\alpha\gamma}$	Diffusive flux of $m_{\alpha\gamma}$
$\underset{\sim}{J}^{d\gamma}$	$\underset{\sim}{J}^m_{\alpha\gamma}$
$\underset{\sim}{J}^h$	Conductive heat flux of α-phase
$\underset{\approx}{k}_\alpha$	Permeability of α-phase
m_α	Mass of α -phase
$m_{\alpha\gamma}$	Mass of γ-component of α-phase
N_{Re}	Reynolds number
n	Porosity
nw	Subscript for nonwetting phase
p	Pressure
$\underset{\sim}{q}_\alpha$	Specific discharge of α-phase
$\underset{\sim}{q}_{r\alpha}$	Specific discharge of α-phase relative to solid
$\underset{\sim}{q}^{m_\alpha}_r$	Mass weighted specific discharge of α-phase relative to solid
R	Radius of spherical REV
$S_{\alpha\beta}$	Surface area of interface between α - phase and all other phases in REV (e.g., $S_{\alpha s}$)
S_o	Surface area of sphere surrounding (U_o)
$S_{\alpha\alpha}$	Surface area of α-phase on (S_o)
$S_{o\alpha}$	Total surface area of $(U_{o\alpha})$
s	Subscript indicating solid phase
s_ν	Length measured along $\underset{\sim}{\nu}_\alpha$
$\underset{\sim}{t}'$, $\underset{\sim}{t}''$	Unit vectors in plane tangent to α-s surface
$\underset{\approx}{T}^*$	$\frac{1}{U_{o\alpha}} \int_{(S_{\alpha\alpha})} \underset{\sim}{x}\underset{\sim}{\nu}_\alpha \, dS$
T_α	Temperature of α-phase (also T_f and T_s)
$\underset{\sim}{u}$	Velocity of $(S_{\alpha\beta})$
U_o	Volume of REV
$U_{o\alpha}$	Volume of α-phase in (U_o) (also U_{os}, U_{of})
$\underset{\sim}{V}_\alpha$	Volume weighted velocity of α-phase (also $\underset{\sim}{V}_s$)
$\underset{\sim}{V}^{m_\alpha}$	Mass weighted velocity of α-phase

$\underset{\sim}{V}^{m\alpha\gamma}$	Mass weighted velocity of γ-component of α-phase
w	Subscript for wetting phase
$\underset{\sim}{x}$	Position vector of a point
$\underset{\sim}{x}_o$	Position vector of centroid of REV
z	Vertical coordinate (positive upward)
α	A phase; also as subscript indicating a phase
β	Subscript for all other phases (except α) in REV
γ	A component of α-phase
$\underset{\approx}{\delta}$	Kronecker's delta (components δ_{ij})
Δ	Characteristic distance from solid wall to interior of the phase
θ_α	Volumetric fraction of α-phase
θ_α^s	Fraction of α area on S_o (= $S_{\alpha\alpha}/S_o$)
λ_α	Thermal conductivity of α-phase (also λ_f, λ_s)
$\underset{\approx}{\lambda}_\alpha^*$	Thermal conductivity of α-phase in a porous medium
$\underset{\approx}{\Lambda}$	Thermal conductivity of a porous medium
μ_α	Dynamic viscosity of fluid α-phase
$\underset{\sim}{\nu}_\alpha$	Unit normal vector on ($S_{\alpha\beta}$) or ($S_{\alpha\alpha}$) pointing outwar
ρ_α	Mass density of α-phase
$\Sigma'_{\alpha\beta}$	$S_{\alpha s}/U_o$ = Specific area of α-phase
$\underset{\approx}{\tau}_\alpha$	Viscous stress in α-phase
$\overset{o}{(\ )}$	Deviation of ( ) from intrinsic phase average of ( ) over REV
$\overline{(\ )}$	Phase average of ( )
$\overline{(\ )}^\alpha$	Intrinsic phase average of ( )
$\tilde{(\ )}^{\alpha\beta}$	Average of ( ) over ($S_{\alpha\beta}$)
$D^{m\alpha}(\ )/Dt$	Material derivative ($\equiv \partial(\ )/\partial t + \underset{\sim}{V}^{m\alpha}.\nabla(\ )$)

PORE SCALE PHYSICAL MODELING OF TRANSPORT PHENOMENA IN POROUS MEDIA

Richard A. Dawe, Eric G. Mahers, and John K. Williams

PORE SCALE PHYSICAL MODELING OF TRANSPORT PHENOMENA IN POROUS MEDIA

Richard A. Dawe, Eric G. Mahers, and John K. Williams

Mineral Resources Engineering Department
Imperial College
London SW7 2AZ
United Kingdom

ABSTRACT

The flow of fluids through natural reservoir bodies is complicated, particularly for multiphase processes and especially if there is mass transfer. Physical modeling using visual techniques can give some of the necessary descriptions leading to the proper formulation of mathematical models for predicting reservoir performance. This chapter describes the micromodel techniques developed at Imperial College, highlighting particularly those involving pore scale events which depend on network and pore morphology.

1. INTRODUCTION

Fluid flow in porous media, with or without mass transfer, needs to be understood for many applications, including petroleum reservoir engineering, especially enhanced oil recovery, groundwater hydrology, soil science and waste water disposal. An accurate description of reservoir characteristics is therefore required on many length scales, ranging from 1-100 μm for pore-scale features, through 1-100 cm for core samples to 1-100 km for reservoir bodies; a range of 10^{11}. Such a description can only be achieved through a thorough understanding of the geological setting combined with geological, geophysical, and petrophysical data from well tests, logs and cores.

Single phase bulk fluid transfer in porous media at a continuum level is described by Darcy's law, and is fully discussed elsewhere (e.g., 3, 10, 12). Displacement processes, when there are differences in physical properties, such as viscosity, density or interfacial tension, (e.g., water displacing oil, salt water intruding into potable water) cannot be described so easily, and when the matrix itself has non-uniform properties the processes become very complex. However, physical descriptions must be sought before mathematical models can be devised and the equations solved to predict the behavior of some specific field operation be it a potential aquifer or hydrocarbon reservoir depletion plan, nuclear waste site or other application.

1.1. The Scaling Problem

Physical experiments must be performed in order to gain the understanding of the mechanisms of flow, displacement and entrapment at both the qualitative and quantitative level. Initially the objectives are to identify mechanistic processes using simple models, using an ever increasing range of fluids with diverse physical properties, and ultimately to develop scaled models representing as realistically as possible the various scaling groups relating model to prototype obtained through dimensional or inspectional analysis (3).

Clearly, caution must always be exercised to ensure that experimental conditions, such as flow rates, pressure gradients, interfacial tensions, wettability and flow regimes are similar and do not invalidate any scaled conclusions, especially as normally some scaling criteria have to be relaxed. The physical processes occurring need to be understood at the scale one lower than that required to be predicted, as well as at the actual level and possibly even one level higher, although here the properties can usually be modeled directly through the choice of equation coefficients, (an example can be found in Haldorsen and Lake (14) and Begg, Chang and Haldorsen (4)). Care must always be taken,as emphasized throughout this chapter, that the physical processes are being properly described and averaged in the scale-up.

Although it has been recognized for some time that transport through porous media depends on the microstructure of the pore space, the macroscopic effects cannot, as yet, be interpretted in terms of simple cause and effect relationships due to the microscopic events. Correlating a change in one variable with a change in another is a long way from demonstrating that one is the cause of the other. Because of such difficulties, studies on simple model systems are particularly valuable in clarifying our understanding. However, any one model can often highlight only a few components although a good model will emphasize the most important features and ignore inessential detail. Therefore a series of models are needed, each focussing on a number of key features, and at a range of length scales. Clearly the scale of detail explicit in the pore scale cannot be approached in the larger scale, and a whole hierarchy of models are needed to handle the description of phenomena on different length scales.

The motivation for our work can be summarized by the vital questions:

1) Where is the oil in the pore structure?
2) How does the oil move to the well-bore (mobilization)?
3) Why does the oil stop moving (entrapment)?
4) Can the oil be remobilized (enhanced oil recovery)?

In this chapter, some of our microscopic, pore level modeling studies (1-1000 μm) carried out at Imperial College will be outlined, (fuller details are referenced later.) Our objectives have been to gain a better understanding of the basic parameters affecting the mechanisms of displacment of immiscible and miscible systems with and without mass transfer. This is of particular relevance in our case to petroleum reservoir engineering especially for improving oil recovery. Consequently the rest of this chapter refers to oil recovery, although most of our observations are applicable to other processes. We shall describe the micromodeling techniques used (Sections 2 and 3) and give a few examples of qualitative work with emulsions (Section 4.1), a quantitative study of diffusional mass transfer processes using micromodel (Section 4.2) and live-fringe holographic methods (Section 4.3) and an indication of the effects of pore structure in displacements, such as snap-off and entrapment due to pore space morphology, and hydrodynamic instabilities due to mass transfer and capillary pressure changes (Section 5).

2. MICROSCOPIC BEHAVIOR

The pore structure, when examined under the Scanning Electron Microscope, is seen to be extremely complicated (31) as can be seen in Figure 1. The pores are microscopic and are considered to be on average no more than a few micrometers in size (perhaps 2-100 μm). If the reservoir drawdown area has a radius of some 1 km, the fluids from the outer boundaries will have passed through some tens of millions of pores. Since there are some 10^9 pores in a 1 cm^3 sample and within the pore space, oil, water and sometimes gas distribute themselves, it therefore seems pertinent to understand fully the behavior of multiphase fluids within the network system.

In multiphase systems capillary forces will dominate the fluid distributions and control fluid flow behavior with the capillary number ($V\mu/\gamma$) - the ratio of viscous to interfacial forces - being an important parameter. (Various other definitions of capillary number are used in the literature (5, 19, 20, 35).) Unfortunately even today, a piece of reservoir rock is essentially a black box when it comes to observing the detail of multiphase flow. One can measure input and output flow rates and compositions and pressure profiles along the core, but the exact distribution of fluids cannot

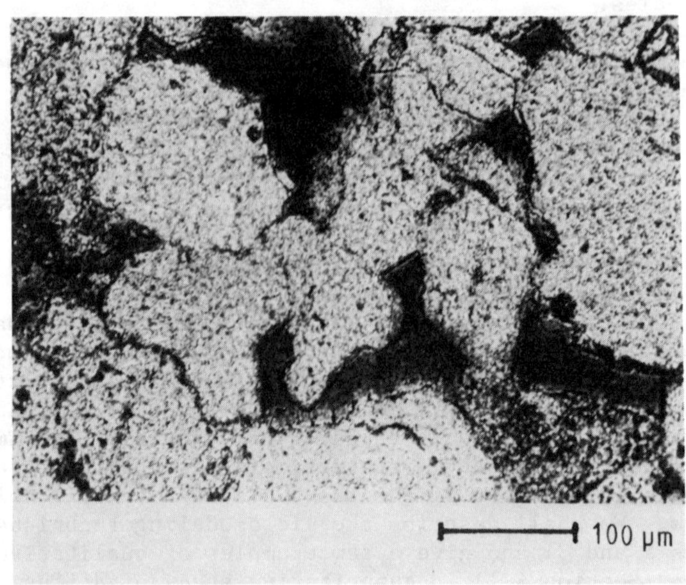

Figure 1. Micrograph of a North Sea Sandstone showing the pores as areas impregnated with dark resin. Note the irregularities in the shape.

be determined except in a few cases, (e.g., filling cores with plastics and solids and then destructive examination of them by sectioning (6) or tomography experiments using "brain scanners", (e.g., 9, 15, 16, 34), although here the resolution is still insufficient, or perhaps certain radioactive tracer methods, (2).) A fuller discussion of pore structure has been given recently by Quiblier (32). More specifically the need is to understand the mechanism of fluid flow and tracer transport in order to have a physical basis for interpolation and extrapolation (e.g. to perhaps 3 phases, oil, water, gas or a second liquid phase) of the limited experimental data. Often the measurements have to be made on small samples and the results scaled upwards towards reservoir dimensions.

It is therefore essential to model the porous system using simplified networks and fluids which imitate the reservoir fluids, so that saturation patterns can be followed, and displacement sequences understood (28, 30). The pore space is treated as an assemblage of pore segments and transport is governed by rules incorporating pore level mechanics which determine where the fluids go during a transport process and which, when suitably averaged, give rise to the calculation of the macroscopic properties. The geometry of the network and its degree of interconnection are important in determining the transport within the matrix. Individual segments of the network vary in size; those which are larger are termed "pore bodies", and these are connected by numerous smaller passageways termed "pore throats". The sequence of bodies and throats give a transport path which has a converging-diverging character. The distribution of the multiple phase within the porous matrix is usually important in determining the nature and magnitude of the transport. Here, wettability strongly affects the distribution but this itself depends on the actual fluids present in the porous medium and how they arrived there. Transport relevant to multiphase flow with mass transfer, including condensate recovery (excluding thermal effects) requires an understanding of:

1) the flow of each phase,
2) the transport of each chemical species within each phase,
3) mass transfer at the interface.

The complexity of the pore geometry makes it difficult to scale effects observed in one pore to those occurring at reservoir dimensions, but the randomness of the pore structure can sometimes allow one to statistically average over a continuous volume that is large with respect to the size of an individual pore yet small with respect to the size of the sample, and large compared to the length scale of the phenomenon being studied. This enables one to scale from the microscopic behavior to a macroscopic average (1, 3, 12).

Clearly there are also a number of size scales even at the micro-level. One can examine in detail the morphology of the clay infillings within a pore (e.g., 29); or observe diffusion, (24), or surfactant transfer mobilizing residual oil within a pore as is described later, (8), or events over a few pores, such as ganglion backflow (26), or over a few thousand pores to obtain average saturations, (27).

3. PORE LEVEL PHYSICAL MODELS - THE MICROMODEL

Micromodels are 2-D flow cells which represent idealized porous media. They have a network of flow channels and are constructed in transparent material to enable the fluids to be directly observed and recorded as displacement and mass transfer occur.

Etched network models have been used in our work because of the advantageous control over network design and pore geometry (size and shape), since these determine the interfacial curvatures and hence the interfacial forces, i.e., capillary pressure. There are several ways micromodels and their results may be used: (1) as a purely visual aid to gain insight into the physics of displacement within porous media, (2) to measure the volume average properties such as fluid saturation, permeability and dispersion coefficients, and relate these to network parameters, (3) to study pore level events, such as the mechanics of oil ganglia and fluid snap-off, in terms of local pore topology and imposed boundary conditions (velocit and pressure fields).

The first of these ways has proved to be valuable to us and is also the forerunner of any more quantitative studies. The essence of micromodeling is not only to seek the answers to questions of fluid flow, but also to pose the questions which need to be answered.

3.1. Micromodel Construction

The networks are produced by etching into silica glass or photoetching into nylon from which replicas in epoxy resin are cast. The glass models have a surface chemistry similar to that of clean sandstone and are water-wet when clean. The resin models demonstrat a mixed wettability: decane completely wets them, whereas water show a finite contact angle. For decane-water-alcohol systems the aqueous phase tends to wet the resin completely, but not spontaneous when the surfaces are initially contacted by decane.

Figure 2 shows the complete method of producing glass micromodels. It has been developed from that described by McKellar and Wardlaw (27) and involves six stages: (1) photographing a hand-drawn or computer-drawn pore network, (2) coating a glass base-plate with a photoresist, (3) projecting the photographed network design onto the casting with UV light, (4) washing away the unpolymerised (unexposed to UV) sections of the coating, (5) etching the design into the base-plate with HF, (6) sealing on a cover plate by heating in a furnace.

The epoxy resin model production procedures have been described fully previously (22,26). In our very early work we performed flow experiments directly with the etched nylon film but these were found to suffer from significant absorption of dyes and solvents which ruined the models. Consequently we now use the etchings as patterns for silicone rubber moulds, from which rigid non-absorbent epoxy resin replicas are cast. The casts accurately reproduce the microstructure of the nylon model. A flat epoxy resin film is sealed on top of the casting to produce the 2-D micromodel. Inlet and outlet ports are drilled into the model and fine tubing sealed in place with epoxy cement. The arrangement of valves helps to eliminate fluid mixing in the entry tube thereby ensuring injection of uncontaminated fluid. The model is mounted on the stage of a microscope and fluids pumped through it using microsyringe pumps at typical reservoir rates (typically less than 1 m/day). The flow mov ments can be observed through a microscope and recorded in colour on videotape or still photographs.

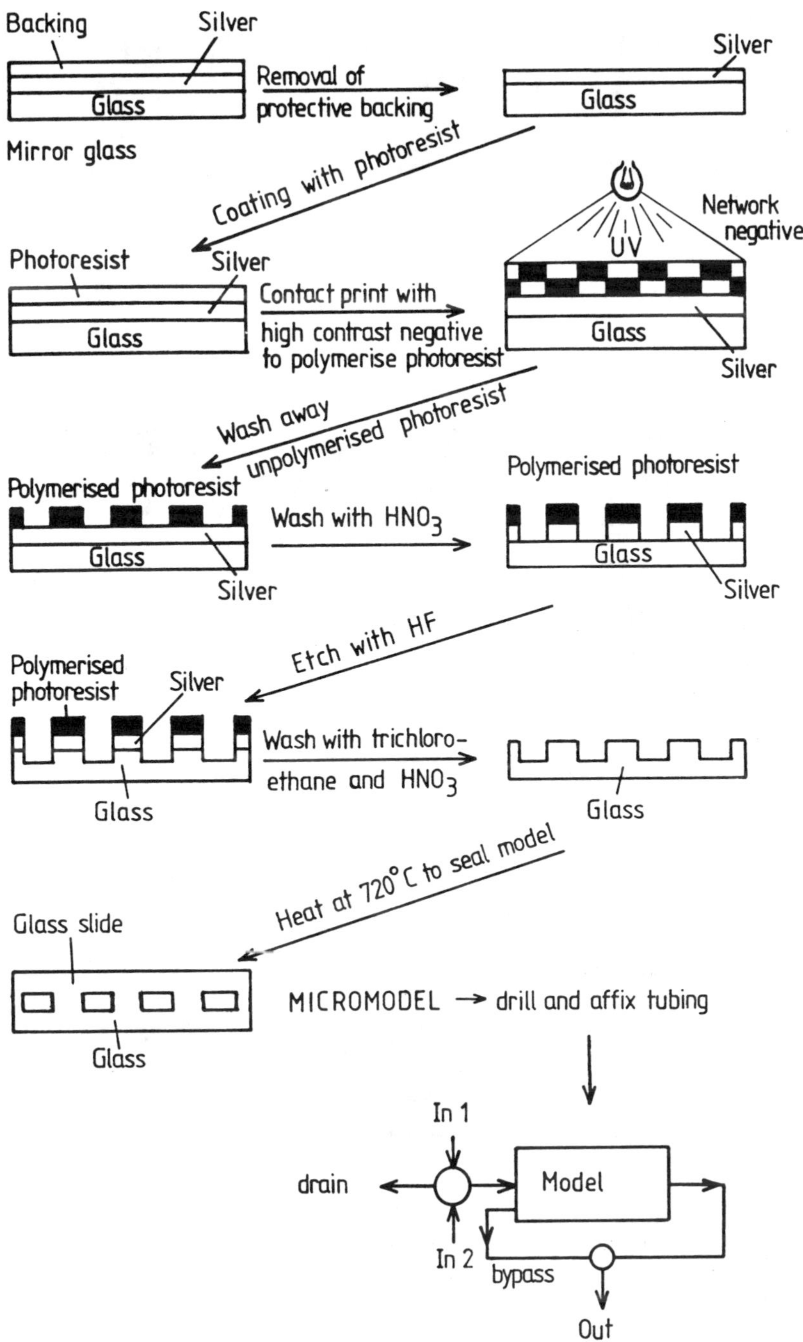

Figure 2. Preparation of glass micromodel.

(a) Hand-drawn parallel layer model with serial heterogeneities. This was designed to yield realistic and predictable relative permeability and capillary pressure functions.

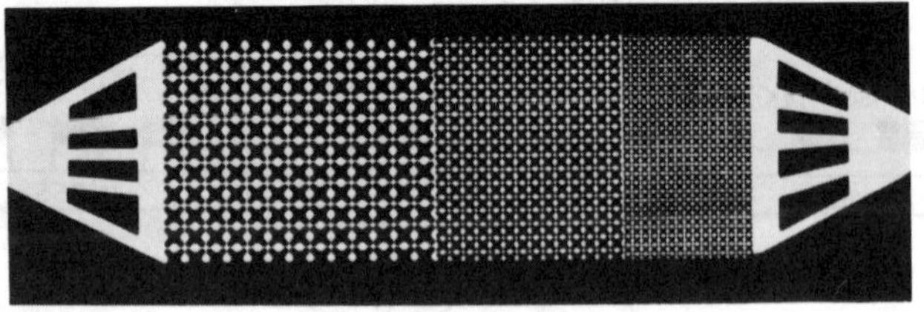

(b) Computer drawn regular network of curved channels with reducing pore throat sizes, to investigate fines movement and entrapment.

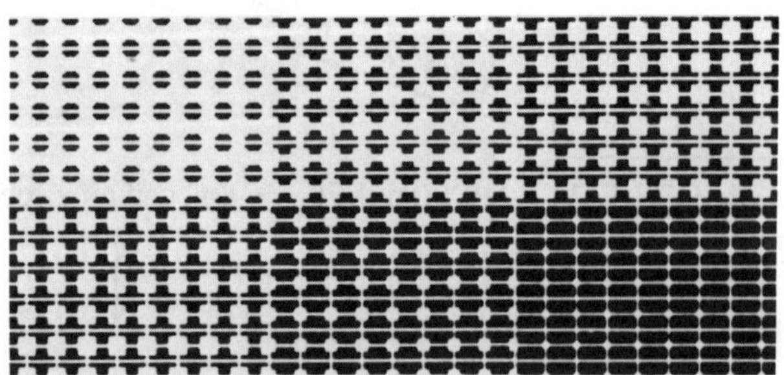

(c) Computer drawn doublet network with a variation of pore parameters and having abrupt pore necks. This was designed to demonstrate the effects of two pore sizes in parallel and in series.

Figure 3. A selection of micromodel networks.

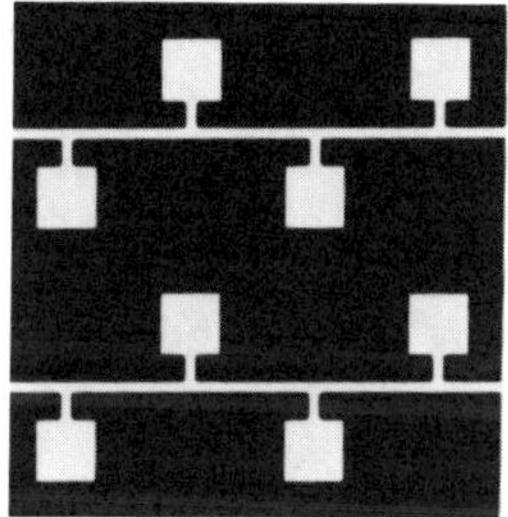

(d) Part of a computer-drawn model having dead-end pores (sometimes known as ink bottle pores) in high conductivity channels. This was designed to study the effects of diffusion into stagnant regions by holography. It is also a possible model of some carbonate rocks.

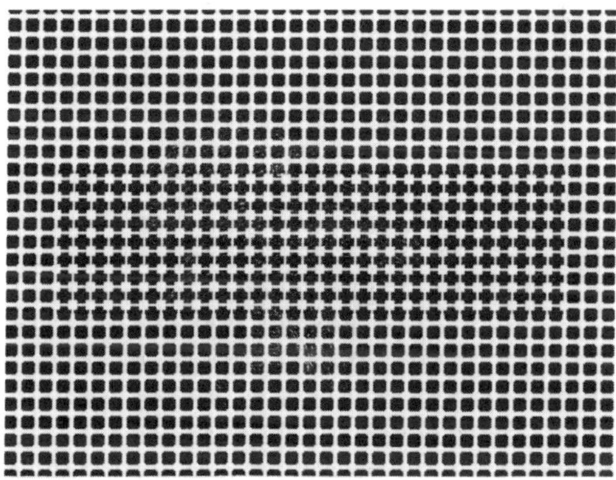

(e) Heterogeneous network to study bypassing and entrapment. The middle section has a different pore throat/body size from the surrounding matrix and therefore different capillary pressure and permeability.

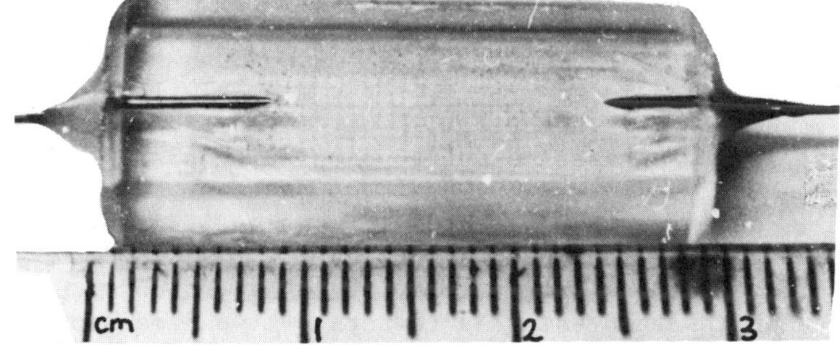

(f) A complete micromodel of the smaller variety.

Figure 3. A selection of micromodel networks (Cont.)

Sensitive pressure transducers can be attached to the micromodels to allow the fluid movements and pressure events to be related. Often one of the fluids is dyed so that it can be distinguished. Dyes can also be used to follow dispersion patterns within a single phase although they are surface active and their effects on wettability must be taken into account when analysing results.

3.2. Network Design

The photomasks used to control the etching procedures can be produced either by hand-drafting or by computer graphics and microfilm facilities, allowing both design flexibility and control over the network parameters. Figure 3 gives a number of examples. This degree of control over network parameters, such as pore body and throat size distributions and network connectivity, offers the opportunity to test theoretical models on a microscopic scale to an extent not yet possible in a real porous system and to obtain wherever necessary quantitative data to confim the principles observed. By these methods pore networks have been fabricated with well controlled geometries down to pore throat dimensions of about 15 μm.

The pore dimensions can be made realistically small to ensure that capillary forces are of the correct order of magnitude as occur in reservoir rocks (they decrease as pore size increases) and the neck to pore cavity ratios can be representative of real porous media (i.e., 1:2 to 1:10). It is important to ensure that the capillary number and other dimensionless groups are of the correct order of magnitude as found in the field. We also vary pore size and shape, pore wall irregularity and roughness to represent the various aspects of porous media which affect oil movement and entrapment. The effects of heterogeneities, where areas of oil may be bypassed, and methods of contacting and remobilising this oil can also be studied. Models incorporating random distributions of pore bodies and pore throats or defined non-random heterogeneities are now being developed with pore roughness and network heterogeneity based on a fractal approach (18, 33, 36).

The two dimensional nature of the models will affect some global aspects of the behavior, in particular the simultaneous continuity of two phases is not topologically possible in two-dimensions but is in three dimensions. However this is not a major problem if one is more interested in the local pore-level physics. The global aspects of the behavior can be understood in terms of invasion percolation theory, (e.g., 19, 35), which should help to determine which features are artefacts of two dimensions and which are not. For instance at breakthrough in two dimensions the displaced phase will no longer be continuous. Care must also be exercised when considering local grain contact effects, for instance there are difficulties in modeling "pendular" rings of wetting phase in micromodels. The shape may be totally different. Also there is the problem of any "unseen" curvature between the top and bottom of the model, and which could in some circumstances dominate the curvature seen in the plane. It is therefore erroneous to measure micromodel permeability, dispersion coefficients and fluid saturations, and to treat these as absolute quantities which can be compared directly with values derived from real, three-dimensional media. Measurements of, for instance, residual oil saturations can be valuable in a relative rather than absolute sense.

The actual shape of the pore may have some effect. For instance our glass micromodels are sealed by sintering which tends to round-the corners, whereas our resin models tend to have sharp corners. This allows the resin models to have a "groove effect" for the wetting phase to bypass residual phase or transport chemical to the interface. Such phenomena have been discussed by Lenormand and Zarcone (21) although they suggest some of the effects are due to a surface roughness factor.

Nevertheless, in spite of these limitations we feel that many microscopic aspects of fluid flow in reservoir rocks can be realistically modeled.

4. MICROMODEL STUDIES

In this and the following section a few examples will be given of our microscopic visualisation studies which demonstrate the scope of the method.

4.1. Qualitative Studies

Our initial objectives were to understand the microscopic mechanics of miscible displacements, then the immiscible displacement of water flooding, the usual method of secondary oil recovery. Following then to study the unproven but promising techniques of enhanced oil recovery, particularly those utilising low interfacial tension and miscible processes (23). Recently discovered surfactant mixtures can lower the oil/water interfacial tension some four orders of magnitude to values below 10^{-3} mNm^{-1}. In miscible processes the solvent either dissolves the oil/water interfaces on first contact or after a period of mass transfer involving diffusion and interfacial instability (Marangoni effects) and other solubilizing events. Model fluid systems are used to simulate reservoir behaviour as hydrocarbon reservoirs are at high temperatures and pressures (often near 100°C and 500 bar) experiments at real conditions are difficult and expensive. Alcohols are suitable fluids since the wide variety allows a spectrum of single to multiple-contact-miscible to immiscible systems to be studied with variations in such properties as viscosity, density, interfacial tension, solubility and diffusion coefficient.

In these studies (23,25), not only were the phase effects studied, but also the pore network geometry and its effect on capillary pressure, displacement and entrapment by varying size distribution, pore shape and throat sizes, and the connectivity of the pores. Further discussion of this work is given in Section 5. Another interesting phenomenon was observed with a study of displacement from dead end pores. We found that oil can be displaced by water moving into the pore along the wetting film. This model was preferentially water-wet.

4.2. Quantitative Studies

There are many facets of reservoir behaviour which must be studied and quantified, apart from the purely visual description, which is how micromodels have been used previously, such as

1) the effects of network and pore structure (topology, connectedness, the shape of the diverging-converging connection, roughness and irregularities of the pores,
2) the effects of shapes and sizes of the pore elements (pore bodies and throats),
3) the effects of flow rate,
4) the effects of viscosity,
5) the effects of density differences,
6) the investigation of wetting preferences of immiscible phases,
7) the study of one, two or perhaps even three mobile phases,
8) the effects of low interfacial tensions.

The phenomena become even more challenging when the system is close to a critical point, such as is found in retrograde condensate systems (11, 37) or when the phases are in the form of an emulsion (7, 8). This is an important area of study for when the oil is mobilized, emulsions are frequently formed, especially if surfactants or thermal methods are being used. Figure 4 shows a still photograph taken from a video-taped sequence of an oil-in-water emulsion (interfacial tension about 10^{-2} mNm^{-1}) flowing through a curved pore channel network. This study demonstrated beyond doubt that the interconnectivity of the pores affects the flowing properties of the emulsion. It showed that the previous work on emulsion movement through single straight capillary tubes may not be readily extended to porous media. We have shown that the emulsion droplets do not all move at the same velocity; some slip or stick against the pore walls whereas those in the centre of the pore move far more readily. This affects the rheology as well as the creaming and coalescence properties of the emulsions (7).

Figure 5 shows photographs taken from the video monitor showing

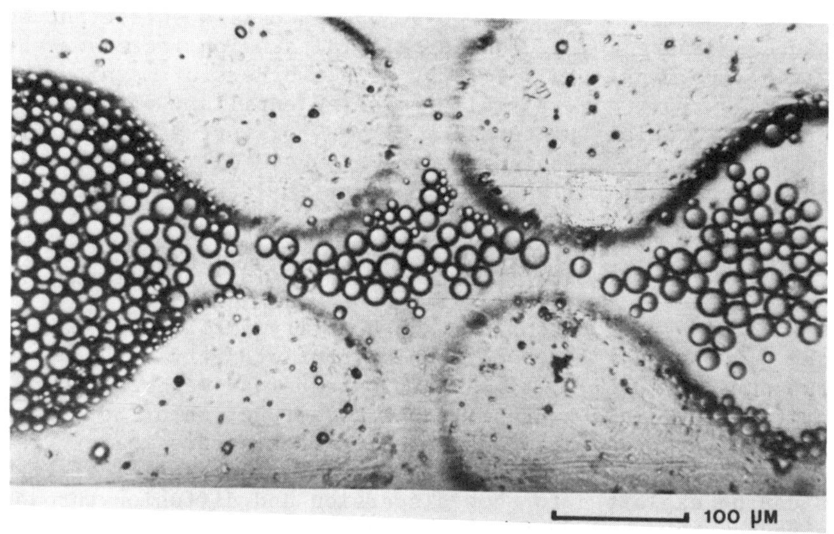

Figure 4. Oil-in-water emulsion flowing (left to right) within a curved pore channel.

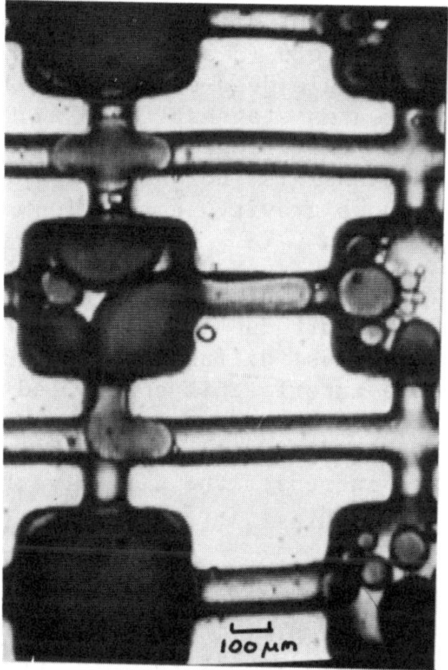

Figure 5a. Detail of square-grid model after a surfactant solution (white) has begun to displace oil (black) by breaking it into smaller droplets (flow is left to right).

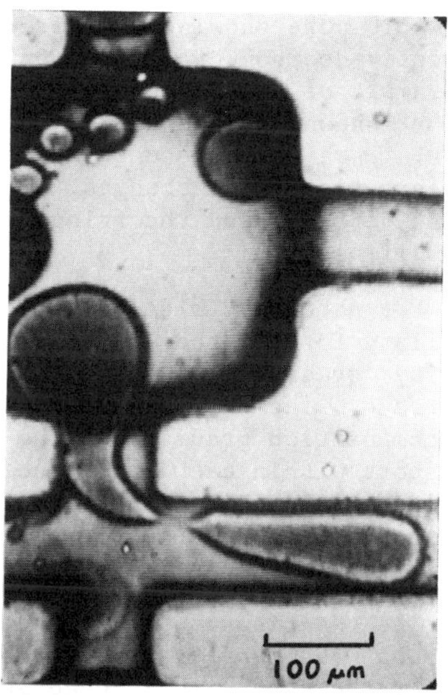

Figure 5b. Close-up of a moving oil droplet at the instant of snap-off from a trapped oil blob. The elongated forms of the oil-aqueous surfactant interfaces are due to the very low interfacial tension (about 10^{-3} mNm^{-1}) between the oil and surfactant.

the interfacial shapes that can be obtained as a surfactant solution contacts entrapped oil. The interfacial tension increases the ratio of viscous to capillary forces by some 4 orders of magnitude because there is now only a very small resisting capillary pressure. The trapped oil droplet's interfaces can now distort and manoeuvre through the pores; this can lead to improved oil recovery. Interfacial forces can still however be very influential in such low tension systems.

4.3. Mass Transfer Quantification Using Holographic Interferometry

A novel approach to study and quantify diffusional mass transfer at the pore scale is by holographic interferometry using a micromodel as the porous medium, (24). By this technique we are able to follow quantitatively fluid concentrations as a function of both position and time. The speed at which mass transfer occurs is very influential in the recovery of oil by chemical agents. Absorption of light by dyes can be exploited to show dispersion and diffusion through the pore system but the quantitative effects of the tracers may be different to those of the fluids. By using live fringe holographic interferometry, which exploits the refractive index differences of the liquids it is possible to map lines of constant composition over the whole pore showing the direction and rate of diffusion and any convective component of mass transfer. Figure 6 (a and b) illustrates an example of diffusion effects within a single dead-end pore in the network shown in Figure 3d. The model creates stagnant fluid in the pores. Here we can see the development of the fringe pattern as the concentration field changes. The flow channel is inclined at 45° to the vertical and the fringe pattern shows gravity segregation plus diffusion.

For partially miscible systems, where the interface is held by capillary forces, the convective mass transport due to unstable density gradients is limited because mass must diffuse across the interface. The diffusive flux across the interface is controlled by the composition gradient at the interface in both phases. Convective transport within each phase due to buoyancy forces increases the gradients at the interface and therefore increases the diffusive flux across it. Thus the mass transfer coefficient is larger for the gravity unstable case.

The contribution of convective transport will become even more apparent when mass transfer occurs into systems of interconnected pores rather than single dead-end pores. As a result calculations of mass transfer using only diffusion coefficients measured in bulk fluids, and pore size parameters will be erroneous when density effects are present.

These experiments demonstrate the possible importance of the influence of gravity on the microscopic mass transfer between fluids od different density, previously always assumed to be

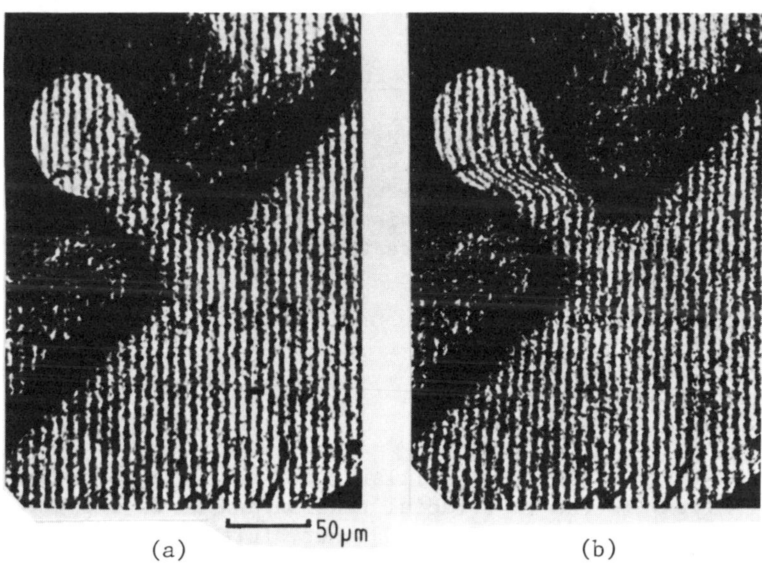

Figure 6. Live fringe holographic interferogram in a single dead-end pore; the model was held vertical with the channels inclined at 45°. Flow within the wide channel is from bottom left to top right.
a. Model filled with oil.

b. After introducing solvent of greater density and refractive index than the oil; the displacement is complete in the channel while the curved fringes indicate a diffusion zone within the dead-end pore.

negligible. Full details have been given in Mahers (22) and Mahers and Dawe (24).

5. CAPILLARY PRESSURE, MASS TRANSFER AND HYDRODYNAMIC INSTABILITY IN DISPLACEMENT

In this section we show an example of the quantitative scope of the micromodel demonstrating the effects of pore geometry (23, 25, 26).

Hydrodynamic instability during displacement processes can occur due to dynamic capillary pressure phenomena where there is mass transfer across interfaces. It is most pronounced when the residual oil distribution is not uniform and relatively large volumes of "continuous oil" are present: this can occur for instance in a heterogenous zone of the pore network. The chemical solute may only contact part of the oil/water interface, thereby setting up concentration gradients across the interface.

5.1. Hydrodynamic Instability - Haines Jumps - Drainage Displacement

Figure 7 illustrates the displacement of water by oil in a water wet pore. The oil-water interface in the right hand pore is a head meniscus, and any pore constrictions are termed pore throats or necks. In order for the head meniscus to pass through the throat, the capillary pressure must be greater than the threshold value of the throat:

$$P_c = P_o - P_w = \frac{2\gamma}{R_T} \tag{5.1}$$

where P_c, P_o and P_w are the capillary, oil and water pressures respectively. γ is the interfacial tension and R_T is the throat radius, defined as the harmonic mean curvature of the largest ellipse that will pass through the pore neck. The pore body radius, R_B, is defined as the harmonic mean radius of curvature of the largest ellipsoid that will fit into the pore.

As the meniscus passes through the throat and enters the pore, its interfacial curvature decreases (R_2 increasing), reducing the capillary pressure and thereby lowering the pressure in the oil in the right hand pore. The water pressure at the meniscus is increased and the resulting pressure gradients accelerate the meniscus, termed a *Haines Jump*. When the radius of curvature R_2 equals R_B and more oil then enters the pore, R_2 must begin to decrease and the capillary pressure correspondingly rises. The meniscus becomes stationary when the capillary pressure equals the local difference in pressure between the two phases, and further displacement produces an increase in capillary pressure.

5.2. Imbibition Displacement

Imbibition is illustrated by Figure 8. As the interface is displaced from point A to point B the interfacial curvature increases, increasing the capillary pressure and thus increasing the pressure gradients in both phases near the meniscus, so accelerating the meniscus; we now have a Haines jump in imbibition mode. As the meniscus passes point B, the capillary pressure will decrease, and meniscus will become stationary when the pressure difference equals the capillary pressure.

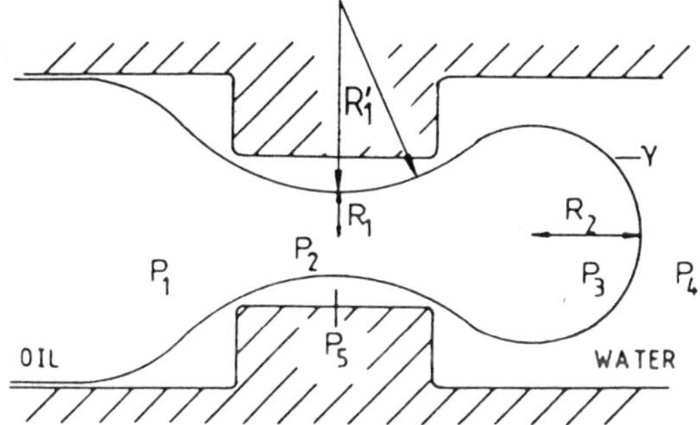

Figure 7. Oil invading a water-wet pore.

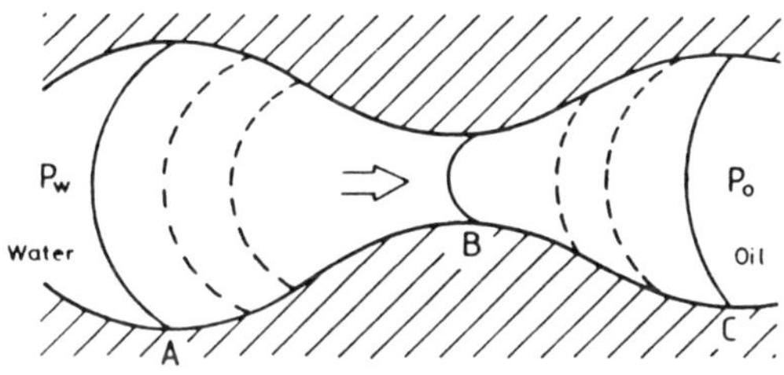

Figure 8. Haines jump in imbibition mode.

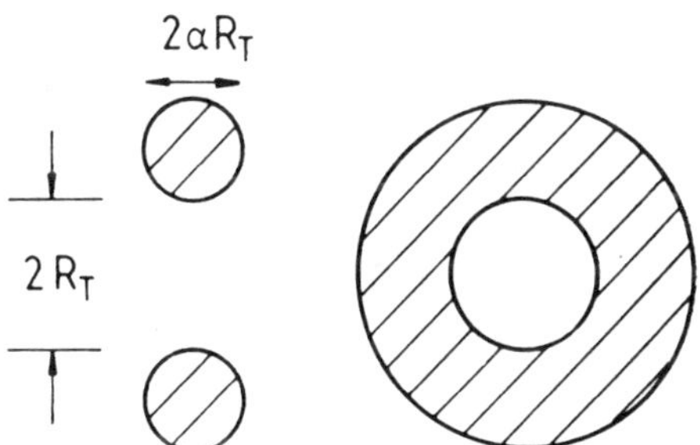

Figure 9. Toroidal pore throat.

5.3. Interfacial Instability - Snap-off

Let us now consider the stability of an oil-water interface during displacement, and further examine the configuration shown in Figure 7. If we let the pressure in the non-wetting oil phase be approximately constant then

$$P_1 \cong P_2 \cong P_3 \tag{5.2}$$

Eq. (5.1) gives $P_w = P_o - P_c$, and therefore,

$$P_4 = P_3 - \frac{2\gamma}{R_2} \tag{5.3}$$

$$P_5 = P_2 - \frac{\gamma}{R_1} + \frac{\gamma}{R_1'} = P_2 - \frac{2\gamma}{AR_T} \tag{5.4}$$

where

$$\frac{1}{A} = \frac{R_T}{2} \left(\frac{1}{R_1} - \frac{1}{R_1'} \right)$$

Subtracting eqn. (5.3) from eqn. (5.4) gives

$$P_5 - P_4 = 2\gamma \left(\frac{1}{R_2} - \frac{1}{AR_T} \right) \tag{5.5}$$

since $P_2 \cong P_3$. The oil filament within the pore throat will become unstable if $P_5 - P_4 < 0$, because the water will flow into the throat. Thus, if

$$\frac{1}{R_2} - \frac{1}{AR_T} < 0, \quad ; R_2 > AR_T, \tag{5.6}$$

snap-off may occur. The limiting value of R_2 is R_B, the pore body radius. Therefore for snap-off,

$$R_B / R_T > A. \tag{5.7}$$

R_B/R_T is the pore aspect ratio, and therefore A can be defined as the critical pore aspect ratio. For a long, straight throat, $R_1 \cong R_T$ and $R_1' = \infty$, so A = 2. For the toroidal throat shape shown in Figure 9,

$$\frac{2}{AR_T} = \frac{1}{R_T} - \frac{1}{\alpha R_T} , \tag{5.8}$$

and

$$A = 2\alpha\ /(\alpha - 1)$$

where α is the throat length to width ratio. If $\alpha = 2$ then $A = 4$; also $A \to 2$ as $\alpha = \infty$ and $A \to \infty$ as $\alpha \to 1$.

The snap-off analysis is valid for both drainage and imbibition displacements, however for drainage, snap-off occurs during the rapid Haines jump and therefore does not always take place. For imbibition, snap-off occurs during the slow displacement step, although at very high flow rates there may again be insufficient time for snap-off.

Further details of displacement mechanisms and residual oil formation within networks in drainage and imbibition modes and the ganglion stability towards remobilization are given in Mahers (22) and Mahers and Dawe (23, 25).

5.4. Instabilities due to Mass Transfer and Interfacial Tension Non-uniformities

If we consider a large discrete volume of oil occupying several pores, which is anistropically contacted by a solute which changes the interfacial tension, then the interfacial curvature must change to maintain constant capillary pressure, as shown in Figure 10. In Figure 10a, the fluid boundary layer surrounding the oil is water and the interfacial tension (IFT) will be constant and equal to γ. The capillary pressure is constant and therefore the interfacial curvature is also constant. The end lobe of the oil is hemispherical, and if assumed to be as shown, the capillary pressure is:

$$P_c = 2\gamma/R_B \tag{5.9}$$

If the IFT is changed only in the end pore by $\delta\gamma$, then, to

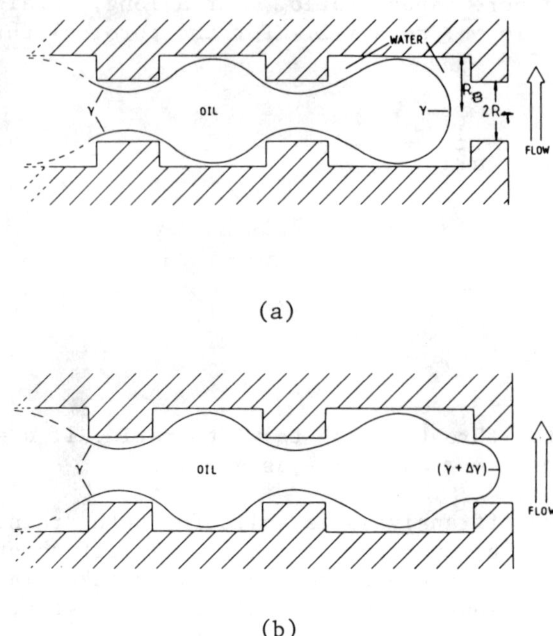

Figure 10. Instability due to non-uniform interfacial tension.

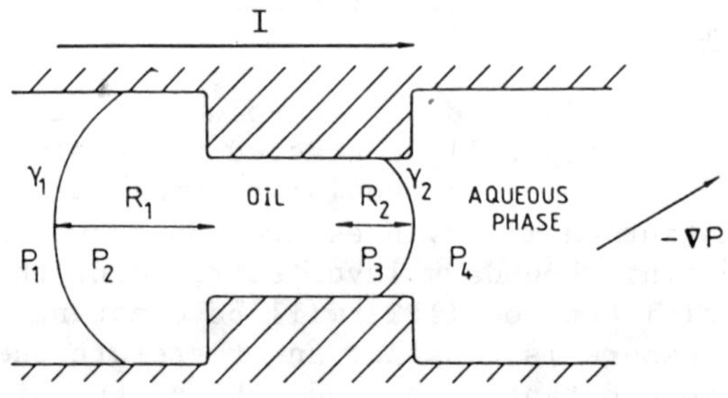

Figure 11. Solute contacting ganglion.

maintain constant capillary pressure, the curvature must increase. The limiting curvature is when the oil invades the pore throat (Figure 10b) and:

$$P_c = 2(\gamma + \delta\gamma)/R_T \tag{5.10}$$

If P_c is maintained constant, i.e., stable structure, then Eqs. (5.9) and (5.10) may be equated and

$$\gamma/R_B = (\gamma + \delta\gamma)/R_T \tag{5.11}$$

Thus the fractional change required in the interfacial tension for invasion of the next pore is:

$$- \delta\gamma/\gamma > 1 - R_T/R_B \tag{5.12}$$

To conserve mass when invasion of the pore occurs, oil must retract elsewhere. This may be from adjacent pores, where the interfacial tension may also be changing. Retraction will increase interfacial curvature, thereby maintaining capillary pressure. Although this mechanism depends on the structure of the immobile phase and the local velocity field. It is a feasible method for mobilization of some of the bypassed oil.

Figure 11 illustrates hydrodynamic instability of relatively small residual oil ganglia powered by interfacial tension gradients. This work was described by Mahers, Wright and Dawe (26) with illustrations of more simple oil-water displacements. The IFT gradients are due to the concentration gradients of the solvent, propan-1-ol, in the continuous aqueous phase. The ganglia moved in a direction opposite to the direction of flow, i.e., in the direction of the concentration gradient.

From Figure 11,

$$P_3 - P_2 = \nabla P.I + 2\gamma_2/R_2 - 2\gamma_1/R_1 \tag{5.13}$$

if,

$$P_3 - P_2 = 0, \tag{5.14}$$

the ganglion will be immobile, but if,

$$P_3 - P_2 < 0, \tag{5.15}$$

the ganglion will move forward, and if,

$$P_3 - P_2 > 0, \tag{5.16}$$

the ganglion will move backward. If the ganglion is initially immobile and $\gamma_1 = \gamma_2 = \gamma$, then from equation (5.13),

$$2\gamma(1/R_1 - 1/R_2) = \nabla P.\underset{\sim}{I} \tag{5.17}$$

$\underset{\sim}{I}$ is the vector length of the ganglion. If the interface is subsequently contacted by solute, then, substituting for $\nabla P.\underset{\sim}{I}$ from Eq. (5.17), Eq. (5.13) becomes:

$$P_3 - P_2 = 2\gamma(1/R_1 - 1/R_2) + 2\gamma_2/R_2 - 2\gamma_1/R_1 \tag{5.18}$$

if $\gamma_2 < \gamma_1$, the ganglion moves backwards until the interfacial curvature changes sufficiently to balance the IFT effect. If interface 1 is then at a pore throat, assuming the contact angle between the fluid interface and the pore surface is zero, and interface 2 has a radius of curvature equal to R_1, then:

$$P_3 - P_2 = 2\gamma(1/R_1 - 1/R_2) + 2\gamma_2/R_1 - 2\gamma_1/R_2 \tag{5.19}$$

For further backflow, the radius of curvature of interface 1 must be equal to R_T and from Eqs. (5.13) and (5.19),

$$\gamma(1/R_1 - 1/R_2) > \gamma_1/R_T - \gamma_2/R_1 \tag{5.20}$$

If we let $\gamma_2 = \gamma_1 + \delta\gamma$, then Eq. (5.20) gives:

$$\delta\gamma/\gamma > (R_1/R_2 - 1) + (R_1/R_T - 1)\gamma_1/\gamma \tag{5.21}$$

If initially $R_2 = R_T$, then,

$$\delta\gamma/\gamma > (1 + \gamma_1/\gamma).(R_1/R_T - 1) \qquad (5.22)$$

The maximum value for $\delta\gamma/\gamma$ is 1, with $\gamma_1 = 0$ and $\gamma_2 = \gamma$, then Eq. (5.22) reduces to:

$$R_1/R_T < 2 \qquad (5.23)$$

If initially $R_1 = R_B$, then the maximum aspect ratio for backflow would be $R_B/R_T = 2$. However, obviously R_1 can be less than R_B and R_2 greater than R_T, thus allowing backflow at larger aspect ratios.

This programme of work shows that oil droplets will move along the local pressure gradient, which may not be in the general flow direction and is created by non-uniform interfacial tensions. We have demonstrated that pore sizes and shapes, especially the pore/body aspect ratio and network geometries are important and snap-off and residual oil formation are partially controlled by these factors.

6. CONCLUDING REMARKS

It is clear that micromodel studies are a very useful and necessary tool in identifying and describing the microscopic mechanisms controlling the various transport phenomena within the pore structure and in assessing the influence of pore space morphology. This paper has given just a few of our studies of imbibition and drainage displacements, residual oil formation, emulsion flow and mass transfer effects on the pore scale. (There are many more projects underway). The observations point out clearly that many of the descriptions and mechanistic assumptions previously accepted are too simple. The control over network parameters and design flexibility now available in our micromodel manufacture enable critical studies of the effects of pore space morphology and matrix-fluid interactions such as wettability on these processes. For instance, as shown in Section 5, the capillary pressure and the pore geometry, especially the pore/throat aspect ratio play a major role in the physics of immiscible displacement. For low aspect ratios snap-off processes do not occur whereas in high aspect ratio networks they do, and residual oil is found in most pores as small ganglia stretching over only one or two pore bodies.

The quantitative experiments with emulsions and condensate fluids mentioned in Section 4.2 are demonstrating clearly the importance of the influence of the pore space structure on the rheological, creaming and coalescence properties of the fluid mixture. Previous studies of emulsion movements through straight capillary tubes are totally inadequate for modeling the behaviour in real porous media. The holographic interferometry techniques used in the studies to quantify mass transfer on the pore scale, the first time such techniques have been used on such a microscopic scale, show that buoyancy forces play a greater role in the transport at the pore scale than previously thought. Also we find that dynamic and local non-equilibrium interfacial forces are very important in remobilising residual phases.

However, although giving much information and occasionally beautiful demonstrations,it must always be remembered that micromodels must not be considered in isolation. The problems associated with the two-dimensional nature of the models must be confronted and methods of scaling-up the microscopic observations, including the changes of the balance of the forces at the different scale lengths, must be developed to provide the macroscopic descriptions sought by industry. This will be achieved by pursuing micromodel studies in conjuction with larger scale models, laboratory experiments and computer studies which model other key features. There are of course a multitude of different and essential factors to be examined.

Hopefully this intimate description of the rock-fluid behaviour provided by our and other micromodel studies will be integrated with the other approaches to yield a more satisfying picture of transport processes in porous media.

7. ACKNOWLEDGEMENTS

We thank all who have helped in this successful development of micromodel work, and J. Leney for secretarial assistance. We are grateful to SERC, British Gas (LRS), British Petroleum (Sunbury) and U.K. DoE for their support.

8. REFERENCES

1. Bachmat, Y., and Bear, J., "On the Concept and Size of an REV," *Advances in Transport Phenomena in Porous Media*, J. Bear and M. Y. Corapcioglu, eds., Martinus Nijhoff Publishers BV, The Hague, The Netherlands, 1986.
2. Bailey, N. A., Rowland, P. R., and Robinson, D. P., "Nuclear Measurements of Fluid Saturations in EOR Flood Experiments," *Proceedings of the European Symposium on Enhanced Oil Recovery*, F. J. Fayers, ed., Elsevier,Bournemouth,U.K., 1981, pp. 483-98.

3. Bear, J., *Dynamics of Fluids in Porous Media,* American Elsevier Publishing Co., Inc., New York, N.Y., 1972.
4. Begg, S., Chang, D.M., and Haldorsen, H. H., "A Simple Statistical Method for Calculating the Effective Vertical Permeability of a Reservoir Containing Discontinuous Shales," presented at the, 1985, Society of Petroleum Engineers 60th Annual Technical Conference, held at Las Vegas, Nevada (Preprint SPE 14271).
5. Chatzis, I., and Morrow, N. R., "Correlation of Capillary Number Relationships for Sandstone," *Society of Petroleum Engineers Journal,* Vol. 24, No. 5, Oct., 1984, pp. 555 - 562.
6. Craig, F. F., "The Reservoir Engineering Aspects of Waterflooding," *Monograph 3,* Society of Petroleum Engineers, Dallas, 1971, p. 16.
7. Coucoulas, L., "Some Interfacial and Rheological Properties of Emulsions Influencing Flow in Porous Media," thesis presented to the University of London (Imperial College), at London, U.K., in 1986, in partial fulfillment of the requirements for the degree of Doctor of Philosophy.
8. Coucoulas, L., and Dawe, R. A., *Emulsion Flow in Porous Media-Commentary to Micromodel Demonstration Tape,* Imperial College, London, Apr., 1983, p. 21, (available from Author).
9. Cromwell, V., Kortum, D. J., and Bradley, D.J., "The Use of a Medical Computer Tomography (CT) System to Observe Multiphase Flow in Porous Media," presented at the, 1984, Society of Petroleum Engineers 59th Annual Technical Meeting, held at Houston, Texas (Preprint SPE 13098).
10. Dake, L.P., *Fundamentals of Reservoir Engineering,* Elsevier Scientific Publishing Co., Amsterdam, 1978, pp.443.
11. Dawe, R. A., "Condensate Fluid Behaviour in Porous Media-The Microscale," presented in the October, 1984, at A One Day Seminar on Condensate Reservoir Studies at British Gas, held at London Research Station.
12. Dullien, F. A. L., *Porous Media:Fluid Transport and Pore Structure,* Academic Press, New York, N.Y., 1979,
13. Greenkorn, R. A., *Flow Phenomena in Porous Media: Fundamentals and Applications in Petroleum, Water and Food Production,* Marcel Dekker, New York, N. Y., 1984.; also "Steady Flow Through Porous Media", *American Institute of Chemical Engineers Journal,* Vol. 27, 1981, pp. 529-545.
14. Haldorsen, H.H., and Lake, L.W., "A New Approach to Shale Management in Field Scale Models," *Society of Petroleum Engineers Journal,* Vol. 24, No. 4, Aug., 1984, pp. 447-452.
15. Honarpour, M. M., Cromwell, V., Satchwell, R., and Hutton, D., "Reservoir Rock Descriptions Using Computerized Tomography (CT) Scannings," presented at the 1985, Society of Petroleum Engineers 60th Annual Technical Conference, held at Las Vegas, Nevada (Preprint SPE 14272).

16. Hove, A., Ringen, J.K. and Read, P. A., "Visualisation of Laboratory Corefloods with the Aid of Computerized Tomography of X-rays," presented at the 1985 Society of Petroleum Engineers 55th California Meeting (Preprint 13654).
17. Jennings, D.A., and Dawe, R. A., "Two Dimensional Steady State Relative Permeabilities - An Experimental Study by Micromodel," (in preparation).
18. Katz, A. J., and Thompson, A. H., "Fractal Sandstone Pores: Implications for Conductivity and Pore Formation," *Physical Review Letters*, Vol. 54, No. 12, 1985, pp. 1325 - 1329.
19. Koplik, J., Wilkinson, D., and Willelmsen, J. F., *Percolation and Capillary Fluid Displacement*, Lecture Notes in Mathematics, No. 1035, Hughes, B.O. and Ninham, B.W., eds., Springer-Verlag, Berlin, 1983, pp. 169-183.
20. Larson, R. G., Davis, H. T., and Scriven, L. E., "Displacement of Residual Non-Wetting Fluid from Porous Media," *Chemical Engineering Science*, Vol. 36, 1981, pp. 75-85.
21. Lenormand, R., and Zarcone, C., "Role of Roughness at Edges During Imbibition in Square Capillaries," presented at the 1984, Society of Petroleum Engineers 59th Annual Meeting, held at Houston, Texas (Preprint SPE 13264).
22. Mahers, E.G., "Mass Transfer and Interfacial Phenomena in Oil Recovery," thesis presented to the University of London (Imperial College), at London, in 1983 in partial fulfillment of the requirements for the degree of Doctor of Philosophy.
23. Mahers, E. G., and Dawe, R.A., "The Role of Diffusion and Mass Transfer Phenomena in the Mobilisation of Oil During Miscible Displacement," presented at the 1982, European Symposium on Enhanced Oil Recovery, Editions Technip., held at Paris, France, pp. 279-288.
24. Mahers, E. G., and Dawe, R. A., "Quantification of Diffusion Inside Porous Media for Enhanced Oil Recovery Processes by Micromodel and Holography," presented at the 1984, SPE/DOE, Fourth Symposium on Enhanced Oil Recovery, held at Tulsa, Oklahama (Preprint SPE 12679); revised as "Gravity Effects During Diffusional Mass Transfer at the Pore Scale," *Society of Petroleum Engineers, Formation Evaluation Journal*, April 1986.
25. Mahers, E.G., and Dawe, R. A., "Visualisation of Microscopic Displacement Processes Within Porous Media in Enhanced Oil Recovery - Capillary Pressure Effects," presented at the April, 1985, AGIP 3rd European Meeting on Improved Oil Recovery, held at Rome, Italy.
26. Mahers, E. G., Wright, R. J., and Dawe, R. A., "Visualisation of the Behaviour of Enhanced Oil Recovery Reagents in Displacements in Porous Media," presented at the 1981, European Symposium on Enhanced Oil Recovery, held at Bournemouth, U.K., F. J. Fayers, ed., Elsevier, pp. 511-526.

27. McKellar, M., and Wardlaw, N. C., "A Method of Making Two-Dimensional Glass Micromodels of Pore Systems," *Journal of Canadian Petroleum Technology*, Vol. 21, No. 4, 1982, pp. 39-41.
28. Mohanty, K. K., and Salter, S. J., "Advances in Pore Level Modeling of Flow Through Porous Media," presented at the 1983, AIChE Annual Fall Meeting, held at Washington, D.C.
29. Neasham, J. W., "The Morphology of Dispersed Clays in Sandstone Reservoirs and its Effect on Sandstone Shaliness. Pore Space and Fluid Flow Properties," presented at the 1977, Society of Petroleum Engineers 52nd Annual Fall Meeting, held at Denver, Colorado (Preprint SPE 6858)
30. Payatakes, A.C., and Dias, M. M., "Immiscible Microdisplacement and Ganglion Dynamics in Porous Media," *Reviews in Chemical Engineering*, Vol. 2, No. 2, April 1984, pp. 85-174.
31. Pittman, E.D., "The Pore Geometries of Reservoir Rocks," *Proceedings of Physics and Chemistry of Porous Media*, American Institute of Physics, D.L. Johnson and P.N. Sen, eds., Conf. Proc. No. 107, 1984, pp. 19.
32. Quiblier, J. A., "A New Three Dimensional Modeling Technique for Studying Porous Media," *Journal of Colloid and Interface Science*, Vol. 98, No. 1., 1984, pp. 84-102.
33. Stanley, H. E., "Application of Fractal Concepts to Polymer Statistics and to Anomalous Transport in Randomly Porous Media," *Journal of Statistical Physics*, Vol. 36, No. 5, 1984, pp. 843-860.
34. Wang, S. Y., Ayral, S., Castellana, F. S., and Gryte, C. C., "Reconstruction of Oil Saturation Distribution Histories During Immiscible Liquid-Liquid Displacement by Computer Aided Tomography," *American Institute of Chemical Engineering Journal*, Vol. 30, No. 4, 1984, pp. 642-649.
35. Wilkinson, D., "Percolation Model of Immiscible Displacement in the Presence of Buoyancy Forces," *Physical Review A*, Vol. 30, No. 1, July 1984, pp. 520-531.
36. Williams, J. K., and Dawe, R. A., "Physical Modeling of Porous Media - Networks, Fractals and Pores," *Transport in Porous Media*, in press, 1986.
37. Williams, J. K., and Dawe, R. A., "Critical Behavior of Phase Separating Mixtures in Porous Media, *Journal of Colloid and Interface Science*, submitted (1986).

9. LIST OF SYMBOLS

A	Critical pore aspect ratio
g	Acceleration due to gravity (ms^{-2})
$\underset{\sim}{l}$	Vector length of ganglion (m)
N_c	$\mu V/\gamma$ - Capillary number
P_c	Capillary pressure (Pa)
P_o	Pressure in oil phase (Pa)
P_w	Pressure in aqueous phase (Pa)

R_1, R_2	Radii of curvature of fluid interfaces (m)
R_B	Pore body radius
R_T	Pore throat radius (m)
R_B/R_T	Pore aspect ratio
V	Average interstitial velocity (ms^{-1})
α	Pore throat length to width ratio for toroidal throat shape
γ	Interfacial tension (Nm^{-1})
μ	Fluid viscosity (Pa.s)

NATURAL CONVECTION IN POROUS MEDIA

Serge Bories

NATURAL CONVECTION IN POROUS MEDIA

Serge Bories

Institut de Mécanique des Fluides de l'Institut National
Polytechnique de Toulouse - Laboratoire Associé au CNRS N° 0005
2, rue Charles Camichel
31071 Toulouse Cedex, France

ABSTRACT

When the temperature of the saturating fluid phase in a porous medium is not uniform, some flows induced by buoyancy effects may occur. Commonly called free or natural convective movements, these flows depend on density differences due to temperature gradients and boundary conditions. Generally speaking, convective movements which tend to homogenize the whole fluid volume where they take place have two main effects: produce a non-uniform insitu temperature distribution characterized by hot and cold zones, and increase the overall heat transfer.

Due to its numerous applications in geophysics and energy-related engineering problems, natural convection in porous media has been receiving increased interest over the last few decades (1,2).

In this review, we deal mainly with the presentation of fundamental results obtained through the study of this phenomena in dispersed saturated porous media. Beginning with the formulation of basic equations and boundary conditions, we then successively review:

- first, the results concerning natural convection in homogeneous isotropic porous layers of wide lateral extent in horizontal or inclined positions,
- second, the studies on natural convection in confined porous media, i.e., when the lateral extent of the layer is of the same order of magnitude so that the thickness and the lateral thermal boundary effects are taken into account,
- and finally, the problems related to natural convection in more complex configurations, such as anisotropic porous layers or porous layers saturated by a fluid of non-constant properties.

1. INTRODUCTION: THE DIFFERENTIAL EQUATIONS AND BOUNDARY CONDITIONS

The analysis of flow and heat transfer is usually based on the transport equations resulting from the differential balance laws. Prediction of global effects such as flow resistance or heat flux from a given object requires detailed information of the surrounding velocity and temperature fields. For a continuous medium, this information is extracted from the solution of the associated microscopic transport equations subject to pertinent boundary conditions.

When a flow through a complex structure such as a porous medium is involved, these local or microscopic equations are generally still valid within the pores. However, the geometric complexity of the internal solid surfaces that bound the flow domain inside the porous medium prevents general solution of the detailed velocity and temperature field.

To overcome these difficulties, physical phenomena in porous media are generally described by "macroscopic" equations valid at the level of a block of porous medium: the Representative Elementary Volume (REV) containing many pores. "Macroscopic" equations are either established using "a priori", an equivalence between the heterogeneous porous medium and a fictitious continuum, or rigorously derived from microscopic equations by means of a volume averaging technique (3,4,5,6,7). These equations are assumed to be representative on the REV of average values of microscopic quantities.

Such is the case, for instance, of the two most frequently used quantities, the porosity ε and the filtration velocity V. These quantities are the mean values in the REV of parameters that are different from zero only in the pore space, and respectively equal to 1 and to the local microscopic velocity.

As far as the thermal behavior of a porous medium is concerned, having a given thermal and hydrodynamic state with a moving or motionless fluid phase, for any geometrical point and its associate representative elementary volume, we can define two average temperatures, T_s for the solid phase and T_f for the fluid one. T_s and T_f characterize the thermal state of each phase in the same elementary volume. In the mathematical modeling for the heat transfer, two alternate methods are used, depending on the difference between T_s and T_f (8).

In the first method, the difference $T_s - T_f$ is assumed to be negligible, and the thermal behavior is described by a single equation for the average temperature $T = T_s = T_f$.

This approach, which is the most commonly used, is valid when the flow velocity is not too high, and if both phases, solid and fluid, are well dispersed.

The second method applies when it is not possible to assume that $T_s - T_f$ is negligible. Then it is necessary to distinguish the two phases and to explicitly define the interphase heat transfer. The medium is considered as equivalent to two continua, and two equations are used.

The "dividing line" between the two models has been recently explored (9). This has led to a series of constraints that must be satisfied if the homogeneous model is to be used with confidence.

1.1. Basic Equations

Due to the general complexity of heat transfer phenomena in porous media, most studies are based on simplified mathematical model in which it is assumed that:

- the solid matrix is homogeneous, non-deformable, and chemically inert with respect to the fluid,
- the fluid is single phase and Newtonian; its density does not depend on pressure variations, but only on variations of temperature,
- no heat sources or sinks exist in the fluid; thermal radiation and viscous dissipation are negligible.

Under these conditions, filtration velocity $\vec{V}$ and temperature T distributions are described by the following set of equations:

Mass conservation equation:

$$\varepsilon \frac{\partial \rho}{\partial t} + \nabla.(\rho \vec{V}) = 0 \qquad (1.1)$$

Momentum conservation equation:

$$\frac{\rho}{\varepsilon} \frac{\partial \vec{V}}{\partial t} + \frac{\rho}{\varepsilon^2} (\nabla . \vec{V})\vec{V} = \nabla P + \rho \vec{g} - \frac{\mu}{\bar{\bar{K}}} \vec{V} \qquad (1.2)$$

State equation:

$$\rho = \rho_o(1 - \alpha(T - T_o)) \qquad (1.3)$$

and energy equations for the solid and the fluid phases when a distinction is made between the average temperatures of the solid phase, T_s, and the moving fluid phase T_f are

$$(1 - \varepsilon)(\rho C)_s \frac{\partial T_s}{\partial t} = \nabla.(\bar{\bar{\lambda}}_s^* . \nabla T_s) + h(T_f - T_s) \quad (a) \qquad (1.4)$$

$$\varepsilon(\rho C)_f \frac{\partial T_f}{\partial t} = \nabla . (\overline{\overline{\lambda}}_f^* . \nabla T_f) - (\rho C)_f \vec{V} . \nabla T_f + h(T_s - T_f) \quad \text{(b)}$$

When $T_s = T_f$, the energy equation for the fictitious continuum medium equivalent to the real dispersed medium is

$$(\rho C)^* \frac{\partial T}{\partial t} = \nabla . (\overline{\overline{\lambda}}^* . \nabla T) - (\rho C)_f \vec{V} . \nabla T \quad (1.5)$$

In these equations, α, ρ, μ are respectively the volumetric thermal expansion coefficient, the density, and the dynamic viscosity of the fluid; g is the gravitational acceleration; P is the pressure; $\overline{\overline{K}}$ and ε are the permeability tensor and the porosity of the porous media; and T_o is a reference temperature level for which ρ is equal to ρ_o.

For the energy equations, $(\rho C)_s$ and $(\rho C)_f$ are the heat capacities of the solid and the fluid phases for constant pressure; $\overline{\overline{\lambda}}_f^*$ and $\overline{\overline{\lambda}}_s^*$ are the equivalent thermal conductivity tensors of the dispersed structures of the solid and the fluid phases; h is the heat transfer coefficient between the two phases, and $(\rho C)^*$ and $\overline{\overline{\lambda}}^*$ are the heat capacity and the equivalent thermal conductivity tensor for the saturated porous media. Due to the addition of equations (1.4.a) and (1.4.b), the following relations may be derived from the hypothesis $T_s = T_f$:

$$(\rho C)^* = (1 - \varepsilon)(\rho C)_s + \varepsilon(\rho C)_f$$

$$\overline{\overline{\lambda}}^* = \overline{\overline{\lambda}}_s^* + \overline{\overline{\lambda}}_f^*$$

As we can see, a mathematical model of heat transfer based on equations (1.4) is rather difficult to apply because we need an estimate for three unknown quantities, $\overline{\overline{\lambda}}_s^*$, $\overline{\overline{\lambda}}_f^*$ and h. Despite the existence of formal descriptions for $\overline{\overline{\lambda}}_s^*$, $\overline{\overline{\lambda}}_f^*$ and h, computation of these coefficients, which depend on thermal conductivity λ_s and λ_f of the phases, on porosity, on structural properties of porous media, and on hydrodynamic dispersion is generally impossible and their experimental determination very difficult (10). Fortunately, in most actual situations, the approximation $T_s = T_f$ is valid, and it is possible to use the simple model described by Eq. (1.5) in which only the value of $\overline{\overline{\lambda}}^*$ is needed. As indicated in Eqs. (1) and (11), two methods may be used to determine the components of this coefficient: theoretical estimation through a model or physical measurement. Like $\overline{\overline{\lambda}}_s^*$ and $\overline{\overline{\lambda}}_f^*$, this coefficient is also a complicated function of λ_s, λ_f, of the porosity, of the structural properties, and of the hydrodynamic dispersion.

For the momentum equation (1.2), a generalized form of the experimental Darcy's law is used, pending a complete and rigorous development from theoretical studies in progress.

Boussinesq's approximation and some other standard assumption:

Due to the complexity of the previous set of equations, other approximations or assumptions are commonly added to facilitate the theoretical approach of convection. Satisfied in numerous practical cases, these assumptions and approximations are:

- the thermophysical properties of the saturating fluid, ρ, μ, α, are assumed to be constant, except in the buoyancy term ρg where variation in fluid density clarifies the real cause of thermal convection (Boussinesq assumption), and
- the thermal physical characteristics of the porous medium, $\bar{\bar{\lambda}}^*$ and $(\rho C)^*$, are also assumed to be constant with $\bar{\bar{\lambda}}^*$ and $\bar{\bar{K}}$ isotropic.

With these assumptions, equation sets (1.1), (1.2), (1.3) and (1.5) yield:

$$\nabla \cdot \vec{V} = 0 \tag{1.6}$$

$$\frac{\rho}{\varepsilon}\frac{\partial V}{\partial t} + \frac{\rho}{\varepsilon^2}(\nabla.\vec{V})\vec{V} = -\nabla P + \rho\vec{g} - \frac{\mu}{K}\vec{V} \tag{1.7}$$

$$(\rho C)^* \frac{\partial T}{\partial t} = \lambda^* \nabla^2 T - (\rho C)_f \vec{V} \cdot \nabla T \tag{1.8}$$

$$\rho = \rho_o(1 - \alpha(T - T_o)) \tag{1.9}$$

1.2. Inspectional Analysis

Equations (1.6) to (1.9) may be rendered dimensionless by use of the following reference parameters: H for the length scale, $(\rho C)^* \frac{H^2}{\lambda^*}$ for the time scale, ΔT for the temperature scale, $\lambda^*/(\rho C)_f H$ for the velocity scale, and $\lambda^*\mu/K(\rho C)_f$ for the pressure scale. The dimensionless equations for an incompressible fluid are therefore employing the same symbolic representation as before:

$$\nabla\vec{V} = 0 \tag{1.10}$$

$$Pr^{*-1}\, MF\left(\frac{1}{\varepsilon}\frac{\partial V}{\partial t} + \frac{1}{\varepsilon^2}(\vec{V} \cdot \nabla)\vec{V}\right) = \nabla\bar{P} - \vec{V} + Ra^* \vec{k}\, T \tag{1.11}$$

$$\frac{\partial T}{\partial t} = \nabla^2 T - \vec{V} \cdot \nabla T \tag{1.12}$$

where $\vec{k} = \frac{\vec{g}}{||\vec{g}||}$ is the unit accelerator vector and $\bar{P} = P + \rho_o gz$.

The following dimensionless terms appear in the equations:

$$Ra^* = g \frac{\alpha(\rho C)_f}{\nu} \frac{K}{\lambda^*} \Delta T \, H \; ; \quad Pr^* = \frac{(\rho C)_f \nu}{\lambda^*} \; ; \quad F = \frac{K}{H^2}$$

$$M = \frac{(\rho C)_f}{(\rho C)^*} \tag{1.13}$$

where Ra^* is the filtration Rayleigh number, Pr^* the equivalent Prandtl number for the porous media, $F = \frac{K}{d^2} \frac{d^2}{H^2}$ with $\bar{d}$ the mean diameter of the pore or grain size of the material making up the porous medium, and d/H is a scale factor which characterizes the fineness of the medium. As this ratio is generally very small, $Pr^{*-1} \, MF \, (\frac{1}{\varepsilon} \frac{\partial \vec{V}}{\partial t} + \frac{1}{\varepsilon^2} (\vec{V} \cdot \nabla)\vec{V})$ may be neglected in the momentum equation. Natural convection in porous media appears as only dependent on the filtration Rayleigh number, Ra^*, and on the boundary conditions.

The equations we have used for this inspectional analysis are consistent only in a specific area of validity, i.e., a unique heat transfer equation, Boussinesq's approximation, and fluid and solid matrix properties are assumed to be constant.

More extensive analyses are possible from more thorough descriptions of the phenomena which yield additional dimensionless numbers. This is the case, for instance, when the dimensionless description of heat transfer is derived from Eq. (1.4):

$$(1 - \varepsilon M)(1 + \Lambda) \frac{\partial T_s}{\partial t} = \nabla^2 T_s - \Lambda \chi (T_s - T_f) \tag{1.14}$$

$$\varepsilon M (\frac{1 + \Lambda}{\Lambda}) \frac{\partial T_f}{\partial t} = \nabla^2 T_f - (\frac{1 + \Lambda}{\Lambda}) \vec{V} \cdot \nabla T - \chi (T_f - T_s) \tag{1.15}$$

in which three complementary dimensionless numbers appear:

$$\varepsilon \; ; \quad \Lambda = \frac{\lambda_s^*}{\lambda_f^*} \; ; \qquad \chi = \frac{K \, H^2}{\lambda_f^*} \tag{1.16}$$

1.3. Boundary Conditions and Physical Configurations

The context of free thermal convection study is defined by hydrodynamic boundary conditions, thermal boundary conditions, and the shape of the volume containing the porous medium.

For hydrodynamic boundary conditions, two situations can be found:

- impervious surface on which the normal component of the filtration velocity is equal to zero, $\vec{V} \cdot \vec{n} = 0$ ($\vec{n}$ unit vector normal to the surface), or
- free surface on which the pressure is constant, P = cte.

Two extreme cases of thermal boundary conditions are also possible:

- isothermal boundary, i.e., uniform temperature on the surface, or,
- and adiabatic or perfectly insulating boundary, i.e., heat flux density $-\lambda^* \vec{n} \cdot \nabla T = 0$ on the surface.

Between these two extreme cases, thermal boundary condition is formulated by stating the continuity of heat flux density through the limiting surfaces of the porous medium and the external parts.

The number of natural or laboratory configurations studied theoretically as well as experimentally has continually increased during the last few decades. After the simple case of an homogeneous layer with constant thickness and large lateral extent, researchers are now interested in more sophisticated configurations, such as confined and heterogeneous porous medium, or porous medium saturated by a fluid of non-constant properties.

From a dimensionless standpoint, the geometrical configurations are characterized by aspects ratios, for instance L_x/H, L_y/H, L_z/H, with L_x, L_y, L_z dimensions of the porous medium along three orthogonal axes.

1.4. Brief Look at the Existence of an Equilibrium

If we look for the conditions required for the existence of an equilibrium state of the saturating fluid ($\vec{V} = 0$), the equation of motion (1.2) gives:

$$- \nabla P + \rho \vec{g} = 0 \tag{1.17}$$

and by taking the curl of each term of Eq. (1.17)

$$\nabla \rho \wedge g = 0 \tag{1.18}$$

Moreover, the equation of state of the fluid yields:

$$\nabla\rho = \frac{\partial\rho}{\partial T} \cdot \nabla T \quad (1.19)$$

Then combining Eq. (1.18) and (1.19):

$$\nabla T \wedge \vec{g} = 0 \quad (1.20)$$

Hence, the condition required for equilibrium is defined by the fact that in the entire volume concerned, the temperature gradient and the body force are colinear.

As the condition expressed by Eq. (1.20) is only a necessary condition for equilibrium, a distinction has to be made between:

- the configurations for which there is no motionless state satisfying Eq. (1.20) and for which convective movements always exist, for instance, in sloped layers or around heated surfaces embedded in an infinite porous medium, and
- the configurations for which a motionless state satisfying Eq. (1.20) exist. In this last case, it is not possible to conclude with the previous simple analysis, and the stability conditions must be derived from an extensive theoretical study based on the Eqs. (1.10), (1.11), (1.12) or (1.14), (1.15) associated with the boundary conditions imposed on the medium. Such is particularly the case of the horizontal porous layer.

Specific theoretical approaches have been used to study the criterion for the onset of natural convection in horizontal porous layers as well as the stability of any convective movements, or the transition criterion. Based on the well-known Hydrodynamic Stability Theories, they have been respectively:

- the linear stability analysis founded on the study of the reactions of the saturating fluid phase to perturbation of small amplitude, allowing the linearization of the equations (12), and
- the non-linear stability analysis founded on the study of finite amplitude perturbation, using variational technique or perturbation expansion for weakly non-linear convection, or numerical simulation for fully non-linear convection (13), (14).

In the following paragraphs, we shall see the possibilities of using these methods to determine criteria for the onset of the convection, the form of convective movements, and the mean heat transfer.

2. NATURAL CONVECTION IN HOMOGENEOUS AND ISOTROPIC POROUS LAYER OF WIDE LATERAL EXTENT

Numerous studies have been completed during recent years describing natural or free convection in porous media. Most have been devoted to the case of homogeneous and isotropic porous layer of uniform thickness H and of large lateral extent $L >> H$ and $W >> H$, bounded by impervious surfaces maintained at different temperatures T_1 for the upper cold boundary and $T_2 = T_1 + \Delta T$ with $\Delta T > 0$ for the other boundary (Fig. 1).

2.1. Horizontal Porous Layer

2.1.1. Linear theory: onset of natural convection

This situation satisfying the condition given by (1.20), a motionless state of the saturating fluid, may exist. For the dimensionless boundary conditions:

$$\left.\begin{array}{l} T = 1 \\ \vec{V} \cdot \vec{n} = 0 \end{array}\right| \text{ at } z = 0 \qquad \left.\begin{array}{l} T = 0 \\ \vec{V} \cdot \vec{n} = 0 \end{array}\right| \text{ at } z = 1 \qquad (2.1)$$

it is defined by $\vec{V}_o = 0$; $T_o = 1 - z$; $P_o = Ra^*(z - \frac{z^2}{2}) + cte$ (2.2) and corresponds to an equilibrium for which the heat transfer is essentially due to the conduction.

The first uses of the linear theory applied to the study of the stability of fluid saturating a porous layer are attributable to Horton and Rogers (17), Lapwood (16) to Katto and Masuoka (15).

Extensively described by Chandrasekhar (12), the mathematical treatment of a problem of instability generally proceeds along the following lines:

a) the solutions corresponding to the initial flow, representing a stationary state, are disturbed by perturbations of infinitesimal amplitude;

b) the new solutions, i.e., disturbed initial solutions, are put into the governing equations of the phenomena, and by using the initial solutions, give the perturbation equations;

c) the perturbation equations are linearized by neglecting all products and powers (higher than the first) of the perturbations; and

d) the perturbation equations are finally resolved by expressing an arbitrary disturbance as a superposition of certain possible basic modes and examining the stability of the system with respect to each of these modes.

The use of this procedure to study the stability of the saturating fluid of an horizontal porous layer heated from below yields the following perturbation equations:

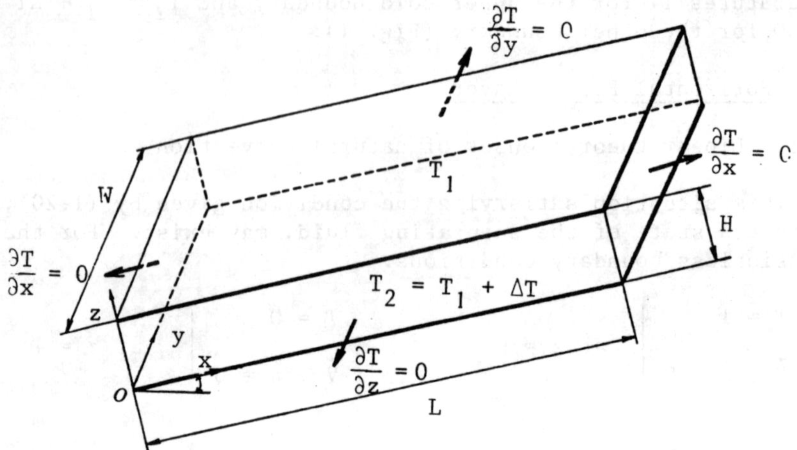

Figure 1. The porous layer.

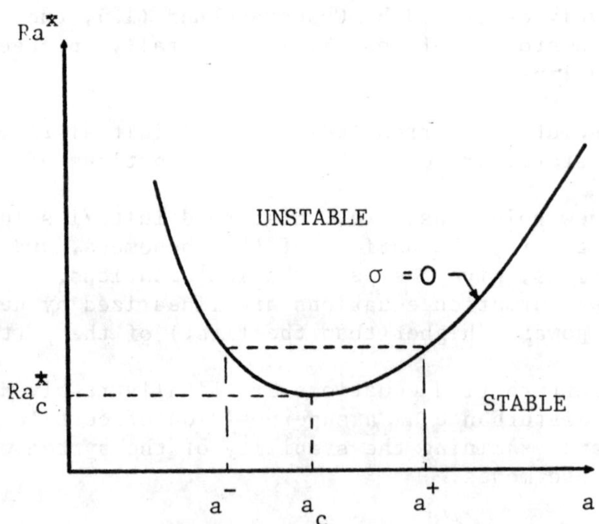

Figure 2. Neutral stability curve.

$$\nabla \vec{V} = 0 \tag{2.3}$$

$$- \nabla \pi + Ra^* \vec{k}\theta - \vec{V} = 0 \tag{2.4}$$

$$\frac{\partial \theta}{\partial t} - w = \nabla^2 \theta \tag{2.5}$$

where $\vec{V}$, θ and π are respectively the velocity, the pressure, and the temperature perturbations of the initial state such as:

$$\vec{V} = 0 + \vec{V} \quad ; \quad T = T_o + \theta \quad ; \quad P = P_o + \pi \quad ; \quad \vec{V} = \vec{i}u + \vec{j}v + \vec{k}w$$

By eliminating the pressure in Eq. (1.20), and by looking for θ and w, a two-dimensional perturbation such as

$$\theta = \theta(z) \exp\,(i(lx + my) + \sigma t) \tag{2.6}$$

$$w = w(z) \exp\,(i(lx + my) + \sigma t) \tag{2.7}$$

we obtain for Eqs. (2.4) and (2.5) with $D = d/dz$:

$$- Ra^* a^2\theta + (D^2 - a^2)w = 0 \tag{2.8}$$

$$(D^2 - a^2 - \sigma)\theta + w = 0 \tag{2.9}$$

associated to the boundary conditions:

$$\theta = w = 0 \quad \text{at } z = 0 \text{ and } \quad z = i \tag{2.10}$$

Equations (2.8) and (2.9) define an eigenvalue problem the solution of which is simplified by the so-called principle of exchange of stabilities. This principle states that when $Re\{\sigma\} = 0$, then also $I_m(\sigma) = 0$. Thus, the neutral curve defining the transition between the convective regimes is given by $\sigma = 0$, and the eigenvalue problem requires only real-valued arithmetic.

The solutions obtained for $\sigma = 0$ corresponding to the existence of a stationary convective flow ($\vec{V} \neq 0$) of the saturating phase are:

$$\theta(z) = A\sin(s\pi z) \quad ; \quad w(z) = B\sin(s\pi z) \tag{2.11}$$

when $$Ra^* = \frac{(a^2 + s^2\pi^2)^2}{a^2} \tag{2.12}$$

s is an integer, $a = (l^2 + m^2)^{\frac{1}{2}}$, and A and B are respectively the wave number, and the amplitudes of the perturbations left undetermined by the linear analysis.

Equation (2.12) determines the critical filtration Rayleigh number as a function of the convective mode s, and the wave number a. The variations of Ra^* corresponding to the first mode s = 1 are displayed in Figure 2. All the points located below this curve are related to a stable situation ($\sigma < 0$, no convection), and the minimum value $Ra_c^*(a_c)$ obtained for $(dRa^*/da)_{s=1} = 0$, correspons to the neutral stability. In other words, the transition between the conductive state and the convective state is equal to $4\pi^2$ for $a_c = \pi$. This value, which is the lowest value of Ra^*, is commonly called critical value for the onset of the convection for an horizontal layer of wide lateral extent, i.e., the occurrence of convection is defined by $Ra^* \geqq Ra_c^* = 4\pi^2$.

In principle, the curve of Figure 2 means that for supercritical Ra^* number, the entire range of wavenumbers are possible. In fact, when non-linear analysis gives a curve which is inside the neutral stability curve, then the range of allowed, a, is restricted (in infinite geometry). The presence of lateral boundaries in the layer restricts again the allowed range of, a, (case of finite geometry).

Another method in which not only small but also arbitrary disturbances may be considered has been used for the computation of the stability criterion (14), (19). Based on the study of the temporal evolution of a linear combination of kinetic and thermal energies of perturbations when $t \to \infty$:

$$\frac{1}{2}\frac{d}{dt} \langle\theta^2\rangle = Ra^* \langle(\lambda^{-\frac{1}{2}} - \lambda^{\frac{1}{2}})W\hat{\theta} - \langle |\nabla\hat{\theta}|^2 + |\vec{V}|^2\rangle \qquad (2.13)$$

(where $\langle . \rangle = \int_\Omega d\omega / \int_\Omega d\omega$, λ is a coupling parameter such as $\theta = \hat{\theta}/\lambda^{\frac{1}{2}}$, and Ω the volume of the porous medium.

The stability is defined by the condition $\frac{d}{dt} \langle\theta^2\rangle \leq 0$ when $t > 0$. For the problem in which we are interested, this variational technique called global stability analysis gives the same result as the linear theory.

The Physical Reasons of a Critical Threshold Ra_c^*

Consider a spherical portion (radius $r \cong K^{\frac{1}{2}}$) of the fluid submitted to the constant gradient of temperature $\Delta T/H$. If we move this sphere of fluid from a warm to a cold region, for example, the relaxation time of its temperature will be:

$\tau = \frac{r^2}{\lambda^*}(\rho C)_f \cong \frac{K}{\lambda^*}(\rho C)_f$. This means that if this sphere moves with a constant velocity w at a given instant, the temperature of this portion of fluid is that of its surroundings at an earlier instant of time $t-\tau$. Thus, at time t, the temperature difference δT between the sphere and the surrounding porous media is:

$$\delta T = \frac{\Delta T}{H} w\tau \qquad \text{or :} \qquad \delta T = \frac{\Delta T}{H} w \frac{K}{\lambda^*} (\rho C)_f \tag{2.14}$$

This produces a buoyancy force:

$$Fa = \rho_o \alpha \delta T r^3 g = \rho_o g \alpha r^3 \frac{\Delta T}{H} w \frac{K}{\lambda^*} (\rho C)_f \tag{2.15}$$

On the other hand, the viscous drag is:

$$Fd = - 6\pi \mu r w \tag{2.16}$$

If the buoyancy force Fa overcomes the drag force Fd, the motion tends to amplify and the system becomes unstable. The ratio Fa/Fd clearly increases with r, and roughly, we can say that the instability of the layer will begin with fluid portions of the maximum size, i.e., H the distance of horizontal boundaries of the layer.

So, the instability criterion Fa > Fd is:

$$g \frac{\alpha(\rho C)_f}{\nu} \frac{K}{\lambda^*} \Delta T\, H > cte \quad (\text{or: } Ra^* > cte = Ra^*_c) \tag{2.17}$$

This simple analysis confirms that the dimensionless number which controls the stability of the layer is the Rayleigh number, and that the balance between destabilizing effects and stabilizing ones produces the existence of a critical threshold for the appearance of fluid motion. If $Ra^* < Ra^*_c$ in spite of the thermal gradient applied to the porous layer, the fluid remains at rest.

Shape of convective movements:

When the filtration Rayleigh number is marginally higher than Ra^*_c, the perturbations are of finite amplitude and a steady state convective flow exists. From the result giving the horizontal wave number, a, two kinds of frequently encountered configurations may be deduced:

- the first one corresponds to two-dimensional contrarotative rolls, i.e., to the value of l or m = 0 given the reduced size of each roll l/H = 1,
- the second consists of a juxtaposition of polyhedral cells, the solution of which is given by Christopherson (12) and corresponds to hexagonal cells of side l' such as l'/H = 1.33 (Fig. 3). As in Busse and Riahi (20), the preference for hexagonal pattern would be the result of non-symmetrical boundary condition with respect to the midplane.

As shown in theoretical studies devoted to the problem of non-linear thermal convection at small amplitude in a horizontal porous layer with finite conducting boundaries (18), (20), (21), critical Rayleigh number, horizontal wave number, and shape of convective movements largely depend on thermal boundary conditions. An example of neutral curves for different conductivity ratios λ_b/λ^* (λ_b = thermal conductivity of the lower and upper surfaces) is presented in Figure 4.

2.1.2. Non-linear Theory - Mean Heat Transfer

When amplitude A increases for supercritical Rayleigh numbers, the non-linear self interactions of the first order mode becomes important and cannot be neglected in the perturbation equations if we want to keep a satisfying description of the phenomena. Such is the case, for instance, when we want to compute the influence of the convective flow on the mean heat transfer.

Generally described by means of a dimensionless number Nu^* (Nusselt number) equal to the ratio of the mean heat flux density when convective movements exist on the mean heat flux density due to the thermal conduction alone, this number is equal to 1 for $Ra^* \leq 4\pi^2$ (conductive regime), then increases for $Ra^* > Ra_c^*$ (convective regime).

The analytical expression of the mean temperature gradient $d\bar{T}/dz$ computed from the complete heat transfer perturbation equation yields (1):

$$\frac{d\bar{T}}{dz} = -1 + \overline{\theta w} - \int_0^1 \overline{\theta w}\, dz \qquad (2.18)$$

and for the Nusselt number : $Nu^* = - \int_\Sigma \left.\frac{\partial T}{\partial z}\right|_{z=0} d\sigma = - \left.\frac{d\bar{T}}{dz}\right|_{z=0}$

$$Nu^* = 1 + \int_0^1 \overline{w\theta}\, dz \qquad (2.19)$$

where $\bar{T}$ and $\overline{W\theta}$ are average values on horizontal planes of surface Σ.

Equation (2.18) exhibits how a linear temperature gradient in the absence of motion is modified by the presence of motion and, as a consequence, the relative influence of conductive and convective mechanisms on the heat transfer in the different parts of the layer.

The typical vertical temperature profiles through the porous layer respectively for conductive and convective state deduced from Eq. (2.18) are presented on Figure 5.

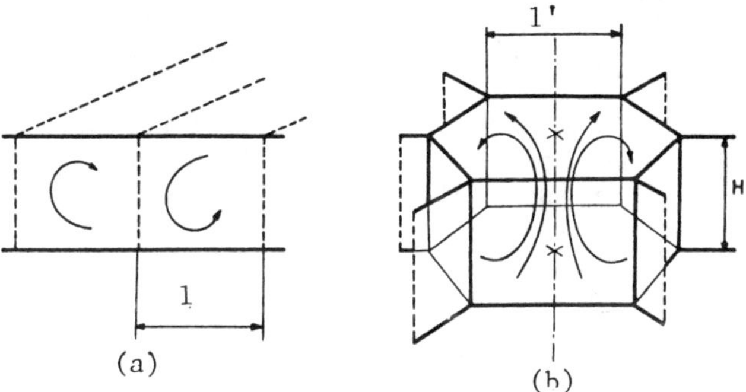

Figure 3. (a) Contratotative rolls;
(b) Polyhedral cell

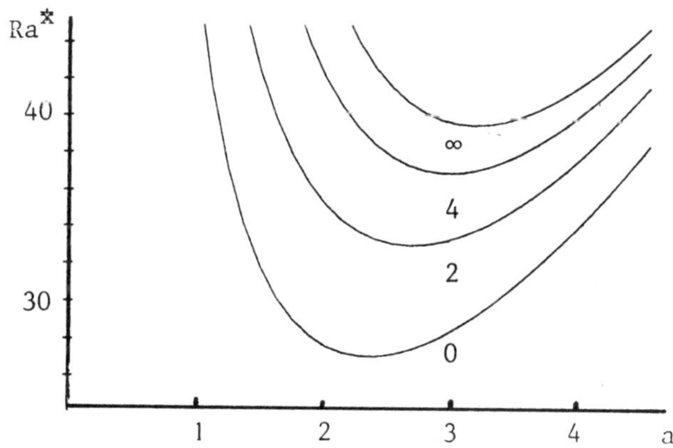

Figure 4. Neutral curves for different conductivity of the lower boundary (after Riahi (21))

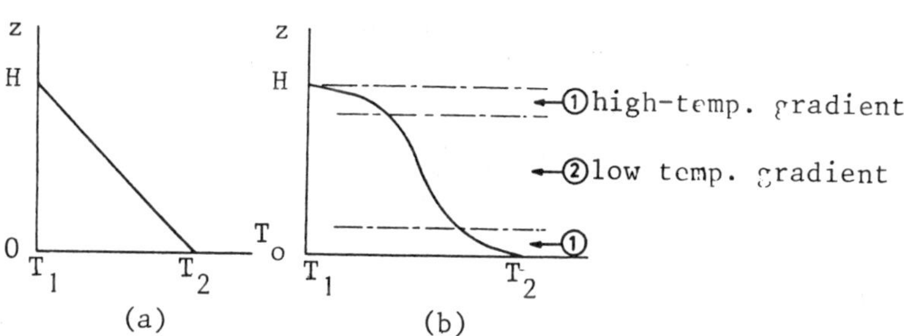

Figure 5. Vertical temperature profiles - (a) conductive state; (b) convective state.

As there is no phase shifting (see Eq. (2.4)) between θ and W, $<\theta.W>$ is at a maximum and $d\bar{T}/dz$ is at a minimum in the horizontal midplane $z = 1/2$. On the contrary, when $<\theta W>$ is low, i.e., along the isothermal boundaries, $|d\bar{T}/dz|$ is at a maximum.

This explains the cause for the development of high temperature gradients along the impervious isothermal surfaces, and the non linear term $\vec{V} \cdot \nabla\theta$ is the reason this zone becomes unstable when Ra^* increases.

Weakly non-linear analysis

Since any product of disturbance components is omitted by the basic assumption, the linear theory is unable to describe the evolution of the temperature profile through the porous layer. Nor is it able to describe the increase in the mean heat transfer or the component l and m of the horizontal wave number.

The only systematic method for analyzing the numerous three-dimensional non-linear steady solutions of (1.10), (1.11), (1.12) is the perturbation approach based on the amplitude A of convection as small parameter. This approach is particularly appropriate in the case of convection because the instability occurs in the form of infinitesimal disturbance. Obviously, the perturbation expansion is of limited usefulness when the filtration Rayleigh number is increased much beyond its critical value. In this case, direct numerical methods must be used to solve the problem of fully non-linear convection.

Among the existing different weakly non-linear approaches (22,23,24), the Malkus-Veronis which is the one we used (1) for the study of natural convection in porous media is based on the properties of integral relations governing the steady convection. Obtained by multiplying the equations of motion and the heat transfer equation by the disturbance components, these integrals are respectively:

$$\int_0^1 (\bar{u}^2 + \bar{v}^2 + \bar{w}^2)\, dz = Ra^* \int_0^1 \overline{\theta . w}\, dz \tag{2.20}$$

$$\int_0^1 \overline{\theta . W}\, dz + \int_0^1 \overline{\theta \nabla^2 \theta}\, dz = \int_0^1 \overline{\theta W^2}\, dz - \left(\int_0^1 \overline{\theta W}\, dz \right)^2 \tag{2.21}$$

Due to its nonhomogeneity versus the disturbance components, the equation (2.21) derived from the energy equation can be used to compute the amplitude, provided that analytical form of θ and W is known. Assuming that W and θ are the solutions of the linearized equations, the amplitude of the perturbations is:

$$A = (Ra^* - Ra^*_{cs})^{\frac{1}{2}} \tag{2.22}$$

(where Ra^*_{cs} is the critical Rayleigh number corresponding to the successive convective modes, s), and the Nusselt number computed by Eq. (2.19) is:

$$Nu^* = 1 + \sum_{s=1}^{\infty} k_s \left(1 - \frac{Ra^*_{cs}}{Ra^*}\right) \qquad \text{Fig. 6} \tag{2.23}$$

$$\text{with } k_s = \begin{cases} 2 & \text{for } Ra^* > Ra^*_{cs} \\ 0 & \text{for } Ra^* < Ra^*_{cs} \end{cases} \qquad \text{and} \quad Ra^*_{cs} = 4s^2\pi^2$$

Numerical computation

Different numerical techniques such as the finite difference method, the finite elements method, or the spectral method have been used to resolve the governing equation of natural convection in porous media (28 - 35).

The spectral method (36) based on the well-known Galerkin method consists of developing the temperature and velocity solutions using a set of linearly independent trial functions:

$$T = (1 - z) + \sum_{l=o}^{L} \sum_{m=o}^{M} \sum_{n=o}^{N} a_{lmn}(t) \cos l\pi x \cos m\pi y \sin n\pi z \tag{2.24}$$

$$u = - A^2 \sum_{l=o}^{L} \sum_{m=o}^{M} \sum_{n=o}^{N} b_{lmn} \, ln\pi^2 \sin l\pi x \sin m\pi y \cos n\pi z \tag{2.25}$$

$$v = - B^2 \sum_{l=o}^{L} \sum_{m=o}^{M} \sum_{n=o}^{N} b_{lmn} \, mn\pi^2 \cos l\pi x \sin m\pi y \cos n\pi z \tag{2.26}$$

$$w = \sum_{l=o}^{L} \sum_{m=o}^{M} \sum_{n=o}^{N} b_{lmn} (l^2 + m^2)\pi^2 \cos l\pi x \cos m\pi y \sin n\pi z \tag{2.27}$$

satisfying boundary conditions:

$$T = 1 \quad w = 0 \quad \text{for } z = 0; \quad \frac{\partial T}{\partial x} = 0, \ u = 0 \ \text{for } x = 0 \text{ and } x = A$$

$$T = 0 \quad w = 0 \quad \text{for } z = 1; \quad \frac{\partial T}{\partial y} = 0, \ v = 0 \ \text{for } y = 0 \text{ and } y = B$$

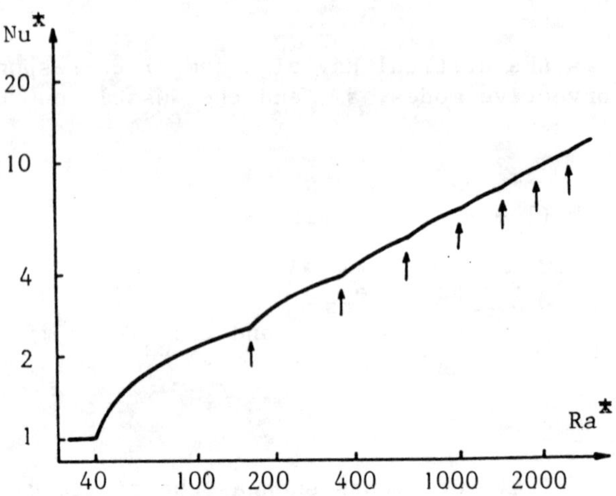

Figure 6. A theoretical relationship $Nu^*(Ra^*)$

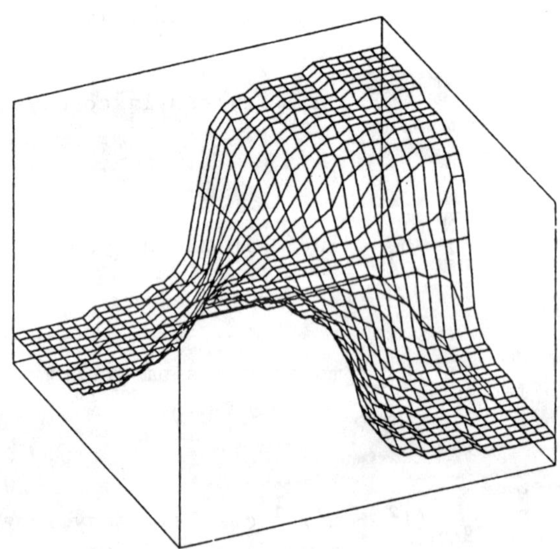

Figure 7. Three-dimensional flow
$Ra^* = 100$ T = 0.5 (1,1,1) mode
(after Caltagirone and Meyer (31))

where A and B are the aspect ratios respectively in the x and y directions : $A = L/H$; $B = W/H$.

The method consists of finding the remainder by means of a trial function on integrating over the whole volume. After eliminating the pressure term by applying the divergence theorem and by taking into account the continuity equation, the motion equation allows the determination of the explicit relationship between the b_{ijk} and a_{ijk} coefficients.

If solutions given by Eqs. (2.24) and (2.27) are introduced into the energy equation, the following differential system is obtained:

$$\frac{da_{ijk}}{dt} = - (\frac{i^2}{l^2} + \frac{j^2}{B^2} + k^2)\pi^2 \; a_{ijk} + (i^2 + j^2)\pi^2 \; b_{ijk} - N(a,b) \tag{2.28}$$

where $N(a,b)$ is a non-linear operator corresponding to the convective term $\vec{V}.\nabla T$. The initial conditions are represented by the temperature coefficients $a_{ijk}(0)$.

Numerical results obtained from the computation of the non-linear differential system (2.28) show different types of evolution according to the value of the filtration Rayleigh number (31):

- for $Ra^* < 4\pi^2$, the perturbation induced by initial conditions decreases and the system tends to the pure conduction solution,

- for $4\pi^2 \leq Ra^* < 300$, the initial perturbation develops to give a stable convergent solution (which does not depend on the intensity or nature of this perturbation), and different stable tridimensional convective flows are observed : contrarotative rolls (2D), superposition of contrarotative rolls (3D), and polyhedral cells (3D) (Fig. 7). To these flows, among which polyhedral cells appear as the less stable, correspond the Nusselt-Rayleigh correlation given on Figure 16. This correlation has been computed for a two-dimensional roll of reduced size H/1, starting from the Malkus's hypothesis that a flow evolves to a steady configuration that maximizes the heat transport (1), i.e., Nu^* given for the maximum of the curve $Nu^*(H/L)_{Ra^*}$.

Unfortunately, there has been no rigorous demonstration of the validity of this proposal, and some numerical simulations do not support it (38).

- for $Ra^* \geq 240\text{-}300$, a stable regime cannot be reached. Described as a fluctuating convective state, this situation is characterized by a continuous fluctuating in situ temperature and velocity distribution inside the porous layer, and by a relative increase of heat transfer compared to the state we previously described.

Caused by the instability of the thermal boundary layer at horizontal boundaries, the existence of such a state has been deduced from a stability analysis of the finite amplitude two-dimensional solutions (38) (Figure 8). It has been interpreted (28,33) as continuous creations and disappearances of convective cells, even in the thermal steady state, in the area of highest temperature gradients (Figure 9) (30).

<u>Sophisticated model with the heat transfer coefficient</u>

In order to more accurately explain the experimental results, as well as to examine the influence of the parameter Λ and χ on the convection, the sophisticated heat transfer model (1.14) and (1.15) has been also used (1). With this model, all aspects of convection appears as affected by the three numbers Ra^*, Λ and χ, and particularly the Nusselt number that in this case is defined by:

$$Nu^* = f(Ra^*, \Lambda , \chi) \tag{2.29}$$

$$\text{with} \quad \Lambda = \lambda_f^*/\lambda_s^* \quad \text{and} \quad \chi = (h \, . \, H^2)/\lambda_f^* \tag{2.30}$$

<u>Influence of the parameters on heat transfer</u>. As in the case of the simple model (1.15) for given Ra^*, Λ and χ values, the variation of Nu^* versus H/L shows the existence of a maximum. The H/L value, for which this maximum is reached, is not very different from (1.15) and, to specify the influence of Λ and χ on the heat transfer, we look in particular at the case $Ra^* = 200$ with H/L = 1 (Figure 10).

For a given value of Λ , Nu^* is an increasing function of χ which tends, when $h \to \infty$, toward the value computed with the simple numerical model. Indeed, since Λ is constant, when the conditions for the heat transfer between the solid and fluid phases are improved, the porous medium tends to behave like a single continuum.

The influence of Λ, when χ is maintained constant, may be explained by considering the relative contributions of the solid and fluid phases to the overall heat transfer. If Λ increases, i.e., if the contribution of heat conduction by the solid phase becomes negligible, then Nu^* tends toward the value given by the simple model. On the contrary, when heat conduction throughout the solid phase is very large, the Nusselt number decreases ($Nu^* \to 1$ if $\Lambda \to 0$.

<u>Preferential zones for heat transfer between solid and fluid phases</u>. The computed temperature distributions reveal the existence of two specific zones where the difference between the temperatures of the

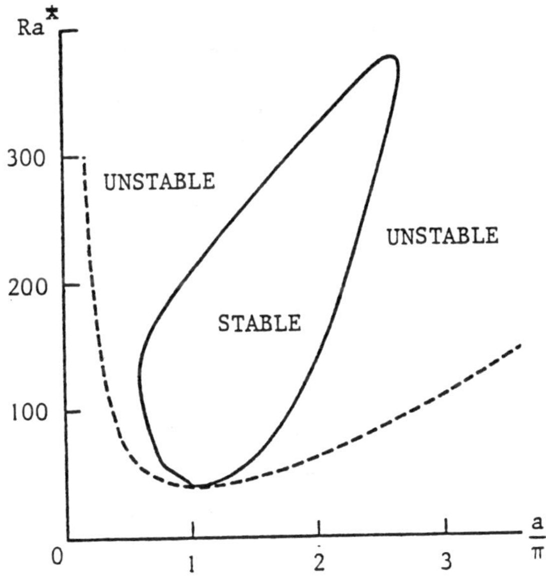

Figure 8. The region of stable two-dimensional solutions. Only within the closed region are finite amplitude two-dimensional solutions stable (after Straus and Schubert (38)).

solid and fluid phases is at a maximum; these zones are the upper part of the upward current and the lower part of the downward current (Figure 11).

This result, which can be intuitively forecast, explains the main role of the solid phase as a heat exchanger in those areas in which the assumption of the equivalence of the real porous medium with a single continuum may thus be questionable.

Influence of texture of the porous or cracked medium. Through the influence it has on both parameters Λ and χ, the texture of the porous or cracked medium, as well as the thermal characteristics of the solid and fluid phases, strongly influences the phenomenon. However, the influence of texture is mainly appreciable by studying the variation of χ. Let $\bar{d}$ be a characteristic length of the porous structure, for instance, the average bead diameter for an unconsolidated aquifer, or the average length of a matrix block for a cracked medium. Parameter χ can be written :

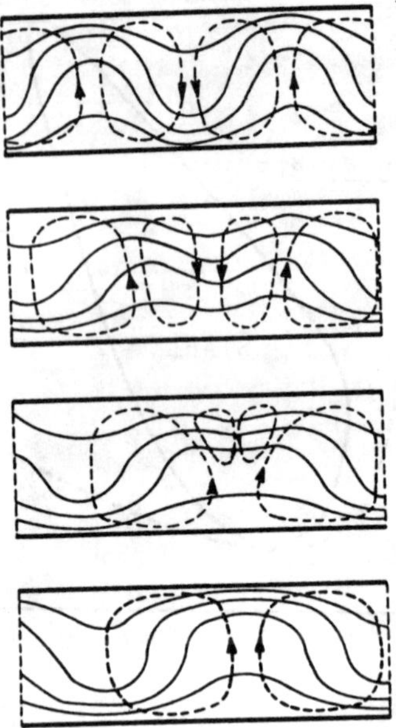

Figure 9. Fluctuating convective state. Evolution of two-dimensional rolls (after Caltagirone and Cloupeau (37)).

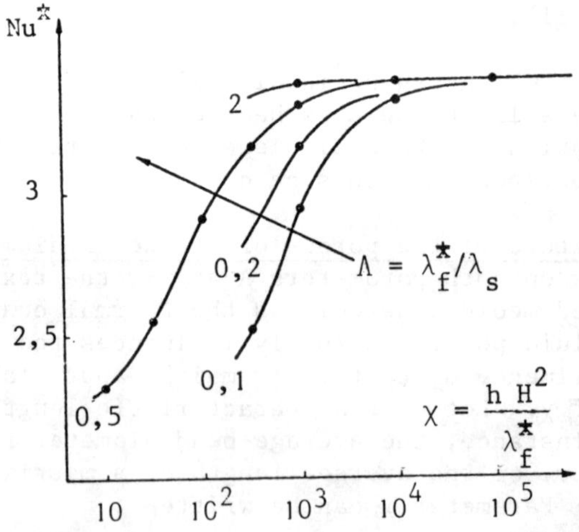

Figure 10. Influence of χ and Λ on mean heat transfer.

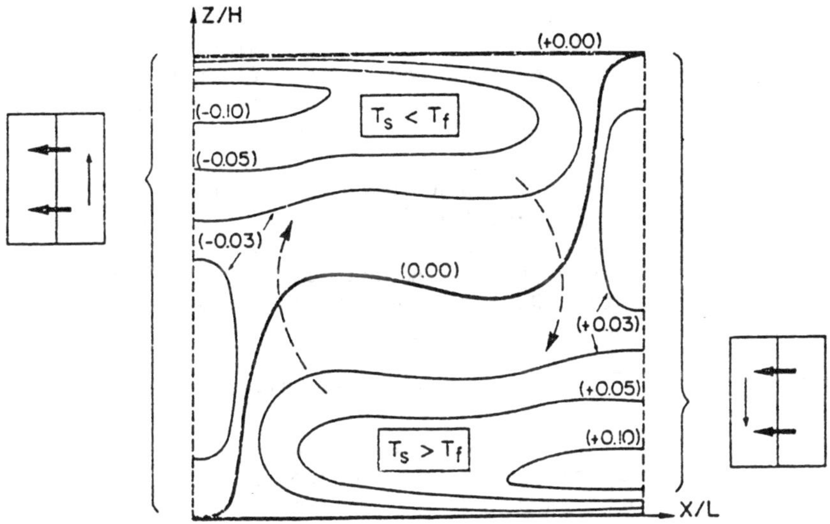

Figure 11. Dimensionless temperature difference between solid and fluid phases for a stable two-dimensional solution $Ra^* = 200$ $\quad \Lambda = 0.5$ $\quad \chi = 1000$ (after Combarnous and Bories (1)).

$$\chi = \frac{h \cdot H^2}{\lambda_f^*} = \frac{h \cdot K}{\lambda_f^*} \cdot \frac{H^2}{K} = \frac{h \cdot d^2}{\lambda_f^*} \left(\frac{H}{d}\right)^2 \tag{2.31}$$

The two factors in the right term of this equation take into account, respectively:

- the influence of heat transfer on the local scale, pore scale, or block scale,
- the influence of a scale factor H^2/K or $(H/d)^2$ which describes the extent of division of the structure compared with the vertical extent of the layer. When the scale factor is high, the porous medium can be considered as a very thorough blend of solid and fluid phases. When it is low, then the porous medium is more heterogenous.

The main result obtained by using the sophisticated model, with two heat transfer equations, explains the influence of this scale factor on the mean heat transfer. Let us assume that h does not depend on the filtration velocity and is affected solely by the thermal characteristics of the constituting phases as well as the texture of the medium. We will find, from the numerical results, that the mean heat transfer which, in addition to Λ, depends solely on $\chi = (h \, . \, H^2)/\lambda_f^*$, is affected by the height H of the layer. The higher the scale factor is, the better the description is of the porous medium as a single continuum by the simple numerical model with a sole heat transfer equation. The influence of these different parameters on the fluctuating regime has been also studied (39).

2.2. Sloped Porous Layer

It is well known that in a sloped, saturated, porous layer bound by isothermal planes, the fluid phase is always moving, and the basic flow which develops is of a unicellular two-dimensional type. The structure of this flow and its stability are defined not only by the filtration Rayleigh number and the slope ϕ, but also by the parameter L and W (39,40,51,52).

2.2.1. Stability of the flow in an infinite extension layer

If the porous layer is of infinite extension in the x and y directions, the solution corresponding to the basic unicellular flow can be readily found, and leads to the following expressions for the temperature and the velocity fields:

$$T_o = 1 - z \; ; \quad U_o = Ra^* \sin\phi(\frac{1}{2} - z) \quad ; \quad V_o = 0 \; , \; W_o = 0 \qquad (2.32)$$

Equations of perturbation relative to this flow deduced from the (1.10), (1.11), (1.12) system become:

$$\nabla^2\theta - Ra^* \sin\phi(\frac{1}{2} - z) \frac{\partial\theta}{\partial x} + W = \frac{\partial\theta}{\partial t} \qquad (2.33)$$

$$\nabla^2 w + Ra^*(\sin\phi \frac{\partial^2\theta}{\partial x \partial z} - \cos\phi(\frac{\partial^2\theta}{\partial x^2} + \frac{\partial^2\theta}{\partial y^2})) = 0 \qquad (2.34)$$

Developing the perturbations into complex exponential functions of the spatial coordinates x, y and of time t and eliminating w, Eqs. (2.33) and (2.34) are reduced to one equation in :

$$\begin{aligned}(D^2 - a^2)\theta - \sigma(D^2 - a^2)\theta - Ra^* \cos\phi a^2\theta \\ - ilRa^* \sin\phi((\frac{1}{2} - z) (D^2 - a^2)\theta + D\theta) = 0 \qquad (2.35)\end{aligned}$$

in which l represents the component of the wave number $a = (l^2 + m$ $a = (l^2 + m^2)^{\frac{1}{2}}$ of the perturbation in the direction of the slope and $D = d/dz$.

The problem of eigenvalues (Eq. (2.35)) corresponding to the stationary solution, i.e., $\sigma = 0$, can be solved by means of the Galerkin method developed in the following form:

$$\theta = \sum_{k=1}^{N} a_k \sin k\pi z$$

By subtitution, Eq. (2.35) can be written in the form:

$$\sum_{s=1}^{N} \{ Ra^* \cos\phi - \left(\frac{j^2\pi^2 + a^2}{a^2}\right)^2 \delta_{js} - i\, Ra^* \sin\phi \frac{1}{a^2}\left(\frac{4js}{(j^2 - s^2)^2}\right)$$

$$\left(\frac{2a^2}{\pi^2} + j^2 + s^2\right)\delta_{j+s,2p-1} \}\, a_s = 0 \qquad (2.36)$$

Let us say L(a) = 0 gives a homogeneous linear system accepting only a non-zero solution for a particular value of Ra^* such that det(L)=0. The analysis of three-dimensional linear stability enables three flow domains to be distinguished in the (Ra^*,ϕ) plane:

- for Ra^* and ϕ such that $Ra^*\cos\phi<4\pi^2$, only the basic two dimensional unicellular flow T_o, U_o remains;
- when the Ra^* - ϕ couple is such that $Ra^*\cos\phi>4\pi^2$, the $a \neq 0$ three-dimensional flow becomes steady, and the $Ra^*\cos\phi = 4\pi^2$ conditions (39,40,41) correspond to the appearance of a flow in a longitudinal coil of wavelength $2\pi/1 = 2H$;
- when Ra^* and ϕ gives representative points located over the transition defined by Eq. (2.36), the flow can be steady with transverse rolls: $a = 1 \neq 0$ and generally with polyhedral cells $a = (1^2 + m^2)^{\frac{1}{2}}$ with $1 \neq 0$ and $m \neq 0$ corresponding to the superimposition of several groups of rolls.

When the angle ϕ increases, at a fixed Ra^*, the slope for which the transition between polyhedral cells or transverse and longitudinal rolls appear is theoretically found equal to $\phi_t = 31^o48'$ (Fig. 12). When $\phi < \phi_t$, it has been observed that initial conditions have a great influence on the setting up of the stationary flow (47).

2.2.2. Mean heat transfer

Concerning the mean heat transfer, as for the horizontal layer, the Malkus technique may be used and produces a relationship similar to Eq. (2.23) in which Ra^* is replaced by $Ra^*\cos\phi$.

2.3. Experimental Observations

In the case of porous layers of wide lateral extent bounded by isothermal planes, numerous experimental results are now available that describe the criterion between the different configurations of flow, the convective movements, and the mean heat transfer.

Horizontal layers

- the criterion for the onset of natural convection, i.e., $Ra^* \geqq 4\pi^2$, is well confirmed by experimental results presented in (41) through (47), starting with the evolution of the correlation $Nu^*(Ra^*)$ (Figure 13).

- the convective flows deduced through in situ temperature measurements in the medium plane of the layer or by visualization tests were proven to be consistent with the linear theory predictions, i.e., convective steady state is characterized by adjacent polyhedral cells or two-dimensional rolls, the reduced size of which are respectively $l'/H = 1.33$ and $l/H = 1$ (Figure 14).

Also, the horizontal extent of convective cells was observed to be a slightly decreasing function of Ra^* (Figure 15).

When Ra^* is higher than a critical value which lies in the range 240-300 (depending on the porous medium), the fluctuating convective state, characterized by a continuously fluctuating in situ temperature distribution inside the porous layer and by a relative increase of the mean heat transfer (Figure 13), has been proven to exist (30).

Concerning the mean heat transfer due to convection, where the standard mathematical study leads, as for the case of a fluid layer, to a unique relationship between the Rayleigh and Nusselt numbers, experimental data have shown that the mean heat transfer does not depend solely on the Rayleigh number, but also on the thermal characteristics of the constituting phases, the solid matrix, and the saturating fluid (Figure 16).

As we can see on Figure 16 where a comparison between experimental, numerical, and theoretical $Nu^*(Ra^*)$ correlation is presented, numerical results are in good agreement with the average experimental results. However, the results are unable, contrary to the more sophisticated model, to explain the influence of the porous medium characteristics on the heat transfer. The theoretical prediction is restricted to a small range of variation of Ra^* near Ra^*_{c1}.

Sloped layers

In the case of sloped porous layers whose extension is greater than their thickness, published experimental studies are not very numerous (39,40,41,49,50,51,52).

The experimental heat transfer results put in the form of a relation between Nu^* and $Ra^*\cos\phi$ (Figure 17) show that the correlation $Nu^*(Ra^*\cos\phi)$ is quite good, and that despite the finite extent

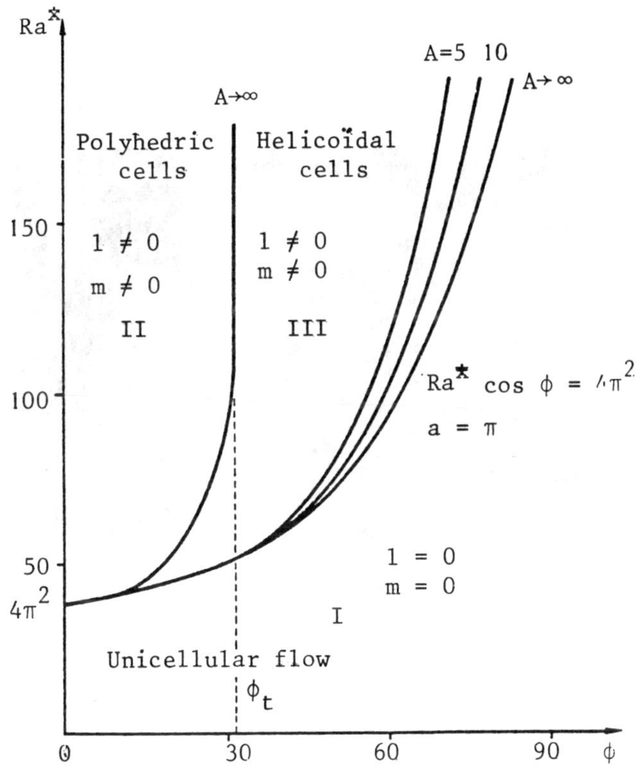

Figure 12. Criteria for transition between the different types of flow (after Caltagirone and Bories (51)).

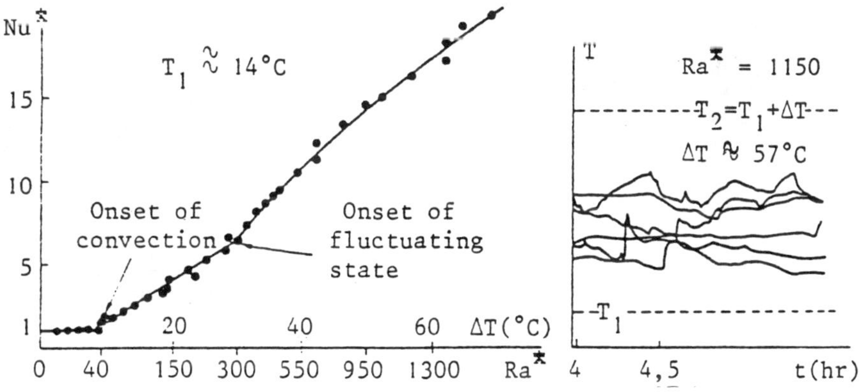

Figure 13. Evolution of Nu^* and in situ temperature in the fluctuating convective state (after Combarnous (48)).

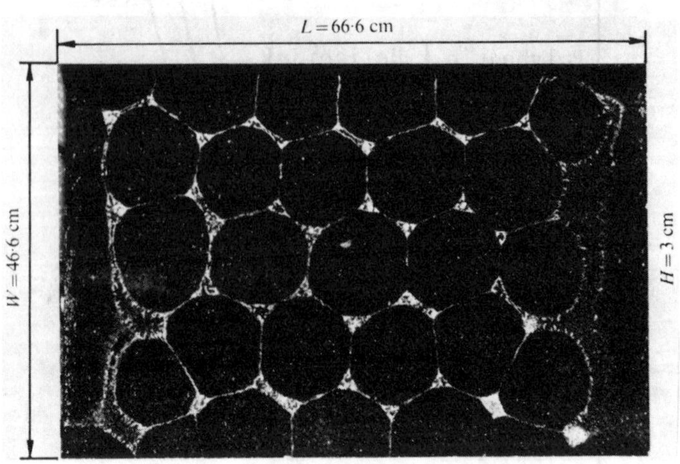

A picture of the upper free surface of a horizontal porous layer with polyhedral cell ; the lower boundary is impervious and isothermal.

(a)

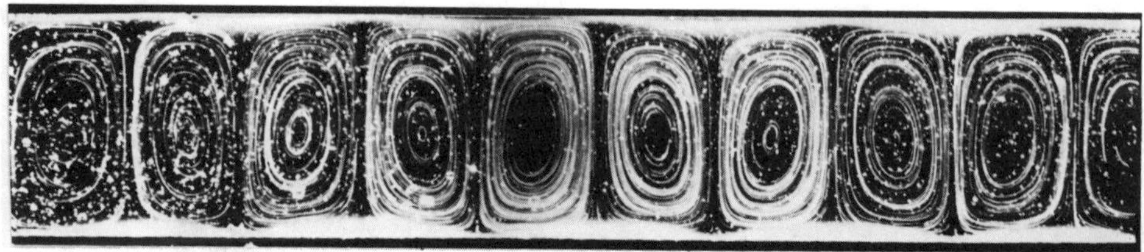

A picture of the streamlines in convective stable cells appearing in a vertical two-dimensional Hele-Shaw model bounded by isothermal impervious boundaries.

(b)

Figure 14 (a), (b).

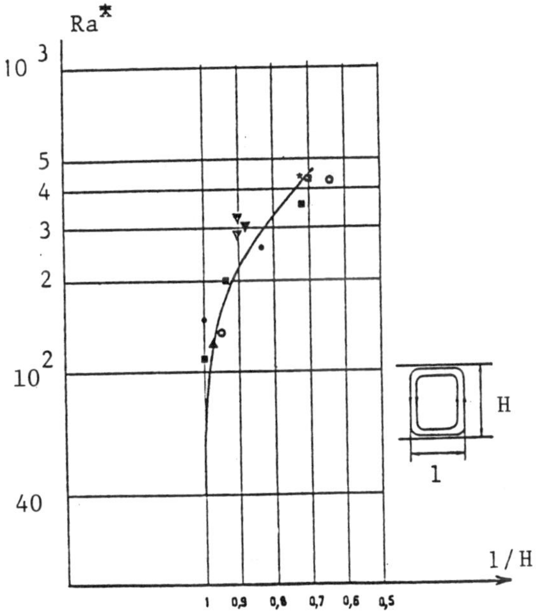

Figure 15. Evolution of wavelength for two-dimensional convection (after Jaffrennou and Bories (60)).

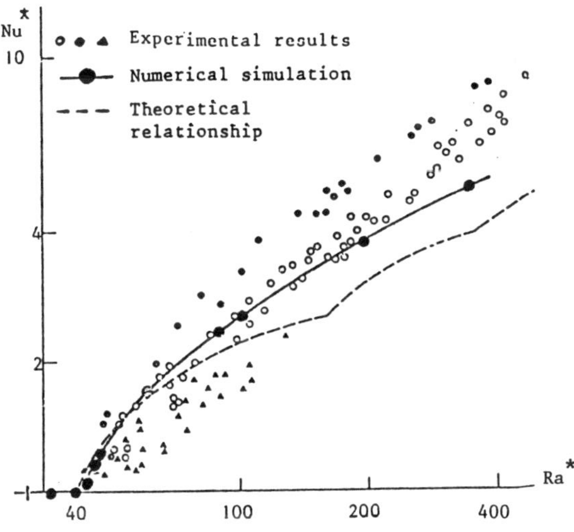

Figure 16. Nusselt-Rayleigh correlation (after Combarnous and Bories (1)).

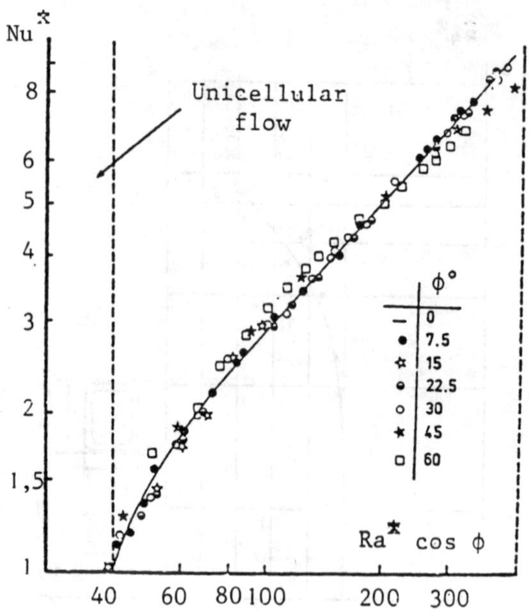

Figure 17. Correlation between Nu^* and $Ra^*\cos\phi$ (after Bories and Monferran (39)).

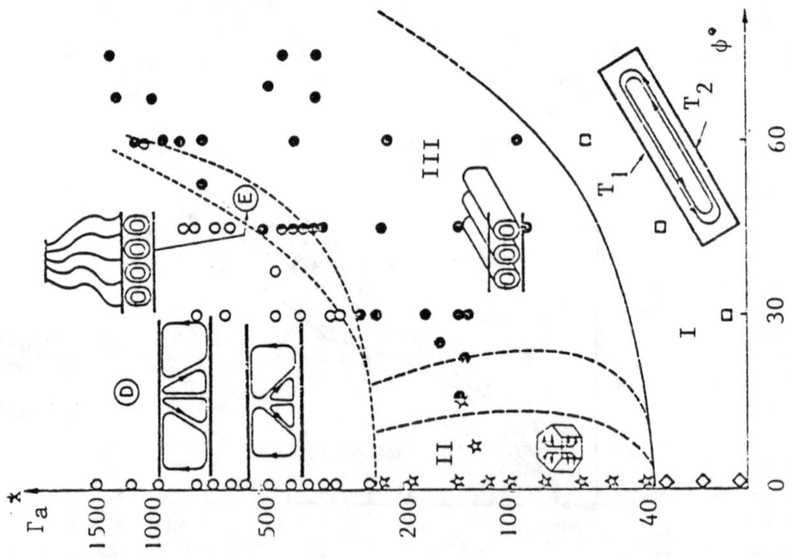

Figure 18. Different types of convective flows in a sloped layer (after Bories and Combarnous (27)).

of the experimental model, the criterion for transition between the two-dimensional unicellular flow and the three-dimensional flows is well defined by relationship $Ra^{*}\cos\phi = 4\pi^2$.

The experimental observation, deduced through visualization and in situ temperature measurements, for the convective movements is presented in a synoptic manner in Figure 18. These observations confirm the theoretical and numerical computations , except for the value of the transition angle between the structures of type II and III, where $\phi_{exp} \cong 15^{o} < \phi_t$.

3. NATURAL CONVECTION IN CONFINED POROUS MEDIA

When the porous layer is not of large lateral extent, i.e., when all dimensions are of the same order of magnitude, convective movements are influenced by geometrical dimensions and lateral thermal boundary conditions. This influence is appreciable for the convection criterion, the organization of convective movements, and for the overall heat transfer. It has also has been investigated for two basic configurations, i.e., horizontal or inclined porous boxes.

3.1. Natural Convection in Horizontal Porous Boxes

When the horizontal porous layer is laterally bounded by a solid material, complementary conditions corresponding to the impermeability of the lateral surface, $\vec{n} \cdot \vec{V} = 0$, and to the thermal conditions at the interface porous material vertical boundaries, have to be taken into account. Three cases of lateral thermal boundary conditions have been investigated:

- perfectly insulating wall, $\lambda^{*}\,\vec{n}.\nabla T = 0$ (35,52,53,54,55),
- perfectly conducting wall, $T = T(z)$ (55,56);
- and finally, imperfectly conducting wall for which the heat conduction in the region bounding the porous zone must be taken into account (55,57).

In this last case, the equation govern the heat conduction inside the wall is:

$$(\rho C)_p \frac{\partial T_p}{\partial t} = \lambda_p \nabla^2 T_p \tag{3.1}$$

where p characterizes a parameter defined in the wall, associated to the boundary conditions:

$$\lambda^{*}\,\vec{n}.\nabla T = \lambda_p\,\vec{n}\,.\,\nabla T \tag{3.2}$$

at the interface between the porous material and the side wall,

and : $\lambda_p\,\vec{n}\,.\,\nabla T_p = \vec{n}\,.\,\vec{\phi}$ (3.3) at the interface between the wall and the outside ($\vec{n} \cdot \vec{\phi}$) characterizes the heat transfer between the wall and the surrounding), has to be simultaneously solved with the governing equations of the phenomena inside the porous medium.

Such a problem was studied in (58) for the case of a porous material confined in a vertical circular cylinder horizontally bounded by two impervious isothermal surfaces and laterally by an impervious wall of given thermal conductivity and thickness (Figure 19). For this physical system, linearized steady state perturbation equations and supplemented boundary conditions become:

$$\nabla^2\theta + w = 0 \qquad \text{in the porous medium} \tag{3.4}$$

$$\nabla^2 w - Ra^* \nabla_1^2 \theta = 0 \quad \text{with} \quad \nabla_1^2 = \nabla^2 - \frac{\partial}{\partial z^2} \tag{3.5}$$

$$\nabla^2 \theta_p = 0 \qquad \text{in the wall} \tag{3.6}$$

$$\text{with } \theta = \theta_p = 0 \text{ at } z = 0 \text{ and } z = 1 \tag{3.7}$$

$$\theta = \theta_p, \quad \Lambda \frac{\partial \theta}{\partial \tau} = \frac{\partial \theta_p}{\partial r} \text{ at } r = \frac{1}{2R_o} \quad ; \quad R_o = \frac{H}{D} \; ; \quad \Lambda = \frac{\lambda^*}{\lambda_p} \tag{3.8}$$

$$\frac{\partial \theta_p}{\partial r} = \vec{n} \cdot \vec{\phi} \quad \text{at } Re = \frac{H}{D_e} \tag{3.9}$$

$$\vec{V} \cdot \vec{n} = 0 \quad \text{on the impervious surface} \tag{3.10}$$

The eigenvalue problem associated to this set of equations was solved to determine the influence of the aspect ratios R_o, R_e and of the conductivity ratio $\Lambda = \lambda^*/\lambda_p$ upon the mean features of natural convection in confined porous media (55), (56), (58).

3.1.1. Criterion for the onset of the convection

Results obtained through the linear stability theory (55), (58) for two extreme values of Λ, i.e., $\Lambda = 0$ (perfectly conductive wall) and $\Lambda \to \infty$ (perfectly insulating wall) and for an infinite thickness of the wall $Re \to \infty$, are presented in Figure 20.

These results, like those previously obtained (52), show that it is only for relative tall and slender cavities $R_o \gg 1$ that the lateral walls have much effect, such as tending delay the appearance of natural convection. This lack of influence of the walls on Ra_c^* for $R_o \leq 1$ is to be expected since, unlike the corresponding case of

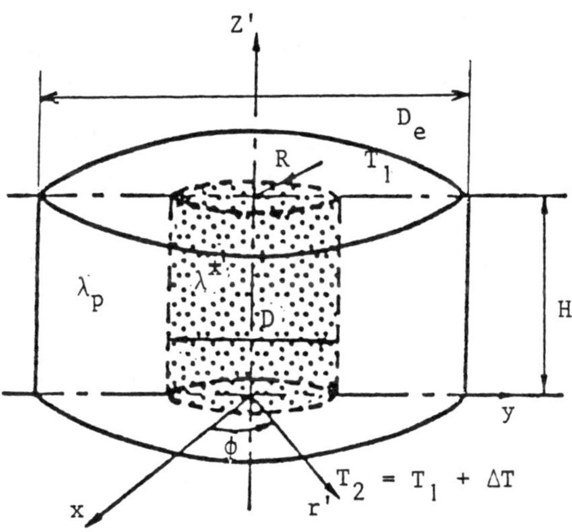

Figure 19. The porous cavity (after Bories and Deltour (55)).

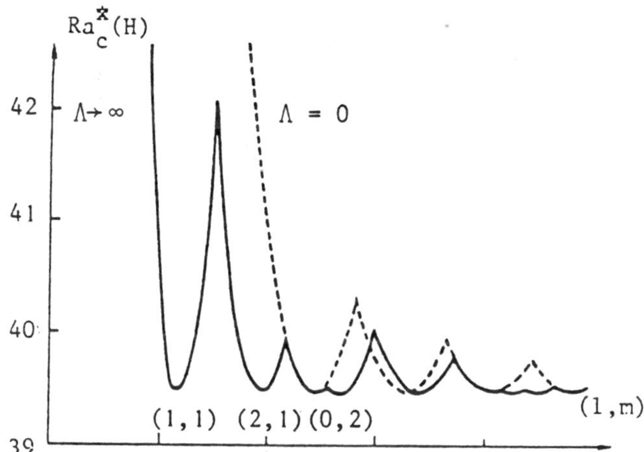

Figure 20. Critical Rayleigh number and preferred convective modes (1,m) (after Deltour (58)).

a continuous fluid, there is no viscous dissipation at the lateral walls. This result justifies the choice of aspect ratios marginally higher or lower than 1 generally used for the numerical simulation of convective movements in a large extent porous layer.

In confined porous media, the only influence of the wall is thus to select the cellular modes, i.e., the shape of the convective structures.

For high values of R_o, an estimate of the order of magnitude of each term appearing in the energy equation shows that $\partial^2T/\partial z^2 \ll \partial^2T/\partial r^2$; hence, the representative length scale of the convection is not H but D. As a consequence of the simplifications derived from this inequality, the critical value of Ra^* built on the diameter of the cavity appears as independent of H when $R_o > 6$ (Figure 21).

As for the influence of the thermal conductivity λ_p,a similar effect for the confinement has been observed, i.e., a tendency to stabilize the fluid by damping the perturbations of temperature on the side wall, when λ_p is increasing.

At low filtration Rayleigh number, this result, as well as the following concerning the mean heat transfer and the shape of convective movements has been found in good agreement with experimental observations (58).

3.1.2. Mean heat transfer

The weakly non-linear analysis based on the Malkus technique has also been used to derive the relationship between the Nusselt number and the filtration Rayleigh number in confined horizontal porous layer (54), (57), (58). Since the influence of the aspect ratio and of the lateral thermal boundary condition is only to select the cellular modes, it has been proven (54), (55), (58) that the analytical expression $Nu^*(Ra^*)$ is the same as for the large extent porous layer, i.e., :

$$Nu^* = 1 + \sum_{s=1}^{\infty} k_s(1 - \frac{Ra^*_{cs}}{Ra^*} \tag{3.11}$$

$$\text{with} \quad k_s = \begin{cases} 2 \quad \text{for} \quad Ra^* > Ra^*_{cs} \\ 0 \quad \text{for} \quad Ra^* < Ra^*_{cs} \end{cases} \tag{3.12}$$

and Ra^*_{cs} which is the critical value of appearance of the selected modes depending on R_o, R_e and Λ.

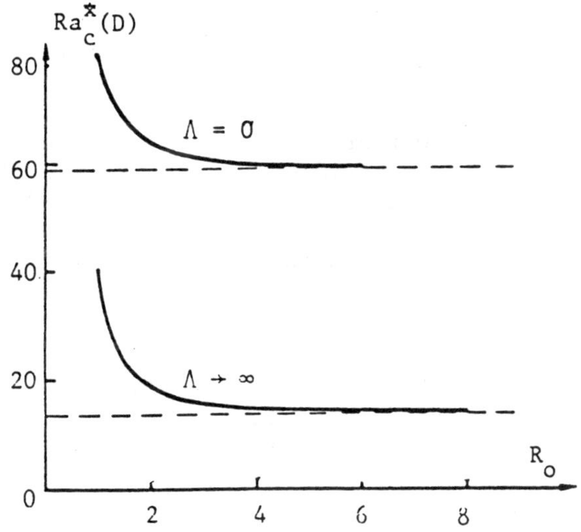

Figure 21. Criterion for the onset of convection (after Bories and Deltour (55)).

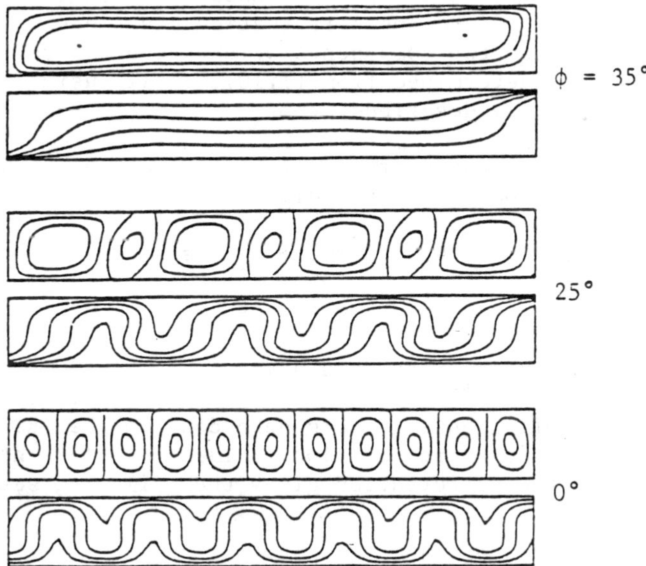

Figure 22. Isotherms and streamlines for various ϕ values $Ra^* = 100$ $A = 8$ (after Caltagirone and Bories (51)).

3.1.3. The convective structure

In the case of a porous media confined in a circular cylinder, the preferred convective modes for the two lateral boundary conditions, i.e., $\Lambda = 0$ and $\Lambda \to \infty$, are indicated (l,m) on Figure 20.

These results show that in the case of slender cavities, the convective modes are nonaxisymmetric, and that for low R_o values, they tend to become polycellular. As thoroughly described in (59), (52), between these extreme cases, different two- or three-dimensiona solutions exist, depending on the geometry of the lateral walls and on initial conditions (57). This question is very delicate and not yet well understood.

3.2. Sloped Porous Layer of Finite Lateral Extent Bounded by Perfectly Insulating Wall (51)

3.2.1. Stability of the unicellular flow

According to the modifications introduced to the basic flow by the variations of the lateral dimension L, the stability criterion $Ra^*\cos\phi = 4\pi^2$ is no longer satisfied when the extension of the layer in the direction of the inclination takes finite values. In order to determine the influence of this parameter upon the stability of unicellular flow, it is then necessary to know the new field T_o, $\vec{V}_o$ to be introduced into the perturbation equations. Noting that in this case $\vec{V}_o$ has two components which are not zero, say: $\vec{V}_o = U_o\vec{i} + W_o\vec{k}$ this has been determined from (1.10), (1.12) using the Galerkin method previously described.

Restricting the approximation rank to N = 2, the solution of the basic flow can be stated as follows:

$$T_o = (1 - z) + a_{o2}\sin 2\pi z + a_{11}\cos\pi x \sin\pi z \qquad (3.13)$$

$$U_o = - Ra^*\pi^2 A^2 b_{11} \sin \pi x \cos \pi z \qquad (3.14)$$

$$W_o = Ra^* \pi^2 b_{11} \cos \pi x \sin \pi z \qquad (3.15)$$

where the coefficients a_{ij} are dependent on the (Ra^*,ϕ) couple. Substituting the basic solution (3.13) to (3.15) into the linearized equation of the perturbation gives the following set of equation:

$$\frac{1}{A^2}\frac{\partial u}{\partial x} + \frac{1}{B^2}\frac{\partial v}{\partial y} + \frac{\partial w}{\partial z} = 0 \qquad (3.16)$$

$$- \nabla\pi + Ra^* \vec{k}\,\theta - \vec{V} = 0 \qquad (3.17)$$

$$\frac{\partial\theta}{\partial t} - \left(\frac{1}{A^2}\frac{\partial^2\theta}{\partial x^2} + \frac{1}{B^2}\frac{\partial^2\theta}{\partial y^2} + \frac{\partial^2\theta}{\partial z^2}\right) + \frac{1}{A^2}\left(U_o\frac{\partial\theta}{\partial x} + u\frac{\partial T_o}{\partial x}\right)$$

$$+ \frac{1}{B^2}\left(U_o\frac{\partial\theta}{\partial y} + u\frac{\partial T_o}{\partial y}\right) + w\frac{\partial T_o}{\partial z} + W_o\frac{\partial\theta}{\partial z} = 0 \qquad (3.18)$$

and associated boundary conditions:

$v = 0$ at $y = 0$ and $y = 1$
$w = 0$ at $z = 0$ and $z = 1$
$\theta = 0$ at $z = 0$ and $z = 1$

$$\frac{\partial\theta}{\partial x} = \frac{\partial\theta}{\partial y} = 0 \quad \text{at} \quad x = y = 0 \quad \text{and } x = y = 1 \qquad (3.19)$$

corresponding to the restating of Eqs. (1.10), (1.12), using an orthogonal reference system with differently distorted coordinates.

Since the transition we are searching for corresponds to the change from a unicellular regime to longitudinal coils, the perturbation is taken as two-dimensional, and then the x component of the velocity perturbation is equal to zero ($U = 0$).

The solution of the eigenvalue problem carried out by adopting the following forms:

$$\theta = \cos m\pi z \sum_{s=1}^{N} A_s \sin s\pi z$$

$$v = - Ra^* \pi^2 B^2 m \sin m\pi y \sum_{s=1}^{N} s B_s \cos s\pi z \qquad (3.20)$$

$$w = Ra^* \pi^2 m^2 \cos m\pi y \sum_{s=1}^{N} B_s \sin s\pi z$$

where A_s and B_s coefficients are indeterminate using linear theory, gives the results shown in Figure 18 for different values of L. It can be observed that with a given angle, the critical Rayleigh number is always greater than $Ra^*\cos\phi = 4\pi^2$ corresponding to an ($L \to \infty$) infinite aspect ratio.

This result, the analytical relation of which has been previously given (60):

$$Ra^* \cos\phi = 4\pi^2 + 3M^2 \tag{3.21}$$

where M is a decreasing function of L, ($M \to 0$ when $L \to \infty$), emphasizes the stabilizing role of longitudinal confinement upon the unicellular flow.

As for the case of the inclined fluid layer, experimental and theoretical studies are still necessary in order to provide a good understanding of the influence of small aspect ratios and high Rayleigh numbers on both convective flows and mean heat transfer.

3.2.2. Two- and three-dimensional flows

Numerical computation based either on the Galerkin method or on the finite difference method (51) has been used to predict the structures of the convective flows and their evolution as functions of the inclination and aspect ratio L.

Two examples respectively for two- and three-dimensional flows are shown on Figure 22 and 23. Figure 22 illustrates the different transitions explaining the number rolls decreasing when the angle ϕ increases for a two-dimensional flow. Figure 23 shows the existence of solutions corresponding to longitudinal coils for L and W >> H, but finite in the domain III on Figure 18.

A few other numerical results (53,62,63) are also available in the literature describing natural convection in inclined porous boxes, such as $L \cong H$.

3.3. Vertical Porous Layers

Due to its fundamental importance in thermal insulation engineering, two-dimensional steady natural convection in rectangular porous cavities bounded by vertical walls at different temperatures and adiabatic horizontal walls has been extensively studied during the last two decades.

Theoretical work reported on this problem includes boundary layer solutions (66,67,68), integral analysis (67,69,70), and numerical results (71,72,73,74,75,76,77,78). Based on these studies, different flow regimes, temperature profiles, and Nusselt number correlations have been presented as functions of the Rayleigh number and aspect ratios.

For the observed convective flows, a distinction has to be made between the tall cavities A >> 1 in which the convective flow is always unicellular and the shallow cavities A << 1 in which the convective flow may be multicellular when the Rayleigh number greatly increases (78). In spite of these structural differences at

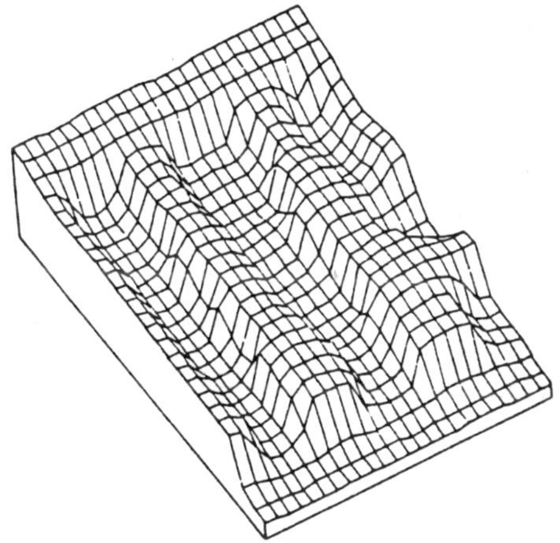

Figure 23. Three-dimensional flow: longitudinal rolls isotherm T = 0.5 Ra^* = 100 A = 6 B = 4 $\phi = 30^o$ (after Caltagirone and Bories (51)).

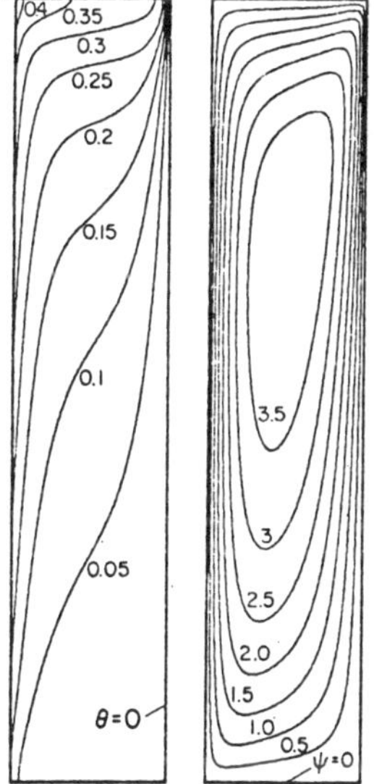

Figure 24. Isotherms and streamlines for a porous cavity for Ra^* = 1000 and A = 5 (after Prasad and Kulacki (78)).

high Rayleigh numbers, the same flow patterns are evident for both configurations, i.e., :

- the conduction regime in which the isotherms are almost parallel to the vertical walls for $Ra^* \to 0$ and A is finite or Ra^* is finite and $A \to \infty$ or $A \to 0$,
- and the asymptotic and boundary layer states characterized by a stratified core in which the temperature gradient is quite modest, surrounded by thin thermal layers on the cavity walls when A is finite Ra^* increases (Figure 24).

Numerous correlations for the mean transfer function, a synthesis of which can be found in (70), have been proposed to describe the evolution of the average Nusselt number when Ra^* and A vary. For boundary layer regimes, the most generally used correlations are, respectively, Weber's correlation for tall enclosures, $Nu^* = (3)^{\frac{1}{2}}(Ra^*)^{\frac{1}{2}}(A)^{\frac{1}{2}}$ and the Walker-Homsy correlations for shallow enclosures, $Nu^* = \frac{1}{120}(Ra^*)^2(A)^4$. Between these two extreme cases, numerical results show that Nu^* goes through maxima. Depending on Ra^*, these maxima are obtained for A varying from 1 to 30 when the Rayleigh number is varying from 20 to 300 (71,72,78) (Figure 25).

This influence of the reduced extension on the Nusselt number is due to the fact that heat transfer caused by the displacement of the saturating fluid is mainly concentrated on the vertical or horizontal ends respectively in tall or shallow cavities.

A thorough description of the convective flows, temperature fields, and overall heat transfer for a large range of aspect ratios and filtration Rayleigh numbers can be found in papers recently published (77,78). Among the main features of these numerical studies, we shall mention:

- the definition of criteria for flow regimes in tall and shallow cavities;
- the testing, applicability, and accuracy of the analytic correlation given to estimate the heat transfer;
- the finding of solutions corresponding to multicellular flow in shallow cavities; and
- the study of the effect of a constant heat flux on one vertical wall compared to the case of a cavity with two vertical walls at constant temperatures.

Free convective heat transfer in cylindrical annuli filled with a saturated porous medium has been also studied. For an annuli whose inner wall is heated at constant temperature and outer wall is isothermally cooled, the top and bottom being insulated (82,83, 84,85), heat transfer results have been obtained for a wide range

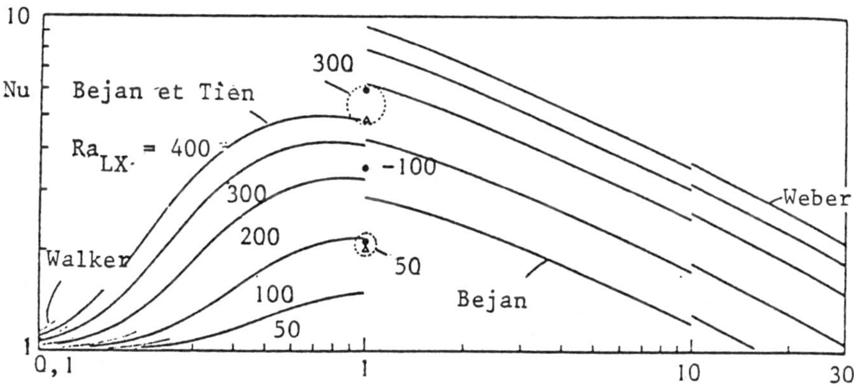

Figure 25. Summary of heat-transfer theories (after Bejan (72)).

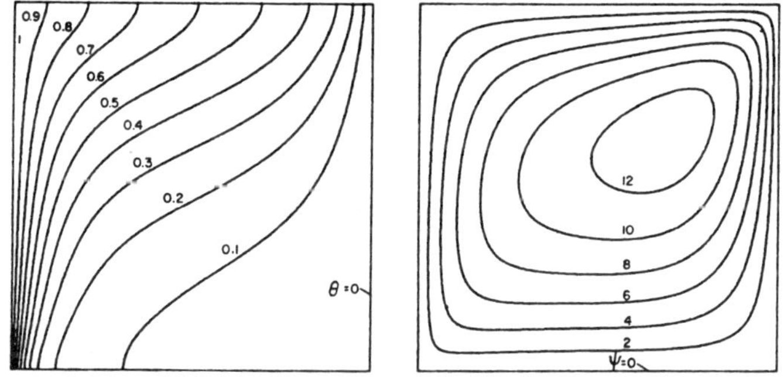

Figure 26. Streamlines and isotherms in a vertical porous annuli: $Ra^* = 100$, A = 1, C = 6 (after Prasad (82)).

of Rayleigh numbers, aspect (height to gap width) ratios, A, and radius ratios, C. Results obtained through a numerical study show that the curvature effects are significant, and completely disturb the centrosymmetrical nature found in the vertical cavity case (Figure 26). Though the effect of the Rayleigh number and the aspect ratio are qualitatively similar to what has been observed for the vertical cavity, the correlations for the average Nusselt number requires modification in order to include the influence of the curvature C.

4. NATURAL CONVECTION IN MORE COMPLEX CONFIGURATIONS

In the previous sections, we have presented the main aspects of thermal convection occurring in porous media which was considered to be homogeneous and isotropic. Likewise, the saturating fluid was considered as obeying the Boussinesq assumption. If the study of these simple cases is an essential step toward a better understandin of the phenomena, it is not sufficient to analyze the effect of convection in more complex cases like those generally encountered in natural or industrial situations.

For the simple geometrical configuration of an horizontal porou layer of large lateral extent, for instance, it is often necessary to consider that the porous material is not homogeneous or isotropic and that the physical properties of the saturating fluid are depende upon the temperature. Such is the case in the modeling of geotherma fields and in the study of the insulating material submitted to larg temperature gradients.

4.1. The Multilayered Porous Medium

The multilayered system studied by several authors is assumed to comprise n separately homogeneous layers of total thickness H saturated by a fluid obeying the Boussinesq's assumption. Beneath layer 1, the system is bound by an impermeable isothermal surface at temperature $T_2 = T_1 + \Delta T$, where T_1 is the temperature of the isothermal top surface which is considered to be either impermeable or at constant pressure. The porous material contained in layer i of thickness H_i, has a permeability tensor $\bar{\bar{K}}_i$ and an equivalent thermal conductivity tensor $\bar{\bar{\lambda}}_i$. Within each layer, the usual equations of conservation of mass, momentum, and energy hold, and appropriate continuity considerations for the temperature, the vertical component of the velocity, of the heat flux, and of the pressure determine boundary conditions at the interface between the layers (Figure 27).

For this physical system, the formalism required to determine the criterion for the onset of convection has been developed using a straightforward stability analysis (86,87). The post-onset

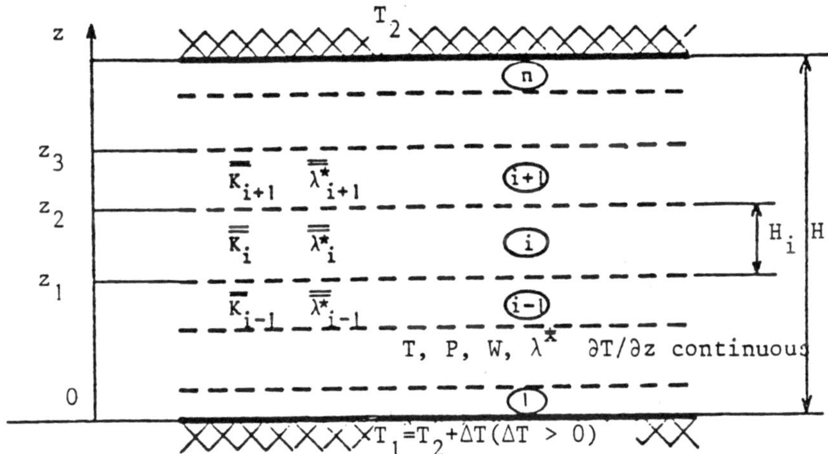

Figure 27. Schematic diagram of a layered porous medium.

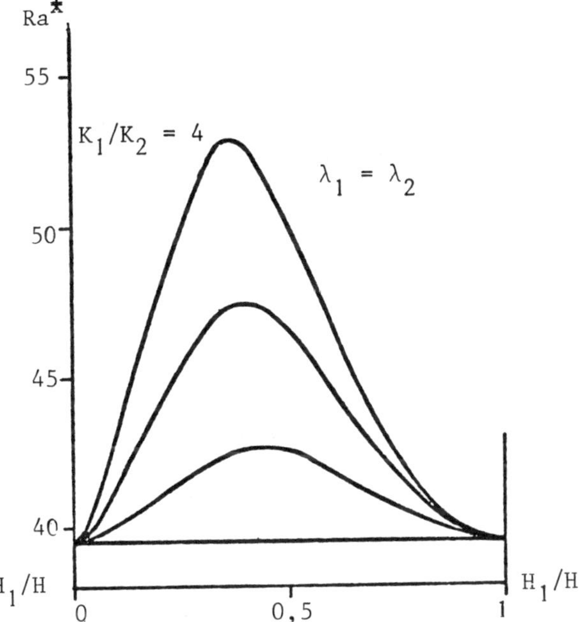

Figure 28. Criterion for the onset of convection in a two-layered porous media (after Richard (87), and Richard and Combarnous (88)).

behavior provides an estimate of the heat transported by convection, for slightly supercritical Rayleigh numbers, by means of the perturbation approach based on the amplitude A of convection as small parameter (88) (weakly non-linear analysis).

The method proceeds by introducing the series expansion:

$$Ra_i^* = Ra_{ic}^* + A\, Ra_i^{*(1)} + A^2\, Ra_i^{*(2)} \tag{4.1}$$

$$\theta_i = A\, \theta_i^{(1)} + A^2\, \theta_i^{(2)} + \dots \tag{4.2}$$

(where i = 1 ... n characterizes every layer), and analogous expressions for V_i and P_i into the equations of the perturbations, and by solving successively the linear equations corresponding to each power of A associated with the boundary conditions at each interface. Since only steady solutions are considered, the $\partial/\partial t$ terms vanish and in the order A, the problem becomes identical to the linear problem. In the higher-order, inhomogeneous linear system equations determine the amplitude of the perturbation, hence the Nusselt number A second order approximation for the Nusselt number yields:

$$Nu^* = 1 + K\left(1 - \frac{Ra_c^*}{Ra^*}\right) \tag{4.3}$$

where K depends on the amplitude of the perturbation and Ra^* is a Rayleigh number defined in terms of the thickness and temperature drop of the whole system and the conductivity and permeability of layer 1. As shown in (87,88,89,90), A and Ra_c^* are dependent upon the number of layers, layer depths, layer permeability and conductivity ratios and cell width.

An extensive study of this problem has been developed (see (87 -92)), for multilayered porous media, the layers of which have different thickness, permeability, and conductivity ratios. For two-dimensional convection patterns, a wide variety of possible configurations and values of parameters has been studied.

Some results(87) studying the evolution of Ra^* function of K_1/K_2 and H_1/H with $\lambda_1^* = \lambda_2^*$ for a two-layer system of isotropic porous media are shown in Figure 28, and the streamlines corresponding to a four-layer system of isotropic porous media are in Figure 29 (89).

These results show clearly that the presence of layers of different permeability can have great influence on the convective flow in porous media heated from below, and that the modeling of

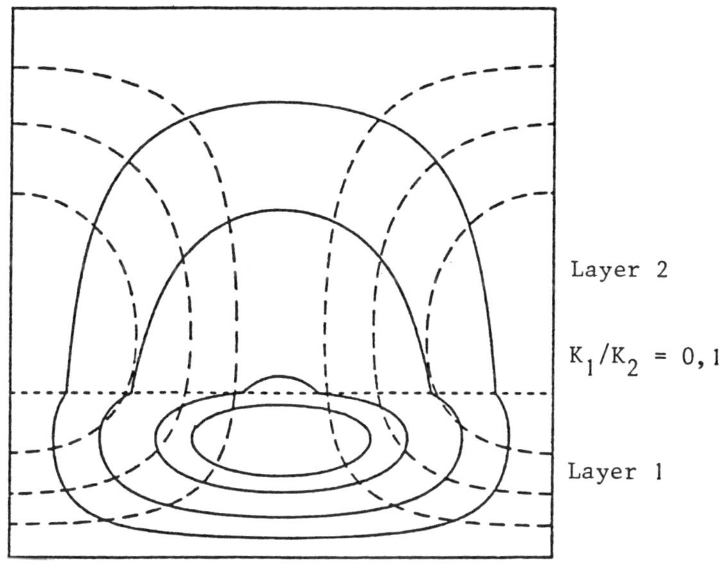

Figure 29. Streamlines (———) and isotherms (-----) at onset of convection (after McKiabin and O'Sullivan (86)).

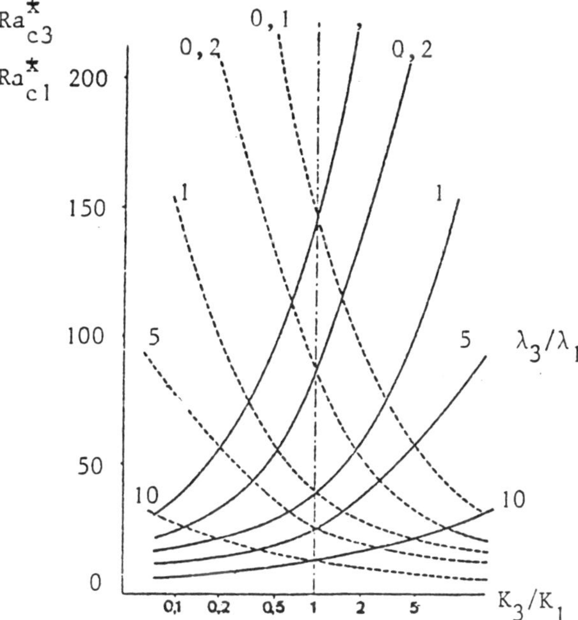

Figure 30. Criteria for onset of convection Ra^*_{c3}(——) Ra^*_{c1}(---).

such systems by a homogeneous layer may give quite erroneous predictions as far as convection and heat flux are concerned.

4.2. The Homogeneous Anisotropic Porous Layer (93,94)

If $\bar{\bar{K}}$ and $\bar{\bar{\lambda}}^*$ are respectively the dimensionless tensors of permeability and thermal conductivity such as:

$$\bar{\bar{K}} = k_1 = \vec{i}.\vec{i} + k_2\, \vec{j}.\vec{j} + \vec{k}.\vec{k} \tag{4.4}$$

$$\bar{\bar{\lambda}}^* = \lambda_1\, \vec{i}.\vec{i} + \lambda_2\, \vec{j}.\vec{j} + \vec{k}.\vec{k}$$

with:

$$k_1 = \frac{K_1}{K_3} \;;\quad k_2 = \frac{K_2}{K_3} \;;\quad \lambda_1 = \frac{\lambda_1^*}{\lambda_3^*} \;;\quad \lambda_2 = \frac{\lambda_2^*}{\lambda_3^*} \tag{4.5}$$

where K_1, K_2, K_3, λ_1^*, λ_2^*, λ_3^* are the principal components of $\bar{\bar{K}}$ and $\bar{\bar{\lambda}}^*$, the Rayleigh number for the onset of convection is found to be:

$$Ra^* = \frac{1 \quad m^2 + k_2\, 1^2 + \pi^2}{k_1\, m^2 + k_2\, 1^2} \;(\lambda_1\, m^2 + \lambda_2\, 1^2 + \pi^2) \tag{4.6}$$

with $a^2 = 1^2 + m^2$ horizontal wave number and:

$$Ra^* = \frac{g\ \alpha\ \Delta T\ H\ K_3 (\rho C)_f}{\nu\ \lambda_3^*}$$

defined in terms of the vertical components of $\bar{\bar{K}}$ and $\bar{\bar{\lambda}}^*$.

Minimizing Eq. (4.6) with respect to 1 and m, yields the critical Rayleigh number:

$$Ra_c^* = \pi^2 (\min \{ (\frac{\lambda_1}{k_1})^{\frac{1}{2}} \quad , \quad (\frac{\lambda_2}{k_2})^{\frac{1}{2}} \} + 1)^2 \tag{4.7}$$

Three cases can be considered:

$\lambda_1/k_1 < \lambda_2/k_2$ which gives rolls aligned in the y direction:

$$1 = \pi(\lambda_1 k_1)^{\frac{1}{4}}, \; m = 0 \tag{4.8}$$

$\lambda_1/k_1 > \lambda_2/k_2$ which gives rolls aligned in the x direction:

$$l = 0, \quad m = \pi(\lambda_2 \, k_2)^{-\frac{1}{4}} \tag{4.9}$$

$\lambda_1/k_1 = \lambda_2/k_2$ which gives the critical wave number vector: $\vec{a} = l\vec{i} + m\vec{j}$ such as:

$$(\lambda_1 \, k_1)^{\frac{1}{2}} \, l^2 + (\lambda_2 \, k_2)^{\frac{1}{2}} \, m^2 = \pi^2 \tag{4.10}$$

In the case of horizontal isotropy with : $k_1 = k_2 = k$; $\lambda_1 = \lambda_2 = \lambda$, we obtain:

$$Ra_c^* = \pi^2((\lambda/k)^{\frac{1}{2}} + 1)^2 \quad \text{and } a = \pi(\lambda \, k)^{-\frac{1}{4}} \tag{4.11}$$

For this configuration, two relations can be used for the filtration Rayleigh number: Ra_3^* defined in terms of K_3 and λ_3^* or $Ra_1^* = Ra_2^*$ defined in terms of $K_1 = K_2$ and $\lambda_1^* = \lambda_2^*$. Criterion for onset of corresponding convection, i.e., Ra_{c3}^* and Ra_{c1}^*, are presented in Figure 30 for different values of the ratios K_3/K_1 and λ_3^*/λ_1^*.

For supercritical conditions, the steady non-linear problem has been investigated both numerically and analytically. Regions of stable wave numbers and Rayleigh numbers have been found for two-dimensional motion. The results obtained show that the Nusselt number and the stability regions depend on the anisotropic parameters only through the ratios k_1/λ_1 and k_2/λ_2.

Some experiments performed in anisotropic porous layer (95,96), seem to confirm the validity of the relations giving the critical Rayleigh number and the mean heat transfer. However, this problem needs further experimental studies.

4.3. Natural Convection in a Porous Layer Saturated by a Fluid of Non-Constant Properties

All the studies reviewed until now assumed that fluid properties such as thermal expansion α, viscosity μ, and specific heat remain constant. In addition, each also involved the Boussinesq approximation that the fluid density ρ is constant, except in so far as it affects the buoyancy forces. Valid when the differences of temperature in the porous layer are small, these assumptions are not relevant at high temperature differences.

The influence of variable viscosity due to large temperature differences at the onset of natural convection and horizontal platform of the cellular motion has been studied in (114), and both the variations of viscosity and density have been taken into account in (115,116,117). For these two cases, solutions derived from the linear stability analysis show that the critical Rayleigh number is lower than $4\pi^2$, and depends on the coefficient of variation of the viscosity with the temperature and on the state equation of the fluid.

For a porous medium saturated with an ideal gas at constant pressure such as:

$$\mu = \mu_m(1 + \gamma(T - T_m)) \quad \text{and} \quad \rho_m T_m = \rho_1 T_1 \tag{4.12}$$

where γ is a constant and μ_m and ρ_m, the dynamic viscosity and the volumetric mass at mean temperature $T_m = (T_1 + T_2)/2$, it was shown that the critical filtration Rayleigh number and the wave number of the perturbation are dependent upon γ, ΔT, T_1 : $Ra_c^*(\gamma, \Delta T, T_1)$; $a_c(\gamma, \Delta T, T_1)$ with:

$$Ra^* = \frac{g\ \alpha(T_m)\ \rho^2(T_m) C_p\ \Delta T\ K\ H}{\mu(T_m)\ \lambda^*} \tag{4.13}$$

based on the physical properties of the saturating gas at mean temperature. An example of the correlation $Ra_c^*(\Delta T/T_1)$ is given on Figure 31 in the case of a dry air-saturated porous layer. In the range of temperature varying from 80 to 300°K, Ra_c^* tends toward the maximum $4\pi^2$ when ΔT tends toward zero, i.e., when the Boussinesq assumption becomes relevant.

A few other results have also been published on natural convection near 277°K in water-saturated porous media. For the preceding cases, due to the non-linear relationship between water volumetric mass and temperature near 4°C the linear Boussinesq approximation is not applicable (118).

4.4. Free Convection Around Surfaces and Concentrated Heat Sources in Infinite Porous Media (External Convective Flows)

Steady free convection around heated surfaces or concentrated heat sources embedded inside an infinite porous media i.e., external convective flows, also plays an important role in numerous geophysical and engineering applications (108).

Among the most frequently studied configurations, we shall mention horizontal or inclined surfaces (97 - 108) and concentrated heat sources (109 to 113).

Based on the boundary layer approximations derived from the scale analysis of the problem, similarity solutions, i.e., solutions giving a similar temperature profile from one location, x, on the

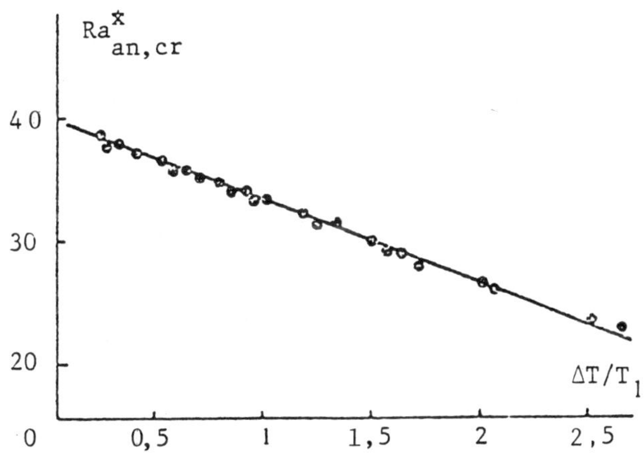

Figure 31. Criterion for the onset of convection in a porous layer saturated by a perfect gas (after Epherre et al. (117)).

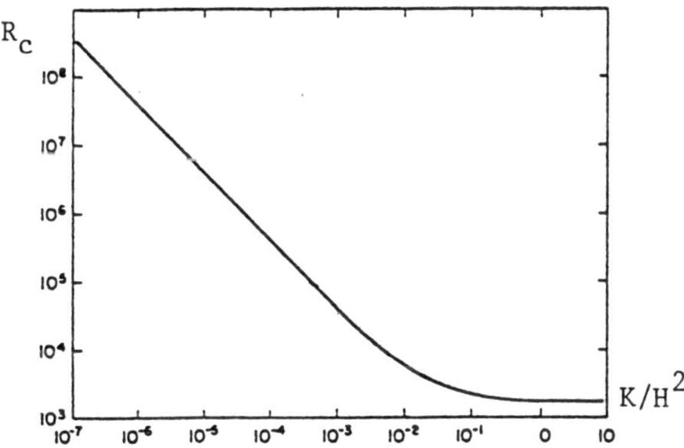

Figure 32. The eigenvalue R_c as a function of dimensionless permeability ($R_c = 4\pi^2H^2/K$) (after Walker and Homsy (121)).

surface to another, have been obtained for two-dimensional free convection on a heated vertical cylinder (98), and on a horizontal and inclined heated surface (see (99 - 108)).

In the case of a horizontal heated surface, a linear stability analysis has been also made to determine the conditions of appearance of secondary flow in the boundary layer when the prescribed wall temperature is a power function of distance (101). Starting from a basic external flow corresponding to the steady two-dimensional buoyancy-induced boundary layer flow, disturbance equations have been solved numerically in order to obtain the critical Rayleigh number governing the stability of the boundary layer and the shape of perturbations. From these computations, it has been proved that the transverse rolls correspond to the most stable solution and that the wall temperature plays an important role on the stability of the layer; the existence of a temperature gradient along the surface has a stabilizing effect.

Natural convection in an infinite porous medium with concentrated heat sources has been studied recently (112,113). For a unique heat source point which is continuous in time and is suddenly embedded in a infinite fluid-saturated porous medium, the transient time-dependent temperature and flow pattern around the source and the steady regime obtained as time approaches infinity was investigated (112) using a standard perturbation analysis.

Solutions presented as a function of the Rayleigh number: $Ra^* = g \frac{\alpha(\rho C)_f}{\nu} \frac{K}{\lambda^{*2}} Q$ based on the source strength Q (rate of energy release) and the permeability of the medium, show that the transient flow pattern consists of an expanding vortex ring situated in the horizontal plane containing the source, and that the steady state of an upward flow pattern becomes very intense near the source. Due to the approximations chosen in the expansion series (first order for the transient state and third order for the steady state), solutions are only valid for small Rayleigh numbers.

For the study of a variety of isolated heat sources (113), thermal and flow fields have been approximately determined through an analytical approach based on the superpositions of solutions. Limited to the linear system derived from the first term of the expansion series, the solutions are only valid for $Ra^* \to 0$.

4.5. The Boundary Effects on Natural Convection in Porous Media

Due to increasing use of high porosity media in thermal insulating techniques, a thorough understanding of boundary layer effects, either in the vicinity of an impermeable surface or in the transition zone between a porous material and a fluid, has become more and more important.

As the Darcy's law is unable to describe the hydrodynamics phenomena occurring on these interfaces, another momentum equation has been proposed to study these problems. Known as the Brinkman model (119), this momentum equation may be considered as an extension of the Darcy's law. Generally written:

$$\frac{\mu}{K}\vec{V} = -\nabla P + \rho\vec{g} + \mu\nabla^2\vec{V} \tag{4.14}$$

it includes the viscous forces term, and consequently, is able to satisfy the no-slip conditions on an impermeable surface or the continuity in stress on the interface between a porous media and a fluid. In a few cases, inertial terms $\frac{\partial\vec{V}}{\partial t} + (\vec{V}\cdot\nabla)\vec{V}$ were taken into account (120,121).

When the Brinkman model is used instead of the Darcy's law, it appears that natural convection does not only depend on the Rayleigh number, but also on the ratio $K/H^2 = D_a$ known as the Darcy-Brinkman number.

This dimensionless number emphasizes both the influence of the porosity and grain size, or pore diameter, on the phenomena. As shown in a study concerning the onset of free convection in an horizontal porous layer of large lateral extent bounded by isothermal impermeable surfaces, this influence is negligible as far as $D_a \lesssim 10^{-4}$, (Figure 32).

Among the results recently published on the use of the Brinkman model to study convective heat transfer, we mention:

- natural convection in vertical porous enclosures (122) where a thorough analysis of the influence of D_a and A, both on the boundary layer and mean heat transfer, is developed. For a given value of A, these results confirm the influence of D_a previously derived in (121), i.e., a satisfactory approximation of the phenomena by means of the Darcy's law for $D_a \lesssim 10^{-4}$;

- natural convection on a semi-infinite vertical flat plate in a porous medium (123), where it is found that the no-slip boundary conditions have a lesser effect on the heat transfer than on the vertical velocity profile; and

- natural convection in a vertical fluid cavity divided by a permeable porous layer (124), where a numerical simulation of both the hydrodynamics and the heat transfer is developed by means of a sole conservation equation of the Brinkman type with inertial terms added. In this case, the transition from the fluid to the porous medium is realized via the permeability which is the chosen function of the spatial coordinate. Thus, the need to state a boundary condition at the interface of the two media is avoided. As for the preceding cases, the limit of the validity of Darcy's law is also confirmed for $D_a \leq 10^{-4}$.

5. CONCLUDING REMARKS

This review surveys a set of papers that have been published during the last decade on natural convection in saturated porous media.

Despite the great number and the variety of studied configurations, further experiments and theoretical analysis are still necessary to improve the most fundamental aspects of the problem.

Focusing our attention on the sole thermal phenomena, many questions have to be solved in order to give an undeniable predictive character for the use of mathematical formulation.

Of general interest for studies of heat transfer in saturated porous media, these questions especially concern: first, the definition of conditions that must be satisfied in order to use with confidence the homogeneous model based on a unique heat transfer equation, and second, how to accurately estimate the coefficients of equivalent thermal conductivity tensors $\bar{\bar{\lambda}}^*$, $\bar{\bar{\lambda}}_s^*$, $\bar{\bar{\lambda}}_f^*$ and the heat transfer coefficient, h.

Nevertheless, when the porous medium is a very thorough blend of solid and fluid phases, and the equivalent thermal conductivity is known, results obtained on natural convection confirm the validity of the approach based on the homogeneous model.

6. ACKNOWLEDGEMENTS

The author is grateful to Miss Marie-Luz Tison for material fulfilment of this chapter, to Australian friend Greame Clark for his active contribution in translation of this paper.

7. REFERENCES

1. Combarnous M.A., and Bories S.A., "Hydrothermal Convection in Saturated Porous Media," *Advances in Hydrosciences*, Vol. 10, Academic Press, New York, N.Y., 1975, pp. 231-307.
2. Klarsfeld S., "Champs de température associés aux mouvements de convection naturelle dans un milieu poreux limité," *Revue Générale de Thermique*, Vol. 9, 1970, pp. 1403-1424.

3. Marle C., "From the Pore Scale to the Macroscopic Scale: Equation Governing Multiphase Fluid Through Porous Media," *Flow and Transport in Porous Media,* A.A. Belkema, ed., 1981, pp. 57-61.
4. Matheron G., *Elements Pour Une Théorie des Milieux Poreux,* Masson et Cie Editeurs, Paris VIe, 1967.
5. Withaker, S., "Simultaneous Heat, Mass and Momentum Transfer in Porous Media. A Theory of Drying," *Advances in Heat Transfer,* Vol. 13, 1977, pp. 119-203.
6. Bear, J., *Dynamics of Fluids in Porous Media,* American Elsevier, 1972.
7. Hassanizadeh M., and Gray, W.G., "General Conservation Equations for Multiphase Systems: 3. Constitutive Theory for Porous Media," *Advances in Water Resources,* Vol. 3, 1980, pp. 25-40.
8. Slattery, J.C., *Momentum Energy and Mass Transfer in Continua,* Mc Graw Hill Book Company, 1972.
9. Whitaker,S., "Local Thermal Equilibrium: An Application to Packed Bed Catalytic Reactor Design," to be published in *Chemical Ingr. Science.,* 1984.
10. Jaguaribe,E., and Beasley, D., "Modeling of the Effective Thermal Conductivity and Diffusivity of a Packed Bed with Stagnant Fluid," *Journal of Heat and Mass Transfer,* Vol.27, No. 3, 1984, pp. 399-407.
11. Bear, J., and Bachmat, Y., "Transport Phenomena in Porous Media. The Basic Equations," *Review NATO Advanced Study Institute on Mechanics of Fluids in Porous Media,* Newark, DE, July 18-27, 1982.
12. Chandrasekhar, S., *Hydrodynamic and Hydromagnetic Stability,* Oxford at the Clarendon Press, 1961.
13. Westbrook, D.R., "Stability of Convective Flow in a Porous Medium," *Physics of Fluids,* Vol. 12, 1969, pp. 1547-1551.
14. Joseph, D.D., "Stability of Fluid Motions," *Springer Tracts in Natural Philosophy,* Vol. 27 and 28, Springer, Berlin, Heidelberg, New York, 1976.
15. Katto, Y., and Masuoka, T., "Criterion for Onset of Convective Flow in a Fluid in a Porous Medium," *International Journal of Heat and Mass Transfer,* Vol. 10, 1967, pp. 297-309.
16. Lapwood, E.R., "Convection of a Fluid in a Porous Medium," *Proceedings of Cambridge Pil. Society,* Vol. 44, 1948 pp. 508-521.
17. Horton, C.W., and Rogers, F.T., "Convection Currents in a Porous Medium," *Journal of Applied Physics,* Vol. 16, 1945, pp. 367-370.
18. Nield, D.A., "Onset of Thermohaline Convection in a Porous Medium," *Water Resources Research,* Vol. 4, 1968, pp. 553-560.
19. Caltagirone, J.P., "Stability of a Saturated Porous Layer Subject to a Sudden Rise in Surface Temperature: Comparison Between the Linear and Energy Methods," *The Quarterly Journal of Mechanics and Applied Mathematics,* Vol. 23, 1980, pp. 47-58.

20. Busse, F.H., and Riahi, N., "Non Linear Convection in a Layer With Nearly Insulating Boundaries," *Journal of Fluid Mechanics*, Vol. 96, 1980, pp. 243-256.
21. Riahi, N., "Non Linear Convection in a Porous Layer with Finite Conducting Boundaries," *Journal of Fluid Mechanics*, Vol. 129, 1983, pp. 153-171.
22. Malkus, V.W.R., and Veronis, G., "Finite Amplitude Convection," *Journal of Fluid Mechanics*, Vol. 4, 1958, pp. 225-260.
23. Gorkov, L.P., "Stationary Convection in a Plane Liquid Layer Near the Critical Heat Transfer Point," *English Transl. Sov. Phys. JETP*, Vol. 6, 1957, pp. 311-315.
24. Edwards, D.K., and Catton, I., "Prediction of Heat Transfer by Natural Convection in Closed Cylinder Heated From Below," *International Journal of Heat & Mass Transfer*, Vol. 12, 1969, pp. 23-30.
25. Schluter, A., and Lortz, D., and Bosse, F.H., "On the Stability of Steady Finite Amplitude Convection," *Journal of Fluid Mechanics*, Vol. 23, 1965, pp. 129-144.
26. Bories, S.A., "Comparaison des prévisions d'une théorie non Linéaire et des résultats expérimentaux en convection naturelle dans une couche poreuse saturée horizontale," *Comptes-rendus Acadēmie des Sciences B*, Vol. 271, 1970, pp. 269-272.
27. Bories, S.A., and Combarnous, M.A., "Natural Convection in a Sloping Porous Layer," *Journal of Fluid Mechanics*, Vol. 57, 1973, pp. 63-79.
28. Caltagirone, J.P., "Thermoconvective Instabilities in a Porous Layer," *Journal of Fluid Mechanics*, Vol. 72, 1975, pp. 269-287.
29. Caltagirone, J.P., "Thermoconvective Instabilities in a Porous Medium Bounded by Two Concentric Horizontal Cylinders," *Journal of Fluid Mechanics*, Vol. 76, 1976, pp. 337-362.
30. Caltagirone, J.P., and Combarnous, M., and Mojtabi, A., "Natural Convection Between Two Concentric Spheres: Transition Towards a Multicellular Flow," *Num. Heat Transfer*, Vol. 3, 1980, pp. 107-114.
31. Caltagirone, J.P., and Meyer, G., and Mojtabi, A., "Structurations Thermoconvectives tridimensionnelles dans une couche poreuse horizontale," *Journal de Mécanique 20*, Vol. 2, 1981, pp. 219-232.
32. Horne, R.N., "Three Dimensional Natural Convection in a Confined Porous Medium Heated From Below," *Journal of Fluid Mechanics*, Vol. 92, 1979, pp. 751-766.
33. Horne, R.N., and Caltagirone, J.P., "On the Evolution of Thermal Disturbances During Natural Convection in a Porous Medium," *Journal of Fluid Mechanics*, Vol. 100, 1980, pp. 385-395.
34. Schubert, G., and Straus, J.M., "Three Dimensional and Multicellular Steady and Unsteady Convection in Fluid Saturated Porous Media at High Rayleigh Number," *Journal of Fluid Mechanics*, Vol. 94, 1979, pp. 25-38.

35. Straus, J.M., and Schubert, G., "Three-dimensional Convection in a Cubic Box of Fluid Saturated Porous Material," *Journal of Fluid Mechanics*, Vol. 91, 1979, pp. 155-165.
36. Gottleib, D., and Orstag, S.A., "Numerical Analysis of Spectral Methods: Theory and Applications," *NSF - CBMS Monograph 26*, Society of Industrial and Applied Mathematics, Philadelphia, 1978.
37. Caltagirone, J.P., and Cloupeau, M., and Combarnous, M., "Convection Naturelle Fluctuante dans Une Couche Poreuse Horizontale," *Compte-Rendu á l'Académie des Sciences, B* Vol. 273, 1971, pp. 833-836.
38. Straus, J.M., "Large Amplitude Convection in Porous Media," *Journal of Fluid Mechanics*, Vol.64, 1974, pp. 51-63.
39. Bories, S.A., and Monferran L., "Conditions de Stabilité et Échanges Thermiques Par Convection Naturelle Dans Une Couche Poreuse Inclinée de Grande Extension," *Compte-rendu á l'Académie des Sciences, B*, Vol. 274, 1972, pp. 4-7.
40. Bories, S.A., and Combarnous, M.A., and Jaffrennou, J.Y., "Observation des différentes formes d'écoulements convectifs dans une couche poreuxe inclinée," *Compte-Rendu à l'Académie des Sciences, A*, Vol. 275, 1972, pp. 857-860.
41. Schneider, K.H., "Investigation of the Influence of Free Thermal Convection on Heat Transfer Through Granular Material," *Proceedings of International Congress on Refrigiration, 11th*, Munich, 1963, pp. 11-14.
42. Elder, J.W., "Steady Free Convection in a Porous Medium Heated from Below," *Journal of Fluid Mechanics*, Vol. 27, 1967, pp. 29-48.
43. Steen, P.H., "Pattern Selection for Finite Amplitude Convection States in Boxes of Porous Media," *Journal of Fluid Mechanics*, Vol. 136, 1983, pp. 219-241.
44. Combarnous, M.A., and Lefur, B., "Transfert de Chaleur Par Convection Naturelle Dans Une Couche Poreuse Horizontale," *Compte Rendu à l'Académie des Sciences, B*, 1969, pp. 1009-1012.
45. Bories, S.A., "Sur Les Mécanismes Fondamentaux de la Convection Naturelle en Milieu Poreux," *Revue Générale de Thermique*, Vol. 9, 1970, pp. 1355-1376.
46. Combarnous, M.A., "Convection Naturelle et Convection Mixte Dans Une Couche Poreuse Horizontale," *Revue Générale de Thermique* Vol. 9, 1970, pp. 1335-1355.
47. Hollard, S., "Structures Thermoconvectives Stationnaires Dans Une Couche Poreuse Plane Inclinée de Grande Extension," Thèse de Docteur-Ingénieur, Institut National Polytechnique de Toulouse, France, 1984.
48. Combarnous, M.A., "Convection Naturelle et Convection Mixte Dans Une Couche Poreuse Horizontale," These Université de Paris, Edition Technip, Paris, 1970.
49. Klarsfeld, S., "Champs de Temperature Associés Aux Mouvements de Convection Naturelle Dans Un Milieu Limité," *Revue Générale de Thermique*, Vol. 9, 1970, pp. 1403-1424.

50. Betbeder, J., and Jolas, P., "Influence de la Convection Libre Sur La Conductivité Thermique d'une Couche Verticale d'isolant Poreux," *International Journal of Heat & Mass Transfer*, Vol.24, No. 3, 1972, pp. 496-506.
51. Caltagirone, J.P., and Bories, S., "Solutions and Stability Criteria of Natural Convective Flow in an Inclined Porous Layer," in press in *Journal of Fluid Mechanics*, 1985 (see also "Solutions Numériques Bidimensionnelles et Tridimensionnelles de l'écoulement de Convection Naturelle Dans Une Couche Poreuse Inclinée," *Compte-Rendu à l'Académie des Sciences*, B, Vol. 190, 1980, pp. 197-200).
52. Beck, J.L., "Convection in a Box of Porous Material Saturated With Fluid," *Physics of Fluids*, Vol. 15, 1972, pp. 1377-1383.
53. Holst, P.H., and Aziz, K., "Numerical Simulation of Three-Dimensional Natural Convection in Porous Media, *Preprint S.P.E. 2805*, 1970.
54. Zebib, A., and Kassoy, D.R., "Three-dimensional Natural Convection Motion in a Confined Porous Medium," *The Physics of Fluids*, Vol. 21, 1978, pp. 1-3.
55. Bories, S., and Deltour, A., "Influence des Conditions Aux Limites Sur La Convection Naturelle Dans Un Volume Poreux Cylindrique," *International Journal of Heat & Mass Transfer*, Vol. 23, No. 6, 1980, pp. 63-79.
56. Bau, H.H.,and Torrance, K.E., "Low Rayleigh Number Thermal Convection in a Vertical Cylinder Filled with Porous Material and Heated from Below," *International Journal of Heat & Mass Transfer*, Vol. 104, No. 1, 1982, pp. 166-172.
57. Lowel, R.P., and Hernandez, H., "Finite Amplitude Convection in a Porous Container with Fault-like Geometry: Effect of Initial and Boundary Conditions," *International Journal of Heat & Mass Transfer*, Vol. 5, No. 6, 1982, pp. 631-641.
58. Deltour, A., "Convection Naturelle Au Sein d'un Milieu Poreux Saturé Confiné Dans un Domaine Cylindrique Vertical," Thèse Institut National Polytechnique Toulouse, France, 1982.
59. Horne, R.N., "Three-dimensional Natural Convection in a Confined Porous Medium Heated from Below," *Journal of Fluid Mechanics*, Vol. 92, 1979, pp. 751-766.
60. Jaffrennou, J.Y., and Bories, S., "Convection Naturelle Dans Une Couche Poreuse Inclinée," *Rapport Interne, G.E. 14*, Institut de Mécanique des Fluides de Toulouse, 1974.
61. Caltagirone, J.P., "Convection in Porous Medium. In Convective Transport and Instability Phenomena," *G. Braun Editor Karlsruhe* 1981, pp. 199-232.
62. Vlasuk, M.P., "Transfert de Chaleur Par Convection Dans Une Couche Poreuse," (in Russian), presented at the 1972, 4th All Union Heat & Mass Transfer Conference, held at Minsk.
63. Kaneko, T., "An Experimental Investigation of Natural Convection in Porous Media," thesis presented to the University of Calgary, at Canada, in 1972, in partial fulfillment of the Master of Science.

64. Walch, J.P., and Dulieu, B., "Natural Convection in a Slightly Inclined Rectangular Box Filled With a Porous Medium," *International Journal of Heat & Mass Transfer*, Vol. 22, No. 12, 1979, pp. 1607-1612.
65. Gill, A.E., "The Boundary Layer Regime for Convection in a Rectangular Cavity," *Journal of Fluid Mechanics*, Vol. 26, 1966, p. 515.
66. Weber, J.E., "The Boundary Layer Regime for Convection in a Vertical Porous Layer," *International Journal of Heat & Mass Transfer*, Vol. 18, 1975, pp. 569-573.
67. Walker, K.L., and Homsy, G.M., "Convection in a Porous Cavity," *Journal of Fluid Mechanics*, Vol. 87, 1978, p. 449.
68. Bejan, A., "On the Boundary Layer Regime in a Vertical Enclosure Filled with a Porous Medium," *International Journal of Heat & Mass Transfer*, Vol. 6, No. 2, 1979, pp. 93-102.
69. Simpkins, P.G., and Blythe, P.A., "Convection in a Porous Layer," *International Journal of Heat & Mass Transfer*, Vol. 23, No. 6, 1980, pp. 881-887.
70. Bejan, A., and Tien, C.L., "Natural Convection in a Horizontal Porous Medium Subjected to an end-to-end Temperature Difference," *Journal of Heat Transfer*, Vol. 100, No. 2, 1978, pp. 191-198.
71. Chan, B.C.K., and Ivey, C.M., and Barry, J.M., "Natural Convection Enclosed Porous Media with Rectangular Boundaries," *Journal of Heat Transfer*, Vol.2, 1970, pp. 21-27.
72. Bejan, A., "A Synthesis of Analytical Results for Natural Convection Heat Transfer Across Rectangular Enclosures," *International Journal of Heat & Mass Transfer*, Vol. 2, No. 2, 1980, pp. 723-726.
73. Aziz, K., and Kaneko, T., and Mohtadi, M.F., "Natural Convection in Confined Porous Media," presented at the 1972, 1st Pacific Chemical Engineering Congress, held at Kyoto, Japan.
74. Bankvall, C.G., "Natural Convection in Vertical Permeable Space," *Stoffubertragung*, Vol. 7, 1974, pp. 22-30.
75. Burns, P.J., and Chow, L.C., and Tien, C.L., "Convection in a Vertical Slot Filled with Porous Insulation," *International Journal of Heat & Mass Transfer*, Vol. 20, No. 9, 1977, pp. 919-926.
76. Hickox, C.E., and Gartlink, D.K., "A Numerical Study of Natural Convection in Horizontal Porous Layer Subjected to an end-to-end Temperature Difference," *Journal of Heat Transfer*, Vol. 103, No. 4,1981, pp. 797-802.
77. Prasad, V., and Kulacki, F.A., "Convective Heat Transfer in a Rectangular Porous Cavity - Effect of Aspect Ratio on Flow Structure and Heat Transfer," *Journal of Heat Transfer*, Vol. 106,1984, pp. 158-165.
78. Prasad, V., and Kulacki, F.A., "Natural Convection in a Rectangular Porous Cavity with Constant Heat Flux on one Vertical Wall," *Journal of Heat Transfer*, Vol. 106, 1984, pp. 152-157.

79. Seki, N., Fukusako, S., and Inaba, H.K., "Heat Transfer in a Confined Rectangular Cavity Packed with Porous Media," *Journal of Heat and Mass Transfer*, Vol. 21, No. 7, 1978, pp. 985-989.
80. Inaba, H., and Seki, N., "An Experimental Study of Heat Transfer Characteristics in a Porous Layer Enclosed Between Two Opposing Vertical Surfaces with Different Temperatures," *International Journal of Heat & Mass Transfer*, Vol. 24, No. 11, 1981, pp. 1854-1857.
81. Bejan, A., "Natural Convection Heat Transfer in a Porous Layer With Internal Flow Obstructions," *International Journal of Heat and Mass Transfer*, Vol. 26, No. 6, 1983, pp. 815-822.
82. Prasad, V., "Natural Convection in Porous Media. An Experimental and Numerical Study for Vertical Annular and Rectangular Enclosures," thesis presented to the University of Delaware, at Newark, De., in 1983, in partial fulfillment of the requirements for the degree of Doctor of Philosophy.
83. Prasad, V., and Kulacki, F.A., "Natural Convection in a Vertical Porous Annulus," *International Journal of Heat & Mass Transfer*, Vol. 27, No. 2, 1984, pp. 207-219.
84. Hickox, C.E., and Gartling, G.K., "A Numerical Study of Natural Convection in a Vertical, Annular Porous Layer," presented at the 1982, ASME, 21st National Heat Transfer Conference (Paper No. 82-HT 68).
85. Havstad, M.A., and Burns, P.J., "Convective Heat Transfer in Vertical Cylindrical Annuli Filled with a Porous Medium," *International Journal of Heat & Mass Transfer*, Vol. 25, No. 11, 1982, pp. 1755-1766.
86. Mckibbin, R.,and O'Sullivan,M.,"Onset of Convection in a Layered Porous Medium Heated from Below," *Journal of Fluid Mechanics*, Vol. 96, No. 2, 1980, pp. 375-393.
87. Richard, J.P., "Critère d'apparition de la convection naturelle Dans Des Couches Poreuses Stratifiées," presented at the 1979, 15th Congrès International du Froid, Communication B1-92.
88. Richard, J.P., and Combarnous, M., "Critère d'apparition de la Convection Naturelle Au Sein d'une Couche Poreuse Horizontale," *Compte-rendu Académie des Sciences, Paris, B*, Vol. 285, 1977, pp. 73-75.
89. McKibbin, R., and O'Sullivan, M.J., "Heat Transfer in a Layered Porous Medium Heated From Below," *Journal of Fluid Mechanics*, Vol. 111, 1981, pp. 141-173.
90. McKibbin, R., and Tyvand, P.A., "Anisotropic Modeling of Thermal Convection in Multilayered Porous Media," *Journal of Fluid Mechanics*, Vol. 118, 1982, pp. 315-339.
91. McKibbin, R., and Tyvand, P.A., "Thermal Convection in a Porous Medium Composed of Alternating Thick and Thin Layers," *International Journal of Heat and Mass Transfer*, Vol. 26, No.5, 1983, pp. 761-780.

92. McKibbin, R., and Tyvand, P.A., "Thermal Convection in a Porous Medium with Horizontal Cracks," *International Journal of Heat and Mass Transfer*, Vol. 27, No. 7, 1984, pp. 1007-1023.
93. Epherre, J.F., "Critère d'apparition de la Convection Naturelle Dans Une Couche Poreuse Anisotrope," *Revue Générale de Thermique*, Vol. 168, 1975, pp. 949-950.
94. Kvernvold, O., and Tyvand, P.A., "Non Linear Thermal Convection in Anisotropic Porous Media," *Journal of Fluid Mechanics*, Vol. 90, 1979, pp. 609-624.
95. Epherre, J.F., "Convection Naturelle Dans Une Couche Poreuse Anisotrope Saturée Par Un Gaz Parfait," Thèse Université de Bordaux I, 1976.
96. Castinel, G., and Combarnous, M., "Critère d'apparition de la Convection Naturelle Dans Une Couche Poreuse Horizontale Anisotrope," *Compte-rendu Académie des Sciences, Paris B*, Vol. 278, 1974, pp. 701-704.
97. Minkowycz, W.J., and Cheng, P., "Free Convection About a Vertical Cylinder Embedded in a Porous Medium," *International Journal of Heat & Mass Transfer*, Vol. 19, No. 7, 1976, pp. 805-813.
98. Cheng, P., "Similarity Solutions for Mixed Convection from Horizontal Impermeable Surfaces in Saturated Porous Media," *International Journal of Heat and Mass Transfer*, Vol. 20, 1977, pp. 893-898.
99. Cheng, P., "Combined Free and Forced Convection Flow About Inclined Surfaces in Porous Media," *International Journal of Heat and Mass Transfer*, Vol. 20, 1977, pp. 807-814.
100. Johnson, C.H., and Cheng, P., "Possible Similarity Solutions for Free Convection Boundary Layers Adjacent to Flat Plates in Porous Media," *International Journal of Heat & Mass Transfer*, Vol. 21, 1978, pp. 709-718.
101. Hsu, C.T., and Cheng, P., and Homsy, G.M., "Instability of free Convection Flow Over a Horizontal Impermeable Surface in Porous Medium," *International Journal of Heat and Mass Transfer*, Vol. 21, 1978, pp. 1221-1228.
102. Hsu, C.T., and Cheng, P., "The Onset of Longitudinal Vortices in Mixed Convective Flow Over an Inclined Surface in a Porous Medium," *Journal of Heat Transfer*, Vol. 102, 1980, pp. 544-549.
103. Joshi, Y., and Gerbhart, B., "Vertical Natural Convection Flows in Porous Media. Calculations of Improved Accuracy," *International Journal of Heat & Mass Transfer*, Vol. 21, No. 1, 1984, pp. 69-75.
104. Pop, I., and Cheng, P., "The Growth of a Thermal Layer in a Porous Medium Adjacent to a Suddenly Heated Semi-infinite Horizontal Surface," *International Journal of Heat & Mass Transfer*, Vol. 26, No. 10, 1983, pp. 1574-1576.

105. Rees, D.A., and Riley, D.S., "Free Convection Above a Near Horizontal Semi-Infinite Heated Surface Embedded in a Saturated Porous Medium," *International Journal of Heat and Mass Transfer*, Vol. 28, No. 1, 1985, pp. 183-190.
106. Bejan, A., and Anderson, R., "Natural Convection at the Interface Between a Vertical Porous Layer and an Open Space," *Journal of Heat Transfer*, Vol. 105, 1983, pp. 124-129.
107. Chandrasekhara, B.C., and Namboodiri, P.M.S., "Influence of Variable Permeability on Combined Free and Forced Convection About Inclined Surfaces in Porous Media," *International Journal of Heat & Mass Transfer*, Vol. 28, No. 1, 1985, pp. 199-206.
108. Cheng, P., "Heat Transfer in Geothermal Systems," *Advances in Heat Transfer*, Vol. 14, 1978, pp. 1-105.
109. Hozanski, J.M., and Huet, S., "Etude Théorique et Expérimentale de la Convection Naturelle en Milieu Poreux Saturé Produite Par Une Source de Chaleur Concentrée," Symposium de l'A.I.R.H. sur les Transferts de Chaleur et de Masse en Milieux Poreux, Toulouse, France, 1980.
110. Bejan, A., "Natural Convection in an Infinite Porous Medium with a Concentrated Heat Source," *Journal of Fluid Mechanics*, Vol. 89, 1978, pp. 97-107.
111. Hickox, C.E., and Watts, H.A., "Steady Thermal Convection from a Concentrated Source in a Porous Medium," *Journal of Heat Transfer*, Vol. 102, No. 2, 1980, p. 248.
112. Hickox, C.E., "Thermal Convection at Low Rayleigh Number from Concentrated Sources in Porous Media," *Journal of Heat Transfer*, Vol. 103, 1981, pp. 232-236.
113. Poulikakos, D., "On Buoyancy Induced Heat and Mass Transfer from a Concentrated Source in an Infinite Porous Medium," *International Journal of Heat & Mass Transfer*, Vol. 28, No. 3, 1985, pp. 621-629.
114. Zebib,A., and Kassoy, D.R., "Onset of Natural Convection in a Box of Water-Saturated Porous Media with Large Temperature Variation," *The Physics of Fluids*, Vol. 20, No. 1, 1977, pp. 4-9.
115. Saatdjian, E., "Natural Convection in a Porous Layer Saturated with a Compressible Ideal Gas," *International Journal of Heat & Mass Transfer*, Vol. 23, 1980, pp. 1681-1683.
116. Epherre, J.F., "Convection Naturelle au Sein d'une Couche Poreuse Anisotrope Saturée par un Gaz Parfait," Thèse Université de Bordeaux I, Bordeaux, France , 1976.
117. Epherre, J.F., and Combarnous, M., and Klarsfeld, S., "Critère d'apparition de la Convection Naturelle Dans des Couches Poreuses Anistropes Saturées d'air Soumises à de Grandes Différences de Température," *Commission Bulletin*, Institut. Int. du Froid, Washington, 1976, pp. 55-62.
118. Blake, K.R., and Bejan, A., and Poulikakos, D., "Natural Convection Near 4°C in a Water Saturated Porous Layer Heated From Below," *International Journal of Heat & Mass Transfer*,

Vol. 27, No. 2, 1984, pp. 2355-2364.

119. Brinkman, H.C., "On the Permeability of Media Consisting of Closely Packed Porous Interface, *Aplied Science Research.*, Vol. A1, 1947, pp. 27-34.

120. Vafai, K., and Tien, C.L., "Boundary and Intertia Effects on Convective Mass Transfer in Porous Media," *International Journal of Heat and Mass Transfer*, Vol. 25, No. 8, 1982, pp. 1183-1190.

121. Walker, K., and Homsy, G.M., "A Note on Convective Instabilities on Boussinesq Fluids and Porous Media," *Journal of Heat Transfer*, Vol. 99, 1977, pp. 338-339.

122. Tong, T.W., and Subramanian, E., "A Boundary Layer Analysis for Natural Convection in Vertical Porous Enclosures - Use of the Brinkman Extended Darcy Model," *International Journal of Heat and Mass Transfer*, Vol. 28, No. 3, 1985, pp. 563-571.

123. Hsu, C.T., and Cheng, P., "The Brinkman Model for Natural Convection About Semi-Infinite Vertical Flat Plate in a Porous Medium," *International Journal of Heat & Mass Transfer*, Vol. 28, No. 3, 1985, pp. 683-697.

124. Rudraiah, N., and Veerappa B., and Balachandra Rao S., "Effects of Non Uniform Thermal Gradients and Adiabatic Boundaries on Convection in Porous Media," *Journal of Heat Transfer*, Vol.102, 1980, pp. 254-260.

125. Nield, D.A., "The Boundary Correction for the Rayleigh Darcy Problem: Limitations of the Brinkman Equation," *Journal of Fluid Mechanics*, Vol. 128, 1983, pp. 37-46.

126. Caltagirone, J.P., and Arquis, E., "Natural Convection in a Vertical Fluid Cavity Divided by a Permeable Porous Layer," in press.

8. LIST OF SYMBOLS

A	Aspect ratio in x direction
a	Wave number
a_{ijk}	Coefficient
$a_{ijk}(0)$	Initial condition
B	Aspect ratio in the y direction
b_{ijk}	Coefficient
Da	Darcy-Brinkman number
d	Characteristic length of the porous structure
Fa	Buoyancy force
Fd	Viscous drag force
g	Gravitational acceleration
H	Reference parameter for length scale; thickness of the layer
h	Heat transfer coefficient between two phases
$\bar{\bar{K}}$	Permeability tensor
K_1, K_2, K_3	Principal components of permeability tensor $\bar{\bar{K}}$

L_x	Dimension of porous medium along x axis
L_x/H	Aspect ratio in x direction
L_y	Dimension of porous medium along y axis
L_y/H	Aspect ratio in y direction
L_z	Dimension of porous medium along z axis
L_z/H	Aspect ratio in z direction
$N(a,b)$	Non linear operator
Nu^t	Nusselt number
n	Unit vector normal to the surface
P	Pressure
Pr^*	Equivalent Prandtl number
Q	Rate of energy release
Ra	Aspect ratio, H/D
Ra^*	Filtration Rayleigh number
Ra^*_c	Critical Rayleigh number
Ra^*_{cs}	Critical Rayleigh number corresponding to the successive convective modes
r	Radius of spherical portion of fluid
T	Temperature
T_f	Temperature for fluid phase
T_o	Reference temperature
T_s	Temperature for solid phase
$\bar{T}$	Average values of temperature on horizontal planes of surfaces
$\bar{\bar{\lambda}}_i$	Equivalent thermal conductivity tensor
t	Time
V	Filtration velocity
α	Volumetric thermal coefficient
γ	Constant
ε	Porosity
θ	Temperature perturbations
θ_p	Temperature perturbation in the wall
Λ	Thermal conductivity ratio
λ	Coupling parameter
λ_b	Thermal conductivity of lower and upper surface
λ_f	Thermal conductivity of liquid phase
λ_s	Thermal conductivity of solid phase
$\bar{\bar{\lambda}}^*$	Equivalent thermal conductivity tensor for saturated porous media
$\bar{\bar{\lambda}}_f$	Equivalent thermal conductivity tensor for the dispersed structure of the fluid phase

$\overline{\overline{\lambda}}_s^*$	Equivalent thermal conductivity tensor for the dispersed structure of the solid phase
μ	Dynamic viscosity of the fluid
μ_m	Dynamic viscosity at mean temperature
Π	Pressure perturbations
ρ	Density of fluid
ρ_o	Reference density
ρ_m	Volumetric mass at mean temperature
ϕ	Slope; ϕ_t=theoretical slope
Ω	Volume of the porous media
ω	Velocity of the sphere(constant)

HEAT AND MASS TRANSPORT IN GEOTHERMAL RESERVOIRS

Sveinbjorn Bjornsson and Valgardur Stefansson

HEAT AND MASS TRANSPORT IN GEOTHERMAL RESERVOIRS

Sveinbjorn Bjornsson

Science Institute
University of Iceland
Dunhagi 3
107 Reykjavik, Iceland

Valgardur Stefansson

National Energy Authority
Grensasvegur 9
108 Reykjavik, Iceland

ABSTRACT

Geothermal reservoirs are generally more complex than reservoirs of groundwater or petroleum. Physical states of the hydrothermal fluid fall into four categories: vapor-saturated, two-phase boiling, liquid-saturated and supercritical. Liquid-saturated reservoirs and liquid-dominated or vapor-dominated reservoirs of the two-phase boiling type are the most common types exploited so far. There is growing interest in submarine geothermal systems and heat extraction from hot rock or magma bodies, where the hydrothermal fluid circulates at supercritical temperatures and pressures. Meteoric water dominates in continental systems and ocean water in submarine systems. The contribution of magmatic water is small at upper levels in the crust, but may increase as magma bodies are approached. The larger fumarolic fields have magma as a heat source. The rate of heat transfer required to sustain the intense heat output of such fields remains problematic, unless an intimate contact between circulating fluids and hot boundary rock of the magma is maintained over the lifetime of the activity. Convective downward migration of fluid along existing fractures and water penetration by thermal cracking of hot rock are important processes in this respect. Two-phase convection is of major importance in geothermal reservoirs. The phase change instability mechanism induces convection prior to the onset of ordinary buoyancy-driven thermal convection. Mathematical modelling of geothermal systems has greatly advanced the understanding of the dynamic nature of geothermal reservoirs and their response to exploitation.

1. INTRODUCTION

Geothermal reservoirs have many features in common with groundwater and petroleum reservoirs, and geothermal technology has naturally drawn on the experience gathered within these disciplines. This applies to drilling technology, as well as theoretical attempts to define reservoir properties and estimate production capacity. Geothermal reservoirs are, however, more complex than their counterparts in the other disciplines. Efficient utilization of geothermal resources requires understanding the physics of fluid flow and heat transport in fractured rocks. The aim is to extract heat, but the fluid serves as carrier for the heat energy to be mined. The first systems to be exploited were vapor-dominated systems yielding steam for the generation of electricity and hot groundwater systems delivering water for space heating. The most common type of reservoirs exploited in recent years is the liquid-dominated type, which under utilization develops into a boiling reservoir. Interest is growing in experiments to extract heat from hot impermeable rocks by controlled hydraulic fracturing and injection of fluid to carry the heat to the surface. Similar ideas are developing towards heat extraction from magma bodies. Geopressured geothermal reservoirs have been found in association with petroleum reservoirs. Submarine geothermal systems have been discovered, rich in metals and minerals. Although these systems will hardly be exploited for heat energy, their investigation may cast light on the processes that govern the formation of metalloferous deposits in the roots of continental geothermal systems. This great variety in geothermal phenomena illustrates that the physical processes of interest are not limited to subcritical temperatures and pressures, but may range to magmatic temperatures and lithostatic pressures at a depth of about 10 km.

This chapter discusses evidence on the source of fluid and heat in geothermal systems. Geothermal reservoirs are classified according to the physical state of the reservoir fluid using pressure, density, and volume saturation of phases as parameters. Conceptual models of the dynamic natural state of geothermal reservoirs are described by examples. The importance of considering the effects of temperature on the physical properties of the fluid is emphasized, as well as the physics of convection in geothermal systems where boiling occurs. The chapter concludes with an outline of recent developments in mathematical modelling of geothermal systems and the application of these models to obtain quantitative descriptions of the natural state of geothermal reservoirs and to study the response of reservoirs to exploitation.

This chapter draws heavily on a related review by Stefansson and Björnsson (95). Useful discussions were also found in the reviews of Mercer and Faust (69), Garg and Kassoy (49), Donaldson and Grant (35) and the recent textbook by Grant, Donaldson and Bixley (52).

2. SOURCE OF FLUID

In geothermal reservoirs, heat is mainly transported by the hydrothermal fluid. The fluid consists of liquid water with dissolved solids, water vapor, and gases dissolved in the liquid and free in the vapor. Generic types of water as a hydrothermal fluid [White (102,103,108), Ellis and Mahon (42)] are defined as:

Meteoric water. Water recently involved in atmospheric circulation.

Ocean water. Water penetrating into the crust of ocean floor spreading centers.

Juvenile water. "New" water from mantle-derived magma and which has not previously been part of the hydrosphere.

Magmatic water. Water derived from magma, but not necessarily juvenile water, since magma may incorporate meteoric or ocean water of deep circulation, or water from sedimentary material.

Connate water. "Fossil" water incorporated in sediments of the time of deposition.

Metamorphic water. Modified connate water, derived from hydrous minerals during their recrystallization to less hydrous minerals during metamorphic processes.

Systematic studies of stable oxygen and hydrogen isotopes in geothermal water (2,28,29,30) have established meteoric water to be the dominant source of fluid in most active continental geothermal systems. Evidence for this origin of thermal waters was further strengthened by Ellis and Mahon (40,41), and Mahon (62), who showed experimentally that the chemical composition of most waters could be attained by the solvent action of hot water on the local volcanic rocks.

The role of ocean water, and the possibility of submarine geothermal systems on oceanic ridges, was first pointed out by Elder (39). Geothermal systems on land where ocean water is the main source of fluid have been described by Bjornsson et al. (8,9), Arnorsson (3), and Kjaran et al. (57). Submarine geothermal systems have recently been discovered at a number of sites near ocean spreading centers [reviews by Rona and Lowell (83) and White and Guffianti (109), Spiess et al. (92)]. It is now well established that hydrothermal circulation plays a major role in the thermal balance of ocean ridges where ocean water is the dominant fluid source.

Although the contribution of magmatic or juvenile water appears to be minor, there is growing evidence for magmatic influence on thermal fluids (31,58,64,94).

Distinction between ocean water and meteoric water might seem to be of little importance, but the chemical composition of ocean water has a large effect on the solvent action of the geothermal fluid. When seawater is heated within the rock matrix, the removal of Mg from the seawater generates acidity which maintains heavy metals in solution at moderate temperatures (about 300°C)(88). This acidity, and the higher hydrostatic pressure, influences the chemical output of submarine geothermal systems towards much higher metal concentration as compared to geothermal systems fed by meteoric water (7,87).

In geothermal systems in the Imperial Valley, U.S.A., fluids of both high salinity and high temperature (350°C) are found. Metallic concentration is unusually high in these fluids (71). In the Krafla geothermal reservoir in Iceland, the fluid is of very low salinity, but high metallic (mainly Fe) concentrations have been encountered in some wells. These high concentrations result from a very low pH value of the thermal fluid due to intermittent flow of volcanic gases (SO_2, Cl_2) into the hydrothermal system (1,4). Ore deposits in fossil hydrothermal brine systems are suggestive of a brine fluid at the time of deposition (105,106,110). One of the early signs of submarine hydrothermal systems was the observation of the metallic content of sediments near the East Pacific Rise (23,24). The significance of these observations was not generally recognized until the physical evidence for submarine hydrothermal systems became commonly known.

3. MAGMA AS HEAT SOURCE

A recent and extensive review of the literature on the heat source of geothermal systems is given by Stefansson and Bjornsson (95). The following discussion is based on that review.

Of the numerous speculations about the nature of thermal activity, the work of Einarsson (38) is the first quantitative treatment. He contended that the hot springs of Iceland were not physically different from ordinary cold springs, except for the greater depth of penetration of the water. The heat comes simply from the conductive heat flux from the interior of the earth. Bodvarsson (11,13,14) elaborated this concept further. He agreed with Einarsson on the nonvolcanic origin of low-temparature fields, but concluded on the basis of energy balance considerations, that the conduction process involved in the heating must be of a transient nature. Bodvarsson (15) suggested that the deglaciation of Iceland has generated the hydrological and elastomechanical impulses that activated the hydrothermal circulation. Noting that there is a strong positive correlation between temperature and the mass flow of the systems,

Bodvarsson (16) concluded further that convective downward migration of fracture spaces along the walls of mafic dykes appeared to be a dominant thermomechanical process in the development of the low temperature systems. The mechanism of this process involves concepts suggested earlier (12) for the case of the high temperature systems in Iceland, and by White (107). Convective fluid motion in open vertical fracture spaces is associated with the withdrawal of heat from the formation at the lower boundary, resulting in thermoelastic contraction of the adjacent rock and opening of additional fracture spaces at the bottom. A single fracture or system of fractures harboring such convective fluid motions can therefore migrate downward by the process. Since the walls of dykes are not welded to the country rock, the downward migration process does not involve thermoelastic fracturing of solid rock.

The intense heat output of the high temperature fumarolic fields, which is of the order of 100-1000 MW thermal over periods of 10^4 years, cannot be explained by the normal conductive heat flux of the earth. The concentrated source of heat must be magma or hot, recently solidified rock. Limited surface area and slow thermal conduction through solid rock require intimate contact between the rock and the percolating fluid.

Lister (59,60,61) has presented a conceptual model of the downward penetration of water into hot rocks by a process of cooling and thermal cracking. Evidence in support of water penetration into hot rock boundaries of solidifying magma is reported by Bjornsson et al. (10). Watering of a molten lava flow demonstrated a heat extraction efficiency of 40 KW/m^2. The authors conclude that this process of heat extraction is required to explain the sustained heat output of 5000 MW of the subglacial Grimsvötn geothermal area in Iceland. Hydrothermal vents jetting out water at 380°C have been discovered on the axis of the East Pacific Rise at a water depth of 2500 to 2900 m (92). These submarine systems may also be regarded as evidence for convective downward migration of water into the oceanic crust.

White (102,107) found that a magma supply of at least 10^2-10^3 km^3 was required to support the Steamboat Springs system through its life of 10^5 - 10^6 years. He considered that a batholith intruded into the shallow crust, and then remaining static as it cools and crystallizes, is not a satisfactory model, unless the fissure system controlling the circulating water can gradually extend deeper into the batholith as stored heat is removed at higher levels by circulating water. As an alternative for the heat-flow problem, he suggested convection within the magma chamber to maintain magmatic temperatures near the base of the hydrothermal circulation (107).

Irvine (55) has studied the relation between temperature in a magma body and a crystallization mechanism where crystal fractionatio is a major process. He described a convective process in the magma body where crystals are accumulated in the lower part of the intrusion, but the temperature near the top remains close to or above the liquidus temperature of the magma. This process allows higher rate of heat loss and solidification than would occur if the crystals were frozen to the roof of the magma chamber. Convective processes of this nature appear to be capable of providing sufficient heat transfer to the hydrothermal fluid to explain the heat output of most geothermal systems.

Although magma bodies are considered to be a common heat source of geothermal systems, direct evidence on the existence of these bodies and their relationship to the geothermal systems is rather scarce. S-wave shadows have indicated a small magma body at 3 to 7 km depth beneath the Krafla geothermal system in the axial rift zone of NE-Iceland (37). Similar conditions are apparently found in the geothermal system of the Puna district on Hawaii (48) and its relation to the summit magma chamber of Kilauea. Gravimetric, magnetic, and seismic data have shown anomalous zone under the Avachinsky volcano on the Kamchatka Peninsula. This is suspected to be a peripheral magma chamber (45).

Teleseismic P-wave delays have been used extensively to infer velocity structure at several geothermal systems. At the Geysers, a molten chamber about 14 km in diameter is inferred with its top about 7 km beneath the volcanic field (56). At the Coso Geothermal area, an intense low-velocity body, which coincides with the surface expressions of late Pleistocene rhyolitic volcanism, high heat flow, and hydrothermal activity, is resolved between 5 and 20 km depth (82

Eroded central volcanoes are widely distributed within the Tertiary basalt formations in Iceland (99,100). The volcanic centres are places of unusually vigorous volcanic activity. This is shown by the great concentration of dykes and intrusive sheets. Walker (100) estimates that the intrusions amount to at least 50% of the rock in some of the complexes. Each centre has a down-sagged core region. The hydrothermal alteration , typically found in the collapsed core, bears witness of ancient geothermal activity, which is attributable to the basic sheet swarm in the core. The zones of intense hydrothermal alteration appear to have hosted large hydrothermal reservoirs, each with a volume of the order of 100 km^3.

To summarize this discussion, one may conclude that the heat source of the major geothermal systems is magma. The intense loss of heat observed in some geothermal fields is, however, difficult to explain unless the heat transfer from the magma to the hydrothermal fluid is caused either by downward penetration of water and cracking of the hot intrusion, or that convection is taking place within the magma body in such a way that the boundary between the magma and the hydrothermal fluid remains relatively thin for a considerable time.

4. PHYSICAL STATES OF GEOTHERMAL SYSTEMS

There have been many attempts to define basic types of geothermal systems. The classification chosen in this review is by the physical state of the reservoir fluid as discussed by Stefansson and Bjornsson (95). Geothermal systems are generally of a complex nature and contain zones representing different physical states. Figure 1 uses pressure and density of the reservoir fluid to recognize four regions of physical states. These are the *vapor-saturated* region, the two-phase *boiling* region, the *liquid-saturated* region, and the *supercritical* region. These regions are separated by the Clapeyron curves for saturated vapor and liquid, the isothermal T_{crit} for $P > P_{crit}$, and the isobar P_{crit} for $T > T_{crit}$. In three of the regions, the thermal fluid occurs as a single-phase. The *liquid-saturated* region contains the class of geothermal systems where the temperature never reaches boiling. Most important of these are hydrothermal systems in the ocean crust. Hydrothermal circulation at hydrostatic pressures exceeding the critical pressure for the fluid will not reach boiling, unless it is induced by the release of volatiles from geothermal fluid. These pressures are found beneath oceans of 2.2 km depth. Hydrothermal circulation in the ocean crust is thus generally a single-phase convection of seawater. Geothermal systems belonging to the *vapor-saturated* region are found on active volcanoes and low pressure superheated steam is common at shallow depth in geothermal fields. Exploitation also leads to dry-out of water in vapor-dominated rocks near production wells. *Supercritical* conditions are expected in geothermal systems that penetrate deep into the crust to supercritical pressures, where young igneous intrusions have generated supercritical temperatures. On land, these conditions could be found below 3.5 km depth in the crust, assuming boiling conditions in the hydrostatic fluid above. On the sea floor, the hydrostatic head of the ocean may exceed the critical pressure, and supercritical temperatures can therefore exist at shallow depth beneath the floor. The presence of dissolved salts in geothermal fluids has an important effect on phase transitions. The temperature and pressure of the critical point increase with increased salinity (91)(see also Fig. 21), displacing the conditions for supercritical fluid to greater depth. Secondly, the saline fluid has lower vapor pressure (53). This effect delays the initiation of boiling in an ascending saline fluid.

The most important region for the discussion of geothermal systems is the two-phase boiling region enveloped by the Clapeyron curves at subcritical temperatures and pressures. To uniquely define

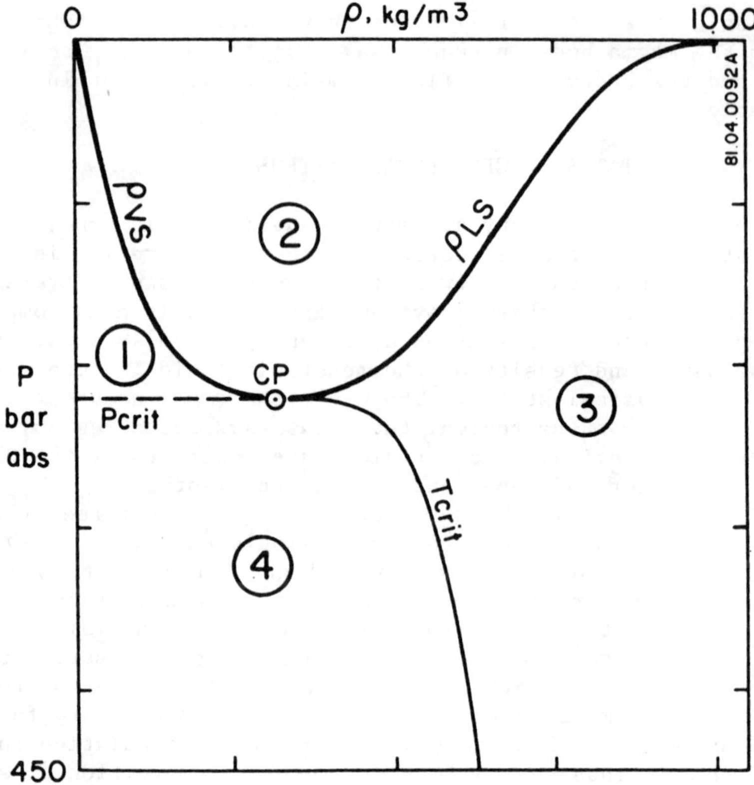

Figure 1. Physical states in hydrothermal systems [After (95)]

1. *Vapor-saturated* region, where the density is less than the density of saturated steam (ρ_{vs}) and the pressure is equal to or less than the critical pressure (P_{crit}).

2. *Boiling* region, where two phases are present. The region is enveloped by the Clapeyron curves for saturated vapor (ρ_{vs}) and saturated liquid (ρ_{ls}) at subcritical temperatures and pressures.

3. *Liquid-saturated* region, where the temperature is less than the critical temperature (T_{crit}) and the density is greater than the density of saturated liquid (ρ_{ls}) for all $P \leq P_{crit}$ and greater than $\rho(T_{crit}, P)$ for all $P > P_{crit}$.

4. *Supercritical* region, where both temperatures and pressures exceed the critical point values (T_{crit}, P_{crit}).

physical conditions within this region, an additional parameter such as the water saturation, i.e., the volume fraction of water in the fluid, is needed. In the three single-phase regions, the vertical pressure gradient is proportional to the density of the respective fluid. In the two-phase region, the situation is different. Physical states with an intermediate mixture of liquid and vapor are not stable, although they may exist as transient phenomena. Gravity segregation of the phases leads to separation of the reservoir into a lower zone with high water saturation and a pressure gradient dominated by liquid density, and an upper zone with low water saturation where vapor density controls the vertical pressure gradient.

Geothermal reservoirs are often referred to as either liquid-dominated or vapor-dominated, depending on the phase which controls the vertical pressure gradient (Fig. 2). The liquid-dominated reservoirs are either liquid-saturated or boiling with a small vapor saturation. The vapor-dominated reservoirs are either vapor-saturated or boiling with a high vapor saturation. The majority of known geothermal reservoirs is of the liquid-dominated type, although vapor-dominated fields have been favored for the generation of electricity.

5. CONCEPTUAL MODELS OF GEOTHERMAL SYSTEMS

The aim of exploratory surveys and exploratory drilling is to gather information on the physical state and nature of a prospective reservoir. The evidence brought forward by different disciplines is combined in a descriptive and qualitative model of the geothermal system. This conceptual model incorporates the essential features of the system and guides further exploratory and appraisal studies. It also provides the basis for numerical modelling of the natural state of the reservoir before withdrawal of fluid affects the physical state. A frequently cited model of large-scale circulation of fluid in the natural state of a geothermal system was presented by White (104) as shown in Figure 3. Cold groundwater percolates down faults, dykes, and fissures to considerable depth, where it picks up heat in permeable hot rocks. The density difference between the cold and the hot water results in buoyancy imbalance, which drives the hot water back to the surface along permeable channels. The heat source may be a magma chamber at greater depth, or just the general heat flow from the interior of the earth. The latter case represents, e.g., the so called low temperature systems where the temperature at the base of circulation is generally below 150°C. A few samples of temperature profiles in such systems in Iceland are shown in Figure 4. The profiles for Laugaland and Reykjavik are typical for upflow zones. If the conventional model of Figure 5 is applied, the base of circulation appears to lie below 2000 m, but the inflow temperature is only about half of that predicted by the regional geothermal gradient. The dyke convector model (Fig. 6),

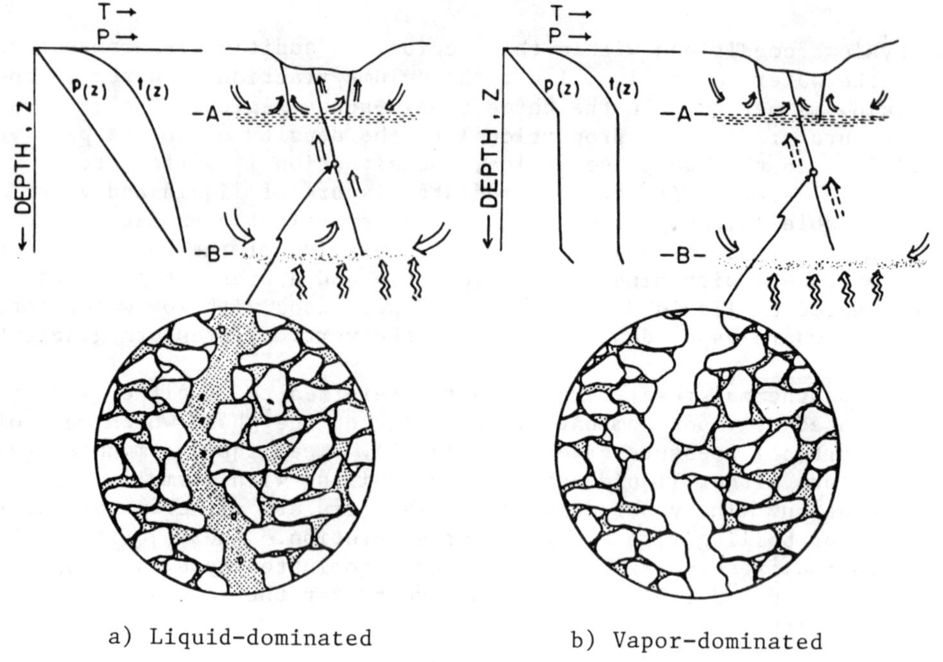

a) Liquid-dominated

b) Vapor-dominated

Figure 2. Fluid distribution in pores and fractures [After (46)]

a) Liquid dominates in the open channels although bubbles of steam and gas are present.

b) Vapor dominates in the open channels, but liquid fills most of the intergranular pore space.

proposed by Bodvarsson (16), resolves this discrepancy by assuming that the recharge enters the dyke along some relatively shallow path, but sinks through cracks or fractures along the walls of the dyke where it takes up heat from the hot adjacent rock. Convection within the dyke transports heat from lower lying rocks and delivers excess heat to the upper layers. In this way, the convection equalizes the temperature within the convector. As water in a liquid-saturated reservoir ascends to lower pressure, it eventually reaches saturation pressure and begins to boil (Fig. 3). Below the boiling level, the temperature is practically constant and equal to the base temperature. Above the boiling level, the temperature and pressure are related by the Clapeyron curve for saturated liquid. Ignoring the dynamic pressure drop caused by the upflow, usually less than 10% of the static gradient (35), the temperature and pressure in the rising column may be found approximately by summing up the static

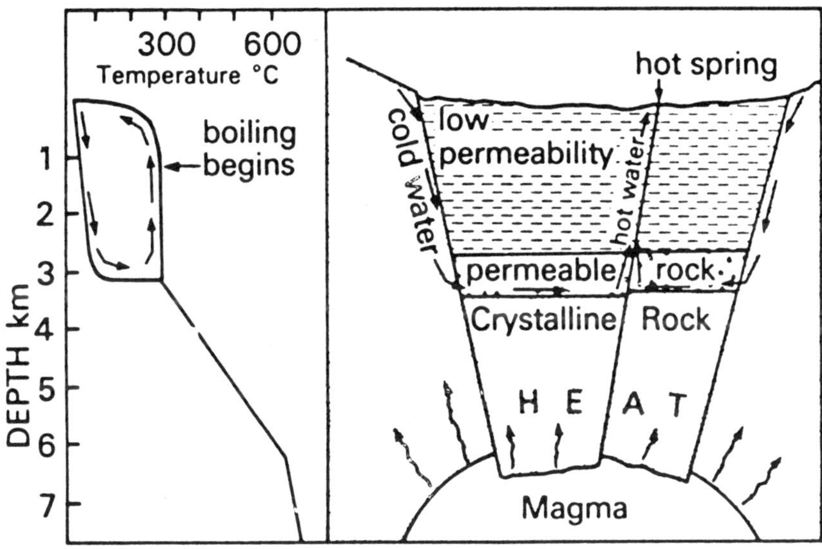

Figure 3. Model by White (104) of the large-scale circulation of fluid in the natural state of geothermal system.

weight of a column of water whose temperature is everywhere at saturation for the local pressure. Examples of such boiling point curves are given in Figs. 7 and 8. If noncondensible gas is present, its partial pressure adds to the vapor pressure and initiates boiling at greater local pressure than for pure water (Curve E in Fig. 7). A boiling column of reservoir fluid implies vertical upflow of steam and, in most cases, also that of water. Structural control and regional groundwater flow usually impose some component of lateral flow.

Exploration of geothermal systems is usually limited to the exploitable part of the reservoir which is the region of upflow. Most conceptual models therefore describe only that part of the system. Examples of conceptual models of upflow regions in several reservoirs are given in Figs. 9 to 13. These models illustrate that the natural state and the initial fluid distribution in the reservoir are controlled by a dynamic balance of mass and heat flow.

The examples presented above are all of the liquid-dominated category of reservoirs. A general conceptual model of vapor-dominated reservoirs in their natural state was presented by White et al. (110) as shown in Fig. 14. The main part of the reservoir is dominated by vapor ascending from a layer of boiling convecting brine. At the top of the reservoir, impermeable cap rock prevents escape of the steam. Heat is lost to surface and boundaries by conduction. This heat loss is balanced by condensation of steam. The condensate

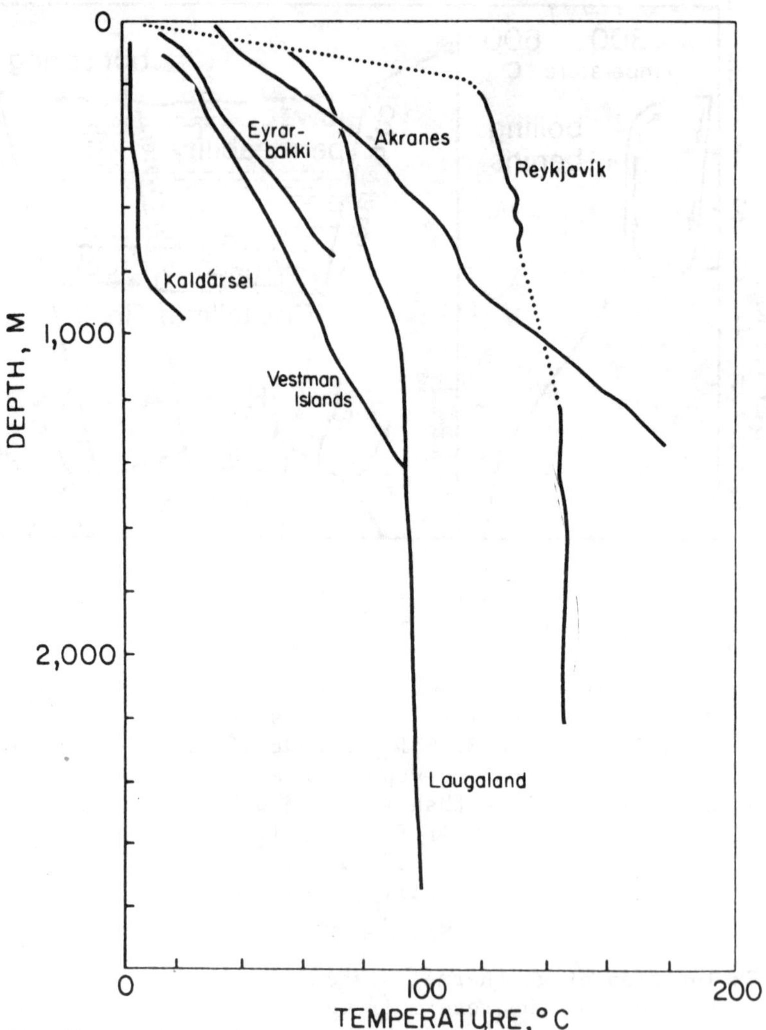

Figure 4. A few samples of temperature profiles in boreholes in Iceland [After Bodvarsson (16), based on data by Palmason (74)]. Regional conductive gradient in Reykjavik is near 100°C/km but about 60°C/km near Laugaland.

trickles against the rising steam down to the brine. In this counter flow of steam and condensate, the steam occupies the wider channels but the water favors small pores and channels because of its high surface tension. D'Amore and Truesdell (32) suggested a modified model as shown in Fig. 15. The natural upflow of steam from the

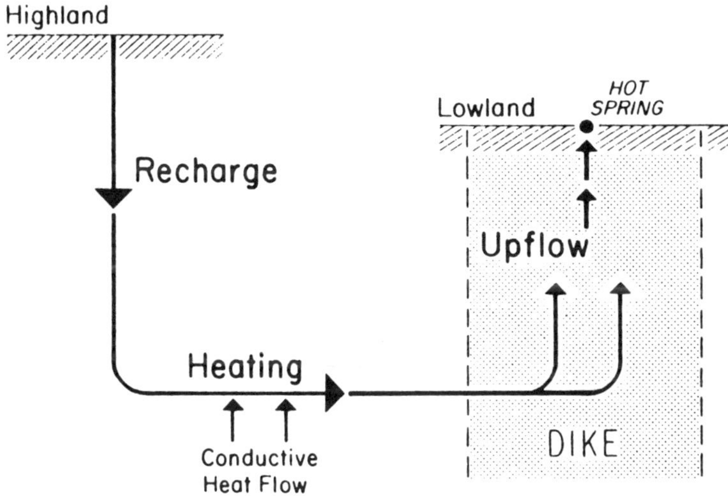

Figure 5. The conventional model for low temperature systems [After Bodvarsson (16)].

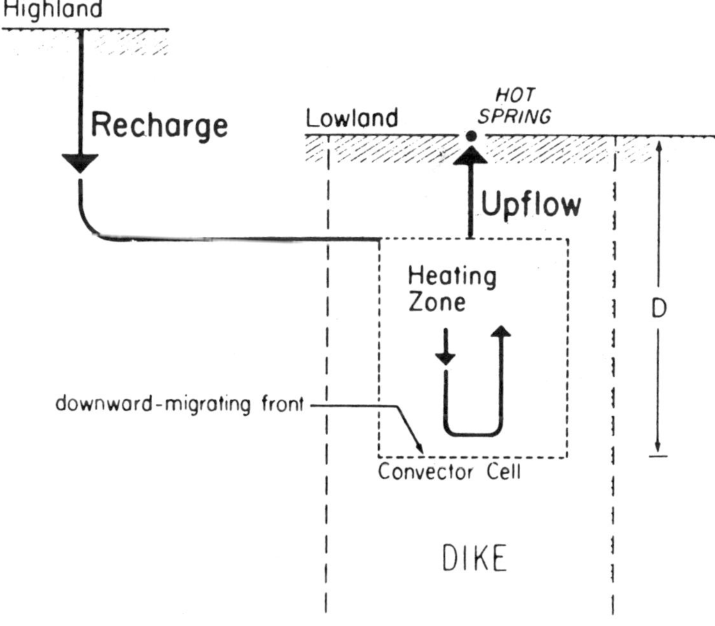

Figure 6. Dyke convector model proposed by Bodvarsson (16).

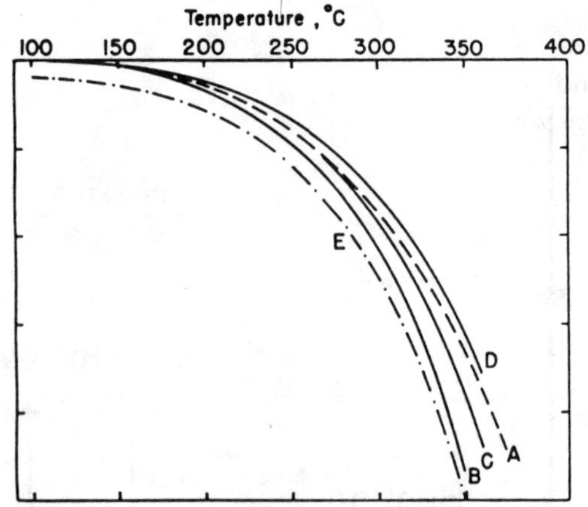

Figure 7. Depth-temperature relations for boiling solutions [After Fournier (47)]. Depth-pressure relations for curve A fixed by the weight per unit area of a free-standing column of cold water extending to the surface. Depth-pressure relations for curves B to E fixed by the weight per unit area of free-standing columns of the given solutions everywhere at their boiling temperatures, and extending to the surface. Curves A and B for pure water, curve C for 10 weight percent aqueous NaCl, curve D for 20 weight percent aqueous NaCl, and curve E for water plus a partial pressure of CO_2 of 10 bars.

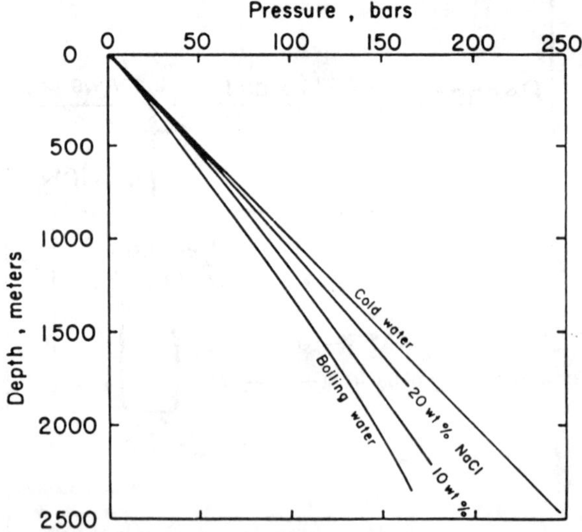

Figure 8. Depth-pressure relations for boiling and cold columns of pure water and boiling aqueous NaCl solutions. [After Fournier (47)].

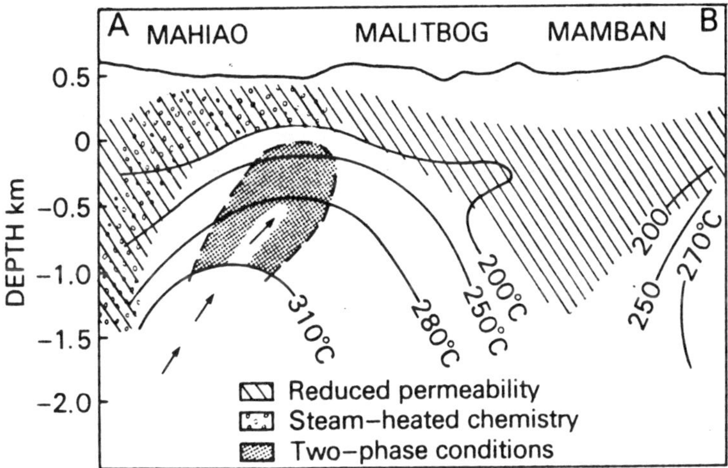

Figure 9. Section through Tongonan reservoir, The Philippines [After Grant, Donaldson and Bixley (52), based on Grant and Studt (51)]. Temperature contours are based on downhole data. An outer region containing liquid water encompasses most of the field. Temperatures in the two-phase region are close to saturation. A zone of reduced permeability above the reservoir is identified by conductive temperature profiles and by the presence of heavily silicified rock.

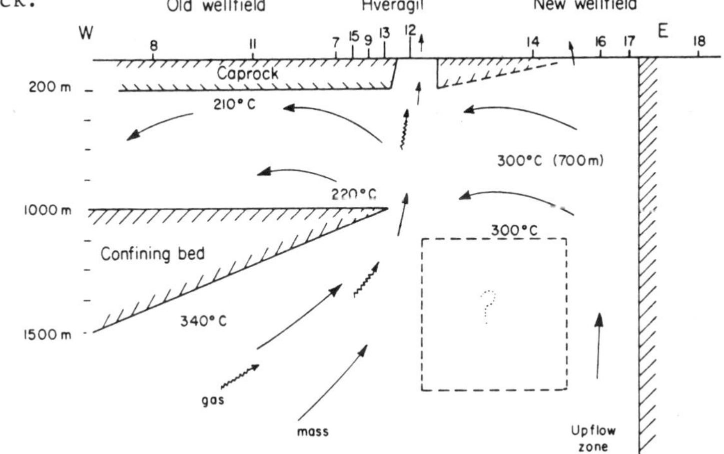

Figure 10. Conceptual model of the Krafla reservoir, Iceland [After Bodvarsson et al. (18)]. A gas rich steam-water mixture of at least 340°C temperature flows from the west in the lower reservoir and rises through a fracture zone where it mixes with 300°C fluid from an eastern upflow zone. The boiled and degassed fluid flows west in an upper reservoir at a temperature about 200°C.

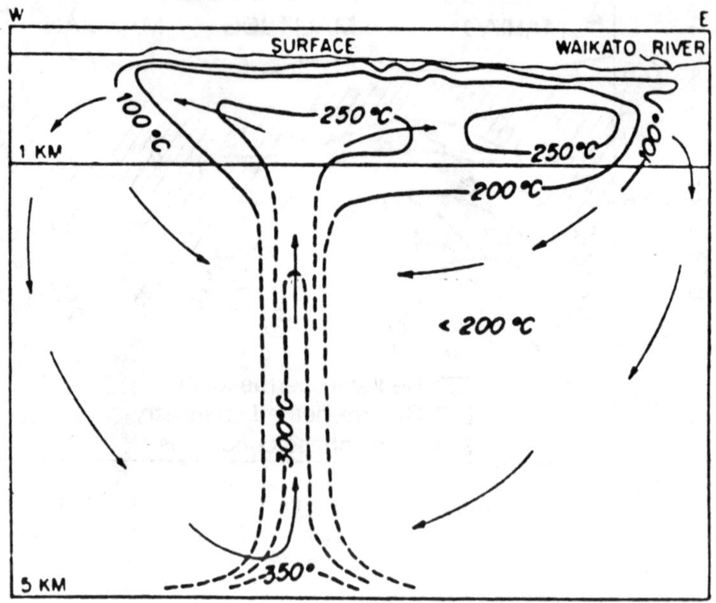

Figure 11. Cross section through the geothermal system at Wairakei, New Zealand [After Fournier (47) based on Elder (39)]. Solid lines are isotherms derived from borehole data, dashed lines are estimated isotherms. The approximate flow lines of meteoric water are shown by arrows.

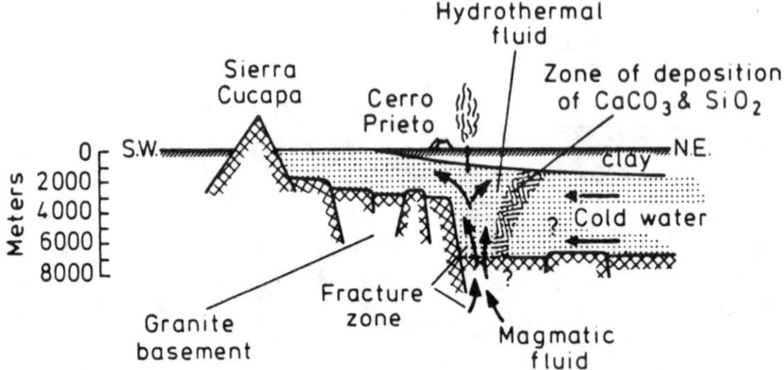

Figure 12. Schematic cross section of the Cerro Prieto field, Mexico [After Ellis and Mahon (42), based on Mercado (66)]. The system is capped by some 700 m of plastic clay, forcing the hot fluids to flow horizontally away from the fractured upflow zones mainly toward west. Interaction of the hot fluids with cold water in the east causes precipitation of calcite and silica.

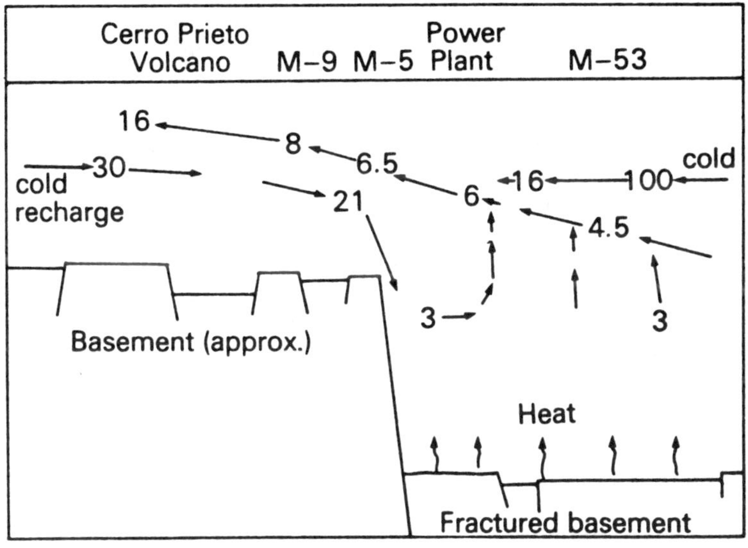

Figure 13. Conceptual model of the natural flow in Cerro Prieto [After Grant, Donaldson and Bixley (52), based on Mercado (67)]. Numbers are Na/K - ratios in the reservoir fluid. Low ratio indicates high temperature.

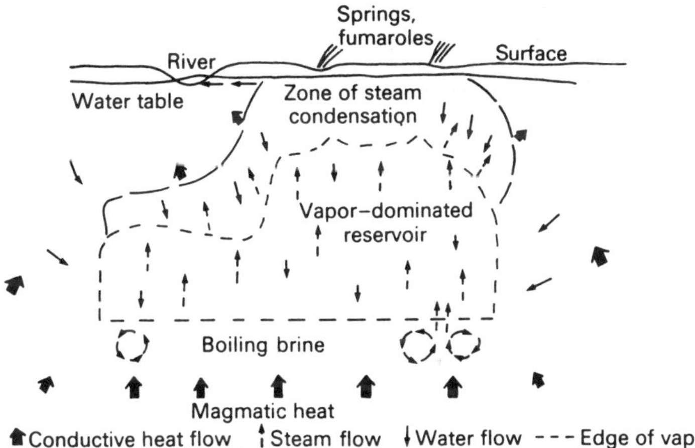

Figure 14. Conceptual model of the fluid flow in the natural state of a vapor-dominated reservoir. [After Grant, Donaldson and Bixley (52), after White et al. (110)].

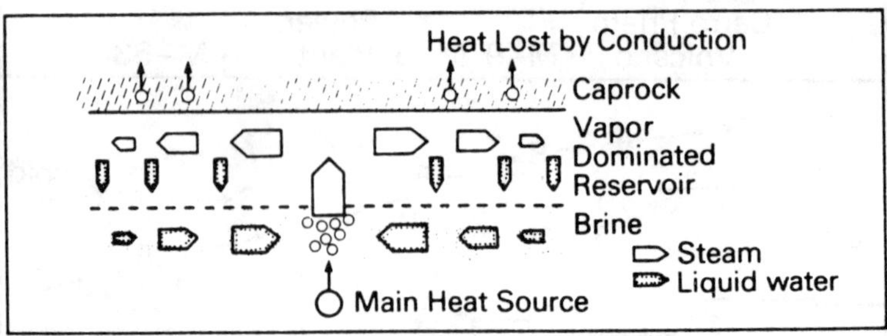

Figure 15. Conceptual model of the natural state of a vapor-dominate reservoir [After Grant, Donaldson and Bixley (52). and D'Amore and Truesdell (32)].

boiling brine occurs in a limited area, from which the steam spreads laterally through the reservoir. This model fits well with observed variations in steam chemistry in the Lardarello and the Geysers reservoirs. Figure 16 presents a model for the geothermal system beneath the Lassen volcano [Muffler et al. (70)]. A relatively shallow vapor-dominated reservoir is underlain by 240°C liquid-dominated reservoir rich in chloride. A local 240°C steam zone of a similar water reservoir in Kenya is illustrated by Figure 17. Lateral flow of the boiling fluid to lower pressures leads to an increased thickness of the steam zone.

White et al. (110) suggested that phorphyry copper mineralization may occur in the zone of boiling brine below the vapor-dominated systems. Porphyry copper deposits occur in tertiary and older orogenic-volcanic belts around the world. Isotope and fluid inclusion studies have shown that in a number of deposits, the development of the characteristic ore alteration pattern involved the interaction of meteoric groundwaters with saline fluids evolved from magma.

Henley and McNabb (54) considered the nature of the interaction between a buoyant thermal plume of a low density and salinity magmatic vapor carrying copper and other ore components and cooler groundwater in an essentially hydrostatic environment. They presented the model shown schematically in Figure 18 to explain porphyry copper emplacement. Condensation of the vapor creates a high salinity brine with dissolved acids. A neutral, chloride rich solution, of moderate salinity and dominated by meteoric water evolves above and at the sides of the acid-altered rock and hot gas region. Interaction between the two fluid systems leads to copper ores and alteration patterns. The acid alteration and precipitation of quartz from the hydrothermal fluid above, due to retrograde solubilit

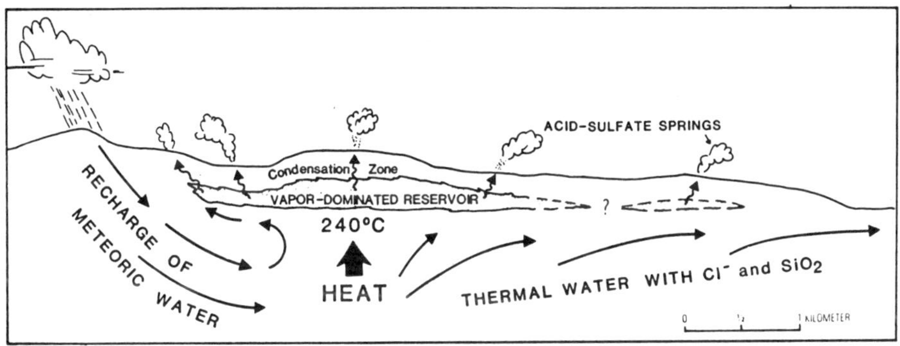

Figure 16. Schematic cross section of the Lassen geothermal system [After Fournier (47), based on Muffler et al. (70)].

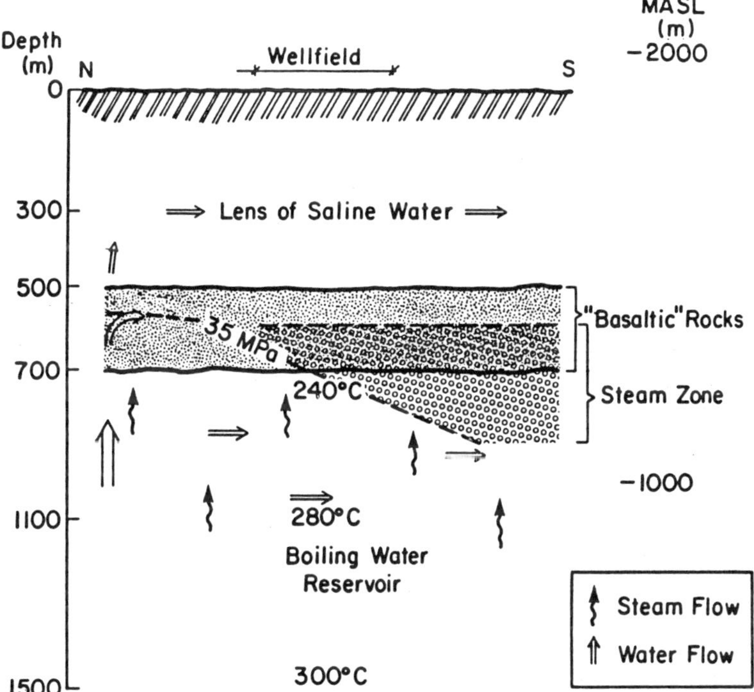

Figure 17. Conceptual model of the East Olkaria geothermal reservoir, Kenya [After Bodvarsson et al. (21)].

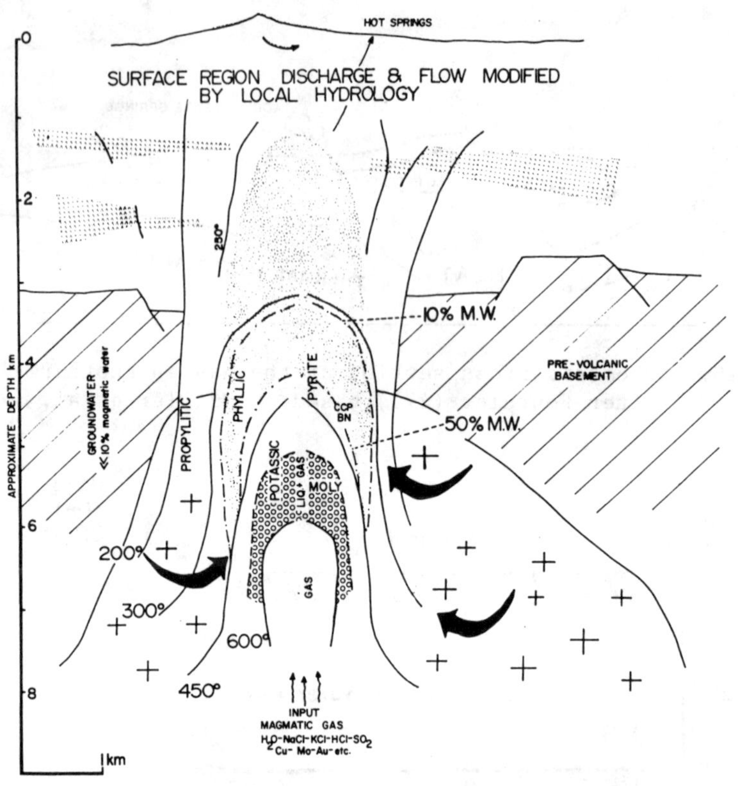

Figure 18. Thermal plume model for a developing porphyry copper deposit [After Henley and McNabb (54)]. Low salinity magmatic vapor carries metals into a hydrostatic environment of cooler groundwater. A small portion of the vapor condenses into a high-salinity liquid. The high salinity phase is diverted toward the margins of the two-phase region where refluxing occurs. Groundwater is progressively entrained into the magmatic vapor plume (% M.W: Percentage of magmatic water).

at temperatures above 350-400°C (Fig. 19), may lead to a self-sealed envelope [Fournier (47)] in the country rock close to the magmatic intrusion (Fig. 20). Fluid within the envelope is then at lithostatic pressure, whereas the hydrothermal system above circulates at hydrostatic pressures.

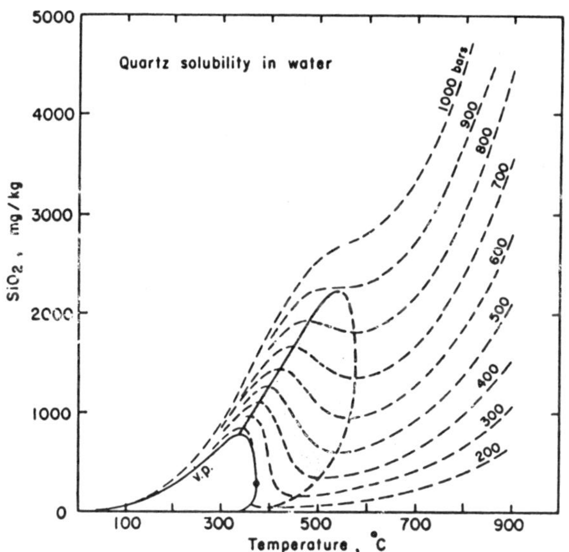

Figure 19. Solubilities of quartz in water up to 900°C at the indicated pressures. The shaded area emphasizes a region of retrograde solubility [After Fournier (47)].

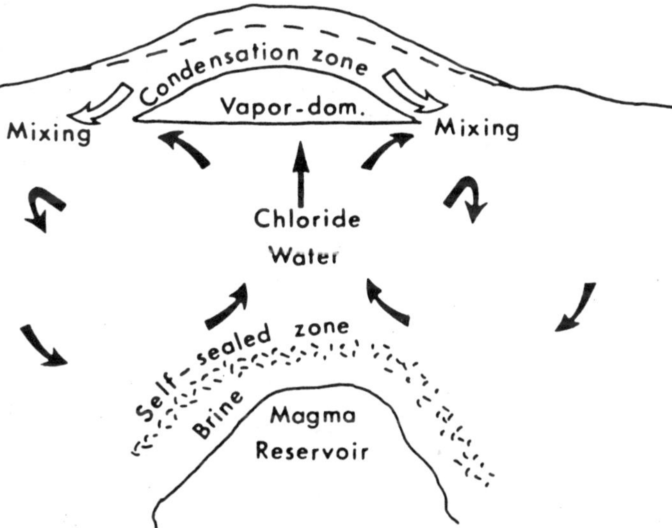

Figure 20. Schematic cross section of a hydrothermal system with a self-sealed envelope separating geopressured fluid from a hydrostatically pressured convecting fluid [After Fournier (47)].

6. CONVECTION AND BOILING

Thermal convection is a fundamental process of heat transport in hydrothermal systems. Although single-phase convection of water has received most attention in the literature, two-phase convection of water, steam and, gases is probably the dominant mode of convection in most geothermal reservoirs. The following section reviews some important studies on single-phase and two-phase convection in geothermal systems [See also Stefansson and Bjornsson (95)].

6.1. Single Phase Convection of Water

A thorough review of the basic characteristics of free convection of a single-phase fluid in porous media is included in Whitherspoon et al. (111). A linear stability analysis shows that thermal convection in a liquid-saturated porous layer is initiated when a critical value of the Rayleigh number, Ra, is exceeded. In a horizontal layer of thickness H and a temperature difference ΔT across, the Rayleigh number can be expressed,

$$R_a = \frac{\alpha . g . \Delta T . H . \rho^2 . c_p . K}{\mu . \lambda}$$

where:

α = coefficient of thermal expansion of the fluid
g = acceleration of gravity
ρ = fluid density
c_p = specific heat of the fluid at constant pressure
K = permeability of the rock
μ = dynamic viscosity of the fluid
λ = thermal conductivity of saturated rock

The most common approximation in the extensive literature on the subject of thermal convection is to consider the viscosity, the permeability, the thermal conductivity, the thermal expansivity, and the specific heat as constant values in the convection process [Boussinesq (25)]. Variations in density are included in the buoyancy term of the vertical balances of forces, but otherwise density is assumed to be constant. Straus and Schubert (96) studied the conditions for the onset of thermal convection in a water-saturated porous layer. They used an accurate representation of the equation of state for liquid water and an empirical formula for the viscosity of water as a function of temperature and pressure. The properties of water considered were:density ρ, thermal expansion coefficient α, isothermal compressibility β, specific heat at constant pressure

c_p, adiabatic temperature gradient $\alpha gT/c_p$, and dynamic viscosity μ. By allowing for variations in the thermal properties of water, Straus and Schubert determined the critical Rayleigh number for the onset of convection for various thicknesses of the porous layer, as well as for various thermal gradients in the layer. They found that the permeability necessary for convection is seriously overestimated when the thermal properties of water are assumed to be constant values. Due to the effects of variable water properties, convection can occur for smaller vertical temperature differences in rock of a given permeability or for smaller permeability at given temperature difference. The primary reasons for increased tendency to initiate convection are the substantial increase of thermal expansivity and the decrease of viscosity with increased temperature. Variations in the specific heat, the adiabatic temperature gradient, and the compressibility were found of minor importance in the cases considered by Straus and Schubert (96).

6.2. Two-Phase Convection

In many geothermal systems, the flowing water reaches the saturation pressure due to release of pressure, and boiling is initiated. The fluid becomes a two-phase mixture of steam and water with thermodynamic and transport properties different from those of liquid water. Where steam and water are in thermodynamic equilibrium, the fluid temperature and pressure are uniquely related by the Clapeyron equilibrium equation which determines the boiling (Clapeyron) curve separating the steam and water phases on a p-T diagram. The thermodynamic properties of each phase are unique functions of temperature (or pressure) only. The two-phase flow is generally assumed to be laminar. This assumption, however, might not be valid where rapid boiling occurs. Darcy's law is generally applied separately to the steam and water phases, introducing relative permeability factors to account for the restricted flow of each phase in the presence of the other. The relative permeability factors are expressed as functions of the volume fraction of each phase, S denoting the volume saturation of water and 1-S that of the steam. These functions are still poorly defined, however, and important values such as the water saturation at which the water becomes immobile are inadequately known. In view of the difficulties met in defining the relative permeability, the use of complex relations is hardly warranted. Many authors simply assume that the relative permeability factor for each phase is equal to the saturation value of the respective phase.

The Darcy flow of either phase is driven by the pressure gradient in excess of the static gravity gradient of each phase. Donaldson (34) considered boiling processes within a one-dimensional steady upflow of hot water. He found that a two-phase zone of steam and water formed for vertical massflow rates above a threshold value.

The steam ascended more rapidly than the water, and was condensed at the upper boundary of the two-phase zone. Sheu et al. (89) extended the model of Donaldson (34) to include a more complete energy equation and more realistic thermodynamic properties. For flow rates below a critical value, only liquid water existed at all depths. Above the critical value, three zones exist, consisting of a near-surface water layer, an underlying two-phase zone of water undergoing pressure release boiling , and a deeper zone of liquid water.

Schubert and Straus (84) studied the conditions for the onset of convection, and the nature of that convection, in a three-dimensional porous medium containing a steam-water mixture or water at saturation temperature at all depths. The steam-water mixtures were described by a homogeneous model, in which a single Darcy-velocity specifies the mass flow of the mixture, and the thermodynamic and transport properties of the mixture depend only on the properties of the individual phases and their relative amounts. The tendency of these fluids to convection is quite different from that governing the instability of an ordinary single-phase fluid. The ordinary Rayleigh instability is driven by buoyancy forces which cause relatively hotter, lighter fluid elements to rise and relatively colder, heavier fluid parcels to sink. The two-phase convection proceeds by way of a phase change instability mechanism associated with the requirement that the fluid temperature and pressure always lie on the equilibrium Clapeyron curve. Temperature variations are directly responsible for the pressure gradients which drive convection. A perturbed hotter region of the fluid is also at somewhat higher pressure than its surroundings, and fluid will flow horizontally away from the hot spot. Conservation of mass then requires that the horizontal divergence of fluid out of the hotter region be balanced by a vertical influx of fluid. Condensation and boiling occur to achieve a balance of forces in the vertical. The most striking aspects of this type of convection are the small lateral dimensions of the cells and the concentration of the flow, phase changes, and temperature variations toward the bottom of the porous layer. The saturated liquid convection cells are only about half as wide as those of ordinary buoyancy-driven convection in water, two-phase cells are still narrower, and the flow more concentrated toward the bottom. This phase change instability mechanism induces convection prior to the onset of ordinary buoyancy-driven thermal convection [Schubert and Straus (84)]. Although buoyancy-driven convection has been assumed to dominate in many hydrothermal systems, the real geometry of convection cells has never been observed in nature. One reason for this might be that phase change driven convection dominates the convective pattern with narrower cells and concentration of the convective flow near the bottom of the system.

Many properties of the fluid change rapidly as the thermodynamic critical point is approached. For pure water, the critical point is found at 374°C and at 221 bar pressure. Increasing salinity of the fluid raises the critical point both in temperature and pressure (Fig. 21). Theoretically, as the critical point is approached, the specific heat at constant pressure, coefficient of volume expansion, compressibility, and thermal conductivity become infinite (93). Straus and Schubert (96) analyzed the onset of natural convection in porous media near the critical state, which they found significantly influenced by large property variation as the critical point is approached. Norton and Knight (72) calculated heat flux and fluid movement in hydrothermal systems surrounding cooling plutons. The fluid circulation was found to be effectively controlled by expansion coefficient , specific heat, and viscosity variation near the critical point. Dunn and Hardee (36) presented data from laboratory experiments with natural convection in a permeable medium which showed the expected significant increase in heat transfer rates near the critical point.

The presence of gases in geothermal fluids may greatly enhance convective instability. In geothermal systems where chemical equilibrium is attained for all major components incorporated in alteration minerals, the concentration of gases (CO_2, H_2S, H_2 and CH_4) is directly related to the temperature of the geothermal fluid. The concentration of CO_2 in geothermal systems near 300°C is sufficiently large to profoundly influence the physical state of the system (5, 6,50,63,94). Boiling in water-CO_2 fluid occurs where the sum of the partial pressures of CO_2 and steam exceeds the ambient pressure in the fluid. Boiling refers thus to the phenomenon in which CO_2 and vapor create a gaseous phase in equilibrium with the liquid phase at the ambient pressure. The boiling temperature of the water-CO_2 fluid is well below the saturation temperature for pure water at the same ambient pressure (Fig. 7). The ascending fluid initiates boiling at greater depths than pure water. The boiling point curve within a water-CO_2 geothermal system is displaced to progressively greater depths as the partial pressure of the CO_2 increases with temperature and depth. Straus and Schubert (97) showed that the buoyancy of the geothermal fluid depends critically on the presence of CO_2, because of the large volume changes that occur when CO_2 enters or leaves the solution and forces water to simultaneously change phase. While the presence of CO_2 can, in principle, enhance or inhibit convection in geothermal fluids, the general effect is to strongly enhance convection for most temperatures and pressures of interest.

6.3. Steam Water Counterflow In Vapor-Dominated Systems

In the discussion of single-phase and two-phase convection above, it has implicitly been assumed that water is the continuous

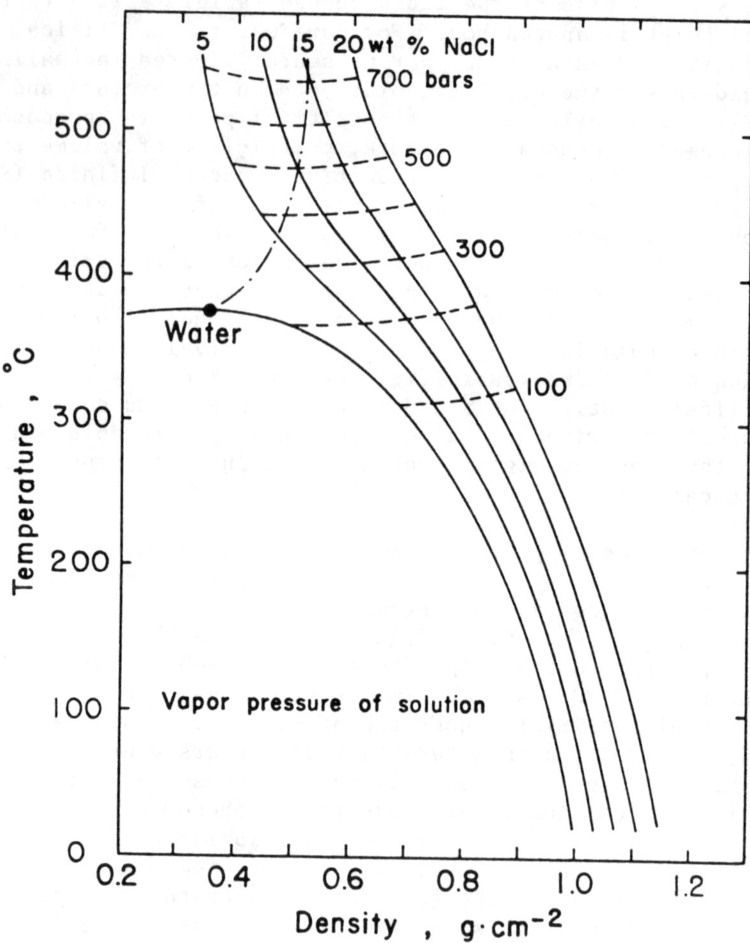

Figure 21. Density-temperature relations in the system NaCl-H_2O at the vapor pressures of the solutions. The dashed lines are isobars. The dotted-dashed line is the critical curve [After Fournier (47)].

phase throughout the system and thus provides control. Pressure in such systems is near hydrostatic values. This appears to be the most common state of geothermal reservoirs. There are however important exceptions, such as the vapor-dominated reservoirs, where the steam is the continuous, pressure-controlling phase, although liquid water is also present.

An essential feature of a two-phase vapor-dominated system is the counterflow of rising steam and descending water, which has been termed one-dimensional convection as opposed to single-phase convection which can only occur in two or three dimensions. The large difference in density between steam and water provides the driving force that tends to segregate the two phases. Martin et al. (65) demonstrated that one-dimensional, vertical two-phase, steady state, convective, and conductive heat flow is theoretically possible in geothermal reservoirs (See also Schubert and Straus (85)). Martin et al. assumed steam to be generated at depth by heat conducted from below. The steam flows upward and an equal mass of hot water flows downward within the geothermal reservoir. At the top of the geothermal reservoirs, the steam condenses into hot water which then flows downwards. Above the reservoir the heat flow is again only conductive. Martin et al. (65) found that in many cases two water/steam volume ratios satisfy the same heat flow rate. One is a high ratio in which liquid water is the principal mobile phase. The pressure gradient is approximately that of hydrostatic water. The other ratio corresponds to a high vapor saturation in which steam is the principal mobile phase and the pressure gradient is approximately that of static steam.

One characteristic of both liquid- or vapor-dominated two phase geothermal systems is that some wells produce superheated steam (86,94,98). As first pointed out by Bodvarsson (12) in a consideration of the thermodynamic behavior of the Lardarello system, the superheat can be explained by assuming that steam flowing to producing wells receives heat from the reservoir rock. In this early study, Bodvarsson also concludes that the steam produced at Lardarello must originate from liquid water in the reservoir, and that the heat capacity of the reservoir rock contributes significantly to the energy withdrawn from the system. These features of the Lardarello system are found in many later models (See e.g. 33,79,98).

7. MATHEMATICAL MODELLING OF GEOTHERMAL SYSTEMS

Mathematical modelling related to geothermal systems has advanced rapidly in the last decade. Empirical methods fitting analytical functions to data, and analytical methods including lumped-parameter models are being replaced by distributed-parameter models (numerical simulators). A useful review of the development of numerical simulators and the status of numerical modelling of geothermal systems is given by Bodvarsson (17). Other valuable references on geothermal reservoir modelling are, e.g., Witherspoon et al. (111), Faust and Mercer (43,44), Mercer and Faust (68,69), Castanier and Sanyal (26), Cheng (27), Pinder (75), Wang, Sterbentz and Tsang (101), Pruess et al. (81), Pruess and Truesdell (80), Pruess and Narasimhan (78), Garg and Kassoy (49), Donaldson and Grant (35), and Grant, Donaldson and Bixley (52). The present status in numerical modelling of geothermal systems is presented in a recent review by Bodvarsson, Pruess and Lippmann (22). The discussion here will therefore be brief.

To account for spatial and temporal variations in the physical properties of the fluid and the reservoir rocks as well as the effects of gravity segregation of steam and water, a general model must be a distributed-parameter model with the capability of three-dimensional simulation. The model must not only accommodate the transient flow of single-phase fluid, but also allow for phase changes and temperature changes. The general governing equations consist of mass, momentum, and thermal energy balances for each phase (Table 1) and sets of constitutive relationships between variables. For general two-phase reservoir applications, the constitutive relationships concern thermodynamics, capillary pressure relative permeability, viscosity, reservoir consolidation, thermal exchange between phases, and thermal dispersion-conduction. Spacing of joints and fractures is an important variable in the mathematical treatment of flow through fractured rock. In geothermal systems, fracture geometry is usually poorly known. The spacing of discontinuities is, however, small in comparison to the size of the reservoir being considered and, consequently, the fractured rock can in many cases be treated as continuous medium with anisotropic permeability. The fissured rock mass is then represented by an equivalent porous media of anisotropic and spatially varying permeability. There are, however, situations in which the fractured nature of the rock cannot be ignored, such as thermal changes during cold water advance and heat transfer in fractured reservoirs with boiling. A double-medium theory is in development, which models the fractures and the rock matrix as two interpenetrating media with fluid and heat transfer between them (76,78). Another simplification in most studies is to treat the pore fluid as a pure water substance. Effects of noncondensible gases and dissolved solids are thereby neglected, although they may be significant (73). Coupled equations for consolidation, fluid flow, and heat transport in geothermal reservoirs may be derived and solved, but consolidation is not the process of concern in many studies. The movement of steam and water is in most cases sufficiently slow to assume that phases of the fluid and the rock matrix are in local thermal equilibrium.

In order to solve the governing equations, hydrodynamic and thermal boundary and initial conditions must be specified. The hydrodynamic boundary conditions are generally: (1) constant pressure, (2) impermeable surfaces, and (3) specified mass flux. The corresponding thermal boundary conditions are: (1) constant temperature, (2) insulated surfaces, and (3) specified heat flux. When sources or sinks are present, their pressures or massflow rates must be specified. Initial conditions specify the physical state of the reservoir before the simulation. Numerical models for the simulation of geothermal systems are mainly of two categories:

(1) studies of the natural conditions prevailing in the reservoir prior to exploitation. (This study aims at estimates of heat flow, recharge rates, and the initial distribution of fluid in the system), and (2) studies of the response of the reservoir to exploitation. Well test data are used to infer the distribution of permeability and porosity in the reservoir. The limited data generally available require sensitivity studies of these parameters and also the reservoir dimensions, initial distribution of the reserves, and possible recharge. The simulation considers different exploitation alternatives and provides preliminary estimates of the generating capacity of a field, appropriate well spacing, and production depths. As data on the production performance accumulate, a simulation, matched for the production data of individual wells, yields predictions of future field performance and the longevity of the field according to a given exploitation scheme. Simulation models are also found useful to study different alternatives for fluid injection into producing reservoirs.

Numerical modelling of geothermal systems has provided insight into the processes of mass and heat transfer that govern the physical state of geothermal reservoirs. Quantitative models have replaced qualitative conceptual models. Numerical simulation has also become a valuable strategic tool for the optimum exploitation of geothermal systems. Examples of the results obtained in various fields are given by Garg and Kassoy (49), Donaldson and Grant (35), and Grant, Donaldson and Bixley (52). More recent studies include Sigurdsson et al. (90) that presents a summary of reservoir engineering studies of low-temperature reservoirs in Iceland and Bodvarsson et al. (18, 19,20), Pruess et al. (77), and Bodvarsson et al. (21) that describe extensive modelling studies of the boiling reservoirs of the Krafla Field in Iceland and the Olkaria Field in Kenya.

In line with Mercer and Faust (69), we may conclude that the mathematical tools that have been developed for numerical modelling are capable of solving the most difficult problems encountered in geothermal reservoirs, but the conceptual basis of these mathematical models is still weak and awaits further study.

Table 1. Governing Equations for the Physical Processes in Geothermal Systems [After Mercer and Faust (69)]

Mass Balance

$$-\frac{\partial}{\partial x_i}(\underset{\sim}{v}_s \rho_s) + q_s + d_v = \frac{\partial(\phi S_s \rho_s)}{\partial t} \qquad (7.1)$$

$$-\frac{\partial}{\partial x_i}(\underset{\sim}{v}_w \rho_w) + q_w - d_v = \frac{\partial(\phi S_w \rho_w)}{\partial t} \qquad (7.2)$$

Momentum Balance

$$\underset{\sim}{v}_s = -\frac{\underset{\approx}{k} k_{rs}}{\mu_s}\left(\frac{\partial p_s}{\partial x_j} - \rho_s \underset{\sim}{g}\right) \qquad (7.3)$$

$$\underset{\sim}{v}_w = -\frac{\underset{\approx}{k} k_{rw}}{\mu_w}\left(\frac{\partial p_w}{\partial x_j} - \rho_w \underset{\sim}{g}\right) \qquad (7.4)$$

Energy Balance

$$\frac{\partial(\phi S_s \rho_s U_s)}{\partial t} + \frac{\partial}{\partial x_i}(\rho_s U_s \underset{\sim}{v}_s) + \frac{\partial \underset{\sim}{\lambda}_s}{\partial x_i} + p_s \frac{\partial \underset{\sim}{v}_s}{\partial x_i}$$

$$+ q_s h_s + Q_{ws} + Q_{rs} = 0, \qquad (7.5)$$

$$\frac{\partial(\phi S_w \rho_w U_w)}{\partial t} + \frac{\partial}{\partial x_i}(\rho_w U_w \underset{\sim}{v}_w) + \frac{\partial \underset{\sim}{\lambda}_w}{\partial x_i} + p_w \frac{\partial \underset{\sim}{v}_w}{\partial x_i}$$

$$+ q_w h_w + Q_{sw} + Q_{rw} = 0 \qquad (7.6)$$

$$\frac{\partial[(1-\phi)\rho_r U_r]}{\partial t} + \frac{\partial \underset{\sim}{\lambda}_r}{\partial x_i} + Q_{sr} + Q_{wr} = 0 \qquad (7.7)$$

8. REFERENCES

1. Armannsson, H., Gislason, G., and Hauksson, T., "Magmatic Gases in Well Fluids Aids the Mapping of the Flow Pattern in a Geothermal System," *Geochimica et Cosmochimica Acta*, Vol. 46, 1982, pp. 167-177.
2. Arnason, B., "Groundwater Systems in Iceland Traced by Deuterium", *Rit Societas Scientiarum Islandica*, Vol. 42, 1976, p. 236.
3. Arnorsson, S., "Major Element Chemistry of the Geothermal Seawater at Reykjanes and Svartsengi, Iceland", *Mineralogical Magazine*, June 1978, pp. 209-220.
4. Arnorsson, S., "Mineral Deposition from Icelandic Geothermal Waters:Environmental and Utilization Problems," *Journal of Petroleum Technology*, Vol. 33, Jan. 1981, pp. 181-187.
5. Arnorsson, S., "Gas Pressures in Geothermal Systems," *Chemical Geology*, Vol. 49, 1985, pp. 319-328.
6. Arnorsson, S., Gunnlaugsson, E., and Svavarsson, H., "The Chemistry of Geothermal Waters in Iceland. III, Chemical Geothermometry in Geothermal Investigations," *Geochimica et Cosmochimica Acta*, Vol. 47, 1983, pp. 567-577.
7. Bischoff, J.L., and Seyfried, W., "Hydrothermal Chemistry of Seawater from 25° to $350^{o}C$," *American Journal of Science*, Vol. 278, 1978, pp. 838-860.
8. Bjornsson, S., Arnorsson, S., and Tomasson, J., "Exploration of the Reykjanes Brine Area," *Geothermics*, Special Issue 2, 1970, pp. 1640-1650.
9. Bjornsson, S., Arnorsson, S., and Tomasson, J., "Economic Evaluation of Reykjanes Thermal Brine," *Bulletin of the American Association of Petroleum Geologists*, Vol. 56, 1972, pp. 2380-2391.
10. Bjornsson, H., Bjornsson, S., and Sigurgeirsson, Th., "Penetration of Water Into Hot Rock Boundaries of Magma in Grimsvotn," *Nature*, Vol. 295, 1982, pp. 580-581.
11. Bodvarsson, G., "Geophysical Methods in the Prospecting for Hot Water in Iceland," (in Danish), *Journal of the Iceland Association of Engineers*, Vol. 35, 1950, pp. 48-59.
12. Bodvarsson, G., "Report on the Hengill Thermal Area," (In Icelandic with a Summary in English), *Journal of the Iceland Association of Engineers*, Vol. 36, 1951, pp. 1-48.
13. Bodvarsson, G., "Physical Characteristics of Natural Heat Resources in Iceland," *Jokull*, Vol. 11, 1961, pp. 29-38.
14. Bodvarsson, G., "Elastomechanical Phenomena and the Fluid Conductivity of Deep Geothermal Reservoirs and Source Regions," presented at the Dec., 1979, Fifth Workshop on Geothermal Reservoir Engineering, held at Stanford University, Stanford, California.
15. Bodvarsson, G., "Glaciation and Geothermal Processes in Iceland," *Jokull*, Vol. 32, 1982, pp. 21-28.

16. Bodvarsson, G., "Temperature/Flow Statistics and Thermomechanics of Low-Temperature Geothermal Systems in Iceland," *Journal of Volcanology and Geothermal Research,* Vol. 19, 1983, pp. 255-280.
17. Bodvarsson, G.S., "Mathematical Modelling of the Behavior of Geothermal Systems under Exploitation," *LBL-13937,* Lawrence Berkeley Laboratory, University of California, Berkeley, California, Jan. 1982, 353pp..
18. Bodvarsson, G.S., Benson, S.M., Sigurdsson, O., Stefansson, V., and Eliasson, E.T., "The Krafla Geothermal Field, Iceland:1. Analysis of Well Test Data," *Water Resources Research,* Vol. 20, No. 11, Nov. 1984, pp. 1515-1530.
19. Bodvarsson, G.S., Pruess, K., Stefansson, V., and Eliasson, E.T "The Krafla Geothermal Field, Iceland: 2. The Natural State of the System," *Water Resources Research,* Vol. 20, No. 11, Nov. 1984, pp. 1531-1544.
20. Bodvarsson, G.S., Pruess, K., Stefansson, V., and Eliasson, E.T., "The Krafla Geothermal Field, Iceland: 3. The Generating Capacity of the Field," *Water Resources Research,* Vol. 20, No. 11, Nov. 1984, pp. 1545-1559.
21. Bodvarsson, G.S., Pruess, K., Stefansson, V., Bjornsson, S., and Ojiambo, S.B., "A Summary of Modeling Studies of the East Olkaria Geothermal Field, Kenya," presented at the August 26-30, 1985, Geothermal Resources Council International Symposium on Geothermal Energy, held at Kailua-Kona, Hawaii.
22. Bodvarsson, G.S., Pruess, K., and Lippmann, M.J., "Modeling of Geothermal Systems," presented at the March 27-29, 1985, Society of Petroleum Engineers California Regional Meeting, held at Bakersfield, California (LBL-18268)
23. Bostrom, K., and Peterson, M.N.A., "Precipitates From Hydrothermal Exhalations of the East Pacific Rise," *Economic Geology* Vol. 61, 1966, pp. 1258-1365.
24. Bostrom, K., and Peterson, M.N.A., "The Origin of Aluminium Ferromanganoan Sediments in Areas of High Heat Flow on the East Pacific Rise," *Marine Geology,* Vol. 7, 1969, pp. 427-447.
25. Boussinesq, J., *Theorie Analytique de la Chaleur,* Gauthier-Villars, Paris, Vol. 2, 1903, p. 172.
26. Castanier, L.M., and Sanyal, S.K., "Geothermal Reservoir Modeling - A Review of Approaches," *Transactions,* Geothermal Resources Council, Vol. 4, Sept. 1980, pp. 313-314.
27. Cheng, P., "Heat-Transfer in Geothermal Systems," *Advances in Heat Transfer,* Vol. 14, 1978.
28. Craig, H., "The Isotopic Geochemistry of Water and Carbon in Geothermal Areas," *Nuclear Geology on Geothermal Areas,* E. Tongiorgi, ed., Consiglio Nazionale Delle Ricerche, Laboratorie di Geologica Nucleare, Spoleto,Pisa, 1963, pp. 17-53.

29. Craig, H., "The Isotopic Composition and Origin of the Red Sea and Salton Sea Geothermal Brines," *Science*, Vol. 154, 1966, pp. 1544-1548.
30. Craig, H., Boato, G., and White, D.E., "Isotopic Geochemistry of Thermal Waters," *Nuclear Processes in Geological Settings, Series Report 19*, National Research Council Communications on Nuclear Science, 1956, pp. 29-38.
31. Craig, H., Lupton, J.E., Welham, J.A., and Poreda, R., "Helium Isotope Ratios in Yellowstone and Lassen Park Volcanic Gases," *Geophysical Research Letters*, Vol. 5, 1978, pp. 897-900.
32. D'Amore, F., and Truesdell, A.H., "Models for Steam Chemistry at Lardarello and the Geysers," *Proceedings of the Tenth Workshop on Geothermal Reservoir Engineering*, Stanford University, Stanford, California,Jan. 22-24, 1985, pp. 113-121.
34. Donaldson, I.G., "The Flow of Steam and Water Mixtures Through Permeable Beds. A Simple Simulation of a Natural Undisturbed Hydrothermal Region," *New Zealand Journal of Science*, Vol. 11, 1968, pp. 3-23.
35. Donaldson, I.G., and Grant, M.A., "Heat Extraction from Geothermal Reservoirs," *Geothermal Systems: Principles and Case Histories*, L. Rybach and L.J.P. Muffler, eds., Wiley Interscience, New York, N.Y., 1981, pp. 145-174.
36. Dunn, J.C., and Hardee, H.C., "Superconductive Geothermal Zones," *Journal of Volcanology and Geothermal Research*, Vol. 11, 1981, pp. 189-201.
37. Einarsson, P., "S-wave Shadows in the Krafla Caldera in NE-Iceland, Evidence for a Magma Chamber in the Crust," *Bulletin Volcanologique*, Vol. 41, 1978, pp. 1-9.
38. Einarsson, T., "Ueber das Wesen der Heissen Quellen Islands," *Rit Societas Scientiarum Islandica*, Vol. 26, 1942, p. 91.
39. Elder, J.W., "Physical Processes in Geothermal Areas," *Terrestrial Heat Flow*, W.H.K. Lee, ed., American Geophysical Union Monograph 8, 1965, pp. 211-239.
40. Ellis, A.J., and Mahon, W.A.J., "Natural Hydrothermal Systems and Experimental Hot-Water/Rock Interactions," *Geochimica et Cosmochimica Acta*, Vol. 28, 1964, pp. 1323-1357.
41. Ellis, A.J., and Mahon, W.A.J., "Natural Hydrothermal Systems and Experimental Hot-Water Rock Interactions (Part II)," *Geochimica et Cosmochimica Acta*, Vol. 31, 1967, pp. 519-538.
42. Ellis, A.J., and Mahon, W.A.J., *Chemistry and Geothermal Systems*, Academic Press, New York, 1977.
43. Faust, C.R., and Mercer, J.W., "Geothermal Reservoir Simulation 1: Mathematical Models for Liquid- and Vapor-Dominated Hydrothermal Systems," *Water Resources Research*, Vol. 15, 1979, pp. 23-30.
44. Faust, C.R., and Mercer, J.W., "Geothermal Reservoir Simulation 2: Numerical Solution Techniques for Liquid- and Vapor-Dominated Hydrothermal Systems," *Water Resources Research*, Vol. 15, 1979, pp. 31-46.

45. Fedotov, S.A., Balesta, S.T., Droznin, V.A., Masurenkov, Y.P., and Sugrobov, V.M., "On a Possibility of Heat Utilization of the Avachinsky Volcanic Chamber," *Proceedings of the Second United Nations Symposium on the Development and Use of Geothermal Resources*, Vol. 1, 1976, pp. 363-369.

46. Fournier, R.O., "Application of Water Geochemistry to Geothermal Exploration," *Geothermal Systems: Principles and Case Histories*, L. Rybach and L.J.P. Muffler, eds., John Wiley and Sons, Inc., New York, N.Y., 1981, pp. 109-143.

47. Fournier, R.O., "Active Hydrothermal Systems as Analogues of Fossil Systems," *The Role of Heat in the Development of Energy and Mineral Resources in the Northern Basin and Range Province, Special Report No. 13*, Geothermal Resources Council, Davis, California, 1983, pp. 263-284.

48. Furumoto, A.S., "The Relationship of a Geothermal Reservoir to the Geological Structure of the East Rift of Kilauea Volcano, Hawaii," *Transactions 2*, Geothermal Resources Council, 1978, pp. 199-201.

49. Garg, S.K., and Kassoy, D.R., "Convective Heat and Mass Transfer in Hydrothermal Systems," *Geothermal Systems: Principles and Case Histories*, L. Rybach and L.J.P. Muffler, eds., Wiley-Interscience, New York, N.Y., 1981, pp. 37-76.

50. Grant, M.A., "Broadlands - A Gas - Dominated Geothermal Field," *Geothermics*, Vol. 6, 1977, pp. 9-29.

51. Grant, M.A., and Studt, F.E., "A Conceptual Model of the Tongonan Geothermal Reservoir," presented at the 1981, Fourth GEOSEA Conference.

52. Grant, M.A., Donaldson, I.A., and Bixley P.F., *Geothermal Reservoir Engineering*, Academic Press, New York, N.Y., 1982, p. 369.

53. Haas, J.L., Jr., "The Effect of Salinity on the Maximum Thermal Gradient of a Hydrothermal System at Hydrostatic Pressure," *Economic Geology*, Vol. 66, 1971, pp. 940-946.

54. Henley, R.W., and McNabb, A., "Magmatic Vapor Plumes and Ground-Water Interaction in Phorphyry Copper Emplacement," *Economic Geology*, Vol. 73, No. 1, Jan.-Feb., 1978, pp. 1-20.

55. Irvine, T.N., "Heat Transfer During Solidification of Layered Intrusions. I. Sheets and Sills," *Canadian Journal of Earth Sciences*, Vol. 7, 1970, pp. 1031-1061.

56. Iyer, H.M., Oppenheimer, D.H., and Hitchcock, T., "Large Teleseismic P-Wave Delays in the Geysers - Clear Lake Geothermal Area, California," *Science*, Vol. 204, 1979, pp. 495-497.

57. Kjaran, S.P., Halldorsson, G.K., Thorhallsson, S., and Eliasson, J., "Reservoir Engineering Aspects of Svartsengi Geothermal Area," *Transactions*, Geothermal Resources Council, Vol. 3, 1979, pp. 337-339.

58. Kononov, V.I., and Polak, B.G., "Indicators of Abyssal Heat Recharge of Recent Hydrothermal Phenomena," *Proceedings of Second United Nations Symposium on the Development and Use of Geothermal Resources*, Vol. 1, 1976, pp. 767-773.
59. Lister, C.R.B., "On the Penetration of Water into Hot Rock," *Geophysical Journal of the Royal Astronomical Society*, London, Vol. 39, 1974, pp. 465-509.
60. Lister, C.R.B., "Qualitative Theory on the Deep End of Geothermal Systems," *Proceedings of the Second United Nations Symposium on the Development and Use of Geothermal Resources*, Vol.1, 1976, pp. 459-463.
61. Lister, C.R.B., "Qualitative Models of Spreading - Center Processes, Including Hydrothermal Penetration," *Tectonophysics*, Vol. 37, 1977, pp. 203-218.
62. Mahon, W.A.J., "Natural Hydrothermal Systems and the Reaction of Hot Water with Sedimentary Rocks," *New Zealand Journal of Science*, Vol. 10, 1967, pp. 206-221.
63. Mahon, W.A.J., and Finlayson, J.B., "The Chemistry of the Broadlands Geothermal Area, New Zealand," *American Journal of Science*, Vol. 272, 1972, pp. 48-68.
64. Mamyrin, B.A., Tolstikhin, I.N., Anufriev, G.S., and Kamensky, I.L., "Isotopic Composition of Helium in Thermal Springs of Iceland," *Geokhimiya* (USSR), No. 11, 1972, p. 1396.
65. Martin, J.C., Wagner, R.E., and Kelsey, F.J., "One-Dimensional Convective and Conductive Geothermal Heat Flow," *Summaries of the Second Workshop on Geothermal Reservoir Engineering*, Stanford University, Stanford, California, Dec. 1976, pp. 251-261, (SGP-TR-20).
66. Mercado, S., "Geoquimica Hidrothermal en Cerro Prieto, B.C., Mexico," *Commision Federal de Electricidad*, Mexicali, Mexico, 1967.
67. Mercado, S., "Movement of Gecthermal Fluids and Temperature Distribution in the Cerro Prieto Geothermal Field, Baja California, Mexico," *Proceedings of the Second United Nations Symposium on the Development and Use of Geothermal Resources*, Vol. 1, 1976, pp. 487-494.
68. Mercer, J.W., and Faust, C.R., "Geothermal Reservoir Simulation 3: Application of Liquid- and Vapor- Dominated Hydrothermal Modeling Techniques to Wairakei, New Zealand," *Water Resources Research*, Vol. 15, 1979, pp. 653-671.
69. Mercer, J.W., and Faust, C.R., "The Physics of Fluid Flow and Heat Transport in Geothermal Systems," *Sourcebook on the Production of Electricity from Geothermal Steam*, J. Kestin, ed., United States Department of Energy, Washington, D.C., 1980, pp. 121-135.
70. Muffler, L.J.P. Nehring, N.L., Truesdell, A.H., Janik, C.J., Clynne, M.A., and Thompson, J.M., "The Lassen Geothermal System," *Proceedings of the Pacific Geothermal Conference 1982*, incorporating the *Fourth New Zealand Geothermal Workshop*,

Part 2, 1982, pp. 349-356.
71. Muffler, L.J.P., and White, D.E., "Active Metamorphism of Upper Cenozoic Sediments in the Salton Sea Geothermal Field and the Salton Trough, Southeastern California," *Geological Society of America, Bulletin*, Vol. 80, 1968, pp. 157-182.
72. Norton, D., and Knight, J., "Transport Phenomena in Hydrothermal Systems: Cooling Plutons," *American Journal of Science*, Vol. 277, 1977, pp. 937-981.
73. O'Sullivan, M.J., Bodvarsson, G.S., Pruess, K., and Blakeley, M.R., "Fluid Flow in Gas-Rich Geothermal Reservoirs," presented at the 1983, Society of Petroleum Engineers 58th Annual Technical Conference and Exhibition, held at San Francisco, California.
74. Palmason, G., "Heat Flow and Hydrothermal Activity in Iceland," *Geodynamics of Iceland and the North Atlantic Area*, L. Kristjansson, ed., Reidel Publishing Company, Dordrecht, Netherlands, 1974, pp. 297-307.
75. Pinder, G.F., "State-of-the-Art Review of Geothermal Reservoir Modeling," *Geothermal Subsidence Research Management Program, LBL-9093*, Earth Sciences Division, Lawrence Berkeley Laboratory, University of California, Vol. 5, March 1979, p. 144.
76. Pruess, K., "Heat Transfer in Fractured Geothermal Reservoirs with Boiling," *Water Resources Research*, Vol. 19, 1983, pp. 201-208.
77. Pruess, K., Bodvarsson, G.S., Stefansson, V., and Eliasson, E.T., "The Krafla Geothermal Field, Iceland: 4. History Match and Prediction of Individual Well Performance," *Water Resources Research*, Vol. 20, No. 11, Nov. 1984, pp. 1561-1584.
78. Pruess, K., and Narasimhan, T.N., "A Practical Method for Modeling Fluid and Heat Flow in Fractured Porous Media," presented at the 1982, Society of Petroleum Engineers 6th Symposium on Reservoir Simulation, held at New Orleans, La.
79. Pruess, K., and Narasimhan, T.N., "On Fluid Reserves and the Production of Superheated Steam From Fractured Vapor-Dominated Geothermal Reservoirs," *Journal of Geophysical Research*, Vol. 87, No. B11, 1982, pp. 9329-9339.
80. Pruess, K., and Truesdell, A.H., "A Numerical Simulation of the Natural Evolution of Vapor-Dominated Hydrothermal Systems," *Summaries of the Sixth Workshop on Geothermal Reservoir Engineering*, Stanford University, Stanford, California, Dec. 1980.
81. Pruess, K., Zerzan, J.M., Schroeder, R.C., and Witherspoon, P.A., "Description of the Three-Dimensional Two-Phase Simulator SHAFT 78 for Use in Geothermal Reservoir Studies," presented at the 1979, Society of Petroleum Engineers AIME Fifth Symposium on Reservoir Simulation, held at Denver, Colorado, (SPE 7699).
82. Reasenberg, P., Ellsworth, W.L., and Walter, A., "Teleseismic Evidence for a Low-Velocity Body Under the Coso Geothermal Area," *Journal of Geophysical Research*, Vol. 85, 1980,

pp. 2471-2483.

83. Rona, P.A., and Lowell, R.P., "Hydrothermal Systems at Ocean Spreading Centers," *Geology*, Vol. 6, 1978, pp. 299-300.

84. Schubert, G., and Straus, J.M., "Two-Phase Convection in a Porous Medium," *Journal of Geophysical Research*, Vol. 82, 1977, pp. 3411-3421.

85. Schubert, G., and Straus, J.M., "Steam-Water Counterflow in Porous Media," *Journal of Geophysical Research*, Vol. 84, 1979, pp. 1621-1628.

86. Sestini, G., "Superheating of Geothermal Steam," *Geothermics, Special Issue 2*, Vol. 2, Part 1, 1970, pp. 622-648.

87. Seyfried, W., and Bischoff, J.L., "Hydrothermal Transport of Heavy Metals by Seawater: The Role of Seawater/Basalt Ratios," *Earth and Planetary Science Letters*, Vol. 34, 1977, pp. 71-77.

88. Seyfried, W.E., and Bischoff, J.L., "Experimental Seawater-Basalt Interaction at 300°C, 500 bars, Chemical Exchange, Secondary Mineral Formation and Implications for the Transport of Heavy Metals," *Geochimica et Cosmochimica Acta*, Vol. 45, 1981, pp. 135-147.

89. Sheu, J.P., Torrance, K.E., and Turcotte, D.L.,"On the Structure of Two-Phase Hydrothermal Flows in Porous Media," *Journal of Geophysical Research*, Vol. 84, 1979, pp. 7524-7532.

90. Sigurdsson, O., Kjaran, S.P., Thorsteinsson, Th., Stefansson, V., and Palmason, G., "Experience of Exploiting Icelandic Geothermal Reservoirs," presented at the August 26-30,1985, Geothermal Resources Council's International Symposium on Geothermal Energy, held at Kailua-Kona, Hawaii.

91. Sourirajan, S., and Kennedy, G.C., "The System H_2O-NaCl at Elevated Temperatures and Pressures," *American Journal of Science*, Vol. 260, 1962, pp. 115-141.

92. Spiess, F.N., MacDonald, K.C., Atwater, T., Ballard, R., Garranza, A., Cordoba, D., Cox, C., Diaz Garcia, V.M., Francheteau, J., Guerrero, J., Hawkins, J., Haymon, R., Hessler, R., Juteau, T., Kastner, M., Larson, R., Luyendyk, B., MacDougall, J.D., Miller, S., Normark, W., Orcutt, J., Rangin, C., "East Pacific Rise; Hot Springs and Geophysical Experiments,' *Science*, Vol. 207, 1980, pp. 1421-1433.

93. Stanley, H.E., *Introduction to Phase Transitions and Critical Phenomena*, Oxford University Press, New York, N.Y. 1971.

94. Stefansson, V., "The Krafla Geothermal Field, Northeast Iceland," *Geothermal Systems and Case Histories*, L. Rybach and L.P.J. Muffler, eds., John Wiley and Sons, New York, N.Y., 1981, pp. 273-294.

95. Stefansson, V., and Bjornsson, S., "Physical Aspects of Hydrothermal Systems," *Continental and Oceanic Rifts*, G. Palmason, ed., Geodynamic Series, Vol. 8, American Geophysical Union, Washington, D.C., 1982, pp. 123-145.

96. Straus, J.M., and Schubert, G., "Thermal Convection of Water in a Porous Medium: Effects of Temperature and Pressure-Dependent Thermodynamic and Transport Properties," *Journal of Geophysical Research*, Vol. 82, 1977, pp. 325-333.
97. Straus, J.M., and Schubert, G., "Effect of CO_2 on the Buoyancy of Geothermal Fluids," *Geophysical Research Letters*, Vol. 6, 1979, pp. 5-8.
98. Truesdell, A.H., and White, D.E., "Production of Superheated Steam From Vapor-Dominated Geothermal Reservoirs," *Geothermics*, Vol. 2, 1973, pp. 154-175.
99. Walker, G.P.L., "The Breiddalur Central Volcano, Eastern Iceland, *Quarterly Journal of the Geological Society*, London, Vol. 119, 1963, pp. 29-63.
100. Walker, G.P.L., "Acid Volcanic Rocks in Iceland," *Volcanologiqu* Vol. 29, 1966, pp. 375-406.
101. Wang, J.S.Y., Sterbentz, R.A., and Tsang, C.F., "The State-of-the-art of Numerical Modeling of Thermohydrologic Flow in Fractured Rock Masses," *LBL-10524*, Lawrence Berkeley Laboratory Berkeley, California, Feb. 1980.
102. White, D.E., "Thermal Waters of Volcanic Origin," *Geological Society of America, Bulletin*, Vol. 68, 1957, pp. 1637-1658.
103. White, D.E., "Magmatic, Connate, and Metamorphic Waters," *Geological Society of America, Bulletin*, Vol. 68, 1957, pp. 1659-1682.
104. White, D.E., "Some Principles of Geyser Activity, Mainly From Steamboat Springs, Nevada," *American Journal of Science*, Vol. 265, 1967, pp. 641-684.
105. White, D.E., "Mercury and Base-Metal Deposits with Associated Thermal and Mineral Waters," *Geochemistry of Hydrothermal Ore Deposits*, H.L. Barnes, ed., Holt, Rinehart and Winston, Inc., New York, 1967, 575-631.
106. White, D.E., "Environments of Generation of Base-Metal Ore Depth," *Economic Geology*, Vol. 63, 1968, 301-335.
107. White, D.E., "Hydrology, Activity, and Heat Flow of the Steamboat Springs Thermal System, Washoe County, Nevada," *U.S. Geological Survey Professional Paper 458-C*, 1968, p. 109.
108. White, D.E., "Diverse Origin of Hydrothermal Ore Fluids," *Economic Geology*, Vol. 6, 1974, pp. 954-973.
109. White, D.E., and Guffianti, M., "Geothermal Systems and Their Energy Resources," *Reviews, Geophysics and Space Physics*, Vol. 17, 1979, pp. 887-902.
110. White, D.E., Muffler, L.J.P., and Truesdell, A.H., "Vapor-Dominated Hydrothermal Systems Compared With Hot-Water Systems, *Economic Geology*, Vol. 66, 1971, pp. 75-97.
111. Witherspoon, P.A., Neuman, S.P., Sorey, M.L., and Lippmann, M.J., "Modeling Geothermal Systems," presented at the March 3-5, 1975, Academia Nazionale dei Lincei International Meeting on Geothermal Phenomena and its Applications, held at Rome, Italy, (LBL-3263).

9. LIST OF SYMBOLS

CP	Critical Point
c_p	Specific heat at constant pressure
d_v	Rate of vaporization
g	Gravitational acceleration
$\underset{\sim}{g}$	Gravitational acceleration vector
H	Thickness of layer
h	Enthalphy per unit mass
K	Permeability of the rock
$\underset{\approx}{k}$	Local intrinsic permeability tensor
k_r	Relative permeability
P	Pressure
P_{crit}	Critical pressure
p	Pressure
Q	Interphase heat transfer term
q	Source term
Ra	Rayleigh number
S	$= S_w =$ Volume saturation, where $S_w + S_s = 1$
T	Temperature
T_{crit}	Critical temperature
U	Internal energy per unit mass
$\underset{\sim}{v}$	Phase average velocity
z	Depth
α	Coefficient of thermal expansion of the fluid
β	Isothermal compressibility
Δ	Difference
λ	Thermal conductivity of saturated rock
$\underset{\sim}{\lambda}$	Combined conduction-dispersion vector
μ	Dynamic viscosity
ρ	Average density
ϕ	Porosity

Subscripts and Superscripts:

ls	Saturated liquid
vs	Saturated vapor
r,s,w	Rock,Steam,Water

THERMOHYDRAULICS OF AN AQUIFER THERMAL ENERGY STORAGE SYSTEM

Chin-Fu Tsang

THERMOHYDRAULICS OF AN AQUIFER THERMAL ENERGY STORAGE SYSTEM

Chin-Fu Tsang

Earth Sciences Division
Lawrence Berkeley Laboratory
University of California
Berkeley, California 94720, USA

ABSTRACT

The thermohydraulics of an aquifer thermal energy storage system is reviewed. The storage of hot or chilled water in an aquifer involves three major physical processes: (1) buoyancy flow, (2) forced convection, (3) thermal conduction and thermal dispersion. This chapter is divided into two parts. The first part presents an analysis of buoyancy flow causing the tilting of the interface (thermal front) between hot and cold water. The second part describes a numerical approach that studies the thermohydraulics of an aquifer thermal energy system where all three processes are taken into account.

1. INTRODUCTION

The need for energy storage arises from the disparity between energy demand and production. The development of viable storage methods will play a significant role in our ability to implement alternative energy technologies and use what is now waste heat. The ability to provide heat at night and during inclement weather is a key factor in the development of solar energy. Conversely, winter cold, in the form of melted snow or water cooled to winter air temperatures, can be used as a coolant or for air-conditioning.

Practical storage systems would also allow us to capture the heat that occurs as a by-product of industrial processes and power production. Industrial plants and electric utilities generate tremendous amounts of waste heat which are usually dissipated through an expensive system of cooling towers or ponds to avoid thermal pollution. Because periods of heat demand do not generally coincide with electricity generation or industrial production, a viable storage method is essential if this heat is to be used. Such a method would not only utilize what is now waste heat, but would significantly decrease the necessary investment in cooling and backup heating systems.

In recent years, aquifers have been actively studied as a very promising means of long-term, large-scale thermal energy storage(18, 22). Aquifers are physically well-suited to thermal energy storage because of their low heat conductivities, large volumetric capacities and in the case of confined aquifers, their ability to contain water under high pressure.

A basic aquifer thermal energy storage system is illustrated in Figure 1. The aquifer is penetrated by two wells some distance apart. These two wells are connected by a closed hydraulic arrangement so that water pumped from one well is injected into the other with the net withdrawal of groundwater being kept at zero. Waste heat from a power plant is transferred to the groundwater by means of a heat exchanger. When heat is to be stored, the flow follows the open arrow in the figure. The hot water is stored around the

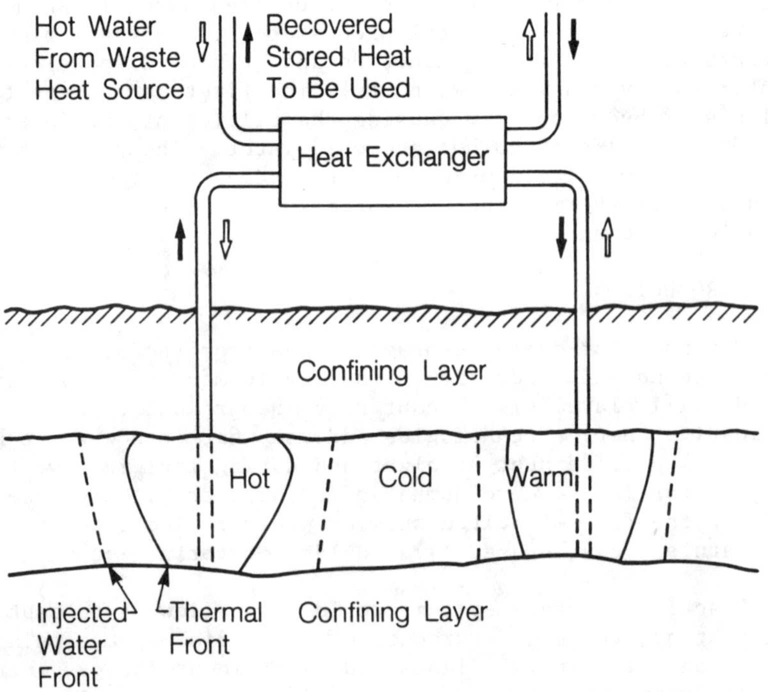

Figure 1. Basic concept of an aquifer thermal energy storage system

hot water well. The broken line in the figure indicates the location of the front of injected fluid, but the thermal front is retarded due to heat transfer to the porous rock medium. After a period ranging from several days to a few months, the hot water is withdrawn (indicated by the solid arrow) when its use for space heating or process heat is required.

In this application, it is critical to know not only the thermal behavior of the groundwater system, but also the amount of energy that can be recovered during withdrawal, and the temperature variations of the produced water.

Figure 2 is a schematic diagram of the physical processes operating in the groundwater system during injection or withdrawal of hot water. The well is represented by the double line at the left of the figure, with radial symmetry about this line. The injected hot water is shown with a tilted front at an angle α with the vertical. The tilting of the front is due to the lower density of the hot water with the resulting buoyancy flow.

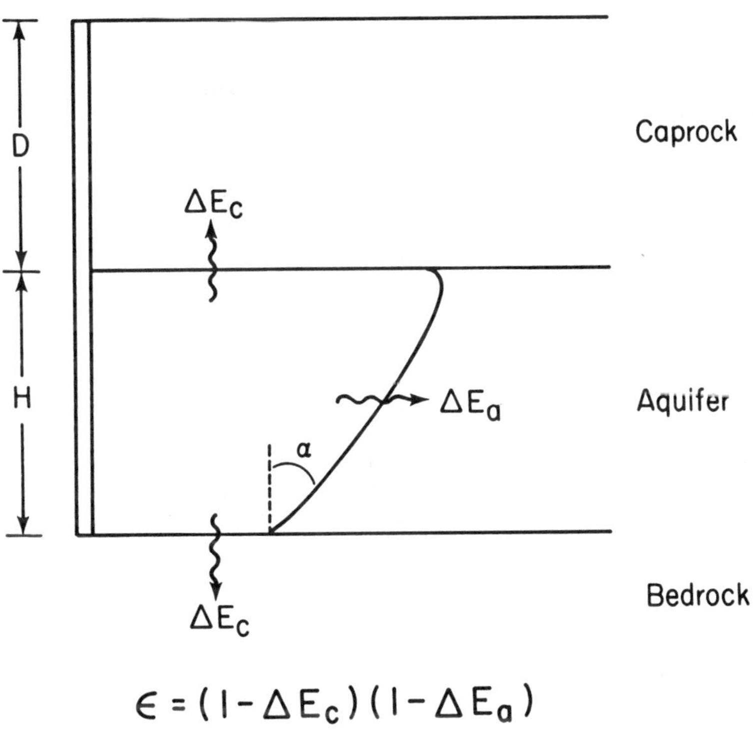

Figure 2. Physical processes involved in an aquifer thermal energy storage system.

The physical processes involved are: (1) buoyancy flow, (2) thermal conduction and thermal dispersion, and (3) forced convection. Forced convection occurs whenever hot water is injected or withdrawn from the aquifer. Thermal conduction from the hot water to the native cold water and to the confining layers is a primary cause of heat loss from the stored heat. Dispersion is present due to the tortuosity of the fluid flow paths in the porous medium. This result in an additional mixing of the stored hot water with the surrounding cold water. Buoyancy flow causes the hot water to move above the cold, so that during withdrawal, both hot water from upper layers of the aquifer and cold water from the lower layers are produced.

Many hydrogeological factors have to be studied when considering aquifer storage:formation geometry, aquifer permeability and storativity, inhomogeneity, confining layer leakage, regional groundwater flow, and geochemistry of injected and native waters in the aquifer.

The paper is divided into two parts. In the first part an analysis of the motion of the interface between hot and cold water in an anisotropic porous medium is presented. This is commonly referred to as the thermal front tilting problem. In the second part a numerical study of an aquifer thermal energy storage system involving the three physical processes listed above is described. Detailed application to a series of field experimental studies is presented. The materials in this paper are based primarily on work performed in collaboration with Buscheck, Claesson, Doughty, and Hellstrom over the last few years (3,7,11,12,20,21).

2. THERMAL FRONT TILTING PROBLEM

In this section, we present the methodology to calculate the motion of the interface that occurs when fluid of one density and viscosity is injected into an aquifer stratum containing fluid of another density and viscosity. The interface region constitutes a transition zone between aquifer regions with different fluids. Initially, the interface is primarily vertical. This situation is intrinsically unstable due to the difference in density between the two fluids. Buoyancy will cause the fluid of lower density to flow towards the upper part of the aquifer. The two-fluid interface will gradually tilt. Forced convection will act on the differences in viscosity, and hence the differences in flow resistance, along different flow paths, and thereby influence the tilting. Depending on the situation, the forced convection may either reinforce or counteract the pure buoyancy tilting. It is often of great interest to be able to predict the rate at which the two-fluid interface tilts.

2.1. Thermohydraulic Equations

The coupled groundwater and heat flow process in the aquifer stratum is governed by two partial differential equations. The volumetric ground water flow $\bar{q}$ is related to the pressure gradient and the gravity force through the empirical law of Darcy:

$$\bar{q} = -\frac{k}{\mu}(\nabla P + \rho g \hat{z}) \tag{2.1}$$

The intrinsic permeability is k. The water density ρ and the viscosity μ are temperature dependent.

Equation (2.1) assumes isotropic permeability in the aquifer. In this paper we will also consider cases where the aquifer has different permeabilities in the horizontal (x,y) and vertical (z) directions. Horizontal and vertical permeabilities are denoted k and k', respectively. We then have:

$$q_x = -\frac{k}{\mu}\frac{\partial P}{\partial x}, \qquad q_y = -\frac{k}{\mu}\frac{\partial P}{\partial y}, \qquad q_z = -\frac{k'}{\mu}\left(\frac{\partial P}{\partial z} + \rho g\right) \tag{2.2}$$

Compressibility effects are neglected, and the divergence of the groundwater flow $\bar{q}$ is then zero at each point:

$$\nabla.\bar{q} = \nabla.\left[-\frac{k}{\mu}(\nabla P + \rho g \hat{z})\right] = 0 \tag{2.3}$$

The aquifer temperature satisfies the equation for convective-diffusive heat transfer:

$$C\frac{\partial T}{\partial t} = \nabla.(\lambda \nabla T - TC_w \bar{q}) \tag{2.4}$$

where C and C_w are the volumetric heat capacities for the aquifer (matrix plus water) and water, respectively. The thermal conductivity is denoted by λ.

The convective heat flow is given by $TC_w\bar{q}$. The thermal velocity is

$$\bar{v}_T = \frac{C_w}{C}\bar{q} \tag{2.5}$$

which represents the convective displacement of the temperature field.

The aquifer stratum is confined by impermeable layers at which the perpendicular groundwater flow component vanishes.

2.2. Buoyancy Flow

A non-uniform temperature field in the aquifer gives a variable fluid density and an ensuing buoyancy flow in the aquifer. We will in particular consider the situation where the aquifer is divided int a warm region ($T = T_1$) and a cold region ($T = T_0$). These regions are separated by a thermal front zone, through which the temperature falls from T_1 to T_0. The idealization of an infinitely thin or sharp thermal front will also be considered. This is often quite a useful approximation.

Let Γ be any closed curve in an isotropic aquifer ($k' = k$). The line integral along Γ of the pressure gradient is automatically zero. Darcy's equation (1) then gives:

$$\int_\Gamma \frac{\mu}{k}\,\bar{q}.d\bar{r} = -\,g\hat{z}.\int_\Gamma \rho d\bar{r} \tag{2.6}$$

The right-hand term represents a net driving force due to density variations along Γ. The left-hand side gives an integral of the tangential component of the flow $\bar{q}$ along Γ. The flow is weighted with the flow resistance coefficient μ/k. The right-hand side is known when the temperature, and hence the density field, is given. Equation (2.6) provides some information on the magnitude of the flow velocities.

Figure 3 shows a case when the curve Γ crosses a sharp thermal front. The density and the viscosity on the warm and cold sides are denoted ρ_1, μ_1 and ρ_0, μ_0 respectively. The vertical distance between the two points where Γ crosses the thermal front is H. It follows from equation (2.6) that

$$\int_{\Gamma_1} \frac{\mu_1}{k}\,\bar{q}.d\bar{r} + \int_{\Gamma_o} \frac{\mu_0}{k}\,\bar{q}.d\bar{r} = (\rho_0 - \rho_1)gH \tag{2.7}$$

Let L_Γ denote the arc length of Γ, and q_Γ a suitable mean tangential component of $\bar{q}$ along Γ. Equation (2.7) may then be written:

$$q_\Gamma = \frac{k(\rho_0 - \rho_1)g}{\mu_0 + \mu_1} \cdot \frac{2H}{L_\Gamma} \tag{2.8}$$

The first factor will appear often in the following. We will call it the characteristic buoyancy flow q_0:

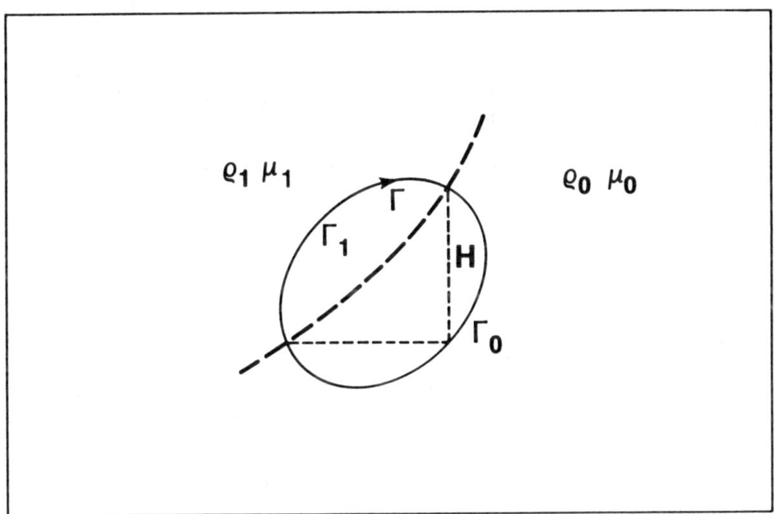

Figure 3. Closed curve Γ in an aquifer with a sharp thermal front (dashed line).

$$q_0 = \frac{k(\rho_0 - \rho_1)g}{\mu_0 + \mu_1} \tag{2.9}$$

2.3. Analytical Solutions

Based on equations (2.1 - 2.3), it is possible to derive explicit expressions for the pressure distribution and the buoyancy flow pattern in some idealized situations. Figures 4a-h show the eight cases, A-H, explained in detail below. The permeability may be different in the horizontal (k) and the vertical (k') directions in all cases except in case H. We will use the notation

$$\kappa = \sqrt{k'/k} \tag{2.10}$$

Case A is a sharp, vertical thermal front in an infinite aquifer bounded by two impermeable horizontal planes. The thickness of the aquifer stratum is H. A vertical cross-section through the aquifer becomes an infinite strip. The expressions for the pressure distribution and the flow field are derived in Appendix A. The flow field is shown in Figure 5. A solution of a limited version of this problem for two fluids with different density, but equal viscosity (μ_o=μ_1) and isotropic permeability (k'=k) has previously been given by de Josselin de Jong (6). Verruijt (23) solved the problem with different viscosities for the two fluids in an isotropic porous medium.

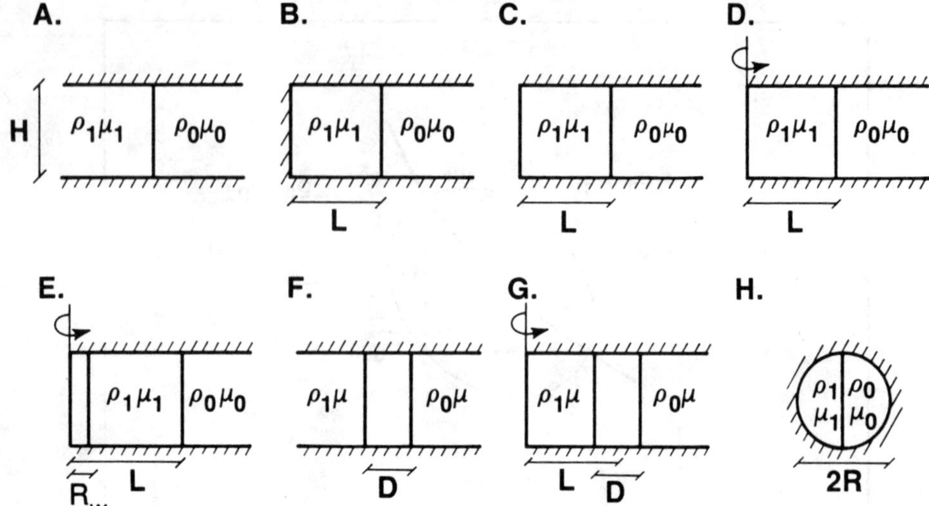

Figure 4. Cases considered for analytical solutions: (a) Infinite strip, (b) Semi-infinite strip with impermeable left boundary, (c) Semi-infinite strip with constant-head left boundary, (d) Cylindrical case with no horizontal flow at the inner boundary, (e) Cylindrical case with constant-head inner boundary, (f) Infinite strip with thermal front thickness D, (g) Cylindrical case with thermal front thickness D, (h) Circular disc.

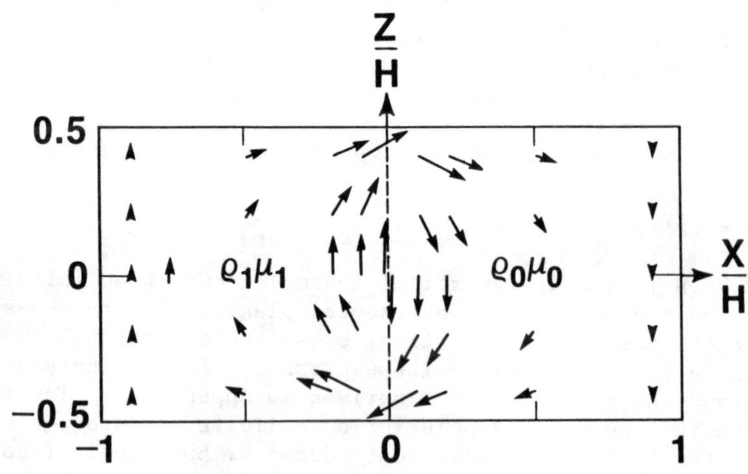

Figure 5. Velocity flow field for case A and isotropic medium.

The cross-section of the aquifer stratum is a semi-infinite strip in cases B and C. There is a sharp, vertical thermal front. The warmer region to the left has a horizontal thickness L. The left boundary is impermeable in case B. In case C the hydrostatic pressure $P = - \rho_1 gz$ prevails along the left, vertical boundary.

Cases D and E consider an infinite aquifer with cylindrical symmetry bounded by two horizontal planes. The warmer region occupies a cylindrical volume with radius L. There is no horizontal flow at the inner boundary in case D. In case E there is hydrostatic pressure $P = - \rho_1 gz$ at the inner boundary at radius $r = R_w$.

In case F we have, as in case A, an infinite aquifer bounded by two horizontal planes. The thermal front has a thickness D. The viscosity in this case is assumed to be constant, i.e., $\mu=\mu_o=\mu_1$. The density is ρ_1 in the warm region and ρ_o in the cold region. The density is assumed to increase linearly through the thermal front region.

Case G is similar to case D with cylindrical symmetry and no horizontal flow at the inner boundary, but with a diffuse thermal front of thickness D. The viscosity is constant, i.e., $\mu=\mu_o=\mu_1$, and the density varies logarithmically through the thermal front region.

In case H the aquifer is an infinite circular cylinder. A vertical cut perpendicular to the cylinder axis becomes a circular disk (Figure 4h). The permeability must be isotropic (κ=1) in this case.

The analytical solutions for these cases are derived in Appendices A-H. The given expressions are only valid at the moment when the thermal front is vertical.

The motion of the thermal front is determined by the magnitude of groundwater flow across the front. The vertical coordinate is denoted by z, and z=0 is the mid-point of the aquifer. The horizontal groundwater flow across the front is denoted $q_f(z)$. The following expressions are obtained for the considered cases:

A. Infinite strip:

$$q_f(z) = \kappa q_0 \cdot \frac{1}{\pi} \ln \left(\frac{1+\sin(\frac{\pi z}{H})}{1-\sin(\frac{\pi z}{H})} \right) \tag{2.11}$$

B. Semi-infinite strip; impermeable left boundary:

$$q_f(z) = \kappa q_0 \cdot \frac{4}{\pi} \sum_{n=0}^{\infty} \frac{(-1)^n}{2n+1} \cdot \frac{\sin\left(\frac{(2n+1)\pi z}{H}\right)}{\frac{\mu_0}{\mu_0+\mu_1} + \frac{\mu_1}{\mu_0+\mu_1}\coth\left(\frac{(2n+1)\pi\kappa L}{H}\right)} \tag{2.12}$$

C. Semi-infinite strip; hydrostatic pressure conditions at the left boundary:

$$q_f(z) = \kappa q_0 \cdot \frac{4}{\pi} \sum_{n=0}^{\infty} \frac{(-1)^n}{2n+1} \cdot \frac{\sin\left(\frac{(2n+1)\pi z}{H}\right)}{\frac{\mu_0}{\mu_0+\mu_1} + \frac{\mu_1}{\mu_0+\mu_1}\tanh\left(\frac{(2n+1)\pi\kappa L}{H}\right)} \tag{2.13}$$

D. Cylindrical case; no horizontal flow at inner boundary:

$$q_f(z) = \kappa q_0 \cdot \frac{4}{\pi} \sum_{n=0}^{\infty} \frac{(-1)^n}{2n+1} \cdot \frac{\sin\left(\frac{(2n+1)\pi z}{H}\right)}{\frac{\mu_1}{\mu_0+\mu_1} \cdot \frac{I_0(\theta_n)}{I_1(\theta_n)} + \frac{\mu_0}{\mu_0+\mu_1} \cdot \frac{K_0(\theta_n)}{K_1(\theta_n)}} \tag{2.14}$$

where

$$\theta_n = \frac{(2n+1)\pi\kappa L}{H} \tag{2.15}$$

Here we have used the modified Bessel functions K_n and I_n.

E. Cylindrical case; hydrostatic pressure conditions at inner boundary:

$$q_f(z) = \kappa q_0 \cdot \frac{4}{\pi} \sum_{n=0}^{\infty} \frac{(-1)^n}{2n+1} \cdot \sin\left(\frac{(2n+1)\pi z}{H}\right) \tag{2.16}$$

$$\times \frac{1 + \frac{K_1(\theta_n^L)}{I_1(\theta_n^L)} \cdot \frac{I_0(\theta_n^R)}{K_0(\theta_n^R)}}{\frac{\mu_0-\mu_1}{\mu_0+\mu_1} \cdot \frac{K_0(\theta_n^L)}{I_1(\theta_n^L)} \cdot \frac{I_0(\theta_n^R)}{K_0(\theta_n^R)} + \frac{\mu_1}{\mu_0+\mu_1} \cdot \frac{I_0(\theta_n^L)}{I_1(\theta_n^L)} + \frac{\mu_0}{\mu_0+\mu_1} \cdot \frac{K_0(\theta_n^L)}{K_1(\theta_n^L)}}$$

where

$$\theta_n^R = \frac{(2n+1)\pi\kappa R_w}{H} \tag{2.17}$$

$$\theta_n^L = \frac{(2n+1)\pi\kappa L}{H} \tag{2.18}$$

F. Infinite strip with diffuse thermal front:

$$q_f(z) = \kappa q_0 \cdot \frac{4}{\pi} \sum_{n=0}^{\infty} \frac{(-1)^n}{2n+1} \cdot \frac{1 - e^{-\frac{(2n+1)\pi\kappa\frac{D}{2}}{H}}}{\frac{(2n+1)\pi\kappa\frac{D}{2}}{H}} \cdot \sin\left(\frac{(2n+1)\pi z}{H}\right) \tag{2.19}$$

The flow $q_f(z)$ refers to the middle of the thermal front region. For large values of D/H, equation (2.19) becomes:

$$q_f(z) \cong 2q_0 \cdot \frac{z}{D} \tag{2.20}$$

G. Cylindrical case with diffuse thermal front:

$$q_f(z) = \kappa q_0 \cdot \frac{2}{\ln\left(\frac{(L+D/2)}{(L-D/2)}\right)} \cdot \frac{4}{\pi} \sum_{n=0}^{\infty} \frac{(-1)^n}{2n+1} \cdot \tag{2.21}$$

$$\left(\Phi(\theta_n^0)\,\frac{I_1(\theta_n^0)}{I_0(\theta_n^0)} - \Phi(\theta_n^+)\,\frac{I_1(\theta_n^0)}{I_0(\theta_n^+)}\right.$$

$$\left. + \Phi(\theta_n^0)\,\frac{K_1(\theta_n^0)}{K_0(\theta_n^0)} - \Phi(\theta_n^-)\,\frac{K_1(\theta_n^0)}{K_0(\theta_n^-)}\right) \cdot \sin\left(\frac{(2n+1)\pi z}{H}\right)$$

where the function Φ is defined by:

$$\Phi(\theta_n) = \frac{1}{\theta_n} \cdot \frac{1}{\dfrac{I_1(\theta_n)}{I_0(\theta_n)} + \dfrac{K_1(\theta_n)}{K_0(\theta_n)}} \tag{2.22}$$

and

$$\theta_n^+ = \frac{(2n+1)\pi\kappa(L+D/2)}{H} \tag{2.23}$$

$$\theta_n^0 = \frac{(2n+1)\pi\kappa L}{H} \tag{2.24}$$

$$\theta_n^- = \frac{(2n+1)\pi\kappa(L-D/2)}{H} \tag{2.25}$$

The flow $q_f(z)$ refers to the middle of the thermal front region $(r = L)$.

H. Circular disk:

$$q_f(z) = q_0 \cdot \frac{1}{\pi}\left(\left(1+\left(\frac{R}{z}\right)^2\right)\cdot \ln\left(\frac{1+\frac{z}{R}}{1-\frac{z}{R}}\right) - 2\cdot\frac{R}{z}\right) \tag{2.26}$$

The flows $q_f(z)$ are all odd functions of z. Further results of these analyses are given by Hellstrom et al. (12).

2.4. Tilting of a Thermal Front

The buoyancy flow will cause a thermal front to tilt. A quantitative measure of the rate of tilting is of great interest. For the case when there is no forced convection, the tilting rate may be defined as follows.

Consider a straight, vertical thermal front at a time t. The total water flow across the upper half of the thermal front is called the tilting flow. The same amount passes in the other direction through the lower half of the front. The tilting flow Q_t is defined by:

$$Q_t = \int_0^{H/2} q_f(z)dz = -\int_{-H/2}^{0} q_f(z)dz \tag{2.27}$$

Figure 6 illustrates the tilting of a vertical front. Each point on the front is displaced a length $\bar{v}_T dt$ during a small time increment. The displacement in the normal direction of the front

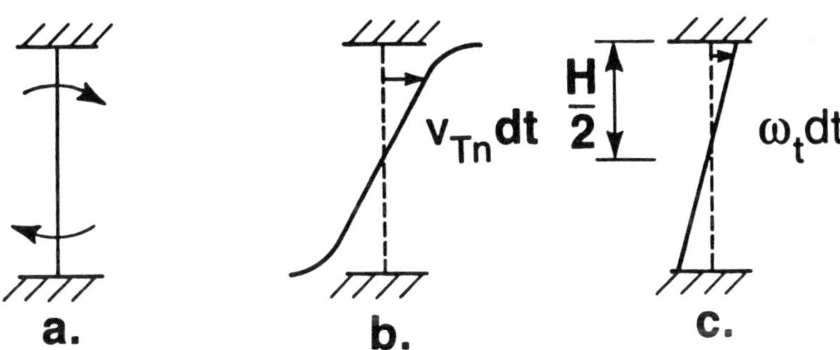

Figure 6. Definition of angular tilting rate ω_t: (a) thermal front at time t; (b) thermal front at time ttdt; (c) linear approximation of situation; (d) with the same total flow.

is $v_{Tn} \cdot dt$, where v_{Tn} is the thermal velocity component perpendicular to the thermal front. The curved thermal front (Figure 6b) is approximated by an appropriate straight line (Figure 6c). The front is tilted an angle $\omega_t dt$ during the time increment dt. A heat balance for the thermal front gives:

$$\frac{1}{2} \cdot \frac{H}{2} \cdot \frac{H}{2} \tan(\omega_t dt) = \frac{C_w}{C} Q_t dt \tag{2.28}$$

The time increment is small, so $\tan(\omega_t dt) = \omega_t dt$. The tilting rate then becomes:

$$\omega_t = \frac{8}{H^2} \cdot \frac{C_w}{C} Q_t \tag{2.29}$$

The tilting flow Q_t is obtained by integrating $q_f(z)$ over the interval $0 \leq z \leq H/2$. The integration of each term in the different series of equations (2.11-2.14, 2.16, 2.19, 2.21, 2.26) is straightforward.

The tilting flow for case A is found to be:

$$Q_t = \frac{4G}{\pi^2} \kappa q_0 H \qquad (G = 0.915\ldots \text{ Catalan's constant}) \tag{2.30}$$

The corresponding rate of angular tilting is:

$$\omega_0 = \frac{32G}{\pi^2} \cdot \kappa \cdot \frac{C_w q_0}{C} \cdot \frac{1}{H} \qquad \left(\frac{32G}{\pi^2} \cong 3.0\right) \tag{2.31}$$

The corresponding tilting time t_0 is then:

$$t_0 = \frac{1}{\omega_0} = \frac{HC}{\kappa C_w k} \; \frac{\pi^2(\mu_0+\mu_1)}{32G(\rho_0-\rho_1)g} \tag{2.32}$$

The second factor on the right hand side is a function only of the temperatures T_0 and T_1.

In case H, the circular disk, it is possible to obtain an analytical solution for the case when the straight thermal front is tilted at an angle α from the vertical direction (see Appendix H). The tilting flow and the tilting rate in case H become:

$$Q_t = \frac{2}{\pi} q_0 R \cdot \cos(\alpha) \tag{2.33}$$

$$\omega_t = \frac{8}{\pi} \cdot \frac{C_w q_0}{C} \cdot \frac{1}{2R} \cdot \frac{1}{\cos(\alpha)} \qquad \left(\frac{8}{\pi} \cong 2.5\right)$$

The tilting rate is thus reduced in the proportion 2.5/3.0, when an infinite strip (case A with κ=1) is compared to a corresponding circular disk with a vertical thermal front.

3. NUMERICAL THERMOHYDRAULIC STUDIES

An analytic approach is useful for providing physical insight and functional dependence for a given process. However, in general, a study of the coupled effects of buoyancy flow, forced convection and thermal conduction in an aquifer thermal energy storage system requires a numerical approach. In this section we introduce the numerical code PT (2), which has been extensively used at Lawrence Berkeley Laboratory for such studies. The code is briefly described and some generic results are presented. Then, several field applications are given to demonstrate the approach and methodology used in applying a numerical code to study the thermohydraulics of such a system.

3.1. Numerical Code

The three-dimensional computer code PT, developed at LBL (2), is capable of calculating coupled liquid and heat flows in a water-saturated porous or fractured-porous medium. The governing equations for PT consist of the conservation equations for mass and energy, and Darcy's law for fluid flow. Pressure and temperature are dependent variables. One-dimensional consolidation of the rock matrix can be considered as well, using the theory of Terzaghi. The mass and energy conservation equations are coupled through the fluid flow in the convection term of the energy equation and the pressure and temperature dependent fluid and rock properties. The rock matrix and fluid are considered to be in local thermal equilibrium at all times. Energy changes due to fluid compressibility, acceleration, and viscous dissipation are neglected.

The following physical effects are included in PT calculations: (1) heat convection and conduction; (2) regional groundwater flow; (3) multiple heat and/or mass sources and sinks; (4) constant pressure or temperature boundaries; (5) hydrologic or thermal barriers; (6) gravitational effects (buoyancy); (7) complex geometries due to heterogeneous materials; and (8) anisotropic permeability and thermal conductivity.

PT carries out the spatial discretization of the flow regime using the Integral-Finite-Difference method (8,17). This method treats one-, two-, or three-dimensional problems equivalently. An efficient sparse solver is used to solve the linearized mass and energy matrix equations. The equations are solved implicitly to allow for large time steps. PT adjusts the time step automatically, so that the temperature or pressure change in any node during one time step is within user-specified limits. Mass and energy balances are calculated for each node at every time step.

PT has been verified against the following analytical solutions:

(1) Theis problem (19);
(2) Cold water injection into a hot reservoir (1);
(3) The temperature variation at a production well due to cold water injection (10);
(4) Radial conduction outside a constant temperature cylinder (4);
(5) Two-node problem, transident conduction heat transfer between two adjacent blocks (4);
(6) The rate of thermal front tilting when hot water is injected into a cold reservoir (11);
(7) Pressure response in a well intercepting a finite conductivity vertical fracture (5);
(8) Pressure response in a well intercepting a (uniform flux) horizontal fracture (9).

3.2. Some Generic Results

To give some physical insight into the thermohydraulic behavior of an aquifer thermal energy storage system, results of calculations using PT on a few hypothetical cases are given here. This particular set of calculations is for the case of high-temperature hot water storage in a deep, low-permeability aquifer. The temperature of the injected water is assumed to be 220°C (in liquid phase under pressure) and the instrinsic permeability of the deep aquifer is assumed to be $10^{-13} m^2$.

With injection and production ranges equal to 10^6 kg/day for a 90-day injection, 90-day storage and 90-day production cycle, the energy balance is calculated as shown in Table 1. Figures 7,8 and 9 show the thermal front advancement and diffusion at the end of the injection period for an inhomogeneous aquifer (Figure 7), an aquifer with a clay lens (Figure 8), and an aquifer with both an injection-production well and a supply well (Figure 9). These figures show the buoyancy flow effects and in the last case, the influence of a neighboring well.

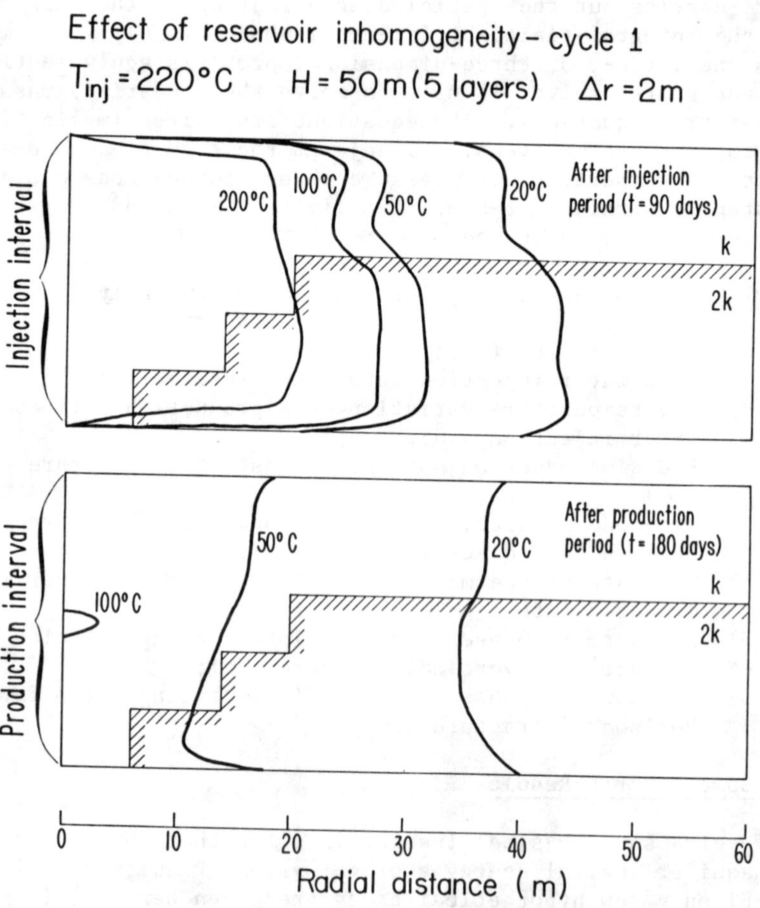

Figure 7. Temperature contours at the end of 90-day injection period and at the end of 90-day production period for an inhomogeneous two-layer aquifer.

Table 1. Computed Energy Balance for a Low-Permeability Aquifer*

Cycle	Energy injected ($J \times 10^{13}$)	Energy recovered ($J \times 10^{13}$)	Energy loss from aquifer ($J \times 10^{13}$)	Energy diffused to heat aquifer ($J \times 10^{13}$)	Energy recovered (%)	Prod. temp. at end of cycle (^{o}C)
1	5.71	4.96	0.053	0.71	86.8	124
2	5.71	5.09	0.068	0.55	89.2	139
3	5.71	5.14	0.077	0.49	90.0	147
4	5.71	5.18	0.084	0.45	90.7	151
5	5.71	5.20	0.091	0.42	91.1	155

*Calculations are for the first five cycles for the case of 90-day injection (flow rate of 10^6 kg/d) and 90-day production periods.

Injection and ambient temperatures are 220 and 20°C, respectively; the aquifer is 100 m thick.

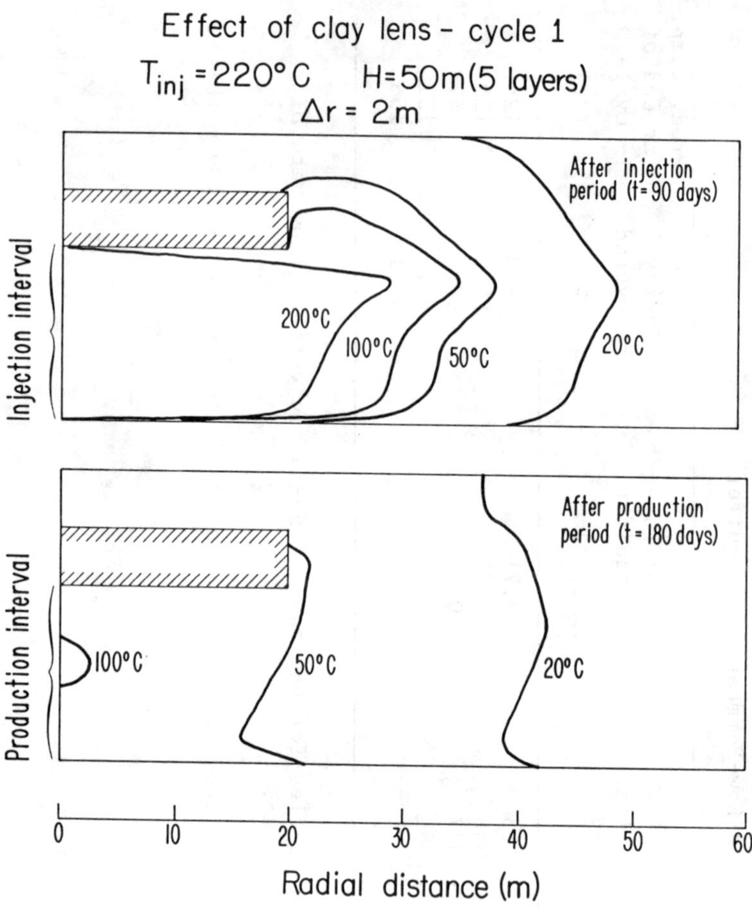

Figure 8. Temperature contours at the end of 90-day injection period and at the end of 90-day production period for an aquifer with a clay lens.

3.3. Field Study Using a Numerical Code

Application of the numerical code PT to a set of field experiments has the goal of studying the thermohydraulics of a practical field situation and of verifying the validity of the numerical code against field data. The field experiments by Auburn University described below were chosen for this purpose. First, PT was used to do a history match of the first two cycles of the field experiment. All data from the experiment were available to the modelers. Second, PT was used to make a double-blind prediction of the next two cycles of the field experiment. Only the design parameters of

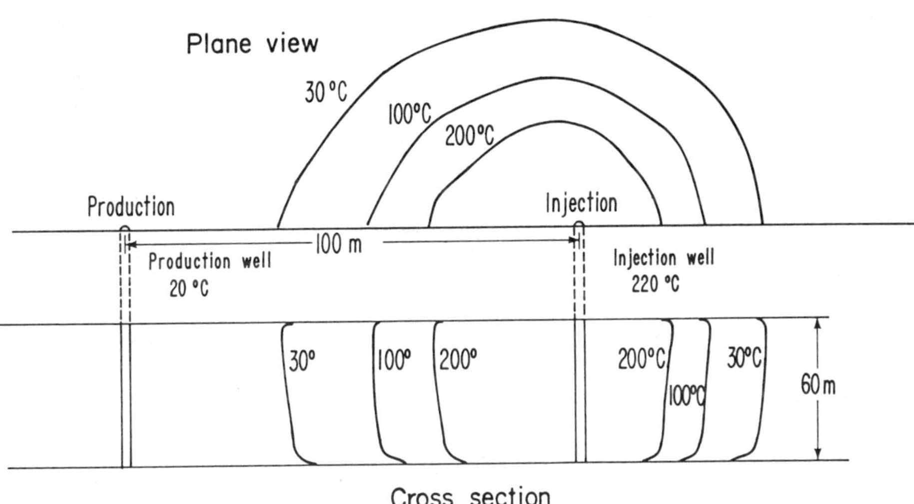

Figure 9. Temperature contours for a two-well system after 90-days of injection.

the experiment were available to the modelers, not the results. Third, PT was used to do optimization design studies for a planned cycle of the field experiment; subsequently, results of the actual cycle were compared with the PT calculations. Progressing from each stage to the next provided a more stringent test of the numerical model of the thermohydraulics of the system, as the code was used more and more as a predictive tool. At each stage of the calculation, a number of parameter sensitivity studies were made to determine which parameters affected the results of the experiment most significantly. Study of the discrepancies between the calculated and observed field results gave insight into possible physical processes not included in the numerical model, and provided direction for future work.

3.3.1. History match

The Water Resources Research Institute of Auburn University initiated a two-cycle injection-storage-production field experiment in a shallow aquifer in northeastern Mobile County, Alabama in 1978 (13,14,15,16). A single injection/production well was screened in the upper half of a confined 21-m-thick aquifer. The aquifer matrix consists primarily of medium to fine sand, with approximately 15 percent interstitial silt and clay. The aquifer is located from about 40 to 61 m below the land surface and is capped by a 9-m-thick clay sequence; it is bounded below by another clay sequence of undetermined thickness. Above the upper clay unit lies another aquifer, which provided the injection water. A number of observation wells were located around the injection/production well; each was completed with thermistors that measured temperature at six depths in the aquifer. Each injection-storage-production cycle lasted six months and involved the injection and recovery of about 55,000 m^3 of water heated to an average temperature of 55°C. The ambient wate temperature of the supply and storage aquifers was 20°C. A convenie quantitative measure of each cycle is the recovery factor, defined as the ratio of the energy produced to the energy injected, with energies measured relative to the ambient groundwater temperature. The first-cycle recovery factor was 0.66 and the second-cycle recove factor was 0.76.

Well tests were done to determine the hydraulic properties of the aquifer, and laboratory tests were made on samples to determine thermal properties of the aquifer and clay layers. Several of the material properties needed for the numerical calculation were not provided; in these instances reasonable values from the literature were used. Whenever possible, sensitivity studies were done to examine the effect of the variation of such parameters. Table 2 summarizes the material properties used for the different layers. Field measurements indicated very small regional groundwater flow, so an axisymmetric model was devised for the calculation. The spatial discretization for a model considering combined heat and fluid flow from a central well must be done with care. For the

Table 2. Parameters Used in the History Match Calculation

Formation thickness	Aquifer	21 m
	Aquitard	9 m
Thermal conductivity	Aquifer	2.29 J/m s °C
	Aquitard	2.56 J/m s °C
Heat capacity of rock		1.81×10^{6} J/m^3 °C
Density of rock		2600 kg/m^3
Aquifer horizontal permeability		0.53×10^{-10} m^2 (53 darcies)
Vertical to horizontal permeability ratio		0.10
Aquitard to aquifer permeability ratio		10^{-5}
Porosity	Aquifer	0.25
	Aquitard	0.15
Storativity	Aquifer	6×10^{-4}
	Aquitard	9×10^{-2}

pressure calculation, the size of the nodes should logarithmically increase with increasing radial distance from the injection well; for the temperature calculation, the size of the nodes should decrease. A compromise of equally spaced nodes within the region of thermal influence, about 60 m, was used. Beyond this region, the size of the nodes steadily increased out to a distance of 20 km, where there was a constant pressure boundary. The vertical spacing of the nodes was made fine wherever sharp gradients in temperature were expected, such as at the bottom of the well screen, and in the clay layers adjacent to the aquifer. To assure that the spatial discretization (the calculational mesh) was adequate for the problem, alternative meshes were devised and used for parts of the first cycle calculation, and results compared to results from the primary mesh. Figure 10 shows a vertical section of the central portion of the primary mesh.

After the calculation for each cycle was carried out, the calculated temperature distributions in the aquifer at various times were compared to measured temperatures (20). The overall match was very good; however, the calculated temperature profiles, shown in Figure 11, appeared to be sharper than the observed ones, indicating that the mathematical model underpredicted thermal diffusion. This is because the model did not include the heterogeneities of the real aquifer that caused fingering, leading to a diffuse front. By comparing the calculated temperatures with temperatures from observation wells located in different directions from the injection/production well, some deviation from axisymmetry was noted. However, the calculated temperature of the produced water agreed very closely with the observed data, as shown in Figures 12 and 13, since the production temperature is the average temperature of water produced from all directions around the injection/production well. The time-average of the production temperature is proportional to the recovery factor. PT calculated recovery factors of 0.68 and 0.78 for the first and second cycles, respectively, as compared to experimental values of 0.66 and 0.76. This excellent agreement indicated that the small heterogeneties of the system tended to balance out, and that on the whole an axisymmetric model of the system was appropriate.

One of the properties of the aquifer not determined by the well tests was the permeability anisotropy, the ratio of vertical to horizontal permeability in the aquifer. A value of 0.10 was chosen for the model, based on previous modeling studies done at this site. A sensitivity study was carried out for the first cycle using values of 1.0 and 0.02 for permeability anisotropy. For the smaller value of anisotropy (i.e., smaller vertical permeability) there was less buoyancy flow of the injected water than in the original first cycle calculation, resulting in a more compact hot plume with a lower surface-to-volume ratio. This led to smaller

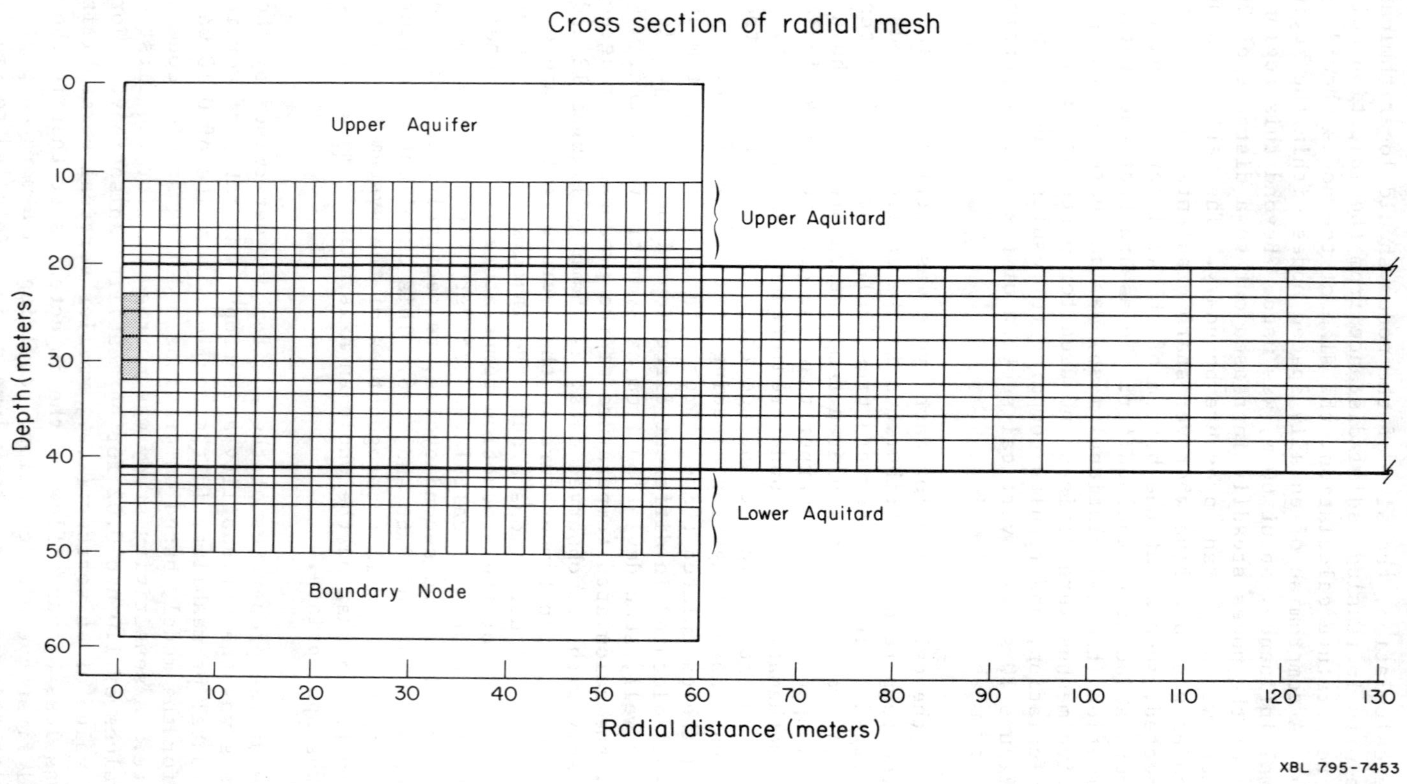

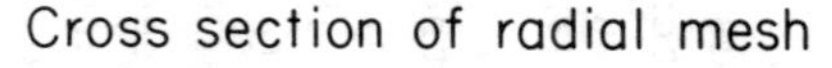

Figure 10. A typical mesh used in the numerical calculation. This mesh was used in the calculation of first and second cycle with the results shown in Figures 11 and 12.

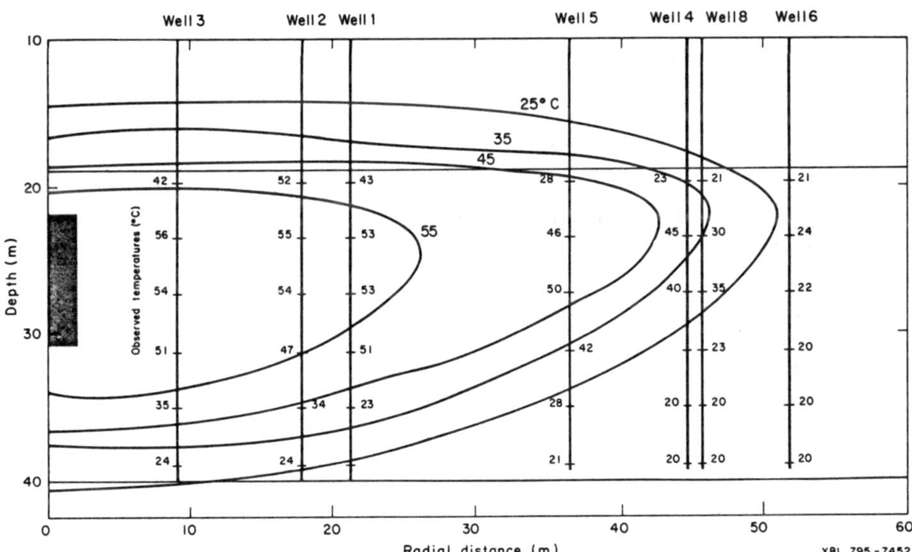

Figure 11. Temperature contours in an aquifer after first cycle injection. Temperature values at observation wells are also indicated.

conductive heat losses, hence a higher recovery factor -- 0.71. For the larger value of anisotropy, buoyancy flow was increased, creating a more elongated plume with larger heat losses, leading to a recovery factor of 0.57. The large variation in recovery factor indicated that the permeability anisotropy is an important parameter.

In summary, the history match indicated that the numerical code PT and an axisymmetric model could match the results of the injection-storage-production cycles very well. Detailed comparisons with individual wells showed some discrepancies, but they tended to cancel out when integrated results such as production temperature and recovery factor were considered. A parameter study indicated that the permeability anisotropy is a very important parameter affecting the results of the experiment. The mesh variation demonstrated the range of mesh spacing appropriate for this particular problem, and showed that numerical dispersion may mimic physical dispersion caused by aquifer heterogeneities.

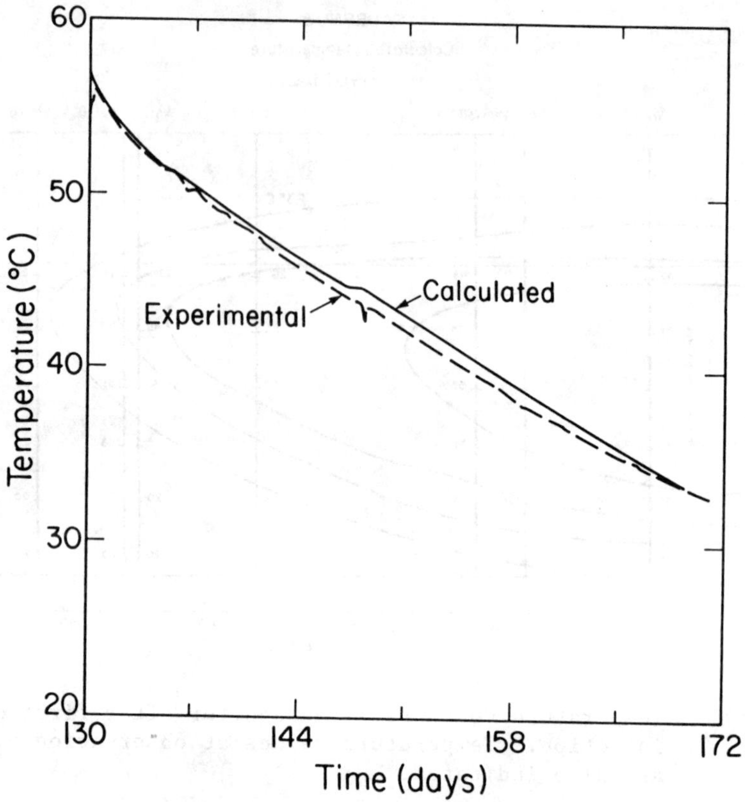

Figure 12. First cycle production temperature.

3.3.2. Double-Blind prediction

In contrast to the history match, in which all the experimental results were available to us throughout the course of the modeling study, in the double-blind prediction, we were provided with only the basic geological, well test, injection flow rate and injection temperature data, and the planned production flow rate. Numerical simulations were conducted to predict the outcome of each cycle before its conclusion. During the course of the study, we were not informed of the experimental observations and the experimenters were not informed of our calculated results. Thus we call this a "double-blind" prediction. It was only after both parties concluded their work that detailed comparisons between the calculated and experimenta recovery factors, production temperatures and in situ temperature distributions were made. Our double-blind prediction studies were carried out in the following fashion.

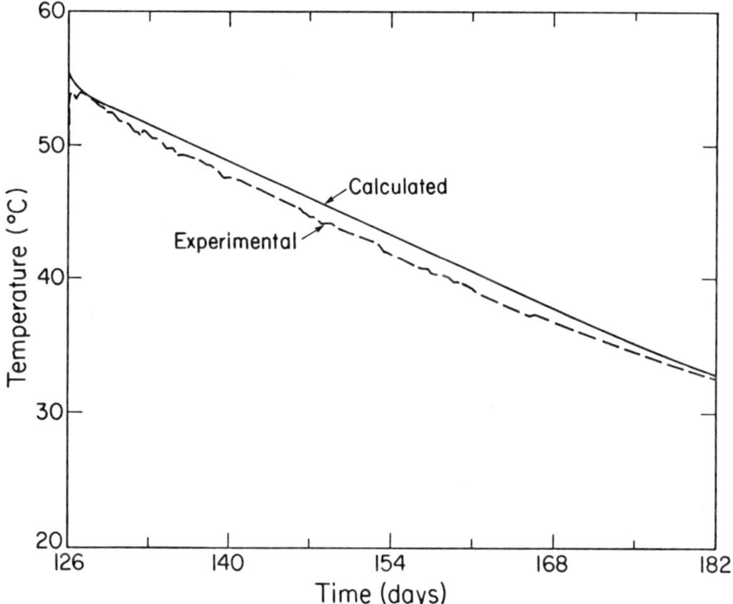

Figure 13. Second cycle production temperature.

The third, fourth and fifth cycles of the Auburn experiments were conducted in 1981 and 1982 in a new area of the aquifer, located about 120 m from the site of the first and second cycles. A fully penetrating injection well was used. Rather than using water from the overlying aquifer, a supply well penetrated the storage aquifer itself, creating an injection-supply doublet. During the third cycle, 25,000 m^3 of water at an average temperature of 59°C was injected over a period of one month. The water was then stored for one month and subsequently produced. During the fourth cycle, a total of 58,000 m^3 of water at an average temperature of 82°C was injected over a period of 4.5 months, then stored for one month. Production began using a well screen open to the full aquifer thickness. After two weeks production stopped and the well screen was modified to withdraw water from only the upper half of the aquifer. Production then resumed and continued until the total water volume produced equaled the volume injected.

Parameter studies done during the first- and second-cycle history match indicated that the temperature field was not very sensitive to the pressure field in the aquifer. Therefore, in the development of a numerical grid for the later cycles, emphasis was placed upon accurate calculation of the temperature distribution. An estimate of the radial extent of the hot region in the aquifer around the injection well, i.e., the thermal radius, can be made based on conservation of energy. The thermal radius was calculated to be about 25 m for the third cycle and 38 m for the fourth cycle. These values are small compared to the doublet spacing (240 m), and large compared to displacement caused by regional flow. Therefore, an axisymmetric model of the aquifer system centered at the injection/production well was used for the calculation.

The wellbore was modeled by a column of nodes 0.1 m wide with a porosity of 1 and a very high vertical permeability. Injection and production were accounted for by a source or sink element connected to the top wellbore node. Well tests conducted prior to the third cycle included a test to determine the vertical permeability of the aquifer. A value of 0.15 was determined for the permeability anisotropy, and used in the mathematical model. Other material properties remained similar to those shown in Table 2.

The third-cycle calculation predicted a recovery factor of 0.61 as well as the production temperature curve and calculated temperature distributions in the aquifer as a function of time. Subsequently, the experimental recovery factor was found to be 0.56. Although this is somewhat below the calculated value, it is an acceptable prediction. However, the experimental temperature distributions in the aquifer at the end of the injection period appeared rather different from the calculated results, as shown in upper part of Figure 14, where two experimental plots show perpendicular cross sections through the aquifer. Apparently, there is a high-permeability layer in the middle of the aquifer into which the injected fluid preferentially flows. After some parameter studies, we decided to use a three-layer-aquifer model in which the middle layer has a permeability 2.5 times that of the upper and lower layers. This three-layer-aquifer model reproduced the experimental temperature distributions and production temperature quite well, as shown in Figures 14 and 15, and predicted a recovery factor of 0.58, much closer to the experimental value than the previous one-layer-aquifer model. This is significant because layering is difficult to detect through conventional well test analysis, which typically gives a single average permeability value for a heterogeneous medium.

The fourth-cycle predictive calculation was made with the three-layer-aquifer model also. This cycle involved injection of much hotter water (82°C) than had been used before. Calculated results (3) indicated that for this large temperature, buoyancy effects were very large, and over-shadowed the preferential flow into the high permeability layer. Based on the original production plan, which called for a fully penetrating production well, the recovery factor was calculated to be 0.40. However, due to low production temperatures, the experimenters modified the production well during

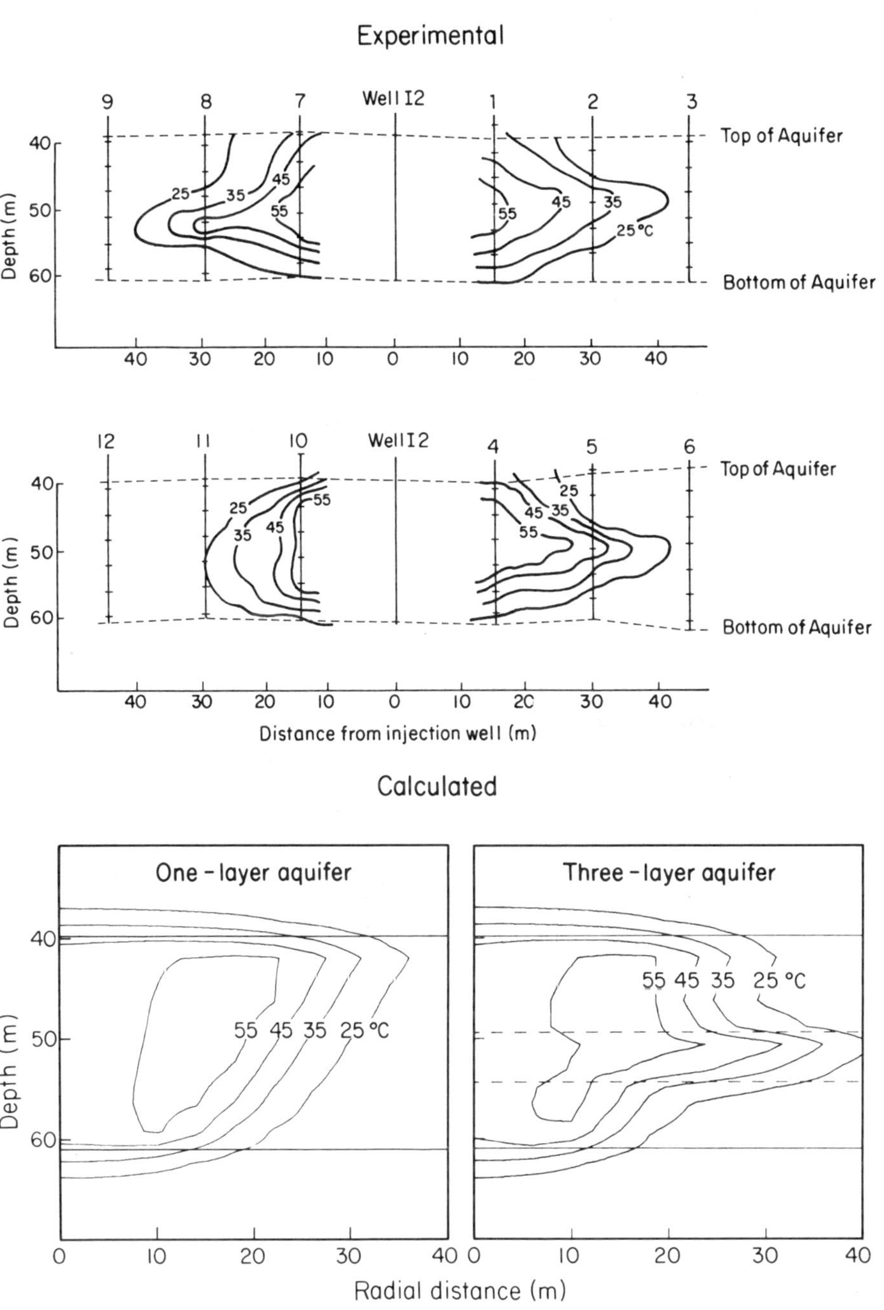

Figure 14. Experimental and calculated temperature contours after the third cycle injection period.

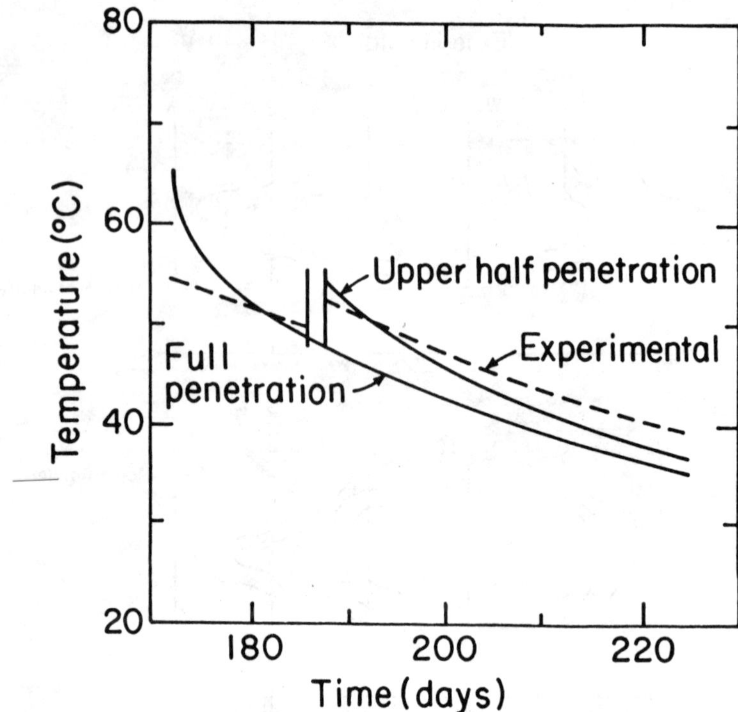

Figure 15. Third cycle production temperature.

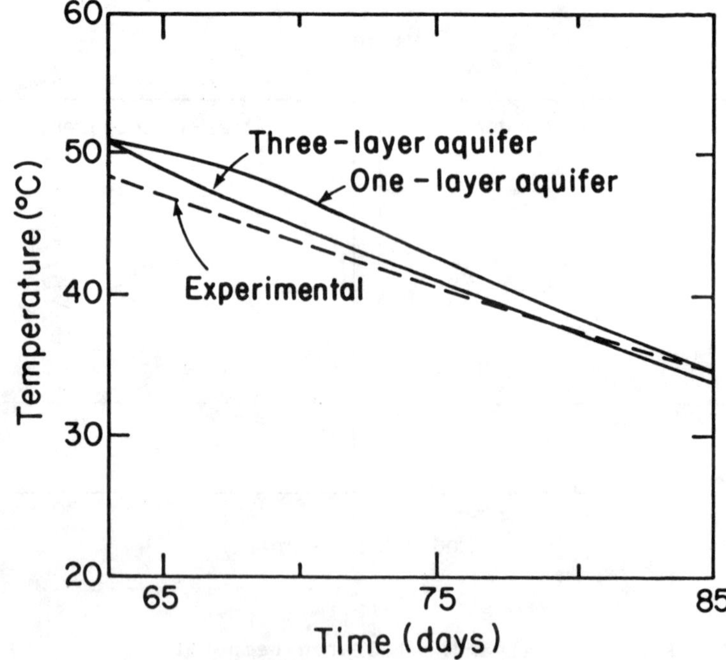

Figure 16. Fourth cycle production temperature.

the production period to eliminate production from the lower half of the aquifer. By following this procedure, the calculated recovery factor was 0.42, as compared to the experimental value of 0.45. This agreement was acceptable, but a comparison of the experimental and calculated production temperature curves, shown in Figure 16, indicated a moderate discrepancy. The calculated production temperature started about 10°C higher than the experimental value, but decreased much more rapidly, so that by the time of the well modification it underpredicted the experimental temperature. When production resumed after the two days of well modification, the calculated result again overpredicted the experimental value, although by just 2°C. Again the calculated temperature decreased more rapidly than the experimental curve, and ended up underpredicting it. This discrepancy of production temperatures was most noticeable for the fourth cycle, but the pattern of early over-prediction followed by late underprediction was evident in the third-cycle production temperature curves as well. It was much smaller for the first and second cycles, suggesting that it might be related to the wellbore model,which was first incorporated in the model for the third-cycle calculation.

In summary, the double-blind prediction made using the code PT yielded reasonable results. The third-cycle comparison of one-layer and three-layer aquifers indicated the importance of aquifer layering. The higher injection temperature of the fourth cycle caused buoyancy flow to be a dominant effect in the aquifer. PT can adequately model the fourth cycle, although there is a larger discrepancy between the calculation and the experiment than for the earlier cycles.

3.3.3. Optimization design studies

Because of the decrease in recovery factor from the third to fourth cycles (0.56 to 0.45) corresponding to the increase in injection temperature, an optimization design study was done before the fifth cycle of the field experiment in an attempt to design an experiment that would yield an optimal recovery factor for an 80°C, three-month cycle. A series of injection-production schemes using different well-screen open intervals were simulated. Each assumed a constant injection flow rate of 112 gpm (7 kg/sec) at 82°C. The three-layer-aquifer model developed for the third-cycle calculation was used. Two variations in cycle design were considered: the first assumed injection, storage, and production periods of one month each; the second assumed a two-month injection period (resulting in double the volume of hot water injected), no storage period, and a one-month production period (at double the injection flow rate). Making use of the results of the fourth cycle calculation, which indicated that buoyancy flow had a dominant effect on the system, three general approaches were taken in the design studies (Fig. 17):

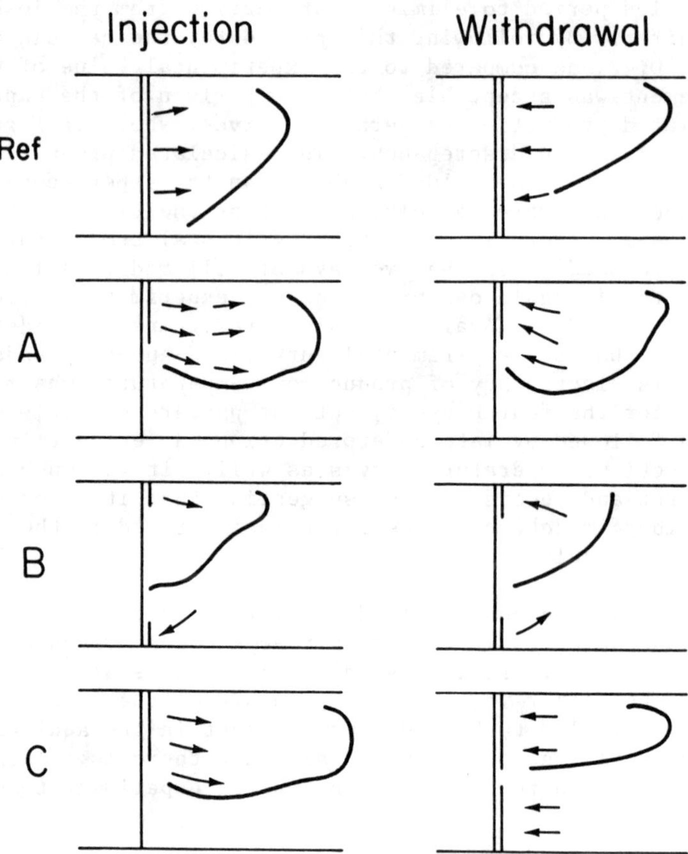

Figure 17. Schematic diagram showing design study alternatives.

(a) simply inject into and produce from the upper portion of the aquifer where most of the hot water would naturally flow because of buoyancy effects; (b) inject hot water at the top of the aquifer while simultaneously producing water from the bottom in order to have a vertical flow field with minimal buoyancy effect; and (c) inject into the upper portion of the aquifer, then while producing from the upper portion, produce (and discard) colder water from the lower portion of the aquifer through a "rejection well" located next to the injection/production well, thus eliminating any upward flow of cool water that would lower production temperature. Table 3 summarizes the results of the calculations. Cases A,B and C correspond to the three approaches listed above. The reference case considered an injection/production well screened over the entir aquifer thickness. For a cycle consisting of one month each of injection, storage, and production, the maximum recovery factor was

Table 3. Fifth-cycle Design Studies.
$T_1 = 82^{o}C$, $Q = 112$ gpm

			Wall Screen Interval		
			Injection	Production	Recovery Factor
Case I.	One month each,	Ref.	Full	Full	0.404
	injection,	A1	Upper 40%	Upper 40%	0.448
	storage,	A2	Upper 40%	Upper 20%	0.501
	production	B1	Upper 20%	Upper 20%	0.516
	$V = 18,300\ m^3$		Lower 20%		
		B2	Upper 20%	Upper 20%	0.487
			Lower 20%	Lower 20%	
		C1	Upper 40%	Upper 40%	0.500
				Lower 55%	
		C2	Upper 40%	Upper 20%	0.521
				Lower 55%	
Case II.	Two months	A1	Upper 40%	Upper 40%	0.609
	injection,	C1	Upper 40%	Upper 40%	0.629
	one month		Lower 55%		
	production	C3	Upper 40%	Upper 40%	0.631
	$V = 36,600\ m^3$			Lower 20%	
	$Q_p = 2Q_i$	C4	Upper 40%	Upper 20%	0.661
				Lower 20%	

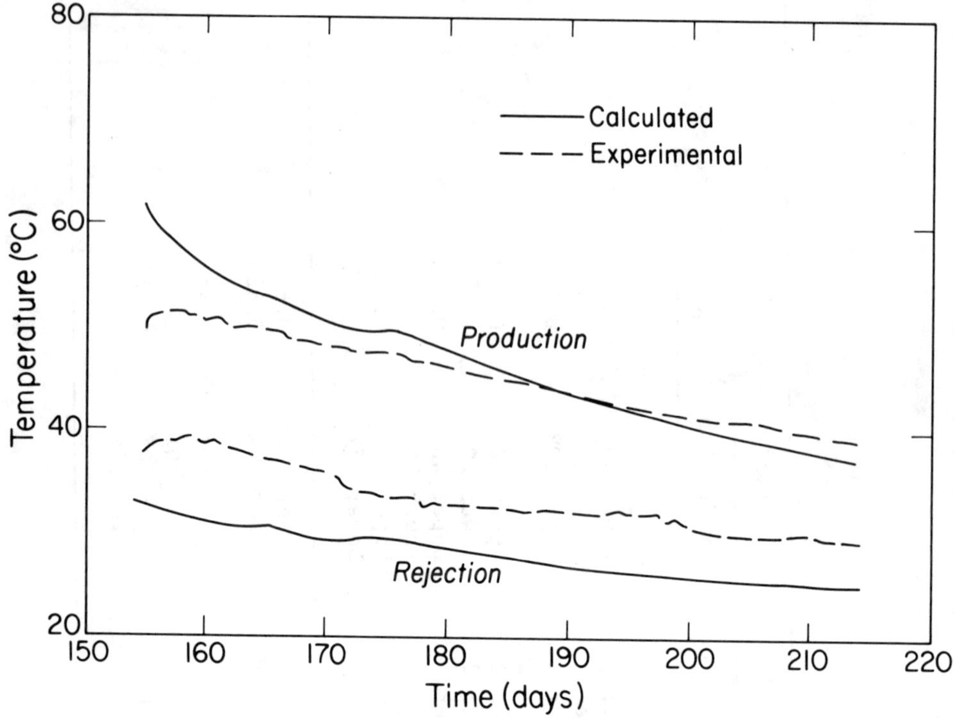

Figure 18. Fifth cycle production temperature.

about 0.52, representing an improvement of about 0.12 over the reference case. For the larger injected volume a recovery factor of 0.66 was possible. Hence, for this system, the volume of fluid injected is as important as the manner in which it is injected and produced. In general, the third method appears to be most successful in yielding a high recovery factor.

After the optimization design studies were completed, the fifth cycle was carried out, using 80°C water and an injection-production scheme patterned after case C. The injection/production well was screened over the upper 9 m of the aquifer and the rejection well, located less than 2 m away, was screened over 9 m in the lower half of the aquifer. Instead of a three-month cycle storing 18,000 or 36,000 m^3, as in the design studies, the fifth cycle lasted seven months, and 56,700 m^3 of water was injected, making a direct check of the design study calculations impossible. The recovery factor for the fifth cycle was 0.42. A history match calculation yielded a recovery factor of 0.44. As in the case of the fourth cycle, the calculated production temperature, shown in Figure 18, initially overpredicted the experimental value, then decreased more rapidly, and finally underpredicted it. The calculated temperature from the rejection well consistently underpredicted the experimental value, indicating that the model may have somewhat overpredicted buoyancy flow. This finding was consistent with the larger production -temperature discrepancy noted for higher temperature (greater buoyancy flow) cycles.

In summary, the optimization design studies investigated a variety of possible injection-production schemes, and indicated the range of recovery factors for them. Although the actual fifth cycle was quite different than the design studies, the relative results of the design studies proved to be useful in the choice of the fifth-cycle injection-production scheme.

4. SUMMARY

The thermohydraulics of an aquifer thermal energy storage system are reviewed and discussed. Effects of thermal conduction, buoyancy flow and forced convection are studied and calculated. Both an analytic approach and a numerical method are presented to familiarize the readers with these techniques.

5. APPENDIX A: ANALYTICAL SOLUTION FOR A SHARP, VERTICAL THERMAL FRONT IN AN INFINITE STRIP

An analytical expression for the pressure distribution in case A, which is shown in Figure 4a, will be derived in this appendix. The aquifer stratum occupies the region $-\infty < x < \infty$, $-H/2 < z < H/2$. The thermal front is located at $x = 0$, $-H/2 < z < H/2$.

Let $P(x,z)$ denote the pressure distribution in the aquifer. In region 1, $x < 0$, $-H/2 < z < H/2$, the pressure satisfies:

$$\frac{\partial}{\partial x}\left(\frac{k}{\mu_1}\frac{\partial P}{\partial x}\right) + \frac{\partial}{\partial z}\left[\frac{k'}{\mu_1}\left(\frac{\partial P}{\partial z} + \rho_1 g\right)\right] = 0 \qquad \text{(A1)}$$

In region 0, $x > 0$, $-H/2 < z < H/2$, we have:

$$\frac{\partial}{\partial x}\left(\frac{k}{\mu_0}\frac{\partial P}{\partial x}\right) + \frac{\partial}{\partial z}\left[\frac{k'}{\mu_0}\left(\frac{\partial P}{\partial z} + \rho_0 g\right)\right] = 0 \qquad \text{(A2)}$$

The upper and lower boundaries are impermeable:

$$\frac{\partial P}{\partial z} + \rho_1 g = 0 \qquad z = \pm\frac{H}{2}, \quad -\infty < x < 0 \qquad \text{(A3)}$$

$$\frac{\partial P}{\partial z} + \rho_0 g = 0 \qquad z = \pm\frac{H}{2}, \quad 0 < x < \infty \qquad \text{(A4)}$$

Hydrostatic conditions prevail far away from the thermal front:

$$P \to -\rho_1 gz \qquad x \to -\infty \tag{A5}$$

$$P \to -\rho_0 gz \qquad x \to +\infty \tag{A6}$$

The pressure and the groundwater flow are continuous at the thermal front:

$$P(-0,z) = P(+0,z) \qquad -\frac{H}{2} < z < \frac{H}{2} \tag{A7}$$

$$-\frac{k}{\mu_1} \cdot \frac{\partial P}{\partial x} = -\frac{k}{\mu_0} \cdot \frac{\partial P}{\partial x} \qquad x = 0, \quad -\frac{H}{2} < z < \frac{H}{2} \tag{A8}$$

We start with the following expressions:

$$x < 0: \qquad P(x,z) = -\rho_1 gz + \sum_{n=0}^{\infty} a_n u_n(x,z) \tag{A9}$$

$$x > 0: \qquad P(x,z) = -\rho_0 gz + \sum_{n=0}^{\infty} b_n u_n(x,z) \tag{A10}$$

where

$$U_n(x,z) = \sin\left(\frac{(2n+1)\pi z}{H}\right) \cdot e^{-\frac{(2n+1)\pi\kappa|x|}{H}} \tag{A11}$$

It is not difficult to verify that these expressions satisfy (A1-A6) for any choice of the coefficients a_n and b_n. The coefficients are determined by the two remaining conditions (A7) and (A8):

$$a_n = -\frac{q_0 H\mu_1}{k} \cdot \frac{k}{\pi^2} \cdot \frac{(-1)^n}{(2n+1)^2} \tag{A12}$$

$$b_n = -\frac{\mu_0}{\mu_1} \cdot a_n \tag{A13}$$

In particular we have for the flow across the thermal front:

$$q_f(z) = -\frac{k}{\mu_1} \cdot \frac{\partial P}{\partial x} = \kappa q_0 \cdot \frac{4}{\pi} \sum_{n=0}^{\infty} \frac{(-1)^n}{2n+1} \cdot \sin\left(\frac{(2n+1)\pi z}{H}\right) \tag{A14}$$

The series may be expressed in the simpler form of equation (11).

6. APPENDIX B: ANALYTICAL SOLUTION FOR A SHARP, VERTICAL THERMAL FRONT IN A SEMI-INFINITE STRIP: IMPERMEABLE LEFT BOUNDARY

An analytical expression for the pressure distribution in case B, which is shown in Figure 4b, will be derived in this appendix. The aquifer stratum occupies the region $0 < x < \infty$, $-H/2 < z < H/2$. The thermal front is located at $x = L$, $-H/2 < z < H/2$.

Let $P(x,z)$ denote the pressure distribution in the aquifer. In region 1, $0 < x < L$, $-H/2 < z < H/2$, the pressure satisfies equation (A1). In region 0, $x > L$, $-H/2 < z < H/2$, we have equation (A2).

The upper and lower boundaries are impermeable:

$$\frac{\partial P}{\partial z} + \rho_1 g = 0 \qquad z = \pm\frac{H}{2}, \quad 0 < x < L \tag{B1}$$

$$\frac{\partial P}{\partial z} + \rho_0 g = 0 \qquad z = \pm\frac{H}{2}, \quad L < x < \infty \tag{B2}$$

Hydrostatic conditions far away from the thermal front give (A6). The left boundary is impermeable:

$$\frac{\partial P}{\partial x} = 0 \qquad x = 0, \quad -\frac{H}{2} < z < \frac{H}{2} \tag{B3}$$

The pressure and the groundwater flow are continuous at the thermal front:

$$P(-L,z) = P(+L,z) \qquad -\frac{H}{2} < z < \frac{H}{2} \tag{B4}$$

$$-\frac{k}{\mu_1} \cdot \frac{\partial P}{\partial x} = -\frac{k}{\mu_0} \cdot \frac{\partial P}{\partial x} \qquad x = L, \quad -\frac{H}{2} < z < \frac{H}{2} \tag{B5}$$

We start with the following expressions:

$$0 < x < L: \qquad P(x,z) = -\rho_1 g z + \sum_{n=0}^{\infty} a_n \sin\left(\frac{(2n+1)\pi z}{H}\right) \cdot \cosh\left(\frac{(2n+1)\pi\kappa x}{H}\right) \tag{B6}$$

$$x > L: \qquad P(x,z) = -\rho_0 g z + \sum_{n=0}^{\infty} b_n \sin\left(\frac{(2n+1)\pi z}{H}\right) \cdot e^{-\frac{(2n+1)\pi\kappa(x-L)}{H}} \tag{B7}$$

These expressions satisfy (A1-A2), (B1-B3), and (A6) for any choice of the coefficients a_n and b_n: The coefficients are determined by the two remaining conditions (B4) and (B5).

$$a_n = -\frac{q_0 H\mu_1}{k} \cdot \frac{1}{\frac{\mu_0}{\mu_0+\mu_1}\sinh(\theta_n) + \frac{\mu_1}{\mu_0+\mu_1}\cosh(\theta_n)} \cdot \frac{4}{\pi^2}\frac{(-1)^n}{(2n+1)^2} \tag{B8}$$

$$b_n = -\frac{\mu_0}{\mu_1} \cdot \sinh(\theta_n) \cdot a_n \tag{B9}$$

where

$$\theta_n = \frac{(2n+1)\pi\kappa L}{H} \tag{B10}$$

Finally we obtain the flow across the thermal front (Eq. (2.12)) by differentiation of Eqs.(B6) or (B7).

7. APPENDIX C: ANALYTICAL SOLUTION FOR A SHARP, VERTICAL THERMAL FRONT IN A SEMI-INFINITE STRIP; HYDROSTATIC PRESSURE CONDITIONS AT THE LEFT BOUNDARY

An analytical expression for the pressure distribution in case C, which is shown in Figure 4c, will be derived in this appendix. The aquifer stratum occupies the region $0 < x < \infty$, $-H/2 < z < H/2$. The thermal front is located at $x = L$, $-H/2 < z < H/2$.

Let $P(x,z)$ denote the pressure distribution in the aquifer. In region 1, where $0 < x < L$, $-H/2 < z < H/2$, the pressure satisfies equation (A1). In region 0, where $x > L$, $-H/2 < z < H/2$, we have equation (A2).

The upper and lower boundaries are impermeable, which implies the boundary conditions (B1-B2). Hydrostatic conditions far away from the thermal front give (A6). Hydrostatic conditions prevail at the left boundary:

$$P(0,z) = -\rho_1 g z \qquad -\frac{H}{2} < z < \frac{H}{2} \tag{C1}$$

The pressure and the groundwater flow are continuous at the thermal front as expressed by (B4) and (B5). We start with (B7) for $x > L$ and the following expression for $0 < x < L$:

$$P(x,z) = -\rho_1 g z + \sum_{n=0}^{\infty} a_n \sin\left(\frac{(2n+1)\pi z}{H}\right) \cdot \sinh\left(\frac{(2n+1)\pi\kappa x}{H}\right) \tag{C2}$$

Expressions (C2) and (B7) satisfy (A1-A2), (B1-B2), (C1), and (A6) for any choice of the coefficients a_n and b_n: the coefficients are determined by the two remaining conditions (B4) and (B5).

$$a_n = -\frac{q_0 H \mu_1}{k} \cdot \frac{1}{\frac{\mu_0}{\mu_0+\mu_1}\cosh(\theta_n) + \frac{\mu_1}{\mu_0+\mu_1}\sinh(\theta_n)} \cdot \frac{4}{\pi^2}\,\frac{(-1)^n}{(2n+1)^2} \tag{C3}$$

$$b_n = -\frac{\mu_0}{\mu_1} \cdot \cosh(\theta_n) \cdot a_n \tag{C4}$$

where θ_n is defined by (B10).

The flow across the thermal front (Eq. (2.13)) is obtained by differentiation of Eqs. (C2) or (B7) with a_n and b_n given by Eqs. (C3) and (C4) respectively.

8. APPENDIX D: ANALYTICAL SOLUTION FOR A SHARP, VERTICAL THERMAL FRONT IN THE CYLINDRICAL CASE; NO FLOW AT THE INNER BOUNDARY

An analytical expression for the pressure distribution in case D, which is shown in Figure 4d, will be derived in this appendix. Cylindrical coordinates, r and z, are used. The aquifer stratum occupies the region $0 < r < \infty$, $-H/2 < z < H/2$. The thermal front is located at $r = L$, $-H/2 < z < H/2$.

Let $P_1(r,z)$ denote the pressure distribution in region 1, $0 < r < L$, $-H/2 < z < H/2$. The pressure satisfies:

$$\frac{1}{r}\frac{\partial}{\partial r}\left(\frac{k}{\mu_1} r \frac{\partial P_1}{\partial r}\right) + \frac{\partial}{\partial z}\left[\frac{k'}{\mu_1}\left(\frac{\partial P_1}{\partial z} + \rho_1 g\right)\right] = 0 \tag{D1}$$

In region 0, $r > L$, $-H/2 < z < H/2$. We have for the pressure $P_0(r,z)$:

$$\frac{1}{r}\frac{\partial}{\partial r}\left(\frac{k}{\mu_0} r \frac{\partial P_0}{\partial r}\right) + \frac{\partial}{\partial z}\left[\frac{k'}{\mu_0}\left(\frac{\partial P_0}{\partial z} + \rho_0 g\right)\right] = 0 \tag{D2}$$

The upper and lower boundaries are impermeable:

$$\frac{\partial P_1}{\partial z} + \rho_1 g = 0 \qquad z = \pm\frac{H}{2}, \quad 0 < r < L \tag{D3}$$

$$\frac{\partial P_0}{\partial z} + \rho_0 g = 0 \qquad z = \pm\frac{H}{2}, \quad L < r < \infty \tag{D4}$$

At the inner boundary, r=0, symmetry requires that:

$$\frac{\partial P_1}{\partial r} = 0 \qquad -\frac{H}{2} < z < \frac{H}{2} \tag{D5}$$

Hydrostatic conditions prevail far away from the thermal front:

$$P_0 \to -\rho_0 gz \qquad r \to \infty \tag{D6}$$

The pressure and the groundwater flow are continuous at the thermal front:

$$P_1(L,z) = P_0(L,z) \qquad -\frac{H}{2} < z < \frac{H}{2} \tag{D7}$$

$$-\frac{k}{\mu_1} \cdot \frac{\partial P_1}{\partial r} = -\frac{k}{\mu_0} \cdot \frac{\partial P_0}{\partial r} \qquad r = L\ , \ -\frac{H}{2} < z < \frac{H}{2} \tag{D8}$$

We start with the following expressions:

$$P_1(r,z) = -\rho_1 gz + \sum_{n=0}^{\infty} a_n \sin\left(\frac{(2n+1)\pi z}{H}\right) \cdot I_0\left(\frac{(2n+1)\pi\kappa r}{H}\right) \tag{D9}$$

$$P_0(r,z) = -\rho_0 gz + \sum_{n=0}^{\infty} b_n \sin\left(\frac{(2n+1)\pi z}{H}\right) \cdot K_0\left(\frac{(2n+1)\pi\kappa r}{H}\right) \tag{D10}$$

Here we make use of the modified Bessel functions I_n and K_n. These expressions satisfy (D1-D6) for any choice of the coefficients a_n and b_n. The coefficients are determined by the two remaining conditions (D7) and (D8).

$$a_n = \frac{q_0 H\mu_1}{k} \cdot \frac{1}{\dfrac{\mu_1}{\mu_0+\mu_1} \cdot \dfrac{I_0(\theta_n)}{I_1(\theta_n)} + \dfrac{\mu_0}{\mu_0+\mu_1} \cdot \dfrac{K_0(\theta_n)}{K_1(\theta_n)}} \cdot \frac{1}{I_1(\theta_n)} \cdot \frac{4}{\pi^2} \frac{(-1)^n}{(2n+1)^2} \tag{D11}$$

$$b_n = -\frac{\mu_0}{\mu_1} \frac{I_1(\theta_n)}{K_1(\theta_n)} \cdot a_n \tag{D12}$$

where θ_n is given by (B10).

Finally we obtain the flow across the thermal front (Eq. (2.14)) by differentiation of Eqs. (D9) or (D10).

9. APPENDIX E: ANALYTICAL SOLUTION FOR A SHARP, VERTICAL THERMAL FRONT IN THE CYLINDRICAL CASE; HYDROSTATIC PRESSURE CONDITIONS AT THE INNER BOUNDARY

An analytical expression for the pressure distribution in case E, which is shown in Figure 4e, will be derived in this appendix. Cylindrical coordinates, r and z, are used. The aquifer stratum occupies the region $R_w < r < \infty$, $- H/2 < z < H/2$. The thermal front is located at $r = L$, $- H/2 < z < H/2$.

Let $P_1(r,z)$ denote the pressure distribution in region 1, where $R_w < r < L$, $- H/2 < z < H/2$. The pressure P_1 satisfies equation (D1). For the pressure $P_0(r,z)$ in region 0, $r > L$, $- H/2 < z < H/2$, we have equation (D2).

The upper and lower boundaries are impermeable, which implies the boundary conditions (D3-D4). Hydrostatic conditions far away from the thermal front give (D6). We also have hydrostatic pressure conditions at the inner boundary, $r = R_w$:

$$P_1(R_w,z) = - \rho_1 gz \qquad - \frac{H}{2} < z < \frac{H}{2} \tag{E1}$$

Pressure and groundwater flow are continuous at the thermal front as expressed by (D7) and (D8).

We start with the following expressions:

$$P_1(r,z) = - \rho_1 gz + \sum_{n=0}^{\infty} a_n u_n(r,z) + \sum_{n=0}^{\infty} b_n v_n(r,z) \tag{E2}$$

$$P_0(r,z) = - \rho_0 gz + \sum_{n=0}^{\infty} c_n v_n(r,z) \tag{E3}$$

where

$$u_n(r,z) = \sin\left(\frac{(2n+1)\pi z}{H}\right) \cdot I_0\left(\frac{(2n+1)\pi\kappa r}{H}\right) \tag{E4}$$

$$v_n(r,z) = \sin\left(\frac{(2n+1)\pi z}{H}\right) \cdot K_0\left(\frac{(2n+1)\pi\kappa r}{H}\right) \tag{E5}$$

These expressions satisfy (D1-D4) and (D6) for any choice of the coefficients a_n, b_n, and c_n. The coefficients are determined by the three remaining conditions (E1), (D7), and (D8).

$$a_n = \frac{q_0 H \mu_1}{k} \cdot \frac{1}{I_1(\theta_n^L)} \cdot \frac{4}{\pi^2} \cdot \frac{(-1)^n}{(2n+1)^2} \cdot \tag{E6}$$

$$\frac{1}{\frac{\mu_0-\mu_1}{\mu_0+\mu_1} \cdot \frac{K_0(\theta_n^L)}{I_1(\theta_n^L)} \cdot \frac{I_0(\theta_n^R)}{K_0(\theta_n^R)} + \frac{\mu_1}{\mu_0+\mu_1} \cdot \frac{I_0(\theta_n^L)}{I_1(\theta_n^L)} + \frac{\mu_0}{\mu_0+\mu_1} \cdot \frac{K_0(\theta_n^L)}{K_1(\theta_n^L)}}$$

$$b_n = - \frac{I_0(\theta_n^R)}{K_0(\theta_n^R)} \cdot a_n \tag{E7}$$

$$c_n = - \frac{\mu_0}{\mu_1} \cdot \frac{I_1(\theta_n^L)}{K_1(\ _n^L)} \cdot \left[1 + \frac{K_1(\theta_n^L)}{I_1(\ _n^L)} \cdot \frac{I_0(\theta_n^R)}{K_0(\ _n^R)}\right] \cdot a_n \tag{E8}$$

where

$$\theta_n^R = \frac{(2n+1)\pi\kappa R_w}{H} \tag{E9}$$

$$\theta_n^L = \frac{(2n+1)\pi\kappa L}{H} \tag{E10}$$

Finally we obtain the flow across the thermal front (Eq.(2.16)) by differentiation of Eqs. (E4) or (E5).

10. APPENDIX F: ANALYTICAL SOLUTION FOR AN INFINITE STRIP WITH DIFFUSE THERMAL FRONT

An analytical expression for the pressure distribution in case F, which is shown in Figure 4f, will be derived in this appendix. The aquifer stratum occupies the region $-\infty < x < \infty$, $-H/2 < z < H/2$. The thermal front region, which has a thickness D, is located at $-D/2 < x < D/2$, $-H/2 < z < H/2$. The viscosity is constant in this case ($\mu=\mu_0=\mu_1$). The density is ρ_1 in region 1, where $x < -D/2$, $-H/2 < z < H/2$, and ρ_0 in region 0, where $x > D/2$, $-H/2 < z < H/2$. The density in the thermal front region varies linearly with x between ρ_1 and ρ_0

$$\rho(x) = \frac{1}{2}(\rho_0+\rho_1) + \frac{x}{D}(\rho_0-\rho_1) \tag{F1}$$

The pressure distribution $P_1(x,z)$ in region 1 satisfies (A1) with $\mu_1=\mu$. For the pressure $P_0(x,z)$ in region 0, we have equation (A2) with $\mu_0=\mu$. The pressure $P_D(x,z)$ in the thermal front region is the solution of:

$$\frac{\partial}{\partial x}\left(\frac{k}{\mu}\frac{\partial P_D}{\partial x}\right) + \frac{\partial}{\partial z}\left[\frac{k'}{\mu}\left(\frac{\partial P_D}{\partial z} + \rho(x)g\right)\right] = 0 \tag{F2}$$

The upper and lower boundaries are impermeable, so that

$$\frac{\partial P}{\partial z} + \rho g = 0 \qquad z = \pm\frac{H}{2} \tag{F3}$$

for the different regions. Hydrostatic conditions far away from the thermal front give (A5-A6).

The pressure and the groundwater flow are continuous at the interfaces between the thermal front region and the surrounding regions:

$$P_1(-D/2,z) = P_D(-D/2,z) \qquad -\frac{H}{2} < z < \frac{H}{2} \tag{F4}$$

$$P_0(D/2,z) = P_D(D/2,z) \qquad -\frac{H}{2} < z < \frac{H}{2} \tag{F5}$$

$$-\frac{k}{\mu}\frac{\partial P_1}{\partial x} = -\frac{k}{\mu}\frac{\partial P_D}{\partial x} \qquad x = -D/2,\ -\frac{H}{2} < z < \frac{H}{2} \tag{F6}$$

$$-\frac{k}{\mu}\frac{\partial P_0}{\partial x} = -\frac{k}{\mu}\frac{\partial P_D}{\partial x} \qquad x = D/2,\quad -\frac{H}{2} < z < \frac{H}{2} \tag{F7}$$

We start with the following expressions:

$$P_1(x,z) = -\rho_1 gz + \sum_{n=0}^{\infty} a_n u_n(x,z) \tag{F8}$$

$$P_0(x,z) = -\rho_0 gz + \sum_{n=0}^{\infty} b_n v_n(x,z) \tag{F9}$$

$$P_D(x,z) = -\rho(x)gz + \sum_{n=0}^{\infty} c_n u_n(x,z) + \sum_{n=0}^{\infty} d_n v_n(x,z) \tag{F10}$$

where

$$u_n(x,z) = \sin\left(\frac{(2n+1)\pi z}{H}\right) \cdot e^{\frac{(2n+1)\pi\kappa x}{H}} \tag{F11}$$

$$v_n(x,z) = \sin\left(\frac{(2n+1)\pi z}{H}\right) \cdot e^{-\frac{(2n+1)\pi\kappa x}{H}} \tag{F12}$$

These expressions satisfy (A1-A2), (F2-F3), (A5-A6) for any choice of the coefficients a_n, b_n, c_n, and d_n. The coefficients are determined by the four remaining conditions (F4-F7):

$$a_n = \frac{q_0 H\mu}{2k} \cdot \frac{4}{\pi^2} \cdot \frac{(-1)^n}{(2n+1)^2} \cdot \frac{H}{(2n+1)\pi\kappa\frac{D}{2}} \tag{F13}$$

$$\cdot \left(e^{-\frac{(2n+1)\pi\kappa\frac{D}{2}}{H}} - e^{\frac{(2n+1)\pi\kappa\frac{D}{2}}{H}} \right)$$

$$b_n = - a_n \tag{F14}$$

$$c_n = \frac{q_0 H\mu}{2k} \cdot \frac{4}{\pi^2} \cdot \frac{(-1)^n}{(2n+1)^2} \cdot \frac{H}{(2n+1)\pi\kappa\frac{D}{2}} \cdot e^{-\frac{(2n+1)\pi\kappa\frac{D}{2}}{H}} \tag{F15}$$

$$d_n = - c_n \tag{F16}$$

In particular we have for the flow across a vertical cut in the middle of the thermal region:

$$q_f(z) = - \frac{k}{\mu} \cdot \frac{\partial P_D}{\partial x} \qquad x = 0 \tag{F17}$$

The result may be expressed as the series of Eq. (2.19).

11. APPENDIX G: ANALYTICAL SOLUTION FOR THE CYLINDRICAL CASE WITH DIFFUSE THERMAL FRONT

An analytical expression for the pressure distribution in case G, which is shown in Figure 4g, will be derived in this appendix. The aquifer stratum occupies the region $0 < r < \infty$, $- H/2 < z < H/2$. The thermal front region, which has a thickness D, is located at $L\text{-}D/2 < r < L\text{+}D/2$, $-H/2 < z < H/2$. The thermal front region must not extend into the well, i.e. $D < 2L$. The viscosity is constant in this case ($\mu=\mu_0=\mu_1$). The density is ρ_1 in region 1, $0 < r < L\text{-}D/2$, $-H/2 < z < H/2$, and ρ_0 in region 0, $r > L\text{+}D/2$, $-H/2 < z < H/2$. The density in the thermal front region varies with r between ρ_1 and ρ_0 according to:

$$\rho(r) = \frac{\rho_1 \ln\left(\frac{(L+D/2)}{r}\right) + \rho_0 \ln\left(\frac{r}{(L-D/2)}\right)}{\ln\left(\frac{(L+D/2)}{(L-D/2)}\right)} \tag{G1}$$

The pressure distribution $P_1(r,z)$ in region 1 satisfies (D1) with $\mu_1 = \mu$. For the pressure $P_0(r,z)$ in region 0, we have equation (D2) with $\mu_0 = \mu$. The pressure $P_D(r,z)$ in the thermal front region is the solution of:

$$\frac{1}{r}\frac{\partial}{\partial r}\left(\frac{k}{\mu} r \frac{\partial P_D}{\partial r}\right) + \frac{\partial}{\partial z}\left[\frac{k'}{\mu}\left(\frac{\partial P_D}{\partial z} + \rho(r)g\right)\right] = 0 \qquad \text{(G2)}$$

The upper and lower boundaries are impermeable, so that

$$\frac{\partial P}{\partial z} + \rho g = 0 \qquad z = \pm\frac{H}{2} \qquad \text{(G3)}$$

for the different regions. There is no horizontal flow at the inner boundary, r = 0, and hydrostatic conditions far away from the thermal front give (D5-D6).

The pressure and the groundwater flow are continuous at the interfaces between the thermal front region and the surrounding regions:

$$P_1(L-D/2,z) = P_D(L-D/2,z) \qquad -\frac{H}{2} < z < \frac{H}{2} \qquad \text{(G4)}$$

$$P_0(L+D/2,z) = P_D(L+D/2,z) \qquad -\frac{H}{2} < z < \frac{H}{2} \qquad \text{(G5)}$$

$$-\frac{k}{\mu}\cdot\frac{\partial P_1}{\partial r} = -\frac{k}{\mu}\cdot\frac{\partial P_D}{\partial r} \qquad r = L-D/2, \quad -\frac{H}{2} < z < \frac{H}{2} \qquad \text{(G6)}$$

$$-\frac{k}{\mu}\cdot\frac{\partial P_0}{\partial r} = -\frac{k}{\mu}\cdot\frac{\partial P_D}{\partial r} \qquad r = L+D/2, \quad -\frac{H}{2} < z < \frac{H}{2} \qquad \text{(G7)}$$

We start with the following expressions:

$$P_1(r,z) = -\rho_1 gz + \sum_{n=0}^{\infty} a_n u_n(r,z) \qquad \text{(G8)}$$

$$P_0(r,z) = -\rho_0 gz + \sum_{n=0}^{\infty} b_n v_n(r,z) \qquad \text{(G9)}$$

$$P_D(r,z) = -\rho(r)gz + \sum_{n=0}^{\infty} c_n u_n(r,z) + \sum_{n=0}^{\infty} d_n v_n(r,z) \qquad \text{(G10)}$$

where the functions $u_n(r,z)$ and $v_n(r,z)$ are given by (E4) and (E5) respectively.

These expressions satisfy (D1-D2), (G2-G3), (D5-D6) for any choice of the coefficients a_n, b_n, c_n, and d_n. The coefficients are determined by the four remaining conditions (G4-G7):

$$a_n = \frac{2q_0 H\mu}{k} \cdot \frac{1}{\ln\left(\frac{(L+D/2)}{(L-D/2)}\right)} \cdot \frac{4}{\pi^2} \cdot \frac{(-1)^n}{(2n+1)^2} \cdot \left(\frac{\Phi(\theta_n^+)}{I_0(\theta_n^+)} - \frac{\Phi(\theta_n^-)}{I_0(\theta_n^-)}\right) \quad \text{(G11)}$$

$$b_n = \frac{2q_0 H\mu}{k} \cdot \frac{1}{\ln\left(\frac{(L+D/2)}{(L-D/2)}\right)} \cdot \frac{4}{\pi^2} \cdot \frac{(-1)^n}{(2n+1)^2} \cdot \left(\frac{\Phi(\theta_n^+)}{K_0(\theta_n^+)} - \frac{\Phi(\theta_n^-)}{K_0(\theta_n^-)}\right) \quad \text{(G12)}$$

$$c_n = \frac{2q_0 H\mu}{k} \cdot \frac{1}{\ln\left(\frac{(L+D/2)}{(L-D/2)}\right)} \cdot \frac{4}{\pi^2} \cdot \frac{(-1)^n}{(2n+1)^2} \cdot \frac{\Phi(\theta_n^+)}{I_0(\theta_n^+)} \quad \text{(G13)}$$

$$d_n = -\frac{2q_0 H\mu}{k} \cdot \frac{1}{\ln\left(\frac{(L+D/2)}{(L-D/2)}\right)} \cdot \frac{4}{\pi^2} \cdot \frac{(-1)^n}{(2n+1)^2} \cdot \frac{\Phi(\theta_n^-)}{K_0(\theta_n^-)} \quad \text{(G14)}$$

where the function Φ is defined by:

$$\Phi(\theta_n) = \frac{1}{\theta_n} \cdot \frac{1}{\frac{I_1(\theta_n)}{I_0(\theta_n)} + \frac{K_1(\theta_n)}{K_0(\theta_n)}} \quad \text{(G15)}$$

and

$$\theta_n^+ = \frac{(2n+1)\pi\kappa(L+D/2)}{H} \quad \text{(G16)}$$

$$\theta_n^- = \frac{(2n+1)\pi\kappa(L-D/2)}{H} \tag{G17}$$

In particular we have for the flow across a vertical cut in the middle of the thermal region:

$$q_f(z) = -\frac{k}{\mu} \cdot \frac{\partial P_D}{\partial r} \qquad r = L \tag{G18}$$

The result may be expressed as the series of Eq. (2.21).

12. APPENDIX H: ANALYTICAL SOLUTION FOR A SHARP THERMAL FRONT IN A CIRCULAR REGION

An analytical expression for the pressure distribution in case H, which is shown in Figure 4h, will be derived in this appendix. The aquifer has the shape of an infinite circular cylinder with an horizontal symmetry axis. A vertical cut through the cylinder becomes a circular disk with radius R. We first consider the case with a vertical thermal front. Both polar coordinates (r,ϕ) and cartesian coordinates (x,z) are used. Here ϕ denotes the angle with respect to the upward vertical direction, which is denoted $\hat{z}$.

Let $P_1(r,\phi)$ denote the pressure distribution in the left part of the circular region, $0 < r < R$, $-\pi < \phi < 0$. In the right part, $0 < r < R$, $0 < \phi < \pi$, the pressure is $P_0(r,\phi)$. The pressure P_1 and P_0 both satisfy:

$$\frac{1}{r}\frac{\partial}{\partial r}\left(r\frac{\partial P}{\partial r}\right) + \frac{1}{r^2}\frac{\partial^2 P}{\partial \phi^2} = 0 \tag{H1}$$

The periphery of the disk is impermeable. Let $\hat{r}$ and $\hat{z}$ denote unit vectors in the radial and vertical direction respectively. Then we have:

$$\frac{\partial P_1}{\partial r} + \rho_1 g\hat{z}.\hat{r} = 0 \qquad r = R, \quad -\pi < \phi < 0 \tag{H2}$$

$$\frac{\partial P_0}{\partial r} + \rho_0 g\hat{z}.\hat{r} = 0 \qquad r = R, \quad 0 < \phi < \pi \tag{H3}$$

The pressure and the groundwater flow are continuous at the thermal front:

$$P_1(r,\phi) = P_0(r,\phi) \qquad 0 < r < R\,, \quad \phi = 0 \quad \text{and } \pm\pi \tag{H4}$$

$$-\frac{k}{\mu_1} \cdot \frac{\partial P_1}{\partial x} = -\frac{k}{\mu_0} \cdot \frac{\partial P_0}{\partial x} \qquad 0 < r < R,\ \phi = 0 \text{ and } \pm\pi \tag{H5}$$

where x is a horizontal coordinate.

We start with the following expressions:

$$P_1(r,\phi) = -\alpha r\cos(\phi) + \sum_{n=1}^{\infty} a_n \left(\frac{r}{R}\right)^{2n} \cdot \sin(2n\phi) \tag{H6}$$

$$P_0(r,\phi) = -\alpha r\cos(\phi) + \sum_{n=1}^{\infty} b_n \left(\frac{r}{R}\right)^{2n} \cdot \sin(2n\phi) \tag{H7}$$

These expressions satisfy (H1) for any choice of the coefficients α, a_n, and b_n. The coefficients are determined by the four remaining conditions (H2-H5):

$$\alpha = \frac{(\mu_0\rho_1 + \mu_1\rho_0)g}{\mu_0 + \mu_1} \tag{H8}$$

$$a_n = -\frac{q_0 R\mu_1}{k} \cdot \frac{4}{\pi} \cdot \frac{1}{(2n+1)(2n-1)} \tag{H9}$$

$$b_n = \frac{\mu_0}{\mu_1} \cdot a_n \tag{H10}$$

The pressure in the aquifer is now:

$$P_i(r,\phi) = -\frac{(\mu_0\rho_1 + \mu_1\rho_0)gz}{\mu_0+\mu_1} - \frac{q_0 R\mu_i}{k}\, \tilde{P}\left(\frac{r}{R},\phi\right) \tag{H11}$$

where $i = 0$ for $0 < \phi < \pi$ and $i = 1$ for $-\pi < \phi < 0$. We have introduced a dimensionless pressure:

$$\tilde{P}(r',\phi) = -\frac{4}{\pi}\sum_{n=1}^{\infty} \frac{1}{(2n+1)(2n-1)} \cdot (r')^{2n} \cdot \sin(2n\phi) \tag{H12}$$

This series may be expressed in closed form with the use of the complex number

$$w = r'\cos(\phi) + ir'\sin(\phi) = (z+ix)/R \tag{H13}$$

The dimensionless pressure (H12) may then be written:

$$\tilde{P} = -\frac{1}{\pi}\,\mathrm{Im}\left[\left(w - \frac{1}{w}\right)\ln\left(\frac{1+w}{1-w}\right)\right] = -\frac{1}{\pi}\,\mathrm{Im}[f(w)] \tag{H14}$$

The symbol Im denotes the imaginary part. An evaluation of the complex function f(w) gives:

$$\tilde{P} = -\frac{1}{\pi}\left[(r' - \frac{1}{r'})\cos(\phi)\cdot\arctan\left(\frac{2r'\sin(\phi)}{1-(r')^2}\right)\right. \tag{H15}$$

$$\left. + \frac{1}{2}(r' + \frac{1}{r'})\sin(\phi)\cdot\ln\left(\frac{1+(r')^2+2r'\cos(\phi)}{1+(r')^2-2r'\cos(\phi)}\right)\right]$$

In particular we have for the flow across the thermal front:

$$q_f(z) = - q_0 R \cdot \frac{\partial\tilde{P}}{\partial x} = \frac{q_0}{\pi} \cdot \mathrm{Re}[\frac{df}{dw}] \tag{H16}$$

Here Re denotes the real part. The result is given in Eq. (2.26).

In this particular case it is possible to solve the problem with a straight thermal front which is tilted an angle α from a vertical position. By making the substitution:

$$\phi' = \phi - \alpha \tag{H17}$$

we find that equations (H1-H7) remain unchanged except that the gravitational constant g is replaced by g.cos(α). This means that all pressures and flows are reduced by the factor cos(α) when the thermal front is tilted an angle α.

13. ACKNOWLEDGEMENTS

The cooperation and assistance from J. Claesson and G. Hellstrom of Lund University, Sweden and T. Buscheck, C. Doughty and E. Klahn of Lawrence Berkeley Laboratory are gratefully acknowledged. This work was supported by the Assistant Secretary for Energy Research, Office of Basic Energy Sciences, Division of Engineering and Geosciences through U.S. Department of Energy Contract No. DE-AC03-76SF00098.

14. REFERENCES

1. Avdonin, N.A., "Some Formulas for Calculating the Temperature Field of a Stratum Subjected to Thermal Injection," *Neft'i Gaz* Vol. 3, 1964, p. 3741.
2. Bodvarsson, G. S., "Mathematical Modeling of the Behavior of Geothermal Systems under Exploitation," thesis presented to the University of California, at Berkeley, California, in Jan. 1982, in partial fulfillment of the requirements for the degree

of Doctor of Philosophy, *Rep LBL-13937*, pp. 14-48.
3. Buscheck, T.A., Doughty, C., and Tsang, C.F., "Prediction and Analysis of a Field Experiment on a Multilayered Aquifer Thermal Energy Storage System with Strong Buoyancy Flow," *Water Resources Research*, Vol. 19, Oct., 1983, pp. 1307-1315.
4. Carslaw, N.W., and Jaeger, J.C., *Conduction of Heat in Solids*, Second Edition, Oxford University Press, Oxford, Great Britain, 1959, p. 335.
5. Cinco-Ley, H., Samaniego, V. F., and Dominquez, A. N., "Transient Pressure Behavior for a Well with a Finite-Conductivity Vertical Fracture," *Society of Petroleum Engineers Journal*, August 1978, pp. 253-264.
6. de Josselin de Jong, G., "Singularity Distributions for the Analysis of Multiple-Fluid Flow Through Porous Media, *Journal of Geophysical Research*, Vol. 65, pp. 3739-3758, 1960.
7. Doughty, C., Hellström, G., Tsang, C.F., and Claesson, J., "A Dimensionless Parameter Approach to the Thermal Behavior of an Aquifer Thermal Energy Storage System, *Water Resources Research*, Vol. 18, No. 3, pp. 571-587, 1982.
8. Edwards, A.L., "TRUMP: A Computer Program for Transient and Steady-State Temperature Distributions in Multidimensional Systems," *Rep. UCRL-14754 Rev.3*, Lawrence Livermore Laboratory, Livermore, California, September 1972.
9. Gringarten, A.C., "Unsteady-State Pressure Distribution Created by a Well with a Single Horizontal Fracture, Partial Penetration or Restricted Entry," thesis presented to the Stanford University, at Palo Alto, California, in 1971, in partial fulfillment of the requirements for the degree of Doctor of Philosophy.
10. Gringarten, A.C., and Sauty, J.P., " A Theoretical Study of Heat Extraction from Aquifers with Uniform Regional Flow," *Journal of Geophysical Research*, Vol. 80, pp. 4956-4962, 1975.
11. Hellstrom, G., Tsang, C.F., and Claesson, J., "Heat Storage in Aquifers: Buoyancy Flow and Thermal Stratification Problems," *Rep. LBL-14246*, Lawrence Berkeley Laboratory, Berkeley, California, October 1979.
12. Hellstrom, G., Tsang, C.F., and Claesson, J., "Motion of a Two-Fluid Interface in a Porous Medium: 1. Analytic Studies," submitted to *Water Resources Research*, March 1986.
13. Molz, F.J., Melville, J.G., Guven, O., and Parr, A.D., "Aquifer Thermal Energy Storage: An Attempt to Counter Free Thermal Convection," *Water Resources Research*, Vol. 19, pp. 922-930, August 1983.
14. Molz, F.J., Melville, J.G., Parr, A.D., King, D.A., and Hopf, M.T., "Aquifer Thermal Energy Storage: A Well Doublet Experiment at Increased Temperatures," *Water Resources Research*, Vol. 19, pp. 149-160, February 1983.
15. Molz, F.J., Parr, A.D., and Anderson, P.F., "Thermal Energy Storage in a Confined Aquifer: Second Cycle," *Water Resources*

Research, Vol. 17, pp. 641-645, June 1981.

16. Molz, F.J., Parr, A.D., Anderson, P.F., Lucido, V.D., and Warman, J.C., "Thermal Energy Storage in a Confined Aquifer: Experimental Results," *Water Resources Research*, Vol. 15, pp. 1509-1514, December 1979.
17. Narasimhan, T.N., and Witherspoon, P.A., "An Integrated Finite-Difference Method for Analyzing Fluid Flow in Porous Media," *Water Resources Research*, Vol. 12, pp. 57-64, 1976.
18. STES Newsletter, C.F.Tsang, ed., Lawrence Berkeley Laboratory, 1978-1985.
19. Theis, C.V., "The Relationship Between the Lowering of Piezometric Surface and the Rate and Duration of Discharge Using Groundwater Storage," *Transactions*, American Geophysical Union, Vol. 2, pp. 519-524, 1935.
20. Tsang, C.F., Buscheck, T.A., and Doughty, C., "Aquifer Thermal Energy Storage: A Numerical Simulation of Auburn University Field Experiments," *Water Resources Research*, Vol. 17, pp. 647-658, JUne 1981.
21. Tsang, C.F., and Doughty, C., "Detailed Validation of Liquid and Heat Flow Code Against Field Performance," submitted to *Journal of Society of Petroleum Engineers*, Report LBL-18833, Lawrence Berkeley Laboratory, 1985.
22. Tsang, C.F., and Hopkins, D., "Aquifer Thermal Energy Storage-A Survey," in *Recent Trends in Hydrogeology*, T.N. Narasimhan, ed., Geological Society of America, Special paper No. 189, 1981.
23. Verruijt, A., "The Rotation of a Vertical Interface in a Porous Medium," *Water Resources Research*, Vol. 16, No. 1, pp. 239-240, 1980.

15. LIST OF SYMBOLS

C	Aquifer volumetric heat capacity (matrix plus water), J/m^3K
C_w	Volumetric heat capacity of water, J/m^3K
D	Thickness of diffuse thermal front in cases F and G, m
d	Dispersivity, m
f_{bt}	Buoyancy tilting function
f_{ft}	Forced-convection tilting function
f_t	Basic tilting function
f_1	= 0.235
G	Catalan's constant (= 0.915...)
g	Standard gravity, 9.81 m/s^2
H	Thickness of aquifer stratum, m
k	Permeability (horizontal), m^2
k'	Vertical permeability, m^2

L	Horizontal thickness of region left of the thermal front in case B, C, D, E, and G, m
n	Effective porosity of porous medium
P	Pressure, Pa
P_b	Pressure for buoyancy flow part of Eq. (35), Pa.
P_{fc}	Pressure for forced convection part of Eq. (35), Pa.
Q_t	Tilting flow, m^3 H_2O/s.
Q_{bt}	Buoyancy tilting flow, m^3 H_2O/s.
Q_{ft}	Forced-convection tilting flow, m^3 H_2O/s.
Q_1	Forced-convection flow rate through aquifer, m^3 H_2O/s.
$\bar{q}$	Volumetric groundwater flux, m^3 H_2O/m^2s.
$\bar{q}_b$	Volumetric groundwater flux for buoyancy flow part of Eq. (35), m^3 H_2O/m^2s.
$\bar{q}_{fc}$	Volumetric groundwater flux for forced convection part of Eq. (35), m^3 H_2O/m^2s.
q_x	x-component of $\bar{q}$, m^3 H_2O/m^2s.
q_y	y-component of $\bar{q}$, m^3 H_2O/m^2s.
q_z	z-component of $\bar{q}$, m^3 H_2O/m^2s.
q_f	Horizontal buoyancy flow across thermal front, m^3 H_2O/m^2s.
q_o	Characteristic buoyancy flow defined by Eq. (9), m^3 H_2O/m^2s.
R	Radius of circular region in case H, m.
R_s	Retardation factor for solute transport
R_w	Radius at inner boundary in case E, m
r	Radial coordinate, m
S	Tilting function defined by Eq. (82)
s	Tilting parameter equal to $\kappa \tan\alpha$.
T	Temperature, ^oC.
T_o	Temperature of region 0, ^oC; ambient temperature, ^oC.
T_1	Temperature of region 1, ^oC; injection temperature, ^oC.
t	Time, s.
t_o	Characteristic tilting time defined by Eq. (33), s.
$\bar{v}_T$	Thermal velocity equal to $C_w\bar{q}/C$, m/s.
x	Horizontal coordinate, m.
y	Horizontal coordinate, m.
z	Vertical coordinate, m.

α	Tilting angle. Angle between straight thermal front and vertical axis.
α'	Tilting angle for isotropic case ($\kappa = 1$)
β	Viscosity factor equal to μ_0/μ_1
γ	Forced-convection tilting parameter defined by Eq. (77).
ε	Energy recovery factor
κ	Anisotropy factor, equal to $\sqrt{k'/k}$
λ	Thermal conductivity, W/mK
μ	Dynamic viscosity, kg/ms
μ_0	Dynamic viscosity in region 0, kg/ms
μ_1	Dynamic viscosity in region 1, kg/ms
ρ	Density, kg/m^3
ρ_0	Density in region 0, kg/m^3
ρ_1	Density in region 1, kg/m^3
ω_t	Angular tilting rate defined by Eq. (30), rad/s
ω_0	Angular tilting rate for case A, given by Eq. (32), rad/s

MECHANICS OF FLUIDS IN LAYERED SOILS

Arnold Verruijt and Frans B.J. Barends

MECHANICS OF FLUIDS IN LAYERED SOILS

Arnold Verruijt
Dept. of Civil Engineering
University of Delft
2628 CN Delft, The Netherlands

Frans B.J. Barends
Delft Soil Mechanics Laboratory
P.O. Box 69
2600 AB Delft, The Netherlands

ABSTRACT

In this chapter the mechanical behaviour of layered porous, saturated soils, typically consisting of strata of clay and sand is discussed, from the viewpoint of phenomena such as subsidence due to the extraction of fluids from the aquifer. Special attention is paid to the influence of permeability contrasts such as occurring in a soil system consisting of layers of clay and sand, and the relative importance of the compressibilities of the various layers. In many existing models the deformations of the clay layers are disregarded, so that all surface settlements are due to deformations of the sandy aquifers. This may be acceptable for relatively thin layers of stiff clay. In many circumstances, however, such as may occur in delta's of large river systems, it may be necessary to take into account the deformation of the clay layers. For such situations two possible approximate models are presented. The models are compared to a full numerical solution, which can be considered to represent the true solution of the coupled problem.

1. INTRODUCTION

Natural soils often consist of layers of different properties, with layers of high permeability (sand layers) interspersed with layers of low permeability (e.g. clay). For the prediction of the subsidence of such a system due to the withdrawal of groundwater, or due to an external loading of the soil, the compression of all strata must be taken into account.

The problem is of a transient nature, with the propagation of pore water pressure differences being retarded by the combined effect of flow of groundwater and the compression of the soil. The process can be considered to be a generalization of the steady state type problems of groundwater flow in layered soils. For problems of this type two types of approach are available, which can be denoted as the aquifer model and the more general three dimensional Biot model. In the aquifer model, first used by De Glee (7), vertical flow is assumed in the clay layers (the aquitards) and horizontal flow in the sand layers (the aquifers). This leads to relatively simple equations. In the general three dimensional model, which is due to Biot (3) no assumptions with regard to the flow field or the deformation field are necessary, but as a result the system of equations is of a more complicated form. The two types of models can be related by a process of averaging, see e.g. Corapcioglu and Bear (5).

The actual transient behaviour of a soil system is determined on one hand by hydraulic properties, such as permeability, and on the other hand by mechanical properties, such as compressibility and shear resistance. One of the main characteristics of the process is the difference in accuracy of these parameters. Darcy's law, which is the basis of the hydraulic part of the process, is generally considered to be a rather accurate description of the fluid flow in a porous medium. The description of the mechanical behaviour of a soil, by compressibility coefficients and shearing resistance, is of a much more controversial nature, involving uncertainties such as geological history (pre-consolidation, pre-shearing), shear failure, dilatancy, and creep. A review of possible models has been given by Corapcioglu (6).

It is postulated here that a well-balanced description of the physical processes should be of the following general nature. First of all a realistic description of the geometry should be used, taking into account the layered structure of the soil, with different permeabilities and compressibilities. The coupling of the fluid flow and the deformation can be taken into account by an analysis in two stages. In the first stage the pore pressures are calculated, disregarding the feedback from the deformation process, but instead using a simplified unique relation between effective stresses and pore pressures. This is certainly not justified in many problems from soil mechanics practice, in which the fluid flow is generated by external loading of the soil. In groundwater hydraulics, however, the underlying assumption that the total stresses are constant in time, is acceptable, at least as a first approximation. The deformations of the soil due to the pore pressure variations can be calculated in a second stage, using the pore pressures determined from the hydraulic model as input in the deformation model. It may be noted that the same approach is generally used in reservoir engineering (8).

In this chapter the aquifer approach will be used, without going into the possible justification of this model, perhaps involving such refinements as horizontal deformability of the soil layers. A first order model, due to Jacob (10) and Hantush (9), is to take into account the compression of the aquifers only, disregarding the compression of the aquitards. A large number of solutions for this type of model, to be denoted as the first order model, have been obtained, especially by Hantush (9). Physically speaking this may not be a realistic set of assumptions, because clay layers are often more compressible than sand layers, and therefore disregarding their compression is not justified, except in case of clay layers of very small thickness.

In order to account for the deficiencies of the first order model more refined models have been developed, in which the compressibility of the aquitards is taken into account, by Neuman and Witherspoon (11). It follows from this theory that the influence of the compressibility of the aquitards may be considerable. Unfortunately the exact solutions of the problem are mathematically rather inconvenient, and for that reason a simplified approach has been proposed by Barends (1). This approach, to be denoted as the pseudo-exact model, will be presented in this chapter, together with an approximate model, to be denoted as the second order model, in which the pressure variation in the aquitard is represented in a simplified way. It will be shown that these models can be useful tools for the prediction of the progress of subsidence.

A deficiency of most existing transient theories of subsidence is that the deformation properties of the soil are usually highly schematized, in order to keep the system of equations amenable to mathematical analysis. This may mean that the accuracy of the prediction of the ultimate subsidence is less than what might be attained for, taking into account the achievements of non-linear soil mechanics. Thus the long term problem of subsidence at constant pore pressures should be considered separately, taking into account phenomena such as creep.

2. DEFINITION OF THE MODELS

In this section the two approximate theories will be presented. For reasons of simplicity the considerations will be restricted to the simple system of a single aquifer and a single aquitard. The theories can easily be generalized to systems of more layers, however.

The first approximate approach is to consider the flow in the aquifer to be horizontal, and in the aquitard to be vertical, to take into account the compressibility of the aquifer and the aquitard, and then to use the Laplace transform technique with

Schapery's inversion formula to solve the system of equations (1). In this way it is assured that the solution is correct for t=0 and for $t \to \infty$. For smooth processes the Schapery approximation is often very good, and it will be shown that in the present case the approximation is at least reasonably good. This approximation will be denoted as the pseudo-exact model.

As an alternative a physical approximation will be presented, to be denoted as the second order model, in which the flow in the aquitard is represented by a parabolic variation of the head. This can be considered to be analogous to a finite element approximation, using a single second order element to represent the flow and compression in the aquitard. This model will appear to be less accurate for small values of time, but somewhat better for large values of time.

2.1. The Second Order Model

Consider the non-steady flow of groundwater in a layered soil consisting of an aquifer and an aquitard, which separates the aquifer from another layer in which the groundwater head is constant in time and space, (see Fig. 1). This constant will be taken as the reference level for all other heads. The heads are influenced by a certain action in the main aquifer, for instance a well that starts operating at time t=0.

The basic flow mechanism to be considered is the same as in the first order theory: mainly horizontal flow in the aquifer, and vertical flow in the aquitard.

The groundwater head in the aquifer will be denoted by ϕ. Under conditions of steady flow the head in the aquitard would vary linearly from 0 at the top (z=0) to ϕ at the bottom (z=d). In the beginning of the transient state the head in the aquitard will lag

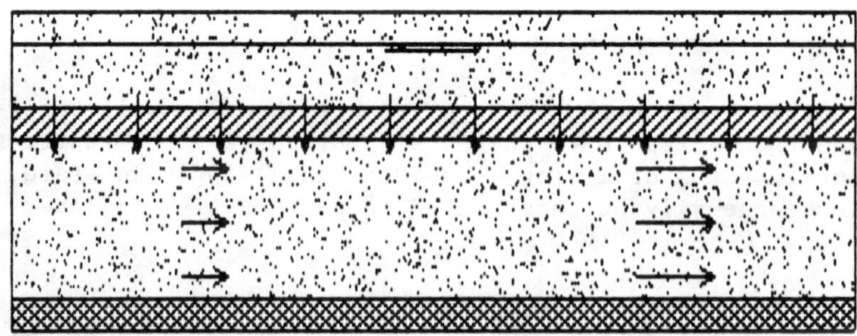

Figure 1. Leaky aquifer.

behind the value in the aquifer, however, and the final state will be reached only after completion of the consolidation process of the clay. In order to describe the head in the aquitard in an approximate way such that the physical process of consolidation as well as the ultimate steady state can be described, it is assumed that at all times this head can be written as

$$h = 4(z/d)(1 - z/d)\psi + (z/d)(2z/d - 1)\phi \tag{2.1}$$

where ψ is the head in the center of the aquitard (for $z = \frac{1}{2}d$). It can be seen from Eq. (2.1) that for $z = 0$: $h = 0$, for $z = \frac{1}{2}d$: $h = \psi$, and that for $z = d$: $h = \phi$. The expression (2.1) actually represents a quadratic function defined by the three values 0, ψ and ϕ, at the top, in the center, and at the bottom, respectively. If $\psi = \frac{1}{2}\phi$ (in the final steady state) the formula reduces to a straight line.

The parabolic variation of the head in the aquitard can be considered to be a first refinement of the classical theory, in which the flow rate in the aquitard is constant at all times, which corresponds to an assumed linear variation of the head. Considering the classical theory due to Jacob (10) as the first order theory the present approach can be denoted as a second order model. Further refinements can be made by assuming a higher order variation of the head.

It follows from Eq. (2.1) that

$$\partial^2 h/\partial z^2 = 4(\phi - 2\psi)/d^2 \tag{2.2}$$

This means that the second derivative is constant over the thickness of the aquitard. Another consequence of Eq. (2.1) is the following expression for the leakage from the aquitard to the aquifer

$$L = - k(\partial h/\partial z)_{z=d} = - (3\phi - 4\psi)/c \tag{2.3}$$

where c is the resistance of the aquitard ($c = d/k$). In the final steady state, when $\psi = \frac{1}{2}\phi$, one obtains $L = - \phi/c$, as in the classical theory.

The equation describing the process of consolidation of the clay layer is (13)

$$\partial h/\partial t = c_v \partial^2 h/\partial z^2 \tag{2.4}$$

where c_v is the consolidation coefficient, defined by

$$c_v = k/(m_v \gamma_w) \tag{2.5}$$

Here m_v is the compressibility of the clay, and γ_w is the volumetric weight of water. For the sake of future convenience a storativity S_c is now defined as

$$S_c = m_v \gamma_w d \tag{2.6}$$

Physically speaking this quantity represents the settlement of the clay layer if the effective stresses in it are uniformly increased by a pressure of 1 meter water, in agreement with the usual definition of storativity.

The consolidation equation (2.4) cannot be satisfied uniformly, but only on the average. This leads to the following condition

$$4\partial\psi/\partial t + \partial\phi/\partial t = 12(\phi - 2\psi)/t_c \tag{2.7}$$

where

$$t_c = \tfrac{1}{2}d^2/c_v = \tfrac{1}{2}cS_c \tag{2.8}$$

This quantity is a measure for the duration of the consolidation process. In the case of uniform consolidation of a layer drained on both sides the consolidation process is usually said to be practically completed (for 99%) if $c_v t/d^2 = 0.5$, or $t = t_c$.

Equation (2.7) is one of the basic equations of the present theory. It describes a relation between the head ϕ in the aquifer, and the head ψ in the (center of the) aquitard.

2.1.1. Flow in the aquifer

The flow in the aquifer can be described by the following equation, which is based on Darcy's law, assuming horizontal flow, and the continuity condition,

$$T(\partial^2\phi/\partial x^2 + \partial^2\phi/\partial y^2) + L = S_a\partial\phi/\partial t \tag{2.9}$$

where T is the transmissivity of the aquifer, assumed to be constant and S_a is the storativity. The leakage from the aquitard is given in Eq. (2.3). Substitution of that expression into Eq. (2.9) gives

$$\partial^2\phi/\partial x^2 + \partial^2\phi/\partial y^2 - (3\phi - 4\psi)/\lambda^2 = (cS_a/\lambda^2)\partial\phi/\partial t \tag{2.10}$$

where λ is the leakage factor,

$$\lambda = \sqrt{(Tc)} \tag{2.11}$$

Equation (2.10) is the second basic equation of the theory presented here. The governing differential equations of the problem considered are Eqs. (2.7) and (2.10), which should be solved together with the appropriate initial and boundary conditions. Some examples will be considered later. First it will be verified whether the present equations are compatible with some well-known limiting cases.

2.2. Rigid Aquitard

If the aquitard is completely rigid $S_c = 0$, and it then follows from Eq. (2.8) that $t_c = 0$, and from Eq. (2.7) that $\psi = \frac{1}{2}\phi$. This means that Eq. (2.10) reduces to

$$\partial^2\phi/\partial x^2 + \partial^2\phi/\partial y^2 - \phi/\lambda^2 = (cS_a/\lambda^2)\partial\phi/\partial t \tag{2.12}$$

This is indeed the familiar differential equation for non-steady flow in a leaky aquifer, disregarding the compression of the aquitard. The coefficient in the right hand side is usually written as (S/T). It can be concluded that the present model is indeed a generalization of the classical theory of Jacob (10). It should be noted that the case of a single aquifer, with no leakage at all, is of course also included in the model. For $\lambda \to \infty$ the third term in Eq. (2.12) vanishes. Thus the solutions of Theis (14) and Hantush (9) are special cases of the second order model.

2.3. Response of Aquitard to Step Function

Another interesting limiting case is the response of the aquitard to a sudden change of the head in the aquifer. This is a standard problem from the theory of consolidation. For the case that the head ϕ jumps at time $t = 0$ from its original zero value to $\Delta\phi$ the solution of the consolidation equation (2.4) can be determined by using the Laplace transform technique. The solution is

$$h = \Delta\phi\{\frac{z}{d} + \frac{2}{\pi}\sum_{k=1}^{\infty}\frac{(-1)^k}{k}\sin(k\pi z/d)\exp[-k^2\pi^2 c_v t/d^2]\} \tag{2.13}$$

The head in the center (for $z = \frac{1}{2}d$) is found to be

$$h_m = \tfrac{1}{2}\Delta\phi\{1 - \frac{4}{\pi}\sum_{j=0}^{\infty}\frac{(-1)^j}{2j+1}\exp[-(2j+1)^2\pi^2 c_v t/d^2]\} \tag{2.14}$$

It can be shown that the value for $t = 0$ is indeed zero, because the sum of the series then is $\pi/4$.

A quantity of special interest is the deformation of the entire layer. In general one may write

$$\frac{\partial \xi}{\partial t} = - m_v \gamma_w \int_0^d \frac{\partial h}{\partial t} \, dz \tag{2.15}$$

where ξ is the displacement of the top of the layer, with respect to its bottom. After performing the necessary differentiations and integrations one obtains, with Eq. (2.6),

$$\xi = - \tfrac{1}{2}\Delta\phi S_c \{1 - \frac{8}{\pi^2} \sum_{j=0}^{\infty} \frac{1}{(2j+1)^2} \exp[-(2j+1)^2 \pi^2 c_v t/d^2]\} \tag{2.16}$$

The negative value obtained for $t \rightarrow \infty$ indicates that the soil expands if the head increases. For $t = 0$ the deformation is zero, because the sum of the series is then precisely $\pi^2/8$.

The analytical results will now be compared with the results obtained from the second order theory presented above. The equation to be solved in this case is Eq. (2.7), where ϕ is the step function defined before. The solution of this problem, which can be obtained most conveniently by using the Laplace transform technique, is

$$\psi = \tfrac{1}{4}\Delta\phi\{2 - 3\exp(-6t/t_c)\} \tag{2.17}$$

For $t \rightarrow \infty$ this indeed approaches the steady state value $\frac{1}{2}\Delta\phi$, but for $t=0$ the solution gives a step value of $-\frac{1}{4}\Delta\phi$, which is perhaps unexpected. This behaviour of the approximate solution is caused by the (implicit) condition in the theory that at the moment of loading no net loss of water can be generated, so that the averag head must be zero. If the head at the lower boundary is increased this must be balanced by a reduction of the head in the center.

The deformation can be calculated from Eq. (2.15) and Eq. (2.4) and (2.2). The result is

$$\xi = - \tfrac{1}{2}\Delta\phi S_c\{1 - \exp(-6t/t_c)\} \tag{2.18}$$

The initial and ultimate values of this expression are both in agreement with the exact solution [Eq. (2.16)]. Apparently the initial step in the head in the center is necessary so that the settlement at $t = 0$ is zero, or that the average head remains zero.

The exact and approximate solutions for the total deformation are shown in Figure 2. The approximate formula [Eq. (2.18)] is indicated as "2nd order". The agreement appears to be reasonably good, except for small values of time.

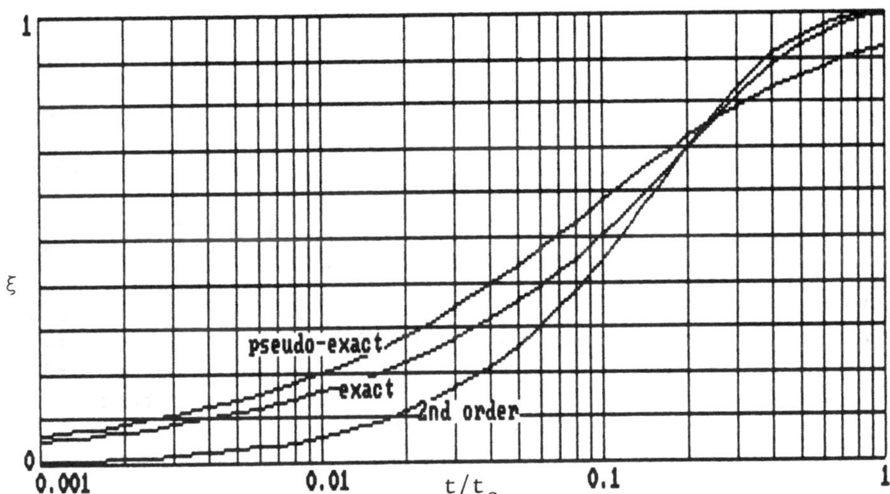

Figure 2. Exact and approximate solutions for the deformation.

As an alternative approximation the results of a pseudo-exact model are also shown in Figure 2. These results have been obtained by following an approach proposed by Barends (1), which is based upon an exact description of the consolidation process in the aquitard. By using the Laplace transformation to transform the two basic differential equations, for the vertical compression of the aquitard, and the horizontal flow in the aquifer, a solution in the domain of the Laplace transform parameter can be obtained. As an approximation to the inversion of this solution Schapery's approximate formula is then used. The result is

$$\xi = -\tfrac{1}{2}\Delta\phi S_c\sqrt{(4t/t_c)}\tanh(\sqrt{(t_c/4t)}) \qquad (2.19)$$

It can be seen from Figure 2 that this approximation is better for small values of time, but somewhat less accurate for large values of time. Both approximations seem to be sufficiently accurate for engineering purposes.

Another interesting quantity is the leakage from the aquitard into the aquifer. The exact solution for this quantity, for a stepwise variation of the head in the aquifer, is

$$L = -(\Delta\phi/c)\{1 + 2\sum_{k=1}^{\infty}\exp[-k^2\pi^2 c_v t/d^2]\} \qquad (2.20)$$

The approximate solution for the second order model is, from Eq. (2.3) and (2.17),

$$L = -\ (\Delta\phi/c)\{1 + 3\exp(-6t/t_c)\} \tag{2.21}$$

The pseudo-exact solution, using Schapery's inversion formula, is

$$L = -\ (\Delta\phi/c)\sqrt{(t_c/t)}\coth(\sqrt{(t_c/t)}) \tag{2.22}$$

A comparison of these exact and approximate solutions for the leakage from the aquitard into the aquifer is shown in Figure 3. Again the approximation by using Schapery's inversion formula is (much) better for small values of time, whereas the approximation based upon a parabolic variation of the head in the aquitard is better for large values of time. The failure of the second order model for small values of time must be due to the fact that at all times the head in the aquitard is approximated by a parabola, see Eq. (2.1), and this is inaccurate immediately after a stepwise variation at one of the boundaries. The leakage into the aquifer, which is infinitely large in the exact and pseudo-exact models, is actually underestimated. The deformation and the average head are approximated much better. Actually the curves shown in Figure 2 can also be considered to represent a comparison of the average head, for which the approximation is reasonably good. It can also be expected that for more gradual changes of the head at the boundary

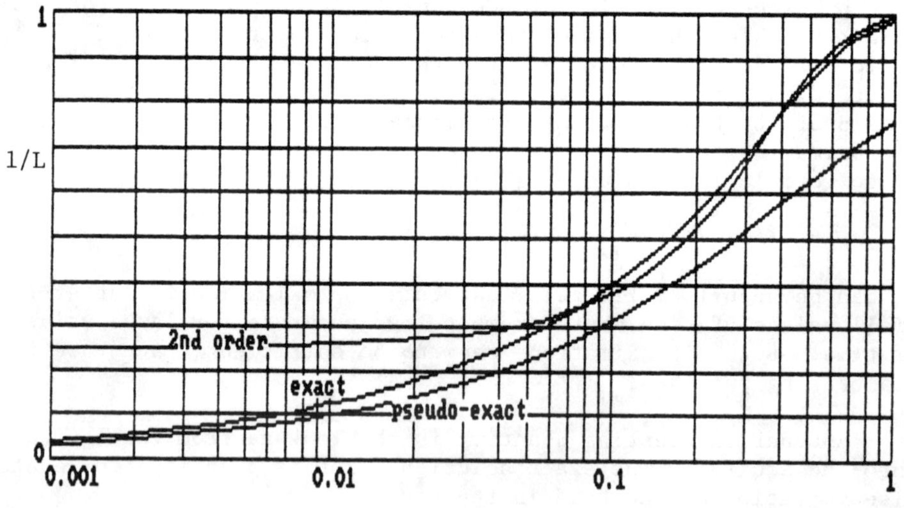

Figure 3. Exact and approximate solutions for the leakage.

a parabolic approximation may be sufficiently accurate. In later sections it will be found, from other comparisons, that the second order model may be a useful tool, despite its inability to correctly describe variations of high frequency.

The two special cases investigated above can be considered as the two extreme situations. In the case of a rigid aquitard the storage in the aquifer is dominant, and in the second case the principal phenomenon considered is the storage in the aquitard. In general it can be concluded that the second order model, based upon a parabolic variation of the head in the aquitard, gives a good approximation of the subsidence, especially after some time. The response immediately after a stepwise variation of the head is approximated somewhat better by the pseudo-exact model, using Schapery's inversion formula.

As a next step in the presentation of the theory the solution of some particular problems will be considered.

3. ONE-DIMENSIONAL FLOW

Perhaps the simplest example is that of one-dimensional flow in a semi-infinite aquifer, bounded by a long canal in which the groundwater head is increased from its initial value 0 to a value ϕ_o at time t=0.

Two special cases of this problem can be used for the purpose of reference, namely the steady state solution, and the solution for a single aquifer in the absence of leakage. The steady state solution is the standard formula for a leaky aquifer,

$$\phi = \phi_o \exp(-x/\lambda) \qquad (3.1)$$

The solution of the problem in the absence of leakage can be obtained from Eq. (2.9), by taking L = 0. This solution is found to be

$$\phi = \phi_o \operatorname{erfc}(x\sqrt{(S/4Tt)}) \qquad (3.2)$$

For t = 0 the value of ϕ is 0, and for $t \to \infty$ the solution approaches ϕ_o, everywhere, as it should.

The solution of the general problem for the approximate model can be investigated by using the Laplace transform technique (4). If the Laplace transforms of ϕ and ψ are denoted as Φ and Ψ, respectively, the first basic equation, the transform of Eq. (2.7), is as follows

$$4s\Psi + s\Phi = 12(\Phi - 2\Psi)/t_c \qquad (3.3)$$

where s is the Laplace transform parameter. It follows from this equation that Ψ can be expressed in terms of Φ,

$$\Psi/\Phi = (12 - st_c)/(24 + 4st_c) \tag{3.4}$$

The Laplace transform of the second basic equation, Eq. (2.10), is

$$d^2\Phi/dx^2 - (3\Phi - 4\Psi)/\lambda^2 = (cS_a/\lambda^2)s\Phi \tag{3.5}$$

The head Ψ in the aquitard can be eliminated with the aid of Eq. (3.4). This gives

$$d^2\Phi/dx^2 = [2st_a + (6 + 4st_c)/(6 + st_c)]\Phi/\lambda^2 \tag{3.6}$$

where t_a is a parameter characteristic for the time scale in the aquifer itself, defined by

$$t_a = \tfrac{1}{2}cS_a \tag{3.7}$$

It is to be noted that this expression is of the same form as Eq. (2.8). The process appears to contain two different time scales, one for the aquifer response and one for the response of the aquitard

The solution of Eq. (3.6) subject to the boundary conditions at infinity and at x = 0 (x = 0: $\phi = \phi_o$) is

$$\Phi = (\phi_o/s)\exp\{-(x/\lambda)\sqrt{[2st_a + (6 + 4st_c)/(6 + st_c)]}\} \tag{3.8}$$

The mathematical problem now remaining is to determine the inverse Laplace transform of Eq. (3.8). Unfortunately this is not a standard inversion, and therefore only an approximation will be presented.

3.1. Schapery Approximation

A simple approximate inversion formula, applicable to non-oscillating phenomena, is the so-called Schapery approximation, already introduced and used in the previous chapter. This approximation is exact for t = 0 and for t = ∞ (12). In the case of Eq. (3.8) the result is

$$\phi/\phi_o = \exp\{-(x/\lambda)\sqrt{[t_a/t + (12t + 4t_c)/(12t + t_c)]}\} \tag{3.9}$$

For t = 0 the head is zero everywhere, as it should be, and for t = ∞ the solution reduces to the familiar result of a simple exponential function, defined by the leakage factor λ.

It is interesting to note that the solution can also be written in the general form

$$\phi/\phi_o = \exp\{-x/\lambda^*\} \tag{3.10}$$

where λ^* is a time-dependent leakage factor, defined by

$$\lambda^* = \lambda/\sqrt{[t_a/t + (12 + 4t_c/t)/(12 + t_c/t)]} \tag{3.11}$$

where, as defined in Eqs. (2.8) and (2.16),

$$t_c = \tfrac{1}{2}cS_c,\ t_a = \tfrac{1}{2}cS_a \tag{3.12}$$

The concept of a transient leakage factor was introduced by Barends (1), using the pseudo-exact approach, as follows. The consolidation of the aquitard is described by the differential equation (2.4). The Laplace transform solution of this equation, satisfying the boundary conditions that $h = 0$ for $z = 0$, and $h = \phi$ for $z = d$, is

$$H/\Phi = \sinh[z\sqrt{(s/c_v)}]/\sinh[d\sqrt{(s/c_v)}] \tag{3.13}$$

This means that the leakage into the aquifer is, with Eq. (2.8),

$$L/\Phi = -\ (1/c)\sqrt{(2st_c)}\coth(\sqrt{(2st_c)}) \tag{3.14}$$

Substitution into the Laplace transform of the differential equation (2.9) for the head in the aquifer gives

$$d^2\Phi/dx^2 = \Phi/(\lambda^{\#})^2 \tag{3.15}$$

where $\lambda^{\#}$ is a modified leakage factor, depending upon the Laplace transform parameter s. After solution of the differential equation and inverse transformation by Schapery's approximation the result is again of the simple form of Eq. (3.10), with now the time dependent leakage factor defined as

$$\lambda^{\#} = \lambda/\sqrt{[t_a/t + \sqrt{(t_c/t)}\coth(\sqrt{(t_c/t)})]} \tag{3.16}$$

Formula (3.16) is an alternative to Eq. (3.11). Both expressions tend toward λ if $t \to \infty$, and they vanish if $t = 0$. Some more insight in the behaviour of the solutions can be obtained by plotting the expressions (3.11) and (3.16), (see Figure 4 and Figure 5). In these figures the transient leakage factor is plotted for 5 values of t_c/t_a namely 0.1, 1, 10, 100 and 1000 (from left to right in the figures). Values of t_c/t_a smaller than 0.1 (down to 0) give the same result as for $t_c/t_a = 0.1$. This means that the classical solution is correct if the storativity of the aquitard is less than 10% of the storativity of the aquifer, or, as indicated by the figure, even if the two storativities are of the same order of

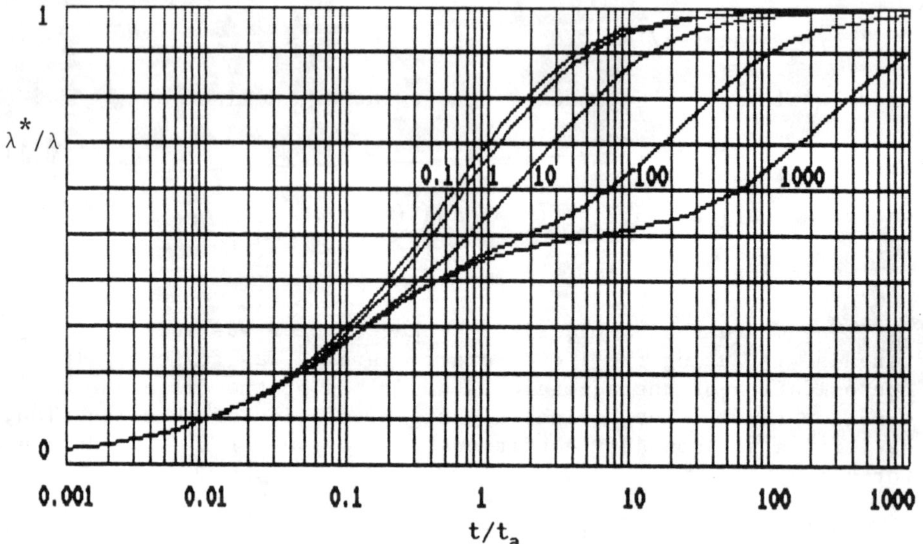

Figure 4. Transient leakage factor, Eq. (3.11)

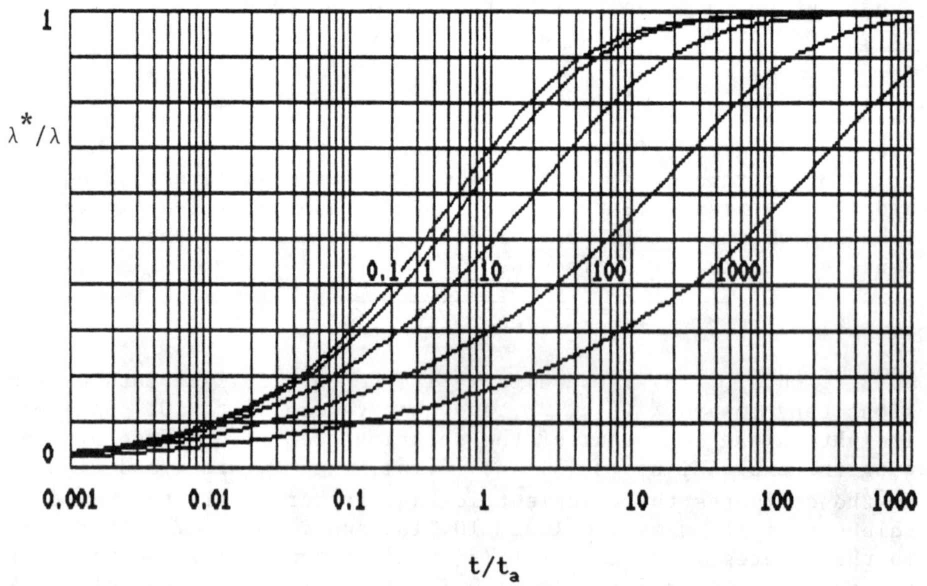

Figure 5. Transient leakage factor, Eq. (3.16).

magnitude. If the storativity of the aquitard is large compared to the storativity of the aquifer the classical solution, disregarding the storage in the aquitard, is no longer applicable.

For $t = 0$ the value of the transient leakage factor is indeed zero, which indicates that for small values of time the drawdown is restricted to the immediate vicinity of the disturbance at $x = 0$. For larger values of time the transient leakage factor increases, indicating a growing region of influence. For large values of time the transient leakage factor approaches the steady state value λ.

For very large values of the parameter t_c/t_a (that is for relatively stiff aquifers) the solution (3.11) seems to have a plateau at the level $L = \frac{1}{2}$. Actually this is confirmed by the behaviour of the original transformed solution. If the aquifer is completely rigid one obtains $t_a = 0$, and then the transformed solution (3.8) reduces to the following form

$$\Phi = (\phi_o/s)\exp\{-(x/\lambda)\sqrt{[(6 + 4st_c)/(6 + st_c)]}\} \qquad (3.17)$$

As is well known (see e.g. 4) the behaviour of a function for small values of time can be obtained by taking the transform parameter s very large. It follows from Eq. (3.17) that for large values of s

$$\Phi \cong (\phi_o/s)\exp\{-2x/\lambda\} \qquad (3.18)$$

Inverse transformation now shows that for small values of time the solution is approximately,

$$\phi \cong \phi_o\exp\{-2x/\lambda\} \qquad (3.19)$$

This is the steady state solution, with λ replaced by $\frac{1}{2}\lambda$.

It should be noted that the particular behaviour of the second order model, with the plateau at half the maximum level, is a mathematical property, which is caused by the particular type of approximation of the head in the aquitard, namely the parabolic variation, and has no physical significance. The parabolic approximation underestimates the leakage at the beginning of the process. It has been found that a higher order approximation, involving a third order approximation, leads to a better approximation for small values of time. As compared to the direct Schapery approximation for an aquitard of finite thickness, which is more accurate for small values of time, the second order model has the advantage that it consists of a complete set of differential equations, and thus can easily be extended to nonhomogeneous soil layers. This model also admits a numerical solution, which will be presented below.

4. NUMERICAL SOLUTION

As the analytical approach followed above did not lead to a closed form solution, even for the simple one dimensional case, but only lead to an approximate solution, the accuracy of which remains unknown, it may be illuminating to attempt a numerical solution of the problem. On the basis of the results obtained above it seems reasonable to distinguish between two types of problems: one for rigid or almost rigid aquifers, in which case there is an immediate response followed by a gradual adjustment, and one for systems in which the storativity of the aquifer is of the same order of magnitude, or even larger, than the storativity of the aquitard. In the latter case it can be expected that the response of the aquitard is semistatic.

The system of equations to be solved is, for the one-dimensional case, see Eqs. (2.7) and (2.10),

$$4\partial\psi/\partial t + \partial\phi/\partial t = 12(\phi - 2\psi)/t_c \qquad (3.20)$$

$$\partial\phi/\partial t = (\lambda^2/2t_a)\partial^2\phi/\partial x^2 - (3\phi - 4\psi)/2t_a \qquad (3.21)$$

The simplest way to approximate this system of equations is by an explicit finite difference scheme. An explicit expression for $\partial\psi/\partial t$ can be obtained by elimination of $\partial\phi/\partial t$ from the two equations, and then both the increment of ψ and of ϕ can easily be calculated, using a central finite difference for the second order spatial derivative. The main disadvantage of such an approximation is that the time steps must be taken rather small in order to maintain stability. Therefore a somewhat better approximation is to use a central finite difference in time or a backward difference, in which case an implicit system of equations is obtained.

An elementary computer program performing the numerical solution by a fully implicit scheme, using a constant spatial finite difference, is reproduced below. The program has been written in BASIC, with input entered interactively, with the program asking for values of λ, the space step Δx, the two time parameters t_c and t_a, the time step Δt, and some output parameters. Output consists of a list on the screen or the printer, of the head in the aquifer and the head in the aquitard, both expressed as a ratio of the final steady state value of the head in the aquifer, $\exp(-x/\lambda)$, in a single point of the system. The data calculated are also stored in a datafile, for later processing, for instance the construction of a graph. The program itself gives a suggestion for the magnitude of the first time step, which is based upon the stability criterion for the explicit process, and which has been derived in the usual way, by requiring that all possible distributions of errors are damped by the numerical process. After each time step the magnitude of the

Table 1. Computer Program for One Dimensional Flow

```
100 CLS:PRINT"Aquifer-Aquitard":PRINT:DEFDBL A-H,O-Z:DEFINT I-N
110 DIM F(1000),G(1000),DF(1000),DG(1000):NN=1000
120 DIM TJ(1000),FJ(1000),GJ(1000):NT=1000
130 INPUT"Leakage factor .............. ";Z
140 INPUT"Space step .................. ";DX:IF DX>Z/2 THEN DX=Z/2
150 INPUT"Characteristic time clay ..... ";TC
160 INPUT"Characteristic time sand ..... ";TA
170 N=INT(5*Z/DX+.5):IF N>NN THEN N=NN
180 A$="####.####":B=1+DX*DX*(7+72*TA/TC)/(4*Z*Z)
190 T1=2*TA*DX*DX/(B*Z*Z):B=1+7*DX*DX/(4*Z*Z):T2=TA*DX*DX/(B*Z*Z)
200 PRINT"Suggestion 1 for time step ... ";:PRINT USING A$;T1
210 PRINT"Suggestion 2 for time step ... ";:PRINT USING A$;T2
220 INPUT"Time step .................... ";DT
230 PRINT"Output on printer (Y/N) ...... ? ";:GOSUB 480
240 INPUT"Output for node number ....... ";J
250 INPUT"Name of datafile ............. ";D$
260 CLS:T=0:FF=EXP(-J*DX/Z):FOR I=0 TO N:F(I)=0:G(I)=0:NEXT I:F(0)=1
270 IF P$="N" THEN 300
280 LPRINT"Leaky aquifer, tc/ta = ";:LPRINT USING A$;TC/TA
290 LPRINT"x/L = ";:LPRINT USING A$;J*DX:LPRINT
300 FOR I=0 TO N:DF(I)=0:DG(I)=0:NEXT I
310 AC=DT/TC:AA=DT/TA:A=Z*Z/(DX*DX):A2=AA/2:C=(1+(1.5+A)*AA):D=4+24*AC
320 K=K+1:FOR IT=1 TO N:FOR I=1 TO N-1
330 B=A2*(A*(F(I+1)-2*F(I)+F(I-1)+DF(I+1)+DF(I-1))-3*F(I)+4*(G(I)+DG(I)))
340 DF(I)=B/C:B=12*AC*(F(I)+DF(I)-2*G(I))-DF(I):DG(I)=B/D
350 NEXT I:NEXT IT:FOR I=1 TO N:F(I)=F(I)+DF(I):G(I)=G(I)+DG(I):NEXT I
360 T=T+DT:B=G(J)/F(J):TJ(K)=T/TA:FJ(K)=F(J)/FF:GJ(K)=G(J)/FF
370 PRINT" t/ta = ";:PRINT USING A$;T/TA;
380 PRINT"   f/ff = ";:PRINT USING A$;F(J)/FF;
390 PRINT"   g/ff = ";:PRINT USING A$;G(J)/FF;
400 PRINT"   g/f = ";:PRINT USING A$;B:IF P$="N" THEN 450
410 LPRINT" t/ta = ";:LPRINT USING A$;T/TA;
420 LPRINT"   f/ff = ";:LPRINT USING A$;F(J)/FF;
430 LPRINT"   g/ff = ";:LPRINT USING A$;G(J)/FF;
440 LPRINT"   g/f = ";:LPRINT USING A$;B
450 DT=1.2*DT:E=1-F(J)/FF:IF E>.0001 AND K<NT  THEN 310
460 OPEN "O",#1,D$:PRINT#1,K:FOR I=1 TO K:PRINT#1,USING A$;TJ(I)
470 PRINT#1,USING A$;FJ(I):PRINT#1,USING A$;GJ(I):NEXT I:CLOSE #1:END
480 P$=INPUT$(1):IF P$="Y" OR P$="y" THEN P$="Y":PRINT"Yes":RETURN
490 IF P$="N" OR P$="n" THEN P$="N":PRINT"No":RETURN
500 GOTO 480
```

time step is multiplied by a given factor 1.2, in order to accelerate the process. Experience with the program, on an IBM-PC, has shown that it is working reasonably well, although it may be slow, especially when using small space steps. Of course the calculation speed can be improved by using a compiler, or an arithmetic coprocessor.

For four values of t_c/t_a namely 0.1, 1, 10 and 100, the results of the numerical solution are shown in Figure 6. Figure 7 shows the approximate solutions obtained for the second order model by using Schapery's inversion formula, and Figure 8 shows the results obtained from the pseudo-exact model, again using Schapery's inversion formula, but now on the basis of the exact expression for the Laplace transform of the solution. All results apply to the point $x = \lambda$.

The numerical solutions shown in Figure 6 agree reasonably well with the approximate solutions shown in Figure 7, which indicates that the simple approximation by Schapery's formula is sufficiently accurate. Again for large values of the ratio of the response times of aquitard and aquifer a solution consisting of two waves (at least on the semi-logarithmic scale used) is observed. This is probably unrealistic, as can be seen from Figure 8, which shows the Schapery approximation of the exact solution. Actually a better approximation to the true solution might be obtained by using a more refined numerical inversion scheme, involving a series of terms rather than the single term used here. The advantages of a simple analytical result is somewhat lost then, however.

5. COMPLETE NUMERICAL SOLUTION

In order to compare the approximate results presented above, in the Figures 7 and 8, with the exact solution a fully numerical model for the complete system of aquifer and aquitard has been developed. This model is again based upon the usual assumptions that the flow in the aquifer is horizontal, and that the flow in the aquitard is strictly vertical. The head in the aquifer is approximated by a finite element scheme, and then in each node of the network a vertical inflow (leakage) is entered from a vertical column. The soil in this column (representing the aquitard) is consolidating under the influence of the boundary condition on its lower boundary. The column is subdivided into a number of one dimensional elements (say 10). This means that the number of unknown values of the head equals the number of nodes in the finite element mesh, multiplied by the number of points in the columns. This may be a rather large number, but because of the simple geometrical structure of the system (see Figure 9), with horizontal connections only in the lower plane only, the system matrix can be set up

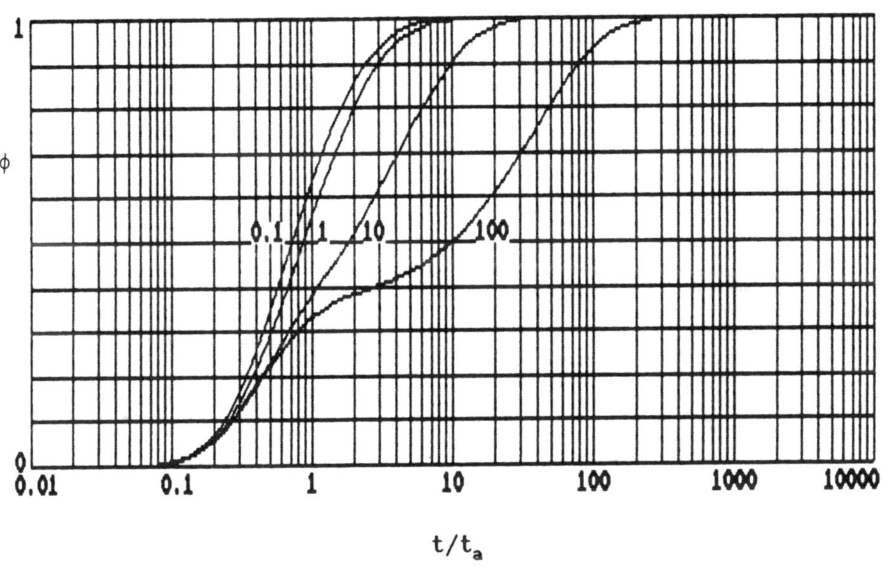

Figure 6. Second order model, numerical solution.

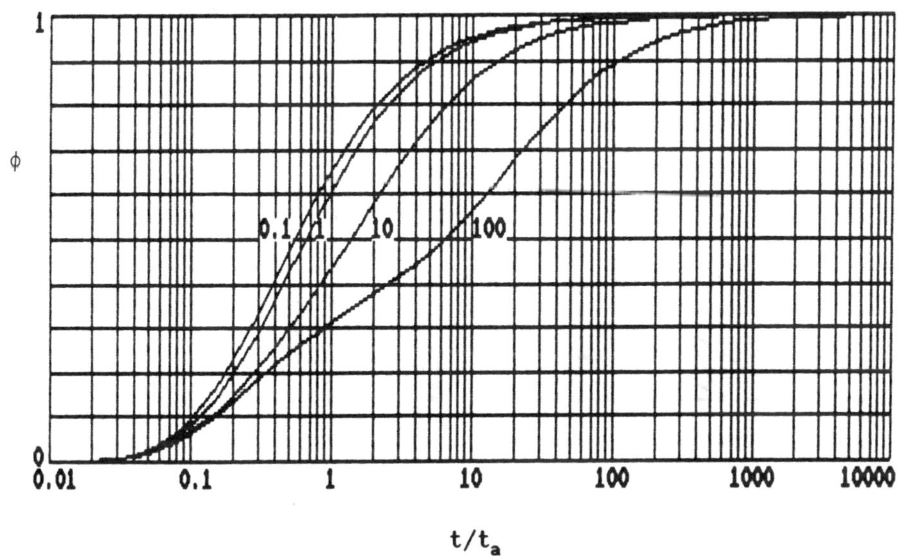

Figure 7. Second order model, approximate solution.

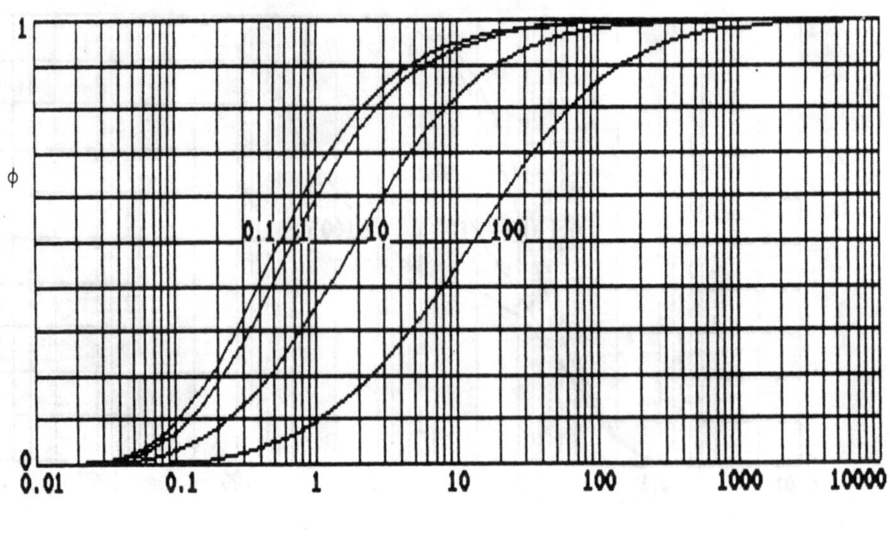

Figure 8. Pseudo-exact solution.

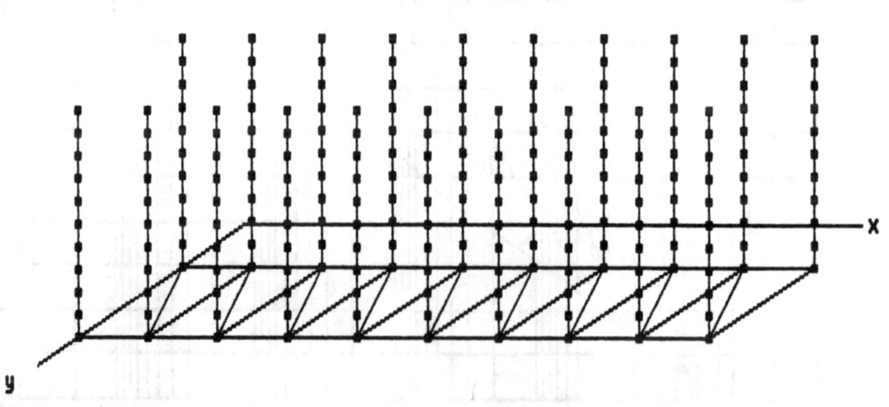

Figure 9. Nodal points in complete numerical model.

in such a way that solution of the system of linear equations takes advantage of that structure.

A numerical model for this problem has been written, again in BASIC for the IBM-PC, using a finite element approximation in the x,y-plane and a fully implicit finite difference approximation in the vertical columns. The one dimensional problem considered in the previous section can be analyzed by using a simple network of a line of elements (as shown schematically in Figure 9). The results are shown in Figure 10, again for the point $x = \lambda$. The numerical solution shown in Figure 10, which can be considered to be a close approximation of the "true" solution, compares reasonably well with the approximate solutions shown in Figures 7 and 8. It again appears that the classical first order model is justified if the storativity of the aquitard is less than the storativity of the aquifer.

6. FLOW TOWARDS A WELL

An important problem is the case of radial flow towards a well. In this case the simplest transient solution known is the solution of Theis (14) for a well in a completely confined aquifer. This solution can be considered to be a limiting case of the more general system considered here, applicable for situations in which there is no leakage, or when the permeability of the aquitard is so small that it can hardly contribute to the flow in the aquifer. The Theis solution is

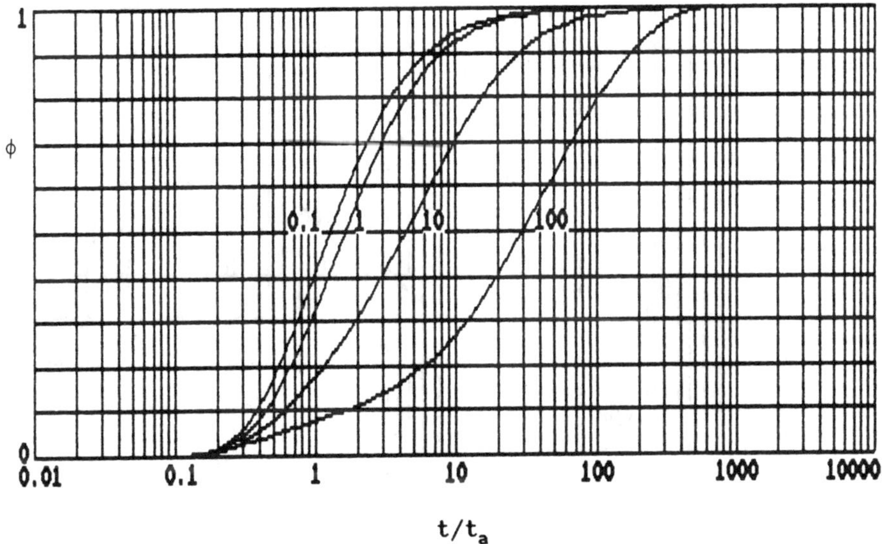

Figure 10. Aquifer-aquitard, numerical solution.

$$\phi = (Q/4\pi T)E_1(S_a r^2/4Tt) \tag{6.1}$$

where $E_1(x)$ is the exponential integral, and Q is the discharge of the well.

For the case of a leaky aquifer the first order solution, in which the storage in the aquitard is disregarded, is due to Hantush (9),

$$\phi = (Q/4\pi T)W(S_a r^2/4Tt, r/\lambda) \tag{6.2}$$

The function W(u,x) has been tabulated by Hantush, or can be calculated by an appropriate numerical subroutine, which can easily be programmed.

The approximate solution obtained from the second order or pseudo-exact theory is of the form

$$\phi = (Q/2\pi T)K_o(r/\lambda) \tag{6.3}$$

where $K_o(x)$ is a (modified) Bessel function of order zero, and λ is the transient leakage factor, defined either by Eq. (3.11) for the second order model, or Eq. (3.16) for the pseudo-exact model.

For four values of t_c/t_a, namely 0.1, 1, 10 and 100, the approximate solutions are shown in the Figures 11 and 12. These results apply to the point $r = \lambda$.

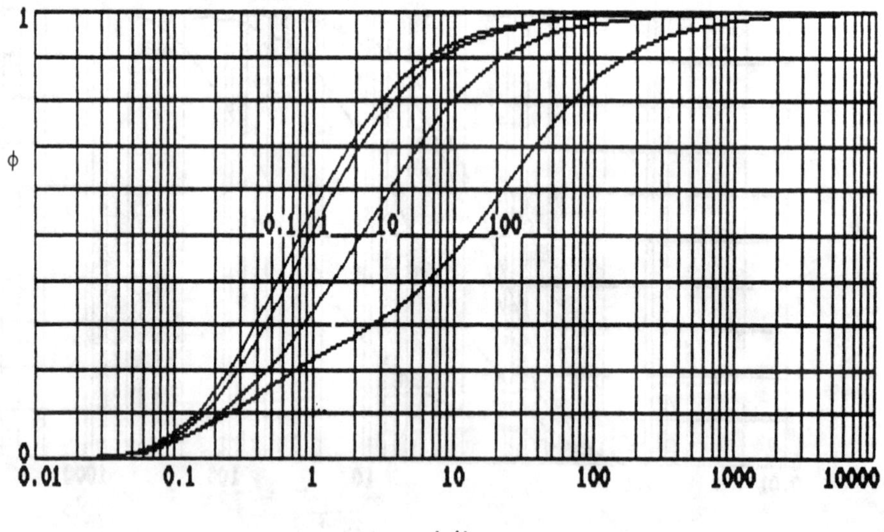

Figure 11. Drawdown in radial flow, second order model.

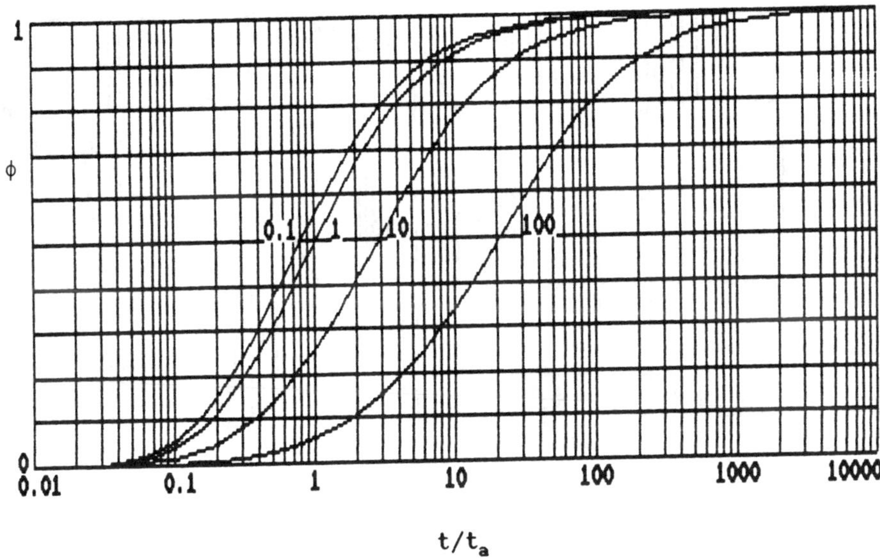

Figure 12. Drawdown in radial flow, pseudo-exact model.

A good approximation of the complete solution can again be obtained by the numerical model outlined in the previous section, using a wedge shaped mesh of finite elements. The results of these calculations are shown in Figure 13.

In Figure 13, the limiting solution for small values of the storativity of the aquitard (Hantush's solution) is also shown, indicated by the value "0". This indeed seems to be a limiting curve of the numerical results.

The general shape of the approximate solutions is similar to the shape of the numerical solution. This means that the approximate solutions might well be used as a good indication of the behaviour of the system.

It should be noted that the simple character of the approximate solutions presented in this paper enables the application of superposition of solutions, for instance for systems involving a large number of wells, each perhaps with its own discharge function.

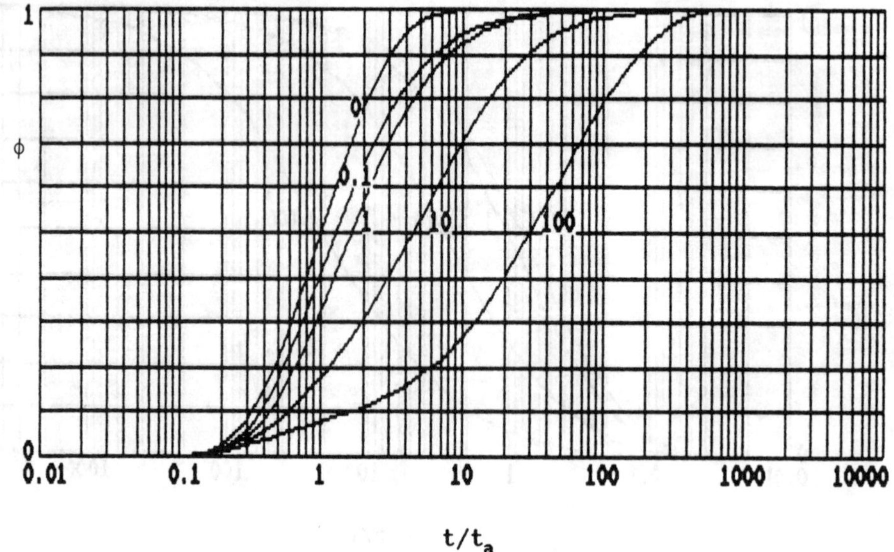

Figure 13. Drawdown in radial flow, complete numerical model.

7. CONCLUSION

Two approximate approaches for the analysis of non-steady groundwater flow in a layered soil, consisting of a permeable aquifer and an aquitard of low permeability, have been presented. Although these approximations suffer from the unavoidable deficiency that they are unable to describe the behaviour of the systems in all its details, the solutions are an improvement on the classical first order theory, in which the deformation of the aquitard is completely ignored. Compared to the exact solutions available for simple systems (11) or to a full numerical solution using a three-dimensional finite element method, the present solutions are much simpler to use. The concept of a transient leakage factor may be particularly useful.

8. REFERENCES

1. Barends, F.B.J., "Transient flow in leaky aquifers," *Proceedings of the Symposium on Modern Approach to Groundwater Management*, Capri, 1983, pp. 151-161.
2. Bear, J., *Hydraulics of Groundwater*, McGraw-Hill Publishing Co. New York, N.Y., 1979.
3. Biot, M.A., "General Theory of Three-Dimensional Consolidation," *Journal of Applied Physics*, Vol. 12, 1941, pp. 155-164.

4. Churchill, R.V., *Operational Mathematics*, 2nd ed., McGraw-Hill Publishing Company, New York, 1958.
5. Corapcioglu, M.Y., and Bear, J., "Land Subsidence - B, A Regional Mathematical Model for Land Subsidence Due to Pumping," *Fundamentals of Transport Phenomena in Porous Media*, J. Bear and M.Y. Corapcioglu, eds., Martinus Nijhoff Publishers BV, The Hague, The Netherlands, 1984, pp. 445-497.
6. Corapcioglu, M.Y, "Land Subsidence - A, A State-of-the-art Review," *Fundamentals of Transport Phenomena in Porous Media*, J. Bear and M.Y. Corapcioglu, eds., Martinus Nijhoff Publishers BV, The Hague, The Netherlands, 1984, pp. 369-444.
7. De Glee, G.J., *Over Grondwaterstromingen bij wateronttrekking door middel van putten*, Waltman, Delft, 1930.
8. Geertsma, J., "The Effect of Fluid Pressure Decline on Volumetric Changes of Porous Rocks,"*Transactions*, American Institute of Mining Engineers, Vol. 210, 1957, pp. 331-343.
9. Hantush, M.S., "Hydraulics of Wells," *Advances in Hydroscience*, Vol. 1, 1964, pp. 281-431.
10. Jacob, C.E., "Radial Flow in a Leaky Artesian Aquifer," *Transactions*, American Geophysical Union, Vol. 27, 1946, pp. 198-208.
11. Neuman, S.P., and Witherspoon, P.A., "Applicability of Current Theories of Flow in Leaky Aquifers," *Water Resources Research*, Vol. 5, 1969, pp. 817-829.
12. Schapery, R.A., "Approximate Methods of Transform Inversion in Viscoelastic Stress Analysis," *Proceedings of 4th U.S. National Congress of Applied Mechanics*, American Society of Mechanical Engineers, Vol. 2, June 1961, pp. 1075-1085.
13. Terzaghi, K., *Theoretical Soil Mechanics*, Chapman & Hall, London, 1943.
14. Theis, C. V., "The Relation Between the Lowering of the Piezometric Surface and the Rate and Duration of Discharge of a Well Using Groundwater Storage," *Transactions*, American Geophysical Union, Vol. 16, 1935, pp. 519-524.
15. Verruijt, A., *Theory of Groundwater Flow*, 2nd ed., Macmillan Co., London, 1982.

9. LIST OF SYMBOLS

c	d/k, resistance of aquitard
c_v	$k/m_v\gamma_w$, consolidation coefficient of aquitard
d	Thickness of aquitard
h	Head in aquitard
k	Hydraulic conductivity of aquitard
r	Radial coordinate
S_a	Storativity of aquifer
S_c	$m_v\gamma_w d$, storativity of aquitard
s	Laplace transform parameter

T	Transmissivity
t	Time
t_a	$\frac{1}{2}cS_a$, response time of aquifer
t_c	$\frac{1}{2}cS_c$, response time of aquitard
x	Linear coordinate
z	Vertical coordinate
λ	$\sqrt{cT}$, leakage factor
ϕ	Head in aquifer
ψ	Head in center of aquitard
ξ	Vertical settlement

PART 2

PARTICLE TRANSPORT IN POROUS MEDIA

GOVERNING EQUATIONS FOR PARTICLE TRANSPORT IN POROUS MEDIA

M. Yavuz Corapcioglu, Nelly M. Abboud and A. Haridas

GOVERNING EQUATIONS FOR PARTICLE TRANSPORT IN POROUS MEDIA

M. Yavuz Corapcioglu, Nelly M. Abboud and A. Haridas

Department of Civil Engineering
City College of New York
New York, NY 10031 USA

ABSTRACT

The migration and capture of particles, such as colloidal materials and microorganisms, through porous media occur in fields as diverse as water and wastewater treatment , well drilling, and in various liquid-solid separation processes. In liquid waste disposal projects, suspended solids can cause the injection well to become clogged, and groundwater quality can be endangered by suspended clay and silt particles migrating to the formation adjacent to the wellbore. In addition to reducing the permeability of the soil, mobile particles can carry groundwater contaminants adsorbed onto their surfaces. Furthermore, as in the case of contamination from septic tanks, the particles themselves may be pathogens, i.e., bacteria and viruses.

In this chapter, the equations governing the transport and capture of suspended solid particles have been studied in two categories. The first category includes transport and deposition of particles in an established porous medium. In this category, following the review of governing equations and various capture mechanisms in deep bed filters, the transport equation for microbial particles has been studied. For microbial particles, the governing equation for bacterial transport is coupled with a transport equation for the bacterial nutrient present in the suspension. The deposition and declogging mechanisms are incorporated into the model as a rate process for bacteria and as an equilibrium partitioning for viruses.

Formation of a cake by deposition of solid particles on a filter cloth or on a previous cake constitutes the second category. Following a literature survey, a governing equation for cake thickness is obtained by averaging the conservation of mass equation for solid particles along cake thickness. Then, the resulting equation is solved with known average porosity functions. In addition to the balance equation for solid particles, the fluid flow equation has been averaged and solved simultaneously to obtain an expression for cake thickness. Furthermore, temporal and spatial variation of pore liquid pressure across filter cake is obtained with a variable total stress expression.

1. INTRODUCTION

The transport of suspended particles in a liquid through porous media has great importance from the view point of engineering practice and industrial applications. The migration and capture of particles, such as powders and microorganisms through porous media, occur in fields as diverse as water purification (101), wastewater treatment, activated sludge processes (4), oil and water well drilling (35,55), sugar and paper pulp drying, and in many other industrial liquid-solid separation processes (102). In addition to the filtration processes which separate solids in a liquid slurry the retention of suspended particles in drilling fluids in water and oil well drilling operations is another important phenomenon (55). The accumulation of these suspended particles on perforated well screens causes a pressure drop, and sometimes causes the shutdown of the well (60,118). By a similar mechanism, unwanted perforations in a well case can be closed by squeeze cementing operations (10). As noted by Avogadro and others (6,86), colloids are released to geologic environments from radioactive waste materials. In petroleum reservoirs, the use of diverting agents, which are fine ground polymer or resin particles, requires the employment of a wellbore model to simulate the behaviour of these agents (42).

In addition to suspended solid particles, microbial particles, such as bacteria and viruses, can be introduced to soils and groundwater from septic tanks and cesspools or by land application of municipal wastewater. Although some authors found the microbial mass transport negligible for granular filters and soil-microbial mass systems (e.g., Wollum and Cassel, 128; Sykes et al., 100), many others considered the transport of microorganisms the most significant, and used bacteria and viruses to trace groundwater movement in much the same manner as chemical tracers are used. A review presented by Keswick et al (58) finds that bacterial viruses appear to be the microorganisms most suited as a microbial tracer because of their size, ease of assay, and lack of pathogenicity.

The transport and capture of suspended solid particles in a porous medium can be studied in two categories. The first category is deep bed filters made of granular materials. Suspended particles in a slurry are accumulated onto the grains of an established porous medium. The deposited particles decrease the available pore volume, and change the geometry and the structure of the medium and the nature of the grain surfaces. In constant velocity filtration operations, pressure drop across the filterbed increases due to the loss of permeability, and eventually, filtration efficiency is reduced. Usually, this type of filtration technique is used for liquid suspension with a particle concentration in the range of 100 ppm. In constant headloss operations, the liquid flux decreases during filtration. Formation of a cake by deposition of solid particles on a filter cloth or on a previous medium constitutes the second category. This type of filter media is known as filter cakes. These cakes are compressible, and during filtration, the cake is compacted while fresh solid particles are laid down on the cake surface, thereby gradually increasing the thickness. This type of mechanism which is used for concentrated slurries, causes a time-dependent cake build-up. When the cake is deposited at the surface, it has a high porosity and large liquid content. As a new filter cake is built up, the previous cake surface passes into the cake interior, and the liquid is squeezed out as the cake is compressed during filtration. In this case, the cake is relatively less permeable to permit the build-up of a head of slurry, and the liquid flow through the filter cake does not approach steady-state conditions due to changes in compaction and cake thickness. A schematic representation of these two types of filtration is given in Figure 1.

The main objective of this primarily theoretical study is to review and present mathematical statements of particle transport and capture in porous media. We will consider both types of porous media noted above. We should note that one possibility in modeling particle transport in porous media is to treat the filter bed as an assemblage of individual collectors instead of description by phenomenological equations. The former leads to expressions for isolated collectors which can be integrated to obtain expressions for the entire assemblage. For a detailed discussion of this approach, the reader is referred to Yao et al. (130) or Spielman (96). On the other hand, phenomenological methods to be reviewed in this chapter would yield partial differential equations either at the microscopic or macroscopic level depending on the size of the differential element of filter volume. Microscopic level equations can be transformed to macroscopic ones by volume averaging over the representative elementary volume. All governing equations developed in this study are at the macroscopic level. For a microscopic level treatment of filtration equation, the reader is referred either to the paper by Willis (126) or to Bear and Bachmat (8) in the previous volume.

2. PARTICLE TRANSPORT THROUGH A FIXED BED (DEEP BED FILTRATION)

Deep bed filtration has been studied by various researchers. Ives (48,50), Tien and Payatakes (103), Tien (102), Rajagopalan and Tien (78), O'Melia (75), Adamczy et al. (1), McDowell-Boyer et al. (69), Sakthivadivel and Irmay (85), Spielman (96), Irmay (45) and Herzig et al. (41) provide detailed reviews of various aspects of deep bed filtration studies that include theoretical considerations.

As noted earlier, when a liquid (filtrate) carrying suspended particles flows through an established porous medium, the particles are transported to the surface of the filter grains (collectors). Particles are captured on collectors by mechanisms caused by the action of fluid-mechanical forces along with other forces acting between the particles and collectors (96). Therefore, we will first review these mechanisms.

2.1. Governing Mechanisms

To describe the dynamic behavior of deep bed filtration at the macroscopic level, we make use of the conservation of mass equation for suspended solid particles in a liquid flowing through a porous medium, the conservation of mass equation for captured particles on solid grains composing the filter bed, and particle capture relationships for the deposition of suspended particles onto the grain surfaces.

2.1.1. Transport mechanisms

The conservation of mass equation for *suspended solid particles* in a single phase fluid flowing through a saturated *fixed bed* can be expressed as

$$\frac{\partial(nC)}{\partial t} + R_a = -\nabla.[C\vec{q} - nD\nabla C - nD^*\nabla C] + S \qquad (2.1)$$

where n is the porosity, C is the mass of suspended particles per unit volume of liquid (*filtrate*), R_a is the rate of deposition of particles on grains by various particle capture mechanisms, $\vec{q}$ is the specific discharge vector, D is the coefficient of convective dispersion, and D^* is the coefficient of molecular diffusion. Both D and D^* are second rank tensors. Eq. (2.1) is known in the literature as the equation of hydrodynamic dispersion. The term $C\vec{q}$ in Eq. (2.1) represents the convective transport. The dispersive flux represented by $nD\nabla C$ exists only at the macroscopic level, and is obtained by volume averaging of microscopic level equations. We assume that the dispersive flux is expressed as a Fickian type law as given in Eq. (2.1). The term S denotes the growth or decay of

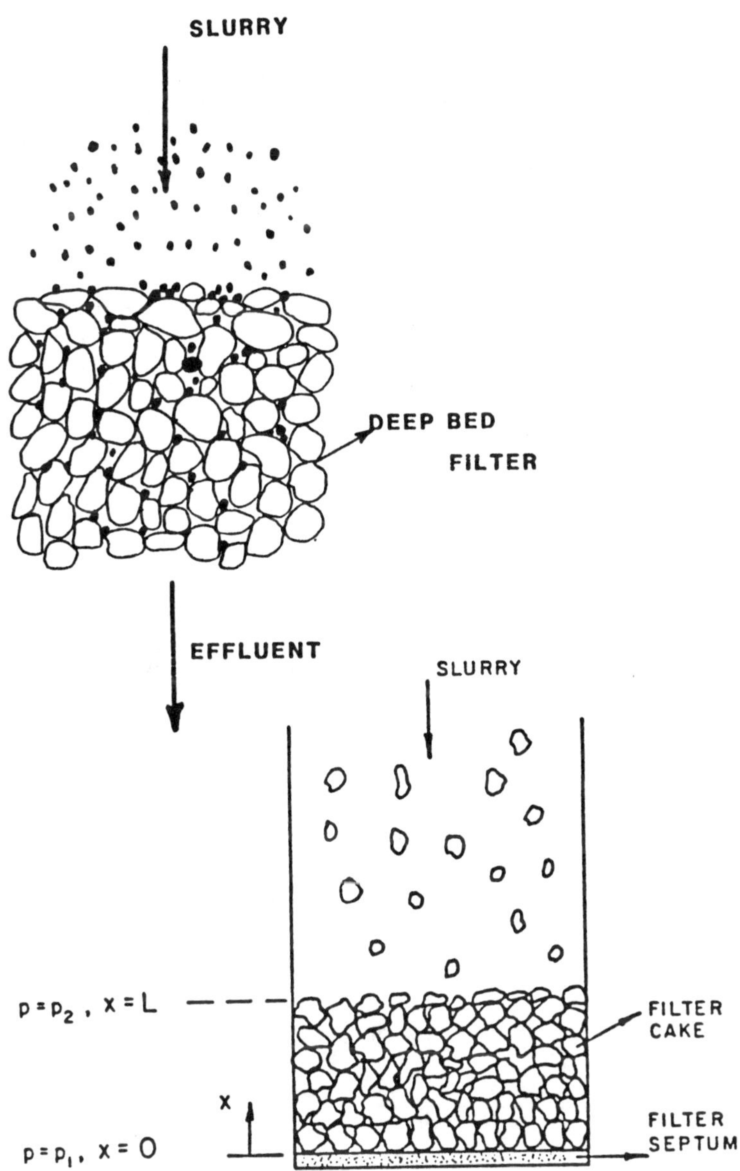

Figure 1. Schematic representation of particle capture. (a) in deep bed filter, and (b) in filter cakes.

suspended particles. One possibility is the decay of radioactive colloids or the death or growth of microbial particles.

The most common types of fixed bed filters used in engineering practice are rapid granular-medium filters, deep bed filters, and slow sand filters. The first two types of filters are essentially the same except for the depth of the filter bed and the size of the filtering media. The latter one has a depth of 1-3 meters. Slow sand filters have a shorter depth with a removable sand bed which is replaced when clogged. The filtration rate is in the range of 2-5 liters/m^2.min with a head loss from 0.05 m initially to 1.25 m when clogged (101). The range of filtration rate in all three types of filters justifies the assumption of *plug flow*. Furthermore, Herzig, et al. (41) has noted that particle diffusion is negligible when the particle size is larger than 1 micrometer (see Fig. 2). With these assumptions, Eq. (2.1) would reduce to

$$\frac{\partial (nC)}{\partial t} + R_a = - \nabla . C\vec{q} + S \tag{2.2}$$

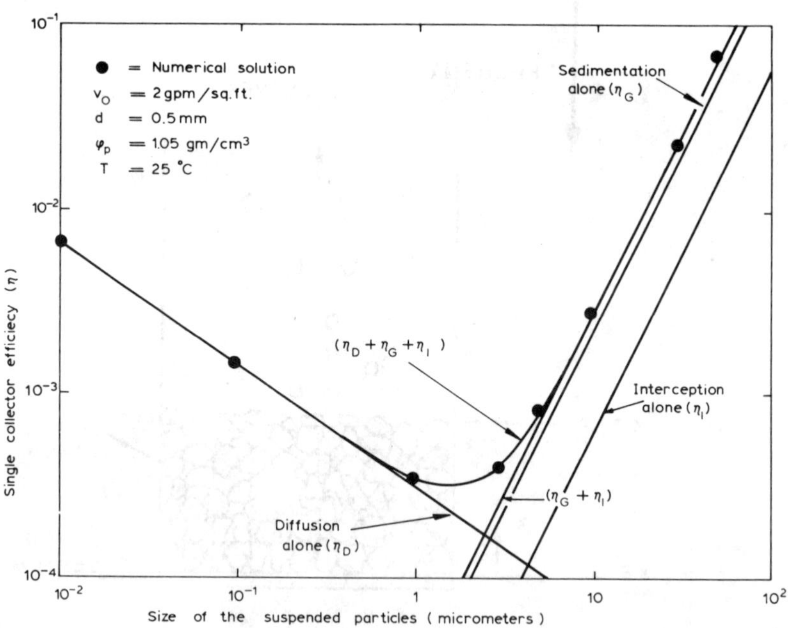

Figure 2. Comparison of various capture mechanisms [after Yao et al. (130)].

At this point, we can introduce the conservation of mass equation for *captured solid particles*. Assuming that the bed and deposited particles are completely rigid, and there is no diffusion in the solid phase, we can write

$$\frac{\partial(\rho_s \sigma)}{\partial t} - R_a = S_c \tag{2.3}$$

where ρ_s is the density of the particles and σ is the volume of particles per unit volume of filter bed. It is also referred as the "absolute specific deposit". S_c denotes the growth or decay rate of deposited particles. As a special case, if we assume constant particle density and neglect the growth or decay term, and combine Eqs. (2.2) and (2.3), we obtain

$$\frac{\partial(nC)}{\partial t} + \rho_s \frac{\partial \sigma}{\partial t} = - \nabla . C\vec{q} \tag{2.4}$$

This equation is the widely accepted filtration equation in the literature (e.g., Tien(102,103), Rajagopalan and Tien (78)). The first term at the left hand side of Eq. (2.4),which describes the rate of change of mass of suspended particles in a filtrate,can be neglected due to the fact that "in a flowing process [through a filter bed] the quantity of liquid contained within the bed is usually small compared with the volume of liquid passing through the bed" (101). In other words, moving particles are neglected in comparison to captured particles. Gruesbeck and Collins (37) and Deb (25) have included this term in their formulation to study migration of fine particles in petroleum reservoirs. The assumption stated earlier would be invalid in such an environment. As an alternative to total omission of this term, Tien (102,103) has introduced the concept of *corrected time variable*, θ' in an axial flow filter as

$$\theta' = t - \int_o^z \frac{n}{q_z} dz$$

where z is the vertical coordinate and q_z/n is the superficial velocity of filter bed in the z-direction. Herzig et al. (41) names θ' as "retention age". Then, (nz/q_z) is the time required for suspension to reach the bed depth z to replace the clear liquid initially filling the porous bed. Then in terms of variables z and θ', Eq. (2.4) becomes

$$q_z \frac{\partial C}{\partial z} + \rho_s \frac{\partial \sigma}{\partial \theta'} = 0 \tag{2.5}$$

Tien (103) has noted that "the difference between t and θ' is usually small [due to the length of filtration operations which takes several hours]. However, it may become important in the interpretation of data from small experimental filters". For the sak of obtaining a solution, as seen in Eq. (2.5), either C should be expressed in terms of σ or vice versa. Such a relation would actually quantify the particle capture mechanism. The conservation of mass equation given by either Eq. (2.1) or Eq. (2.5) with various assumptions stated earlier is totally independent of particle capture mechanisms. However, the relationship between C and σ is a function of physics of particle capture mechanisms and surface properties of collectors,including the chemistry of surfaces and the type of filtrate.

2.1.2. Capture mechanisms

As noted by Spielman (96), the capture mechanism of suspended particles on grain surfaces of a porous medium is governed by the combined effect of various forces of "fluid-mechanical origin", in addition to forces acting between the suspended particle and grain which acts as a collector. Various researchers adopted the collector approach and studied elementary mechanisms of particle capture by utilizing idealized geometrical models of collectors (i.e., spherical, cylindrical and constricted tube). Among them, Payatakes et al. (76), Spielman and Fitzpatrick (97), Yao et al. (130), Herzig et al. (41),and Ives (46) can be given as representative studies. For a review of particle capture mechanisms, the reader is referred to Ives (49,50), Spielman (96), McDowell-Boyer et al. (69), and O'Melia (75).

The mechanisms of particle capture can be listed as straining, sedimentation, interception, Brownian diffusion, inertial impaction and hydrodynamic action. Once a suspended particle is brought to the vicinity of a solid surface by one or a combination of these forces, London-van der Waals forces and electrical forces at interfaces which both act between the particle and collector contribute to the attachment of particles colliding with collectors. Furthermore, when the flow velocity is increased either locally by microscopic changes or throughout the medium by filter operation, the deposited particles can be detached from grain surfaces and returned to suspension in the flow. This process is known as detachment or reentrainment or decolmatage or scouring. At this point, we should note that the particle capture process is also known in the literature as colmatage or clogging. Following Corapcioglu and Haridas (21), we will review these elementary capture mechanisms briefly.

Straining in the contact zones of adjacent pores: Straining takes place when a particle in suspension flowing through a pore is larger than the pore opening, resulting in the accumulation of suspended particles on grains. Theoretically, a particle of any diameter may wedge in a void between two grains; this is no longer valid if the ratio of suspended particle diameter to grain diameter is small, since it can be assumed that the particle lies on a surface site due to some other mechanisms (e.g., surface forces) (41). Although this process is not important in many filtration problems, it has been reported to be one of several limitations for bacteria traveling through soils (Krone et al. (61), Krone (62) and Gerba et al. (33)). To estimate the significance of this effect, Herzig et al. (41) gave the following expression for the volume of deposited particles with uniform shape per unit volume of total porous medium based on purely geometric considerations:

$$\sigma = \frac{1}{2}\,(1-n_o)\pi Z(d/d_g)^2[(1+d/d_g)^2-1]^{\frac{1}{2}} \tag{2.6}$$

where n_o is the initial porosity; d and d_g are suspended particle and grain mean diameters, respectively; and Z is the coordination number which indicates the interconnectedness in the network of a porous medium. Herzig et al. (41) have shown that for $n_o = 0.40$ and $Z = 7.0$, the retention by this mechanism is important if $d/d_g \geq 0.05$. For bacteria with $d = 1$ µm and silt with a mean grain diameter, 0.01 mm, Eq. (2.6) would give $\sigma = 3.02\%$ which is not a negligible amount. For very small particles such as viruses, the limit could hardly be reached. For a polio virus with a mean diameter of 0.01 µm in the same soil, it would give $\sigma = 3.10^{-5}\%$, which is practically negligible. Therefore, for large particles, the effect of straining should be taken into consideration, but for colloidal particles, the effect of straining can be neglected.

Sedimentation in the pores: *Gravitational deposition* on grains can occur if the particles have a density different from that of the liquid. Due to their extremely small size, viruses and some bacteria are neutrally buoyant and therefore do not tend to settle. Then, any term in the conservation of mass equation characterizing the effects of gravitational settling can be neglected. But, Gerba et al. (33) reported that the sedimentation could be a mechanism of removal for some large bacteria. Yao et al. (130) noted that gravitational settling plays a significant part only in the capture of relatively large particles (> 5 micrometers) (Fig. 2); for these particles the removal efficiency is proportional to d^2. The gravitational velocity as expressed by Yao et al. (130) can be used as a criterion to measure the significance of sedimentation:

$$v_g = (1-\rho_w/\rho_s)(m_d g/3\pi\mu_w d) \tag{2.7}$$

where ρ_s, d and m_d are the density, diameter and mass of the particles, respectively; and ρ_w and μ_w are the density and viscosity of the water, respectively.

Interception: Even with exactly the same density as the fluid, some suspended particles, owing to their large size, would not be able to follow the smallest tortuosities of the fluid streamlines, and they will thus collide with the walls of the convergent areas of the pores.

Brownian Diffusion: Colloids, like bacteria and viruses, also partially rely on Brownian motion for their movement. Brownian motion is a random motion, caused by the thermal motion of molecules following their collision with other molecules or with colloids. The mass discharge of colloids, J_B, by Brownian motion is expressed by

$$\vec{J}_B = - D_B \, n\nabla C \tag{2.8}$$

where C is the concentration of the particles and D_B is the diffusio coefficient of the suspended particles, which could be estimated by the Stokes-Einstein equation:

$$D_B = k_b T/3\pi\mu_w d \tag{2.9}$$

where k_b is the Boltzman constant (energy per degree); T is the absolute temperature; μ_w is the fluid viscosity; and d is the diameter of the particles. Smaller particles are collected more efficiently due to their greater Brownian motion. Yao et al. (130) have shown that for suspended particles smaller than 1 μm, removal efficiency increases with decreasing particle size which is accomplished by Brownian diffusion. Many bacteria (7-0.2 μm) and viruses (0.5-0.01 μm) in soils are within this range. In deep bed filtration, the diffusion process is neglected when the particle size is larger than 1 μ. McDowell-Boyer et al. (69) calculates $D_B = 4.3\times10^{-9}$ cm²/sec for 1 μm diameter particles in water at 20°C. They conclude that although it is a small number,it can be significant within soil pores. Avogadro and deMarsily (6) have noted that due to Brownian diffusion colloidal particles may move with an average velocity that is faster than that of water in ground water. This is known as *hydrodynamic chromatography*.

Inertial Impaction: If the particles are massive enough, the inertia will force them to collide with the grain instead of following the flow streamline. This type of mechanism is determined by the Stokes number

$$S_t = m_d U/6\pi a_p \mu_w a \tag{2.10}$$

where U is the superficial velocity, a is the collector radius, and a_p is the particle radius. For liquid-borne particles, S_t is very small because μ_w is large (Spielman, (96)). Therefore, inertial impaction is important for gas-borne particles only.

Hydrodynamic Action: This type of capture mechanism is caused by a non-uniform drag force on particles due to varying shear field. Ives (49) has noted that this phenomenon has been observed for different Reynolds numbers in a flow field. As noted by O'Melia (75), in many studies hydrodynamic retardation was neglected based on the assumption of balance between the hydrodynamic drag increase and Van der Waals forces. But when interception and gravitational deposition are important, hydrodynamic retardation reduces the filter bed removal efficiency by 10% for suspended particles with a radius of 1 μm (75).

Surface forces, which include London-van der Waals and electrical forces, contribute to the attachment of particles to the collector surfaces. The van der Waals forces, also called secondary bonds or intermolecular forces, are the result of the mutual interaction of electrons and nuclei of molecules or atoms. Although Van der Waals forces are always attractive, the electrical forces can be either repulsive or attractive depending on the surface charges. Sharma et al. (88) has shown that there is an excellent correlation between surface charge alterations and bacteria transportability in sandpack columns.

When suspended particles accumulate on the grain surface, the straining effect increases with particle accumulation and eventually captured particles behave like a filter and remove finer particles. When the accumulations grow and become unstable, clusters break off which are transported by the flow and may be removed by straining and sedimentation. When the clusters break off, particles saturate the straining sites below, and the saturated front progresses (61). The rate of removal depends on the flow rate, size and density of the particle clusters. The deposition of the suspended particles by various mechanisms and sloughing off clusters (declogging) are usually simultaneous and caused by changes in the flow field.

In practice, the particle capture mechanism takes place with more than one particular mechanism dominating the filtration process. The superimposed equations of motion of the mechanisms listed above will yield equations predicting the particle trajectory at the microscopic level. However, such an approach is theoretically more rigorous; its usefulness is limited in solving a macroscopic equation such as Eq.(2.5). Instead, empirical equations can be used to express σ in terms of C or vice versa, and the parameters of such a relation can be determined by functional relations obtained by theoretical considerations. Such an approach is similar to using mathematical relationships to describe the sorption of reactive solutes by soil grains. A survey of mathematical sorption relationships can be found in Travis and Etnier (115); their review of equilibrium models of adsorption processes include the linear, Freundlich, Langmuir and others. First order kinetic sorption models include reversible linear, reversible nonlinear, kinetic product, and several others.

Equilibrium sorption isotherm developed by Langmuir to quantify the adsorption of gases by solids has very limited use in particle capture mechanism in porous media. Various studies such as Drewry and Eliassen (29), Filmer, et al. (31), and Burge and Enkiri (13) have shown that data on capture of viruses by soils were found to fit a Freundlich isotherm, and were not describable by a Langmuir isotherm. Despite this general consensus, Cookson (16) has utilized the Langmuir isotherm in his study for virus removal through packed beds. Saltelli et al. (86) utilized the following form of Langmuir isotherm to study the filtration of microcolloids such as anionic $Am(CO_3)_2^-$ or cationic $Am(HCO_3)_2^+$ in glauconitic sand columns.

$$\bar{C} = \frac{\alpha_1 C}{1+\alpha_2 C} \tag{2.11}$$

where $\bar{C}$ is the dimensionless concentration of captured microcolloids, α_2 is a measure of the bond strength holding the microcolloids on the grain surface, and the ratio (α_1/α_2) is the maximum amount of microcolloids that can be captured by the grains. The Langmuir isotherm given by Eq. (2.11) has been employed to explain the observed early adsorption of Americium colloids, in addition to a second order kinetic model to explain capture mechanism. In other words, Eq. (2.4) would yield to

$$\frac{\partial(nC)}{\partial t} + \rho_s \left(\frac{\partial\sigma}{\partial t} + \frac{\partial\bar{C}}{\partial t} \right) = - \nabla . C\vec{q} \tag{2.12}$$

It was assumed that the parameters of Eq. (2.11) are independent of the parameters of $\partial\sigma/\partial t$.

The use of equilibrium models like Langmuir isotherm requires that the two-phase system (liquid-solid) is at equilibrium after a sufficient time period so that the concentrations C and σ are constant. However, equilibrium is not easily obtained in filtration columns, or when non-colloidal suspended particles migrate in natural soils. In this case, it is much more appropriate to use a kinetic model to describe the particle capture mechanisms.

Sorption-desorption relationships for reactive solutes in soil are usually represented by first order reversible linear models which, in general, can be expressed as

$$\frac{\partial \sigma}{\partial t} = k_1 \frac{\theta}{\rho_s} C - k_2 \sigma \tag{2.13}$$

where k_1 and k_2 are sorption (clogging) and desorption (declogging) coefficients, θ is the volumetric water content (θ=n when the soil is saturated with water), and ρ_s is the particle density. In terms of particle capture mechanism, Eq. (2.13) states that the rate of particle capture is proportional "to the difference between what can be [captured] at some concentration and what has already been" [captured] (115). In sorption studies, usually k_1 and k_2 are constants. But, in filtration studies, not only the state of the collector surface, but also the concentration of suspended particles participate in the deposition reaction. Therefore, the rate of deposition is faster than the one expressed by a first order linear model. Then, in a most general way, we can write a higher order model as

$$\frac{\partial \sigma}{\partial t} = \psi(f_1, f_2, f_3, \ldots\ldots, f_n) \tag{2.14}$$

where f_1 .... f_n are the variables governing the capture mechanism, for example;

$$f_1 = C \tag{2.15}$$

$$f_2 = \sigma \tag{2.16}$$

$$f_3 = q_z \tag{2.17}$$

$$f_4 = n \tag{2.18}$$

$$f_5 = \rho_s \tag{2.19}$$

$$f_6 = \theta \tag{2.20}$$

By any means, this list is not comprehensive. Any other variable which governs the capture mechanism can be included, provided that a functional form of ψ can be obtained either experimentally or theoretically. Obviously, the dependence of ψ on C and σ is fundamental due to the change in particle concentration during filtration and change in grain and pore characteristics by deposition.

By rewriting Eq. (2.5), we obtain

$$-\frac{\rho_s}{q_z}\frac{\partial\sigma}{\partial\theta'} = \frac{\partial C}{\partial z} \tag{2.21}$$

In 1937, Iwasaki (52) for slow sand filters and later, Ives (46) for rapid sand filters have assumed that the amount of captured particles within the filter at a depth of z is proportional to particle concentration. This may be written as

$$\frac{\partial C}{\partial z} = -\lambda C \tag{2.22}$$

The proportionality parameter λ is known as the filter coefficient in the literature. It is usually assumed that λ is a function of σ and independent of C due to change in pore geometry. Tien (103) notes that the change of C with z is initially linear, and as time increases, a non-linear behavior is observed. Iwasaki (52) assumes that

$$\lambda = \lambda_o + A\sigma \tag{2.23}$$

where A is a constant and λ_o is the initial clean filter coefficient. λ_o is a function of local velocity, grain size, and density of particles. An empirical clean filter expression assuming additive particle capture mechanisms by Brownian motion, interception, and gravitational sedimentation is available in the literature (69).

$$\lambda_o = \frac{3}{2}\frac{(1-n)}{d_g}\left[4\Lambda^{1/3}\left(\frac{Ud_g}{D_B}\right)^{-2/3} + 0.56\Lambda\left(\frac{A^*}{\mu d^2 U}\right)^{1/8}\left(\frac{d}{d_g}\right)^{15/8} + 2.4\times10^{-3}\Lambda\left(\frac{V_s}{U}\right)^{1.2}\left(\frac{d}{d_g}\right)^{-0.4}\right] \tag{2.24}$$

where A^* is the Hamaker's constant, V_s is the Stokes settling velocity, and

$$\Lambda = \frac{1-(1-n)^{5/3}}{1-\frac{3}{2}(1-n)^{1/3}+\frac{3}{2}(1-n)^{5/3}-(1-n)^2}$$

A general nonlinear relation could be written as (103)

$$\lambda = \lambda_o \phi(\sigma, Q_i) \tag{2.25}$$

A general form of ϕ has been suggested by Ives (40) as

$$\phi = (1+B\beta^* \frac{\sigma}{n_o})^{\beta_1} (1-\beta^* \frac{\sigma}{n_o})^{\beta_2} (1- \frac{\sigma}{\sigma_{max}})^{\beta_3} \tag{2.26}$$

where β_1, β_2, β_3 are constants, n_o is the initial filter porosity, β^* is the bulking factor, and B is the packing constant. Ives (50) stated that "The first term is based on the changes in geometry of a spherical grain due to accumulating deposit, and accounts for the initial rise in filter efficiency [ripening state] which is observed in practice [due to larger surface area, see also the initial increase of λ with σ in Fig. 3]. The second term is based on the internal coating of a cylindrical capillary, reducing the surface area. The third term is derived when the interstitial velocity reaching [increasing] a critical value at σ_{max} when no further deposition takes place." [rate of detachment equals that of adherence]. Existence of critical velocity above which no particle deposition occurs has been experimentally verified by Maroudas and Eisenklam (64,65). The bulking factor β^* is defined as

$$\beta^* = 1/(1-n_d) \tag{2.27}$$

where n_d is the porosity of deposited particles. Then, the porosity of the filter bed would be

$$n = n_o - \beta^* \sigma \tag{2.28}$$

as the pores become clogged with deposited particles. Herzig et al. (41) suggest that since particle concentratiions in deep filtration are quite low, n in Eq. (2.4) can be replaced by n_o.

When σ reaches σ_{max}, the filter coefficient reduces to zero, then complete shutoff of the filter is reached. Usually σ_{max} varies from $0.2n_o$ to $0.4n_o$ (72). Various functional forms of λ reported in the literature are listed in Table 1. A graphical comparison of these expressions are given in Figure 3. Wright(129) notes that as seen in Fig. 3, even though the filter coefficient increases initially, it will decrease after some time due to particle deposition.

Sakthivadivel (84) has proposed an expression for λ, considering also the dependence on C

$$\lambda = \lambda_o[1 + a^*(\sigma+nC/\rho_s)] \tag{2.29}$$

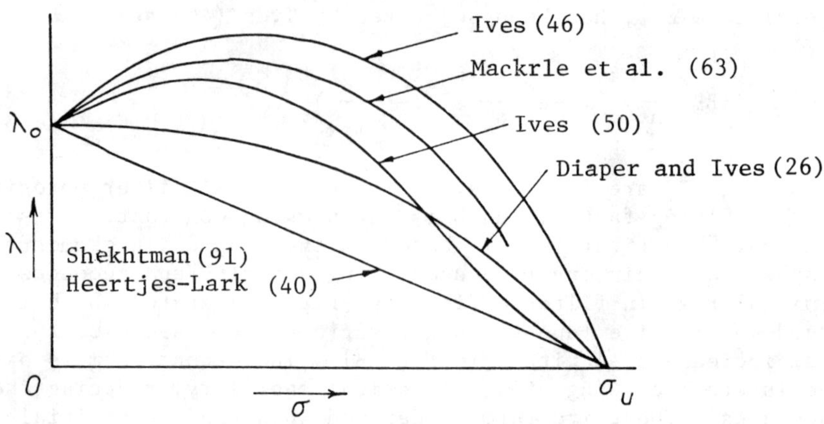

Figure 3. Variation of the filter coefficient, λ according to various researchers [after Ives (48)].

Table 1. Functional Forms of λ as Compiled from Ives (48)

Researcher	Expression for λ
Ives(46)	$\lambda_o + a\beta^*\sigma - b\beta^{*2}\sigma^2(n_o-\beta^*\sigma)$
Mackrle et al. (63)	$\lambda_o[1 + a'\beta^*\sigma/n_o]^{\alpha_1}[1 - \beta^*\sigma/n_o]^{\alpha_2}$
Diaper and Ives (26)	$\lambda_o - a''\beta^*\sigma^2$
Maroudas and Eisenklam (64)	$\lambda_o[1 - \sigma/\sigma_{max}]$
Shekhtam (91) Heertjes & Lerk (40)	$\lambda_o[1 - \beta^*\sigma/n_o]$

where a^* is a constant. Eq. (2.29) is erroneous since only captured particles can affect the filter coefficient due to the change in pore geometry (41). Suspended particles would not have any substantial effect on this phenomenon.

As noted earlier, if the declogging (colmatage) and clogging mechanisms are simultaneous, the rate equation proposed by Mints (70) can be utilized

$$\frac{\partial \sigma}{\partial t} = q_z \lambda_o \, C/\rho_s - \alpha^* \sigma \tag{2.30}$$

where α^* ($=k_2$ in Eq. (2.13)) is the scour coefficient. The first term is the particle capture rate, whereas the second one represents the declogging rate. When the limit condition is reached, i.e., $\sigma = \sigma_{max}$, $\partial\sigma/\partial t = 0$ and $C = C_o$ is the inlet concentration at the filter bed surface. Then

$$\alpha^* = \frac{q_z \lambda_o C_o}{\rho_s \sigma_{max}}$$

This implies that α^* is constant. However, Ives (40) notes that the experimental evidence is against such an implication.

Although head loss has an effect on the rate of deposition, with the exception of Adin and Rebhun (3), it has been generally neglected in filter coefficient expressions. They proposed

$$\frac{\partial \sigma}{\partial t} = k_1' \, q_z (\sigma_{max} - \sigma) C/\rho_s - \alpha^* \sigma \, \frac{\partial H}{\partial z} \tag{2.31}$$

where k_1' is a deposition coefficient, $\partial H/\partial z$ is the hydraulic gradient along the filter bed and approximately a linear function of σ(50). Adin and Rebhun note that hydrodynamic shear forces acting on captured particles are represented by the hydraulic gradient.

Wnek et al. (127) present a model which contains no empirical factors to be determined from filter runs. The effect of forces between the collector and the particle due to electrical double layer and Van der Waals interactions was introduced to the formulation by treating the collector surface with first order reaction kinetics. The rate constant is taken as a function of the stability ratio of colloid chemistry.

Gruesbeck and Collins (37) introduced a conceptual partitioning of the pore space into two classes which they call plugging and non-plugging pathways. For the volume of deposited particles in the non-pluggable pathways, σ_{np}

$$\frac{\partial \sigma_{np}}{\partial t} = - \alpha_1^* (U_{np} - U_c) \sigma_{np} + \beta_1^* C \tag{2.32}$$

where U_{np} and U_c are the volumetric flux density of fluid flowing through non-pluggable pathways and the critical volumetric flux respectively. α_1^* and β_1^* are dimensional constants. The first term on the right hand side is zero for $U_{np} < U_c$. In the pluggable pathways they assumed that the volume of plug deposits, σ_p increase as the rate of deposit increases.

$$\frac{\partial \sigma_p}{\partial t} = (\gamma_1^* + \delta_1^* \sigma_p) U_p C \tag{2.33}$$

where U_p is the volumetric flux density of fluid flowing through the pluggable pathways. γ_1^* and δ_1^* are dimensional constants. We should note that

$$n_o \sigma = n_o [f^* \sigma_p + (1-f) \sigma_{np}] \tag{2.34}$$

where f^* is the dimensionless fraction of pore space containing pluggable pathways.

Maroudas (67) has shown that based on data obtained earlier (66), two different modes of deposition lead to entirely different forms of λ. The blocking mode of deposition,which is assumed to be applicable to granular beds and to suspensions of particles under comparable conditions of shape and size range,results in the blocking of flow paths; the ratio between the porosity, n, and the surface area available for deposition per unit volume of bed, s remains constant. Then,

$$\frac{\partial C}{\partial z} = - (K_1 \frac{s}{n}) \frac{1}{q_z} C \tag{2.35}$$

where K_1 is the volume fraction of particles depositing in unit time per unit area. The quantity within the paranthesis is constant. However, if the deposition results in the gradual constriction of flowpaths rather than in blocking, the product of the porosity and velocity remains constant during the run,while the surface area available for deposition may vary, Then,

$$\frac{\partial C}{\partial z} = - (\frac{K_1}{n q_z}) s C \tag{2.36}$$

The connection between the macroscopic equations considered in this chapter and the microscopic properties of individual filter bed elements is introduced by Rajagopalan and Tien (78). They assumed that, since the physical dimension of each unit bed element, ℓ^*, is always small, λ is constant over this distance. Hence, they obtained

$$\lambda = - [3(1-n)/2d_c] \ln(1-\eta) \tag{2.37}$$

where d_c is the collector diameter and η is the filter efficiency, which is the ratio of amount of particles captured to the amount in the feed. η is usually in the order of 10^{-3}.

Various researchers have used statistical techniques to model particle transport in porous media. Donaldson et al. (28) developed a *random walk model* using Poiseuille's capillary flow equation and the actual pore size distribution to calculate the pressure drop across the core. Particles are selected using a random number generator and the actual particle size distribution. Travis and Nuttall (114) and Nuttall (73) have solved the *population balance equation* with log-normal population density distribution for the mass concentration of colloids.

Khilar et al. (59) developed a capillary model to predict *piping* and *plugging* of clay particles. Their model included mass balance equations for eroding particles in water and in solid phase. Khilar et al. considered convective transport, rate of erosion, rate of capture, and rate of change of mass of suspended clay particles. The rate of erosion term was assumed to be proportional to the difference of flow rate and shear stress at a pore wall. The capture term is proportional to suspended particle concentration through a coefficient, which is in turn proportional to the flow rate.

2.1.3. Solutions of filtration equations

There are various attempts to simultaneously solve the conservation of mass equation for suspended particles with the rate equation. In some cases, a close form analytical solution is possible for simple rate equations. For others, numerical techniques can be applied.

The simplest solution can be obtained by integrating Iwasaki's (52) equation

$$\frac{\partial C}{\partial z} = - \lambda C \tag{2.22}$$

The solution for a constant $\lambda = \lambda_o$ and $C(z=0)=C_o$ would be

$$C = C_o \exp(-\lambda_o z) \tag{2.38}$$

Shekhtman's (91) and Heertjes and Lerk's (40) model

$$\frac{\partial C}{\partial z} = \lambda_o (1-\beta^* \sigma/n_o) C \tag{2.39}$$

can be simultaneously solved (48,50) with

$$\frac{\partial C}{\partial z} + \frac{\rho_s}{q_z} \frac{\partial \sigma}{\partial t} = 0 \tag{2.40}$$

Eq. (2.40) is identical to Eq. (2.5) except the fact that θ' is replaced by t

$$\frac{C}{C_o} = \frac{\exp(\lambda_o \beta^* q_z C_o t/(n_o \rho_s))}{\exp(\lambda_o z) + \exp(\lambda_o \beta^* C_o q_z t/n_o \rho_s) - 1} \tag{2.41}$$

$$\frac{\sigma}{n_o} = \frac{\exp(\lambda_o \beta^* q_z C_o t/(n_o \rho_s))}{\exp(\lambda_o z) + \exp(\lambda_o \beta^* q_z C_o t/(n_o \rho_s)) - 1} \tag{2.42}$$

Diaper and Ives (26) obtained various analytical solutions which are reviewed in Ives (50) for different conditions. Furthermore, as shown by Ives (48), the Mint's Equation [Eq. (2.30)] yields to

$$\frac{\partial^2 C}{\partial t \partial z} + \lambda_o \frac{\partial C}{\partial t} + \alpha^* \frac{\partial C}{\partial z} = 0 \tag{2.43}$$

by differentiating Eq. (2.30) and substituting Eq. (2.40). The exact solution of Eq. (2.43) is given by Ives (48) as

$$\frac{C}{C_o} = \exp -(\lambda_o z + \alpha^* t) \sum_{i=o}^{\infty} \left(\frac{\alpha^* t}{\lambda_o z}\right)^{\frac{1}{2}} I_i [(\lambda_o z \alpha^* t)^{\frac{1}{2}}] \tag{2.44}$$

where I_i is the modified Bessel function of the first kind of order i. Ives (48) has shown that Mint's (70) solution is an approximation to Eq. 2.44. Hall (38) assumes that some particles deposited on grains creep slowly over the grain surfaces at a rate q' under the combined effect of pressure gradient and shear stresses. This causes the progress of saturated front deeper into the filter bed. Hall's approach is an alternative representation of declogging mechanism in the filtration equation instead of rate equation (Eq. 2.30). Then, Hall proposed the following equation

$$\rho_s \frac{\partial \sigma}{\partial t} + q_z \frac{\partial C}{\partial z} + \rho_s q' \frac{\partial s}{\partial t} = 0 \qquad (2.45)$$

where the last term denotes the creep flow.

Sakthivadivel and Irmay (85) provide a detailed review of Shekhtman's (91) development of a governing equation for σ

$$\frac{\partial^2 \sigma}{\partial z \partial t} + \frac{n_o}{q_z} \frac{\partial^2 \sigma}{\partial t^2} + \frac{\beta^*}{n_o} \frac{\partial \sigma}{\partial t} \frac{\partial \sigma}{\partial x} + \frac{A^*}{\beta^* (q_z/n_o)^2} \frac{\partial \sigma}{\partial t} = 0 \qquad (2.46)$$

where A^* is a dimensional constant. Eq. (2.46) is a non-linear hyperbolic one, and solved by Shekhtman by the method of characteristics.

Irmay (45) notes that Sakthivadivel (84) also obtains a hyperbolic nonlinear equation for σ

$$\lambda \frac{\partial^2 \sigma}{\partial z \partial t} + \lambda^2 \frac{\partial \sigma}{\partial t} = a \frac{\partial \sigma}{\partial t} \frac{\partial \sigma}{\partial x} \qquad (2.47)$$

where a is a dimensional constant. Sakthivadivel (84) obtained a numerical solution agreeing fairly well with experimental data.

Herzig et al. (41) obtained a rate equation based on probabilistic analysis of the filtration mechanism

$$\frac{\partial \sigma}{\partial t} = K_1^* \; n \; C/\rho_s - K_2^* \sigma \qquad (2.48)$$

where K_1^* and K_2^* are dimensional probabilities of clogging and declogging respectively.

$$K_1^* = \frac{1}{n \; C/\rho_s} \left(\frac{\partial \sigma}{\partial t} \right) \quad ; \quad K_2^* \; - \frac{1}{\sigma} \left(\frac{\partial \sigma}{\partial t} \right) \qquad (2.49)$$

Herzig et al. shows the existence of a clogging front moving at a velocity of $q_z C/(\rho_s \sigma)$

Since the governing equations of filtration processes are hyperbolic, the method of characteristics has been employed to solve them by various researchers. Ring (81) has employed a simple numerical technique to solve resulting first order equations. Among others, Hsieh et al. (43) and Adin (2) can be noted. A general trend of resulting solutions can be represented graphically as shown by Ives (50) in Figure 4.

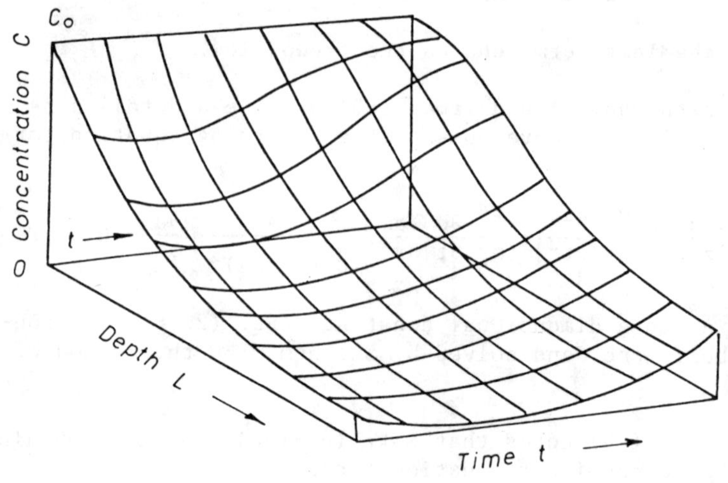

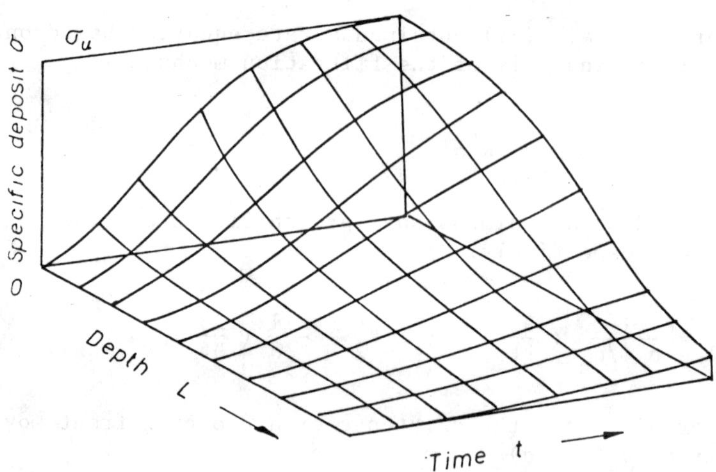

Figure 4. Variation of suspended particle concentration, and amount of deposited particles with depth and time [After Ives (50)]

The solution of filtration equation critically depends on the type of rate equation to be used. Wright (129) shows that in general rate equations can be represented by

$$\frac{\partial \sigma}{\partial t} = m_1 C + m_2 \sigma C + m_3 \sigma + m_4 f(\sigma) C \tag{2.50}$$

Wright notes four possible cases in use (i) $m_1 > 0$, $m_2 = 0$, $m_3 < 0$, $m_4 = 0$, (ii) $m_1 > 0$, $m_2 < 0$, $m_3 < 0$, (iii) $m_1 \gtrless 0$, $m_2 \lessgtr 0$, $m_3 \gtreqless 0$, and (iv) $m_1 > 0$, $m_2 > 0$, $m_3 = 0$, $m_4 < 0$. The first three correspond to reversible rate equations (Eqs. (2.30), (2.31), (2.48)), and the last one represents an irreversible reaction (e.g., Eq. (2.35) or (64),(91), and (40)).

2.1.4. Pressure drop due to deposition and permeability reduction

Due to clogging, the permeability of the filter bed reduces, and pressure drop Δp builds up, reducing the efficiency of the filter. Various empirical formulas are reported in the literature to calculate the pressure drop as a function of deposited particles. They all alter the porosity terms in the permeability expression of the Kozeny-Carman equation. Ives and Pienvichitr (51) present a theoretical development and proposes the expression

$$\frac{\Delta p}{\Delta p_o} = (1-\eta_1 t)^{\eta_2} \tag{2.51}$$

where Δp_o is the initial pressure drop along the filter bed, and η_1 and η_2 are constants; η_2 can take either sign. A review of various expressions are given by Ives and Pienvichitr (51).

The permeability reduction due to clogging by captured particles is the subject matter of numerous studies. In petroleum engineering, bacteria and/or colloidal particles commonly found in water injected into oil-bearing formations to increase recovery might cause plugging of the adjacent formation (7,79). Sharma et al. (88) used a statistical approach and general population balance equations to model the entrapment of fines at pore throats. The equations are solved for open pore densities and size distributions and based on these solutions, the permeability has been calculated. Swartzendruber and Uebler (99) have obtained an equation for hydraulic conductivity as an exponential function of the volume of suspension inflow, with a clogging coefficient based on the data obtained by Uebler and Swartzendruber (116). Since Uebler and Swartzendruber's data are obtained with constant difference in hydraulic head, the expression derived is valid under this condition. Their expression is given by

$$\bar{K} = K_s \exp[-CV/A(H+L)] \tag{2.52}$$

where K is the hydraulic conductivity, A is the cross sectional area, V is the cumulative volume of flowing suspension, H is the depth of water on the top of the sand column, L is the length of column, and C is the clogging coefficient. When V=0, $K=K_s$.

Gruesbeck and Collins (37) assumed different approximate forms for permeability in the pluggable (k_p) and non-pluggable (k_{np}) pathways, which are expressed by Eqs. (2.32) and (2.33)

$$k_p = k_{pi} \exp(-a_1^* \sigma_p^4) \tag{2.53}$$

$$k_{np} = k_{npi}/(1+b_1^* \sigma_{np}) \tag{2.54}$$

where k_{pi}, k_{npi}, a_1^* and b_1^* are phenomenological constants to be specified.

Wright (129) has suggested a cubic expression to estimate the hydraulic conductivity in terms of specific deposit σ and theoretical filter capacity, F, denoting the amount of retained material per unit volume of filter bed that could clog the pores completely

$$K=K_o(1 - \sqrt{\sigma/F}\ )^3 \tag{2.55}$$

where K_o is the initial value of hydraulic conductivity.

2.2. Transport of Bacterial and Viral Particles

Microbial particles such as bacteria and viruses enter soil and groundwater through various ways, such as by land application of wastewater or through the septic system. Rain infiltrating through sanitary landfills and artificial recharge of groundwater aquifers by treated sewage water are additional sources. While natural processes can, in some cases, help to reduce the pollution, some biological contaminants can travel considerably through the earth.

A literature survey presented by Corapcioglu and Haridas (21) shows evidence that microbial contamination of groundwater does occur when human wastes enter into the soil, and that microbes under proper conditions can travel long distances in groundwater. Romero (82) reviewed various case studies of microbial groundwater pollution until 1970. Butler et al. (14), Hagedorn (39), Vaughn et al. (117), and Smith et al. (95) studied the underground movement of bacteria and viruses in soil columns and in pilot scale field studies. Gerba (34) and Keswick and Gerba (57) investigated the factors affecting the migration and survival of viruses in groundwater. They conclude that bacteria may travel little more than

5 feet in moist or dry fine soil, but will travel much further through such means as root channels and rodent holes. Zyman (131) studied the migration of organisms in sludge-soil mixture columns, and reported that heavy rainfall rates appear to promote significant vertical migration of viable indicator organisms to the bottom of the 20.3 cm columns. Zyman also noted an increase in die-off rates due to an increase in desiccation.

The transport of bacteria through porous geological materials has received attention from various researchers due to its significance in *microbial enhanced oil recovery* (53,54,90). This tertiary recovery process is achieved through injection with nutrient,followed by a period of static incubation during which cells multiply and migrate. Jang et al. (54) has found that bacteria can migrate 1 ft/day through the sandpack columns saturated with nutrient broth. Jang et al. (53) have shown that the presence of oil in the sandstone core can facilitate bacterial penetration. Furthermore, they have shown that certain types of bacteria (e.g., B.subtilus) due to the phenomenon called chemotaxis, can migrate in nutrient saturated Berea sandstone cores without applying any pressure gradient.Laboratory experiments have shown that the adsorption of bacteria becomes an important factor in bacterial transport, provided that the rock has a high permeability and the inflow bacterial concentration is low. Otherwise, a filter cake develops at the inlet surface with a decrease in effluent bacterial concentration. Sharma et al. (88) found out that the use of polyanionic species alters the surface charges on sand grains as well as bacteria. This charge alteration causes facilitated transport of organisms through a porous medium.

Although the problem has great practical importance, the mathematical statement of the phenomenon has been attempted only in a few recently published studies. The first conceptual model for bacterial movement in soils was presented by Matthess and Pekdeger (68). Later, Sykes et al. (100) presented a model to predict the concentrations of leachate organics, measured as chemical oxygen demand in groundwaters below sanitary landfills. Simultaneous substrate utilization and microbial mass production equations, with convection and dispersion included for the former, are used for modeling biodegradation. The only processes considered for the latter are microbial growth and decay. None of these modeling studies presents a complete picture of the phenomenon,that is, complete coupling of microbial and substrate mass conservation equations, transient conditions, convective transport of microbial population, etc. The migration of viruses in soils and groundwater has been studied by Vilker(119) and Grosser(36), by using the conventional solute transport equation with a retardation factor due to viral adsorption. Although Filmer et al. (31) have used an equation similar to Eq. (2.5) to simulate viral transport in soils, Filmer and Corey (30) have utilized a diffusion equation. Corapcioglu and Haridas (22) recently presented a coupled mathematical model for the transport and fate of bacteria and viruses in soils and groundwater in the presence of a substrate. The model is designed to predict how long a given population of microorganisms can live in soils and how far they can spread while they are alive.

2.2.1. Biomass transport

The details of the following mathematical model can be found in Corapcioglu and Haridas (22). Here we will repeat the basic steps of the derivation.

We start with the macroscopic mass conservation equation for suspended microbial particles in porous media,

$$R_a + \frac{\partial(\theta C)}{\partial t} = -\nabla.\vec{J} + R_{d_f} + R_{g_f} \tag{2.56}$$

where C is the concentration of suspended particles (bacteria or viruses), R_a is the rate of deposition of particles on grains, R_{d_f} and R_{g_f} are the decay and growth terms of the suspended particles, respectively, and θ denotes the volume occupied by the flowing suspension per unit total volume. Some of the removal mechanisms of bacteria and the transport processes are summed up in the term denoted by J, which is the specific mass discharge of suspended particles.

2.2.2. Microbial capture

The capture of suspended microbial particles from water passing through soil are dominated by mechanisms discussed in section (2.1.2). The important ones for the capture of bacterial particles are straining and sedimentation. Due to the very small size of viruses and microcolloids such as Americium particles, adsorption is the major removal mechanism.

The accumulation of bacteria on grain surfaces forms clusters called dendrides. The straining effects increase with dendridic growth, resulting in further growth, until the clusters become unstably large and break off. The rate of removal depends on the flow rate and the size and density of the bacterial clusters (61). If the deposition (clogging) of the bacteria by various mechanisms (straining and sedimentation) and sloughing off of clusters (declogging) are simultaneous, the conservation equation for the deposited material may be written as

$$\frac{\partial \rho\sigma}{\partial t} = R_a + R_{d_s} + R_{g_s} \tag{2.57}$$

where R_{g_s} and R_{d_s} are the growth and decay terms respectively in the deposited state, ρ is the density of bacteria, and σ is the volume of deposited bacteria per unit volume of bulk soil. The term R_a can be expressed by a kinetic equation

$$R_a = k_c(n-\sigma)C - k_y\rho\sigma^h \tag{2.58}$$

where k_c and k_y are the clogging and declogging rate constants respectively, and h is a constant. Eqs. (2.57) and (2.58) are similar to Eq. (2.30) proposed by Mints (70).

In the case of viruses, since adsorption is the major removal mechanism, Eq. (2.57) should be replaced by

$$\frac{\partial \rho \bar{C}}{\partial t} = R_a + R_{d_s} \tag{2.59}$$

where $\bar{C}$ is the mass of adsorbed phase per unit mass of the solid part of the porous medium and is related to C by an equilibrium isotherm. Note that $R_{g_s} = 0$ for viruses, since viruses reproduce only inside an appropriate host cell.

Adsorption of viruses relies heavily on various factors (33): (a) the physical and chemical nature of viruses and (b) the pH of the solution, (c) the characteristics of the flow, and (d) the degree of saturation. Soil type, ionic strength of soil solution, amount of organic matter and humic substances are all considered in the first category. High salt content in groundwater would increase the adsorption due to double layer compression. Also, it is usually agreed that fine-textured soils like clay retain more viruses (11, 12,29,33) and bacteria (39) than do sandy soils. Increasing adsorption occurs with the reduction of pH below 8.0 and with the addition of cations, especially the divalent species (11). It was also concluded that retention mechanisms by soil were due to an adsorption mechanism which increases with decreasing soil moisture. Bitton et al. (12) critically examined the various methods frequently used to assess soils' potential to retain viruses.

As noted earlier, various studies (29,31,13) have shown that adsorption data of viruses to soils were found to fit a Freundlich isotherm, and were not describable by a Langmuir isotherm.

In summary, due to surface and electrokinetic forces, R_a is a rate-controlled reaction for bacteria, and an equilibrium controlled reaction for viruses.

2.2.3. Chemotaxis and tumbling of bacteria

The land application by primary treatment effluent or the seepage of raw sewage water from septic tanks can provide enough substrate concentration to support the microbial activity in soil and groundwater. In solutions like wastewater existing substrate concentration gradients stimulate response from microbes. Some microbes move systematically toward a richer food supply, and this motion, induced by the presence of a solute gradient, is termed chemotaxis.

The chaotic, random movement of motile bacteria which was referred to as "tumbling" (56), gives rise to an effective diffusivity or motility coefficient D_T. The random movement may be assumed to be superimposed upon any systematic migration induced by substrate, so the two effects (random and systematic) may be considered to be additive. Note that although this random motion is a sign of vitality, the Brownian motion is exhibited by any particle. Keller and Segel note the additive property of Brownian and chemotactic particle migration by saying that "the chemotactic response of unicellular microscopic organisms is viewed as analogous to Brownian motion. Local assessments of chemical concentrations made by individual cells give rise to fluctuations in path. When averaged over many cells, on a long time interval, a macroscopic flux is derived which is proportional to the chemical gradient." The total flux $\vec{J}_{CT}$, due to chemotactic movement and tumbling, can be expressed by the following equation (24)

$$\vec{J}_{CT} = \theta(C\, k_m \nabla \ln C_F - D_T \cdot \nabla C) \tag{2.60}$$

where k_m is the migration rate constant, C_F is the substrate concentration, and D_T is the motility coefficient. Chemotaxis is reported in nutrient saturated sandstone columns by Jang et al. (53,54). Since the anatomy of viruses is very different than that of bacteria, chemotaxis is irrelevant for viruses.

2.2.4. Decay and growth of microbial particles

Gerba et al. (33) conclude that in most cases, 2 to 3 months is sufficient for reduction of pathogenic bacteria to negligible numbers once they have been applied to the soil. The decay mechanism of viruses is similar to that of bacteria, but certain types which are more resistant to environmental changes might survive longer (1 to 6 months) than their bacterial counterpart. Gerba et al. (33) also conclude that the survival of the enteroviruses in soil is dependent on the nature of the soil, temperature, pH, and moisture.

The death of microorganisms is expressed as an irreversible first order reaction, for bacteria

$$R_{d_f} = - k_d \, \theta C \quad ; \quad R_{d_s} = - k_d \, \rho\sigma \tag{2.61}$$

and similarly for viruses

$$R_d = - k_d(\theta C + \rho\bar{C}) \tag{2.62}$$

where k_d is the specific decay rate, R_{d_f} is the decay term in free state in water, and R_{d_s} is the decay rate in adsorbed state. We assume that k_d is the same in free and adsorbed states.

Bacterial growth occurs with the utilization of the substrate. The growth of bacteria is assumed to follow the Monod equation (71). This equation describes a relationship between the concentration of a limiting nutrient and the growth rate of microorganisms. As stated earlier, nutrients needed for proper biological growth can be present in a sewage water. Bacterial growth in a subsurface environment is slow,and the Monod equation may be safely used. Similar to the decay process, we assume that bacteria can grow at the same rate in the deposited state as well as in the suspension. A generalized Monod equation can then be written as

$$R_{g_f} = \mu\theta C \qquad R_{g_s} = \mu\rho\sigma \tag{2.63}$$

where μ is the specific growth rate and R_{g_f} and R_{g_s} denote the growth terms in free and adsorbed states respectively. The functional relationship between μ and an essential nutrient's concentration C_F was proposed by Monod (71) as

$$\mu = \frac{\mu_m C_F}{K_s + C_F} \tag{2.64}$$

Here μ_m is the maximum growth rate achievable when $C_F >> K_s$ and the concentration of all other essential nutrients is unchanged. K_s is that value of the concentration of the substrate where the specific growth rate has half its maximum value; roughly speaking, it is the division between the lower concentration range where μ is linearly dependent on C_F, and the higher range, where μ becomes independent of C_F.

2.2.5. Substrate transport

The substrate, C_F, which is consumed by the microbes at a rate R_F, is assumed to be transported by various mechanisms. Thus, the mass conservation equation for the concentration of dissolved substrate, C_F may be written as

$$\frac{\partial \theta C_F}{\partial t} + \frac{\partial \rho'_s S_F}{\partial t} = - \nabla.(-D_F \theta \nabla C_F + v_f \theta C_F) + R_F \tag{2.65}$$

where C_F is defined as mass of substrate per unit volume of water and S_F as mass of adsorbed substrate per unit mass of soil grains. Hence ρ'_s is bulk density of dry soil.

Assuming the existence of a stoichiometric ratio, Y, between mass of substrate utilized and microbes formed, the net rate of substrate consumption becomes

$$R_f = - \frac{\mu}{Y} (\rho\sigma + \theta C) \tag{2.66}$$

where Y is called the yield coefficient. Experiments show Y to be constant. An equilibrium isotherm

$$S_F = k_a \; C_F^m \tag{2.67}$$

relates C_F and S_F. k_a and m are experimentally determined constants.

2.2.6. Complete set of governing equations

After substitution of Eq. (2.61) and Eq. (2.63) into the macroscopic mass balance Eq. (2.56), we obtain

$$\frac{\partial \theta C}{\partial t} = - \nabla.\vec{J} + (\mu - k_d)\theta C - R_a \tag{2.68}$$

Based on the earlier discussion, the flux of bacteria, $\vec{J}$, comprises Brownian diffusion, dispersion, convection, chemotaxis and gravitational settling. Therefore

$$\vec{J} = - D\theta\nabla C + u\theta C \tag{2.69}$$

where D is the coefficient of hydrodynamic dispersion, and u is the total velocity. The term R_a in Eq. (2.68) representing the net mass transfer rate is given by Eq. (2.58).

The mass conservation equation for adsorbed bacteria is obtained from Eqs. (2.57), (2.61), and (2.63) as

$$\frac{\partial \rho\sigma}{\partial t} = (\mu - k_d)\rho\sigma + R_a \tag{2.70}$$

where σ is defined as the volume of adsorbed bacteria in unit volume of bulk soil and ρ is the density of bacteria. In a saturated porous medium with deposition of suspended particles on soil grains,

$$\theta = n - \sigma \tag{2.71}$$

where n is the porosity of the medium.

Eq. (2.68) is modified for viruses with equilibrium adsorption and no growth, as discussed previously. Assuming negligible pore volume change due to adsorbed viruses, i.e., $\theta = n$,

$$\frac{\partial \rho \bar{C}}{\partial t} + \frac{\partial nC}{\partial t} = - \nabla . \vec{J} - k_d(nC + \rho \bar{C}) \tag{2.72}$$

where J, the flux of viruses, includes hydrodynamic dispersion and convection. Therefore

$$\frac{\partial \rho \bar{C}}{\partial t} + \frac{\partial nC}{\partial t} = - \nabla . (-Dn\nabla C + v_f nC) - K_d(nC + \rho \bar{C}) \tag{2.73}$$

A special case of Eq. (2.68) may be written for deep bed filtration without leaching substrate, resulting in no growth, i.e., $\mu = 0$ and $u = v_f$

$$\frac{\partial \theta C}{\partial t} = - \nabla . \vec{J} - k_d \theta C - R_a \tag{2.74}$$

where R_a is given by Eq. (2.58) and $J = -D\theta\nabla C + v_f \theta C$. The equation for adsorbed microbes remains as given by Eq. (2.70). Note that Eq. (2.74) is identical to Eq. (2.1), which is given for suspended solid particles.

The model of Matthess and Pekdeger (68) contains some of the terms of Eq. (2.68). The missing features are kinetically controlled deposition and resuspension, sedimentation, growth, and chemotaxis. No solution of the equation is given.

The model of Sykes et al. (100) predicts leachate organic concentration under sanitary landfills. An immobile microbial mass biodegrades the organics. These assumptions would reduce Eq. (2.68) and Eq. (2.65) to

$$\frac{\partial C}{\partial t} = \mu C - k_d C \tag{2.75}$$

$$\frac{\partial C_F}{\partial t} = - \nabla . [-D_F \nabla C_F + v_f C_F] + R_F \tag{2.76}$$

2.2.7. Model parameters

Microbial growth parameters, μ_m, K_s, Y, k_d are available for heterogeneous populations in commercial use (32). Sykes et al. (100) present the following values of parameters for populations of organic leachates under sanitary landfills: μ_m = 0.144 - 0.072 day^{-1}, K_s = 2000 - 4000 mg/l of COD, Y = 0.04, and k_d = 0.015 day^{-1}. The leachate and microbial concentrations were estimated to be 1000 mg/l each in the landfill. This was used as a first kind boundary condition for their model.

Adsorption of substrate onto soil particles depends on parameters k_a and m as given in Eq. (2.67). Estimates from two sources are given as: $k_a = 0$ for landfill leachate by Sykes et al. (100); $k_a = 0.2$ mg/l, $m = 1$ for 204 herbicide on sand, $k_a = 2.0$ mg/l, $m = 1$ for 2-4 herbicide on clay by Selim et al. (87).

Filtration studies are a source for mass transfer coefficients, k_c and k_y as in Eq. (2.58). Experimenting with anaerobic filters composed of crushed stones of diameter 20 mm and 50% porosity, Polprasert and Hoang (77) determined $k_c = 1.06 \times 10^{-5}\ s^{-1}$, $k_y \cong 0$ for fecal coliforms; $k_c = 6.25 \times 10^{-6} s^{-1}$, $k_y \cong 0$ for bacteriophages. Based on experiments with latex suspensions (0.04 μm) filtered through glass beads (diameter 0.397 mm, porosity 0.35), Ring (81) used a similar rate expression to model adhesion and suspension. He obtained $k_c = 6.5 \times 10^{-3} s^{-1}$, $k_y = 4.35 \times 10^{-4} s^{-1}$ for negatively charged latex particles on glass beads.

The other parameters required are density of bacteria ρ, and bulk density of soil ρ_s. Bacteria were assumed neutrally buoyant for the purposes of this model, i.e., ρ = 1 g/ml. ρ'_s was taken as 1.75 g/ml.

2.2.8. An analytical solution

An examination of the governing equations shows the complex nature of the model equations, with a high degree of non-linearity and coupling. It is very difficult, if not impossible, to obtain closed form solutions for C, σ, and C_F, even for a one-dimensional space. Therefore, a numerical solution will be sought for a coupled solution of the governing equations. However, a simplified analytical solution is needed to test the validity of the numerical results. Therefore, we will solve the coupled set of Eqs. (2.68), (2.58) and (2.70) as

$$\frac{\partial C'}{\partial t} + R_a = D\frac{\partial^2 C'}{\partial x^2} - u\frac{\partial C'}{\partial x} + k'C' \tag{2.77}$$

$$\frac{\partial \sigma^*}{\partial t} - R_a = k\sigma^* \tag{2.78}$$

$$R_a = k_c C' - k_y \sigma^* \tag{2.79}$$

where $C' = \theta C$, $\sigma^* = \rho\sigma$ and $k' = \mu - k_d$. The effective porosity, $\theta = n - \sigma$, flow velocity, u, and k are all assumed to be constants. We also assume h = 1 in Eq. (2.58). Eqs. (2.77) - (2.79) are solved for a semi-infinite column ($o \le x \le +\infty$) with boundary and initial conditions.

$$C' = C_0^* \quad \text{at } x = 0 \tag{2.80}$$

$$C' = 0 \quad \text{at } x => \infty \tag{2.81}$$

$$C' = 0 \quad \text{at } t = 0 \tag{2.82}$$

$$\sigma^* = 0 \quad \text{at } t = 0 \tag{2.83}$$

The solution technique is similar to that of Ogata (74). Applying the Laplace transform on t, Corapcioglu and Haridas obtained (22)

$$\frac{C'}{C_0^*} = \frac{2}{\sqrt{\pi}} \exp\left[\frac{ux}{2D} + k't\right] \int_{\frac{x}{2\sqrt{Dt}}}^{\infty} \exp\left\{ -\xi^2 - \left(\frac{ux}{4D\xi}\right)^2 - \frac{x^2 k_c}{4D\xi^2} - k_y\left(t - \frac{x^2}{4D\xi^2}\right)\right\} \cdot \left\{ I_o \left(\sqrt{\frac{x^2 k_c k_y \left(t - \frac{x^2}{4D\xi^2}\right)}{D\xi^2}} \right) + (k_y - k') \exp\left[-(k_y - k')\left(t - \frac{x^2}{4D\xi^2}\right)\right] \cdot \int_0^{\left(t - \frac{x^2}{4D\xi^2}\right)} \exp\left[-(k_y - k')\tau\right] I_0 \left(\sqrt{\frac{x^2 k_c k_y \tau}{D\xi^2}} \right) d\tau \, d\xi \tag{2.84}$$

The solution given by Eq. (2.84) is computed and plotted in Figure 5.

2.2.9. Numerical solution

The Galerkin Finite Element method was used to solve Eq. (2.65) and Eq. (2.68) simultaneously for a one-dimensional soil column. The method involves the approximation of the solution with a known set of basic functions. The rest of the model equations, Eqs. (2.57),

(2.64), (2.70), (2.71), (2.66) and (2.67), were incorporated into the numerical technique as explained in Corapcioglu and Haridas (22). For the model parameters given in Table 2, the largest time step was 60 seconds and the smallest time step was taken as 1 second. There are ten space elements ranging from 0.5 to 2 cm. The scheme was executed using a Fortran computer code. Plots of C, C_F and $(n-\sigma)$ for spatial and temporal variations are given in Figs. 6-8.

The soil column is assumed to be initially free of bacteria and substrate. As seen in these figures, the soil surface (i.e., x=0) is 3% clogged after 1.4×10^6 seconds. At this time, the bacteria are almost totally removed in the upper 7 centimeters of soil, although the substrate in the seeping wastewater travels up to 9 cm. The clogging of the soil is negligible after a depth of 6 cm. Another interesting feature is that the substrate concentration has a peak value at an early time, and then decreases gradually due to bacterial consumption. As shown again in Figure 9, for a smaller declogging rate constant, substrate concentration values will have larger values. Also, when the velocity and dispersion coefficients were taken as a tenth of the values in Table 1, all other parameters being the same, as seen in Fig. 10, a larger time is required to reach the steady state values. Similar results have been obtained for a two-dimensional field by Corapcioglu and Haridas (22).

Table 2. Numerical Model Parameters

Dispersion coefficients	$D=D_f=4\times10^{-2}$ cm^2/s
Density of bacteria and dry soil	$\rho=1$ g/ml; $\rho_s=1.74$ g/ml
Clogging rate constant	$k_c=6.5\times10^{-3}$ s^{-1}
Declogging rate constant	$k_y=4.35\times10^{-4}$ s^{-1}
Specific decay constant	$k_d=1\times10^{-6}$ s^{-1}
Monod half constant	$K_s=2\times10^{-3}$ g/ml
Maximum growth rate	$\mu_m=4.2\times10^{-5}$ s^{-1}
Maximum cell yield	$Y=0.04$
Flow velocity	$u=3\times10^{-2}$ cm/s
Porosity	$n=0.6$
Surface bacteria concentration	$C_0=10^{-3}$ g/ml
Surface substrate concentration	$C_{F_0}=10^{-3}$ g/ml

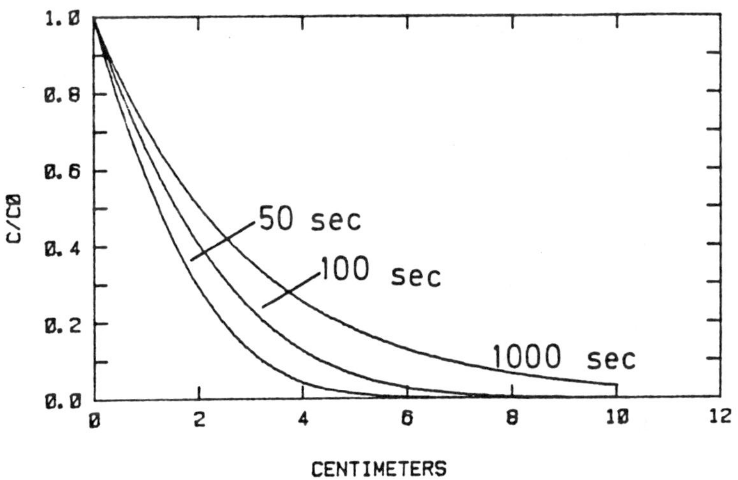

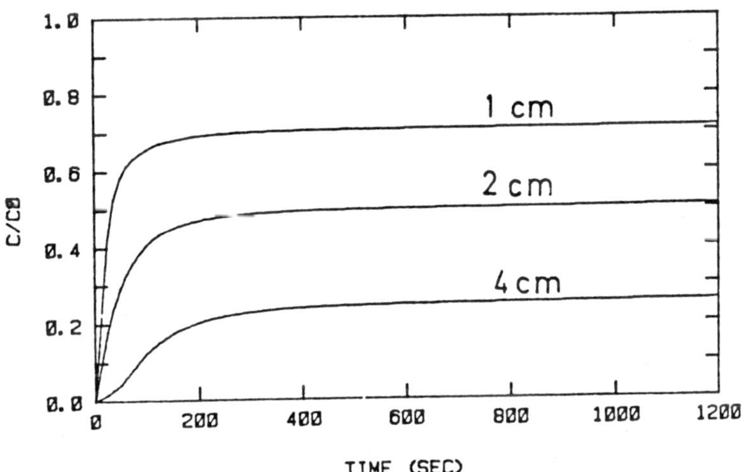

Figure 5. Analytical solution with parameters: D = 0.04 cm^2, u = 0.003 cm/s, k = $-1x10^{-6}s^{-1}$, k_c = $6x10^{-3}s^{-1}$, and k_y = $6x10^{-5}s^{-1}$.

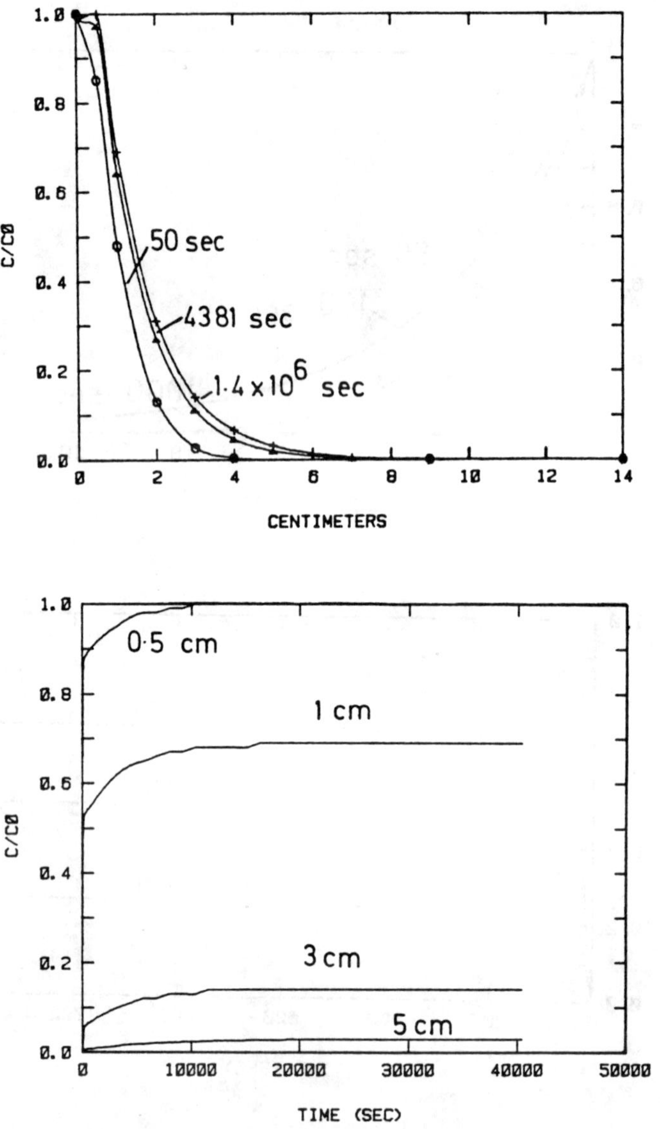

Figure 6. Spatial and temporal variation of bacterial concentration for parameters given in Table 2.

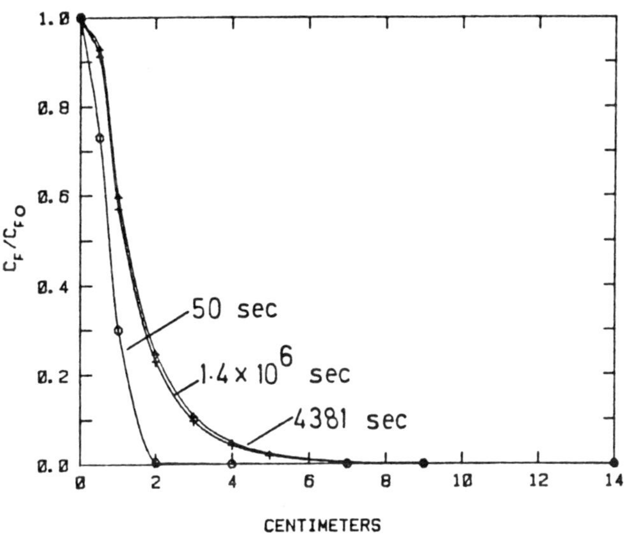

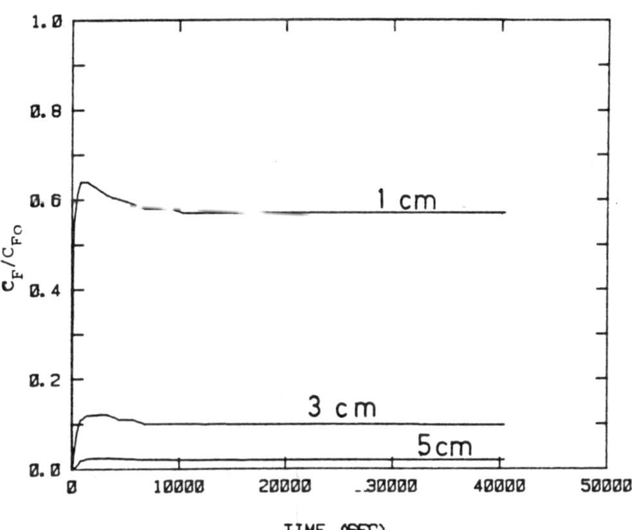

Figure 7. Spatial and temporal variation of substrate concentration for parameters given in Table 2.

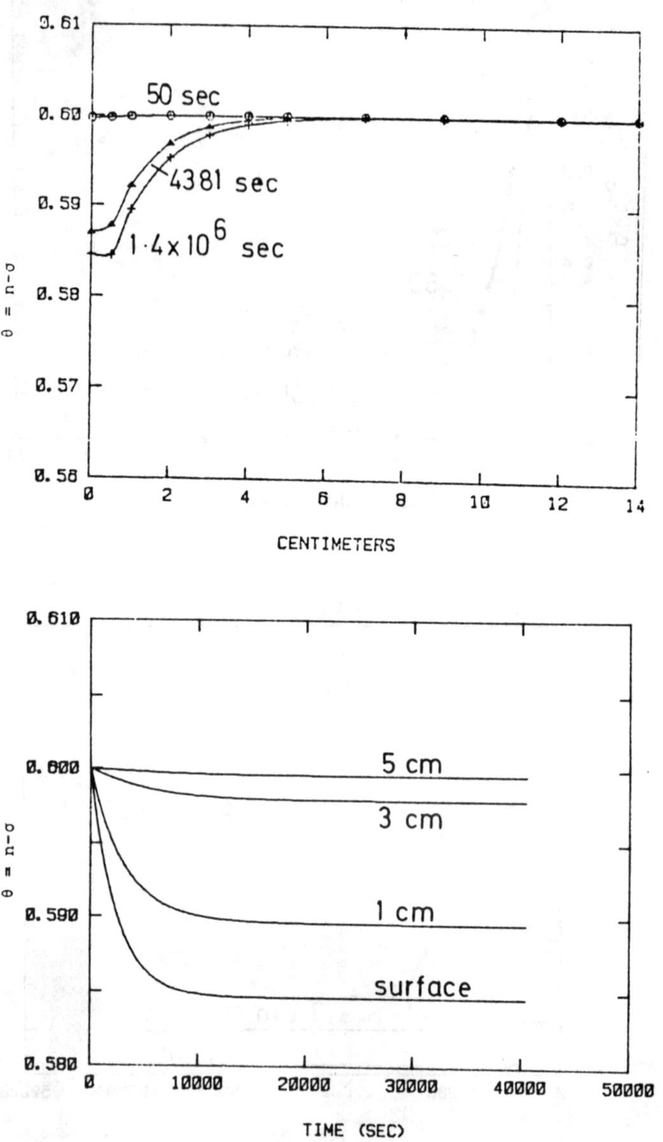

Figure 8. Spatial and temporal variation of effective porosity $\theta = n-\sigma$ due to clogging with parameters given in Table 2.

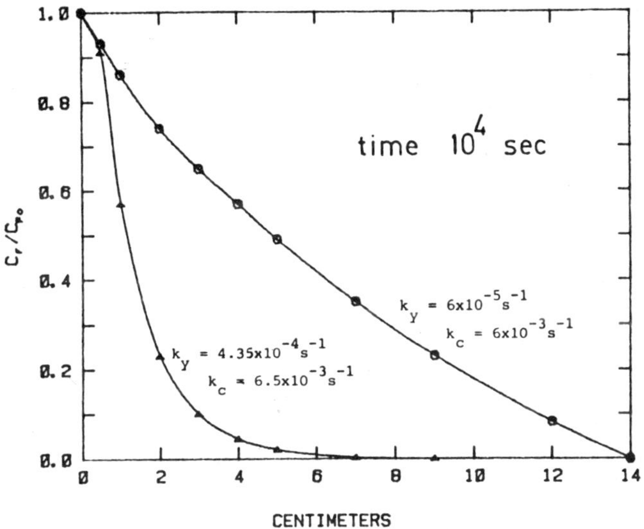

Figure 9. Comparison of the substrate concentrations with two different values of clogging and declogging rate constants.

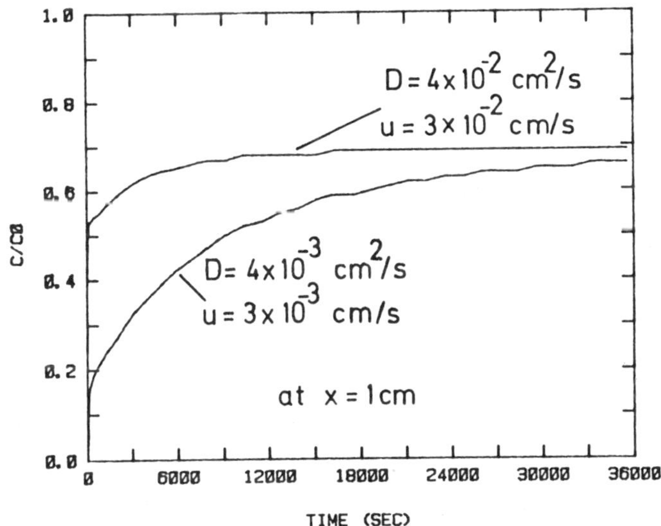

Figure 10. Comparison of bacterial concentration values for two sets of velocity and dispersion coefficients.

3. FORMATION OF A BED (CAKE FILTRATION)

Cake filtration process is used in water treatment process and industrial processes to remove suspended solids. Compressible filter cakes are formed when a liquid with suspended particles are forced through a thin membrane (septum) which allows the liquid transport but retains solid particles by straining and sedimentation mechanisms. During filtration, the filter cake is compacted while new solid particles are laid down on the cake surface (Fig. 1b).

In general, cake filtration is primarily employed for more concentrated slurries. The filtrate from a cake filtration may contain small particles passed through the medium,which must be removed in a polishing step. In addition to industrial filtration processes, the formation of filter cake is important in groundwater and oil wells where drilling fluid contains suspended particles. The accumulation of these suspended particles on perforated well screens causes a pressure drop in the well. This particular problem has been discussed by Binkley et al. (10), Muecke(72), and Kovacs and Ujfaludi (60).

This work is directed at finding mathematical solutions based on a theoretically and physically consistent mathematical model presented by Corapcioglu (18,19). Also, the purpose of this research is to obtain direct results which can be used to predict th cake thickness (L) and the time (t) at different conditions of porosity, particle concentration , and pressure drop.

3.1. Previous Studies on Cake Filtration

Studies governing laws of cake filtration may be found scattered through the literature in different disciplines (e.g., 17,27,92,93,105,106,107,110,121,123). Based on the analysis of various investigators, both cake and filter septum can be considered as porous masses exerting resistance against the moving laminar liquid flow.

Binkley et al. (10) (also in Collins, 17) present the most pertinent features of a mathematical analysis of the factors affecting the rate of deposition of solids by filtration in unfractured perforations during cement slurry injection operations in oil wells. Binkley et al. assume steady state conditions throughout the cake. Their analysis is based on simple volumetric balance of solid and fluid phases (see Eq. (3.7)). In von Engelhardt (120), experiments show that for filtration times up to one hour, the cake thickness varies linearly with time. However, investigations over longer periods (several hours) show that this assumption would not hold as more filtrate is continually collected that can be predicted by a *parabolic function*. Experimental observations by Ruth (83) and Carman (15) imply that the average porosity is constant regardless of cake compressibility. According to Ruth the total filtrate rate can be approximated through an expression which is proportional to the square root of time. The proportionality constant includes the resistivity of the *filter cloth (septum resistance)* which is a source of *non-parabolic* behaviour. Willis et al. (124) present evidence which shows that average porosity is constant throughout a filtration, and the septum clogging determines the extent of deviations from the parabolic relation. Rietema (80) has defined non-compressible filter cakes as the ones whose mean specific resistance does not change with the filtration pressure. All others are called compressible. Rietema has observed a form of "retarded packing compressibility" in which early layers of the cake do not compress gradually until a critical cake thickness is reached. Tosun and Willis (112) conclude that, based on multiphase theory, the classification of filter cakes as compressible and incompressible is unnecessary. Furthermore, they also stated that although parabolic behavior can be achieved by proper septum selection, it is not an optimum condition, due to the need for more energy in comparison to non-parabolic behavior.

Tiller and Cooper (105) pointed out the variation of internal flow rate throughout the cake, and derived a relationship between rate of porosity change and internal flow rate variation. Shirato et al. (92) performed an experiment to determine the liquid pressure drop as a function of the distance through the cake. With the hydraulic pressure p known, the cake compressive pressure p_s can be calculated as $p_s = p_2 - p$, where p_2 is the applied filtration pressure. Knowing p_s, one can estimate the porosity distribution from *compression permeability cell measurements*. Shirato et al. (93) present an analytical method for apparent velocity variations of both liquid and solids through filter cakes, and it has turned out that the effects of the velocity distributions of liquids and solids could not be theoretically neglected, especially for highly concentrated slurries. Tiller and Shirato (106) demonstrated that due to nonuniform flow rate the conventional parabolic relation between filtrate volume and time needs to be modified to include a correction factor, as

$$\frac{dv}{dt} = \frac{g_c\, p}{\mu_w (J\alpha_R w + R_m)} \tag{3.1}$$

where v is the filtrate volume, g_c is the conversion factor, μ_w is the viscosity, w is the total mass of dry solids per unit area, R_m is the medium resistance, α_R is the conventional filtration resistance, and J is the correction factor due to nonuniform flow rate. We should note that these studies assume the validity of the compression-permeability cell tests to determine filter cake resistivity. A study by Atsumi and Akiyama (5), which also utilizes the same test results, describe the transient nature of the problem with a non-linear partial differential equation. A moving boundary condition describing the growing cake surface has been incorporated in Atsumi and Akiyama's model. They solved the following cake filtration equation numerically

$$\frac{\partial e}{\partial t} = \frac{\partial}{\partial w} \{C_p^* \frac{\partial e}{\partial w}\} \tag{3.2}$$

where w is the mass of solid per unit filtering area from the medium e is the void ratio. C_p^* is a variable coefficient associated with permeability and compressibility. The boundary condition at the moving boundary is

$$\left.\frac{\partial e}{\partial w}\right|_{w_i} = \frac{e_o - e_i}{C_p^*(e_i)} \frac{dw_i}{dt} \tag{3.3}$$

where e_i is the void ratio at the cake surface.

Also, Wakeman (121) developed a theoretical analysis for a similar problem recognizing cake filtration as a moving boundary problem and utilizing a variable compressibility coefficient in the form of an exponential function of a normalized porosity. His basic partial differential equation describing liquid movement and cake volume change is

$$\frac{\partial n}{\partial t} = \frac{\partial}{\partial x} \left((1-n) \frac{k}{\mu_w} \frac{\partial p}{\partial n} \frac{\partial n}{\partial x}\right) + \left(\frac{k}{\mu_w} \frac{\partial p}{\partial n} \frac{\partial n}{\partial x}\right)_{x=0} \frac{\partial n}{\partial x} \tag{3.4}$$

where n and k are the porosity and the liquid conductivity (permeability) of the cake and p is the hydraulic pressure. After filtration has started, a cake is deposited on the septum with a porosity varying from n_1 at the septum surface to n_i at the cake/slurry interface. The location of the moving cake/slurry interface is unknown, hence a further condition is required

$$\left.\frac{\partial n}{\partial x}\right|_{x_i} = \frac{n_o - n_i}{1-n_o} \frac{\mu}{k_i} \left.\frac{\partial n}{\partial p}\right|_{x_i} \left.\frac{dx}{dt}\right|_{x_i} \tag{3.5}$$

which is the boundary condition to be satisfied at the cake surface. Wakeman (122) concludes that in all cases the pressure loss across the filter cloth decreases with time. However, at higher filtration pressures, the loss over the cloth decreases more rapidly due to the more rapid formation of a thicker cake with a greater specific resistance. The pressure loss across the cake-forming layer remains reasonably constant after the initial period of filtration.

Von Engelhardt (120) has obtained an expression in his study of filter cake formation in the bore hole during well drilling operations. Drilling mud used to cool and lubricate the drilling bit is a suspension of clay particles in water. Under pressure, a cake of clay particles forms on the wall of the borehold.

$$L = [\frac{2p}{\mu_w} k\ b\ t]^{\frac{1}{2}} \tag{3.6}$$

where μ_w denotes the viscosity of the filtrate, k is the permeability of the cake, and b is the dimensionless specific volume of filter cake (volume of filter cake per cubic centimeter of filtrate). Note that the derivation of Eq. (3.6) assumes constant permeability and porosity of the cake.

As noted earlier, Collins (17) suggested a mathematical description of deposition processes with various assumptions. We should note that Collin's study has been previously published by Binkley et al. (10). Based on steady state conditions and hydrostatic pressure distribution along the cake, Collins (17) obtains an expression for the cake thickness.

$$L = -\frac{k_c^* L^*}{K} + \left[\left(\frac{k_c L^*}{K}\right)^2 + \frac{2k_c^*}{\mu_w} p_a\ w_e t\right]^{1/2} \tag{3.7}$$

where L^* is the thickness of the porous plate septum and K is its permeability. p_a is the applied pressure, t is the time, μ_w is the viscosity of the filtrate, k_c^* is the permeability of the filter cake, and w_e is a factor related to the volume fraction of solids.

Willis and Tosun (123) found experimentally that the discharge is usually a parabolic function of time for a constant pressure filtration. Also, they stated the relation between the cake thickness, L, and time as follows

$$1363L^2 = t \qquad L[\text{cm.}] \text{ and } t[\text{sec.}] \tag{3.8}$$

Tosun and Willis (113) mentioned the derivation of the parabolic filtrate discharge equation, which is restricted to one dimensional filtration. They presented the following equation

$$q_o - q_i = (n_L - \bar{n}) \frac{dL}{dt} - L \frac{d\bar{n}}{dt} \tag{3.9}$$

where q_o is the superficial fluid velocity at the filter septum, q_i is the superficial fluid velocity at the cake surface, L is the cake thickness, $\bar{n}$ is the average porosity, and n_L is the porosity at the cake surface. The two terms on the left represent the difference between the inlet and outlet flow rates, while the first term on the right side is proportional to the change in the cake thickness. The second term on the right hand side is a compaction effect caused by changes in the average porosity.

Tiller et al. (108) have obtained the following expression for the pressure variation along the cake.

$$\frac{p}{p_2} = 1 - \left(1 - \frac{x}{L}\right)^{1/(1-d'-f')} \tag{3.10}$$

where p_2 is the applied filtration pressure. d' and f' are dimensionless empirical constants. A plot of Eq. (3.10) is given in Fig. 11 for different values of (d'+f'). Furthermore, Tiller et al. (109) have proposed the following expression for the porosity variation with cake compressive pressure, p_s

$$1-n = (1-n_L)\left(1 + \frac{p_s}{r'}\right)^{f'} \tag{3.11}$$

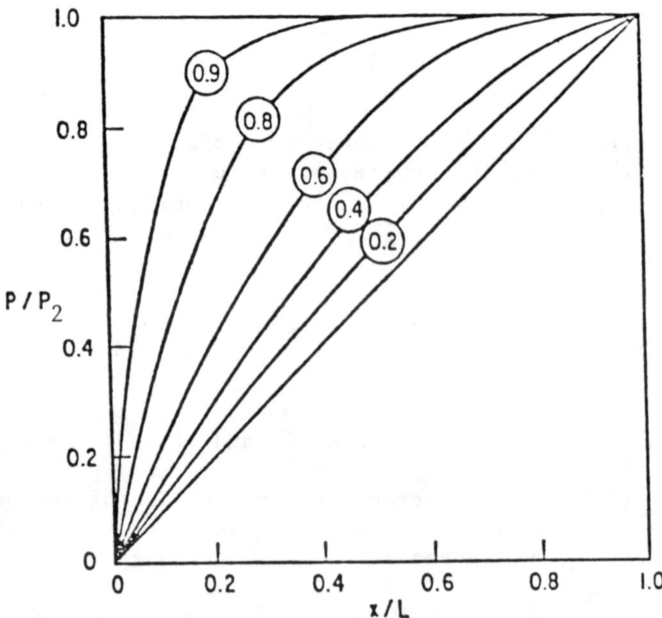

Figure 11. Variation of hydraulic pressure with distance for different compressibility coefficients, [After Tiller et al. (108)].

where r' is an arbitrary parameter. Eq. (3.10) is an alternative to another empirical expression suggested by Tiller in 1953. A review of this work is given by Tiller and Crump (109) as

$$1 - n = r^{*} p_s^{d'} \quad \text{for } p_s \geq p_i \tag{3.12}$$

$$1 - n = r^{*} p_i^{d'} \quad \text{for } p_s \geq p_i \tag{3.13}$$

where r^{*} is a dimensionless empirical constant, and p_i is the pressure below which n is assumed to be constant. The cake permeability, k, is expressed as a function of porosity and *specific cake resistance*, α^{*}, by Tiller (104) as

$$k = \frac{1}{\rho_s \alpha^{*} (1-n)} \tag{3.14}$$

where ρ_s is the true density of solids. The specific cake resistance α^{*} is expressed as

$$\alpha^{*} = \alpha_L^{*} \left(1 + \frac{p_s}{r'}\right)^{d'} \tag{3.15}$$

where α_L^{*} is the specific cake resistance at the cake surface where $p_s = 0$. In his "revised theory", Tiller multiplied the average cake resistance by an exponential function of mass of particles to take "*cake blinding*" into account.

From the review presented in this section, one may conclude that porosity decreases with filtration time, but on the contrary, pressure loss across the deposited cake increases with filtration time. Also, the filtrate discharge is a parabolic function of time for a constant pressure, and it is known that hydraulic pressure variation is linear with distance (94).

Tosun (111) has provided a critical review of Tiller's work. Tosun's review includes other works of Tiller which are not referred to in this study; therefore, the reader is referred to Tosun (111) for a more in-depth discussion of Tiller's work.

As an alternative, Willis and his co-workers (123,126) present a *multiphase theory* which includes conservation of mass and conservation of momentum equations for liquid and solid phases. Willis et al. (124) has shown through a dimensional analysis that deviations from parabolic behavior cannot be attributed to non-Darcian behavior, but rather to septum permeability. They have demonstrated that dominant forces in the conservation of mass equation for the liquid phase are gravitational, pressure and drag forces. Similarly, in the solid phase conservation of momentum equation, the deformation and gravitational forces can be retained. Then, for a Newtonian liquid and non-deformable solid particles, conservation of momentum equations lead to relative Darcy's law and

$$\frac{\partial T_s}{\partial z} = 0$$

where T_s is the intrinsic solid phase stress. Willis et al. (124) has shown that if the slurry concentration is constant, the average cake porosity, $\bar{n}$, is constant at all times and is independent of cake pressure drop. The coupled solution of two conservation equations in terms of porosity, pressure, permeability, and solid velocity would require either additional information such as stress-strain relations between porosity and pressure,and Kozeny-Carman relation between porosity and permeability (not applicable to filter cake studies due to uncertainties and associated restrictions such as pore structure),or experimental determination of any two of these four variables. Willis and Tosun (123) used measured porosity and pressure profiles to calculate internal velocity and permeability distributions. Willis et al. (125) have also presented a "two resistance model" by neglecting the solid's velocity and assuming a conventional Darcy's law instead of a relative one. They fitted the resistance and porosity data as a function of the compressive pressure from compression-permeability cell tests in a form given by Eq. (3.13). The cake thickness is obtained in the form of Eq. (3.10). Willis et al. (125) have shown that this particular model assumes that the porosity is not a function of time,and that the local fluid velocity is uniform throughout the cake at any instant.

Dillingham et al. (27),and later Stephenson and Baumann (98), present governing equations of *diatomite filtration*. In diatomite filtration, diatomite is added to the influent water as a body feed to form a porous incompressible cake. The diatomite filter aid forms a rigid porous mat to accommodate the suspended particles. The filter aid placed on the septum,or filter aid support,before the start of a filter run is called the "*precoat*". The diatomite filter aid fed continuously to the influent during filtration is called the "*body feed*". The filter cake is composed of suspended solid particles and body feed,and is housed over the precoat. Dillingham et al. (27) assumes (1) equal precoat and filter cake porosities, (2) negligible contribution of suspended solids to the mass and the porosity of the filter cake, (3) 100% solid particle capture in the filter cake, and (4) constant flow rate. Following these assumptions, the change in volume of the filter cake V_c is given by

$$\frac{dV_c}{dt} = \frac{C_D \, q \, A}{\rho_D(1 - \varepsilon_p)} \tag{3.16}$$

where C_D is the body feed concentration, A is the cross sectional area, ρ_D is the density of diatomite, and ε_p is the porosity of precoat layer.At the beginning of a filter run, the filter housing is full of clean water from the precoating operation. The mixing of unfiltered influent with the clean water in the filter housing occurs in a transition period till the filter housing contains same quality of water as that of the influent. This transition period of dilution as stated by Dillingham et al. affects V_c and therefore C_D is obtained from

$$\frac{dC_D}{dt} = \frac{qA}{V_f}(C_{D_i} - C_D) \tag{3.17}$$

where C_{D_i} is influent body feed concentration, and V_f is the volume of filter housing. The solution for C_D is substituted to obtain V_c from Eq. (3.16). V_c/A would give the thickness of filter cake which is used to calculate the pressure drop across the cake thickness from the Darcy's law.

3.2. Deposition of Particles on the Cake Surface

In this section, we will obtain an expression to estimate the cake thickness during a solid-liquid separation by cake filtration. As shown in Fig. 1, the slurry is fed from the center with a driving force which produces filtration through the cake while simultaneously increasing its thickness. We will assume that, under the existing pressure conditions during filtration, the liquid is practically incompressible,and that the solid particles are also assumed incompressible. However, the solid matrix as a whole is compressed due to porosity change in time and space. The formulation in this section is based on the theoretical work of Corapcioglu (19), and closely follows that reference.

3.2.1. Formulation by averaging along the cake thickness

Concentration of suspended solid particles in the slurry (or the concentration of suspended particles) per unit volume of the liquid is denoted by C. Under these assumptions, the conservation of mass equation for solid particles is given by Corapcioglu (18,19) as

$$-\frac{\partial[\gamma_s(1-n)]}{\partial t} - \frac{\partial(nC)}{\partial t} = \frac{\partial(Cn\vec{V}_L)}{\partial x} + \frac{\partial[\gamma_s(1-n)\vec{V}_s]}{\partial x} \tag{3.18}$$

where γ_s and n are the specific weight of solid particles per unit volume of solid and the porosity of the cake. $\vec{V}_s$ and $\vec{V}_L$ denote the solid and liquid velocity vectors,and x, which varies with time t denotes the distance measured from the medium to the surface of the cake. The total flux of solids, in Eq. (3.18) given by $\vec{J} = n\vec{V}_L C + \gamma_s(1-n)\vec{V}_s$. Note that Cn is the concentration of suspended particles per unit volume of cake (total volume), and $\gamma_s(1-n)$ is the weight of solids in cake per unit volume of cake. In this study,we define the concentration in terms of weight rather than mass

If we integrate Eq. (3.18) along x from 0 to L,

$$\frac{d[\gamma_s(1-\bar{n})L]}{dt} - \gamma_s(1-n_L)\frac{dL}{dt} + \frac{d(\bar{n}\bar{C}L)}{dt} - n_L C_L \frac{dL}{dt} + [C_L n_L \vec{V}_{L_L}$$

$$+ \gamma_s(1-n_L)\vec{V}_{s_L}].\nabla F_2 - [C_o n_o \vec{V}_{L_o} + \gamma_s(1-n_o)\vec{V}_{s_o}].\nabla F_1 = 0 \tag{3.19}$$

where overbar denotes the averaged quantities, L is the cake thicknes n_L, C_L, $\vec{V}_{s_L}$ and $\vec{V}_{L_L}$ are the surface porosity, surface concentration and the solid and liquid velocity at the cake surface and n_o, C_o, $\vec{V}_s$ and $\vec{V}_{L_o}$ are the porosity, concentration,and the solid and liquid velocity at the filter septum, respectively.

Assuming that all suspended particles are filtered within the cake and a rigid septum, the last term in brackets in Eq. (3.19) would be zero, i.e., $C_o = 0$ and $\vec{V}_{s_o} = 0$. By rearranging Eq. (3.19), we get

$$\frac{dL}{dt}[\gamma_s n_L - \gamma_s \bar{n} + \bar{n}\bar{C} - n_L C_L] + \gamma_s(1-n_L)\vec{V}_{s_L}\cdot\nabla F_2$$

$$+ L[\bar{n}\frac{d\bar{C}}{dt} + \bar{C}\frac{d\bar{n}}{dt} - \gamma_s\frac{d\bar{n}}{dt}] + C_L n_L \vec{V}_{L_L}.\nabla F_2 = 0 \tag{3.20}$$

The solid's velocity at the cake surface can be expressed as

$$\vec{V}_{s_L}\cdot\nabla(x-L) = \vec{V}_{s_L}\cdot\vec{k} = V_{s_L} \tag{3.21}$$

If we re-write Eq.(3.20)

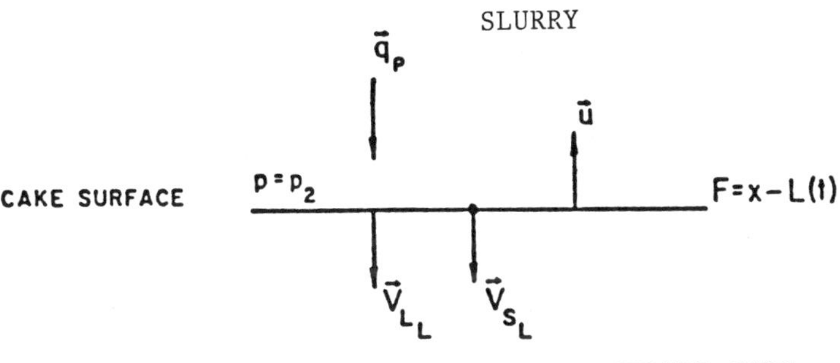

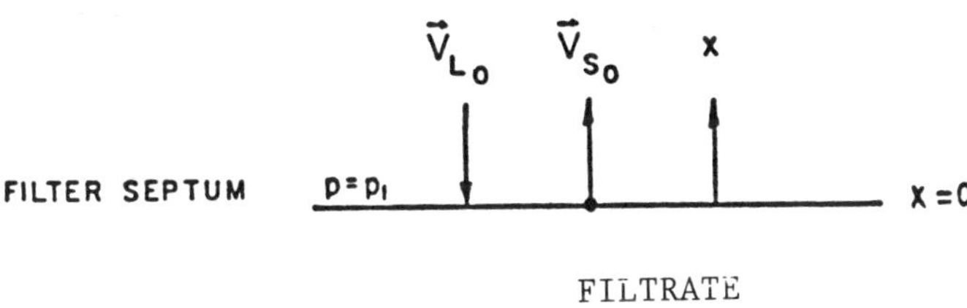

Figure 12. A schematic diagram for nomenculature.

$$[(n_L-\bar{n})\gamma_s + \bar{n}\,\bar{C} - n_L C_L]\,\frac{dL}{dt} + [\bar{n}\,\frac{d\bar{C}}{dt} + \bar{C}\,\frac{d\bar{n}}{dt} - \gamma_s\,\frac{d\bar{n}}{dt}]L$$

$$\gamma_s(1-n_L)\vec{V}_{s_L}.\nabla F_2 + C_L n_L \vec{V}_{L_L}.\nabla F_2 = 0 \qquad (3.22)$$

At this point, we can assume that all suspended particles are totally filtered on the cake surface (i.e., $\bar{C} = 0$). This assumption rules out the possibility of existence of fine particles suspended in the liquid inside the cake. We realize that this may not be true for colloidal particles. Then, Eq. (3.22) becomes

$$[(n_L-\bar{n})\gamma_s - n_L C_L]\,\frac{dL}{dt} - [\gamma_s\,\frac{d\bar{n}}{dt}]L + \gamma_s(1-n_L)V_{s_L} + C_L n_L \vec{V}_{L_L}.\nabla F_2=0 \qquad (3.23)$$

Eq. (3.23) is subject to a general boundary condition at a moving surface F_2, which is defined as

$$F_2 = F = x - L(t) \qquad (3.24)$$

The no-jump condition for the continuity of the mass flux of the solid phase at the cake surface between the filter cake and the slurry can be written in the form

$$[C(\vec{q} - n\vec{u}) + \gamma_s(1-n)(\vec{V}_s - \vec{u})]_{\substack{\text{cake side,} \\ \text{slurry side}}} . \nabla F = 0 \tag{3.25}$$

where n denotes the porosity, C denotes the concentration of solid particles, $\vec{u}$ is the velocity of the moving surface, and $\vec{V}_s$ is the solid particle's velocity. The bracket shows the jump from one side (cake) of the moving surface (cake surface) to the other (slurry) carrying suspended particles. Due to conservation of mass principle with no source or sink terms on the surface, the right hand side of the equation should be equal to zero. Furthermore, $\vec{q}$ denotes the specific discharge and can be determined from Darcy's law. For a deforming porous medium

$$\vec{q} = \vec{q}_r + n\vec{V}_s = n\vec{V}_L \tag{3.26}$$

where $\vec{q}_r$ is the relative discharge, and V_s is the velocity of solid particles. Rewriting Eq. (3.25), we obtain

$$[C(\vec{q} - n\vec{u}) + \gamma_s(1-n)(\vec{V}_s - \vec{u})]_{cake} . \nabla F =$$

$$[C(\vec{q} - n\vec{u}) + \gamma_s(1-n)(\vec{V}_s - \vec{u})]_{slurry} . \nabla F \tag{3.27}$$

The last term at the right-hand side of Eq. (3.27) is equal to zero, since the porosity is equal to unity in the liquid side of the surface. At this point we will assume that solid particles in the slurry are carried by the liquid without any resistance. In other words, we neglect any drag force developing on solid particles in the slurry.

Then

$$\vec{V}_s \cdot \nabla F_2\Big|_{slurry} = \vec{q}_p \cdot \nabla F_2\Big|_{slurry} = - q_p \tag{3.28}$$

Also, for a moving surface

$$\vec{u} \cdot \nabla F + \frac{\partial F}{\partial t} = 0 \tag{3.29}$$

Substitution of Eq. (3.24) into Eq. (3.29) yields

$$\vec{u} \cdot \nabla F = - \frac{d(x-L)}{dt} = \frac{dL}{dt} \tag{3.30}$$

Inserting Eq. (3.28) and (3.30) into Eq. (3.27), and rearranging, we obtain

$$\gamma_s(1-n_L)(\vec{V}_{s_L}-\vec{u}).\nabla F_2 + C_L n_L \vec{V}_{L_L}.\nabla F_2 - C_L n_L \frac{dL}{dt} = C_p \vec{q}_p.\nabla F_2 - \frac{dL}{dt} C_p \tag{3.31}$$

where C_p and $\vec{q}_p$ are the concentration and the flow rate in the slurry side, and n_L, C_L, $\vec{V}_{L_L}$ denote the porosity, concentration and the liquid velocity respectively at the cake surface. That is

$$Cn\vec{u}\Big|_{cake} . \nabla F = C_L n_L \frac{dL}{dt} \tag{3.32}$$

and

$$C_p n_p \vec{u}\Big|_{slurry} . \nabla F = C_p \frac{dL}{dt} \tag{3.33}$$

Then by rewriting Eq. (3.31) and substituting into Eq. (3.23), we obtain

$$[\gamma_s(1-\bar{n}) - C_p] \frac{dL}{dt} - [\gamma_s \frac{d\bar{n}}{dt}]L - C_p q_p = 0 \tag{3.34}$$

where C_p and q_p are the concentration and the flow rate of the injected slurry. This corresponds to the slurry concentration which is a controlled parameter in operation (total weight of suspended particles in a certain volume of liquid). Eq. (3.34) represents the cake filtration equation with the assumption that all material is captured on the surface and the filter septum is rigid. If the slurry concentration is expressed as the mass fraction of particles in the slurry and denoted by s, then $C_p = \gamma_w s$. Furthermore, if we express the slurry flow rate in terms of slurry volume V_p, then $V_p = q_p At$ where A is the filter area.

3.2.2. An analytical solution for the cake thickness with known average porosity functions

Few investigations of porosity variation have been reported in the literature. Hutto (44) concludes that porosity decreases relatively rapidly near the liquid face of the cake, and then continues at a slower, almost linear rate of decrease down to minimum porosity at the filter septum. Tiller and Cooper (105) and Shirato et al. (94) pointed out that the porosity decreases at each point with respect to time. Based on their observations, we can propose an average porosity function decreasing in time. Furthermore, such a linear function can be obtained from the definition of matrix compressibility of a moving solid with an assumption of constant average pressure change in time as shown by Bear, Corapcioglu, and Balakrishna (9).

$$\bar{n} = At + B \tag{3.35}$$

where A is a negative constant and B is a positive constant which is equal to $B = \bar{n}_o - A$ to where $\bar{n}_o$ is the porosity at $t = t_o$.

Equation (3.34) can be written as

$$\frac{dL}{dt} + P(t)L = Q(t) \tag{3.36}$$

where

$$P(t) = \frac{-\gamma_s \frac{d\bar{n}}{dt}}{[\gamma_s(1-\bar{n})-C_p]} \tag{3.37}$$

and

$$Q(t) = \frac{C_p q_p}{[\gamma_s(1-\bar{n})-C_p]} \tag{3.38}$$

Eq. (3.36) is a general first order linear differential equation with variable coefficients and an initial condition $L(t=0)=0$. Then the solution will be

$$L = \frac{C_p q_p t}{[1-\bar{n}]\gamma_s - C_p} \tag{3.39}$$

For $\bar{n} = -0.000178t + 0.864$, the variation of L with t is shown in Fig. 13 (curve a). Other model parameters are given in Table 2. In fact, the solution given in Eq. (3.39) is more general than it implies. For example, for an exponential variation of average porosity in time, we would obtain an identical solution. Wakeman (122) has obtained experimental results showing an exponential decrease of mean cake porosity. If we assume an exponential distribution for $\bar{n}$ as

$$\bar{n} = 0.864 \exp(-0.0002t) \tag{3.40}$$

The temporal variation of L obtained by Eq. (3.40) is shown in Figure 13 (curve b).

Following the results obtained by Willis et al. (124), the average cake porosity can be taken constant. Then for $\bar{n} = 0.70$, the variation of C is illustrated by curve C in Figure 13.

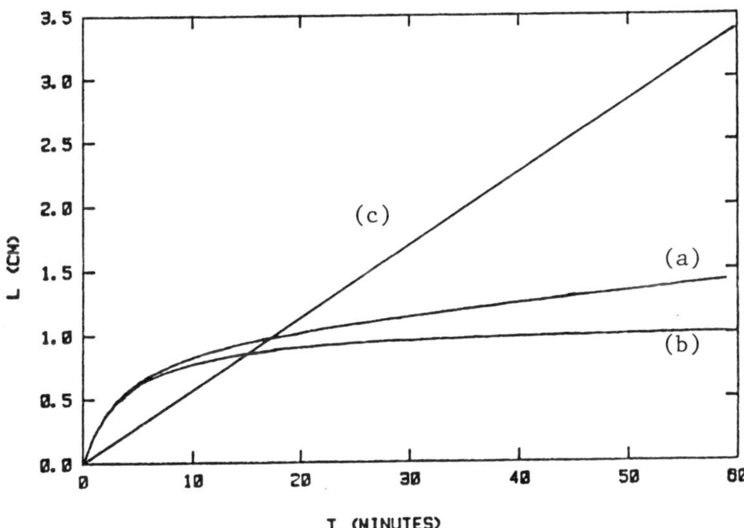

Figure 13. Temporal variation of cake thickness with different functions of average porosity, (a) linear average porosity [Eq. (3.35)], (b) exponential average porosity [Eq. (3.40)], (c) constant average porosity = 0.70.

Table 3. Model Parameters

Concentration of solid particles in the slurry	$C_p = 2x10^{-3} (N/cm^3)$
Flow rate	$q_p = 2x10^{-3} (cm/sec)$
Specific weight of soil	$\gamma_s = 2.1x10^{-2} (N/cm^3)$
Specific weight of water	$\gamma_w = 9.81x10^{-3} (N/cm^3)$
Viscosity of water	$\mu_w = 1x10^{-7} (N\ sec/cm^2)$
Pressure drop	$\Delta p = 15\ (N/cm^2)$
Average hydraulic conductivity of the cake	$\bar{K} = 0.7\ x10^{-6} (cm/sec)$
Average cake porosity	$\bar{n} = 0.7$

3.3. Simultaneous Solution of Conservation of Mass Equations for Liquid and Solid Phases

In Section (3.2.1) we made use of the conservation of mass equation for solid particles in the slurry to obtain an expression for the cake thickness in terms of time. The solution for L given by Eq. (3.39) is obtained in terms of slurry flow rate, slurry concentration and a known average porosity value. In filtration practice, usually the filtrate volume, V_f, is the operational parameter. To obtain a solution in terms of V_f, we will start with the conservation of mass equation for liquid to determine the difference between the volume of slurry filtered, V_p, and filtrate collected. This derivation has been obtained by Corapcioglu (19).

One-dimensional conservation of mass equation for an incompressible liquid in deformable porous media with incompressible solid particles can be written as

$$\frac{\partial(n)}{\partial t} + \frac{\partial \vec{q}}{\partial x} = 0 \tag{3.41}$$

Averaging Eq. (3.41) along x from 0 to L, and introducing the condition at the cake surface as discussed in section (3.2.1), would yield

$$\bar{n}\,\frac{dL}{dt} + L\,\frac{d\bar{n}}{dt} - n_L\,\frac{dL}{dt} + \vec{q}_L \cdot \nabla F_2 - q_o \cdot \nabla F_1 = 0 \tag{3.42}$$

The no-jump condition for the continuity of the mass flux of the incompressible liquid phase at the cake surface between the filter cake and the slurry can be written in the form

$$[\vec{q} - n\vec{u}]_{\substack{\text{cake side,}\\ \text{slurry side}}} \cdot \nabla F = 0 \tag{3.43}$$

If we write Eq. (3.43) at x = L

$$\vec{q}_L \cdot \nabla F_2 - n_L\,\frac{dL}{dt} = \vec{q}_p \cdot \nabla F_2 - \frac{dL}{dt} \tag{3.44}$$

and at x = 0

$$q_o \cdot \nabla F_1 = q_f \cdot \nabla F_1 \tag{3.45}$$

where q_f is the filtrate rate. The substitution of Eqs. (3.44) and (3.45) into Eq. (3.42) would yield

$$q_p = q_f + (\bar{n} - 1)\,\frac{dL}{dt} + L\,\frac{d\bar{n}}{dt} \tag{3.46}$$

If q_p in Eq. (3.39) is expressed by Eq. (3.46) for a cake with constant average porosity

$$L = \frac{C_p q_f t}{[\gamma_s(1-\bar{n}) - \bar{n}C_p]} = \frac{\rho_w s V_F}{A[\rho_s(1-\bar{n}) - \bar{n}\rho_w s]} \tag{3.47}$$

where ρ_f and ρ_s are slurry and particulate densities respectively. A comparison with the data of Willis et al. (124) show that Eq. (3.47) gives very reliable results in predicting cake length in terms of filtrate volume or vice versa (Fig. 14). The linearity of L with V_F indicates the constancy of average porosity throughout the filtration.

3.4. The Variation of Cake Thickness with Cake Pressure Drop

If we rewrite Eq. (3.43) in terms of relative discharges (Eq. (3.26)) and use Eq. (3.30), we obtain

$$\bar{n}\frac{dL}{dt} + L\frac{d\bar{n}}{dt} - n_L\frac{dL}{dt} + \vec{q}_{r_L} \cdot \nabla F_2 + n_L \vec{V}_{s_L} \cdot \nabla F_2 - q_o \cdot \nabla F_1 = 0 \tag{3.48}$$

Noting that in Eq. (3.48), $\vec{V}_{s_o} \cdot \nabla F_1 = 0$ and $\vec{V}_{s_L} \cdot \nabla F_2 = dL/dt$,

$$\bar{n}\frac{dL}{dt} + \vec{q}_{r_L} \cdot \nabla F_2 - \vec{q}_{r_o} \cdot \nabla F_1 = 0 \tag{3.49}$$

The difference between the relative fluxes on the top and bottom of the cake can be approximated in terms of average relative flux. By definition of averaging and Darcy's law as shown by Corapcioglu and Bear (20)

$$\vec{q}_r = \frac{1}{L}\int \vec{q}_r \, dx = \frac{1}{L}\int [-K\nabla\phi] \, dx \tag{3.50}$$

or

$$\vec{q}_r = -\frac{\bar{K}}{L}\left[\frac{\Delta p}{\gamma_w} + L\right] \tag{3.51}$$

where $\bar{K}$ is the hydraulic conductivity of the cake and γ_w is the specific weight of water. $\Delta p = p_2 - p_1$ is the pressure drop. p_2 denotes the applied pressure on the cake surface and p_1 is the pressure at $x = 0$. We realize that the cake thickness L is relatively small, i.e., at the order of a few centimeters. The the gravity terms in Eq. (3.51) are much smaller than the pressure term. Then,

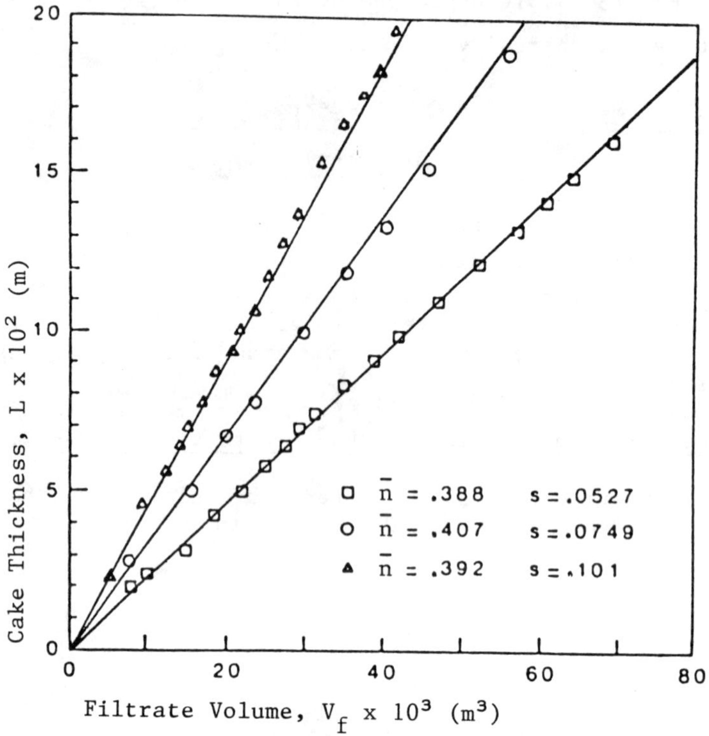

Figure 14. Comparison of theoretical results given by Eq. (3.47) with the experimental data (124). Filtrations of Lucite in water were measured at three different levels of s.

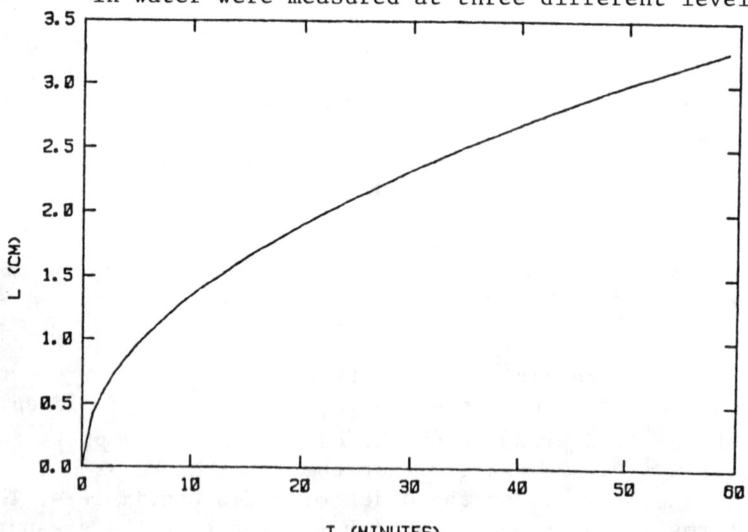

Figure 15. Temporal variation of cake thickness for Eq. (3.54).

$$\vec{q}_r \quad - \frac{\bar{K}}{L}\frac{\Delta p}{\gamma_w} \tag{3.52}$$

To obtain an analytical solution for the cake thickness, we insert Eq. (3.52) into Eq. (3.49)

$$\bar{n}\frac{dL}{dt} - \frac{\bar{K}}{L}\frac{\Delta p}{\gamma_w} = 0 \tag{3.53}$$

The solution of Eq. (3.53) for L(t=0)=0 would be

$$L^2 = \frac{2\bar{K}}{\bar{n}}\frac{\Delta p}{\gamma_w}\,t = \frac{2\bar{k}\,\Delta p}{\bar{n}\mu_w}\,t \tag{3.54}$$

where $\bar{k}$ is the average cake permeability, and μ is the viscosity. This solution is similar to the empirical solution obtained by Collins (17) and by von Engelhardt (120) which are discussed in section (3.1). Using the model parameters given in Table 3, a graphical representation of L and $\bar{n}$ versus t is given in Figure 15.

3.5. Pressure Variation Along the Cake Thickness

To obtain an equation for pressure variations in filter cakes with variable thickness, we follow the procedure given by Corapcioglu et al. (23). As shown by Corapcioglu and others (20,23) in a one-dimensional cake by neglecting the fluid's compressibility, and the gravity, we obtain the following equation for pore pressure p.

$$C_v \frac{\partial^2 p}{\partial x^2} = \frac{\partial p}{\partial t} + \gamma' \frac{\partial L}{\partial t} \tag{3.55}$$

where

$$\gamma' = [n\gamma_w + (1-n)\gamma_s] \tag{3.56}$$

and

$$C_v = \frac{\bar{k}}{\mu_w \alpha'} \tag{3.57}$$

α' is the compressibility coefficient of the cake. With an applied pressure on the cake surface, $p(L,t) = p_2$, a constant pressure at the filter-septum interface, $p(o,t) = p_1$, and a constant initial pressure, Eq. (3.55) would yield a solution for L as expressed by Eq. (3.54)

$$p = \sqrt{t}\left\{\frac{p_2 y}{\sqrt{t}} - \gamma' M(1-y) + \left[\frac{p_1}{\sqrt{t}} + \gamma' M\right]\left\{\exp\left(-\frac{M^2 y^2}{4C_v}\right)\right.\right.$$

$$\left.\left. - y\exp\left(-\frac{M^2}{4C_v}\right) - \frac{yM}{2}\sqrt{\frac{\pi}{C_v}}\left[\operatorname{erf}\left(\frac{M}{2\sqrt{C_v}}\right) - \operatorname{erf}\left(\frac{My}{2\sqrt{C_v}}\right)\right]\right\}\right\} \quad (3.58)$$

where

$$M = \sqrt{\frac{2\Delta p\bar{k}}{\mu}}, \quad y = x/L \quad (3.59)$$

A plot of Eq. (3.58) is given in Fig. 16 for different values of C_v. An examination of Fig. 16 reveals that when C_v decreases, p/p_2 reaches to a linear distribution. This implies that when C_v decreases, the filter cake becomes more and more compressible. As shown in Fig. 11, this phenomenon has been observed by Tiller et al. (108) and has been illustrated by Tosun (111).

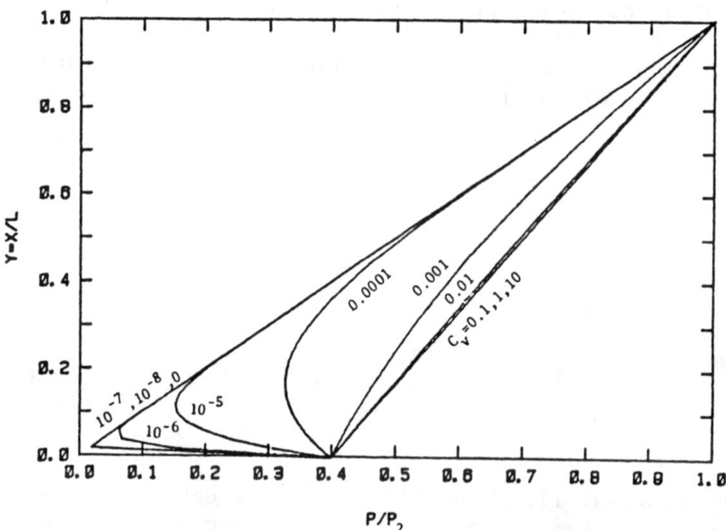

Figure 16. Variation of non-dimensionalized pressure along the cake thickness for different values of C_v (cm^2/sec).

4. SUMMARY AND CONCLUSIONS

In this paper, the equations governing the transport and capture of suspended solid particles have been studied in two categories. The first category includes transport and deposition of particles in an established porous medium. In this category, following a review of governing equations and various capture mechanisms in deep bed filters, the transport equation for microbial particles has been studied. For microbial particles, the governing equation for bacterial transport is coupled with a transport equation for the bacterial nutrient present in the suspension. The deposition and declogging mechanisms are incorporated into the model as a rate process for bacteria,and as an equilibrium partitioning for viruses. While the decay is assumed to be a first order reaction and the growth of bacteria is assumed to follow the Monod equation, the model equations exhibit nonlinearity and coupling. Coupled one-dimensional numerical solutions are obtained at spatial and temporal locations of interest.

Formation of a cake by deposition of solid particles on a filter cloth or on a previous cake,constitutes the second category. This type of filter media is called filter cakes. During filtration, the cake is compacted while new solid particles are laid down. Following a literature survey, a governing equation for cake thickness is obtained by averaging the conservation of mass equation for solid particles along the cake thickness. Then, the first order ordinary differential equation is solved with known average porosity functions,with the assumption of total deposition of solid particles on the cake surface. Results show that the cake thickness changes linearly for constant average cake porosity. In addition to the conservation of mass equation for solid particles, the conservation of mass equation for liquid has been averaged and solved simultaneously with the preceeding one to obtain expressions for the cake thickness in terms of filtrate volume. Excellent agreement has been observed between experimental and theoretical results. The resulting expression shows that the cake thickness is proportional to the square root of time for a constant cake pressure drop. Furthermore, the spatial variation of pore liquid pressure across the filter cake is obtained with a variable total stress expression.

One very important note should be made in regard to the use of these predictive filtration equations. The user should ensure that the model parameters are either experimentally or theoretically available,and that the governing equations represent the system under consideration. This is the only approach to derive reliable results from such studies. Finally, we would like to emphasize that in view of various assumptions and approximations, the results which are obtained by numerical or analytical solutions should be viewed as estimates. These estimates, however, should be useful for assessing the efficiency of a filtration operation or the outcome of a particular wastewater management scheme.

5. ACKNOWLEDGEMENT

The research on which this work is based was financed in part by the United States Department of the Interior as authorized by Grant Number: 14-08-0001-G 833/02.

6. REFERENCES

1. Adamczyk, Z., Dabros, T., Czarnecki, J., and van de Ven, T.G.H., "Particle Transfer to Solid Surfaces," *Advances in Colloid and Interface Science,* Vol. 19, 1983, pp. 183-252.
2. Adin, A., "Solution of Granular Bed Filtration Equations," *Journal of the Environmental Engineering Division,* American Society of Civil Engineers, Vol. 104, No. EE3, June, 1978, pp. 471-483.
3. Adin, A., and Rebhun, M., "A Model to Predict Concentration and Head-Loss Profiles in Filtration," *Journal of the American Water Works Association,* Vol. 69, No. 8, Aug., 1977, pp. 444-453.
4. Anderson, H.M., and Edwards, R.V., "Computational Methods and Control Strategies for Continuous Sedimentation in the Activated Sludge Process," presented at the October 16-17, 1980 International Symposium on Environmental Pollution, held at Atlanta, Georgia.
5. Atsumi, K., and Akiyama, T., "A Study of Cake Filtrations," *Journal of Chemical Engineers of Japan,* Vol. 8, No. 6, 1975, pp. 487-492.
6. Avogadro, A., and de Marsily, G., "The Role of Colloids in Nuclear Waste Disposal," *Proceedings*, Material Research Society, Elsevier, Vol. 26, 1984, pp. 495-505.
7. Barkman, J.H., and Davidson, D.H., "Measuring Water Quality and Predicting Well Impairment," *Journal of Petroleum Technology* July, 1972, pp. 865-873.
8. Bear, J., and Bachmat, Y., "Transport Phenomena in Porous Media-Basic Equations," *Fundamentals of Transport Phenomena in Porous Media,* J. Bear and M.Y. Corapcioglu, eds., Martinuj Nijhoff, Dordrecht, 1984, pp. 3-61.
9. Bear, J., Corapcioglu, M.Y., and Balakrishna, J., "Modeling of Centrifugal Filtration in Unsaturated Deformable Porous Media," *Advances in Water Resources,* Vol. 7, 1984, pp. 150-167.
10. Binkley, G.W., Dumbauld, G.F., and Collins, R.E., "Factors Affecting the Rate of Deposition of Cement in Unfractured Perforations During Squeeze-Cementing Operations," *Petroleum Transactions,* American Institute of Mining Engineers, Vol. 213, 1958, pp. 51-58.

11. Bitton, G., *Introduction to Environmental Virology*, Wiley-Interscience, New York, 1980.
12. Bitton, G., Davidson, J.J., and Farrah, S.R., "On the Value of Soil Columns for Assessing the Transport Pattern of Viruses Through Soils. A Critical Outlook," *Water, Air and Soil Pollution*, Vol. 12, 1979, pp. 449-457.
13. Burge, W.D., and Enkiri, N.K., "Virus Adsorption by Fine Soils," *Journal of Environmental Quality*, Vol. 7, 1978, pp. 73-76.
14. Butler, R.G., Orlob, G.T., McCauhey, P.H., "Underground Movement of Bacterial and Chemical Pollutants," *Journal of the American Water Works Association*, Vol. 46, 1954, pp. 97-111.
15. Carman, P.C., "Fluid Flow Through a Granular Bed," *Transactions*, Institute of Chemical Engineers(London), Vol. 15, 1937, pp. 150-156.
16. Cookson, J.T., "Removal of Submicron Particles in Packed Beds," *Environmental Science and Technology*, Vol. 4, 1970, pp. 128-134.
17. Collins, R.E., *Flow of Fluids Through Porous Materials*, Reinhold, New York, 1961.
18. Corapcioglu, M.Y., "Formation of Filter Cakes," *Filtration and Separation*, Vol. 18, No. 4, 1981, pp. 324-326.
19. Corapcioglu, M.Y., "Deposition of Solids in Drilling Fluids on Bore Hole Walls," (submitted for publication, 1987).
20. Corapcioglu, M.Y., and Bear, J., "A Mathematical Model for Regional Land Subsidence Due to Pumping: 3. Integrated Equations for a Phreatic Aquifer," *Water Resources Research*, Vol. 4, 1983, pp. 895-908.
21. Corapcioglu, M.Y., and Haridas, A, "Transport and Fate of Microorganisms in Porous Media: A Theoretical Investigation," *Journal of Hydrology*, Vol. 72, 1984, pp. 149-169.
22. Corapcioglu, M.Y., and Haridas, A., "Microbial Transport in Soils and Groundwater: A Numerical Model," *Advances in Water Resources*, Vol. 8, No. 6, 1985, pp. 188-200.
23. Corapcioglu, M.Y., Bear, J., and Mathur, S., "One-Dimensional Land Subsidence with Variable Total Stress," *Land Subsidence*, A.I. Johnson, L. Carbognin, and L. Ubertini, eds., International Association of Hydrological Sciences, Wallingford, United Kingdom, Publication No. 151, 1986, pp. 19-28.
24. Dahlquist, F.W., Lovely, P., and Koshland, D.E., "Quantitative Analysis of Bacterial Migration in Chemotaxis," *Nature New Biology*, Vol. 236, 1972, pp. 120-123.
25. Deb, A.K., "Theory of Sand Filtration," *Journal of the Sanitary Engineering Division*, American Society of Civil Engineers, Vol. 95, No. SA3, 1969, pp. 399-422.
26. Diaper, E.W.J., Ives, K.J., "Filtration Through Size-Graded Media," *Journal of the Sanitary Engineering Division*, American Society of Civil Engineers, Vol. 91, No. SA3, 1965, p. 89.
27. Dillingham, J.H., Cleasby, J.L., and Baumann, E.R., "Diatomite Filtration Equations for Various Septa," *Journal of the Sanitary Engineering Division*, American Society of Civil Engineers, Vol. 93, No. SA1, 1967, pp. 41-54.

28. Donaldson, E.C., Baker, B.A., and Carroll, H.B., "Particle Transport in Sandstones," presented at the October 9-12, 1977, Society of Petroleum Engineers 52nd Annual Fall Technical Conference, held at Denver, Colorado, (Paper No. 6905).
29. Drewry, W.A., and Eliassen, R., "Virus Movement in Groundwater," *Journal of Water Pollution Control Federation,* Vol. 40, 1968, pp. 257-271.
30. Filmer, R.W., and Corey, A.T., "Transport and Retention of Virus-sized particles in Porous Media," Sanitary Engineering Papers, No. 1, Colorado State University, Fort Collins, Colorado, 1966.
31. Filmer, R.W., Felton, M., and Yamamoto, T., "Virus Sized Particle Adsorption on Soil, Part 1. Rate of Adsorption," *Proceedings of 13th Conference on Water Quality Control,* University of Illinois, Urbana-Champaign, Illinois, 1971, pp. 75-101.
32. Gaudy, A., and Gaudy, E., *Microbiology for Environmental Scientists and Engineers,* McGraw-Hill Book Company, New York, New York, 1980.
33. Gerba, C.P., Wallis, C., and Melnich, J.L., "Fate of Wastewater Bacteria and Viruses in Soil," *Journal of the Irrigation and Drainage Division,* American Society of Civil Engineers, Vol. 101, No. IR3, 1975, pp. 157-174.
34. Gerba, C.P., "Virus Survival and Transport in Groundwater," *Industrial Microbiology,* Society for Industrial Microbiology, Arlington, Virginia, 1983, pp. 247-251.
35. Grichar, C., "The Mechanical Treatment of Solids in Drilling Fluids," *Water Well Journal,* November, 1983, pp. 46-54.
36. Grosser, P.W., "A One-Dimensional Mathematical Model of Virus Transport," *Proceedings of the Second International Conference on Groundwater Quality Research,* National Center for Groundwater Research, March 26-29, 1984, Tulsa, Oklahoma, 1984, pp. 105-107.
37. Gruesbeck, C., and Collins, R.E., "Entrainment and Deposition of Fine Particles in Porous Media," *Society of Petroleum Engineers Journal,* Vol. 22, No. 6, 1982, pp. 847-856.
38. Hall, W.A., "An Analysis of Some Theories of Sand Filtration," thesis presented to the University of California, at Los Angeles, CA, in 1952, in partial fulfillment of the requirements for the degree of Doctor of Philosophy.
39. Hagedorn, C., "Transport and Fate:Bacterial Pathogens in Groundwater," in "Microbial Health Considerations of Soil Disposal of Domestic Wastewaters," *EPA-600/9-83-017,* Municipal Environmental Research Laboratory, Cincinnati, Ohio, 1983, pp. 153-171.
40. Heertjes, P.M., Lerk, C.F., "The Functioning of Deep Bed Filters, Part II, The Filtration of Flocculated Suspensions," *Transactions,* Institute of Chemical Engineers, Vol. 45, 1967, p. T138.

41. Herzig, J.P., Leclerc, D.M., and Le Golf, P., "Flow of Suspensions Through Porous Media - Application to Deep Filtration," *Flow Through Porous Media*, American Chemical Society, Washington, D.C., 1970, pp. 129-157.
42. Hill, A.D., and Galloway, P.J., "Laboratory and Theoretical Modeling of Diverting Agent Behaviour," paper presented at the Feb., 1983, Society of Petroleum Engineers Production Operation Symposium, held at Oklahoma City, Oklahoma, (SPE 11576).
43. Hsieh, J.S.C., Turian, R.M., and Tien, C., "Multicomponent Liquid Phase Adsorption in Fixed Bed," *American Institute of Chemical Engineering Journal*, Vol. 23, 1976, p. 263.
44. Hutto, F.B., "Distribution of Porosity in Filter Cakes," *Chemical Engineering Progress*, Vol. 53, No. 7, 1957, pp. 328-332.
45. Irmay, S., "The Mechanism of Filtration of Non-Colloidal Fines in Porous Media," *SFB 80/T/123*, Universitat Karlsruhe, 1979, pp. 1-24.
46. Ives, K.J., "Rational Design of Filters," *Proceedings of the Institute of Civil Engineers*, Vol. 16, 1960, p. 189.
47. Ives, K.J., "Simplified Rational Analysis of Filter Behaviour," *Proceedings of the Institute of Civil Engineers*, Vol. 25, 1963, pp. 345-364.
48. Ives, K.J., "Mathematical Models of Deep Bed Filtration," *The Scientific Basis of Filtration*, K.J. Ives, ed., Noordhoff, Leyden, 1975, pp. 203-224.
49. Ives, K.J., "Capture Mechanisms in Filtration," *The Scientific Basis of Filtration*, K.J. Ives, ed., Noordhoff, Leyden, 1975, pp. 183-201.
50. Ives, K.J., "Deep Bed Filter," *Mathematical Models and Design Methods in Solid-Liquid Separation*, A. Rushton, ed., Martinuj Nijhoff, Dordrecht, 1985, pp. 90-149.
51. Ives, K.J., and Pienvichitr, V., "Kinetics of the Filtration of Dilute Suspensions," *Chemical Engineering Science*, Vol. 20, 1965, pp. 965-973.
52. Iwasaki, T.J., "Some Notes on Sand Filtration," *American Water Works Association*, Vol. 29, 1937, pp. 1591-1602.
53. Jang, L.K., Sharma, M.M., Findley, J.E., Chang, P.W., and Yen, J.E., "An Investigation of the Transport of Bacteria Through Porous Media," *Proceedings of the International Symposium on Microbial Enhancement of Oil Recovery*, E.C. Donaldson and J.B. Clark, eds., U.S. Department of Energy, 1983, pp. 60-70.
54. Jang, L.K., Sharma, M.M., and Yen, T.F., "The Transport of Bacteria in Porous Media and Its Significance in Microbial Enhanced Oil Recovery," presented at the April 11-13,1984, Society of Petroleum Engineers California Regional Meeting, held at Long Beach, California, (SPE 12770).
55. Jeremiah's Well, "Drilling with Mud," *Water Well Journal*, Vol. 40, No. 5, 1986, pp. 42-43.

56. Keller, E.F., and Segel, L.A., "Model for Chemotaxis," *Journal of Theoretical Biology,* Vol. 30, 1971, pp. 225-243.
57. Keswick, B.H., and Gerba, C.P., "Viruses in Groundwater," *Environmental Science and Technology,* Vol. 14, NO. 11, 1980, pp. 1290-1297.
58. Keswick, B.H., Wang, D.S., and Gerba, C.P., "The Use of Microorganisms as Groundwater Traces: A Review," *Ground Water,* Vol. 20, 1982, pp. 142-149.
59. Khilar, K.C., Fogler, H.S., and Gray, D.H., "Model for Piping-Plugging in Earthen Structures," *Journal of Geotechnical Engineering,* American Society of Civil Engineers, Vol. 111, No. 7, 1985, pp. 833-846.
60. Kovacs, G., and Ujfaludi, ., "Movement of Fine Grains in the Vicinity of Well Screens," *Hydrological Sciences Journal,* Vol. 28, No. 2, 1983, pp. 247-259.
61. Krone, R.B., Orlob, G.T., and Hodghinson, C., "Movement of Coliform Bacteria Through Porous Media," *Sewage and Industrial Wastes,* Vol. 30, 1958, pp. 1-13.
62. Krone, R.B., "The Movement of Disease Producing Organisms Through Soils," *Municipal Sewage Effluent for Irrigation,* C.W. Wilson and F.E. Beckett, eds., Louisiana Tech. Department of Agricultural Engineering, Ruston, Louisiana, 1968, pp. 75-105.
63. Mackrle, V., Dracka, O., Sved, J., *Hydrodynamics of the Disposal of Low Level Liquid Radioactive Wastes in Soil,* International Atomic Energy Agency Contract Report No. 98, Czechoslovakia Academy of Science, Institute of Hydrodynamics, Prague, 1965.
64. Maroudas, A., and Eisenklam, P., "Clarification of Suspensions: A Study of Particle Deposition in Porous Media, Part 1. Some Observations on Particle Deposition," *Chemical Engineering Science,* Vol. 20, 1965, pp. 867-873.
65. Maroudas, A., and Eisenklam, P., "Clarification of Suspensions: A Study of Particle Deposition in Granular Media, Part 2. A Theory of Clarification," *Chemical Engineering Science,* Vol. 20, 1965, pp. 875-888.
66. Maroudas, A., "Particle Deposition in Granular Filter Media-1," *Filtration and Seperation,* Vol. 2, No. 2, 1965, pp. 369-372.
67. Maroudas, A., "Particle Deposition in Granular Filter Media-2," *Filtration and Seperation,* Vol. 3, No. 2, 1966, p. 115-126.
68. Matthess, G., and Pekdeger, A., "Concepts of a Survival and Transport Model of Pathogenic Bacteria and Viruses in Groundwater," *Proceedings of International Symposium on Quality of Groundwater,* Elsevier, Amsterdam, 1981, pp. 427-437.
69. McDowell-Boyer, L.M., Hunt, J.R., and Sitar, N., "Particle Transport Through Porous Media," *Water Resources Research,* Vol. 22, No. 13, 1986, pp. 1901-1921.
70. Mints, D.M., "Filtration Kinetics of a Low Concentration Suspension in Water of Water Clarification Filters," *Dokl. Akad. Nauk, S.S.S.R.,*(in Russian), Vol.78, 1951, pp. 315-318.

71. Monod, J., *Recherches sur la Croissance des Cultures Bacteriennes,* Hermann et Cie, Paris, 1942.
72. Muecke, T.W,m "Formation of Fines and Factors Controlling Their Movement in Porous Media," *Journal of Petroleum Technology,* Vol. 30, 1979, pp. 144-50.
73. Nuttall, H.E., "Population Balance Model for Colloid Transport," *NNWSI Report/Milestone R318,* Los Alamos National Laboratory, New Mexico, 1986 (also submitted to Water Resources Research)
74. Ogata, A., "Mathematics of Dispersion with Linear Adsorption Isotherm," *U.S. Geological Survey Professional Paper,* 1964, pp. H1-H9.
75. O'Melia, C.R., "Particles, Pretreatment and Performance in Water Filtration," *Journal of Environmental Engineering,* American Society of Civil Engineers, Vol. 111, No. 6, 1985, pp. 874-890.
76. Payatakes, A.C., Rajagopalan, R., and Tien, C., "Application of Porous Media Models to the Study of Deep Bed Filtration," *Canadian Journal of Chemical Engineering,* Vol. 52, 1974, p. 727.
77. Polprasert, C., and Hoang, L.H., "Kinetics of Bacteria and Bacteriophages in Anaerobic Filters," *Journal of Water Pollution Control Federation,* Vol. 35, 1983, pp. 385-391.
78. Rajagopalan, R., and Tien, C., "The Theory of Deep Bed Filtration," *Progress in Filtration and Seperation,* Vol. 1, R.J. Wakeman, ed., Elsevier, New York, 1979, pp. 179-269.
79. Raleigh, J.T., and Flock, D.L., "A Study of Formation Plugging with Bacteria," *Journal of Petroleum Technology,* Feb., 1965, pp. 201-206.
80. Rietema, K., "Stabilizing Effects in Compressible Filter Cakes," *Chemical Engineering Science,* Vol. 2, 1953, pp. 88-94.
81. Ring, T.A., "Bed Filtration of Colloidal Particles," presented at the 1983, American Institute of Mining Engineers Annual Conference, held at Atlanta, Georgia.
82. Romero, J.C., "The Movement of Bacteria and Viruses Through Porous Media," *Ground Water,* Vol. 8, 1970, pp. 37-48.
83. Ruth, B.F., "Studies in Filtration III," *Industrial Engineering Chemistry,* Vol. 27, 1935, p. 708.
84. Sakthivadivel, R., "Theory and Mechanism of Filtration of Non-Colloidal Fines Through a Porous Medium," *Report HEL-15-5,* Hydraulic Engineering Laboratory, University of California, Berkeley, California, 1966.
85. Sakthivadivel, R., and Irmay, S., "A Review of Filtration Theories," *Interim Report HEL-1S-4,* University of California, Berkeley, California, 1966.
86. Saltelli, A., Avagadro, A., and Bidoglio, G., "Americium Filtration in Glauconitic Sand Columns," *Nuclear Technology,* Vol. 67, 1984, pp. 245-254.
87. Selim, H.M., Davidson, J.M., and Rao, P.S., "Transport of Reactive Solutes Through Multilayered Soils," *Soil Science*

Society of America Journal, Vol. 41, 1977, pp. 3-10.

88. Sharma, M.M., Chang, Y.I., and Yen, T.F., "Reversible and Irreversible Surface Charge Modification of Bacteria for Facilitating Transport Through Porous Media," *Colloids and Surfaces*, Vol. 16, 1985, pp. 193-206.
89. Sharma, M.M., and Yortsos, Y.C., "Permeability Impairment due to Fines Migration in Sandstones," presented at the Feb., 26-27, 1986, Society of Petroleum Engineers Seventh Symposium on Formation Damage Control, held at Lafayette, Louisiana (SPE 14819).
90. Sharma, M.M., and Yortsos, Y.C., "Transport of Particulate Suspensions in Porous Media," (submitted for publication)
91. Shekhtman, Y. M., *Filtration of Low Concentration Suspensions*, Publishing House of the USSR Academy of Sciences, Moscow, 1961 (in Russian).
92. Shirato, M., Sambuichi, M., and Okamura, S., "Filtration Behaviour of a Mixture of Two Slurries," *American Institute of Chemical Engineering Journal*, Vol. 9, No. 5, 1963, pp. 599-605.
93. Shirato, M., Sambuichi, M., Kato, H., and Aragaki, T., "Internal Flow Mechanisms in Filter Cakes," *American Institute of Chemical Engineering Journal*, Vol. 15, NO. 3, 1969, pp. 405-490.
94. Shirato, M., Aragaki, T., Ichimura, K., and Ootsuji, N., "Porosity Variation in Filter Cake Under Constant-Pressure Filtration," *Journal of Chemical Engineering of Japan*, Vol. 4, No. 2, 1971, pp. 172-177.
95. Smith, M.S., Thomas, G.W., White, R.E., and Ritonga, D., "Transport of Escherichia Coli Through Intact and Disturbed Soil Columns," *Journal of Environmental Quality*, Vol. 14, No.1, 1985, pp. 87-91.
96. Spielman, L.A., "Particle Capture from Low-Speed Laminar Flows, *Annual Reviews in Fluid Mechanics*, Vol. 9, 1977, pp. 297-319.
97. Spielman, L.A., and Fitz Patrick, J.A., "Theory of Particle Collection Under London and Gravity Forces," *Journal of Colloid and Interface Science*, Vol. 42, 1973, p. 607.
98. Stephenson, R.V., and Baumann, E.R., "Precoat Filtration Equations for Flat and Cylindrical Septa," *Mathematical Models and Design Methods in Solid-Liquid Separation*, A. Rushton, ed., Martinuj Nijhoff, Dordrecht, 1985, pp. 233-256.
99. Swartzendruber, D., and Uebler, R.L., "Flow of Kaolinite and Sewage Suspensions in Sand and Sand-Silt: II. Hydraulic Conductivity Reduction," *Soil Science Society of America*, Vol. 46, 1982, pp. 912-916.
100. Sykes, J.F., Soyupak, S., and Farquhar, G.J., "Modeling of Leachate Organic Migration and Attenuation in Groundwaters Below Sanitary Landfills," *Water Resources Research*, Vol. 18, 1982, pp. 135-145.

101. Tchobanoglous, G., and Schroeder, E.G., *Water Quality*, Addison Wesley, Redding, Mass., 1985.
102. Tien, C., "Deep-Bed Filtration,"*Handbook of Fluids in Motion*, N.P. Cheremisinoff and R. Gupta, eds., Ann Arbor Science, Ann Arbor, Michigan, 1983, pp. 969-1002.
103. Tien, C., and Payatakes, A.C., "Advances in Deep Bed Filtration," *American Institute of Chemical Engineers Journal*, Vol. 25, No. 5, 1979, pp. 737-759.
104. Tiller, F.M., "Compressible Cake Filtration," *The Scientific Basis of Filtration*, K.J. Ives, ed., NATO Advanced Study Institute Series, Noordhoff, Leydem, Vol. 2, 1975, pp. 315-397.
105. Tiller, F.M., and Cooper, H.R., "The Role of Porosity in Filtration: IV - Constant Pressure Filtration," *American Institute of Chemical Engineering Journal*, Vol. 6, No. 4, 1960, pp. 595-601.
106. Tiller, F.M., and Shirato, M., "The Role of Porosity in Filtration: VI - New Definition of Filtration Resistance," *American Institute of Chemical Engineering Journal*, Vol. 10, No. 1, 1964, pp. 61-67.
107. Tiller, F.M., and Green, T.C., "Role of Porosity in Filtration: IV - Skin Effect with Highly Compressible Materials," *American Institute of Chemical Engineering Journal*, Vol. 19, No. 6, 1973, pp. 1266-1269.
108. Tiller, F.M., Shirato, M., and Alciatore, A., *Filtration Principles and Practices*, Part 1, C. Orr, ed., M. Dekker, Inc., New York, New York, 1977.
109. Tiller, F.M., Crump, J.R., and Ville, F., *Proceedings of 2nd World Filtration Congress*, P1, London, 1979.
110. Tiller, F.M., and Crump, J.R., "Recent Advances in Compressible Cake Filtration Theory," *Mathematical Models and Design Methods in Solid-Liquid Separation*, A. Rushton, ed., Martinuj Nijhoff Publishers, Dordrecht, The Netherlands, 1985, pp. 3-24.
111. Tosun, I., "Critical Review of the Cake Filtration Theory," *Belgian Filtration Society* (SBF), 1985.
112. Tosun, I., and Willis, M.S., "Energy Efficiency in Cake Filtration," *Filtration and Seperation*, Vol. 22, No. 2, 1985, pp. 126-127.
113. Tosun, I., and Willis, M.S., "Fluid Velocity Variation in Filter Cakes," *International Journal of Multiphase Flow*, Vol. 8, 1982, pp. 1-3.
114. Travis, B.J., and Nuttall, H.E., "Analysis of Colloid Transport," *Proceedings of the International Symposium on Coupled Processes Affecting the Performance of a Nuclear Waste Repository*, Lawrence Berkeley Laboratory, California, September 18-20, 1985, pp. 205-209.
115. Travis, C.C., and Etnier, E.L., "A Survey of Sorption Relationships for Reactive Solutes in Soil," *Journal of Environmental Quality*, Vol. 10, No. 1, 1981, pp. 8-17.

116. Uebler, R.L., and Swartzendruber, D., "Flow of Kaolinite and Sewage Suspensions in Sand and Sand-Silt: I. Accumulation of Suspension Particles," *Soil Science Society of America Journal*, Vol. 46, 1982, pp. 239-244.
117. Vaughn, J.M., Landry, E.F., and Thomas, M.Z., "The Lateral Movement of Indigenous Enteroviruses in a Sandy Sole-Source Aquifer," *Microbial Health Considerations of Soil Disposal of Domestic Wastewaters, EPA-600/9-83-017*, Municipal Environmental Research Laboratory, Cincinnati, Ohio, 1983, pp. 214-230.
118. Vetter, D.J., Kandarpa, V., and Harouaka, A., "Flow of Particle Suspensions Through Porous Media," U.S. Department of Energy Division of Geothermal Energy, Washington, D.C., 1982.
119. Vilker, V.L., "Simulating Virus Movement in Soils," *Modeling Wastewater Renovation Land Treatment*, I.K. Iskander, ed., Wiley-Interscience, New York, 1981, pp. 223-253.
120. Von Engelhardt, E., "Filter Cake Formation and Water Losses in Deep Drilling Muds," (English Translation) *Circular 191*, Division of the State Geological Survey, Urbana, Illinois,1954.
121. Wakeman, R. J., "A Numerical Integration of the Differential Equation Describing the Formation of and Flow in Compressible Filter Cakes," *Transactions*, Institute of Chemical Engineering, Vol. 56, 1978, pp. 258-265.
122. Wakeman, R.J., "The Formation and Properties of Apparently Incompressible Filter Cakes Under Vacuum on Downward Facing Surfaces," *Transactions*, Institute of Chemical Engineering, Vol. 59, 1981, pp. 260-270.
123. Willis, M.S., and Tosun, I., "A Rigorous Cake Filtration Theory," *Chemical Engineering Science*, Vol. 35, 1980, pp. 2427-2438.
124. Willis, M.S., Collins, R.M., and Bridges, W.G., "Complete Analysis of Non-Parabolic Filtration Behaviour," *Chemical Engineering Research and Design*, Vol. 61, March 1983, pp. 96-109.
125. Willis, M.S., Tosun, I., and Collins, R.M., "Filtration Mechanisms," *Chemical Engineering Research and Design*, Vol.63, May 1985, pp. 175-183.
126. Willis, M.S., "A Multiphase Theory of Filtration," *Progress in Filtration and Separation*, R.J. Wakeman, ed., Vol. 3, Elsevier Scientific Publishing Company, Amsterdam, The Netherlands, 1983, pp. 1-56.
127. Wnek, W.J., Gidaspow, D., and Wasan, D.T., "The Role of Colloid Chemistry in Modeling Deep Bed Filtration," *Chemical Engineering Science*, Vol. 30, 1975, pp. 1035-1047.
128. Wollum II, A.G., and Cassel, D.K., "Transport of Microorganisms in Sand Columns," *Soil Science Society of America Journal*, Vol. 42, 1978, pp. 72-76.
129. Wright, A.M., "Filtration Modeling," thesis presented to the University of California, at Berkeley, California, in 1983,

in partial fulfillment of the requirements for the degree of Doctor of Philosophy.

130. Yao, K.M., Habibian, M.T., and O'Melia, C.R., "Water and Wastewater Filtration: Concepts and Applications," *Environmental Science and Technology*, Vol. 5, 1971, pp. 1105-1112.

131. Zyman, J., "Influence of Rainfall Rates on the Survival and Migration of Selected Indicator Organisms in Sludge Amended Soils," thesis presented to the University of Texas, at Austin, Texas, in December 1986, in partial fulfillment of the requirements for the degree of Master of Science.

7. LIST OF SYMBOLS

A	A constant
A^*	Hamaker's constant
A_3, A_4	Constants
a	A constant
a''	A constant
a^*	A constant
B	A constant
b	A constant
C, C'	Concentration of suspended particles $[M/V_w]$
$\bar{C}$	Concentration of adsorbed phase
C_F	Substrate concentration
C_P	Slurry concentration
C_v	Consolidation coefficient
D	Effective dispersion coefficient
D_B	Brownian diffusion coefficient of particles
D_F	Hydrodynamic dispersion coefficient for substrate
D_T	Diffusivity coefficient due to random tumbling
D^*	Molecular diffusion coefficient
d	Particle diameter
d_g	Grain diameter
e	Void ratio
F	A moving boundary (septum)
F_1	A moving boundary
F_2	A moving boundary (cake surface)

h	The power term in the kinetic equation
$\vec{J}$	Specific mass discharge of particles
$\vec{J}_B$	Specific mass discharge of micro-organisms due to Brownian motion
$\vec{J}_{CT}$	Specific mass discharge of micro-organisms due to chemotactic movement
K	Hydraulic conductivity
K_1^*, K_2^*	Functions [see Eq. (2.48)]
K_s	Monod half constant
k	Cake permeability
k'	First order growth of decay constant in analytical solution
k_a	Adsorption isotherm constant
k_b	Boltzman's constant
k_c	Clogging rate constant
k_d	Specific decay rate
k_m	Migration rate constant (or chemotactic coefficient)
k_y	Declogging rate constant
L	Cake thickness
L^*	Thickness of the porous plate
M	Mass
M	A coefficient
m	The power term in adsorption isotherm
n	Porosity
n_o	Initial porosity
n_d	Porosity of deposited particles
n_L	Surface porosity
$P(t)$	A function of time
$P_1(t)$	A function of time
p	Pore liquid pressure
p_1	Pressure at the interface of medium and cake
p_2	Applied filtration pressure at the cake surface
$Q(t)$	A function of time
$Q_1(t)$	A function of time
Q_i	Parameter vector which determines the mode of deposit morphology

q_r	Specific discharge relative to the moving solid
q_p	Slurry flow rate
q_o	Flow rate at the filter septum
$\bar{q}$	Average flow rate at the medium
$\vec{q}$	Specific discharge vector
q_z	Vertical flow rate
R_a	Rate of deposition of particles $[M/V_T.T]$
R_d	Rate of decay
R_g	Rate of growth
S,S_c	Growth or decay rates
S_F	Mass of adsorbed substrate per unit mass of soil grains
S	Dimensionless slurry concentration $[M/M_w]$
T	Absolute temperature
t	Time
U	Superficial velocity
$\vec{u}$	Flow velocity
u_m	Suspension approach velocity
V_f	Filtrate volume
V_w	Liquid volume
V_s	Solid volume
V_T	Volume of porous medium $(=V_s+V_w)$
$\vec{V}_s$	Solid velocity vector
$\vec{V}_L$	Liquid velocity vector
v_f	Velocity of water flow
v_g	Gravitational settling velocity
Y	Growth yield coefficient, mass of micro-organisms per unit mass of substrate utilized
z,x	Vertical direction
α_1, α_2	Parameters of Langmuir equation
α'	Coefficient of compressibility of a moving solid matrix
α^*	Scour coefficient
β^*	Bulking factor

γ_s	Specific weight of solid particles
γ_w	Specific weight of water
γ'	Specific weight of saturated cake
λ	Filter coefficient
μ	Specific growth rate
μ_w	Viscosity of water
ρ	Density of micro-organisms
ρ_s	Particle density
ρ_s'	Bulk density of dry soil
ρ_w	Density of water $[M/V_w]$
σ	Volume of deposited particles per unit volume of total porous medium
σ_T	Total stress
σ'	Effective stress
η_1	A constant
η_2	A constant
θ	Volume occupied by the suspension per unit volume of total porous medium
θ'	Retention age
τ ξ	Dummy integration variables
Δp	Pressure drop

Superscripts

$-$	(Overbar) indicates averaged variable
$\rightarrow$	Indicates vector

Subscripts

f	Indicates liquid state
o	Indicates initial state
s	Indicates deposited state
1	Indicates component at the filter septum
L,i,2	Indicates component at the cake surface

THEORY OF FILTRATION

Max S. Willis, Steve Bybyk, Ray Collins, Jagadeeshan Raviprakash

THEORY OF FILTRATION

Max S. Willis, Steve Bybyk, Ray Collins, Jagadeeshan Raviprakash

Chemical Engineering Department
The University of Akron
Akron, Ohio 44325 USA.

ABSTRACT

The objective is to describe a fundamental analysis which provides a filtration mechanism that can be used to select filters and filter media.

General multiphase balances are developed from the principles of volume averaging and the new continuum requires constitutive equations which may be different from those of the individual phases. Linear constitutive equations are developed which are function of multiphase variables. For the filtration analysis, governing equations for both non-Newtonian and Newtonian fluids are derived.

Combination of the constitutive and general balance equations is subjected to dimensional analysis to determine that the dominant terms are the pressure and drag forces for both classes of fluids.

Filtrate rate expressions are derived for non-Newtonian and Newtonian fluids, one dimensional cylindrical and circular leaf filters. All are governed by an analogous rate expression.

Experimental data shows the effect of cake pressure drop, filter area, particle size, particulate phases, and filter cake geometry.

1. RESEARCH PERSPECTIVE

Selection of filtration equipment (6,19) relies on (i) duplication of existing operational equipment, (ii) industrial standards when there are few variations from plant to plant, and (iii) pilot plant runs which closely approximate actual operations.

Filter media selection relies on (i) the permeability of the medium as determined from the passage of clear filtrate, (ii) the retention properties relative to the particle size, and (iii) pore size distributions. The data considered most reliable for media selection are obtained from full size equipment in plant tests.

The analysis of filters is largely empirical and consists of correlations in which the design parameters are determined from data obtained from pilot plant or full size equipment. These correlations (16,21,23) normally attribute the dominant resistance to the filter cake and neglect the resistance of the filter media.

The objective here is to describe a fundamental analysis which provides an experimentally verified filtration mechanism and a predictive capability for the selection of filters and filter media that current practice does not have. This fundamental approach (27,28,29) is based on the general multiphase continuum theory and the following generalized design procedure.

1.1 Theory and Experiment in Single Phase Systems

Engineering design problems require integration of theory and experiment and can be placed into one of three levels that depend on the number of independent variables required to describe the problem. Table 1 summarizes this interaction for single phase systems.

Table 1. The Interaction of Theory and Experiment at Three Levels for Single-Phase Engineering Design Problems.

	Theory	Experiment
Microscopic Level ($\underline{x}$,t)	Equations of Change	Constitutive Equations
Macroscopic Level (-,t)	Design Equations	Process Correlations
Equilibrium (-,-)	Thermodynamic Laws	Property Correlations

1.1.1 Theory

In Table 1 under the theory column, the equations of change at the microscopic level are the conservation of mass, momentum, chemical species, energy, and entropy. These equations are spatial and time dependent partial differential equations.

The macroscopic level design equations of Table 1 are obtained from the equations of change by integrating (2) over an arbitrary engineering volume which exchanges mass and energy with the surroundings. For example, when the microscopic level mechanical energy balance is integrated over the arbitrary engineering volume it becomes the macroscopic level Bernoulli equation.

The thermodynamic laws of Table 1 are the conservation of energy and the second law restrictions that apply to spatial and time invariant equilibrium conditions.

1.1.2 Experiment

The microscopic level experimental information shown in Table 1 and termed constitutive equations refers to the additional equations which, when combined with the equations of change, comprise a determinate mathematical system. These constitutive equations include thermo-mechanical equations and equilibrium equations of state. The use of equilibrium property relations at the position and time dependent microscopic level is justified by the assumption of local equilibrium or the continuum hypothesis.

If the required number of constitutive equations is not available, then the mathematical description at the microscopic level is indeterminate and the design procedure then appeals to a process correlation to replace the analytical solution (i.e., the Fanning friction-factor correlation for turbulent tube flow).

All process correlations are limited to a specific geometry, equipment configuration, boundary conditions, and substance. Conversely, the macroscopic design equations are intrinsically independent of these factors. The combination of the process correlation and the macroscopic design equations provides design specification only for those processes which have the same geometry, configuration, boundary conditions, and substance as that for the process correlation.

The property correlations shown in the experiment column of Table 1 are obtained under conditions of thermodynamic equilibrium. Examples are equations of state, such as the ideal gas law, and the caloric equations of state for heat capacity.

This design procedure has evolved to reduce the cost of experimentation. For example, the use of dimensionless numbers, which are obtained from the dimensionless form of the microscopic level equations, reduces the number of independent experimental parameters that must be measured to obtain the process correlation. The experimental effort associated with the search for thermomechanical constitutive equations is reduced by constitutive theory. The kinetic theory of gases provides the temperature and pressure dependence of transport properties with a corresponding reduction in experimental effort. On the other hand, the analysis of multiphase systems has relied entirely on experimental process correlations obtained at the macroscopic level.

1.2 Theory and Experiment in Multiphase Systems

Traditionally, the analysis of multiphase design problems has been done by analogy with single phase concepts where the only essential difference is that the experiments that define a process correlation are executed on multiphase systems. If this analysis of multiphase systems is compared to that for single phase systems, as is done in Table 2, the absence of microscopic multiphase equations of change, constitutive relations, and design equations is apparent.

Table 2. The Interaction of Theory and Experiment at Three Levels for Multiphase Engineering Design Problems.

	Theory	Experiment
Microscopic Level ($\underline{x}$,t)		
Macroscopic Level (-,t)		Process Calculations
Equilibrium (-,-)	Thermodynamic Laws	Property Correlations

1.2.1 Theory

The essential element that is missing in the analysis of multiphase systems, such as filtration, is the microscopic level equations which will permit the determination of a mechanism and the macroscopic design equations. The lack of a multiphase continuum theory has not, however, prevented mechanisms from being postulated for multiphase systems since there is always a need to explain macroscopic observations. Mechanisms postulated solely on the basis of macroscopic measurements must, however, be viewed with caution. This caution can be justified by appeal to the mean value theorem.

$$\langle f \rangle = \int f(x)dx / \int dx \tag{1.1}$$

A given function f(x) has a unique average value, $\langle f \rangle$, but the reverse procedure in which a function is sought which gives a prescribed average value is not unique since there are an infinite number of functions that will give the same average value.

When mechanisms are postulated from only macroscopic level information, it represents only one of an infinite number of plausible mechanisms since, in essence, it is equivalent to using the mean value theorem in reverse. Regardless of the plausibility of the mechanism, it is certainly not unique and the acceptance of such postulated mechanisms depends solely on persuasiveness but the discrimination among these postulated mechanisms requires microscopic level theoretical and experimental information.

The only way to uniquely understand why a multiphase system exhibits the observed macroscopic behavior, is to initiate the analysis at the most general formulation of the fundamental principles at the microscopic level. Proceeding in this fashion will not violate the implications of the mean value theorem and will insure the determination of a unique mechanism.

1.2.2 Experiment

If only microscopic level experimental data is acquired, it is equivalent to having constitutive equations without the equations of change. Such data by itself cannot provide a mechanism because the meaning of this data cannot be discerned due to the absence of microscopic multiphase balance laws. For example, it would not be possible to obtain the laminar flow velocity profile in a tube with only Newton's law of viscosity. Constitutive relations and the microscopic level equations of change are both essential to the design process.

Microscopic level measurements in multiphase systems are not only difficult to measure but also difficult to define. For example, a probe designed to measure a velocity profile in the interstices between 1 μm particles would be extremely difficult to fabricate and manipulate. Larger probes would be easier to use but what interstial velocity are these larger probes measuring?

A similar problem occurs with porosity. The measurement and definition of porosity is no problem at the macroscopic level, but how is a local value of porosity measured and defined such that it is a continuous function of position?

The understanding, analysis, and design of multiphase systems would be considerably improved and the experimental effort reduced if a microscopic level continuum and constitutive theory, that is analogous to that for single phase systems, were also available for multiphase systems. This rationale justifies the effort (12,13,14, 15) that has been expended toward the development and application of a general continuum and constitutive theory for multiphase systems.

2. MULTIPHASE THEORY

The theoretical approaches for describing the thermo-mechanical behavior of homogeneous and heterogeneous mixtures are based on the fundamental work of Truesdell, Toupin, and Noll (24,25) which includes the development of the general balance equations and the constitutive theory. Two approaches that are extensions of this basic framework are the continuum theory of mixtures (1,3) and the principle of local volume averaging (11,12,13,14,15).

In the principle of local volume averaging, phases are viewed as interpenetrating continua, each occupying only part of the space and separated by highly irregular interfaces. Every property is assumed to be continuous throughout each phase but discontinuous at the phase interfaces. The classical single phase equations, together with the appropriate interfacial restrictions, are postulated for each of the individual phases but the arduous task of solving these equations for the complicated geometry of the interstices is circumvented by averaging these single phase equations over a representative element of volume. The averaged results represent an entirely new continuum with its own constitutive equations that are distinct from those of the individual phases.

The theory of mixtures views the mixture as a superposition of overlapping continua and is a better description of multiple components in a single phase mixture rather than multiphase systems. The more realistic conceptualization of the multiphase material is obtained from the volume averaging principle and it is used here to develop a mechanism for filtration.

2.1 Volume Averaging

The averaging volume is shown in Figure 1. The centroid of the volume is located at a position x_i and any point in the averaging volume is located relative to the origin by the position vector r_i and relative to the centroid of the averaging volume by ξ_i such that

$$r_i = x_i + \xi_i \tag{2.1}$$

The phase distribution function is defined as

$$\gamma_\alpha = \gamma_\alpha(r_i,t) = \begin{cases} 1, \text{ if } r_i \in V_\alpha \\ 0, \text{ if } r_i \in V_\beta \end{cases} \tag{2.2}$$

On the interfaces between the two phases γ_α is not defined but its left and right limits exist there. This function not only describes the distribution of the phases, but it also accounts for the points over which the averages are taken.

An average value is obtained by integrating over all the points ξ_i of the averaging volume. For example, the volume of the α-phase in the averaging volume is obtained by

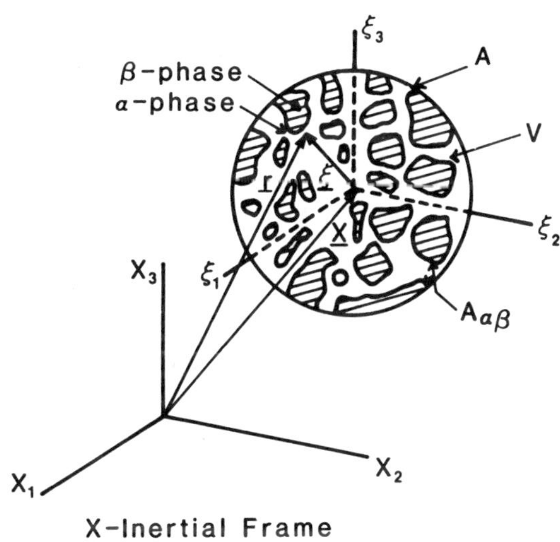

Fig. 1. Reference frames for the averaging volume V with surface area A containing α- and β-phases.

$$V_\alpha(x_i,t) = \int_V \gamma_\alpha(x_i+\xi_i,t)\, d_\xi V \tag{2.3}$$

where $d_\xi V$ indicates an integration over the position variable ξ_i.

Another averaging volume is superimposed over the first and the averaging operation is repeated until each point x_i of the initial averaging volume is filled with a continuum of average values for each of the phases. This procedure is repeated until the entire multiphase body is composed of a continuum of values for each phase at each point x_i in the body.

This averaging procedure then permits the local value of the porosity to be defined as

$$\varepsilon_\alpha(x_i,t) = (1/V) \int_V \gamma_\alpha(x_i+\xi_i,t) d_\xi V \tag{2.4}$$

and a corresponding value ε_β for the β-phase can be also obtained such that

$$\varepsilon_\alpha + \varepsilon_\beta = 1 \tag{2.5}$$

2.1.1 Averaging Operators

For a given averaging volume, the following factors (13) should be taken into account when averaging operators are defined.

First, the intrinsic nature of the property being averaged should be taken into consideration. For example, the stress, which is a surface force, should be averaged over the surface points on which it is acting. The internal energy, on the other hand, should be averaged over the points within the prescribed averaging volume.

Second, the quantities being averaged must be additive. For example, if A is a quantity defined per unit of mass, then A is not additive, AdV is not additive, but ρAdV is additive.

Third, the result of averaging must be a quantity that is observed and can be measured. For example, velocities are mass averaged quantities, therefore, the definition of a volume averaged velocity should also reflect the mass average characteristic normally associated with velocity.

These factors lead to the following definitions for averaging operators:

Volume Average Operator, $\langle\ \rangle_\alpha$

$$\langle f\rangle_\alpha(x_i,t) = (1/V)\int_V f(x_i+\xi_i,t)\,\gamma_\alpha(x_i+\xi_i,t)\,d_\xi V \tag{2.6}$$

Intrinsic Volume Average Operator, $\langle\ \rangle_\alpha^\alpha$

$$\langle f\rangle_\alpha^\alpha(x_i,t) = (1/V_\alpha)\int_V f\gamma_\alpha(x_i+\xi_i,t)d_\xi V \tag{2.7}$$

Mass Average Operator, $\overline{\ }^\alpha$

$$\overline{f}^\alpha(x_i,t) = \frac{\int_V f(x_i+\xi_i,t)\,\gamma_\alpha(x_i+\xi_i,t)\,\rho(x_i+\xi_i,t)d_\xi V}{\int_V \rho(x_i+\xi_i,t)\,\gamma_\alpha(x_i+\xi_i,t)\,d_\xi V} \tag{2.8}$$

Area Average Operator, $\sqcup^\alpha$

$$\overset{\sqcup}{f}^\alpha(x_i,t) = (1/A)\int_A f_j(x_i+\xi_i,t)n_j(x_i+\xi_i,t)\gamma_\alpha(x_i+\xi_i,t)d_\xi A \tag{2.9}$$

2.1.2 Averaging Theorems and Fluctuations

The establishment of balance equations for a volume averaged continuum is achieved by averaging the single phase balance equations over the representative averaging volume. To obtain spatial and time derivatives of averaged quantities, the following theorems (11,13) are used. For these theorems, the averaging volume, V, is constant as it moves throughout the multiphase system.

Average of a Time Derivative

$$\int_V \frac{\partial f}{\partial t}\gamma_\alpha d_\xi V = \frac{\partial}{\partial t}\int_V f\gamma_\alpha d_\xi V - \int_{A_{\alpha\beta}} f\, w_i n_i^{\alpha\beta} d_\xi A \tag{2.10}$$

Average of a Spatial Derivative

$$\int_V \frac{\partial f_j}{\partial r_j}\gamma_\alpha d_\xi V = \frac{\partial}{\partial r_j}\int_V f_j\gamma_\alpha d_\xi V + \int_{A_{\alpha\beta}} f_j n_j^{\alpha\beta} d_\xi A \tag{2.11}$$

The first term on the right side of the average of a spatial derivative, Eq. (2.11), can be written as

$$\frac{\partial}{\partial r_j} \int_V f_j \gamma_\alpha d_\xi V = \int_V \frac{\partial}{\partial r_j} (f_j \gamma_\alpha) d_\xi V \tag{2.12}$$

since the averaging volume is not a function of position. The divergence theorem is used for the term on the right of Eq. (2.12) to obtain an alternate form for the average of a spatial derivative.

Average of a Spatial Derivative (Alternate Form)

$$\int_V \frac{\partial f_j}{\partial r_j} \gamma_\alpha d_\xi V = \int_A f_j \gamma_\alpha n_j d_\xi A + \int_{A_{\alpha\beta}} f_j n_j^{\alpha\beta} d_\xi A \tag{2.13}$$

In the derivation of the averaged balance laws, it is necessary to convert the average of a product to the product of average values. The introduction of deviations, or fluctuations,

$$\hat{f}^\alpha (x_i + \xi_i, t) = f(x_i + \xi_i, t) - \bar{f}^\alpha (x_i, t) \tag{2.14}$$

permits this transformation to be accomplished. These deviations are valid and defined only for those spatial points used to determine the average value.

To determine the average value of deviations, the first step is to establish the set of points, from the phase distribution function, over which the average value $\bar{f}(x_i, t)$ is determined and then for these same set of points, determine the deviation by subtracting the average value from the local values $f(x_i + \xi_i, t)$ in V. There is a deviation for each $f(x_i + \xi_i, t)$ but there is only one average value which remains the same during the averaging of the deviations over the points ξ_i in V. The following identities can then be established.

$$\overline{\hat{f}^\alpha}^\alpha = 0 \tag{2.15}$$

and

$$\overline{\hat{f}^\alpha \bar{g}^\alpha}^\alpha = 0 \tag{2.16}$$

$$\int_V \overline{\rho \hat{f}_i{}^{\alpha} d_\xi V}^{\alpha} = 0 \qquad (2.17)$$

$$\int_V \overline{\rho \hat{f}_i{}^{\alpha} \bar{g}^{\alpha} d_\xi V}^{\alpha} = 0 \qquad (2.18)$$

$$\overline{fg}^{\alpha} = \overline{f}^{\alpha}\overline{g}^{\alpha} + \overline{\hat{f}^{\alpha}\hat{g}^{\alpha}}^{\alpha} \qquad (2.19)$$

2.2 General Balance Laws

These principles and theorems of volume averaging are now applied to the single phase equations which hold for each of the phases in the multiphase system. It is more efficient to volume average a general single-phase equation to obtain a general multiphase balance law. In either case, these general balance equations can be converted to specific balances for mass, momentum, energy, and entropy by proper definition of the general dependent variables.

2.2.1 Single Phase Balance Laws

The general balance law for a single phase system is

$$\frac{\partial}{\partial t}(\rho\Omega) + \frac{\partial}{\partial r_k}(\rho\Omega v_k) - \frac{\partial}{\partial r_k}(q_k) - \rho h = \rho G \qquad (2.20)$$

where Ω is an arbitrary thermodynamic property, q_k is the surface flux of Ω, h is the external supply of Ω, and G is the net production of Ω.

The interactions between phases is important and these interactions are governed by the general interface balance

$$\{\rho\Omega(w_i - v_i) + q_i\}|_\alpha n_i^{\alpha\beta} + \{\rho\Omega(w_i - v_i) + q_i\}|_\beta n_i^{\alpha\beta} = 0 \qquad (2.21)$$

2.2.2 Multiphase Balance Laws

The application of the volume averaging principles to the single-phase equation, Eqs. (2.20), gives the general macroscopic balance law for multiphase systems.

$$\int_V\{\frac{\partial}{\partial t}(\langle\rho\rangle_\alpha\overline{\Omega}^\alpha) + \frac{\partial}{\partial x_j}(\langle\rho\rangle_\alpha\overline{\Omega}^\alpha\overline{v}_j^{\ \alpha})\}d_xV$$

$$- \int_V\{(1/V)\int_{A_{\alpha\beta}} \rho\Omega(w_i - v_i)n_i^{\ \alpha\beta}d_\xi A\}d_xV$$

$$- \int_V\{(1/V)\int_{A_{\alpha\beta}} q_j n_j^{\ \alpha\beta}d_\xi A\}d_xV$$

$$- \int_V \{(1/V)\int_A(q_k - \rho\hat{\Omega}^\alpha\hat{v}_k^{\ \alpha})n_k\gamma_\alpha d_\xi A\}d_xV$$

$$- \int_V\{\langle\rho\rangle_\alpha(\overline{h}^\alpha + \overline{G}^\alpha)\}d_xV = 0 \qquad (2.22)$$

The new terms in the multiphase balance equation are identified with the following notation and physical significance.

The term

$$e^\alpha = (1/\langle\rho\rangle_\alpha V)\int_{A_{\alpha\beta}} \rho\Omega(w_i - v_i)n_i^{\ \alpha\beta}d_\xi A \qquad (2.23)$$

represents the exchange of the property Ω across the interfaces between the α- and β-phases within the averaging volume V due to a phase change. The term 'phase change' is used in a general context that includes interphase transport of chemical species as well as a change of state such as evaporation. The tensorial order of e^α is dependent on that of Ω.

The term

$$F^\alpha = (1/\langle\rho\rangle_\alpha V)\int_{A_{\alpha\beta}} q_j n_j^{\ \alpha\beta}d_\xi A \qquad (2.24)$$

represents the exchange between the α- and β-phases within V due to the surface flux vector q_j of Ω. If the surface flux vector were momentum, then F^α would be the interfacial drag on the particulates due to the fluid motion.

When the flux at the volume averaged level is defined by

$$q_k^\alpha = (q_k - \rho\hat{\Omega}^\alpha\hat{v}_k^{\ \alpha}) \qquad (2.25)$$

then the term which represents the flux through the area A enclosing the averaging volume V can be written as

$$\int_V \{(1/V)\int_A (q_k - \rho\hat{\Omega}^\alpha \hat{v}_k^{\ \alpha})\gamma_\alpha n_k d_\xi A\} d_x V$$

$$= \int_V \frac{\partial}{\partial x_j}\{(1/V)\int_V q_j^{\ \alpha}\gamma_\alpha d_\xi V\} d_x V \tag{2.26}$$

where the divergence theorem has been used for the integral over the area A. As a result, Eq. (2.26) can now be written as

$$\int_V \{(1/V)\int_A (q_k - \rho\hat{\Omega}^\alpha \hat{v}_k^{\ \alpha})\gamma_\alpha n_k d_\xi A\} d_x V = \int_V \frac{\partial}{\partial x_j} \langle q_j^{\ \alpha}\rangle_\alpha \, d_x V \tag{2.27}$$

Since the general macroscopic level multiphase balance, Eq. (2.22), is an integral over an arbitrary volume, it follows that the integrand must be zero and the general microscopic level multiphase balance equation is

$$\frac{\partial}{\partial t}(\langle\rho\rangle_\alpha \overline{\Omega}^\alpha) + \frac{\partial}{\partial x_j}(\langle\rho\rangle_\alpha \overline{\Omega}^\alpha \overline{v}_j^{\ \alpha})$$

$$- \frac{\partial}{\partial x_j}\langle q_j^{\ \alpha}\rangle_\alpha - \langle\rho\rangle_\alpha (e^\alpha + F^\alpha) - \langle\rho\rangle_\alpha \overline{h}^\alpha - \langle\rho\rangle_\alpha \overline{G}^\alpha = 0 \tag{2.28}$$

where the first term represents the accumulation of Ω, the second term represents the convection of Ω, the third term represents the diffusion or conduction of Ω at the volume averaged level, the fourth term represents the transport of Ω across the interfaces $A_{\alpha\beta}$ between the phases, the fifth term represents the body source of Ω, and the last term represents the generation of the property Ω.

The multiphase balance at the interfaces between the phases is obtained by applying the principles of volume averaging to the jump balance, Eq. (2.21), with the result

$$\langle\rho\rangle_\alpha (e^\alpha + F^\alpha) + \langle\rho\rangle_\beta (e^\beta + F^\beta) = 0 \tag{2.29}$$

To obtain specific multiphase balance equations for the conservation of mass and momentum for isothermal flow through a porous media in which there is no transfer of species between phases or phase changes, the general dependent variables are specified as follows,

Mass

$$\overline{\Omega}^\alpha = 1, \quad \langle q_j^\alpha \rangle_\alpha = 0 \quad \overline{h}^\alpha = 0$$

$$\overline{G}^\alpha = 0, \quad F^\alpha = 0, \quad e^{\alpha,\text{mass}} = 0$$

Momentum

$$\overline{\Omega}^\alpha = \overline{v}_i^\alpha \quad \langle q_j^\alpha \rangle_\alpha = t_{ij}^\alpha \quad (q_j - \rho\hat{\Omega}^\alpha \hat{v}_j^\alpha) = (t_{ij} - \rho\hat{v}_i^\alpha \hat{v}_j^\alpha)$$

$$\overline{h}^\alpha = \overline{g}_i^\alpha \quad \overline{G}^\alpha = 0 \quad F^\alpha = F_i^\alpha$$

$$e^{\alpha,\text{mom}} = 0$$

and the multiphase continuity and motion equations are

α-Phase Mass

$$\frac{\partial}{\partial t}(\varepsilon_\alpha \langle \rho \rangle_\alpha{}^\alpha) + \frac{\partial}{\partial x_j}(\varepsilon_\alpha \langle \rho \rangle_\alpha{}^\alpha \overline{v}_j^\alpha) = 0 \tag{2.30}$$

β-Phase Mass

$$\frac{\partial}{\partial t}(\varepsilon_\beta \langle \rho \rangle_\beta{}^\beta) + \frac{\partial}{\partial x_j}(\varepsilon_\beta \langle \rho \rangle_\beta{}^\beta \overline{v}_j^\beta) = 0 \tag{2.31}$$

α-Phase Momentum

$$\frac{\partial}{\partial t}(\varepsilon_\alpha \langle \rho \rangle_\alpha{}^\alpha \overline{v}_i^\alpha) + \frac{\partial}{\partial x_j}(\varepsilon_\alpha \langle \rho \rangle_\alpha{}^\alpha \overline{v}_i^\alpha \overline{v}_j^\alpha) - \frac{\partial}{\partial x_j}(t_{ij}^\alpha) - \varepsilon_\alpha \langle \rho \rangle_\alpha{}^\alpha g_i^\alpha - \varepsilon_\alpha \langle \rho \rangle_\alpha{}^\alpha F_i^\alpha = 0 \tag{2.32}$$

β-Phase Momentum

$$\frac{\partial}{\partial t}(\varepsilon_\beta \langle \rho \rangle_\beta{}^\beta \overline{v}_i^\beta) + \frac{\partial}{\partial x_j}(\varepsilon_\beta \langle \rho \rangle_\beta{}^\beta \overline{v}_i^\beta \overline{v}_j^\beta) - \frac{\partial}{\partial x_j}(t_{ij}^\beta) - \varepsilon_\beta \langle \rho \rangle_\beta{}^\beta g_i^\beta - \varepsilon_\beta \langle \rho \rangle_\beta{}^\beta F_i^\beta = 0 \tag{2.33}$$

where the drag forces between the two phases are related by

$$\varepsilon_\alpha \langle\rho\rangle_\alpha {}^\alpha F_i^\alpha + \varepsilon_\beta \langle\rho\rangle_\beta {}^\beta F_i^\beta = 0 \tag{2.34}$$

Eqs. (2.30)-(2.34) govern the isothermal motion of a multiphase system composed of an α- and β-phase in which the only exchange occurring between the phases is the transfer of momentum. These equations apply to all systems that meet these restrictions. Further identification of the continuum composed of an α-phase and a β-phase requires constitutive equations for t_{ij}^α, t_{ij}^β, and F_i^α. The application of the axioms of constitutive theory are used to develop constitutive relations for these functions.

The notation can be made more compact with the following changes, which will be used henceforth.

$$\langle\varepsilon\rangle_\alpha = \varepsilon_\alpha, \ \langle\rho\rangle_\alpha = \rho_\alpha, \ \bar{g}_i^\alpha = g_i^\alpha, \ \langle\rho\rangle_\alpha^\alpha = \rho^\alpha, \ \bar{v}_i^\alpha = v_i^\alpha$$

3. CONSTITUTIVE THEORY

The results of volume averaging represent an entirely new continuum with its own constitutive equations that are distinct from those of the individual phases.

Without exception, the number of unknowns exceeds the number of fundamental balance equations. The additional relations required to make the problem determinate are called constitutive relations and these relations are determined by combining the constitutive theory with experiment. The basic axioms and principles of constitutive theory (7,18) are generally accepted, although in some details there are still different opinions and approaches (1).

The axioms of constitutive theory require that all of the constitutive equations be functions of the same set of independent variables (i.e., equipresence), that the constitutive function is an absolute invariant relation that is completely independent of observing coordinate frames (i.e., objectivity), that the constitutive function must not violate the entropy inequality, and that the constitutive equation must be consistent with equilibrium conditions represented by the absence of gradients.

The following is an overview (20) of the procedure for the development of constitutive equations.

1. A set of arbitrary independent variables is selected according to the axioms of equipresence and objectivity. Since velocities

and velocity gradients are not objective, they are replaced by such variables as the difference in velocity, v_i^d, and the rate of deformation tensor, d_{ij}^{α}.

2. The entropy inequality is then applied to the constitutive functional. This condition requires that the coefficients of all variables which are not independent variables must vanish to maintain a positive entropy production.

3. The linear constitutive theory approximates the constitutive functions with an appropriate tensorial order polynomial. Material isotropy is normally assumed and, as a result, the coefficients are limited to isotropic tensor forms. The entropy inequality and equilibrium requirements are then applied to the polynomial form of the constitutive equation to determine the algebraic signs of the coefficients in the polynomial.

Constitutive equations for the fluid α-phase are taken to depend on the following independent variables (12,15,20).

$$t_{ij}^{\alpha} = f_{ij}^{\alpha}(\rho^{\alpha}, \varepsilon_{\alpha}, T^{\alpha}, \varepsilon_{\alpha,i}, T^{\alpha}_{,i}, d_{ij}^{\alpha}, v_i^d) \quad (3.1)$$

$$F_i^{\alpha} = h_i^{\alpha}(\rho^{\alpha}, \varepsilon_{\alpha}, T^{\alpha}, \varepsilon_{\alpha,i}, T^{\alpha}_{,i}, d_{ij}^{\alpha}, v_i^d) \quad (3.2)$$

If a constitutive equation were developed for t_{ij}^{β} and used in the momentum balance, Eq. (2.33), the result would be the displacement of the particulate β-phase or, equivalently, the porosity variations. For filtrations, instead of doing this, the porosity profiles are determined directly from electrical conductivity measurements.

Application of the entropy inequality and equilibrium condition (12,15) provide the following modifications to the α-phase constitutive functionals.

$$t_{ij}^{\alpha} = -\varepsilon_{\alpha} P^{\alpha}\delta_{ij} + \tau_{ij}^{\alpha}(\rho^{\alpha}, \varepsilon_{\alpha}, T^{\alpha}, \varepsilon_{\alpha,i}, T^{\alpha}_{,i}, d_{ij}^{\alpha}, v_i^d) \quad (3.3)$$

$$\varepsilon_{\alpha}\rho^{\alpha}F_i^{\alpha} = P^{\alpha}\varepsilon_{\alpha,i} + \pi_i^{\alpha}(\rho^{\alpha}, \varepsilon_{\alpha}, T^{\alpha}, \varepsilon_{\alpha,i}, T^{\alpha}_{,i}, d_{ij}^{\alpha}, v_i^d) \quad (3.4)$$

where τ_{ij}^{α} and π_i^{α} are the dissipative parts of the constitutive relations and vanish at equilibrium.

3.1 Multiphase Stress Tensor

The constitutive function for the multiphase stress is taken to be a second order polynomial (9,10,18,20) and then the assumptions are made that the porous media is isothermal with no temperature gradients, and that the gradient of porosity and the velocity difference between the phases does not affect the multiphase stress tensor. The remaining terms in the polynomial for the multiphase stress tensor are

$$t_{ij}^{\alpha} = (\gamma_o)_{ij} + (\gamma_1)_{ijmn}\, d_{mn}^{\alpha} + (\gamma_2)_{ijmnpq}\, d_{nm}^{\alpha} d_{pq}^{\alpha} \tag{3.5}$$

The multiphase continuum is assumed isotropic and, as a result, the odd order tensor coefficients are all zero and the even order tensor coefficients must be represented by the following isotropic tensor relations (10,18).

$$(\gamma_0)_{ij} = \beta_o \delta_{ij} \tag{3.6}$$

$$(\gamma_1)_{ijmn} = \beta_1 \delta_{ij}\delta_{mn} + \beta_2(\delta_{im}\delta_{jn} + \delta_{in}\delta_{jm}) \tag{3.7}$$

$$\begin{aligned}(\gamma_3)_{ijmnpq} = {} & \beta_3\delta_{ij}\delta_{mn}\delta_{pq} + \beta_4\delta_{ij}\delta_{mp}\delta_{nq} + \beta_5\delta_{ij}\delta_{mq}\delta_{np} + \\ & \beta_6\delta_{im}\delta_{jn}\delta_{pq} + \beta_7\delta_{im}\delta_{jp}\delta_{nq} + \beta_8\delta_{im}\delta_{jq}\delta_{np} + \\ & \beta_9\delta_{in}\delta_{jm}\delta_{pq} + \beta_{10}\delta_{in}\delta_{jp}\delta_{mq} + \beta_{11}\delta_{in}\delta_{jq}\delta_{mp} + \\ & \beta_{12}\delta_{ip}\delta_{jm}\delta_{nq} + \beta_{13}\delta_{ip}\delta_{jn}\delta_{mq} + \beta_{14}\delta_{ip}\delta_{jq}\delta_{mn} + \\ & \beta_{15}\delta_{ip}\delta_{jm}\delta_{nq} + \beta_{16}\delta_{iq}\delta_{jn}\delta_{mp} + \beta_{17}\delta_{iq}\delta_{jp}\delta_{mn}\end{aligned} \tag{3.8}$$

The insertion of Eqs. (3.6)-(3.8) into Eq. (3.5) and the utilization of the equilibrium condition results in the following expression for the fluid α-phase constitutive equation,

$$t_{ij}^{\alpha} = -\varepsilon_{\alpha} P^{\alpha}\delta_{ij} + 2\eta_1\, d_{ij}^{\alpha} + \eta_2\, d_{im}^{\alpha} d_{mj}^{\alpha} \tag{3.9}$$

where the coefficients have the following functional dependence,

$$\eta_1 = \eta_1(\rho^\alpha, \varepsilon_\alpha, T^\alpha, d_{mm}^{\alpha}, d_{mp}^{\alpha} d_{pm}^{\alpha}) \tag{3.10}$$

$$\eta_2 = \eta_2(\rho^\alpha, \varepsilon_\alpha, T^\alpha, d_{mm}^{\alpha}, d_{mp}^{\alpha} d_{pm}^{\alpha}) \tag{3.11}$$

and where the last two independent variables are the first, I_1, and second, I_2, invariants of the rate of deformation tensor d_{ij}.

If the dependence on porosity is removed from these two constitutive parameters then they are converted from multiphase to single phase material properties and if, in addition, a first order approximation is presumed for t_{ij}^{α}, such that,

$$\eta_1 = m(2d_{im}^{\alpha} d_{mi}^{\alpha})^{(n-1)/2} \tag{3.12}$$

then the multiphase stress tensor becomes

$$t_{ij}^{\alpha} = -\varepsilon_\alpha P^\alpha \delta_{ij} + 2\{m(2d_{im}^{\alpha} d_{mi}^{\alpha})^{(n-1)/2}\} d_{ij}^{\alpha} \tag{3.13}$$

3.2 Multiphase Drag

The constitutive equation for the multiphase drag, F_i^{α}, represented by the functional relation given by Eq. (3.2), is determined by the same procedure as that for the multiphase shear stress.

A second order polynomial expansion of Eq. (3.2), combined with the assumptions that the multiphase system is isothermal, results in the following expression for the interfacial drag in multiphase systems

$$F_i^{\alpha} = (\gamma_0)_{im} \varepsilon_{\alpha,m} + (\gamma_1)_{im} v_m^{d} + (\gamma_2)_{imnp} d_{mn}^{\alpha} \varepsilon_{\alpha,p} + (\gamma_3)_{imnp} d_{mn} v_p^{d} \tag{3.14}$$

Utilization of Eqs. (3.6) and (3.7) and the equilibrium condition results in the following expression for the multiphase drag

$$\varepsilon_\alpha \rho^\alpha F_i^{\alpha} = P^\alpha \varepsilon_{\alpha,i} + \lambda_0 v_i^{d} + \lambda_1 d_{ni}^{\alpha} \varepsilon_{\alpha,n} + \lambda_2 d_{ni}^{\alpha} v_n^{d} \tag{3.15}$$

If, as with the multiphase stress tensor, a first degree approximation is presumed, then the multiphase drag becomes

$$\varepsilon_\alpha \rho^\alpha F_i^\alpha = P^\alpha \varepsilon_{\alpha,i} + \lambda_0 v_i^d \tag{3.16}$$

where

$$\lambda_0 = \lambda_0(\rho^\alpha, \varepsilon_\alpha, T^\alpha, d_{mm}^\alpha) \tag{3.17}$$

The residual multiphase entropy inequality (12,15) for the liquid α-phase is

$$\tau_{ij}^\alpha d_{ij}^\alpha - \pi_i^\alpha v_i^d \geq 0 \tag{3.18}$$

The dissipative parts of the multiphase stress and drag can be obtained from Eqs. (3.13) and (3.16) and are, respectively,

$$\tau_{ij}^\alpha = 2\{m(2d_{im}^\alpha d_{mi}^\alpha)^{(n-1)/2}\}d_{ij}^\alpha \tag{3.19}$$

$$\pi_i^\alpha = \lambda_0 v_i^d \tag{3.20}$$

and when these two expressions are substituted into the entropy inequality, Eq. (3.18), then

$$m \geq 0 \tag{3.21}$$

$$(-\lambda_0) \geq 0 \tag{3.22}$$

4. DIMENSIONAL ANALYSIS AND DOMINANT TERMS

The β-phase momentum balance is replaced with a direct experimental determination of the porosity distribution since the combination of a β-phase constitutive equation and the β-phase momentum balance would give the displacement of the particulate β-phase or, effectively, the distribution of the porosity.

However, for the liquid α-phase, the general balance equations, Eqs. (2.30) and (2.33), and the two constitutive equations, Eqs. (3.13) and (3.16), can now be combined to obtain balance equations for that class of multiphase materials that are described by these two constitutive equations.

The accumulation and convection terms in the α-phase motion equation, Eq. (2.32), can be combined with the α-phase mass balance, Eq. (2.30), to obtain

$$\frac{\partial}{\partial t}(\varepsilon_\alpha \rho^\alpha v_i^\alpha) + \frac{\partial}{\partial x_j}(\varepsilon_\alpha \rho^\alpha v_i^\alpha v_j^\alpha) = \varepsilon_\alpha \rho^\alpha \left(\frac{\partial v_i^\alpha}{\partial t} + v_j^\alpha \frac{\partial v_i^\alpha}{\partial x_j}\right) \quad (4.1)$$

and upon insertion of the constitutive equations for t_{ij}^α and F_i^α, Eqs. (3.13) and (3.16), the α-phase motion equation becomes

$$\varepsilon_\alpha \rho^\alpha \left(\frac{\partial v_i^\alpha}{\partial t} + v_j^\alpha \frac{\partial v_i^\alpha}{\partial x_j}\right) + \varepsilon_\alpha \frac{\partial P^\alpha}{\partial x_i}$$

$$- \frac{\partial}{\partial x_j}\{2m(2I_2)^{(n-1)/2} d_{ij}^\alpha\} - \varepsilon_\alpha \rho^\alpha g_i^\alpha - \lambda_0 v_i^d = 0 \quad (4.2)$$

4.1 Dimensional Analysis

Characteristic quantities are defined (5,20,28,29) as follows

L = Characteristic length = L_c

V = Characteristic velocity = $\dot{V}/A\varepsilon_\alpha{}^*$

T = Characteristic time = $L_c A\varepsilon_\alpha{}^*/\dot{V}$

S = Characteristic stress = P_c^o

where the terms on the right are the filtration variables that correspond to the characteristic quantities. The following dimensionless variables can now be defined.

$$x_k = L\, x_k^* \qquad v_k^\alpha = V\, v_k^*$$

$$v_k^d = V\, v_k^{d*} \qquad t = T\, t^*$$

$$P = P^* S + P_o \qquad g_k^\alpha = g\, g_k^*$$

$$\frac{\partial}{\partial x^*_k} = L \frac{\partial}{\partial x_k} \qquad \frac{\partial}{\partial t^*} = T \frac{\partial}{\partial t}$$

4.1.1 Non-Newtonian Fluid

Utilization of these dimensionless quantities permits Eq. (4.2) to be written as

$$\varepsilon_\alpha (Re)\left(\frac{\partial v_i^*}{\partial t^*} + v_j^* \frac{\partial v_i^*}{\partial x_j^*}\right) + \varepsilon_\alpha (Np) \frac{\partial P^*}{\partial x_i^*} -$$

$$\frac{\partial}{\partial x_j^*}\{2(I_2^*)^{(n-1)/2}\, d_{ij}^{\alpha *}\} - \varepsilon_\alpha (Re/Fr) g_i^* + (Nd) v_i^{d*} = 0 \qquad (4.3)$$

where

$$Re = \frac{\rho^\alpha\, \dot{V}^{(2-n)}\, L_c^{\,n}}{m(A\, \varepsilon_\alpha^*)^{(2-n)}} \qquad (4.4)$$

$$Fr = \frac{\dot{V}^2}{gL_c (A\, \varepsilon_\alpha^*)^2} \qquad (4.5)$$

$$Np = \frac{P_c^{\,o}(L_c A \varepsilon_\alpha^*)^n}{m\, \dot{V}^n} \qquad (4.6)$$

$$Nd = \frac{(-\lambda_0) L_c^{\,(n+1)} (A\varepsilon_\alpha^*)^{(n-1)}}{m\, \dot{V}^{(n-1)}} \qquad (4.7)$$

The magnitude of the normalized α-phase momentum balance coefficients reflects the relative importance of the individual terms, and each dimensionless coefficient represents the ratio of forces from which the dominant effects can be deduced. The dimensionless numbers (Re), (Re/Fr),(Np), and (Nd) represent, respectively, the ratios of the inertial, gravity, pressure, and drag forces to the viscous force. Their values are estimated from the data of Christopher and Middleman (8) for the flow a 1% solution of carboxymethylcellulose (CMC) in water through glass beads with diameters between 710 and 840 microns which gave a packed bed with a porosity of 0.370. The results are shown in Table 3.

Neglecting the porosity dependence and assuming the functional dependence given by Eq. (3.12) for η_1 converts this constitutive parameter from that for a multiphase fluid to a single phase fluid and, as a result, it is permissible to use the values of m and n from the data of Christopher and Middleman to calculate the dimensionless numbers shown in Table 3.

Table 3. Dimensionless Numbers for the Flow of 1% CMC in Water through a Packed Tube (Christopher and Middleman, (8)).

$\dot{V}$ cm^3/s	Re	Np	Nd	Re/Fr
1.08	0.028	$6.35x10^3$	$2.11x10^3$	82.8
5.16	0.28	$6.55x10^3$	$4.5\ x10^3$	36.1

Increasing the Reynolds number by a factor of 10 left the pressure force, Np, relatively unchanged, increased the drag force, Nd, and decreased the effect of gravity, Re/Fr. The magnitude of Np and Nd indicate, for flow of CMC in a packed bed of glass beads, that the inertial, gravity, and viscous forces can be neglected.

4.1.2 Newtonian Fluid

If η_1 is assumed to be independent of the invariants, I_1 and I_2, of the multiphase rate of deformation tensor, and the porosity ε_α then

$$\eta_1 = \eta_1(\rho^\alpha, T^\alpha) \tag{4.8}$$

This is equivalent to taking n = 1, m = μ in Eq. (3.12) such that

$$\eta_1 = \mu(\rho^\alpha, T^\alpha) \tag{4.9}$$

where μ now is a single phase Newtonian viscosity rather than a multiphase Newtonian viscosity and it is also now a thermodynamic variable since it is a function of only the pressure and the temperature. The expression for the multiphase stress tensor, Eq. (3.13) becomes

$$t_{ij}^{\ \alpha} = -\varepsilon_\alpha P^\alpha \delta_{ij} + 2\mu d_{ij}^{\ \alpha} \tag{4.10}$$

and the α-phase momentum balance, Eqs. (4.2) becomes

$$\varepsilon_\alpha \rho^\alpha \left(\frac{\partial v_i^\alpha}{\partial t} + v_j \frac{\partial v_i^\alpha}{\partial x_j}\right) + \frac{\partial P^\alpha}{\partial x_i}$$

$$- \frac{\partial}{\partial x_j}(2\mu\, d_{ij}) - \varepsilon_\alpha \rho^\alpha g_i^\alpha - \lambda_0 v_i^d = 0 \qquad (4.11)$$

and the normalized α-phase momentum balance, Eq. (4.3) becomes

$$\varepsilon_\alpha (Re)\left(\frac{\partial v_i^*}{\partial t^*} + v_j^* \frac{\partial v_i^*}{\partial x_j^*}\right) + \varepsilon_\alpha (Np) \frac{\partial P^*}{\partial x_i^*} -$$

$$\frac{\partial}{\partial x_j^*}(2d_{ij}^{\alpha *}) - \varepsilon_\alpha (Re/Fr) g_i^* + (Nd) v_i^{d*} = 0 \qquad (4.12)$$

The dimensionless numbers for n = 1 become

$$Re = \frac{\rho^\alpha \dot{V} L_c}{\mu A \varepsilon_\alpha^*} \qquad (4.13)$$

$$Fr = \frac{\dot{V}^2}{g L_c (A \varepsilon_\alpha^*)^2} \qquad (4.14)$$

$$Np = \frac{P_c^o L_c A \varepsilon_\alpha^*}{\mu \dot{V}} \qquad (4.15)$$

$$Nd = \frac{(-\lambda_0) L_c^2}{\mu} \qquad (4.16)$$

These dimensionless numbers are estimated from filtrations executed with water slurries of Lucite, Solka Floc, Celite, and talc. The filter media used in these filtrations were a fine porosity Whatman No. 3 with a particle retention of 5 microns or larger and a coarse porosity Whatman No. 4 with a particle retention of 25 microns or larger. Porosities of the filter cakes of Lucite, Solka Floc, Celite, and talc were, respectively, 0.391, 0.906, 0.826, 0.616. The results are shown in Table 4.

Again the values of the dimensionless numbers shown in Table 4 reflect the assumption that the viscosity is independent of the porosity and this permits single phase viscosity values to be used in the calculation of the dimensionless numbers shown in Table 4.

It is evident from the relative values of the dimensionless numbers shown in Table 4 that the pressure, drag, and gravity forces dominate the flow and that the inertial and viscous forces can be neglected.

Table 4. Dimensionless Numbers for the Filtration of Water Slurries of Lucite, Solka Floc, Celite, and talc.

Material Media	Re	Np	Nd	Re/Fr
Lucite (5%) No. 3	10.90×10^2	5.2×10^8	5.18×10^8	3.60×10^6
Lucite (5%) No. 4	8.90×10^2	6.7×10^8	7.44×10^8	4.29×10^6
Solka (2%) No. 4	3.78×10^2	13.3×10^8	51.50×10^8	8.43×10^6
Celite (5%) No. 3	1.37×10^2	12.4×10^8	11.00×10^8	19.40×10^6
Celite (5%) No. 4	1.42×10^2	32.4×10^8	11.30×10^8	20.10×10^6
talc (5%) No. 3	$.27 \times 10^2$	19.6×10^8	19.50×10^8	4.05×10^6
talc (5%) No. 4	$.25 \times 10^2$	21.1×10^8	20.70×10^8	4.39×10^6

5. GOVERNING EQUATIONS FOR FILTRATION

In general, the linear constitutive relations for the multiphase stress tensor and the interfacial drag represent material functions for the fluid α-phase in the presence of the particulate β-phase and may not be related to the constitutive equations for the individual phases. However, neglecting the porosity in the development of Eqs. (3.13) and (4.10) eliminates this distinction for these two constitutive equations but this distinction still exists for the multiphase drag coefficient, $(-\lambda_0)$, whose functional dependence is given by Eq. (3.17).

5.1 Multiphase Non-Newtonian Fluid

The dimensional analysis shows, for a fluid α-phases governed by the Eq. (3.13), that the dominant terms in the momentum balance, Eq. (4.2), are the pressure and drag forces. For a one-dimensional axial filtration (20) in which the z-coordinate is located at the cake-media interface and the positive direction is opposite to the fluid motion, the governing balance equations are

$$\frac{\partial}{\partial t}(\varepsilon_\alpha \rho^\alpha) - \frac{\partial}{\partial z}(\varepsilon_\alpha \rho^\alpha v_z{}^\alpha) = 0 \qquad (5.1)$$

$$\frac{\partial}{\partial t}(\varepsilon_\beta \rho^\beta) - \frac{\partial}{\partial z}(\varepsilon_\beta \rho^\beta v_z{}^\beta) = 0 \qquad (5.2)$$

$$\varepsilon_\alpha \frac{\partial P^\alpha}{\partial z} - (-\lambda_0)(v_z{}^\alpha - v_z{}^\beta) = 0 \qquad (5.3)$$

$$\varepsilon_\alpha + \varepsilon_\beta = 1 \qquad (5.4)$$

in which the functional dependence for $(-\lambda_0)$ is

$$\lambda_0 = \lambda_0(\rho^\alpha, \varepsilon_\alpha, T^\alpha, v_{z,z}{}^\alpha) \qquad (5.5)$$

The six unknown functions to be determined are

$$\varepsilon_\alpha,\ \varepsilon_\beta,\ (-\lambda_0),\ v_z{}^\alpha,\ v_z{}^\beta,\ P^\alpha$$

and since there are only four governing equations, two additional relations must be determined experimentally. In filtrations, it is possible to measure the internal porosity and pressure profiles and these profiles, in conjunction with the governing equations, permit the internal profiles of all the unknown functions to be determined.

The particulate β-phase motion equation together with a multiphase constitutive equation would add another equation and provide the displacements of the particulate β-phase. In the formulation as given by Eqs. (5.1)-(5.4), this displacement of the particulate β-phase is replaced by the direct measurement of the local porosity distribution.

These governing equations for multiphase non-Newtonian fluids indicate that the viscous force is small relative to the pressure and drag forces but that the viscous force is sufficient to reduce the effect of the gravity force as reflected in the relatively small value of (Re/Fr). The absence of a viscous force term in the momentum balance also implies that the momentum transfer is predominantly from the fluid α-phase to the particulate β-phase with very little, if any, transfer to the containing walls of the filter cake.

5.2. Newtonian Fluid

The dimensional analysis shows, for a fluid α-phase governed by Eq. (4.10), that the dominant terms in the momentum balance, Eq. (4.11), are the pressure, gravity, and drag forces. For a one dimensional filtration, the governing balance equations are Eqs. (5.1), (5.2) and the momentum balance

$$\varepsilon_\alpha \frac{\partial P^\alpha}{\partial z} - \varepsilon_\alpha \rho^\alpha g_z{}^\alpha - (-\lambda_o)(v_z^\alpha - v_z^\beta) = 0 \tag{5.6}$$

where the functional dependence of λ_o is given by Eq. (5.5) and the six unknown functions are the same as those given previously.

The first invariant of the deformation, $d_{mm}{}^\alpha = v_{z,z}$, in Eq. (5.5) reflects the deformation of the fluid phase at the volume averaged level. If λ_o were independent of the porosity and hence the particulate phase, then the interstitial liquid viscosity could become an independent variable in λ_o by Eq. (4.10) in the form

$$t_{zz}{}^\alpha = -\varepsilon_\alpha P^\alpha + \mu v_{z,z}{}^\alpha \tag{5.7}$$

It is more likely that $v_{z,z}{}^\alpha$ is small and that the viscosity, (a thermodynamic variable) appears as an independent variable in λ_o through the temperature and the Theory of Corresponding States,

$$T^\alpha = T_c T_r (\mu/\mu_c) \tag{5.8}$$

The functional dependence for λ_o, when $v_{z,z}{}^\alpha$ is negligible, is

$$\lambda_o = \lambda_o(\rho^\alpha, \varepsilon_\alpha, \mu, T_c, \mu_c) \tag{5.9}$$

and the permeability, K, can be defined by

$$(-\lambda_o) = \varepsilon_\alpha{}^2 \mu K^{-1} \tag{5.10}$$

where

$$K^{-1} = K^{-1}(\rho^\alpha, \varepsilon_\alpha, T_c, \mu_c) \tag{5.11}$$

The gravity and pressure terms in Eq. (5.6) can be combined by defining

$$P_o^{\alpha} = P^{\alpha} + \rho g^{\alpha} z - P_o \tag{5.12}$$

with the result that Eq. (5.6) becomes

$$\frac{\partial P_o^{\alpha}}{\partial z} - (\varepsilon_{\alpha}\mu/K)(v_z^{\alpha} - v_z^{\beta}) = 0 \tag{5.13}$$

and the governing equations for the one-dimensional axial filtration of a Newtonian liquid α-phase are Eqs. (5.1), (5.2), (5.4), and (5.13).

5.2.1 One-Dimensional Cylindrical Filtration

To obtain the filtrate rate expression for a one-dimensional cylindrical filtration of a Newtonian α-phase liquid, the governing equations, Eqs. (5.1), (5.2), (5.4), and (5.13) are combined with an external macroscopic material balance that equates the difference between the mass of slurry filtered and filtrate to the mass of the filter cake

$$(\rho^{\beta}\varepsilon_{\beta}{*}AL_c)/s - \rho^{\alpha}V = \rho^{\alpha}\varepsilon_{\alpha}{*}AL_c + \rho^{\beta}\varepsilon_{\beta}{*}AL_c \tag{5.14}$$

which can be written as

$$AL_c = GV \tag{5.15}$$

where

$$G = \frac{s\rho^{\alpha}}{\rho^{\beta}\varepsilon_{\beta}{*}(1-s) - \rho^{\alpha}\varepsilon_{\alpha}{*}s} \tag{5.16}$$

Eq. (5.15), which holds at each instant in the filtration, implies that the average porosity is constant if the slurry concentration remains fixed and that this linear relation is also independent of the filter cake geometry.

These observations can be verified experimentally by determining whether or not the cake length is linear with filtrate volume for two filter cake geometries (i.e., cylindrical and leaf filter cakes).

The average porosity is defined by

$$\varepsilon_\alpha^* = (1/L_c)\int_o^{L_c} \varepsilon_\alpha(z,t)dz \tag{5.17}$$

and is constant if

$$\frac{d\varepsilon_\alpha^*}{dt} = 0 \tag{5.18}$$

To meet this requirement $\varepsilon_\alpha^* = \varepsilon_\alpha(\xi)$, where $\xi = z/L_c(t)$, which satisfies Eq. (5.18) since

$$\frac{d}{dt}\int_0^1 \varepsilon_\alpha(\xi)d\xi = 0 \tag{5.19}$$

For the average porosity to be a function of fractional cake volume, ξ, means that the shape of the porosity profile remains the same throughout the filtration.

For a constant density liquid and particulate phase, the change in independent variables from (z,t) to (ξ,t) converts the governing equations for a one dimensional filtration, Eqs. (5.1), (5.2), and (5.13) to

$$\xi\dot{L}_c\frac{d\varepsilon_\alpha}{d\xi} + \frac{\partial}{\partial\xi}(\varepsilon_\alpha v_z{}^\alpha) = 0 \tag{5.20}$$

$$\xi\dot{L}_c\frac{d\varepsilon_\beta}{d\xi} + \frac{\partial}{\partial\xi}(\varepsilon_\beta v_z{}^\beta) = 0 \tag{5.21}$$

$$\frac{\partial P_o^*}{\partial\xi} - (\varepsilon_\alpha \mu L_c/KP_c{}^o)\,(v_z{}^\alpha - v_z{}^\beta) = 0 \tag{5.22}$$

where

$$P_o^* = P_o^{\alpha}/P_c^{o} \tag{5.23}$$

and

$$P_c^{o} = P + \rho^{\alpha} g^{\alpha} L_c - P_o \tag{5.24}$$

in which P is the applied filtration pressure.

The filtrate rate is obtained by evaluating the fluid α-phase momentum balance, Eq. (5.22), at the exit of the filter cake where

$$\xi = 0 \tag{5.25}$$

and

$$v_z^{\beta} = 0 \tag{5.26}$$

This boundary condition defines an interface which is called the "septum" to distinguish it from the filter medium. The septum then is the location in the filter medium where the velocity of the particulate β-phase is zero. The result of applying this boundary condition to Eq. (5.22) is

$$J_o - (\mu L_c / K_o P_c^{o})(\dot{V}/A) = 0 \tag{5.27}$$

where J_o is the ratio of the pressure gradient at the septum to the pressure gradient across the entire filter cake

$$J_o = (\partial P_o^{\alpha}/\partial z)\big|_{\xi=0} (P_c^{o}/L_c)^{-1} \tag{5.28}$$

and

$$(\varepsilon_{\alpha} v_z^{\alpha})\big|_{\xi=0} = \dot{V}/A \tag{5.29}$$

and K_o is the permeability at the septum.

It is common practice to interpret filtration data by plotting the reciprocal rate versus the filtrate volume,

$$\dot{V}^{-1} = (\mu G/A^2 J_o K_o P_c^{\,o})\ V \tag{5.30}$$

where Eq. (5.30) indicates that reciprocal rate is linear in V if J_o, K_o, and P_c^o are all constant.

The objectives of experimental studies are to verify the linearity between the cake length and filtrate volume and to determine the effect of (i) septum permeability, (ii) cake pressure drop, (iii) filter area, (iv) particle size, (v) different β-phases, and (vi) filter cake geometry on Eq. (5.30).

Measurement of $\dot{V}$, V, G, A, μ, and P_c^o permits the determination of the product $K_o J_o$ from Eq. (5.30). Separation of this product requires the measurement of internal pressure profiles and the subsequent calculation of J_o.

The reciprocal rate expression, Eq. (5.30), holds for Newtonian and multiphase non-Newtonian liquids since both classes of fluids are described by momentum balances, Eqs. (5.3) and (5.13), that are the same. The only alteration to Eq. (5.30) would be the substitution of $(-\lambda_0)$ for $(\varepsilon_\alpha^{\,2}\mu/K)$.

5.2.2 Leaf Filtrations

The effect of filter cake geometry (5) is determined from a filtration on a circular leaf on which the shape of the filter cake assumes the form of an approximate oblate spheroid. Previous studies (4,17) have indicated that Darcy's law may not be valid for circular leaf filtrations.

The oblate spheroidal coordinate system that is used for these circular leaf filtrations is shown in Figure 2 and the transformation from rectangular to oblate spheroidal coordinates is

$$x = R((1+\gamma^2)(1-\eta^2))^{1/2}\cos\omega \tag{5.31}$$

$$y = R((1+\gamma^2)(1-\eta^2))^{1/2}\sin\omega \tag{5.32}$$

$$z = R\gamma\eta \tag{5.33}$$

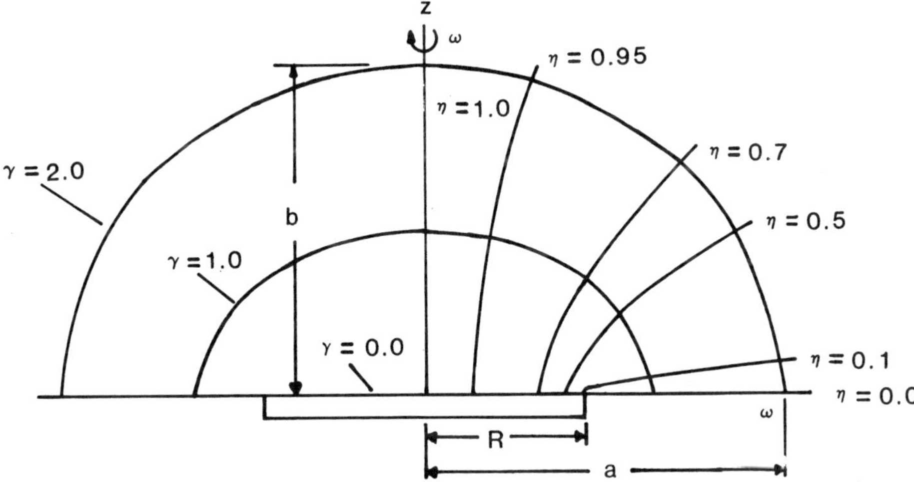

Fig. 2. Oblate spheroidal coordinates.

for $0 \leq \gamma < \infty$, $0 \leq \eta \leq 1$, $0 \leq \omega < 2\pi$. The outer dimensions of the oblate spheroid are given by

$$a = R(1+\gamma_e^2)^{1/2} \tag{5.34}$$

$$b = R\gamma_e \tag{5.35}$$

where R is the radius of the filter media and γ_e is the value of that coordinate on the surface of the filter cake. The scale factors are (16)

$$h_\gamma = R(\gamma^2+\eta^2)^{1/2}/(1+\gamma^2)^{1/2} \tag{5.36}$$

$$h_\eta = R(\gamma^2+\eta^2)^{1/2}/(1-\eta^2)^{1/2} \tag{5.37}$$

$$h_\omega = R((1+\gamma^2)^{1/2}(1-\eta^2))^{1/2} \tag{5.38}$$

The objective is to obtain a filtrate rate expression for circular leaf filters from equations that are analogous to Eqs. (5.20), (5.21), and (5.22) in which the transformation to fractional cake volume, ξ, which is equivalent to a constant average porosity, has been utilized.

The transformation of Eqs. (5.1),(5.2), and (5.13) to oblate spheroidal coordinates gives the governing equations (5)

$$\frac{\partial}{\partial t}(\varepsilon_\alpha \rho^\alpha) - \frac{1}{R(\gamma^2+\eta^2)} \frac{\partial}{\partial \gamma}(\varepsilon_\alpha \rho^\alpha \Omega_\gamma{}^\alpha) = 0 \tag{5.39}$$

$$\frac{\partial}{\partial t}(\varepsilon_\beta \rho^\beta) - \frac{1}{R(\gamma^2+\eta^2)} \frac{\partial}{\partial \gamma}(\varepsilon_\beta \rho^\beta \Omega_\gamma{}^\beta) = 0 \tag{5.40}$$

$$\frac{\partial P_o{}^\alpha}{\partial \gamma} - \frac{\varepsilon_\alpha \mu R(\gamma^2+\eta^2)^{1/2}}{M(1+\gamma^2)^{1/2}} (\Omega_\gamma{}^\alpha - \Omega_\gamma{}^\beta) = 0 \tag{5.41}$$

where M is the permeability and is assumed to be different from that for an axial filtration. The scaled velocities are

$$\varepsilon_\alpha \Omega_\gamma{}^\alpha(\gamma,t) = (\gamma^2+\eta^2)^{1/2}(1+\gamma^2)^{1/2} \; \varepsilon_\alpha v_\gamma{}^\alpha(\gamma,\eta,t) \tag{5.42}$$

$$\varepsilon_\beta \Omega_\gamma{}^\beta(\gamma,t) = (\gamma^2+\eta^2)^{1/2}(1+\gamma^2)^{1/2} \; \varepsilon_\beta v_\gamma{}^\beta(\gamma,\eta,t) \tag{5.43}$$

where the coefficients of $\varepsilon_\alpha v_\gamma{}^\alpha$ and $\varepsilon_\beta v_\gamma{}^\beta$ account for the area change as the fluid moves toward the filter media.

An external macroscopic material balance

$$M_\beta/s - \rho^\alpha V = M_\alpha + M_\beta \tag{5.44}$$

gives a linear relation

$$\Delta = GV \tag{5.45}$$

that is analogous to Eq. (5.15). In this case, the masses of each phase and the filter cake volume are given by

$$M_\alpha = \rho^\alpha \varepsilon_\alpha{*}\Delta \tag{5.46}$$

$$M_\beta = \rho^\beta \varepsilon_\alpha{*}\Delta \tag{5.47}$$

$$\Delta = (2/3)\pi R^3(\gamma_e{}^3+\gamma_e) \tag{5.48}$$

and G is given by Eq. (5.16).

The average porosity

$$\varepsilon_\alpha{*} = (1/\Delta)\int_0^{\gamma_e}\int_0^1\int_0^{2\pi} \varepsilon_\alpha h_\gamma h_\eta h_\omega d\omega d\eta d\gamma \tag{5.49}$$

must be independent of time which means that the local porosity is a function of only fractional cake volume,

$$\varepsilon_\alpha = \varepsilon_\alpha(\xi) \tag{5.50}$$

and where the fractional cake volume for the circular leaf filter cake is given by

$$\xi = (\gamma^3+\gamma)/(\gamma_e{}^3+\gamma_e) \tag{5.51}$$

The conversion of Eqs. (5.39),(5.40), and (5.41) from (γ,t) to (ξ,t) requires the use of the chain rule

$$\frac{\partial}{\partial\gamma} = \frac{(3\gamma^2+1)}{(\gamma_e{}^3+\gamma_e)}\frac{\partial}{\partial\xi} \tag{5.52}$$

$$\frac{\partial\varepsilon_\alpha}{\partial t} = -\frac{\xi\dot{\gamma}_e(3\gamma_e{}^2+1)}{(\gamma_e{}^3+\gamma_e)}\frac{d\varepsilon_\alpha}{d\xi} \tag{5.53}$$

and the elimination of γ by rearranging the definition of fractional cake volume, Eq. (5.51) to obtain the cubic equation

$$\gamma^3 + \gamma - \xi(\gamma_e{}^3+\gamma_e) = 0 \tag{5.54}$$

which has the solution (22)

$$\Upsilon = S + T \tag{5.55}$$

where

$$S(\xi,\Upsilon_e) = \{(\xi\Upsilon_e^3+\xi\Upsilon_e)/2 + \{(4+27\xi^2(\Upsilon_e^3+\Upsilon_e)^2)/108\}^{1/2}\}^{1/3} \tag{5.56}$$

$$T(\xi,\Upsilon_e) = \{(\xi\Upsilon_e^3+\xi\Upsilon_e)/2 - \{(4+27\xi^2(\Upsilon_e^3+\Upsilon_e)^2)/108\}^{1/2}\}^{1/3} \tag{5.57}$$

The final result (5) for the transformation of the Eqs. (5.39), (5.40), and (5.41) to fractional cake volume is

$$\xi\dot{\Upsilon}_e \frac{d\varepsilon_\alpha}{d\xi} + S_c(\xi,\eta,\Upsilon_e) \frac{\partial}{\partial\xi}(\varepsilon_\alpha\Omega_\Upsilon^\alpha) = 0 \tag{5.58}$$

$$\xi\dot{\Upsilon}_e \frac{d\varepsilon_\beta}{d\xi} + S_c(\xi,\eta,\Upsilon_e) \frac{\partial}{\partial\xi}(\varepsilon_\beta\Omega_\Upsilon^\beta) = 0 \tag{5.59}$$

$$\frac{\partial P_o^*}{\partial\xi} - (\varepsilon_\alpha\mu\Upsilon_e/S_m M P_c^o)(\Omega_\Upsilon^\alpha-\Omega_\Upsilon^\beta) = 0 \tag{5.60}$$

where

$$S_c(\xi,\eta,\Upsilon_e) = (3(S+T)^2+1)/\{R((S+T)^2+\eta^2)(3\Upsilon_e^2+1)\} \tag{5.61}$$

$$S_m(\xi,\Upsilon_e) = \{(3(S+T)^2+1)((S+T)^2+1)\}/(R(\Upsilon_e^2+1)) \tag{5.62}$$

$$S_m(0,\Upsilon_e) = 1/(R(\Upsilon_e^2+1)) \tag{5.63}$$

The analogy between the governing equations for axial filtrations, Eqs. (5.20), (5.21), (5.22), and the cylindrical leaf filtrations, Eqs. (5.58), (5.59), (5.60) is apparent.

The filtrate rate expression can be obtained by evaluating Eq. (5.60) at the septum where

$$\xi = 0 \qquad (5.64)$$

$$\gamma = 0 \qquad (5.65)$$

$$\Omega_{\gamma}^{\beta} = 0 \qquad (5.66)$$

$$(\varepsilon_{\alpha}\Omega_{\gamma}^{\alpha})|_{\gamma=0} = \dot{V}/2\pi R^2 \qquad (5.67)$$

$$R(\gamma_e^{3}+\gamma_e) = GV/(2/3)\pi R^2 \qquad (5.68)$$

with the final result (5) for the reciprocal rate expression for circular leaf filtrations

$$\dot{V}^{-1} = \{\mu G/((4/3)A^2 J_o M_o P_o)\}V \qquad (5.69)$$

where M_o is the permeability at the septum for a circular leaf filter. If the filter medium, particulate β-phase, and liquid α-phase are the same, then $M_o = K_o$. This is further evidence that the permeability is a multiphase constitutive property.

Except for the (4/3) in the denominator of Eq. (5.69), the reciprocal rate expressions for the axial cylindrical filtration and the circular leaf filtration, Eq. (5.30), are the same. Although the rate equations are similar, it does not mean that the filtrate rates are the same since the difference in geometries will affect the pressure gradient, J_o, and hence the filtrate rates.

6. EXPERIMENTAL RESULTS

The factors affecting the reciprocal rate expression given by Eq. (5.30)

$$\dot{V}^{-1} = (\mu G/A^2 J_o K_o P_c^{o})\ V \qquad (5.30)$$

are examined with a filtration apparatus (28,29) which is capable of measuring, in addition to the normally acquired macroscopic data of filtrate volume, time, and applied pressure, the filtrate rate, the cake pressure drop, the medium pressure drop, and the microscopic level variables of local porosity and pressure.

The apparatus also is fitted with a pressure controller which enable both constant applied pressure and constant cake pressure drop filtrations to be performed. The local porosities are determined by electrical conductivity probes located axially in the wall of the cylindrical filter chamber. Small (1/16 inch) pressure probes are also located axially inside the filter chamber and connected to transducers which record the internal axial pressure profiles.

6.1 Assumptions and Filtration Mechanism

The assumptions which are associated with Eq. (5.30) are that the viscous and inertial forces are negligible and that both the liquid α-phase and particulate β-phase have constant densities. The viscosity is constant since it is a function of only the temperature and such variations are assumed to be insignificant.

6.1.1 Filtration Mechanism

The mechanism deduced from the theoretical development indicates that the liquid α-phase velocity profile is essentially flat and a function of only the axial coordinate. Momentum is transferred from the α-phase to the particulate β-phase while little, if any, is transferred to the wall of the filter chamber. These conclusions are deduced from the dimensional analysis where it is shown that the viscous force is very small.

Continuity conditions for both phases, Eqs. (5.20) and (5.21), indicate that the porosity distribution determines the liquid α-phase velocity and the particulate β-phase velocity. Pressure profiles are adjusted, via the liquid α-phase momentum balance, Eq. (5.22), to accommodate the mass flow rates and velocities specified by the continuity requirements. This is just the opposite of single phase tube flow where the linear axial pressure profile specifies the flow rate.

The location with the minimum porosity will have the highest local liquid α-phase velocity and in filtrations, this minimum porosity must occur at the septum, which is located somewhere inside the filter media. This then should also be the location of the largest pressure gradient. It is not unexpected then that the filtrate rate depends on the septum permeability, K_o, and the pressure gradient, J_o as indicated in Eq. (5.30).

6.2 Medium and Pressure Gradients

Eq. (5.30) indicates that the pressure of interest in filtration is the pressure drop across the filter cake and not the applied pressure. Experimental data (28,29) indicates that the pressure drop across the filter medium, in some cases, can be as large as the pressure drop across the filter cake.

If it is required to obtain a specific pressure drop over the filter cake, the rate at which the pressure increases to this prescribed pressure is influenced by the pressure gradient across the filter media. A coarse porosity filter paper, such as Whatman No. 4, (retention of 25 μm) has an open structure and a low pressure gradient relative to that of a fine porosity Whatman No. 3 filter paper (retention of 5 μm).

If, for example, an increment of pressure drop, ΔP, is fixed at a specified value, then the filtration with the Whatman No. 4 will reach this prescribed pressure with less filter cake and in a shorter time than that required for the fine porosity Whatman No. 3. Figure 3 illustrates this effect of the medium pressure gradient on the cake pressure gradient for a fixed increment of pressure, ΔP.

6.3 Effect of Pressure on Reciprocal Rate

The initial reciprocal rate is a finite, positive quantity and represents the flow through a clean media. As particulates just begin to deposit, both P_c^o and filtrate volume are zero, but the ratio is finite and proportional to the initial filtrate rate,

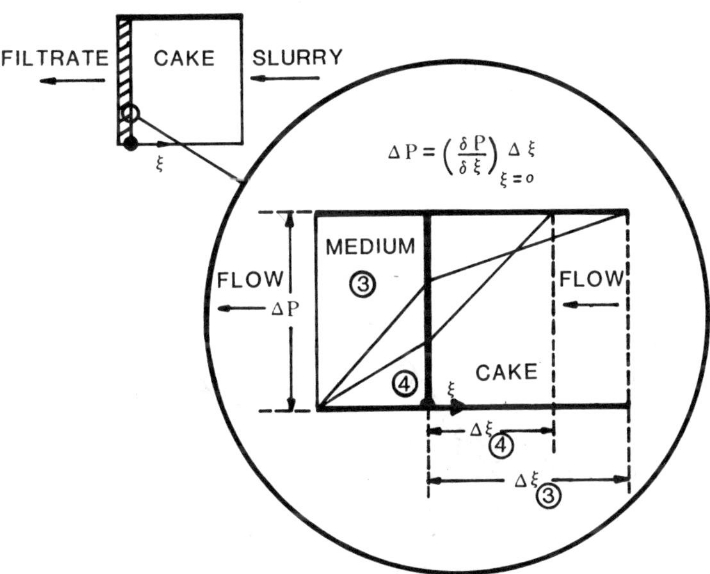

Fig. 3. Interaction of filter medium and cake pressure gradients.

$$\dot{V} = (A^2 K_o J_o/\mu G)(P_c{}^o/V) \tag{6.1}$$

As the filtration progresses from this initial condition, there will be curvature in the reciprocal rate which will persist until all of the quantities, K_o, J_o, and P_c in the slope of Eq. (5.30) become constant.

To isolate the effect of the cake pressure drop, the combination of a Whatman No. 3 filter paper and Lucite 4F molding powder gives a constant value for the product K_oJ_o and this combination is filtered at two values of cake pressure drop. The results are shown in Figures 4 and 5.

The data in Figure 4 shows, for the filtration executed at the higher pressure of 100kPa, that the induced curvature in the reciprocal rate persists for a longer time since it takes longer for the pressure drop across the filter cake, as shown in Figure 5, to reach the specified value of 100kPa.

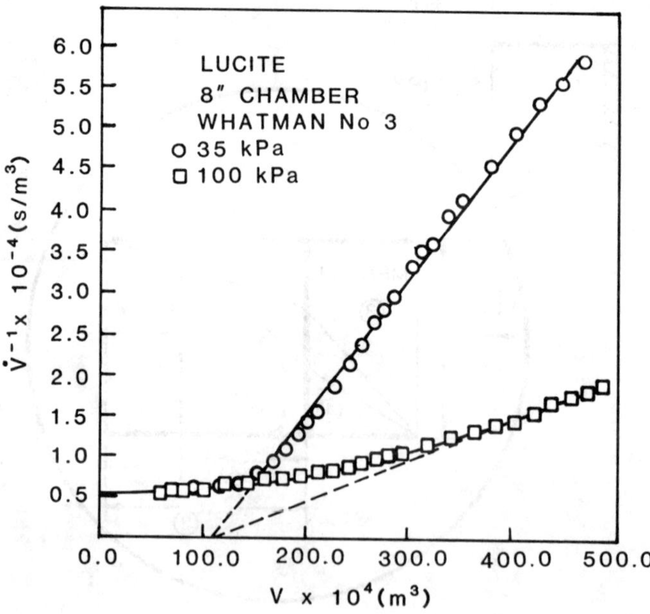

Fig. 4. The effect of cake pressure drop on reciprocal rate.

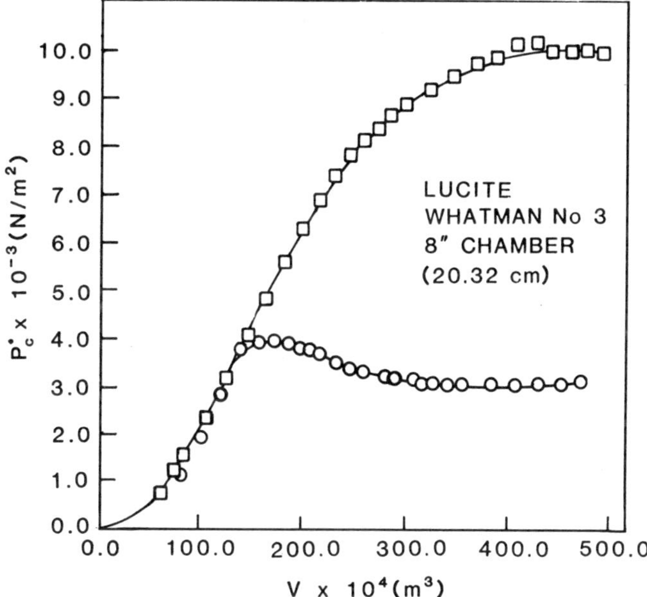

Fig. 5. Cake pressure drops for the reciprocal rate data of Fig. 4.

In addition, the slopes of the reciprocal rates for the two pressure drops vary inversely as the cake pressure drops and the reciprocal rate becomes a linear function, which extrapolates through a displaced origin, when all of the factors in the slope become constant. These observations are all consistent with the behavior predicted by Eq. (5.30) and demonstrate the isolated effect of cake pressure drop on the early part of reciprocal rate data.

The origin of the reciprocal rate is displaced to the right of the measured origin because the filter chamber is initially filled with clear liquid and the concentration of slurry must increase until it reaches that of the incoming slurry. The displaced origin represents a zero value of the filtrate volume which would correspond to a filtration executed with the filter chamber initially filled with slurry rather than clear liquid.

6.4 Pressure, Area, and Mean Porosity

The relation between cake volume and filtrate volume given by Eqs. (5.15) and (5.16) indicates that the slope G is constant if the slurry concentration and average porosity are constant. For a cylindrical axial filtration, the cross sectional area of the filter cake is constant and hence the cake length is linear in filtrate volume.

Filtrations performed in Plexiglas filter chambers of different diameters (D = 4in, D = 8in) and each at two different cake pressure drops (39 kPa, 100 kPa and 30kPa, 100 kPa) indicate that the cake length is a linear function of filtrate volume. The data are shown in Figure 6.

Since these filtrations were executed with a constant slurry concentration, then it must be concluded that the average porosity is constant throughout a filtration and that this conclusion is independent of the cake pressure drop and the size of the filter chamber. The ratio of the slopes of cake length versus filtrate volume shown in Figure 6 is, within experimental accuracy, inversely proportional to the filter areas.

The effect of slurry dilution in the filter chamber at the start of the filtration is again evident. As expected, this dilution effect is greater with the larger diameter filter chamber and, as a consequence, the displacement of the origin is correspondingly greater. In subsequent graphs, this dilution effect, which is approximately equal to the volume of the filter chamber and the liquid in excess of that in the slurry, it subtracted from

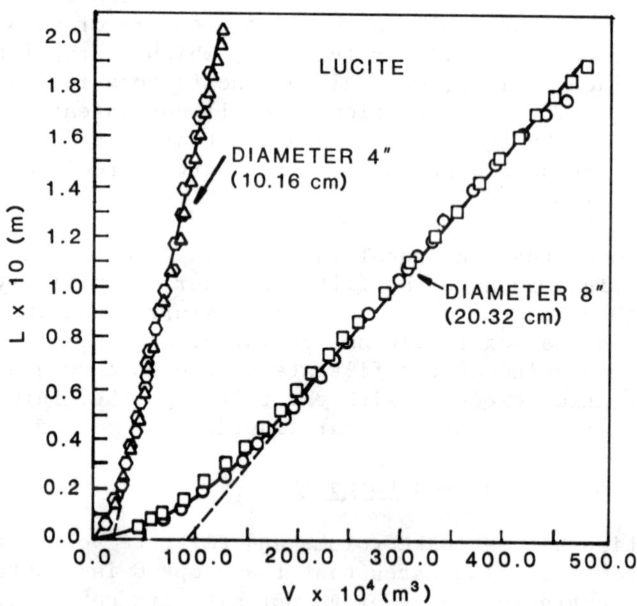

Fig. 6. Cake length versus filtrate volume at constant slurry concentration.

the filtrate volume and the origin is correspondingly translated to compensate for this effect.

6.5 Filter Medium and Septum Permeability

The filtration mechanism and the expression for the reciprocal rate, Eq. (5.30), implies that the location of minimum porosity will have the highest pressure gradient and liquid velocity. To retain particles and form a cake, the smallest openings must occur at the location in the medium where the particulates stop. This interaction between the particulates and the filter medium should be dependent on the relative size and shape of the particulates and the openings in the filter media.

The interfacial permeability, K_o, is determined from the product K_oJ_o which is obtained from the macroscopic measurements dictated by Eq. (5.30) and then this product is separated by determining J_o from the pressure profiles that are directly measured.

The effect on septum permeability is isolated by performing filtrations in which the slurry concentration, average porosity, particulate phase, and cake pressure drop are all maintained, to within experimental error, at the same values. The only difference

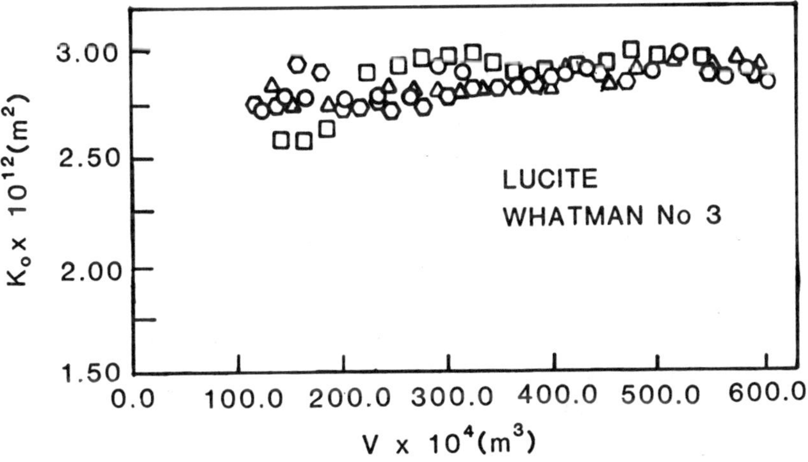

Fig. 7. Septem permeabilities for the Whatman No. 3 filter medium.

in the filtrations are the filter media and two media are examined, a fine porosity Whatman No. 3 (retention of 5μm) and a coarse porosity Whatman No. 4 (retention of 25μm). The results are shown in Figures 7 and 8. Early estimates of the pressure gradient and interfacial permeability are not reliable since only one or two pressure probes are submerged in the filter cake and these early values are correspondingly discounted.

Figure 7 shows replicate runs with water slurries of Lucite at a cake pressure drop of 68.9kPa and a slurry concentration of 7.5% using the fine porosity Whatman No. 3 filter paper. It is evident that the septum permeability resists clogging and maintains a constant, reproducible value throughout each of the four filtrations.

Under identical conditions, the septum permeability of the coarse porosity Whatman No. 4 shows a reproducible decay during the four replicate filtrations shown in Figure 8. Since the only difference between the two filter media are the dimensions of the pores, it must be concluded that the continual penetration of Lucite particulates into the pores of the Whatman No. 4 filter medium accounts for the declining interfacial permeability.

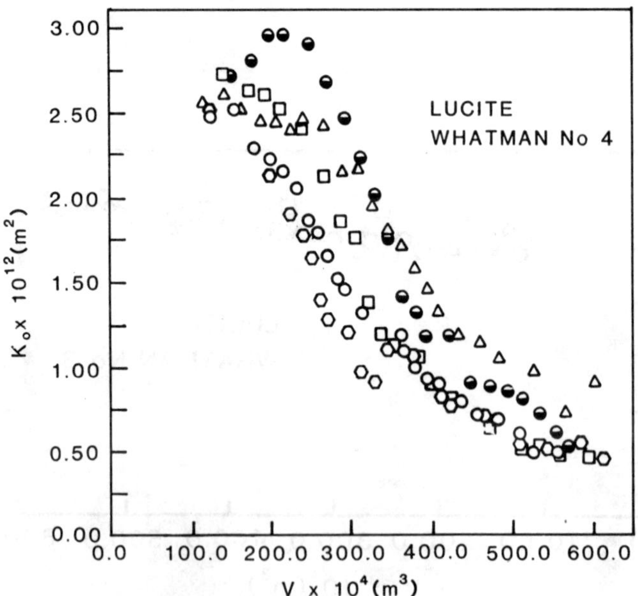

Fig. 8. Septum permeabilities for the Whatman No. 4 filter medium.

A conceptualization of what is occurring within the two media is shown in Figure 9. For the fine porosity Whatman No. 3 filter media, the pores are so small that the location (i.e., septum) where the particulates stop is either at the cake-media interface or only slightly below the surface as shown in the sketch on the left side of Figure 9.

Conversely, for the coarse porosity Whatman No. 4 filter media, shown in the sketch on the right in Figure 9, the location (i.e., septum) where the particulates stop is considerably removed from the cake-media interface and there is a zone in the filter media which is susceptible to further penetration of the Lucite particulates. The net result is a corresponding decrease in the septum permeability, $K_o(t)$ during the course of the filtration.

The evidence thus far indicates that the layer of particles adjacent to the medium accounts for the major resistance to flow, and that the applied pressure increases the stress within the cake but does not alter the porosity.

6.6 Septum Permeability and Reciprocal Rate

Isolation of the effect of septum permeability on reciprocal

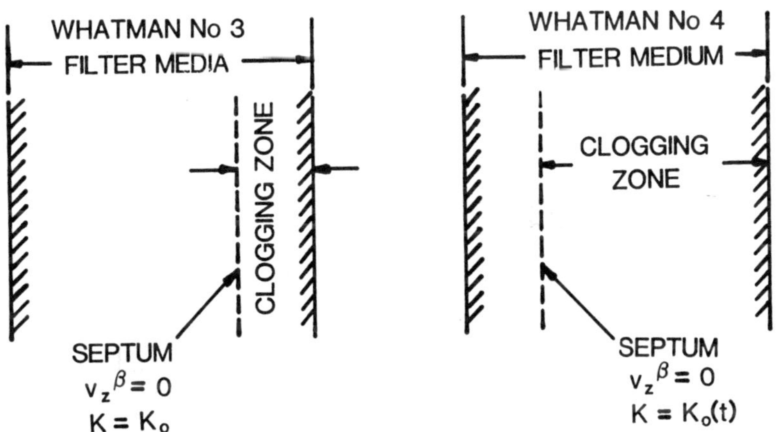

Fig. 9. Location of the septum and clogging zone in Whatman No. 3 and No. 4 filter media.

rate is demonstrated by keeping cake pressure drop (103 kPa), particulate phase (Lucite), and slurry concentration (5%) fixed for filtrations (D = 4in) with the Whatman No. 3 and No. 4 filter media whose characteristics are shown in Figures 7 and 8.

The upper two curves in Figure 10 show the cake pressure drops for the two filtrations. The Whatman No. 3 filter medium rises more slowly to the prescribed cake pressure drop than the Whatman No. 4 filter medium. This is due to the larger pressure gradient across the fine porosity Whatman No. 3 filter media and this effect was discussed previously in regard to Figure 3.

The two lower curves in Figure 10 show the septum permeabilities for the Whatman No. 3 and No. 4 filter media and these curves exhibit the same characteristic form for these two filter media, when they are exposed to Lucite particles, that is shown in Figures 7 and 8.

Reciprocal rate data, shown in Figure 11, reveals that curvature disappears for the Whatman No. 3 filter media as soon as the cake pressure drop becomes constant but persists for the Whatman No. 4 filter media even after the cake pressure drop becomes constant. The persistent curvature must be attributed to the

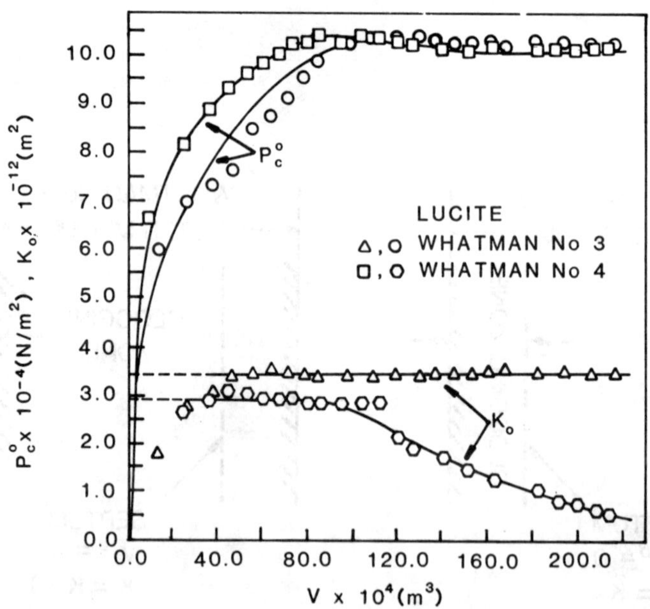

Fig. 10. Cake pressure drops and septum permeabilities for Whatman No. 3 and No. 4 filter media.

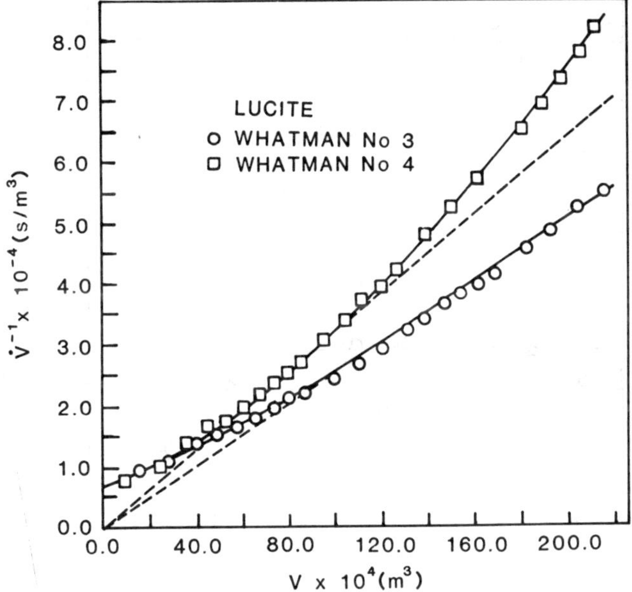

Fig. 11. Reciprocal rates for the Whatman No. 3 and No. 4 septum permeabilities shown in Fig. 10.

decreasing septum permeability as the governing equation for reciprocal rate, Eq. (5.30), indicates.

To summarize, curvature in the initial portion of reciprocal rate data can be attributed primarily to the pressure drop across the filter cake but at later times, persistent curvature is indicative of lower values of the septum permeability due to penetration and clogging of the media with the particulate phase.

6.7 Particle Size and Septum Permeability

The effect of particle size on the septum permeability is determined by separating the spherical Lucite particles into two segments, one of which contains particles in the 125-149 μm range and the other in the 63-90 μm range. These size-graded segments are obtained by using a standard series of screens on an electromagnetic shaker. Photomicrographs of the respective segments revealed that there were no particles in the range of the filter media openings (5 μm to 25 μm) for the particles in the 125-149 μm segment.

Figure 12 is a comparison of the interaction of the septum permeabilities for the two size-graded segments of Lucite with the Whatman No. 3 and No. 4 filter media. The top curve in Figure 12

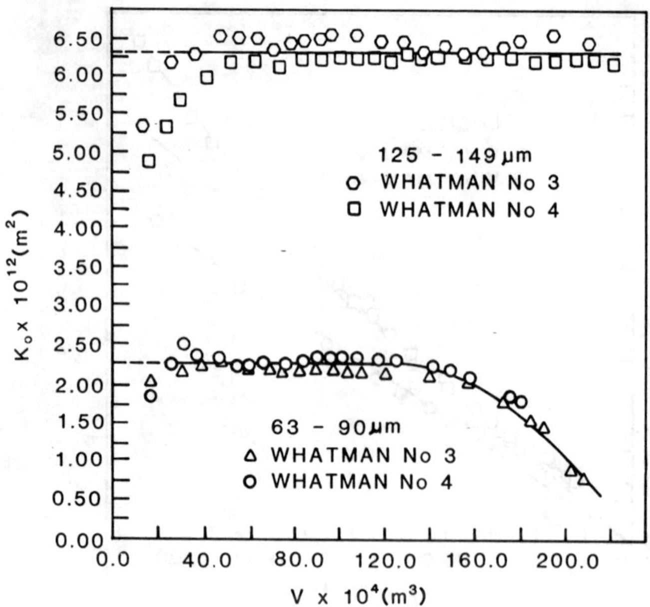

Fig. 12. Effect of Lucite particle size on the septum permeability of Whatman No. 3 and No. 4 filter media.

indicates (i) that limiting the Lucite particles to those which are about five times the pore openings (25 µm) on the Whatman No. 4 filter medium moves the septum closer to the surface of the filter medium and diminishes pore penetration and clogging of the filter medium, and (ii) that this constant septum permeability is about twice that shown in Figure 7 for the unsegmented Lucite.

The lower curve in Figure 12 reveals that the Lucite segment with the smaller particulates (63-90 µm) precipitates pore penetration and clogging in the Whatman No. 3 filter media with the smaller (5 µm) openings and the behavior is indistinguishable from that of the Whatman No. 4 filter media with the larger (25 µm) openings. The decreasing values of septum permeabilities for the 63-90 µm segment of Figure 12 are about the same as those shown in Figure 8 for the unsegmented Lucite.

It is evident that individual characterization of the filter media or particulate phase is of little value since it is the interaction of the particulates with the filter media through the septum permeability, K_o, that affects the filtrate rate.

6.8 Particle Shape and Septum Permeability

It has been demonstrated that the size of the spherical Lucite particulates relative to the opening in the filter media has a measurable effect on the septum permeability, K_o. It is reasonable to expect that particle shape will also influence the values of the septum permeability and three additional shapes are considered, fibers (Solka Floc SW-400), lattices (Celite grade 319), and platelets (Vertal talc grade 77). To maintain a basis of comparison, the same two filter media, Whatman No. 3 and No. 4, are used for the determination of the effect of particle shape on septum permeability.

Septum permeabilities for these substances are determined in the same way as those for the Lucite particulates, that is, the product K_oJ_o is separated by determining J_o from the measured internal pressure profiles.

The upper two curves on Figure 13 show the septum permeabilities for the diatomaceous earth (Celite) which indicates, (i) that Celite exhibits penetration and clogging of both filter media, (ii) that the unique interaction between the Celite and the Whatman No. 3 and No. 4 filter media is characterized by a sharp initial

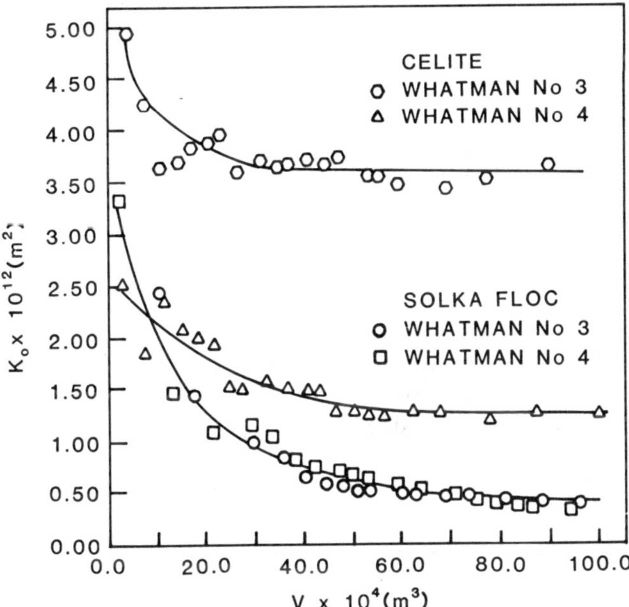

Fig. 13. Effect of particle shape on septum permeability of the Whatman No. 3 and No. 4 filter media.

decline to a relatively constant value of K_o, and (iii) that the permeabilities for the Whatman No. 4 are lower than those for the Whatman No. 3.

The fibrous material (Solka Floc), which is the bottom curve in Figure 13, exhibits a sharp initial decline in the septum permeability which is followed by a continuous, but slower decline, in the latter two-thirds of the filtration. The two types of filter media, however, do not have any effect on the septum permeability for Solka Floc. This can be attributed to the structural similarity between the cellulose fibers of Solka Floc and the paper filter media.

The interaction of submicron platelets (talc) with the two types of filter media is shown in Figure 14. The septum permeabilities with either medium are smaller than all of the previous materials filtered on the Whatman No. 3 and 4 filter media by roughly a factor of ten. As with the Solka Floc, the two different filter media do not have any effect on the septum permeability for talc. The reason, in this case, is that the openings in both filter media are so much larger than the talc particulates that pore penetration and clogging, as evidenced by the decreasing septum permeability, are significant for both the Whatman No. 3

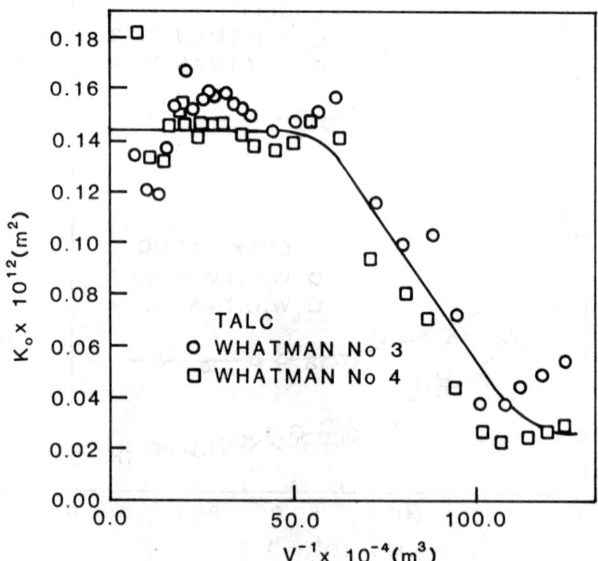

Fig. 14. Septum permeability of talc with Whatman No. 3 and No. 4 filter media.

and No. 4 filter media.

The shape of particulates does not, however, affect the average porosity during the course of the filtration as is illustrated in Figure 15 where the cake length of a function of filtrate volume is plotted for Celite, Solka Floc, and talc.

Under conditions of constant cake pressure drop, the cake pressure drop per unit length decreases as the length of the cake increases. If the medium does not clog, J_o remains at a value of unity indicating that the pressure gradient at the septum inside the filter medium is also decreasing in proportion to the overall pressure gradient. For a material such as talc, filtered with either the Whatman No. 3 or 4 filter media, the clogged interface demands an elevated pressure gradient to satisfy the fluid α-phase continuity condition and, consequently, dissipates a greater fraction of the overall pressure gradient. During this entire process, regardless of the shape or size of the particulates, or the cake pressure drop, or the filter area, the average porosity remains constant.

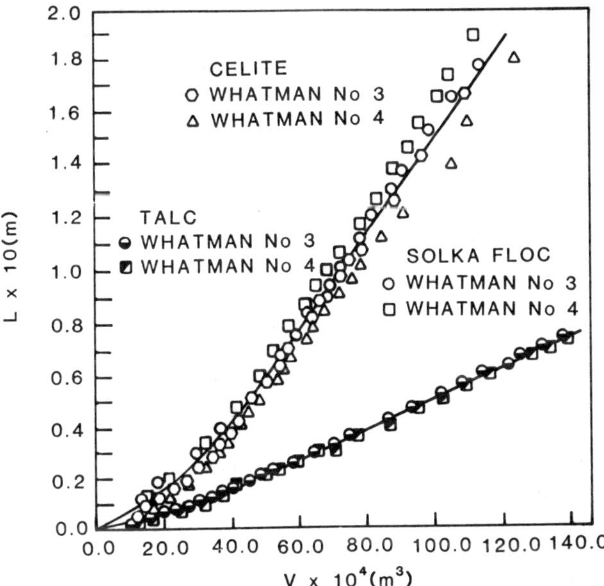

Fig. 15. Effect of particle shape on cake length versus filtrate volume for constant slurry concentration.

6.9 Effect of Filter Cake Geometry

For a cylindrical leaf filter, the factors affecting the reciprocal rate expression

$$\dot{V}^{-1} = \{\mu G/((4/3)A^2 J_o M_o P_o)\}V \tag{5.69}$$

are the same as those of the one-dimensional axial filtration. If the same filter media and particulate phase are used for the cylindrical leaf filtrations as are used in the one-dimensional filtrations, then the filtration mechanism and conclusions, with regard to media and particulate phase, will also apply to the leaf filtrations. The isolated factor, then, is the filter cake geometry.

The experimental procedure involves performing a circular leaf filtration and a one-dimensional axial filtration in which the filter area, filter media, particulate phase, slurry concentration, and cake pressure drop are, within experimental error, the same. The circular leaf filter permits the filter cake to grow in three directions and this is most easily accomplished with a vacuum rather than a pressure filtration. Vacuum filtrations, however, are inherently less reliable than pressure filtrations since the filtrate rates cannot be measured directly due to the degassing and slugging that occur in a rotameter with liquids under a vacuum. As a result, filtrate rates are determined by differentiating the filtrate volume and this is a source of error.

Internal pressure profiles are not measured because it is difficult to locate and maintain the probes along an η-coordinate during the course of the filtration and the probes also distort the shape of the oblate spheroidal filter cake. Hence, the effect of filter cake geometry depends on the interpretation of the behavior of the product $K_o J_o$.

Even with these experimental deficiencies, the results shown in Figure 16 clearly indicate that the product $K_o J_o$ for the leaf filter is significantly larger than the same product for the cylindrical filtration executed under the same conditions. In Figure 17, the reciprocal rates for the leaf filter are less than those for the cylindrical filter or, in terms of filtrate rates, the leaf filter exhibits significantly higher filtration rates.

This behavior can be explained by the filtration mechanism described earlier. Using a Whatman No. 4 filter medium implies that the septum permeability is decreasing during the course of the filtration and this dominance of K_o is evident for the axial filtration. Conversely, the product $K_o J_o$ for the leaf filter is

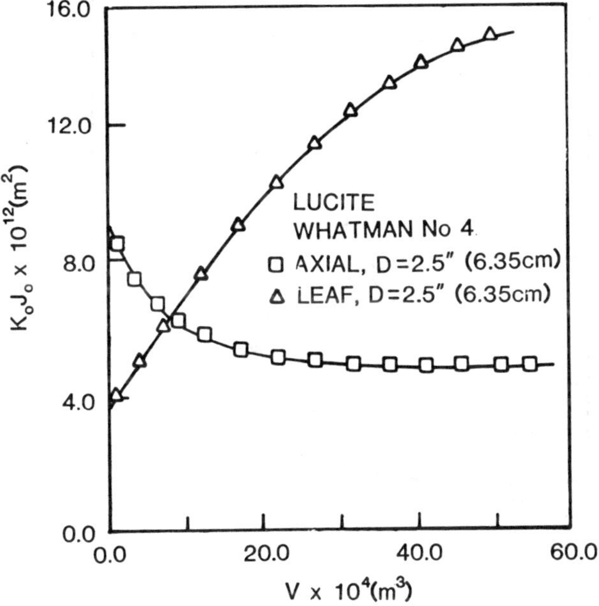

Fig. 16. Comparison of the K_0J_0 product for an axial filtration and a circular leaf filtration.

increasing and this increase must be attributed to the dominant effect of the pressure gradient relative to the decreasing value of the septum permeability.

This dominance of the pressure gradient at the septum for the circular leaf filter is attributed to the significantly larger surface area relative to the filter area. This area difference means that the mass of liquid entering the leaf filter cake must exit through the filter medium, which is a smaller area that has been reduced ever further by the clogging of the filter medium. To assure that the mass flow rate at any instant is the same at both the entrance and exit surfaces requires a proportionate increase in the septum pressure gradient and velocity as dictated by Eq. (5.69).

Filter cake geometry then has considerable influence on filtrate rates. Cake geometries in which the cake surface is larger than the filter surface, other factors being equal, will exhibit higher filtration rates. For example, if inward and outward radial filtrations are to be compared on the basis of maximizing filtrate rates, then an inward radial filtration is superior to an outward radial filtration.

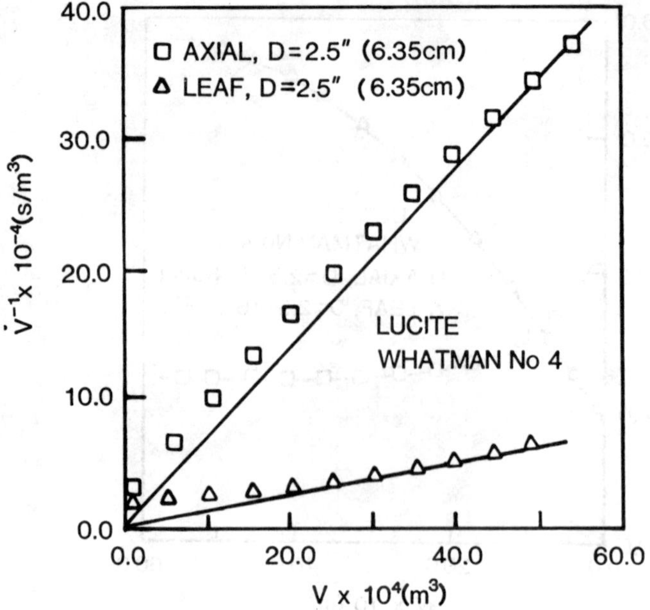

Fig. 17. Comparison of reciprocal rates for an axial filtration and a circular leaf filtration.

7. SUMMARY

Based on the general multiphase continuum and constitutive theory, a filtration mechanism is deduced for the flow of multiphase non-Newtonian and Newtonian fluids in axial and circular leaf filters.

Except for a geometric factor of (3/4), the filtrate rate expressions for axial and circular leaf filters are the same. Non-Newtonian multiphase fluids are also governed by the same rate expression except that the permeability, K, must be properly replaced by $(-\lambda_0)$. These rate expressions assume that the viscous and inertial forces are negligible and that the flow is dominated by the pressure and drag forces.

The mechanism deduced from the governing equations for filtration dictates that the porosity distribution determines the liquid and particulate velocities within the filter cake. Pressure gradients within the filter cake adjust to meet the flow requirements specified by the porosity and velocity distributions.

The location with the highest velocity and pressure gradient occurs in the filter medium at an interface, called the septum, where the solid particulate velocity vanishes. The filtrate rate is controlled by the product of the permeability and pressure gradient at this interface within the filter media.

If the septum is located at the surface of the filter media, where it contacts the particulates of the filter cake, then the filter medium does not clog. If the septum is located at some point within the filter medium, then there is a zone between the septum and the surface of the filter medium in which particulate penetration causes clogging of the filter medium pores and a decline in the septum permeability during the course of the filtration.

Septum permeability is a characteristic property of the filter medium and particulates and this property reflects the interaction of the pore openings in the filter medium and the size and shape of the particulates. However, septum permeabilities alone do not govern filtrate rates. High pressure gradients, J_o, can compensate for low permeabilities, K_o, and it is the product of these two factors that determines the filtrate rate.

The interpretation of data based on this filtration mechanism indicates that initial curvature in the reciprocal filtrate rate is due primarily to the developing pressure drop across the filter cake. At later times, persistent curvature in the reciprocal rate data is indicative of particulate penetration and clogging of the filter medium.

The local porosity is a function of only fractional cake volume and the average porosity is constant and unaffected by the size and shape of the particulate phase or the cake pressure drop and filter area.

Altering the ratio of the surface area to the filter area of a filter cake can have significant effect on filtration rates. An entrance surface which is larger than the exit surface at the filter medium can induce increases in the dimensionless pressure gradient at the septum and thereby enhance filtration rates.

8. ACKNOWLEDGEMENTS

This work has been supported by the National Science Foundation, Grants ENG-7606224 and CPE-8007523, the Chemical Engineering Department, the Research (Faculty Projects) Committee, and the Office of the Coordinator of Research at The University of Akron.

Recognition must also be extended to Miss Jeanine Gray for the typing, Mr. Michael J. Willis for the figures and the following graduate students, Mr. Kenneth Gray, Mr Ming Shen, Dr. Ismail Tosun, Mr. William Bridges, Mr. Raymond Collins, Mr. Steven Bybyk, and Mr. Jagadeeshan Raviprakash.

9. REFERENCES

1. Atkins, F. J., and Craine, R. E., "Continuum Theories of Mixtures: Basic Theory and Historical Development," *Quarterly Journal of Mechanics and Applied Mathematics*, Vol. 29, 1976, pp. 209-244.
2. Bird, R. B., "The Equations of Change and the Macroscopic Mass, Momentum and Energy Balances," *Chemical Engineering Science*, Vol. 6, 1957, pp. 123-130.
3. Bowen, R. M., "Theory of Mixtures," *Continuum Physics*, A. C. Eringen, ed., Academic Press, New York, N.Y., 1976.
4. Brenner, H., "Three-Dimensional Filtration on a Circular Leaf," *A.I.Ch.E. Journal*, Vol. 7, 1961, pp. 666-671.
5. Bybyk, S., "The Effect of Filter Cake Geometry on Multiphase Filtration Theory," thesis presented to The University of Akron, Akron, Ohio, in 1985, in partial fulfillment of the requirements for the degree of Master of Science.
6. Cheremisinoff, N. P., and Azbel, D. S., *Liquid Filtration*, Ann Arbor Science, Woburn, Massachusetts, 1983.
7. Coleman, B. D., and Noll, W., "The Thermodynamics of Elastic Materials with Heat Conduction and Viscosity," *Archive for Rational Mechanics and Analysis*, Vol. 13, 1963, pp. 167-239.
8. Christopher, R. H., and Middleman, S., "Power-Law Flow through a Packed Tube," *Industrial and Engineering Chemistry Fundamentals*, Vol. 4, 1965, pp. 422-426.
9. Eringen, A. C., and Ingram, J. D., "A Continuum Theory of Chemically Reacting Media-I," *International Journal of Engineering Science*, Vol. 3, 1965, pp. 197-212.
10. Eringen, A. C., and Ingram, J. D., "A Continuum Theory for Chemically Reacting Media-II," *International Journal of Engineering Science*, Vol. 5, 1967, pp. 289-321.
11. Gray, W. G., and Lee, P. C. Y., "On the Theorems for Local Volume Averaging of Multi-phase Systems," *International Journal of Multiphase Flow*, Vol. 3, 1977, pp. 333-340.
12. Gray, W. G., "General Conservation Equations for Multi-phase Systems: 4 Constitutive Theory Including Phase Change," *Advances in Water Resources*, Vol. 6, 1983, pp. 130-140.
13. Hassanizadeh, M., and Gray, W. G., "General Conservation Equations for Multi-phase Systems: 1. Averaging Procedure," *Advances in Water Resources*, Vol. 2, 1979, pp. 131-144.
14. Hassanizadeh, M. and Gray, W. G., "General Conservation Equations for Multi-phase Systems: 2. Mass, Momenta, Energy, and Entropy Equations," *Advances in Water Resources*, Vol. 2, 1979, pp. 191-203.
15. Hassanizadeh, M., and Gray, W. G., "General Conservation Equations for Multi-phase Systems: 3. Constitutive Theory for Porous Media Flow," *Advances in Water Resources*, Vol. 3, 1980, pp. 25-40.
16. Happel, J., and Brenner, H., *Low Reynolds Number Hydrodynamics*, Prentice-Hall, Inc., Englewood Cliffs, N.J., 1965.

17. Leonard, J. I., and Brenner, H., "Experimental Studies of Three-Dimensional Filtration on a Circular Leaf," *A.I.Ch.E. Journal*, Vol. 11, 1965, pp. 965-971.
18. Muller, I., "A Thermodynamic Theory of Mixtures of Fluids" *Archive for Rational Mechanics and Analysis*, Vol. 28, 1968, pp. 1-39.
19. Purchas, D., *Solid-Liquid Separation Technology*, Uplands Press, Croydon, England, 1981.
20. Raviprakash, J., "Multiphase Constitutive Theory for Porous Media with Applications to Filtration," thesis presented to The University of Akron, Akron, Ohio, in 1984, in partial fulfillment of the requirements for the degree of Master of Science.
21. Ruth, B. F., "Correlating Filtration Theory with Industrial Practice," *Industrial and Engineering Chemistry*, Vol. 38, 1946, pp. 564-571.
22. Selby, S. M., *Standard Mathematical Tables 18th Edition*, Chemical Rubber Publishing Company, Cleveland, Ohio 1970.
23. Tiller, F. M., Chow, R., Weber, W. and Davies, O., "Clogging Phenomena in the Filtration of Liquified Coal," *Chemical Engineering Progress*, Vol. 25, Dec., 1981, pp. 61-68.
24. Truesdell, C., and Noll, W., "The Nonlinear Field Theories of Mechancis," *Handbuch der Physik*, S. Flugge, ed., Vol. 3, pt. 3, Springer-Verlag, Berlin, West Germany, 1965.
25. Truesdell, C., and Toupin, P., "The Classical Field Theories," *Handbuch der Physik*, S. Flugge, ed., Vol. 3, pt. 1, Springer-Verlag, Berlin, West Germany, 1950.
26. Weber, H. C. and Hershey, R. L., "Some Practical Applications of the Lewis Filtration Equation," *Industrial and Engineering Chemistry*, Vol. 18, 1926, pp. 341-344.
27. Willis, M. S., and Tosun, I., "A Rigorous Cake Filtration Theory," *Chemical Engineering Science*, Vol. 35, 1980, pp. 2427-2438.
28. Willis, M. S., Collins, R. M., and Bridges, W. G., "Complete Analysis of Non-Parabolic Filtration Behavior," *Chemical Engineering Research and Design*, Vol. 61, 1983, pp. 96-107.
29. Willis, M. S., "A Multiphase Theory of Filtration," *Progress in Filtration and Separation*, R. J. Wakeman, ed., Vol. 3, Elsevier Scientific Publishing Company, Amsterdam, The Netherlands, 1983, pp. 1-56.

10. LIST OF SYMBOLS

A	Filter area: surface of the averaging volume
$A_{\alpha\beta}$	Area of the αβ-interface inside the averaging volume
a	Outer dimension of oblate spheroidal filter cake, Eq. (5.34)
b	Height of an oblate spheroidal filter cake, Eq.(5.35)

d_{ij}	$(\partial v_i/\partial x_j + \partial v_j/\partial x_i)$, rate of deformation tensor
e^α	Exchange across the $\alpha\beta$-interface due to a phase change
F^α	Exchange of a property across the $\alpha\beta$-interface
F_i^α	Exchange of momentum across the $\alpha\beta$-interface
Fr	Froude number, ratio of inertial to gravity forces
G	Net production of Ω; ratio defined by Eq. (5.16)
g_i	Gravity
h	External supply of the property Ω
h_γ	Oblate spheroidal scale factor, Eq. (5.36)
h_η	Oblate spheroidal scale factor, Eq. (5.37)
h_ω	Oblate spheroidal scale factor, Eq. (5.38)
I_1	d_{ii}, first invariant of d_{ij}
I_2	$d_{ij}d_{ji}$, second invariant of d_{ij}
J_o	Ratio of the septum to cake pressure gradients, Eq. (5.28)
K	Permeability, Eq. (5.10)
K_o	Permeability at the septum, Eq. (5.27)
L_c	Length of filter cake in a one-dimensional axial filtration
M	Permeability of a circular leaf filter cake, Eq. (5.41)
M_o	Permeability at the septum in a circular leaf filtration
M_α	Mass of the liquid α-phase in a filter cake, Eq. (5.46)
M_β	Mass of the particulate β-phase in a filter cake, Eq. (5.47)
m	Power law constitutive parameter, Eq. (3.12)
Np	Dimensionless ratio of pressure to viscous forces, Eq. (4.6)
Nd	Dimensionless ratio of drag to viscous forces, Eq. (4.7)
n	Power law constitutive parameter, Eq. (3.12)
$n_i^{\alpha\beta}$	Unit normal on the $\alpha\beta$-interface, direction into β-phase
P^α	Mass average α-phase pressure
P_o	Reference pressure for filtrations
P_o^α	Measurable pressure, Eq. (5.12)
P_c^o	Measurable pressure drop across the filter cake, Eq. (5.24)
P_o^*	Dimensionless pressure, Eq. (5.23)

q_k	Flux of the property Ω through A of the averaging volume
R	Radius of circular filter medium in a leaf filtration
Re	Reynolds number, ratio of inertial to viscous forces
r_i	Position vector, Eq. (2.1)
S_c	Oblate spheroidal shape factor, Eqs. (5.58) and (5.59)
S_m	Oblate spheroidal shape factor, Eq. (5.60)
s	Slurry concentration
T^α	Mass average α-phase temperature
T_c	Critical temperature
T_r	Reduced temperature
t	Time
t_{ij}	Flux of momentum through A of the averaging volume
V_α	Volume of the α-phase in the averaging volume
V_β	Volume of the β-phase in the averaging volume
V	Averaging volume: volume of filtrate
v_k	Velocity
v_i^d	Velocity difference, $(v_i^\alpha - v_i^\beta)$
w_i	Velocity of an interface, Eq. (2.23)
x_i	Position vector, Eq. (2.1)
z	Axial coordinate, points from the septum into the cake
Δ	Volume of oblate spheroidal filter cake, Eq. (5.48)
Ω	Arbitrary thermodynamic property
Ω_γ^α	Scaled velocity, Eq. (5.42)
γ_α	Phase distribution function, Eq. (2.2)
δ_{ij}	Kronecker delta
ε_α	α-phase porosity, Eq. (2.4)
ε_α^*	Average filter cake porosity, Eq. (5.17)
η_1	Constitutive parameter, Eq. (3.9)
η_2	Constitutive parameter, Eq. (3.9)
η	Oblate spheroidal coordinate
λ_0	Constitutive parameter, Eq. (3.15)
λ_1	Constitutive parameter, Eq. (3.15)
λ_2	Constitutive parameter, Eq. (3.15)

μ	Single phase viscosity of the liquid phase, Eq. (4.10)
μ_c	Critical viscosity
ξ_i	Position vector, Eq. (2.1)
ξ	z/L_c, Eq. (5.51)
ρ^α	α-phase density
ω	Oblate spheroidal coordinate, Eq. (5.31)

Superscripts

α	α-phase
β	β-phase
$-\alpha$	Mass average operator, Eq. (2.8)
U_α	Area average operator, Eq. (2.9)
$\hat{}\alpha$	Deviation, Eq. (2.14)
$\tilde{}\alpha$	Surface flux, Eq. (2.25)
*	Dimensionless quantity
•	Time derivative

Subscript

$,i$	$\partial/\partial x_i$

PART 3

TRANSPORT PHENOMENA IN FRACTURED ROCKS

TRANSPORT EQUATIONS FOR FRACTURED POROUS MEDIA

Allen M. Shapiro

TRANSPORT EQUATIONS FOR FRACTURED POROUS MEDIA

Allen M. Shapiro

U.S. Geological Survey
Water Resources Division
Reston, Virginia 22092, USA

ABSTRACT

With the advent of analyzing geological settings as possible sites for hazardous waste isolation, the modeling of transport phenomena in fractured rock has been a topic of increasing interest. In studies to date, the means by which transport phenomena in fractured rock have been mathematically visualized has taken two distinct routes. The need for different conceptualizations of fractured rock has arisen due to the diverse nature of fracturing in rock formations. Usually, the length scale of a given transport problem, in relation to the intensity of fracturing, varies from one rock formation to the next. In some instances, there may exist only a few significant fractures (of a given fracture family) over the length scale of the transport problem. In other situations, the length scale of the transport problem may encompass large numbers of interconnected fractures. These observations have led to conceptualizations of fractured rock as either a system of individual and possibly interconnected fractures in a permeable or impermeable host rock, or as one or more overlapping fluid continua, in a manner similar to the mathematical treatment of granular porous materials. The assumptions implicit in the use of each of these conceptualizations are discussed in this chapter along with a selective review of the recent literature. A detailed analysis of the discrete fracture and continuum conceptualizations of fractured rock is provided by developing the appropriate equations of mass, momentum and energy transport for each conceptualization.

1. INTRODUCTION

1.1. Physical Description of Fractured Rock

In our conceptualization of the subsurface, we commonly hypothesize the void space to consist of openings having characteristic lengths and widths of approximately the same dimensions. Under saturated flow conditions, the void space then offers a highly tortuous path to fluid movement through openings in the subsurface which are assumed to be similarly shaped (Figure 1). Usually, we visualize the subsurface to have such characteristics near the ground surface, and in shallow, unconsolidated, granular deposits.

In contrast, indurated deposits exhibit void space, which in addition to having openings such as that described above, possess additional void space that is characterized by openings having lengths which far exceed their associated widths (Figure 2). Depending on their origin, we refer to this additional void space as either fissures, joints, faults, or generically, as fractures. The fractures are usually considered to be a secondary void space of the rock; the primary void space being that associated with the initial formation of the rock itself. Fractures are commonly a result of secondary processes such as tectonic activity, weathering, dissolution of the rock, or other chemically or thermally induced phenomena (17,78). Fractures may also be the result of man-induced phenomena that arise from altering the local stress conditions, as often occurs during quarrying or through hydraulic fracturing in the vicinity of boreholes (43).

Regardless of the rock type, there usually exists more than one family of fractures distributed throughout the formation, each

Figure 1. Schematic drawing of the void space and flow lines in a granular porous medium.

Figure 2. Schematic drawing of the void space and flow lines in an indurated deposit possessing intrinsic void and a single fracture.

fracture family arises from a different secondary process. For example, dissolution channels may constitute one family of fractures, while fractures that arise from different tectonic stresses may constitute additional fracture families. Furthermore, these fracture families need not have the same geometrical characteristics. Microfissures, for example, arising from the cooling of crystalline formations, are not as areally extensive as tectonically induced fractures. Also, fractures arising from different tectonic stresses are known to have unique orientations within a given formation. In Figure 3, several interconnected fracture families are depicted, each having a different areal intensity.

As a consequence of the above discussion and in contrast to our original description of granular porous material, we can usually visualize more than one flow regime within a fractured formation. If the rock exhibits an intrinsic void space, it offers a highly tortuous path to fluid movement in a similar manner to granular porous materials. The existence of fractures, depending on their interconnectivity, constitutes a second, and usually a less tortuous path to fluid movement. In fact, we may wish to subdivide the characteristics of the flow regime even further by considering the influence of each fracture family, or by grouping several fracture families together and considering their combined influence.

The fact that fractured rock is characterized by more than one type of void space that are distinctly dissimilar has led many investigators (3,24,48,52,66,70,87) to question whether the classical

Figure 3. Schematic drawing of several fracture families, each having a different areal intensity.

theories applied to mathematically modeling transport phenomena in (granular) porous materials are equally applicable to the description of such phenomena in fractured rock. That is to say, can we classify all void space in a fractured rock as one, and define bulk parameters that describe the transport processes? Or, is it necessary to formulate alternative mathematical conceptualizations of transport phenomena in fractured rock which account for the multiple characteristics of the void space?

We are unable to supply generic answers to the questions above due to the diversity of physical situations where fractured formations are of hydrologic significance. The geometrical characteristics of fractures in the subsurface differ from one formation to the next; hence, we cannot characterize all fractured rock with one mathematical conceptualization as is done with granular porous materials.

In the recent literature, several mathematical conceptualizations have been employed in the description of transport phenomena in fractured formations. It is the purpose of this article to identify and discuss these various mathematical conceptualizations. At first, our discussion shall be directed toward a general description of these conceptualizations of fractured rock, paying particular attention to the physical situations where each conceptualization is most applicable, and the assumptions that are implicit in their use. In subsequent chapters, a more detailed analysis of these conceptualizations is to be carried out by developing the transport equations for mass, momentum and energy.

1.2. The Continuum Hypothesis - Granular Porous Materials

Mathematical analysis of transport processes in granular porous materials employ properties such as the hydraulic conductivity and porosity. These properties are, in actuality, concepts associated with visualizing the void space and rock matrix of a porous material as continua. The continuum hypothesis is a corner stone of the mathematical modeling of transport phenomena in (granular) porous media. And since fractured rock, in essence, is a porous material (although, not granular in nature), we can visualize the applicability of the continuum hypothesis to fractured rock. In the following paragraphs, we shall provide a brief explanatory discussion of the continuum hypothesis as it is applicable to granular porous materials. The applicability of the continuum hypothesis to fractured media is to be discussed in a subsequent section.

With the continuum hypothesis, instead of describing the actual void space of the medium for the purpose of mathematically defining the physics of transport phenomena, we consider the transport processes on a scale which enables us to employ distributed parameters, such as the hydraulic conductivity and porosity, at points in the medium. These parameters are the manifestation of the geometric and material properties of the medium on a length scale that is much larger than the actual heterogeneities of the porous material, i.e., a length scale much larger than the individual void openings and solid grains.

The continuum conceptualization no longer considers the actual physical structure of the medium. The continuum, in essence, is a fictitious medium where smoothly varying, spatially averaged values are assigned to each mathematical point. Each mathematical point in the continuum is associated with a volume in the actual medium of discontinuous phases (e.g., fluid and rock) over which average values are taken (Figure 4). The explicit mathematical location in the actual medium, which corresponds to the location in the continuum, may lie in any phase; however, associated with this position in the continuum are properties of all phases present in the averaging volume.

We refer to the volumetric dimensions of the scale on which we can invoke the continuum hypothesis as a Representative Elementary Volume (REV) (4). Conceptually, we can visualize the definition of the REV by considering a continuum quantity defined over successively larger averaging volumes about a given point in the actual multi-component medium. For example, considering the porosity (volume of void space per total volume) over successively larger volumes, we anticipate fluctuating values for small averaging volumes due to the disproportionate amount of the void space or rock matrix which is initially encompassed in the volume (Figure 5). As the size of the volume increases these fluctuations are expected to dissipate as more proportionate amounts of the void space and

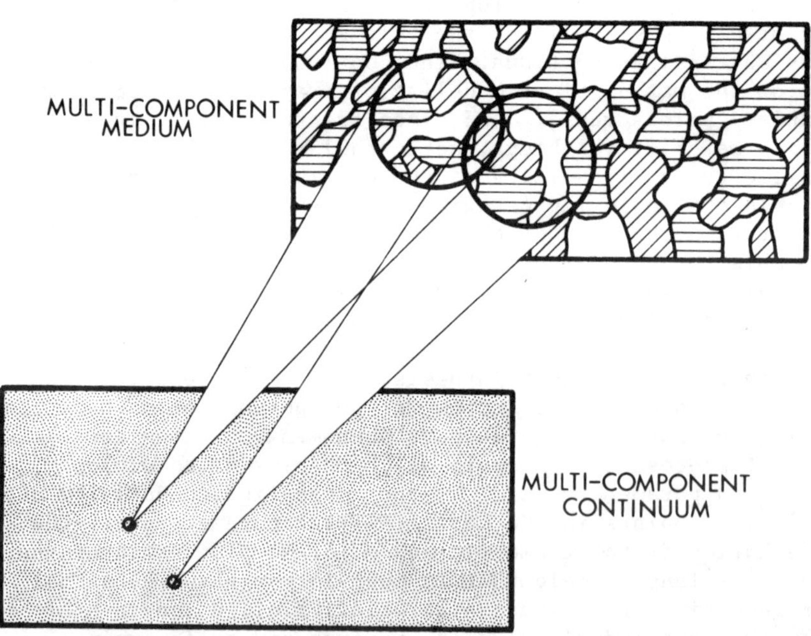

Figure 4. A multi-component medium and its associated continuum representation.

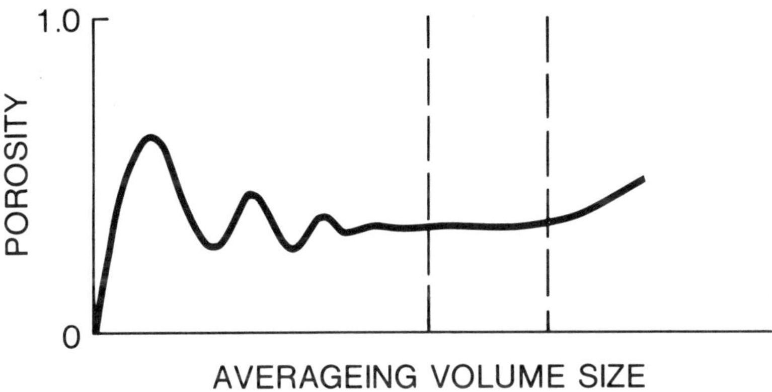

Figure 5. Porosity of a porous medium as a function of the averaging volume size (after Bear (5)).

rock matrix are included. The assumed stationarity of the porosity with respect to the averaging volume size indicates a parameter which is indicative of the continuum scale, and also dictates the size of the REV (54). Extending the size of the averaging volume beyond this point may result in further fluctuations due to heterogeneities on yet a larger scale.

Crucial to the validity of the continuum concept is the assumption that the averaging volume size is independent of the location in the medium, and the parameter that is being averaged. Parameters of all transport processes, e.g., fluid movement, contaminant and thermal migration, etc., must be defined on the same scale due to the coupling of these phenomena. In addition, because the REV acts as the smallest discernible dimensions that are indicative of the continuum scale, the REV concept also has practical implications. If measurements are conducted over dimensions smaller than the REV, they will not be indicative of the continuum since they are biased by the disproportionate amount of the void space or rock matrix which is encompassed in the measurement.

In actuality, we do not measure an REV for a given field situation; the continuum hypothesis is invoked and equations describing transport phenomena are developed based on this assumption. Only through the comparison of field measurements and the predictions from the continuum equations can the validity of the continuum hypothesis be judged. In addition, one must realize that the applicability of the continuum hypothesis is also based on the premise that the scale of the transport phenomenon of interest is orders of magnitude larger than the heterogeneities associated with the individual void openings in the porous material. If the transport problem of interest is of the same length scale as the individual void openings, then we are unable to hypothesize the existence of an REV that will define meaningful continuum parameters. In such instances, we can no longer neglect the geometric intricacies of the void openings, which would greatly influence the transport problem.

For most practical problems in porous materials that are granular in nature, the phenomenon of interest is much larger than the heterogeneities associated with the individual soil grains and void openings. In addition, our measuring instruments sample volumes that are much larger than these heterogeneities. Hence, the application of the continuum hypothesis is viewed as an acceptable assumption from both a theoretical and a practical point of view. In some circumstances, however, we may be interested in phenomena which is significant on a smaller scale, e.g., the specific surface chemistry between the rock matrix and the fluids in the void space. Ultimately, however, it may be necessary to describe this phenomena at the continuum level in order to incorporate such physics into problems on the larger scale where we cannot effectively account for the geometry of the individual soil grains and void openings (see, e.g., James and Rubin (47)).

1.3. Mathematical Conceptualizations of Fractured Rock

In contrast to the treatment of granular porous materials, the mathematical conceptualization of transport phenomena in fractured rock depends highly on the intensity of fracturing in relation to the problem scale. For fractured rock, we may not always have the luxury of assuming the existence of an REV for all fracture families in a given formation. In many instances, the scale of the transport problem may of the same length scale as the (geometric) heterogeneities of one or more fracture families. Hence, conceptualizations other than a continuum conceptualization may be necessary in describing transport in fractured rock.

Furthermore, even if we are able to invoke the continuum hypothesis for all fracture families of a given rock formation, the dissimilarities in the character of the void space may make the equivalent porous medium conceptualization unacceptable with regard to the mathematical description of transport phenomena. In the following sections we shall discuss several conceptualizations of fractured rock which are indicative of a variety of field situations one would anticipate encountering in the description of transport processes in fractured rock. In addition, a selective review of the literature descriptive of each conceptualization accompanies the discussions.

1.3.1. Discrete fracture conceptualization

Under certain geologic circumstances, we may visualize a problem scale for transport phenomena such that only a single fracture, or several interconnected fractures are of significance. For example, if we are interested in identifying the hydrologic responses in the immediate vicinity of a hazardous waste repository, we may encounter relatively few extensive field scale fractures (Figure 6); the site of the repository most likely would be chosen such that these fractures are relatively few in number. Under such circumstances the geometric intricacies of these few fractures will have a direct impact on the transport problem. The sparse number of fractures (of this particular family) prohibits the use of a continuum conceptualization. Thus, in order to accurately describe the transport problem, the geometrical properties of the fractures need to be defined.

Each fracture would be visualized in the rock as an irregularly shaped void space having a characteristic length which exceeds its associated width. In addition, the walls of the rock which constitute the fracture need not be planar and the fracture width, or aperture, would vary from point to point such that at some locations the walls of the fracture come in contact, yielding and obstruction to flow. Usually, more simplified interpretations of the fracture geometry are invoked for modeling purposes. Usually, parallel plates or some distribution of parallel plate apertures within a given fracture is assumed (see, e.g., Snow (77), Witherspoon et al. (91), Neuzil and Tracy (63)).

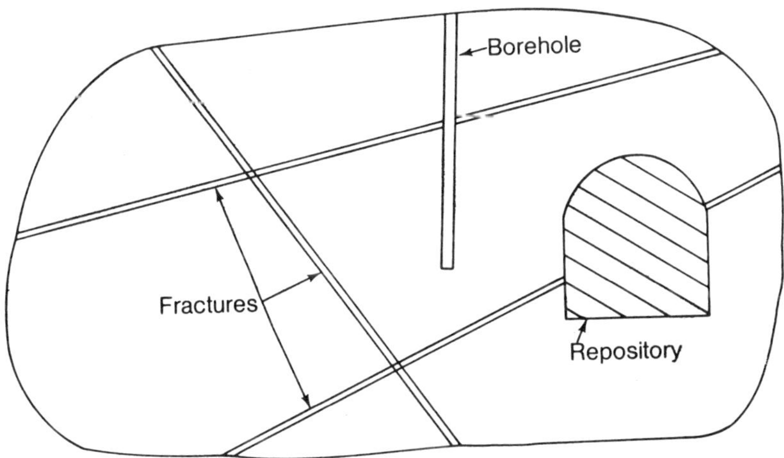

Figure 6. Schematic diagram of a repository site with few field scale fractures in the vicinity.

The rock in which the few significant fractures exist may either be impervious, i.e., there being no void space, or it may have additional void space of a different character than the large fractures. In the latter case, fluid movement would take place in both the fractures and in the adjacent host rock. If there exists void space in the rock adjacent to the fracture, the void space may either be similar to a granular porous medium as we may expect in a sandstone, or the void space in the host rock may result from fracturing on yet a smaller scale, i.e., fracture families whose fractures are less arealy extensive than those few significant fractures. For example, the fracturing which may arise as a result of cooling in crystalline formations would be on a smaller scale and more densely packed than the tectonic fractures. Once again, this is a scale dependent observation. We could, for example, be interested in the scale of transport phenomena associated with these micro-fissures, in which case we would consider the adjacent rock to be impervious. Throughout the present discussion, however, we will assume that if void openings appear in the rock adjacent to the one or more significant fractures at the scale of the transport problem, they occur in such profusity and with such interconnectivity that we can conceptualize the host rock as a continuum in a similar manner to our treatment of granular porous media. Thus, we are invoking the continuum hypothesis (inclusive of the assumptions implicit in its use) with respect to the void space in the host rock.

Employing the discrete fracture conceptualization in some rock formations is attractive since it is descriptive on a scale to which the majority of field instrumentation has been adopted. For example, in rock formations where fractures of a given family are separated by tens of meters or more, field instrumentation would allow hydraulic measurements associated with only one or two fractures of this family, and thus would not be representative of a continuum measurement. However, even though field measurements are adaptable to the discrete fracture conceptualization, this would still not alleviate the problem of identifying the specific geometric characteristics of all significant fractures in the rock formation at the scale of interest, e.g., the aperture and its spatial variation, the areal extent of each fracture, etc. Hence, we can anticipate that large field investigations using the discrete fracture conceptualization are not conceivable unless the fracture geometry is treated statistically (see, e.g., Schwartz et al. (69), Smith and Schwartz (76)), where possibly, we have some information with regard to some, but not all fractures in the domain (see, e.g., Andersson et al. (1)).

In addition, although we consider the fracture aperture as a descriptive parameter of the discrete fracture conceptualization, in actuality, we rarely measure the specific aperture size.

Usually, only the hydraulic properties of a single fracture are measured (see, e.g., Wang et al. (86)), and an effective aperture size is interpreted by an assumed geometric description of the fracture, e.g., parallel fracture walls. Furthermore, in the modeling of transport phenomena in a single fracture, we rarely consider the void space of the fracture in three-dimensions. Usually it is reduced to a two-dimensional surface in the axis of the fracture by assuming the fracture to be relatively narrow. To this two-dimensional surface, the fracture aperture is no longer of significance. Instead, we assign the bulk hydraulic properties which were originally measured. Thus, the added step of defining the fracture aperture is an unnecessary and a potentially confusing procedure since it depends upon our conceptualization of the fracture geometry.

The discrete fracture conceptualization has been widely applied in describing transport phenomena in fractured formations. For the most part, however, analyses that have employed the discrete fracture coneptualization have been directed toward the description of fluid movement. Only recently, with the advent of analyzing rock formations as possible geologic settings for hazardous waste isolation, has emphasis been placed on using the discrete fracture approach in the study of contaminant and thermal migration.

The concepts associated with modeling transport phenomena using the discrete fracture approach are easily conceptualized because they draw upon our knowledge of modeling transport phenomena in porous media (that is, if the host rock is permeable) and also, transport in conduits or parallel plates which are assumed to represent the fractures. Each flow regime is considered separately, and they are coupled through auxiliary conditions imposed at the boundaries that delimit the host rock from the fractures.

In almost all analyses that have employed the discrete fracture conceptualization, fluid movement in the fractures is assumed to be Darcian. With this assumption, the discrete fracture model then resembles a porous medium simulator with bands of high conductivity to represent the individual fractures. Although there has been research conducted into turbulent flow in conduits having porous walls (see, e.g., Whitehead and Davis (89), Chu and Gelhar (14), Munoz Goma and Gelhar (59)) such work has not been extended to the mathematical modeling in discrete fracture networks. This is especially surprising since a majority of discrete fracture models have analyzed fractures in the vicinity of boreholes, or fractures that intersect boreholes, where one would expect high velocity flows.

Early studies which employed the discrete fracture conceptualization used electric analog representations of fractured permeable reservoirs subject to steady-state flow conditions. Among other analyses, studies were made of vertical fissures intersecting wells (16,56), and isolated fractures between pumping and recharge wells (29). Much of this work was initiated in the petroleum industry where the presence of fractures is often of great significance in the management of producing formations. Kiraly (51) also developed an electric analog model with an anisotropic permeability tensor based on fracture geometry in an impermeable host rock. This model parallels a porous media simulator having anisotropic permeability.

Later, analytical techniques offered greater flexibility in analyzing transport phenomena using the discrete fracture conceptualization. Analytical solutions for the cases of vertical and horizontal fractures of finite extent intersecting a well bore were developed by Hartsock and Warren (38), Gringarten and Ramey (32) and Gringarten et al. (33). The applicability of these solutions to field situations was demonstrated by Gringarten et al. (34). Other analytical solutions were developed by de Swann (18) and Boulton and Streltsova (10) who considered the radial flow to a well intersected by an infinite horizontal fracture.

Numerical techniques such as finite difference and finite element methods of approximating solutions to the governing equations made analyses of more intricate fracture geometries and more complicated transport phenomena possible. Russel and Truitt (68), Kazemi (48), Wang et al. (86), Narasimhan and Palen (61), and Narasimhan (60) numerically investigated the responses of fractures intersecting boreholes. In addition, regional aquifer responses using the discrete fracture conceptualization have been made by a number of investigators (see e.g., Wilson and Witherspoon (90), Shapiro and Andersson (73)). Gale et al. (28) considered the combined effect of flow and fracture deformation. Shapiro and Andersson (74) developed a model to treat three-dimensional fracture networks, where fractures are of arbitrary orientation and areal extent in the formation.

Recently, studies that have analyzed contaminant migration in discrete fracture systems have also been conducted. Grisak and Pickens (36) numerically modeled solute transport in a single fracture of uniform thickness to analyze the effect of diffusion into the adjacent permeable matrix. Noorishad and Mehran (64) also numerically solved contaminant movement in a single fracture coupled with a porous medium. Tang et al.(82) and Sudicky and Frind (80,81) developed analytical solutions which investigate contaminant migration in a single fracture and a series of parallel fractures, respectively. Recently, Brown (13) analyzed the stochastic nature of flow and solute transport in a single fracture resulting from a variable fracture aperture.

Contaminant migration for field scale problems using the discrete fracture conceptualization have also been conducted. Rasmuson et al. (67) used numerical methods to solve the advective dispersive equation in the fractures and porous matrix. In contrast, Schwartz et al. (69) and Smith and Schwartz (76) considered an impermeable host rock while using a particle tracking technique to investigate the movement of the contaminant.

1.3.2. Equivalent porous medium conceptualization

The discrete fracture conceptualization is applicable to fractured rock on whatever scale we wish to observe it, that is, provided the geometric information which describes the fractures is well defined. If there exists a large number of interconnected fractures, the use of the discrete fracture conceptualization becomes computationally intractable, as does the possibility of geometrically describing all fractures. In this, we recognize similarities with granular porous materials, where the persistent complex geometry of the void openings makes the solution of the equations governing transport phenomena infeasible. Instead, we assume that the void space and solid matrix are amenable to treatment as continua. Similarly, if we are considering a problem scale in a fractured formation where there are a profuse number of interconnected fractures of all fracture families, then as an assumption, the rock matrix and the void space (inclusive of all fracture families and the intrinsic void space of the rock itself) can be represented as continua. Since all void space is to be characterized as a single continuum, as is assumed in granular porous materials, we refer to this approach of modeling transport phenomena in fractured rock as an equivalent porous medium conceptualization.

By invoking the continuum hypothesis we must assume that the fractured medium satisfies restrictions similar to those discussed in section 1.2. That is, continuum parameters for both the rock matrix and the void space must exhibit stationarity in being averaged over a volume of the fractured formation. We would expect that a sufficient number of fractures from all fracture families must exist in the averaging volume in order for the continuum variable to be independent of the averaging volume size. Otherwise, increasing the averaging volume size would result in fluctuating continuum values as an additional fracture from a given family is included within the averaging volume. Thus, we would likely anticipate two or more plateaus where the continuum properties exhibit stationarity with the averaging volume size (Figure 7). If the host rock possesses an intrinsic void space, the first plateau would be associated with a continuum representation of this void space in a similar manner to the description of a granular porous medium. Enlarging the size of the averaging volume would include different fracture families, and thus, a second plateau is

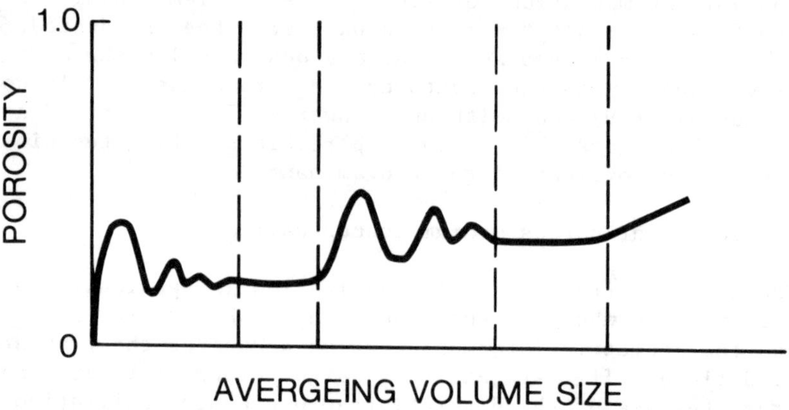

Figure 7. Porosity of a fractured rock as a function of the averaging volume size.

anticipated for the void space of the fractures combined with the intrinsic void space of the host rock. Depending on the frequency of fractures and the nature of the fracture families, additional plateaus of stationarity in the continuum coefficients are also conceivable. The largest averaging volume size that yields stationary continuum coefficients would be indicative of the REV for the equivalent porous medium conceptualization of the fractured medium. The size of the averaging volume depends on the fracture family which is least pervasive. Because the character of fracture families varies from one rock formation to the next, it is impossible to estimate one REV size that would characterize all fractured rock formations. Depending on the character of the fractures, we may anticipate a linear dimension of the REV to be tens of centimeters in some formations, while in others the REV conceivably can be tens of meters or more.

For an example of where one might apply a continuum conceptualization of fractured rock, let us return to the example given in section 1.3.1 where for the discrete fracture conceptualization we considered the immediate vicinity of a waste repository. Carrying this example one step further, we could hypothesize the use of the equivalent porous medium conceptualization, not in the immediate vicinity of the repository, but rather to describe transport processes on the field scale, which may possibly extend for tens of kilometers. Over such dimensions, the relatively infrequent fractures which appeared in the vicinity of the repository, may occur in sufficient numbers on the field scale to justify the continuum hypothesis.

Since the REV is assumed to be the smallest discernible volume which is representative of the continuum, field measurements taken on a smaller scale will not be indicative of continuum behavior. Consequently, measuring devices should encompass a significant number of fractures from all fracture families. In most field situations, it is highly unlikely that a measuring instrument would accommodate such a restriction, especially if the fractures from one family are relatively infrequently spaced. This constitutes a problem in measuring field responses from which we may obtain parameters that describe the transport processes at the continuum level.

There have been attempts to alleviate this problem of parameter identification at the continuum level by theories that express the continuum coefficients in terms of the fracture geometry. This approach has only been adopted for coefficients associated with fluid movement; continuum coefficients defining contaminant and thermal migration have not been considered. Snow (77) defined the tensorial nature of the hydraulic conductivity of a fractured rock solely in terms of the fracture geometry. Shapiro and Bear (75) extended this work by considering the fact that the hydraulic gradient in the individual fractures is not necessarily the same as the gradient of the average hydraulic head, which is the driving force for flow at the continuum level. In contrast, Long et al. (55) evaluated the two-dimensional hydraulic conductivity of a fractured rock by actually solving the problem of steady-state flow through a network of fractures. In this latter work, the assumption of an equivalent porous medium continuum for a network of discrete fractures was evaluated by examining if the hydraulic conductivity tensor could be represented by an ellipse in its rotation.

The equations of fractured rock as an equivalent porous medium are the same as those which have been classically used in analyzing granular porous materials as continua. The values of the material coefficients that are used to describe a fractured medium (e.g., the hydraulic conductivity and porosity), are significantly different from those employed in the modeling of a granular porous medium. Many investigations have been conducted in fractured formations using an equivalent porous medium approach. The majority of these studies, however, consider only single phase fluid movement (see, e.g., Elkins (25), Elkins and Skov (26), Grisak and Cherry (35)). For this transport problem, it has been widely recognized that using an equivalent porous medium conceptualization is acceptable in describing fluid movement in fractured rock. Only in the very early transient responses will the explicit effect of the fractures become evident (31,87). For the most part an equivalent hydraulic conductivity, which is usually anisotropic and accounts for all fracture families and the intrinsic void space of the rock can be employed to describe fluid movement.

There are studies, however, which have been conducted in analyzing contaminant migration in fractured rock using the equivalent porous medium conceptualization (see, e.g., Grove and Beetem (37), Shapiro and Andersson (72), Dougherty (19)). The acceptability of such an approach for contaminant and thermal migration becomes questionable because such processes are highly dependent on the actual fluid velocities in the individual void openings. The equivalent porous medium conceptualization assumes that one continuum velocity is representative of the average of all velocities in the void space of the medium. Although this may be true in granular porous materials, we would anticipate significant variations between the fluid velocity in the intrinsic void space of the host rock and that in the fractures. Although the deviation in velocity between the average continuum value and the velocities in the void space is incorporated into the definition of mechanical dispersion (see, e.g., Bear and Bachmat (6)),it is questionable whether two field parameters alone (longitudinal and transverse dispersivity) are adequate in describing the dispersion process in this complex medium. In addition, we must question the use of the classical dispersion theory since it is based on the assumption of an isotropic porous medium, while rock formations invariably are anisotropically fractured.

1.3.3. Dual porosity conceptualization

Due to the diverse nature of the void space in fractured formations that is anticipated, and the fact that the physics of the transport processes may be significantly different in each flow regime, a natural extension of the equivalent porous medium conceptualization is to consider the fluid in the fractures and the fluid in the intrinsic void space of the host rock as separate continua. Hypothesizing mixtures of fluids and solids as overlapping continua has long been considered in the field of continuum mechanics (see, e.g., Bowen (11)). Barenblatt and Zheltov (2) and Barenblatt et al. (3) were the first to consider the application of this theory in the description of isothermal flow in fractured porous media. In their analysis, continuum equations of fluid mass conservation are written separately for the fluid in the fractures and the fluid in the porous matrix, i.e.,

$$\frac{\partial(\rho_f \varepsilon_f)}{\partial t} + \underset{\sim}{\nabla}\cdot(\rho_f \varepsilon_f \underset{\sim}{V}_f) - \overset{\leftrightarrow}{M}_f = 0 \tag{1.1}$$

$$\frac{\partial(\rho_p \varepsilon_p)}{\partial t} + \underset{\sim}{\nabla}\cdot(\rho_p \varepsilon_p \underset{\sim}{V}_p) + \overset{\leftrightarrow}{M}_f = 0 \tag{1.2}$$

where ρ_α is the mass density per unit volume of the α-fluid continuum, ε_α is the volume fraction (or porosity) of the α-fluid continuum, $\underset{\sim}{V}_\alpha$ is the velocity of the α-fluid continuum, $\overset{\leftrightarrow}{M}_f$ denotes the rate of exchange of fluid mass between the fractures and the porous matrix, and the subscripts f and p denote properties of the fluid in the fractures and properties of the fluid in the porous matrix, respectively. Barenblatt et al. (3) further assumed that the specific discharge for each fluid continuum could be described by a Darcian relationship, i.e.,

$$\varepsilon_\alpha \underset{\sim}{V}_\alpha \equiv q_\alpha = - \underset{\approx}{K}_\alpha \cdot \underset{\sim}{\nabla} \phi_\alpha \tag{1.3}$$

where q_α is the specific discharge of the α-fluid continuum, ϕ_α is the hydraulic head of the α-fluid continuum and $\underset{\approx}{K}_\alpha$ is the hydraulic conductivity tensor of the α-fluid continuum. In addition, the first term in both Eqs. (1.1) and (1.2) can be expanded such that the fluid mass accumulation for each fluid continuum is defined in terms of the time rate of change of the hydraulic head and a storativity coefficient, S_α,

$$\frac{\partial(\rho_\alpha \varepsilon_\alpha)}{\partial t} \equiv \rho_\alpha S_\alpha \frac{\partial \phi_\alpha}{\partial t} \tag{1.4}$$

Introducing Eqs. (1.3) and (1.4) into (1.1) and (1.2) yields a set of two equations in terms of three unknowns, i.e., ϕ_f, ϕ_p and $\overset{\leftrightarrow}{M}_f$. Hence, we require an additional constitutive relationship to describe the exchange of fluid mass between the fracture fluid continuum and teh porous matrix fluid continuum.

In this conceptualization of fractured rock, we have introduced an artificial separation of the fluid occupying the void space. In actuality, for saturated flow conditions, the fluid in the entire void space is continuous. Although this conceptualization is a sophistication in describing the physics of transport (which we may consider as an advantage because it is more representative of the processes in the rock), we are also faced with a disadvantage in that we have introduced an artificial separation of the fluid phase, and we must subsequently describe the interaction between the two fluid continua in terms of quantities that are defined on the continuum scale.

In the discrete fracture conceptualization, the exchange of fluid mass between the fractures and the porous matrix is directly handled by boundary conditions that are imposed at the interface between the two regimes. In the continuum hypothesis, however, boundaries between fractures and porous matrix are no longer recognizable, i.e., we consider transport on a scale that is much larger than the individual geometric heterogeneities of the rock. In the volume over which the average values at the continuum level are defined, a net exchange of fluid mass between fractures and porous matrix is represented by $\overset{\leftrightarrow}{M}_f$ in Eqs. (1.1) and (1.2). For this quantity, Barenblatt et al. (3) assumed the following relationship

$$\overleftrightarrow{M}_f = \beta(\phi_p - \phi_f) \tag{1.5}$$

where β is a coefficient describing the intensity of the rate of fluid mass exchange between the fractures and the porous matrix. Barenblatt et al. (3) assumed that β is proportional to the product of the porous matrix permeability and the specific surface area of the fractures. However, β, eventually will require field evaluation to determine its specific value for a given fractured formation. Methods of evaluating β from pumping test information have been suggested by Warren and Root (87) and Uldrich and Ershaghi (84).

By adding Eqs. (1.1) and (1.2), and assuming that the fluid mixture density, ρ_m, is essentially equal to the densities of the fluid in the fractures and the fluid in the porous matrix, we obtain

$$\frac{\partial(\rho_m n)}{\partial t} + \underset{\sim}{\nabla} \cdot (\rho_m n \underset{\sim}{V}_m) = 0 \tag{1.6}$$

where n is the porosity defined with respect to the entire void space of the rock and $\underset{\sim}{V}_m$ is the velocity of the fluid mixture. The quantities ρ_m and $\underset{\sim}{V}_m$ are defined as

$$\rho_m n \equiv \varepsilon_f \rho_f + \varepsilon_p \rho_p \tag{1.7}$$

$$\underset{\sim}{V}_m \equiv (\varepsilon_f \rho_f \underset{\sim}{V}_f + \varepsilon_p \rho_p \underset{\sim}{V}_p)/n\rho_m \tag{1.8}$$

Eq. (1.6) is a statement of mass conservation for fluid in all void space of the rock. Thus, it is the fluid mass balance equation which is employed in the equivalent porous medium conceptualization discussed in section 1.3.2. Consequently, we can explicitly identify the relationship between the equations describing transport phenomena in the conceptualization of fractured rock described above and that of the previous section.

Since continuum coefficients are assigned to each fluid regime, e.g., conductivities and porosities for the fluid in the fractures and the fluid in the porous matrix, respectively, this approach to modeling fluid movement in fractured rock is usually referred to as the dual porosity conceptualization. In theory, we could extend these ideas to a further subdivision of the flow regime by considering different families of fractures as different fluid continua. However, we have seen that in extending the equivalent porous media conceptualization to two fluid continua there is an added complexity due to the additional continuum coefficients which require specification, e.g., β in Eq. (1.5), and the conductivities and porosities of each flow regime. Considering a further subdivision of the flow regime would introduce additional parameters not only for fluid movement, but also for the other transport processes. Not only would we face the difficulty of formulating field experiments to evaluate the additional continuum coefficients, but also it would be necessary for us to hypothesize constitutive relationships that describe the interaction between each fluid continuum for all transport processes.

The restrictions associated with the continuum hypothesis that were discussed in section 1.2 are also applicable to the dual porosity conceptualization of fractured rock. Here, however, we must assume stationarity of the continuum averages for each flow regime and for the rock matrix. In addition, the averaging volume used to define properties associated with the fluid in the fractures must be the same as that which is used to define properties of fluid in the porous matrix (Figure 8). We must adhere to this restriction since the responses of the two fluid continua must be defined on the same scale due to their interconnected nature. Furthermore, the size of the REV, which one would anticipate for this conceptualization, would be similar to that which was discussed in section 1.3.2. Hence we are faced with similar problems of parameter identification due to field instrumentation (see, e.g., Uldrich and Ershaghi (84)).

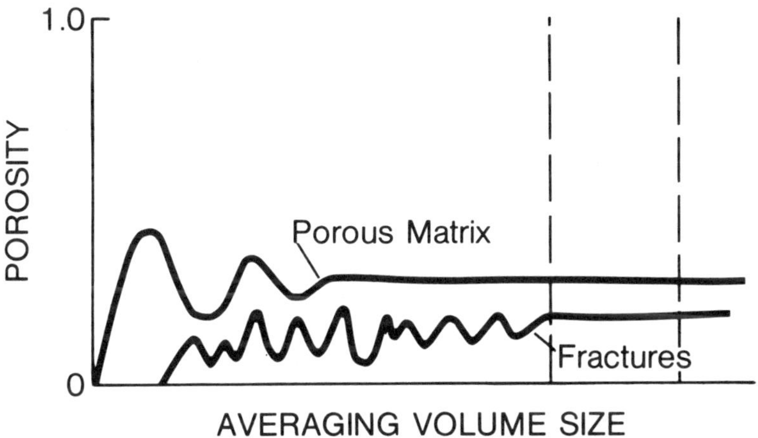

Figure 8. The dependence on the averaging volume size on the porosity of the porous matrix and fractures in a fractured porous medium.

A considerable amount of effort has been directed toward the analysis of the dual porosity conceptualization for fluid movement. Warren and Root (87), Kazemi et al. (49), Streltsova (79) and Dougherty and Babu (20) considered fluid movement using a conceptualization similar to that of Barenblatt et al. (3). Warren and Root (87) analyzed pressure buildup information in a well using an analytical solution to the governing equations. Kazemi et al. (49) developed a numerical simulator for the same purpose. These results indicated that the characteristic behavior of a fractured reservoir is different than an equivalent porous medium in the early pressure responses. At later times an equivalent porous medium conceptualization adequately describes the reservoir behavior. Streltsova (79) analytically solved the governing equations of radial flow to obtain characteristic drawdown functions from which type curves were generated for different parameter values. Dougherty and Babu (20) developed analytical solutions for radial flow to a well partially penetrating a confined fractured formation.

Multi-phase (oil-water) fluid movement in fractured porous media was considered by Bokserman et al. (9) and Verma (85). These analyses proved to be very restrictive since oil movement in only the fractures was considered, and the porous matrix was initially assumed to be saturated with oil. Kazemi et al. (50) considered a more general numerical simulation to this same problem. In addition, Dougherty (19) has demonstrated that an equivalent porous medium conceptualization of multi-phase flow in a fractured porous medium does not adequately represent the physics of fluid motion due to the effect of capillarity and the non-linearity of the fluid mobilities. In addition, Dougherty (19) compared the equivalent porous medium conceptualization with a dual porosity model for contaminant migration under single phase, isothermal conditions.

In analyses of the dual porosity conceptualization, a considerable amount of effort has also been directed toward a rigorous derivation of the equations governing transport phenomena, especially with regard to the development of constitutive relationships to define the exchange of fluid mass, momentum and energy between the porous matrix and the fractures. Braester (12) and Bear and Braester (7) used a hypothetical capillary tube representation to develop a function to describe the exchange of oil and water between fractures and a porous matrix. Duguid and Lee (24) also used a capillary tube model coupled with an analytical solution to describe the exchange of fluid mass under isothermal, single phase flow conditions. In both of these analyses the constitutive relationships were developed from considering the interaction of one fracture (or capillary tube) with an adjacent porous matrix. Thus, it is questionable whether these results truely represent a continuum constitutive function since the averaging volume for the continuum encompasses more than a single fracture. In addition, the constitutive relationships were ultimately defined in terms of parameters such as a characteristic "porous block dimension". Hence, this approach uses attributes of the discrete fracture conceptualization in formulating the continuum equations. Yet, it is questionable if highly idealized shapes of porous blocks such as spheres and cubes actually represent the complex geometry of a fractured medium. This type of approach has also been incorporated into the numerical procedures of Neretnieks and Rasmuson (62) and Huyakorn et al. (44), (45) in analyzing contaminant migration in fractured porous media.

O'Neill (66) analyzed thermal transport in a fractured porous medium. The thermal exchange between the fluid in the fractures and the porous matrix was hypothesized in a form similar to that given in Eq. (1.5), i.e.,

$$\overset{\leftrightarrow}{Q}_f = \sigma(T_p - T_f) \tag{1.9}$$

where $\overset{\leftrightarrow}{Q}_f$ is the rate of energy transfer between the fluid in the fractures and the porous matrix, T_α is the temperature of the α-fluid continuum, and σ is a coefficient describing the intensity of the thermal exchange between the two continua. O'Neill (66) evaluated this coefficient by observing different geometries of a single porous matrix block surrounded by a fracture.

In contrast, Shapiro (70) used mixture theory to define the functional form of the constitutive relationships for the exchange of mass, momentum and energy between the fracture fluid and porous matrix fluid. Terms similar to those given in Eqs. (1.5) and (1.9) appeared in the constitutive relationships; however, additional terms also arose. Through numerical experimentation it was shown that the additional terms in the fluid mass exchange function, $\overset{\leftrightarrow}{M}_f$, were negligible. However, due to differences between fluid mass transport and contaminant and thermal migration (i.e., fluid movement is governed by pressure propagation, a relatively rapid phenomena, while contaminant and thermal migration are governed by the actual fluid velocities), the additional terms which appear in the constitutive function defining thermal and (contaminant) mass exchange between fractures and the porous matrix may be significant. The constitutive functions given in Eqs. (1.5) and (1.9) are usually referred to as a quasi-steady approximations.

2. TRANSPORT EQUATIONS FOR DISCRETE FRACTURE CONCEPTUALIZATIONS

Usually, in the mathematical description of fluid movement in a discrete fracture conceptualization of the rock matrix, the idea of a fracture conductance is introduced. In actuality, fluid movement in an individual fracture is governed by the boundary conditions at the fracture walls and the equations of transport for a single fluid continuum (e.g., the Navier-Stokes equations). In these equations, which describe three-dimensional transport within the void space of an individual fracture, the concept of a fracture conductivity does not appear. The idea of a fracture conductivity arises only in treating the three-dimensional void space of the fracture as an equivalent two-dimensional surface; introducing the conductance of a single fracture corresponds with our ignorance of the geometry of the flow regime, in this case, the specific geometry of the fracture aperture. Treating a single fracture as an equivalent two-dimensional surface is an approximation which is introduced since the aperture of the fracture is usually small in comparison to its areal extent. Hence, transport processes are assumed to be essentially two-dimensional in the hypothesized surface which is descriptive of the fracture geometry. This is also a computational convenience since it reduces the dimensionality (i.e., from three dimensions to two dimensions) of the equations that describe the transport processes in the fracture. Transport process in the rock adjacent to the fracture, in general, retain their three-dimensional character.

In this chapter, the equations of two-dimensional transport in a single fracture are to be developed, paying particular attention to the assumptions that are implicit in their use. Throughout this discussion, we shall restrict our analysis to transport phenomena associated with a single fluid phase which entirely fills the void space of the fractures and the void space in the adjacent host rock. Furthermore, the rock matrix shall be assumed to be non-deformable.

If the rock adjacent to the individual fracture possesses either an intrinsic void space, or fracture families which are amenable to treatment as continua, the equations descriptive of this flow regime are the balance equations which have classically been used in the description of granular porous material. These equations are well known (see, e.g., Bear (5) and shall not be discussed here.

2.1. Mathematical Description of a Single Fracture

Since the aperture of an individual fracture is usually small in comparison to its overall areal extension, the transport processes are assumed to be two-dimensional in the surface of the fracture axis, i.e., the centerline of the aperture with respect to the walls of the fracture (Figure 9). At each point on the surface of the fracture axis, we can define a set of orthogonal coordinates such that the walls of the fracture are defined as

$$F_1(\eta,\xi,\zeta) = \eta - f1(\xi,\zeta) = 0 \quad (2.1)$$

$$F_2(\eta,\xi,\zeta) = \eta - f2(\xi,\zeta) = 0 \quad (2.2)$$

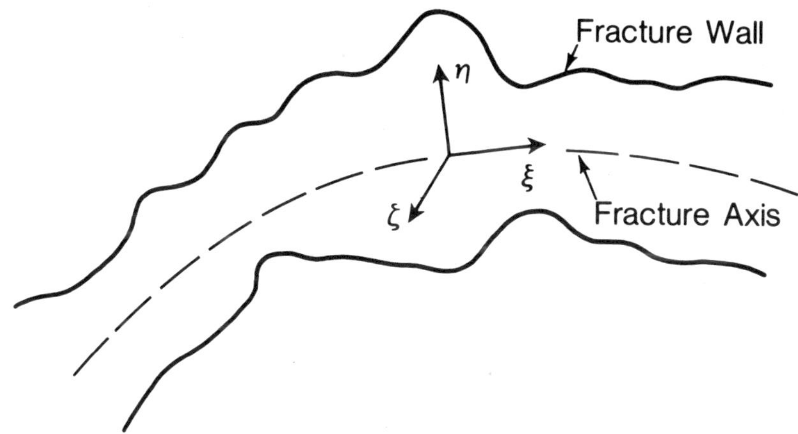

Figure 9. Local coordinate system defined for the axis of a single fracture.

where ξ and ζ are orthogonal coordinates which lie in the surface of the fracture axis, η is the coordinate normal to the fracture surface, and F_1 and F_2 are the functional forms of the two fracture walls where f1 and f2 describe the geometric shape of the fracture walls in the coordinates ξ and ζ. With this definition, the walls of the fracture are surfaces in three-dimensional space that are not necessarily parallel.

From Eqs. (2.1) and (2.2), the unit normal vectors outwardly directed from the fracture walls are defined as

$$\underset{\sim}{n}_i = \underset{\sim}{\nabla} F_i / |\underset{\sim}{\nabla} F_i| \qquad i = 1,2 \tag{2.3}$$

and the fracture aperture, b, which can vary as a function of the position on the fracture axis, is defined as

$$b(\xi,\zeta) = f2(\xi,\zeta) - f1(\xi,\zeta) \tag{2.4}$$

2.2. Balance of Fluid Mass

The two-dimensional equation of fluid movement within an individual fracture is developed by first considering a mass balance written for a single fluid continuum. Assuming the fluid to be incompressible and the mass density to be essentially constant, the equation of mass conservation for a single fluid continuum is (8)

$$\underset{\sim}{\nabla}.\underset{\sim}{v} = 0 \tag{2.5}$$

where $\underset{\sim}{v}$ is the fluid velocity and $\underset{\sim}{\nabla}$ is the gradient operator. The above equation is three-dimensional, and is assumed to be written in a coordinate system which at every point is orthogonal to the centerline of the fracture. Since the mass density is assumed to be constant, Eq. (2.5) can actually be considered an equation of volume balance.

We shall assume that the fracture aperture is narrow in relation to its areal extent and that flow is essentially two-dimensional. A two-dimensional mass balance equation is generated by integrating Eq. (2.5) over the direction normal to the fracture axis, i.e.,

$$\int_{\eta=f1(\xi,\zeta)}^{\eta=f2(\xi,\zeta)} \underset{\sim}{\nabla}.\underset{\sim}{v}\, d\eta = \underset{\sim}{\nabla}.\int_{f1}^{f2} \underset{\sim}{v}' d\eta - \underset{\sim}{v}\Big|_{f2}.\underset{\sim}{\nabla}F_2 + \underset{\sim}{v}\Big|_{f1}.\underset{\sim}{\nabla}F_1 = 0 \tag{2.6}$$

where $\underset{\sim}{v}'$ is the fluid velocity vector in the coordinates which lie in the surface of the fracture axis. The final two terms in (2.6) denote the mass flux normal to the fracture walls. If the adjacent rock is impervious, these quantities are identically equal to zero, otherwise they represent the mass entering or leaving the fracture from the adjacent rock matrix. These terms cannot be directly evaluated since they depend on the responses in the porous matrix. Thus, in order to evaluate the responses in a single fracture, the simultaneous solution of the responses in the adjacent porous matrix is required. With the use of Eq. (2.3), Eq. (2.6) may also be written in the form

$$\underset{\sim}{\nabla}.\int_{f1}^{f2} \underset{\sim}{v}' d\eta - \underset{\sim}{v}\Big|_{f2}.\underset{\sim}{n}_2|\underset{\sim}{\nabla}F_2| + \underset{\sim}{v}\Big|_{f1}.\underset{\sim}{n}_1|\underset{\sim}{\nabla}F_1| = 0 \tag{2.7}$$

We shall define an average velocity over the fracture thickness by

$$\bar{\underset{\sim}{v}}'(\xi,\zeta) = \frac{1}{b}\int_{f1}^{f2} \underset{\sim}{v}'(\eta,\xi,\zeta)\, d\eta \tag{2.8}$$

Introducing Eq. (2.8) into Eq. (2.6) yields the integrated, two-dimensional form of the mass balance expression, i.e.,

$$\underset{\sim}{\nabla}.(b\bar{\underset{\sim}{v}}') - \underset{\sim}{v}\Big|_{f2}.\underset{\sim}{\nabla}F_2 + \underset{\sim}{v}\Big|_{f1}.\underset{\sim}{\nabla}F_1 = 0 \tag{2.9}$$

For fractures that are filled with granular material, a two-dimensional form similar to Eq. (2.9) can also be developed.For such cases, the original three-dimensional mass balance equation would rewritten in terms of the specific discharge, the product of fluid velocity and porosity of the fill material.

2.3. Balance of a Fluid Mass Constituent

The fluid in the fracture may contain several dissolved constituents. We shall assume, however, that the constituents are non-reacting, and that their presence does not effect the fluid density. In the three-dimensional void space of the fracture, the equations of conservation for a single fluid constituent is

$$\frac{\partial c}{\partial t} + \underset{\sim}{\nabla} \cdot (\underset{\sim}{v} c + \underset{\sim}{J}) = 0 \tag{2.10}$$

where c is the concentration of the fluid constituent, the product $\underset{\sim}{v}c$ is the advective flux of the constituent and $\underset{\sim}{J}$ is the diffusive flux, defined as

$$\underset{\sim}{J} = - D \underset{\sim}{\nabla} c \tag{2.11}$$

where D is the coefficient of diffusion (53). Eq. (2.10) is valid for each point in the fluid within a given fracture.

The associated two-dimensional equation of mass conservation for a fluid constituent within an individual fracture shall be developed in a manner analogous to that given in section 2.2, i.e., integrating Eq. (2.10) over the fracture aperture,

$$\int_{f1}^{f2} \left[\frac{\partial c}{\partial t} + \underset{\sim}{\nabla} \cdot (\underset{\sim}{v} c + \underset{\sim}{J}) \right] d\eta = 0 \tag{2.12}$$

Since the limits of integration are independent of time, the first term in the integrand of Eq. (2.12) can be written

$$\int_{f1}^{f2} \frac{\partial c}{\partial t} \, d\eta = \frac{\partial}{\partial t} \int_{f1}^{f2} c(\eta,\xi,\zeta) d\eta = \frac{\partial (b\bar{c})}{\partial t} = b \frac{\partial \bar{c}}{\partial t} \tag{2.13}$$

where $\bar{c}$ is an average concentration over the fracture thickness,

$$\bar{c}(\xi,\zeta) = \frac{1}{b} \int_{f1}^{f2} c(\eta,\xi,\zeta) d\eta \tag{2.14}$$

The second term in the integrand of Eq. (2.12) can be expressed as

$$\int_{f1}^{f2} \underset{\sim}{\nabla}.(\underset{\sim}{v}c + \underset{\sim}{J})d\eta = \underset{\sim}{\nabla}.\int_{f1}^{f2} \underset{\sim}{v}'cd\eta + \underset{\sim}{\nabla}.\int_{f1}^{f2} \underset{\sim}{J}'d\eta$$

$$- (\underset{\sim}{v}c + \underset{\sim}{J})\Big|_{f2} .\underset{\sim}{\nabla}F_2 + (\underset{\sim}{v}c + \underset{\sim}{J})\Big|_{f1} .\underset{\sim}{\nabla}F_1 \qquad (2.15)$$

where $\underset{\sim}{J}'$ denotes the vector components of $\underset{\sim}{J}$ in the ξ and ζ directions. The final two terms on the right hand side of Eq.(2.15) represent the mass flux normal to the fracture walls due to advection and diffusion to or from the rock matrix adjacent to the fracture. Once again, as was noted in section 2.2, if the rock matrix is impervious, the advective flux to the adjacent rock will vanish. However, the diffusive flux may be non-zero if processes such as adsorption to the fracture wall are considered.

The first term on the right hand side of Eq. (2.15) shall be examined by considering the following perturbation expansions for the velocity and concentration,

$$c(\eta,\xi,\zeta) = \bar{c}(\xi,\zeta) + \tilde{c}(\eta,\xi,\zeta) \qquad (2.16)$$

$$\underset{\sim}{v}'(\eta,\xi,\zeta) = \bar{\underset{\sim}{v}}'(\xi,\zeta) + \tilde{\underset{\sim}{v}}'(\eta,\xi,\zeta) \qquad (2.17)$$

where $\tilde{c}$ and $\tilde{\underset{\sim}{v}}'$ are deviations from the average concentration and velocity, respectively, such that when they are added to the average values, the three-dimensional variables are retained. In addition, the deviations $\tilde{c}$ and $\tilde{\underset{\sim}{v}}'$ are subject to

$$\int_{f1}^{f2} \tilde{c}\, d\eta = 0 \qquad (2.18)$$

$$\int_{f1}^{f2} \tilde{\underset{\sim}{v}}' d\eta = 0 \qquad (2.19)$$

Introducing Eqs. (2.16) and (2.17) into the first term on the right hand side of Eq. (2.14) yields

$$\underset{\sim}{\nabla}. \int_{f1}^{f2} \underset{\sim}{v}'cd\eta = \underset{\sim}{\nabla}.(b\bar{\underset{\sim}{v}}'\bar{c}) + \underset{\sim}{\nabla}.\underset{\sim}{J}_d \qquad (2.20)$$

where we have made use of Eqs. (2.18) and (2.19) and the fact that $\bar{\underset{\sim}{v}}'$ and $\bar{c}$ are independent of the direction normal to the fracture axis, and thus can be removed from the integration. The quantity $\underset{\sim}{J}_d$ is a dispersive flux of the constituent that arises due to the averaging procedure,

$$\underset{\sim}{J}_d = \int_{f1}^{f2} \tilde{\underset{\sim}{v}}'\tilde{c}\, d\eta \tag{2.21}$$

Usually, a Fickian relationship is assumed for the dispersive flux, i.e.,

$$\underset{\sim}{J}_d = - \underset{\approx}{D}_d.\underset{\sim}{\nabla}\bar{c} \equiv - \underset{\approx}{\alpha}|b\underset{\sim}{v}'|.\underset{\sim}{\nabla}\bar{c} \tag{2.22}$$

where the dispersion coefficient, $\underset{\approx}{D}_d$, is assumed to be a directionally dependent quantity due to the spatially varying nature of the fracture aperture. The dispersion coefficient is assumed to be functionally dependent on the magnitude of the average specific discharge and a tensorial coefficient $\underset{\approx}{\Omega}$.

There are questions about the significance of the dispersion process within a single fracture. If the fluid velocities are extremely small, the diffusion process may dominate over the dispersion process. On the other hand, if the fluid velocity within the single fracture is large, the deviations from the average velocity may be insignificant; hence, the dispersion could be small. The contaminant would then be transported only through advection.

The second term on the right hand side of Eq. (2.15) (the divergence of the average diffusive flux) is treated in the following manner,

$$\underset{\sim}{\nabla}.\int_{f1}^{f2} \underset{\sim}{J}'d\eta = - D\underset{\sim}{\nabla}.\int_{f1}^{f2} \underset{\sim}{\nabla}c\, d\eta$$

$$= - D\underset{\sim}{\nabla}.\ (b\underset{\sim}{\nabla}\bar{c} + \bar{c}\underset{\sim}{\nabla}b - c\Big|_{f2}\underset{\sim}{\nabla}f2 + c\Big|_{f1}\underset{\sim}{\nabla}f1) \tag{2.23}$$

If we assume that the concentration at the walls of the fracture are essentially equal to the average concentration, then with the use of the definition of b given in Eq. (2.4), Eq. (2.23) reduces to

$$\underset{\sim}{\nabla}.\int_{f1}^{f2} \underset{\sim}{J}'d\eta = - D\underset{\sim}{\nabla}.(b\nabla\bar{c}) \tag{2.24}$$

Introducing Eqs. (2.13), (2.15), (2.20), (2.22) and (2.24) into Eq. (2.12) yields

$$b \frac{\partial \bar{c}}{\partial t} + \nabla.(b\bar{\underset{\sim}{v}}'\bar{c}) - \underset{\sim}{\nabla}.(\underset{\approx}{D}_d.\underset{\sim}{\nabla}\bar{c}) - D\underset{\sim}{\nabla}.(b\underset{\sim}{\nabla}\bar{c})$$

$$- (\underset{\sim}{v}c + \underset{\sim}{J})\Big|_{f2} .\underset{\sim}{\nabla}F_2 + (\underset{\sim}{v}c + \underset{\sim}{J})\Big|_{f1} .\underset{\sim}{\nabla}F_1 = 0 \qquad (2.25)$$

Expanding the second term in the above expression, and using the two-dimensional equation of fluid mass conservation, Eq. (2.9), Eq. (2.25) reduces to

$$b \frac{\partial \bar{c}}{\partial t} + b\bar{\underset{\sim}{v}}'.\underset{\sim}{\nabla}\bar{c} - \underset{\sim}{\nabla}.(\underset{\approx}{D}_d.\underset{\sim}{\nabla}\bar{c}) - D\underset{\sim}{\nabla}.(b\underset{\sim}{\nabla}\bar{c})$$

$$- (\underset{\sim}{v}(c - \bar{c}) + \underset{\sim}{J})\Big|_{f2} .\underset{\sim}{\nabla}F_2 + (\underset{\sim}{v}(c - \bar{c}) + \underset{\sim}{J})\Big|_{f1} .\underset{\sim}{\nabla}F_1 = 0 \qquad (2.26)$$

This two-dimensional equation of mass conservation in an individual fracture must be solved in unison with the equation for mass transport in the adjacent rock matrix due to the terms evaluated at the fracture walls which appear in the above equation.

2.4. Balance of Fluid Momentum

The balance of linear momentum for the two-dimensional interpretation of transport in an individual fracture is developed by integrating the three-dimensional linear momentum balance which is descriptive of transport in a single fluid continuum over the fracture aperture. Assuming the fluid in the fracture to be Newtonian, incompressible and with constant density, the appropriate three-dimensional linear momentum balance is

$$\rho \frac{\partial \underset{\sim}{v}}{\partial t} + \rho\underset{\sim}{\nabla}.(\underset{\sim}{v}\underset{\sim}{v}) + \underset{\sim}{\nabla}p - \mu\nabla^2\underset{\sim}{v} - \rho g = 0 \qquad (2.27)$$

where p is the fluid pressure, μ is the fluid viscosity and $\underset{\sim}{g}$ is the gravitational force, $\underset{\sim}{g} = - gl\underset{\sim}{z}$, with z being the vertical direction which is not necessarily normal to the fracture axis. Introducing into Eq. (2.27), the hydraulic head, ϕ,

$$\phi = \frac{p}{\rho g} + z \qquad (2.28)$$

we obtain

$$\rho \frac{\partial \underset{\sim}{v}}{\partial t} + \rho\underset{\sim}{\nabla}.(\underset{\sim}{v}\underset{\sim}{v}) + \rho g\underset{\sim}{\nabla}\phi - \mu\nabla^2\underset{\sim}{v} = 0 \qquad (2.29)$$

In integrating Eq. (2.29) over the fracture aperture we shall make use of several of the mathematical manipulations that were introduced in section 2.3. The integration of the first term in Eq. (2.29) is expressed by

$$\int_{f1}^{f2} \rho \frac{\partial \underset{\sim}{v}}{\partial t} d\eta = \rho b \frac{\partial \bar{\underset{\sim}{v}}'}{\partial t} \qquad (2.30)$$

The second term in Eq. (2.29) is evaluated as

$$\int_{f1}^{f2} \rho \underset{\sim}{\nabla}.(\underset{\sim}{v}\underset{\sim}{v})d\eta = \rho\underset{\sim}{\nabla}.(\bar{\underset{\sim}{v}}'\bar{\underset{\sim}{v}}'b) + \rho\underset{\sim}{\nabla}.\underset{\approx}{\tau} - \rho\underset{\sim}{v}\underset{\sim}{v}\Big|_{f2}.\underset{\sim}{\nabla}F_2 + \rho\underset{\sim}{v}\underset{\sim}{v}\Big|_{f1}.\underset{\sim}{\nabla}F_1 \tag{2.31}$$

where we have used the definition of $\tilde{\underset{\sim}{v}}'$ in Eq. (2.17) and $\underset{\approx}{\tau}$ is a dispersive momentum flux,

$$\underset{\approx}{\tau} = \int_{f1}^{f2} \tilde{\underset{\sim}{v}}'\tilde{\underset{\sim}{v}}' \, d\eta \tag{2.32}$$

The integration of the third term in Eq. (2.29) is

$$\int_{f1}^{f2} \rho g\underset{\sim}{\nabla}\phi \, d\eta = \rho g\underset{\sim}{\nabla}(b\bar{\phi}) - \rho g\phi\Big|_{f2}\underset{\sim}{\nabla}F_2 + \rho g\phi\Big|_{f1}\underset{\sim}{\nabla}F_1 \tag{2.33}$$

where $\bar{\phi}$ is the average hydraulic head over the fracture thickness,

$$\bar{\phi}(\xi,\zeta) = \frac{1}{b}\int_{f1}^{f2} \phi(\eta,\xi,\zeta) \, d\eta \tag{2.34}$$

If we assume that the hydraulic head at the fracture walls are essentially equal to the average hydraulic head, then by using

$$\int_{f1}^{f2} \rho g\underset{\sim}{\nabla}\phi \, d\eta = \rho g b\underset{\sim}{\nabla}\bar{\phi} \tag{2.35}$$

The previous assumption is equivalent to imposing a hydrostatic condition in the direction normal to the fracture axis. Such an assumption is reasonable since the fracture aperture is small; thus, the effect of potential energy is negligible between the walls of the fracture. In essence, the assumption is equivalent to assuming a uniform pressure distribution over the fracture aperture.

The integration of the final term in Eq. (2.29) is expressed as

$$\int_{f1}^{f2} \mu\nabla^2\underset{\sim}{v} \, d\eta = \mu\nabla^2(b\bar{\underset{\sim}{v}}') - \mu\underset{\sim}{\nabla}.(\underset{\sim}{v}'\Big|_{f2}\underset{\sim}{\nabla}f2 - \underset{\sim}{v}'\Big|_{f1}\underset{\sim}{\nabla}f1)$$

$$- \mu(\underset{\sim}{\nabla}\underset{\sim}{v}\Big|_{f2}.\underset{\sim}{\nabla}F_2 - \underset{\sim}{\nabla}\underset{\sim}{v}\Big|_{f1}.\underset{\sim}{\nabla}F_1) \tag{2.36}$$

If we assume that the fluid velocity at the fracture walls in the

ξ and ζ directions is subject to a no-slip condition, the second term on the right hand side of Eq. (2.36) is identically equal to zero.

$$\int_{f1}^{f2} \mu\nabla^2\underset{\sim}{v}\, d\eta = \mu\underset{\sim}{\nabla}x(\underset{\sim}{\nabla}xb\bar{\underset{\sim}{v}}') - \mu\underset{\sim}{\nabla}(\underset{\sim}{\nabla}.b\bar{\underset{\sim}{v}}') - \mu(\underset{\sim}{\nabla}\underset{\sim}{v}\Big|_{f2}.\underset{\sim}{\nabla}F_2 - \underset{\sim}{\nabla}\underset{\sim}{v}\Big|_{f1}.\underset{\sim}{\nabla}F_1) \tag{2.37}$$

where we have used the following identity to replace the Laplacian of $b\bar{\underset{\sim}{v}}'$ (8),

$$\nabla^2(b\bar{\underset{\sim}{v}}') = \underset{\sim}{\nabla}.\underset{\sim}{\nabla}(b\bar{\underset{\sim}{v}}') = \underset{\sim}{\nabla}x(\underset{\sim}{\nabla}xb\bar{\underset{\sim}{v}}') - \underset{\sim}{\nabla}(\underset{\sim}{\nabla}.b\bar{\underset{\sim}{v}}') \tag{2.38}$$

The final term on the right hand side of Eq. (2.37) represents the drag that the fracture walls impose on the fluid within the fracture.

The two-dimensional, integrated, linear momentum equation within a single fracture is obtained by combining Eqs. (2.30), (2.31), (2.35), and (2.37),

$$\rho b\frac{\partial\bar{\underset{\sim}{v}}'}{\partial t} + \rho\underset{\sim}{\nabla}.(b\bar{\underset{\sim}{v}}'\bar{\underset{\sim}{v}}') + \rho\underset{\sim}{\nabla}.\underset{\approx}{\tau} + \rho g b\underset{\sim}{\nabla}\bar{\phi} - \mu\underset{\sim}{\nabla}x(\underset{\sim}{\nabla}xb\bar{\underset{\sim}{v}}')$$
$$+ \mu\underset{\sim}{\nabla}(\underset{\sim}{\nabla}.b\underset{\sim}{v}') - \rho\underset{\sim}{v}\underset{\sim}{v}\Big|_{f2}.\underset{\sim}{\nabla}F_2 + \rho\underset{\sim}{v}\underset{\sim}{v}\Big|_{f1}.\underset{\sim}{\nabla}F_1$$
$$+ \mu(\underset{\sim}{\nabla}\underset{\sim}{v}\Big|_{f2}.\underset{\sim}{\nabla}F_2 - \underset{\sim}{\nabla}\underset{\sim}{v}\Big|_{f1}.\underset{\sim}{\nabla}F_1) = 0 \tag{2.39}$$

Expanding the second term in Eq. (2.39) and using Eq. (2.9) yields

$$\rho b\frac{\partial\bar{\underset{\sim}{v}}'}{\partial t} + \rho b\bar{\underset{\sim}{v}}'(\underset{\sim}{\nabla}.\bar{\underset{\sim}{v}}') + \rho\underset{\sim}{\nabla}.\underset{\approx}{\tau} + \rho g b\underset{\sim}{\nabla}\bar{\phi} - \mu\underset{\sim}{\nabla}x(\underset{\sim}{\nabla}xb\bar{\underset{\sim}{v}}')$$
$$+ \mu\underset{\sim}{\nabla}(\underset{\sim}{\nabla}.b\bar{\underset{\sim}{v}}') - \rho\underset{\sim}{v}\underset{\sim}{v}\Big|_{f2}.\underset{\sim}{\nabla}F_2 + \rho\underset{\sim}{v}\underset{\sim}{v}\Big|_{f1}.\underset{\sim}{\nabla}F_1 \tag{2.40}$$
$$+ \mu(\underset{\sim}{\nabla}\underset{\sim}{v}\Big|_{f2}.\underset{\sim}{\nabla}F_2 - \underset{\sim}{\nabla}\underset{\sim}{v}\Big|_{f1}.\underset{\sim}{\nabla}F_1) + \rho\bar{\underset{\sim}{v}}'(\underset{\sim}{v}\Big|_{f2}.\underset{\sim}{\nabla}F_2 - \underset{\sim}{v}\Big|_{f1}.\underset{\sim}{\nabla}F_1) = 0$$

The above equations indicate that, in general, two-dimensional flow in the axis of a fracture is rotational. Therefore, a pseudo-potential function defining the average velocity can only be introduced if we ultimately assume that the rotation $\underset{\sim}{\nabla}x(\underset{\sim}{\nabla}xb\bar{\underset{\sim}{v}}')$ is negligible. We cannot neglect this term by assuming the fluid viscosity to be negligible since other terms in the same equation would have to be neglected for similar reasons; it is the effect of the fluid viscosity which creates the frictional drag that ultimately leads to the definition of the fracture conductance. It should be noted, however, that if fluid movement is uni-directional, the rotational term vanishes by definition.

For further analysis of Eq. (2.40), let us consider steady-state, uni-directional flow through parallel fracture walls. If the fracture walls are impervious, the velocity distribution within the fracture will be parabolic and symmetric around the fracture center line (Figure 10), i.e.,

$$v_\xi(\eta) = 6\,\frac{\bar{v}_\xi}{b^2}\,\eta(b - \eta) \qquad 0 \le \eta \le b \tag{2.41}$$

where $\bar{v}_\xi$ is the average velocity over the fracture aperture. With the above assumptions, Eq. (2.40) reduces to

$$\rho g b\,\frac{d\bar{\phi}}{d\xi} + 2\mu\,\frac{dv_\xi}{d\eta}\bigg|_{\eta=0} = 0 \tag{2.42}$$

where evaluating the second term with Eq. (2.41) we obtain

$$\bar{v}_\xi = -\,\kappa\,\frac{d\phi}{d\xi} \tag{2.43}$$

The quantity κ is the fracture transmissivity defined as

$$\kappa = \frac{\rho g b^2}{12\mu} \tag{2.44}$$

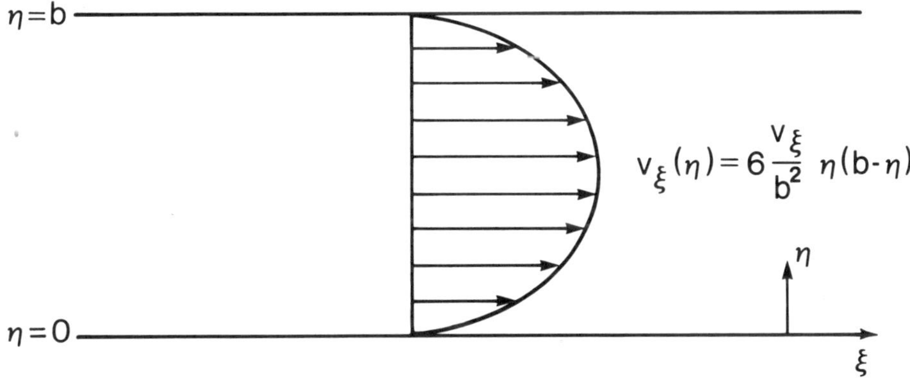

Figure 10. Parabolic velocity profile in a fracture having parallel walls.

which is the classic definition of the fracture transmissivity that is employed in most analyses using the discrete fracture conceptualization.

For the purpose of comparison, let us consider Eq. (2.40) where we assume that the fracture walls are permeable. For uni-directional flow, parallel fracture walls, and assuming that the flux into or out of the fracture is equal on each fracture wall and uniform over the fracture length, Eq. (2.40) reduces to

$$\rho g b \frac{d\bar{\phi}}{d\xi} + 2\rho \bar{v}_\xi (Q-Q^2) + 2\mu \left.\frac{dv_\xi}{d\eta}\right|_{\eta=0} = 0 \tag{2.45}$$

where Q is the volumetric flux leaving the fracture. If we continue to assume that the velocity in the ξ direction is zero at the fracture walls and additionally assume the validity of the parabolic velocity profile, then Eq. (2.45) reduces to

$$\bar{v}_\xi = - \frac{\rho g b^2}{2\rho Q(1-Q) + 12\mu} \frac{d\bar{\phi}}{d\xi} \tag{2.46}$$

This relationship illustrates that if one wishes to define a fracture transmissivity for a permeable host rock, in general, it will be a function of the magnitude of the flux into or out of the fracture. If the volumetric flux, Q, is negligible in comparison to the viscosity, then the fracture transmissivity can be approximated by Eq. (2.44).

2.5. Balance of Energy

In the discrete fracture conceptualization, the two-dimensional, integrated balance of energy for the fluid in a single fracture is analogous to the two-dimensional equation of contaminant migration provided that it is assumed that temperature changes do not appreciably affect the fluid density. The two-dimensional equation of thermal transport in a single fracture is then written

$$\rho C_w b \frac{\partial \bar{T}}{\partial t} + \rho C_w b \bar{\underset{\sim}{v}}'.\underset{\sim}{\nabla}\bar{T} - \nabla.(\underset{\approx}{\lambda}_d.\underset{\sim}{\nabla}\bar{T}) - \lambda\underset{\sim}{\nabla}.(b\underset{\sim}{\nabla}\bar{T}) \tag{2.47}$$

$$- \left.(\rho C_w \underset{\sim}{v}(T-\bar{T})+\underset{\sim}{J}^h)\right|_{f2} .\underset{\sim}{\nabla}F_2 + \left.(\rho C_w \underset{\sim}{v}(T-\bar{T})+\underset{\sim}{J}^h)\right|_{f1} .\underset{\sim}{\nabla}F_1 = 0$$

where $\bar{T}$ is the average temperature over the fracture aperture,

$$\bar{T}(\xi,\zeta) = \frac{1}{b}\int_{f1}^{f2} T(\eta,\xi,\zeta)\, d\eta \tag{2.48}$$

C_w is the specific heat of the fluid in the fracture, λ is the thermal conductivity of the fluid in the fracture, $\underset{\sim}{J}^h$ is the thermal diffusion of energy defined as

$$\underset{\sim}{J}^h = -\lambda \underset{\sim}{\nabla} T \tag{2.49}$$

and $\underset{\approx}{\lambda}_d$ is the coefficient of thermal dispersion. If we assume that the dispersion process is dependent only upon the fracture geometry, then $\underset{\approx}{\lambda}_d$ is related to the dispersion coefficient defined in Eq. (2.22), i.e.,

$$\underset{\approx}{\lambda}_d = \rho C_w \underset{\approx}{\alpha} |b\bar{\underset{\sim}{v}}'| \tag{2.50}$$

As was noted for the two-dimensional equation of solute movement in an individual fracture, the two-dimensional equation of energy transport must also be solved with a similar equation in the adjacent rock matrix. These two regimes are coupled by the terms defined at the fracture walls. If the rock adjacent to the fracture is impervious, the advective flux of energy from the rock matrix to the fracture is identically zero. The diffusive flux of energy, denoted by $\underset{\sim}{J}^h$ in the last two terms on the left hand side of Eq. (2.47), will be non-zero since an impervious boundary to fluid movement does not act as a material surface to energy transport.

3. TRANSPORT EQUATIONS FOR CONTINUUM CONCEPTUALIZATIONS

In field situations where there are numerous interconnected fractures from a variety of fracture families, we are unable to effectively describe transport phenomena on the scale where individual fractures are discernible and modeled as separate geometric entities. Even in those situations where we have a reasonably detailed statistical knowledge of the geometric features of most fracture families, treating highly fractured formations using a discrete fracture conceptualization becomes computationally intractable. In such instances, a continuum conceptualization of the rock matrix and the fluids occupying the void space alleviates the necessity of describing the specific geometry of the fractures.

In sections 1.3.2 and 1.3.3, two continuum conceptualizations of fractured rock were briefly identified along with the assumptions and restrictions implicit in their use. In this chapter we shall discuss the equations that are descriptive of the transport of mass, momentum and energy for the equivalent porous medium and dual porosity conceptualizations.

3.1. Continuum Balance Equations

Both the equivalent porous medium and dual porosity conceptualizations are examples of multi-component continua, that is, a mixture of discontinuous phases that are hypothesized as continua, in this case, a solid phase, and one or more fluid phases. We shall initiate our discussion by briefly examining the general balance equations of mass, momentum and energy that are applicable to any multi-component continuum. Here, we shall discuss two widely used methods of hypothesizing, or formulating, the general continuum balance equations. The application of these balance equations to a specific multi-component continuum (e.g., the equivalent porous medium or dual porosity conceptualizations) arises through restrictions that are imposed on each continuum component and through constitutive relationships that make the general balance equations descriptive of a particular multi-component medium. In subsequent sections we shall discuss the assumptions that are applied to the general balance equations in the application to the two above mentioned continuum conceptualizations of fractured rock.

3.1.1. Mixture theory

One approach to the development of the balance equations for a multi-component medium is to directly hypothesize their form in conjunction with assumptions concerning the nature of the phases which comprise the continuum (see, e.g., Bowen(11), Ingram and Eringen (46)). This is usually referred to as a mixture theory approach. The multi-component continuum may be composed of any number of overlapping continuum phases. The continuum phases are assumed to coexist (or overlap) at each point in the medium; thus, properties of several phases are defined at each mathematical point.

For fractured rock where a single fluid is assumed to saturate the entire void space of the rock, the equivalent porous medium conceptualization is described by only two continuum phases, a rock phase and the fluid phase in all the void space of the rock, regardless of the character of the void space. In contrast, the dual porosity conceptualization is hypothesized by three continuum phases, i.e., the rock matrix, and two distinct fluid continua, for example, the fluid in the fractures and the fluid in the porous matrix of the rock.

Balance equations for mass, momentum and energy shall be written for each phase (or component) of the multi-component medium. We shall express the balance of the thermodynamic properties, mass, momentum and energy, in a general form, where the specific thermodynamic property is referred to as Ψ_α (e.g., mass of the α-phase, momentum of the α-phase, etc.) having a density, ψ_α, defined per unit mass of the α-phase. The hypothesized form of the general balance equation for the property Ψ_α is

$$\frac{\partial(\rho_\alpha\varepsilon_\alpha\psi_\alpha)}{\partial t} + \underset{\sim}{\nabla}\cdot(\rho_\alpha\varepsilon_\alpha\psi_\alpha\underset{\sim}{V}_\alpha + \varepsilon_\alpha\underset{\sim}{J}_\alpha) - \rho_\alpha\varepsilon_\alpha f_\alpha$$

$$= \overleftrightarrow{M}_\alpha\psi_\alpha + \overleftrightarrow{T}_\alpha \qquad \alpha = 1,2,\ldots N \qquad (3.1)$$

where ρ_α is the density of the α-phase defined per unit volume of the α-phase, ε_α is the volume fraction of the α-phase, $\underset{\sim}{V}_\alpha$ is the velocity of the α-phase, $\underset{\sim}{J}_\alpha$ is a non-advective flux of Ψ_α, f_α is the rate of a source of Ψ_α generated internal to the α-phase, $\overleftrightarrow{M}_\alpha$ denotes the rate of exchange of mass between the α-phase and all other phases, $\overleftrightarrow{T}_\alpha$ is the rate of exchange of Ψ_α between the α-phase and all other phases due to mechanical interaction between the phases, and N is the number of phases which comprise the multi-component continuum. The definition of the terms in Eq. (3.1) for the thermodynamic quantities; mass, momentum and energy are listed in Table 1.

Eq. (3.1) is a point balance equation that is assumed to be valid throughout the entire continuum volume. Therefore, it is based on assumptions of smoothness and differentiability in the quantities defined above. In addition, Eq. (3.1) is subject to the following restrictions

$$\sum_{\alpha=1}^{N}\varepsilon_\alpha = 1 \qquad (3.2)$$

$$\sum_{\alpha=1}^{N}(\overleftrightarrow{M}_\alpha\psi_\alpha + \overleftrightarrow{T}_\alpha) = 0 \qquad (3.3)$$

Eq. (3.2) merely implies that volume at a given point is conserved, i.e., the volumetric fractions of all phases must sum to unity. Similarly, Eq. (3.3) insures local (pointwise) conservation of the thermodynamic property Ψ_α at a given point. The quantities $\overleftrightarrow{M}_\alpha\psi_\alpha$ and $\overleftrightarrow{T}_\alpha$ denote exchanges of the property Ψ_α between the α-phase and all other phases. Eq. (3.3) states that the rate at which a thermodynamic property leaves one phase must be equivalent to the rate at which it is gained in the remaining N-1 coexisting phases at the point in question.

3.1.2. Volume averaging

Eq. (3.1) is hypothesized without a specific knowledge or reference to a configuration of the individual phases that comprise the continuum. Although concepts such as an averaging volume or REV are never mentioned in hypothesizing Eq. (3.1), they are implicit in the definition of the continuum quantities. From Eq. (3.1), we gain little physical insight into the meaning and significance of terms such as $\underset{\sim}{J}_\alpha$, $\overleftrightarrow{M}_\alpha$ and $\overleftrightarrow{T}_\alpha$.

Table 1. Definition of the density, flux and internal source of mass, momentum and energy for a multi-component continuum.

	Density	Flux	Internal Supply	Sources Mass Exchange	Mechanical Interaction
Mass	1	0	0	$\overleftrightarrow{M}_\alpha$	0
Mass of a Phase Constituent	ω_α	$\underset{\sim}{J}_\alpha^\omega$	0	$\overleftrightarrow{M}_\alpha \omega_\alpha$	$\overleftrightarrow{J}_\alpha^\omega$
Momentum	$\underset{\sim}{V}_\alpha$	$\underset{\approx}{\tau}_\alpha$	$\underset{\sim}{g}$	$\overleftrightarrow{M}_\alpha \underset{\sim}{V}_\alpha$	$\overleftrightarrow{\underset{\sim}{F}}_\alpha$
Energy	$e_\alpha+\frac{1}{2}V_\alpha^2$	$\underset{\approx}{\tau}_\alpha \cdot \underset{\sim}{V}_\alpha+\underset{\sim}{J}_\alpha^h$	$h_\alpha+\underset{\sim}{g}\cdot\underset{\sim}{V}_\alpha$	$\overleftrightarrow{M}_\alpha(e_\alpha+\frac{1}{2}V_\alpha^2)$	$\overleftrightarrow{\underset{\sim}{F}}_\alpha\cdot\underset{\sim}{V}_\alpha+\overleftrightarrow{Q}_\alpha$

ω_α is the mass fraction of a constituent of the α-phase defined as ρ_ω/ρ_α

ρ_ω is the mass density of the ω-constituent of the α-phase per unit volume of the α-phase

ρ_α is the mass density of the α-phase per unit volume of the α-phase

$\overleftrightarrow{M}_\alpha$ is the rate of mass exchange between the α-phase and all other phases

$\underset{\sim}{J}_\alpha^\omega$ is the non-advective flux of the ω-constituent of the α-phase

$\overleftrightarrow{J}_\alpha^\omega$ is the rate of exchange of the ω-constituent of the α-phase with all other phases due to mechanical interaction between phases

$\underset{\sim}{V}_\alpha$ is the velocity of the α-phase

$\underset{\approx}{\tau}_\alpha$ is the internal stress of the α-phase

$\underset{\sim}{g}$ is the gravitational force

$\overleftrightarrow{\underset{\sim}{F}}_\alpha$ is the rate of mechanical exchange of momentum between the α-phase and all other phases

e_α is the internal energy density of the α-phase

$\underset{\sim}{J}_\alpha^h$ is the non-advective flux of thermal energy

h_α is the internal supply of thermal energy

$\overleftrightarrow{Q}_\alpha$ is the rate of exchange of thermal energy between the α-phase and all other phases

For this reason, a considerable effort has been directed toward a rigorous derivation of the balance equations for the transport of mass, momentum and energy in porous media (see, e.g., Drew (21), Drew amd Segel (22), Bear and Bachmat (6), Hassanizadeh (39), Hassanizadeh and Gray (40,41)). These investigations were directed toward the development of continuum balance equations from a knowledge of the physical processes at the level where individual phases and boundaries between phases are still discernible. This approach, referred to as volume averaging, uses as a point of departure, the balance equations of mass, momentum and energy that are valid at the microscopic level. The continuum balance equations for each phase of the continuum are developed by averaging the microscopic balance equations over the volume of the phase in question within a Representative Elementary Volume.

First attempts at employing such ideas consisted of using specific geometrical models of porous media, e.g., capillary tube models (see, e.g., Duguid and Lee (24), Bear and Braester (7)). More recent investigations consider an arbitrary geometry of the phases within the REV (see, e.g., Hassanizadeh (39), Hassanizadeh and Gray (40,41), Bear and Bachmat (6)). In the following paragraphs, the thoery of volume averaging as applied to the development of the continuum balance equations is outlined for a multi-component continuum.

At the level where we visualize the medium as a conglomerate of discontinuous phases that are separated by boundaries, the general balance equation for the thermodynamic property, referred to as Ψ, at a given point within any phase is

$$\frac{\partial(\rho\psi)}{\partial t} + \underset{\sim}{\nabla}.(\rho\psi\underset{\sim}{v} + \underset{\sim}{j}) - \rho f = 0 \tag{3.4}$$

where ρ is mass density, ψ is the density of the thermodynamic property Ψ, $\underset{\sim}{j}$ is the nonadvective flux of Ψ, f is the rate of an internal source of Ψ and $\underset{\sim}{v}$ is velocity. Equations similar to (3.4) are used in the discrete fracture conceptualization discussed in Section 2.

The continuum balance equations are developed for a given phase by averaging (3.4) over the volume of that phase within the REV, i.e.,

$$\int_{dV_\alpha} \left[\frac{\partial(\rho\psi)}{\partial t} + \underset{\sim}{\nabla}.(\rho\psi\underset{\sim}{v} + \underset{\sim}{j}) - \rho f\right] dv = 0 \tag{3.5}$$

where dV_α denotes the volume of the REV occupied by the α-phase. Due to the assumed arbitrary geometry within the REV, specific averaging rules are applied to Eq. (3.5) in order that spatial and time derivatives can be removed from the integration (see, e.g., Gray and Lee (30)). Consequently, Eq. (3.5) can be placed in a form identical to Eq. (3.1), where as a result of the averaging procedure, we obtain specific definitions of the continuum quantities i.e.,

$$\varepsilon_\alpha = \frac{1}{dV} \int_{dV_\alpha} dv \tag{3.6}$$

$$\rho_\alpha = \frac{1}{dV_\alpha} \int_{dV_\alpha} \rho dv \tag{3.7}$$

$$\psi_\alpha = \frac{1}{\rho_\alpha dV_\alpha} \int_{dV_\alpha} \rho\psi dv \tag{3.8}$$

$$\underset{\sim}{V}_\alpha = \frac{1}{\rho_\alpha dV_\alpha} \int_{dV_\alpha} \rho \underset{\sim}{v} dv \tag{3.9}$$

$$\underset{\sim}{J}_\alpha = \frac{1}{dV_\alpha} \int_{dV_\alpha} (\underset{\sim}{j} + \rho\tilde{\psi}_\alpha \tilde{\underset{\sim}{V}}_\alpha)\; dv \tag{3.10}$$

$$\overleftrightarrow{M}_\alpha = \sum_{\beta \neq \alpha} \frac{1}{dV} \int_{dS_{\alpha\beta}} \rho(\underset{\sim}{w} - \underset{\sim}{v}).\underset{\sim}{n}_{\alpha\beta}\; ds \tag{3.11}$$

$$\overleftrightarrow{T}_\alpha = \sum_{\beta \neq \alpha} \frac{1}{dV} \int_{dS_{\alpha\beta}} (\rho\tilde{\psi}_\alpha(\underset{\sim}{w} - \underset{\sim}{v}) + \underset{\sim}{j}).\underset{\sim}{n}_{\alpha\beta}\; ds \tag{3.12}$$

In the above definitions dV denotes the volume of the REV, $dS_{\alpha\beta}$ denotes the interface between the α and β-phases within the REV having a normal vector $\underset{\sim}{n}_{\alpha\beta}$ outwardly directed from the α-phase, $\underset{\sim}{w}$ is the velocity of the interface between the α and β-phases, and $\tilde{\psi}_\alpha$ and $\tilde{\underset{\sim}{V}}_\alpha$ are deviations from the volume averaged quantities ψ_α and $\underset{\sim}{V}_\alpha$, respectively, such that the microscopic values are defined as

$$\psi = \psi_\alpha + \tilde{\psi}_\alpha \tag{3.13}$$

$$\underset{\sim}{v} = \underset{\sim}{V}_\alpha + \tilde{\underset{\sim}{V}}_\alpha \tag{3.14}$$

The deviations, $\tilde{\psi}_\alpha$ and $\tilde{\underset{\sim}{V}}_\alpha$, are subject to

$$\int_{dV_\alpha} \rho\tilde{\psi}_\alpha \, dv = 0 \tag{3.15}$$

$$\int_{dV_\alpha} \rho\tilde{\underset{\sim}{V}}_\alpha \, dv = 0 \tag{3.16}$$

It can also be shown that Eqs. (3.2) and (3.3) can be rigorously derived through the volume averaging procedure (Hassanizadeh (39), Hassanizadeh and Gray (40)).

In Eqs. (3.6) through (3.12), we see the specific construction of the continuum quantities, which was unrecognizable from merely hypothesizing Eq. (3.1). For example, in Eq. (3.10), the continuum non-advective flux is shown to be the cummulative effect of the volume average of the microscopic non-advective flux and a dispersive flux, denoted by $\tilde{\psi}_\alpha\tilde{\underset{\sim}{V}}_\alpha$. Furthermore, in Eqs. (3.11) and (3.12), the specific form of the terms which define the exchanges between phases are identified. In these definitions the summation indicates that the boundary between the α-phase and all other phases is considered. Furthermore, from Eqs. (3.11) and (3.12) the assumptions that will force these terms to vanish become more evident. For example, if we assume that the microscopic interface between the α-phase and all other phases is a material surface, i.e., $\underset{\sim}{w} = \underset{\sim}{v}$, then $\overset{\leftrightarrow}{M}_\alpha$ is identically zero.

The volume averaging procedure is also significant since it essentially represents the physics of measurement. If a measurement is to be representative of the continuum hypothesis, it should be amenable to a description as a volume or surface average. The validity of continuum measurements are usually questioned (at least in fractured rock formations) due to the fact that a Representative Elementary Volume may not be encompassed by the measuring device, and thus, may not adhere to the definitions given above.

In summary, the volume averaging approach has led to a considerable amount of insight into the actual definition of the continuum fluxes and exchange terms appearing in the continuum balance equations. However, these definitions are actually of little significance in Eq. (3.1) because the continuum quantities defined in Eqs. (3.6) through (3.12) only have meaning in their integrated form; the microscopic quantities within the integration are unrecognizable. Only if restrictions are imposed on the geometric structure of the medium, and the microscopic behavior of the fluid and rock matrix, will the definitions in Eqs. (3.6) through (3.12) be of use.

3.2. Constitutive Relationships

Eq. (3.1), whether obtained from direct hypothesis or through volume averaging, is descriptive of any multi-component continuum since the fluxes and sources of mass, momentum and energy are defined in generic terms only. The balance equations become descriptive of a particular multi-component continuum when the quantities $\underset{\sim}{J}_{\alpha}$, $\overleftrightarrow{M}_{\alpha}$, and $\overleftrightarrow{T}_{\alpha}$ are defined by phenomenological (or constitutive) relationships that are specific to the multi-component continuum under investigation. These constitutive relationships are usually defined in terms of the measurable quantities of the transport problem, e.g., ρ_{α}, ε_{α}, $\underset{\sim}{V}_{\alpha}$, and ψ_{α}. The balance equations of mass, momentum and energy, in conjunction with the constitutive relationships are to provide an equivalent number of equations and unknowns; with boundary and initial conditions, the transport problem is then amenable to a solution.

The constitutive relationships that are to be employed in the balance equations cannot violate any of the balance laws, or the Second Law of Thermodynamics. The Second Law of Thermodynamics acts as a further constraint on the transport processes by dictating the direction that reactions will take. For example, nothing in the balance equations indicates that energy will be transported from domains of higher temperature to domains of lower temperature. It is the Second Law of Thermodynamics which imposes restrictions of this type.

Coleman and Noll (15) employed the Second Law of Thermodynamics as a means of placing restrictions on the functional dependence of the constitutive relationships, and also a means of obtaining their (thermodynamic) equilibrium form. The Coleman and Noll method in conjunction with linear and higher order approximations for the thermodynamic nonequilibrium parts of the constitutive functions has found wide applicability in developing field equations for mixtures of fluids and solids(23,57,58,65,71). Its application to the transport phenomena in porous media has been thoroughly examined by Hassanizadeh (39). Shapiro (70) considered the application of the Coleman and Noll method in the development of transport equations for a dual porosity conceptualization of fractured rock.

For analyses of fractured rock, it is especially important that a systematic and unbiased method of determining constitutive functions, such as that of Coleman and Noll (15), be employed. The reason for this being that there does not exist a wealth of experimental information to allow constitutive relationships to be developed by more heuristic means. Although we may draw upon our physical intuition in the development of constitutive relationships, this alone may not be sufficient since there are different mechanisms of transport which are of importance in fractured rock that are not relevant in granular porous materials, e.g., the exchange of mass, momentum and energy between two fluid continua in the dual porosity conceptualization.

The constitutive relationships are developed by first hypothesizing a general functional dependence of the nonmeasurable quantities. The method of choosing the functional dependence on the measurable quantities is discussed in general by Truesdell and Toupin (83) and Eringen (27), and by Hassanizadeh (39) and Shapiro (70) for subsurface hydrologic phenomena. The Second Law of Thermodynamics restricts the functional dependence of the constitutive functions and determines their thermodynamic equilibrium form. The non-equilibrium part of the constitutive relationships are usually determined by a Taylor series expansion about the equilibrium state. In many instances a linear approximation is suitable to describe the functional form of the constitutive relationships. However, in some instances, a higher order expansion is necessary to describe the physical behavior (see, e.g., Whitaker (88), Shapiro (71)). Continuum coefficients arise in the Taylor series expansion. These are the material coefficients of the medium which ultimately require evaluation through physical experimentation.

In this section we consider the application of the general balance equation to the specific case of the equivalent porous medium and dual porosity conceptualizations of fractured rock. The continuum balance equations for the equivalent porous medium conceptualization are those which have been classically used in models describing transport phenomena in granular porous media (see, e.g., Bear (5), Hassanizadeh and Gray (42), and we shall only dwell briefly on the equations of this conceptualization.

3.2.1. Equivalent porous medium conceptualization

For the equivalent porous medium conceptualization of fractured rock, we shall consider a single fluid occupying the entire void space of the rock matrix. The microscopic interface between the rock and the fluid phases is assumed to be a material surface; thus $\overleftrightarrow{M}_r = \overleftrightarrow{M}_f = 0$. In addition, we shall assume that constituents of the fluid phase do not react with the solid matrix. From these restrictions the balance equations for mass, momentum and energy reduce to the following

Mass

$$\frac{\partial(\rho_\alpha \varepsilon_\alpha)}{\partial t} + \underset{\sim}{\nabla}.(\rho_\alpha \varepsilon_\alpha \underset{\sim}{V}_\alpha) = 0 \qquad \alpha = f,r \qquad (3.17)$$

Mass of a Fluid Constituent

$$\rho_f \varepsilon_f \frac{D^f \omega_f}{Dt} + \underset{\sim}{\nabla} \cdot (\varepsilon_f \underset{\sim}{J}_f^{\omega}) = 0 \tag{3.18}$$

Momentum

$$\rho_\alpha \varepsilon_\alpha \frac{D^\alpha \underset{\sim}{V}_\alpha}{Dt} + \underset{\sim}{\nabla} \cdot (\varepsilon_\alpha \underset{\approx}{T}_\alpha) - \rho_\alpha \varepsilon_\alpha \underset{\sim}{g} = \overset{\leftrightarrow}{\underset{\sim}{F}}_\alpha \qquad \alpha = f, r \tag{3.19}$$

Energy

$$\rho_\alpha \varepsilon_\alpha \frac{D^\alpha e_\alpha}{Dt} + \varepsilon_\alpha \underset{\approx}{T}_\alpha \cdot \underset{\sim}{\nabla} \underset{\sim}{V}_\alpha + \underset{\sim}{\nabla} \cdot \varepsilon_\alpha \underset{\sim}{J}_\alpha^h + \rho_\alpha \varepsilon_\alpha h_\alpha = \overset{\leftrightarrow}{Q}_\alpha \qquad \alpha = f, r \tag{3.20}$$

where

$$\frac{D^\alpha (\ )}{Dt} \equiv \frac{\partial (\ )}{\partial t} + \underset{\sim}{V}_\alpha \cdot \underset{\sim}{\nabla} (\ ) \tag{3.21}$$

and the subscripts f and r denote the fluid and solid phase, respectively. Into Eqs. (3.18), (3.19) and (3.20), we have introduced the mass balance expression, (3.17). In addition, we have subtracted off a balance of mechanical energy to arrive at the form of the energy balance given in Eq. (3.20). This is carried out through a manipulation of the total energy balance and the equation of linear momentum.

The balance equations are also subject to the following restrictions

$$\sum_\alpha \varepsilon_\alpha = 1 \tag{3.22}$$

$$\overset{\leftrightarrow}{\underset{\sim}{F}}_f + \overset{\leftrightarrow}{\underset{\sim}{F}}_r = 0 \tag{3.23}$$

$$\overset{\leftrightarrow}{\underset{\sim}{F}}_f \cdot (\underset{\sim}{V}_f - \underset{\sim}{V}_r) + \overset{\leftrightarrow}{Q}_f + \overset{\leftrightarrow}{Q}_r = 0 \tag{3.24}$$

Usually for analyses in granular porous media, it is assumed that the fluid and the solid matrix are in a local thermodynamic equilibrium, in which case, only one energy equation is required to describe the thermodynamic state of the medium. It is questionable whether such an assumption can be applied to a fractured rock. The time for the equilibration of temperatures between the fluid and the rock matrix would be directly related to the intensity of fracturing, and thus, the size of the averaging volume.

The development of constitutive relationships for the set of equations given above, and their reduction to the classic form used in simulating transport processes in granular porous media is presented in Hassanizadeh (39) and Hassanizadeh and Gray (42). The development of constitutive relationships for the non-advective (dispersive) contaminant and thermal fluxes is discussed by Shapiro (70,71).

3.2.2. Dual porosity conceptualization

Imposing restriction of the solid and fluid phases similar to those stated at the beginning of section 3.2.1, the balance equations of mass, momentum and energy for the dual porosity conceptualization take the form

Mass

$$\frac{\partial(\rho_\alpha \varepsilon_\alpha)}{\partial t} + \underset{\sim}{\nabla} \cdot (\rho_\alpha \varepsilon_\alpha \underset{\sim}{V}_\alpha) = \overset{\leftrightarrow}{M}_\alpha \qquad \alpha = f,p \tag{3.25}$$

$$\frac{\partial(\rho_r \varepsilon_r)}{\partial t} + \underset{\sim}{\nabla} \cdot (\rho_r \varepsilon_r \underset{\sim}{V}_r) = 0 \tag{3.26}$$

Mass of a Fluid Constituent

$$\rho_\alpha \varepsilon_\alpha \frac{D^\alpha \omega_\alpha}{Dt} + \underset{\sim}{\nabla} \cdot (\varepsilon_\alpha \underset{\sim}{J}^\omega_\alpha) = \overset{\leftrightarrow}{J}^\omega_\alpha \qquad \alpha = f,p \tag{3.27}$$

Momentum

$$\rho_\alpha \varepsilon_\alpha \frac{D^\alpha \underset{\sim}{V}_\alpha}{Dt} + \underset{\sim}{\nabla} \cdot (\varepsilon_\alpha \underset{\approx}{T}_\alpha) - \rho_\alpha \varepsilon_\alpha g = \overset{\leftrightarrow}{\underset{\sim}{F}}_\alpha \qquad \alpha = f,p,r \tag{3.28}$$

Energy

$$\rho_\alpha \varepsilon_\alpha \frac{D^\alpha e_\alpha}{Dt} + \varepsilon_\alpha \underset{\approx}{T}_\alpha \cdot \underset{\sim}{\nabla} \underset{\sim}{V}_\alpha + \underset{\sim}{\nabla} \cdot \varepsilon_\alpha \underset{\sim}{J}^h_\alpha + \rho_\alpha \varepsilon_\alpha h_\alpha = \overset{\leftrightarrow}{Q}_\alpha \qquad \alpha = f,p,r \tag{3.29}$$

where the subscripts f, p and r denote properties of the fluid in the fractures, the fluid in the porous matrix and the rock phase, respectively. The equations above are also subject to the following restrictions

$$\sum_\alpha \varepsilon_\alpha = 1 \tag{3.30}$$

$$\overset{\leftrightarrow}{M}_f = - \overset{\leftrightarrow}{M}_p \tag{3.31}$$

$$\overset{\leftrightarrow}{M}_f(\omega_f - \omega_p) + \sum_\alpha \overset{\leftrightarrow}{J}^\omega_\alpha = 0 \tag{3.32}$$

$$\overset{\leftrightarrow}{M}_f(\underset{\sim}{V}^{fr} - \underset{\sim}{V}^{pr}) + \sum_\alpha \overset{\leftrightarrow}{\underset{\sim}{F}}_\alpha = 0 \tag{3.33}$$

$$\frac{1}{2}\overset{\leftrightarrow}{M}_f((V^{fr})^2 - (V^{pr})^2) + \overset{\leftrightarrow}{\underset{\sim}{F}}_f \cdot \underset{\sim}{V}^{fr} + \overset{\leftrightarrow}{\underset{\sim}{F}}_p \cdot \underset{\sim}{V}^{pr} + \sum_\alpha \overset{\leftrightarrow}{Q}_\alpha = 0 \tag{3.34}$$

where $\underset{\sim}{V}^{\alpha r} \equiv \underset{\sim}{V}_\alpha - \underset{\sim}{V}_r$.

The constitutive relationships developed for the exchanges of mass, momentum and energy between the phases are

$$\overset{\leftrightarrow}{M}_f = \beta_1 \rho_f + \beta_2 \rho_p + \beta_3 \underset{\sim}{\nabla} \cdot \underset{\sim}{V}_f + \beta_4 \underset{\sim}{\nabla} \cdot \underset{\sim}{V}_p \tag{3.35}$$

$$\overset{\leftrightarrow}{J}{}^\omega_f = \theta_1 \omega_f + \theta_2 \omega_p + \theta_3 \underset{\sim}{\nabla} \cdot \underset{\sim}{V}_f + \theta_4 \underset{\sim}{\nabla} \cdot \underset{\sim}{V}_p \tag{3.36}$$

$$\overset{\leftrightarrow}{\underset{\sim}{F}}_f = \underset{\approx}{R}_f \cdot \underset{\sim}{V}^{fr} \tag{3.37}$$

$$\overset{\leftrightarrow}{\underset{\sim}{F}}_p = \underset{\approx}{R}_p \cdot \underset{\sim}{V}^{pr} \tag{3.38}$$

$$\overset{\leftrightarrow}{Q}_f = \sigma_1 T_f + \sigma_2 T_p + \sigma_3 T_r + \sigma_4 \underset{\sim}{\nabla} \cdot \underset{\sim}{V}_f + \sigma_5 \underset{\sim}{\nabla} \cdot \underset{\sim}{V}_p \tag{3.39}$$

$$\overset{\leftrightarrow}{Q}_p = \nu_1 T_f + \nu_2 T_p + \nu_3 T_r + \nu_4 \underset{\sim}{\nabla} \cdot \underset{\sim}{V}_f + \nu_5 \underset{\sim}{\nabla} \cdot \underset{\sim}{V}_p \tag{3.40}$$

where T_α is the temperature of the α-phase, and β_i, θ_i, σ_i, ν_i and $\underset{\approx}{R}_\alpha$ are material coefficients.

Eqs. (3.37) and (3.38) when substituted into Eq. (3.28) written for the fluid in the fractures and the fluid in the porous matrix yields a Darcian relationship for the fluid velocity when inertial and internal stress effects are neglected.

The mass exchange function, (3.35), was analyzed by Shapiro (70) where it was demonstrated that $\beta_1 = -\beta_2$ where $\beta_2 > 0$. The significance of β_3 and β_4 was also shown to be negligible due to the rapidity at which fluid pressures equilibrated between the porous matrix and the fractures.

The presence of $\underset{\sim}{\nabla} \cdot \underset{\sim}{V}_f$ and $\underset{\sim}{\nabla} \cdot \underset{\sim}{V}_p$ in Eqs. (3.35), (3.36), (3.39) and (3.40) illustrates that there are additional terms which arise in the development of these constitutive relationships other than those which lead to a quasi-steady exchange function (see Eqs. (1.5) and (1.9)).

The transfer of mass or energy between the porous matrix and fractures is a phenomenon that depends on the previous history of the transport process between the two flow regimes. If we observe the microscopic interface between a single fracture and a porous block, the rate of mass or energy entering the porous matrix from the fractures depends on the distribution of mass or energy initially within the porous matrix. The distribution of mass or energy within the porous matrix, however, depends on the previous history of transport and the geometry of the porous matrix. The quasi-steady approximation is adequate when temperatures or concentrations of the porous matrix and fractures are nearly the same, i.e., small perturbations from equilibrium. For large perturbations from equilibrium, the rock would have to be highly fractured with matrix blocks of small dimensions for equilibration to occur rapidly enough to use the quasi-steady approximation.

The terms $\underset{\sim}{\nabla}\cdot\underset{\sim}{V}_f$ and $\underset{\sim}{\nabla}\cdot\underset{\sim}{V}_p$ may not fully account for the time history of the transport process. Constitutive relationships could also be developed where $\partial\omega_f/\partial t$ and $\partial\omega_p/\partial t$ would appear in Eq. (3.36), and similar terms defined for temperature would appear in Eq. (3.39) and (3.40). The significance of the variable size matrix blocks would arise in the material coefficients.

3.3. Continuum Coefficients

The continuum coeffficients, which arise in the systematic procedure of hypothesizing constitutive relationships, are referred to as material coefficients since they ultimately depend on the specific medium under investigation. The Taylor series expansions define these material coefficients only in a thermodynamic sense. Hence, we gain very little physical insight into the relative significance of these quantities. Only through physical experimentation are we able to determine their relative magnitude so that the significance of the terms appearing in the constitutive relationships can be evaluated. Usually we extract the material coefficients through experimentation by subjecting the medium to a continuum excitation and measuring the induced continuum responses. The material coefficeints are then evaluated by substituting both the responses and the excitation into the field equations that are assumed to describe the physics of transport in the continuum.

In fractured rock, we are faced with many difficulties in evaluating continuum coefficients from physical experiments. The majority of these problems arise because measurements may not be indicative of the continuum hypothesis. For example, the interpretation of measurements for their application in evaluating continuum coefficients is questionable in rock formations where the density of fracturing is sparse in comparison with the length scale of a measuring device. Yet in such formations on a scale much larger than the measuring device, the continuum hypothesis may be applicable. Furthermore, even if a measuring instrument does intersect a significant number of fractures such that it can possibly identify a continuum response, the measurement obtained may not be a representative sample. For example, if we wish to extract a sample of a contaminant from the fluid in the void space through a borehole, the sample will most likely be representative of the fluid in the fractures since this flow regime is more conductive than the porous matrix. If we employ the equivalent porous medium conceptualization, then such a measurement is not indicative of the fluid in all void space. For the dual porosity conceptualization, this measurement may be indicative of the fracture fluid continuum, but we are still left with the problem of identifying the response in the porous matrix.

As a result of these difficulties, investigators have tried to estimate continuum parameters by methods which allow the use of measurements conducted over volumes that are smaller than the REV. If we do have a knowledge of the microstructure of the medium, it is only to our benefit to incorporate this knowledge into the continuum conceptualization. Yet it must be done in a manner that is consistent with the continuum hypothesis and the notion of an averaging volume. Other ways of incorporating the microstructure are more heuristic in nature, and first require a knowledge of the continuum coefficient, e.g., correlating hydraulic conductivity with grain size distribution or fracture intensity.

In the remainder of this section we shall briefly describe the means by which the continuum hydraulic conductivity of a fractured rock can be determined by volume averaging in conjunction with a knowledge of the fracture geometry. Such considerations require assumptions concerning the nature of fluid movement in the fractures; therefore, the applicability of any results must be viewed in the light of these assumptions. Furthermore, it is not the intension of these results to yield exact values of the hydraulic conductivity tensor, but rather they are intended to provide an order of magnitude estimate for this continuum coefficient.

3.3.1. Evaluating hydraulic conductivity from fracture geometry

For the purpose of this discussion, we shall assume that only fractures (from any number of fracture families) constitute the entire void space of the rock. Thus, the host rock is considered to be impervious, and to possess no void space which is characteristic of a granular porous material. In addition, each fracture is assumed to have a variable aperture, however, we also assume that the fractures may be represented by a series of parallel segments (Figure 11). Furthermore, we consider a single fluid to occupy the entire void space of the rock, and the rock to be non-deformable.

With these assumptions, we may draw upon the results presented in Section 2 for the mathematical modeling of fluid movement in a discrete fracture conceptualization. For each parallel fracture segment, the average fluid velocity in the plane of the fracture is given by (see Eqs. (2.43) and (2.44))

$$\underset{\sim}{v} = - \kappa \, \underset{\sim}{\nabla}\bar{\phi} \quad ; \qquad \kappa = \frac{\rho g b^2}{12\mu} \tag{3.41}$$

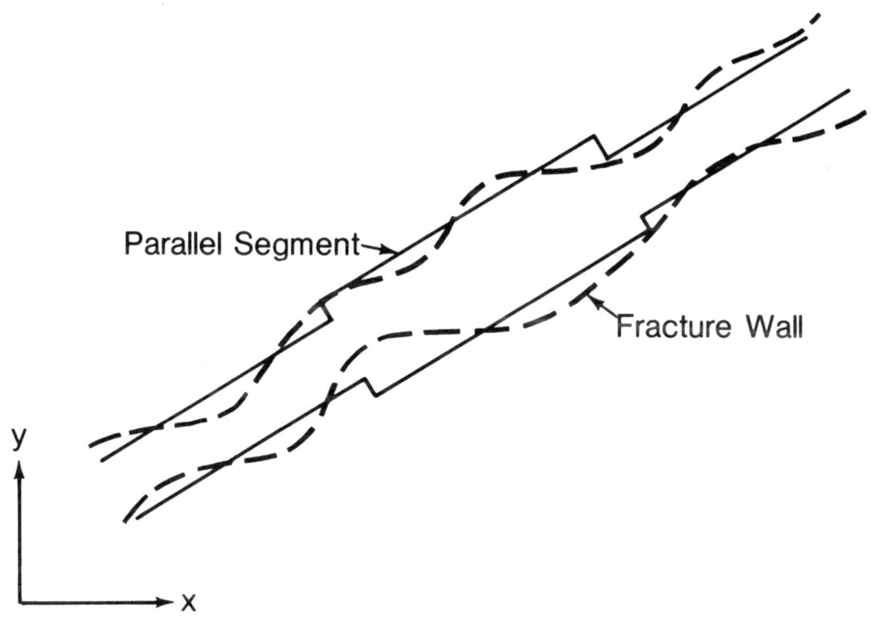

Figure 11. Conceptual model of a fracture as composed of a series of parallel segments.

Here, velocity and the hydraulic head are quantities averaged over the fracture aperture.

The velocity may also be defined with respect to a fixed coordinate system, i.e.,

$$\bar{\underset{\sim}{v}}' = - \underset{\approx}{k} \cdot \underset{\sim}{\nabla} \bar{\phi} \tag{3.42}$$

where $\underset{\approx}{k}$ is a tensor defined as (77)

$$\underset{\approx}{k} = k_{ij} = \kappa(\delta_{ij} - n_i n_j) \quad ; \quad i,j = 1,2,3 \tag{3.43}$$

where δ_{ij} is the Kronecker delta and n_i and n_j are the direction cosines in the i and j directions, respectively, for the vector normal to the fracture axis.

Eq. (3.42) is the result of integrating the equation of linear momentum over the fracture aperture. The fracture network is subsequently described by surfaces in three-dimensional space, where the boundaries of the fractures are defined by lines in three-dimensional space. Since we are considering an impervious host rock, a no-flow condition is imposed at these boundaries, i.e.,

$$\bar{\underset{\sim}{v}}' \cdot \underset{\sim}{\nu} = 0 \qquad \text{on } L_{fr} \tag{3.44}$$

where L_{fr} is the line in three-dimensional space defining the fracture boundary, and $\underset{\sim}{\nu}$ is the outwardly directed unit normal vector to this boundary (Figure 12).

Following the method of volume averaging, we shall integrate Eq. (3.42) over the void space within the REV, and analyze the coefficient that is analogous to the hydraulic conductivity. Since an initial average has been conducted over the fracture aperture, we need to consider the integration of Eq. (3.42) over the fracture surfaces, i.e.,

$$\underset{\sim}{q} \equiv n\underset{\sim}{V} = \frac{1}{dV} \int_{dA} b\, \bar{\underset{\sim}{v}}' da = -\frac{1}{dV} \int_{dA} b\, \underset{\approx}{k} \cdot \underset{\sim}{\nabla} \bar{\phi}\, da \tag{3.45}$$

where the surfaces of all fractures in the REV are denoted by dA, dV is the volume of the REV, $\underset{\sim}{q}$ is the average volumetric flow rate in the fractures per unit volume of the REV, n is the fracture porosity, and $\underset{\sim}{V}$ is the average volumetric flow rate per unit volume of the fractures in the REV.

Since we assume that a sufficient amount of interconnected fractures exist within the REV, in the absence of fluid sources and sinks, it is reasonable to assume that $\underset{\sim}{\nabla}\bar{\phi}$ does not change sign within the REV and is monotonously increasing or decreasing. Using the mean value theorem, we may rewrite Eq. (3.45) as

$$\frac{1}{dV} \int_{dA} b\underset{\approx}{k} \cdot \underset{\sim}{\nabla}\bar{\phi} da = \frac{[\underset{\sim}{\nabla}\bar{\phi}(\underset{\sim}{X})]}{dV} \cdot \int_{dA} b\underset{\approx}{k} da \tag{3.46}$$

where $\underset{\sim}{\nabla}\bar{\phi}(\underset{\sim}{X})$ is a value of $\underset{\sim}{\nabla}\bar{\phi}$ within the REV that yields an exact evaluation of (3.45). We shall approximate $\underset{\sim}{\nabla}\bar{\phi}(\underset{\sim}{X})$ by the average value, such that Eq. (3.45) can be written

$$n\underset{\sim}{V} \cong \left(\frac{1}{dV} \int_{dA} b\underset{\approx}{k} da \right) \cdot \left(\frac{1}{dA} \int_{dA} \underset{\sim}{\nabla}\phi da \right) = \underset{\approx}{\Lambda} \cdot \overline{\underset{\sim}{\nabla}\phi} \tag{3.47}$$

where $\overline{\underset{\sim}{\nabla}\phi}$ is the average of the gradients of the fluid potential over the entire REV, i.e.,

$$\overline{\underset{\sim}{\nabla}\phi} = \frac{1}{dA} \int \underset{\sim}{\nabla}\bar{\phi} da \tag{3.48}$$

and $\underset{\approx}{\Lambda}$ is the average of the fracture conductivities of the parallel fracture segments. The quantity $\underset{\approx}{\Lambda}$ is purely related to the geometric features of the fractures.

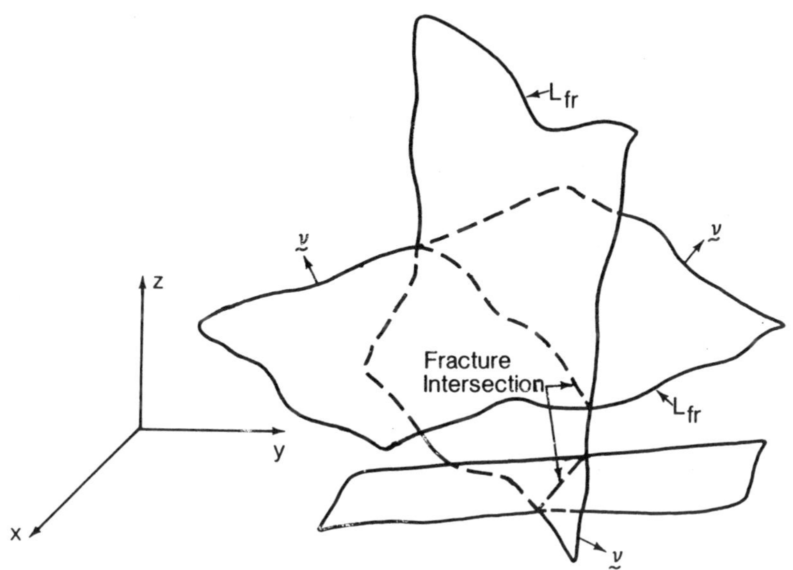

Figure 12. Areal fracture surfaces in three-dimensional space.

Eq. (3.47) is the form that Snow (77) has used to estimate the hydraulic conductivity of a fractured rock, where $\underset{\approx}{\Lambda}$ is considered to be the hydraulic conductivity. Eq. (3.47), however, is not consistent with what we usually consider as a definition for the macroscopic fluid flux. The driving force for fluid movement at the continuum level is $\underset{\sim}{\nabla}\Phi$ and not $\overline{\underset{\sim}{\nabla}\phi}$, where Φ is the fluid potential averaged over the REV, i.e.,

$$\Phi = \frac{1}{dA} \int_{dA} \bar{\phi} \, da \tag{3.49}$$

The correct form of the macroscopic flux is

$$n\underset{\sim}{V} = - \underset{\approx}{K} \cdot \underset{\sim}{\nabla}\Phi \tag{3.50}$$

where $\underset{\approx}{K}$ is the hydraulic conductivity tensor.

In order to arrive at Eq. (3.50), we must develop a relationship between $\overline{\underset{\sim}{\nabla}\phi}$ and $\underset{\sim}{\nabla}\Phi$. Shapiro and Bear (75) demonstrated that $\overline{\underset{\sim}{\nabla}\phi}$ is related to $\underset{\sim}{\nabla}\Phi$ by the following

$$\overline{\underset{\sim}{\nabla}\phi} = \underset{\approx}{\Omega} \cdot \underset{\sim}{\nabla}\Phi \tag{3.51}$$

where $\underset{\approx}{\Omega}$ is a geometric property of the medium defined as

$$\underset{\approx}{\Omega} = \underset{\approx}{I} - \frac{1}{dA} \int_{L_{fr}} \underset{\sim}{x}' \underset{\sim}{\nu} \, dl \tag{3.52}$$

where $\underset{\approx}{I}$ is the unit tensor, $\underset{\sim}{x}'$ the position vector measured from the centroid of the REV and $\underset{\sim}{\nu}$ is the outwardly directed unit normal vector from the lines L_{fr}. The quantity $\underset{\approx}{\Omega}$ is a tensorial coefficient which defines the tortuous path of the fluid within the REV. The hydraulic conductivity tensor of the fractured rock at the point in question is then defined as

$$\underset{\approx}{K} = \underset{\approx}{\Lambda} \cdot \underset{\approx}{\Omega} \tag{3.53}$$

Eq. (3.53) exhibits two effects which contribute to the hydraulic conductivity tensor. The first, $\underset{\approx}{\Lambda}$, is the result of the orientation of the individual fractures within the REV. The quantity, $\underset{\approx}{\Omega}$, in contrast, denotes the effect that arises from the gradients of the fluid potential in the individual fractures that are not necessarily the same as the gradient of the average fluid potential, which is the driving force for fluid movement.

4. SUMMARY AND CONCLUSIONS

Due to the diverse nature of fracturing in rock formations, two distinct means of mathematically conceptualizing transport phenomena in fractured rock have arisen. These are a discrete fracture conceptualization and a continuum conceptualization of rock formations.

The discrete fracture conceptualization is best suited to those situations where there are few significant fractures of a given fracture family within a problem area. To accurately describe transport phenomena under such conditions, the explicit geometrical features of these few fractures must be identified. Because of the difficulty in defining the fracture geometry under field situations, it is only possible to describe these few fractures in a statistical manner.

The fractures constitute a three-dimensional void space within the rock. However, for simplicity in mathematically describing transport phenomena within the fractures, we use the fact that the characteristic length of a fracture is significantly larger than its associated width. The transport of mass, momentum and energy is then assumed to be essentially two-dimensional, in the surface that depicts the center line of the fracture aperture.

Under this assumption, the equations of transport of mass, momentum and energy are developed by integrating over the fracture aperture the three-dimensional transport equations that are valid within a single fracture. The resulting two-dimensional equations are written in terms of quantities averaged over the fracture thickness. From this averaging procedure, properties such as fracture conductivity arise. For parallel fracture walls, the conductivity of the fracture is proportional to the square of the fracture aperture, and the volumetric flux is proportional to the cube of the fracture aperture. If the fracture walls are permeable, the conductivity of the fracture is dependent on the volumetric flux entering or leaving the fracture.

Incorporated into the two-dimensional equations of transport within a single fracture are conditions of transport at the fracture walls. These conditions of mass, momentum, or energy cannot be arbitrarily prescribed at the fracture walls since they depend on the responses in the adjacent host rock. Thus, the equations of transport within a single fracture must be solved in conjunction with similar equations of transport in the host rock.

In contrast to the discrete fracture conceptualization, a continuum conceptualization treats fractured rock in a similar manner to our treatment of granular porous media. The explicit geometry of the void space is neglected, and distributed parameters are employed in the description of the transport processes. These parameters, such as the hydraulic conductivity, are actually the manifestation of averaging the geometric and material properties of the void space over a Representative Elementary Volume of the medium.

A natural extension of the continuum theories applied to granular porous material is to treat fractured formations in a similar manner, i.e., all void space of the rock are treated as one continuum and considered as an equivalent porous medium. Such an approach has proved to be acceptable in the description of single phase, isothermal fluid movement. For contaminant and thermal migration, and multi-phase flow, however, a single continuum velocity may not be descriptive of the diverse character of the void space within a fractured formation, i.e., fractures of different families and even void space that is similar to a granular porous material. An alternative continuum conceptualization of fractured rock is to consider two overlapping fluid continua, for example, the fluid in the fractures and the fluid in the porous matrix of the host rock. Although this continuum conceptualization may better describe the physics of transport in a fractured formation, it offers an added complexity in that additional constitutive relationships and continuum coefficients require evaluation.

The Coleman and Noll (15) procedure provides a systematic and unbiased means of thermodynamically formulating constitutive relationships. The constitutive relationships developed for the (continuum) dual porosity conceptualization illustrate that the exchange of energy and contaminant mass between the fractures and porous matrix may be governed by more than a quasi-steady approximation represented by the difference between the temperatures or concentrations of the two fluid continua.

The constitutive functions developed in Section 3 are only hypothesized since they are based on an assumed functional dependence that was initially imposed on all constitutive functions. The validity of these phenomenological relationships can only be estimated through experimentation. Ideally, the development of constitutive functions should be an iterative process, i.e., a set of constitutive functions are hypothesized, their validity is then checked through experimentation, and the original hypotheses are reevaluated.

For porous materials that are granular in nature, experimental information has been available long before any rigorous attempts were made at developing the field equations for transport phenomena. Hence, the experimental information acted as a guide and a means of verifying any proposed theories. For continuum conceptualizations of fractured rock, however, the quantity of experimental information (except possibly for single phase, isothermal fluid movement) is minute. In addition, the quality of experimental information for the continuum conceptualizations is extremely questionable. Thus, for the verification of the continuum field equations for fractured rock, we do not have the luxury of experimental information. Consequently, unanswered questions still remain concerning the functional forms that describe the exchange of contaminant mass and energy between fractures and a porous matrix, or between fractures of different families. In addition, the assumption of thermal equilibrium between continuum phases (which is an assumption employed in granular porous media) is questionable in its application to fractured rock formations.

The diverse character of fracturing under natural conditions will most likely prohibit us from ever proposing generic answers to these and other questions concerning the continuum conceptualizations of fractured rock. It is more reasonable to assume that we must quantify the physical situations where given assumptions may be applied in the characterization of transport phenomena in fractured rock formations.

5. APPENDIX A: RELATIONSHIP BETWEEN $\overline{\nabla \phi}$ and $\nabla \Phi$

Let us assume that the fracture aperture is essentially uniform over a given fracture surface such that the equation governing fluid movement in the fracture reduces to

$$\nabla^2 \bar{\phi} = 0 \tag{A.1}$$

where $\bar{\phi}$ is the hydraulic head averaged over the fracture aperture. Additionally, let us consider the following integration over the areal fracture surfaces, dA, within the REV,

$$\frac{1}{dA}\int_{dA} \underset{\sim}{x}' \nabla^2\phi \, da = 0 \tag{A.2}$$

where $\underset{\sim}{x}'$ is the position vector measured from the centroid of the REV.

Applying the first form of Green's theorem to Eq. (A.1) yields

$$\frac{1}{dA}\int_{dA} \underset{\sim}{x}'\nabla^2\bar{\phi} \, da = \frac{1}{dA}\int_{L_{ff}+L_{fr}} \underset{\sim}{x}'\underset{\sim}{\nabla}\bar{\phi}.\underset{\sim}{\nu} \, dl - \frac{1}{dA}\int_{dA} \underset{\sim}{\nabla}\bar{\phi}.\underset{\sim}{\nabla}\underset{\sim}{x}' \, da \tag{A.3}$$

where L_{fr} is the line defining the boundaries of the fracture surfaces within the REV, L_{ff} is the line defining the intersection of the fracture surfaces and the surface of the REV and $\underset{\sim}{\nu}$ is the outwardly directed unit normal vector from the lines L_{fr} and L_{ff}. The quantity $\underset{\sim}{\nabla}\bar{\phi}.\underset{\sim}{\nu}$ in the line integrals of Eq. A.1) is proportional to the fluid flux perpendicular to the boundaries L_{ff} and L_{fr}. Because we have assumed that the host rock is impervious, the line integral over L_{fr} vanishes, and Eq.(A.3) reduces to

$$\frac{1}{dA}\int_{dA} \frac{\partial\phi}{\partial x_j} \, da - \frac{1}{dA}\int_{L_{ff}} x_j' \frac{\partial\phi}{\partial x_i} \nu_i \, dl \tag{A.4}$$

where we have made use of the identity

$$\underset{\sim}{\nabla}x' \equiv \frac{\partial x_i}{\partial x_j} = \delta_{ij} \tag{A.5}$$

where δ_{ij} is the Kronecker delta.

The first term in Eq. (A.4) is the average of the microscopic gradients of the fluid potential, $\overline{\underset{\sim}{\nabla}\phi}$, which is defined solely in terms of the fluid flux through the fractures at the boundary of the REV,

$$\overline{\underset{\sim}{\nabla}\phi} = \frac{1}{dA}\int_{L_{ff}} \underset{\sim}{x}' \, \underset{\sim}{\nabla}\bar{\phi}.\underset{\sim}{\nu} \, dl \tag{A.6}$$

Because we had assumed earlier that $\underset{\sim}{\nabla}\bar{\phi}$ is either positive or negative at all points on all fracture surfaces within the REV, $\underset{\sim}{\nabla}\bar{\phi}$ will not change sign on the boundary L_{ff}. Using the mean value theorem, we can then approximate the line integral in Eq. (A.6) by

$$\overline{\underset{\sim}{\nabla}\phi} = \frac{[\underset{\sim}{\nabla}\bar{\phi}(\underset{\approx}{X})]}{dA} \cdot \int_{L_{ff}} \underset{\approx}{x}'\underset{\sim}{\nu}\, dl \tag{A.7}$$

where $\underset{\sim}{\nabla}\bar{\phi}(\underset{\approx}{X})$ is a value of $\underset{\sim}{\nabla}\bar{\phi}$ on the line L_{ff} that yields an exact evaluation of equation (A.7). We shall approximate $\underset{\sim}{\nabla}\bar{\phi}(\underset{\approx}{X})$ by $\underset{\sim}{\nabla}\Phi$, where Φ is the average of the hydraulic head in the fluid within the REV,

$$\Phi = \frac{1}{dA} \int_{dA} \bar{\phi}\, da \tag{A.8}$$

Eq. (A.7) can then be written

$$\overline{\underset{\sim}{\nabla}\phi} = \underset{\approx}{\Omega}.\underset{\sim}{\nabla}\Phi \tag{A.9}$$

where $\underset{\approx}{\Omega}$ is a geometric property of the medium,

$$\underset{\approx}{\Omega} = \frac{1}{dA} \int_{L_{ff}} \underset{\sim}{x}'\underset{\sim}{\nu}\, dl \tag{A.10}$$

The quantity $\underset{\approx}{\Omega}$ can also be related to the boundary L_{fr} through the following identity

$$\underset{\approx}{I} = \frac{1}{dA} \int_{L_{fr}+L_{ff}} \underset{\sim}{x}'\underset{\sim}{\nu}\, dl = \underset{\approx}{\Omega} + \frac{1}{dA} \int_{L_{fr}} \underset{\approx}{x}'\underset{\sim}{\nu}\, dl \tag{A.11}$$

Thus, $\underset{\approx}{\Omega}$ can be defined as

$$\underset{\approx}{\Omega} = \underset{\approx}{I} - \frac{1}{dA} \int_{L_{fr}} \underset{\approx}{x}'\underset{\sim}{\nu}\, dl \tag{A.12}$$

where $\underset{\approx}{I}$ is the unit tensor.

6. REFERENCES

1. Andersson, J., Shapiro, A.M., and Bear, J., "A Stochastic Model of a Fractured Rock Conditioned by Measured Information," *Water Resources Research*, Vol. 20, 1984, pp. 79-88.
2. Barenblatt, G.I., and Zheltov, Yu.P., "Fundamental Equations of Filtration of Homogeneous Liquids in Fissured Rocks," *Soviet Physics - Doklady*, Vol. 5, 1960, pp. 522-525.
3. Barenblatt, G.I., Zheltov, Yu.P., and Kochina, I.N., "Basic Concepts in the Theory of Seepage of Homogeneous Liquids in

Fissured Rocks," *Journal of Applied Mathematics and Mechanics,* Vol. 24, 1960, pp. 1286-1303.

4. Bear, J., *Dynamics of Fluids in Porous Media,* Elsevier, New York, N.Y., 1972.
5. Bear, J., *Hydraulics of Groundwater,* McGraw-Hill, New York, N.Y., 1979.
6. Bear, J., and Bachmat, Y., "Transport Phenomena in Porous Media-Basic Equations," *Fundamental of Transport Phenomena in Porous Media,* J. Bear and M.Y.Corapcioglu, eds., Martinus Nijhoff Publishers, Dordrecht, 1984, pp. 3-61.
7. Bear, J., and Braester, C., "On the Flow of Two Immiscible Fluid in Fractured Porous Media," *Proceedings of the Second Symposium on Fundamentals of Transport Phenomena in Porous Media,* University of Guelph, 1972, pp. 177-202.
8. Bird, R.B., Stewart, W.E., and Lightfoot, E.N., *Transport Phenomena,* John Wiley and Sons, Inc., New York, N.Y., 1960.
9. Bokserman, A.A., Zheltov, Yu.P., and A.A. Kocheshkov, "Motion of Immiscible Liquids in a Cracked Porous Medium," *Soviet Physics - Doklady,* Vol. 9, 1964, pp. 285-287.
10. Boulton, A.S., and Streltsova, T.D., "Unsteady Flow to a Pumped Well in a Fissured Water Bearing Formation," *Journal of Hydrology,* Vol. 35, 1977, pp. 257-269.
11. Bowen, R.M., "Theory of Mixtures," *Continuum Physics,* A.C. Eringen, ed., Vol. III, Academic Press Inc., New York, N.Y., 1976, pp. 2-127.
12. Braester, C., "A Shock Wave in Immiscible Displacement in a Fissured Porous Medium," *Israel Journal of Technology,* Vol. 9, 1971, pp. 433-438.
13. Brown, D.M., "Stochastic Analysis of Flow and Solute Transport in a Variable-Aperture Rock Fracture," thesis presented to the Massachusetts Institute of Technology, Cambridge, Massachusetts, in 1984, in partial fulfillment of the requirements for the degree of Master of Science.
14. Chu, Y., and Gelhar, L.W., "Turbulent Pipe Flow With Granular Permeable Boundaries," *Report No. 148,* Dept. of Civil Engineering, Massachusetts Institute of Technology, 1972.
15. Coleman, B.D., and Noll, W., "Thermodynamics of Elastic Materials With Heat Conduction and Viscosity," *Archive for Rational Mechanics and Analysis,* Vol. 13, 1963, pp. 167-178.
16. Crawford, P.B., and Collins, R.E., "Estimated Effect of Vertical Fractures on Secondary Recovery," *Transactions,* American Institute of Mining Engineers, Vol. 201, 1954, pp. 192-196.
17. Davis, S.N., "Porosity and Permeability of Natural Materials," *Flow Through Porous Media,* R.J.M. DeWiest, ed., Academic Press, New York, N.Y., 1969, pp. 54-89.
18. de Swann O.A., "Analytical Solutions for Determining Naturally Fractured Reservoir Properties by Well Testing," *Transactions,* American Institute of Mining Engineers, Vol. 261, 1976, pp. 117-122.

19. Dougherty, D.E, "On Equivalent Porous Media Modeling of Transport in Fractured Porous Reservoirs," thesis presented to Princeton University, Princeton, N.J., in 1985, in partial fulfillment of the requirements for the degree of Doctor of Philosophy.
20. Dougherty, D.E., and Babu, D.K., "Flow to a Partially Penetrating Well in a Double-Porosity Reservoir," *Water Resources Research*, Vol. 20, 1984, pp. 1116-1122.
21. Drew, D.A., "Averaged Field Equations for Two-Phase Media," *Studies in Applied Mathematics*, Vol. 50, 1971, pp. 133-166.
22. Drew, D.A., and Segel, L.A., "Averaged Equations for Two-Phase Flows," *Studies in Applied Mathematics*, Vol. 50, 1971, pp. 205-231.
23. Drumheller, D.S., and Bedford, A., "A Thermomechanical Theory for Reacting Immiscible Mixtures," *Archive for Rational Mechanics and Analysis*, Vol. 73, 1980, pp. 257-284.
24. Duguid, J.O., and Lee, P.C.Y., "Flow in Fractured Porous Media," *Water Resources Research*, Vol. 13, 1977, pp. 558-566.
25. Elkins, L.F., "Reservoir Performance and Well Spacing, Spraberry Trend Area Field of West Texas," *Transactions*, American Institute of Mining Engineers, Vol. 198, 1953, pp. 177-196.
26. Elkins, L.F., and Skov, A.M., "Determination of Fracture Orientation From Pressure Interference," *Transactions*, American Institute of Mining Engineers, Vol. 219, 1960, pp. 301-304.
27. Eringen, A.C., *Mechanics of Continua*, John Wiley and Sons, Inc., New York, N.Y., 1967.
28. Gale, J.E., Taylor, R.L., Witherspoon, P.A., and Ayatollahi, M.S., "Flow in Rocks with Deformable Fractures," *Finite Element Methods in Flow Problems*, J.T. Oden, O.C. Zienkiewicz, R.H. Gallagher, C. Taylor, eds., University of Alabama Press, Huntsville, Alabama, 1974, pp. 583-598.
29. Givens, J.W., and Crawford, P.B., "Effect of Isolated Vertical Fractures Existing in the Reservoir on Fluid Displacement Response," *Transactions*, American Institute of Mining Engineers, Vol. 237, 1966, pp. 81-86.
30. Gray, W.G., and Lee, P.C.Y., "On the Theorems for Local Volume Averaging of Multi-Phase Systems," *International Journal of Multi-Phase Flow*, Vol. 3, 1977, pp. 333-340.
31. Gringarten, A.C., "Flow Test Evaluation of Fractured Reservoirs," *Recent Trends in Hydrogeology*, Geological Society of America Special Paper 189, T.N. Narasimhan, ed., 1982, pp. 237-263.
32. Gringarten, A.C., and Ramey, H.J., "Unsteady-State Pressure Distribution Created by a Well With a Single Horizontal Fracture Partial Penetration or Restricted Entry," *Transactions*, American Institute of Mining Engineers, Vol. 257, 1974, pp. 413-426.
33. Gringarten, A.C., Ramey, H.J., and Raghavan, R., "Unsteady-State Pressure Distribution Created by a Well With a Single Infinite-Conductivity Vertical Fracture," *Transactions*, American

Institute of Mining Engineers, Vol. 257, 1974, pp. 347-360.

34. Gringarten, A.C., Ramey, H.J., and Raghavan, R., "Applied Pressure Analysis for Fractured Wells," *Transactions*, American Institute of Mining Engineers, Vol. 259, 1975, pp. 887-892.

35. Grisak, G.E., and Cherry, J.A., "Hydrologic Characteristics and Response of Fractured Till and Clay Confining a Shallow Aquifer," *Canadian Geotechnical Journal*, Vol. 12, 1975, pp. 23-43.

36. Grisak, G.E., and Pickens, J.F., "Solute Transport Through Fractured Media: 1. The Effect of Matrix Diffusion," *Water Resources Research*, Vol. 16, 1980, pp. 719-730.

37. Grove, D.B., and Beetem, W.A., "Porosity and Dispersion Constant Calculations for a Fractured Carbonate Aquifer Using Two Well Tracer Method," *Water Resources Research*, Vol. 7, 1971, pp. 128-134.

38. Hartsock, J.H., and Warren, J.E., "The Effect of Horizontal Hydraulic Fracturing on Well Performance," *Journal of Petroleum Technology*, Vol. 13, 1961, pp. 1050-1056.

39. Hassanizadeh, M., "Macroscopic Description of Multi-Phase Systems - A Thermodynamic Theory of Flow in Porous Media," thesis presented to Princeton University, Princeton, N.J., in 1980, in partial fulfillment of the requirements for the degree of Doctor of Philosophy.

40. Hassanizadeh M., and Gray, W.G., "General Conservation Equations for Multi-Phase Systems 1. Averaging Procedure," *Advances in Water Resources*, Vol. 2, 1979, pp. 131-144.

41. Hassanizadeh, M., and Gray, W.G., "General Conservation Equations for Multi-Phase Systems 2. Mass, Momenta, Energy and Entrophy Equations," *Advances in Water Resources*, Vol. 2, 1979, pp. 191-203.

42. Hassanizadeh, M., and Gray, W.G., "General Conservation Equations for Multi-Phase Systems 3. Constitutive Theory for Porous Media Flow," *Advances in Water Resources*, Vol.3, 1980, pp. 25-40.

43. Hubbert, M.K., and Willis, D.G., "Mechanics of Hydraulic Fracturing," *Underground Waste Management and Environmental Implications*, American Association of Petroleum Geologists Memoir 18, 1972, pp. 239-257.

44. Huyakorn, P.S., Lester, B.H., and Mercer, J.W., "An Efficient Finite Element Technique for Modeling Transport in Fractured Porous Media: 1. Single Species Transport," *Water Resources Research*, Vol. 19, 1983, pp. 841-854.

45. Huyakorn, P.S., Lester, B.H., and Mercer, J.W., "An Efficient Finite Element Technique for Modeling Transport in Fractured Porous Media: 2. Nuclide Decay Chain Transport," *Water Resources Research*, Vol. 19, 1983, pp. 1286-1296.

46. Ingram, J.D., and Eringen, A.C., "A Continuum Theory of Chemically Reacting Media - II. Constitutive Equations of Reacting Fluid Mixtures," *International Journal of Engineering*

Science, Vol. 5, 1967, pp. 289-322.

47. James, R.V., and Rubin, J., "Transport of Chloride in a Water-Unsaturated Soil Exhibiting Anion Exclusion," presented at the December 3-7, 1984, American Geophysical Union Fall Meeting, San Francisco, Califormia.
48. Kazemi, H., "Pressure Transient Analysis of Naturally Fractured Reservoirs with Uniform Fracture Distribution," *Society of Petroleum Engineers Journal*, Vol. 9, 1969, pp. 451-462.
49. Kazemi, H., Seth, M.S., and Thomas, G.W., "The Interpretation of Interference Tests in Naturally Fractured Reservoirs with Uniform Fracture Distribution," *Society of Petroleum Engineers Journal*, Vol. 9, 1969, pp. 463-472.
50. Kazemi, H., Merrill, L.S., Potterfield, K.L., and Zelman, P.R., "Numerical Simulation of Water-Oil Flow in Naturally Fractured Reservoirs," *Transactions*, American Institute of Mining Engineers, Vol. 261, 1976, pp. 317-326.
51. Kiraly, L., "Groundwater Flow in Heterogeneous, Anisotropic Fractured Media: A Simple Two-Dimensional Electric Analog," *Journal of Hydrology*, 1971, pp. 255-261.
52. Lewis, D.C., and Burgy, R.H., "Hydraulic Characteristics of Fractured and Jointed Rock," *Groundwater*, Vol. 2, 1964, pp. 4-9.
53. Lightfoot, E.N., and Cussler, E.L., "Diffusion in Liquids," *Selected Topics in Transport Phenomena*, Chemical Engineering Progress Symposium Series, American Institute of Chemical Engineers, No. 58, Vol. 61, 1965.
54. Long, R.R., *Mechanics of Solids and Fluids*, Prentice Hall, Englewood Cliffs, N.J., 1961.
55. Long, J.C.S., Remer, J.S., Wilson, C.R., and Witherspoon, P.A., "Porous Media Equivalents for Networks of Discontinuous Fractures," *Water Resources Research*, Vol. 18, 1982, pp. 645-658.
56. McGuire, W.J., and Sikora, V.J., "The Effect of Vertical Fractures on Well Productivity," *Transactions*, American Institute of Mining Engineers, Vol. 219, 1960, pp. 401-403.
57. Muller, I., "A Thermodynamic Theory of Mixtures of Fluids," *Archive for Rational Mechanics and Analysis*, Vol. 28, 1968, pp. 1-39.
58. Muller, I., "Thermodynamics of Mixtures of Fluids," *Journal de Mecanique*, Vol. 14, 1975, pp. 267-303.
59. Munoz Goma, R.J., and Gelhar, L.W., "Turbulent Pipe Flow With Rough Porous Walls," *Report No. 109*, Dept. of Civil Engineering, Massachusetts Institute of Technology, 1968.
60. Narasimhan, T.N., "Multidimensional Numerical Simulation of Fluid Flow in Fractured Porous Media," *Water Resources Research*, Vol. 18, 1982, pp. 1235-1247.
61. Narasimhan, T.N., and Palen, W.A., "A Purely Numerical Approach for Analyzing Fluid Flow to a Well Intercepting a

Vertical Fracture," presented at the 1979, AIME California Regional Meeting of the Society of Petroleum Engineers, held at Ventura, California.

62. Neretnieks, I. and Rasmuson, A., "An Approach to Modeling Radionuclide Migration in a Medium with Strongly Varying Velocity and Block Sizes Along the Flow Path," *Water Resources Research*, Vol. 20, 1984, pp. 1823-1836.
63. Neuzil, C.E., and Tracy, J.V., "Flow in Fractures," *Water Resources Research*, Vol. 17, 1981, pp. 191-199.
64. Noorishad, J., and Mehran, M., "An Upstream Finite Element Method for Solution of Transient Transport Equation in Fractured Porous Media," *Water Resources Research*, Vol. 18,1982, pp. 588-596.
65. Nunziato, J.W., and Walsh, E.K., "On Ideal Multi-Phase Mixtures With Chemical Reactions and Diffusion," *Archive For Rational Mechanics and Analysis*, Vol. 73, 1980, pp. 285-311.
66. O'Neill,K., "The Transient Three-Dimensional Transport of Liquid and Heat in Fractured Porous Media," thesis presented to Princeton University, Princeton, N.J., in 1977, in partial fulfillment of the requirements for the degree of Doctor of Philosophy.
67. Rasmuson, A., Narasimhan, T.N., and Neretnieks, I., "Chemical Transport in Fissured Rock; Verification of a Numerical Model," *Water Resources Research*, Vol. 18, 1982, pp. 1479-1492.
68. Russel, D.G., and Truitt, N.E., "Transient Pressure Behavior in Vertically Fractured Reservoirs," *Transactions*, American Institute of Mining Engineers, Vol. 231, 1964, pp. 1159-1170.
69. Schwartz, F.W., Smith, L., and Crowe, A.S., "A Stochastic Analysis of Macroscopic Dispersion in Fractured Media," *Water Resources Research*, Vol. 19, 1983, pp. 1253-1265.
70. Shapiro, A.M., "Fractured Porous Media: Equation Development and Parameter Identification," thesis presented to Princeton University, Princeton, N.J., in 1981, in partial fulfillment of the requirements for the degree of Doctor of Philosophy.
71. Shapiro, A.M., "An Alternative Formulation for Hydrodynamic Dispersion in Porous Media," *Flow and Transport in Porous Media*, A. Verruijt and F.B.J. Barends, eds., A.A. Balkema, Rotterdam, 1981, pp. 203-207.
72. Shapiro, A.M., and Andersson, J., "Finite Element Simulation of Contaminant Transport in Fractured Rock Near Karlshamn, Sweden," *Finite Elements in Water Resources*. K.P. Holz, U. Meissner, W. Zielke, C.A. Brebbia, G. Pinder, W. Gray, eds., Springer-Verlag, Berlin, 1982, pp. 12.11-12.20.
73. Shapiro, A.M., and Andersson, J., "Steady-State Fluid Response in Fractured Rock: A Boundary Element Solution for a Coupled Discrete Fracture - Continuum Model," *Water Resources Research*, Vol. 19, 1983, pp. 959-969.

74. Shapiro, A.M., and Andersson, J., "Simulation of Steady-State Flow in Three-Dimensional Fracture Networks Using the Boundary Element Method," *Finite Elements in Water Resources*, J.P. Laible, C.A. Brebbia, W. Gray, G. Pinder, eds., Springer-Verlag, 1984, pp. 713-722.
75. Shapiro, A.M., and Bear, J., "Evaluating the Hydraulic Conductivity of Fractured Rock From Information on Fracture Geometry," presented at the January 7-10, 1985, International Association of Hydrogeologists 17th International Congress on Hydrogeology of Rocks of Low Permeability, held at Tucson, Arizona.
76. Smith, L. and Schwartz, F.W., "An Analysis of the Influence of Fracture Geometry on Mass Transport in Fractured Media," *Water Resources Research*, Vol. 20, 1984, pp. 1241-1252.
77. Snow, D.T., "Anisotropic Permeability of Fractured Media," *Water Resources Research*, Vol. 5, 1969, pp. 1273-1289.
78. Stearns, D.W., and Friedman, M., "Reservoirs in Fractured Rock," *Stratigraphic Oil and Gas Fields*, American Association of Petroleum Geologists Memoir 16, 1972, pp. 82-106.
79. Streltsova, T.D., "Hydrodynamics of Groundwater Flow in a Fractured Formation," *Water Resources Research*, Vol. 12, 1976, pp. 405-414.
80. Sudicky, E.A., and Frind, E.O., "Contaminant Transport in Fractured Porous Media: Analytical Solutions for a System of Parallel Fractures," *Water Resources Research*, Vol. 18, 1982, pp. 1634-1642.
81. Sudicky, E.A., and Frind, E.O., "Contaminant Transport in Fractured Porous Media: Analytical Solutions for a Two-Member Decay Chain in a Single Fracture," *Water Resources Research*, Vol. 20, 1984, pp. 1021-1029.
82. Tang, D.H., Frind, E.O., and Sudicky, E.A., "Contaminant Transport in Fractured Porous Media: Analytical Solution for a Single Fracture," *Water Resources Research*, Vol. 17, 1981, pp. 555-564.
83. Truesdell, C., and Toupin, R.A., "The Classical Field Theories," *Handbuch der Physik*, Vol. III/1, Principles of Classical Mechanics and Field Theory, S. Flugge, ed., Springer-Verlag, Berlin, 1960, pp. 226-793.
84. Uldrich, D.O., and Ershaghi, I., "A Method for Estimating the Interporosity Flow Parameter in Naturally Fractured Reservoirs," *Society of Petroleum Engineers Journal*, Vol. 19, 1979, pp. 324-332.
85. Verma, A.P., "Imbibition in Flow of Two Immiscible Liquids Through a Cracked Porous Medium with Small Viscosity Difference," *Proceedings of the Second Symposium on Fundamentals of Transport Phenomena in Porous Media*, University of Guelph, 1972, pp. 393-402.

86. Wang, J.S.Y., Narasimhan, T.N., Tsang, C.F., and Witherspoon, P.A., "Transient Flow in Tight Fractures," *Report No. LBL-7026*, Lawrence Berkeley Laboratory, Berkeley, California, 1977.
87. Warren, J.E., and Root, P.J., "The Behavior of Naturally Fractured Reservoirs," *Transactions*, American Institute of Mining Engineers, Vol. 228, 1963, pp. 245-255.
88. Whitaker, S., "Diffusion and Dispersion in Porous Media," *American Institute of Chemical Engineers Journal*, Vol. 13, 1967, pp. 420-427.
89. Whitehead, R.E., and Davis, R.T., "Surface Conditions in Slip Flow with Mass Transfer," College of Engineering Report, Virginia Polytechnic Institute, 1969.
90. Wilson, C.R., and Witherspoon, P.A., "Steady-State Flow in Rigid Networks of Fractures," *Water Resources Research*, Vol. 10, 1974, pp. 328-335.
91. Witherspoon, P.A., Wang, J.S.Y., Iwai, K., and Gale, J., "Validity of the Cubic Law for Fluid Flow in a Deformable Rock Fracture," *Technical Report 23*, Lawrence Berkeley Laboratory, Berkeley, California, 1979.

7. LIST OF SYMBOLS

b	Fracture aperture
C_w	Specific heat of the fluid in the fracture
c	Concentration of the fluid within a single fracture
$\bar{c}$	Concentration averaged over the fracture aperture
$\tilde{c}$	Deviation from the concentration averaged over the fracture aperture defined in Eq. (2.16)
D	Diffusion coefficient of a dissolved constituent in a fluid
$\underset{\sim}{D}_d$	Coefficient of dispersion in a single fracture
dA	Surface of all fractures in an REV
$dS_{\alpha\beta}$	Surfaces between α and β-phases within an REV
dV	Volume of REV
dV_α	Volume of the α-phase of REV
e_α	Internal energy density of the α-phase in a multiphase continuum
F_i	Function defining the surface of the fracture wall (i=1,2)
$\overset{\leftrightarrow}{\underset{\sim}{F}}_\alpha$	Rate of momentum exchange between the α-phase and all other phases
f	Rate of a source of the thermodynamic property Ψ within a single continuum
f_α	Rate of a source of the thermodynamic property Ψ_α generated internal to the α-phase of a multiphase continuum

f1	Function defining the shape of the fracture wall
f2	Function defining the shape of the fracture wall
g	Magnitude of the gravitational force
$\underset{\sim}{g}$	Gravitational force
h_α	Rate of a source of thermal energy internal to the α-phase of a multiphase continuum
$\underset{\sim}{J}$	Diffusive flux of a dissolved constituent in a fluid
$\underset{\sim}{J}'$	Vector components of $\underset{\sim}{J}$ in the directions that lie in surface of the fracture centerline
$\underset{\sim}{J}_d$	Dispersive flux of a contaminant arising from averaging over the fracture aperture
$\underset{\sim}{J}^h$	Thermal diffusion within a fluid
$\underset{\sim}{J}_\alpha$	Nonadvective flux of the thermodynamic property Ψ_α in the α-phase of a multiphase continuum
$\underset{\sim}{J}_\alpha^\omega$	Nonadvective flux of the ω-constituent within the α-phase of a multiphase continuum
$\underset{\sim}{J}_\alpha^h$	Nonadvective thermal energy flux within the α-phase of a multiphase continuum
$\overset{\leftrightarrow}{J}_\alpha^\omega$	Rate of mass exchange with all other phases of the ω-constituent of the α-phase
$\underset{\sim}{j}$	Nonadvective flux of the thermodynamic property Ψ within a single continuum
$\underset{\approx}{k}$	Tensorial definition of the transmissivity of a single fracture
$\underset{\approx}{K}_\alpha$	Hydraulic conductivity tensor of the α-phase of a multiphase continuum
L_{ff}	Line in three-dimensional space defining the intersection of fracture surfaces and the surface of the REV
L_{fr}	Line in three-dimensional space defining the boundary of all fractures in an REV
$\overset{\leftrightarrow}{M}_\alpha$	Rate of mass exchange between the α-phase and all other phases
N	Number of phases in a multiphase medium
n	Porosity of all void space in a porous material
$\underset{\sim}{n}_i$	Unit normal vector to the fracture walls (i=1,2)
$\underset{\sim}{n}_{\alpha\beta}$	Normal vector on the surface $dS_{\alpha\beta}$ outwardly directed from the α-phase
p	Fluid pressure
Q	Volumetric flux discharged through one of the fracture walls
$\overset{\leftrightarrow}{Q}_\alpha$	Rate of thermal energy exchange between the α-phase and all other phases
$\underset{\sim}{q}_\alpha$	Specific discharge of the α-fluid phase
$\underset{\approx}{R}_\alpha$	Material coefficient defining the momentum exchange between the α-phase and all other phases
REV	Representative Elementary Volume

S_α	Storativity of the α-phase
t	Time
T_α	Temperature of the α-phase
T	Temperature of the fluid within a single fracture
$\bar{T}$	Temperature of the fluid within a single fracture averaged over the fracture aperture
$\overleftrightarrow{T}_\alpha$	Rate of exchange of the thermodynamic property Ψ_α due to mechanical interaction between the α-phase and all other phases
$\underset{\sim}{V}_\alpha$	Velocity of the α-phase
$\underset{\sim}{V}_m$	Velocity of the fluid in all void space
$\tilde{\underset{\sim}{V}}_\alpha$	Deviations of the average velocity from the point-wise velocity within an averaging volume defined by Eq. (3.14)
$\underset{\sim}{v}$	Fluid velocity within a single fracture or the velocity within any void space
$\underset{\sim}{v}'$	Components of the fluid velocity within a single fracture that lie within the surface of the fracture centerline
$\bar{\underset{\sim}{v}}'$	Fluid velocity averaged over the fracture aperture
$\tilde{\underset{\sim}{v}}'$	Deviation from the fluid velocity averaged over the fracture aperture defined in Eq. (2.17)
v_ξ	Velocity parallel to the fracture centerline
$\bar{v}_\xi$	Velocity parallel to the fracture centerline averaged over the fracture aperture
$\underset{\sim}{x}'$	Position vector measured from the centroid of the REV
$\underset{\approx}{\alpha}$	Material coefficients defining dispersion
β	Material coefficient defining the intensity of the fluid mass exchange between the fractures and the porous matrix
β_i	Set of material coefficients (i=1,2,3,4) that define the fluid mass exchange between the fractures and the porous matrix
δ_{ij}	Kronecker delta
ε_α	Volume fraction of the α-phase
ζ	Coordinate direction within the surface of the fracture axis
η	Coordinate direction normal to the fracture axis
θ_i	Set of material coefficients (i=1,2,3,4) that define the mass exchange of a dissolved fluid constituent

κ	Transmissivity of a single fracture
λ	Coefficient of thermal diffusion in a fluid
$\underset{\approx}{\lambda}_d$	Coefficient of thermal dispersion in a single fracture
$\underset{\approx}{\Lambda}$	Geometric coefficient defining fracture conductivity
μ	Dynamic fluid viscosity
$\underset{\sim}{\nu}$	Outwardly directed normal vector from the boundary L_{fr}
ξ	Coordinate direction within the surface of the fracture axis
ρ	Mass density of any single continuum
ρ_α	Mass density of the α-phase in a multiphase medium
ρ_m	Combined mass density of all fluid phases
σ	Material coefficient defining the intensity of the thermal energy exchange between the fluid in the fractures and the porous matrix
σ_i	Set of material coefficients (i=1,2,3,4,5) that define the thermal energy exchange between the fluid in the fractures and all other phases
$\underset{\approx}{\tau}$	Dispersive momentum flux from averaging over the fracture aperture
$\underset{\approx}{\tau}_\alpha$	Stress tensor of the α-phase in a multiphase continuum
ν_i	Set of material coefficients (i=1,2,3,4,5) that define the thermal energy exchange between the fluid in the porous matrix and all other phases
ϕ	Hydraulic head of the fluid in a single fracture
$\bar{\phi}$	Hydraulic head of the fluid in a fracture averaged over the fracture aperture
ϕ_α	Hydraulic head of the α-phase in a multiphase continuum
Φ	Hydraulic head averaged over the REV
ψ	Density of the thermodynamic property Ψ within a single continuum defined per unit mass
ψ_α	Density of the thermodynamic property Ψ_α defined per unit mass of the α-phase in a multiphase continuum
$\tilde{\psi}_\alpha$	Deviation from the density ψ_α defined by equation (3.13) thermodynamic property of a single continuum such as mass, momentum or energy
Ψ	Thermodynamic property of a single continuum such as mass, momentum or energy
Ψ_α	Thermodynamic property of the α-phase in a multiphase continuum such as mass, momentum or energy
ω_α	Mass fraction of a dissolved constituent in the α-fluid phase
$\underset{\approx}{\Omega}$	Geometric coefficient defining fracture conductivity

Special Symbols

$\frac{D^{\alpha}}{Dt}$	Material time derivative with respect to the velocity of the α-phase
$\underset{\sim}{\nabla}$	Gradient operator
∇^2	Laplacian operator
$\underset{\sim}{\nabla} x \underset{\sim}{P}$	Curl of the vector $\underset{\sim}{P}$

CHEMICAL TRANSPORT IN FRACTURED ROCK

Ivars Neretnieks, Harald Abelin, Lars Birgersson, Luis Moreno, Anders Rasmuson and Kristina Skagius

CHEMICAL TRANSPORT IN FRACTURED ROCK

Ivars Neretnieks, Harald Abelin, Lars Birgersson, Luis Moreno, Anders Rasmuson, and Kristina Skagius

Department of Chemical Engineering
Royal Institute of Technology
S-100 44 Stockholm, Sweden

ABSTRACT

Transport of dissolved species in fractured rock has become an area of special interest in recent years when deep lying crystalline rocks have become potential sites for repositories for nuclear waste. In Sweden, research was started in 1977 to investigate the flow and transport in low permeability crystalline rocks such as granites and gneisses.

About 10 different potential sites have been investigated by surface mapping and by deep (700 m) drillings. The rock even at large depths is fractured. The conductivity of the fractures range from below measurement limit, which is near the permeability of the rock matrix, up to 4-5 orders of magnitude higher values in the more permeable fractures. The fracture frequency is usually 1 to several fractures per meter, but only a few of the visible fractures carry measurable amounts of water. The frequency of permeable fractures is often one in 5 to 10 or more. The range of conductivities of the conductive fractures is very large. The flow in individual fractures seems to take place in permeable sections making up only a minor part of the fracture.

The matrix of the rock is porous and dissolved species can move in and out of the stagnant water in the pores by diffusion. Dissolved species may thus move at a very different rate than the mobile water. Those species which interact by sorption with the large inner surfaces of the matrix are even more retarded in relation to the mobile water.

The sparseness of fractures and very large variability in conductivity cast doubts on the applicability of the Advection-Dispersion equation.

To understand and model the flow and transport in such rock a series of field experiments as well as laboratory experiments have been performed.

Field experiments with tracer migration between injection and with withdrawal boreholes at distances of 10 and up to 51 m have been performed and analysed. Tracer experiments in natural fractures over short distances have been made in the laboratory under well controlled conditions. Tracer tests in the Stripa mine at 360 m depth have been made in 2 natural fractures over distances of 5 and 10 m with sorbing and non-sorbing tracers. A large scale tracer experiment is in progress at the same site where tracers are injected above a 75 m long drift. The water flow and tracers can be collected in more than 350 different sampling sections.

Diffusion experiments in the rock matrix have been performed in laboratory as well as in undisturbed rock at 360 m depth in Stripa granite. Laboratory measurements and in situ measurements of the diffusion of sorbing species have also been performed.

Models describing the transport of dissolved species have been designed and tested against the experiments. The models include such mechanisms as advection, dispersion, channeling, diffusion into stagnant water in the fractures as well as diffusion into the rock matrix with arbitrary geometries and sorption on the inner surfaces.

The models have been used to predict radionuclide transport in crystalline rock. In that context matrix diffusion was shown to be the dominating mechanism for retardation. Channeling was shown to have adverse effects because the fast channels may carry the nuclides at a rate at which they will have less time to decay.

1. BACKGROUND AND INTRODUCTION

In recent years there has been an increasing interest in the area of flow and transport in low permeability fractured rock. The reason for this is that many countries are seriously considering to site final repositories for nuclear waste in such environments, at depths ranging from a few tens of meters for low and intermediate level waste and 500 m to more than a km for high level waste.

There is considerably less information and experience on depths below a few hundred meters than at shallower depths. To assess if a repository is sufficiently isolated, information in several areas is needed. The flow rate and flow distribution at repository depth will strongly influence the rate of dissolution of many radionuclides. The flow paths and velocities will influence their travel time. This will in turn determine the decay of the radionuclides. Axial dispersion will dilute the species in time but also allow a fraction of the nuclides to travel faster. Channeling has the same effect. Transverse dispersion will cause dilution but also exchange species between fast and slow flowpaths.

For those nuclides which sorb on fissure surfaces and or diffuse into the rock matrix, the frequency of water conducting channels and their exposed area directly influence the contact area between flowing water and rock.

Most of the work presented in this chapter has been sponsored by the Swedish nuclear fuel company, SKB, within the project fuel safety "KBS" and by the OECD/NEA stripa project.

2. PRESENT CONCEPTS

2.1. General

Solutes which are dissolved in water will be carried by the moving water but will also move independently by various mechanisms such as diffusion and they will be retarded by interaction with the solids. Small molecules or ions diffuse in a concentration gradient and can move from one "stream tube" to another. Different water volumes move with different velocities and may mix at more or less regular intervals. A regular type of mixing may be described as a random process of the same type as molecular diffusion and is often described as such by what is called Fickian dispersion.

With only advection and dispersion active, the classical advection-dispersion description is obtained. It has been used extensively to describe tracer movement in porous media. It is easily modified to account for instantaneous chemical reaction with linear or nonlinear equilibria and can also easily accomodate reaction rates if the reactions cannot be approximated as instantaneous.

Ideally, hydrologic tracers are not supposed to react chemically with the solid material, but some naturally occurring tracers, e.g. C-14 and H-3 do, and the problem cannot be neglected. Also in cases where it is of interest to describe the movement (and retardation) of reactive species e.g. chemical waste and many radionuclides, the chemical reactions are an integral part of the problem. In many instances in laboratory experiments kinetic effects can be designed out of the experiment. In the field cases they often cannot.

In rocks with a connected matrix porosity the accessible pore volume of the matrix can be very much larger than the mobile water volume in the fracture. The species which has access to the pore water will then have a residence volume and mean residence time determined by the sum of the water volume in fractures with flow and the water volume accessed in stagnant areas. As the stagnant zones are reached by diffusion, the volume of stagnant water accessed is dependent on the residence time of the flowing water. Stagnant zones of water may also exist in fractures with channeling where water volumes between channels do not participate in the flow. This is one of the important mechanisms for long contact times and has been shown to have a dominating effect on the transport of radionuclides from deep geologic repositories in crystalline rock.

The dissolved species may also experience kinetic effects due to physical processes. One such process which has a very large impact for flow in fractured rock is the diffusion in and out of zones with so slowly moving water that it can for practical purposes be assumed to be stagnant. Stagnant water can be expected in fractures with uneven surfaces and with fracture filling materials between zones with channels with flow.

Rock fractures often have preferential channels where the water flows. Channels in the same fracture may not meet and mix their water over considerable distances. The mixing required for the process to become one of hydrodynamic dispersion, in the sense that it may be modeled as a Fickian process, may not be sufficient over the distances of interest. The situation may then be better described as stratified flow or channeling. It may be expected that the mixing will eventually be sufficient to obtain Fickian dispersion. In media where even larger features are encountered when the distance increases this may not happen. The larger pathways may always dominate the flow.

The properties of the medium can vary considerably in crystalline fractured rock. Fracture openings and thus the transport capacity of the individual fractures are known to span many orders of magnitude. Fractures are only open in places where the blocks are not in contact. The open areas may be much smaller than the closed areas. The porosity of and diffusivity in the rock matrix has similar variability. Fracture coating and filling materials also may vary considerably as to type and amount. In fracture zones the block sizes vary from very small particles up to blocks of considerable size. The size of the blocks will strongly influence the amount of stagnant volume accessible at a given contact time.

2.2. Fracture Systems

Recently six potential sites for a spent fuel repository were investigated in Sweden (KBS-3 1983). They cover gneissic and granitic rocks. Aerial photographs and geophysical measurements were used to locate fracture zones over areas on the scale of tens of kilometers. On the more local scale tens of bore holes intersecting the fracture zones were used for mapping them in depth and for obtaining fracture and hydrological data. The widths of the fracture zones range from a few meters to several 100 meters.

It was found that rock blocks between fracture zones of the order of several cubic kilometers could be found in which a repository could be sited. Rock blocks with a horizontal projected area smaller than a square km were not uncommon whereas considerably larger blocks seem to be scarce. Figure 1 shows the interpreted fracture zones around the Kamlunge site and Figure 2 shows a more localized view of the same site. The other sites are similar in regard to fracture zones. Up to 15 deep max. (app. 800 m) diamond drilled holes were investigated in each site in addition to many shallower holes (app. 100 m). Fracture mapping of the cores show fracture frequencies varying between less than 1 and 4 fractures per meter at depth below a few hundred meters. This includes closed fractures. Double packer tests at distances of 5 and 10 m and in some holes at 2-3 m show that only a fraction of the visual fractures are conducting water. At the Finnsjön and Sternö sites the distance between water conducting fracture below 200 m was between 5 and 10 m. This includes all fractures where even the lowest conductivity (above the measurement limit) was obtained.

The distances between fractures with higher conductivity are considerably larger. Figure 3 shows a plot of hydraulic conductivity data from Kamlunge (KBS-IV). The hydraulic conductivities of the fracture zones were about an order of magnitude higher than that of the rock mass.

Within the OECD/NEA Stripa project a study is underway to observe tracer movements in a larger rock mass. For this purpose a drift, 75 m long with two side arms 12.5 m each, has been excavated at the 360 m level in the granitic rock of the rock laboratory. The rock is saturated with water and there is a natural inflow of water to the drift.

The natural flow to the upper half of the drift has been monitored in considerable detail by glueing on approximately 350 plastic collector sheets about 2mx1m in size. The water flow seeping into every sheet is monitored. Figure 4 shows where and how much water seeps into the drift. It is seen that the water flow is unevenly distributed over the 100 m long drift sections.

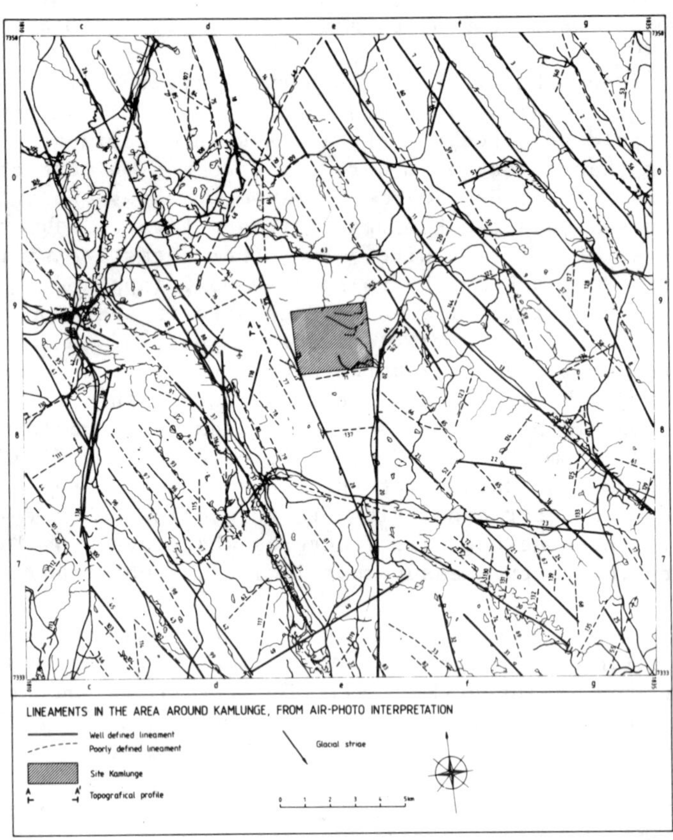

Figure 1. Interpreted lineaments in area around Kamlunge.

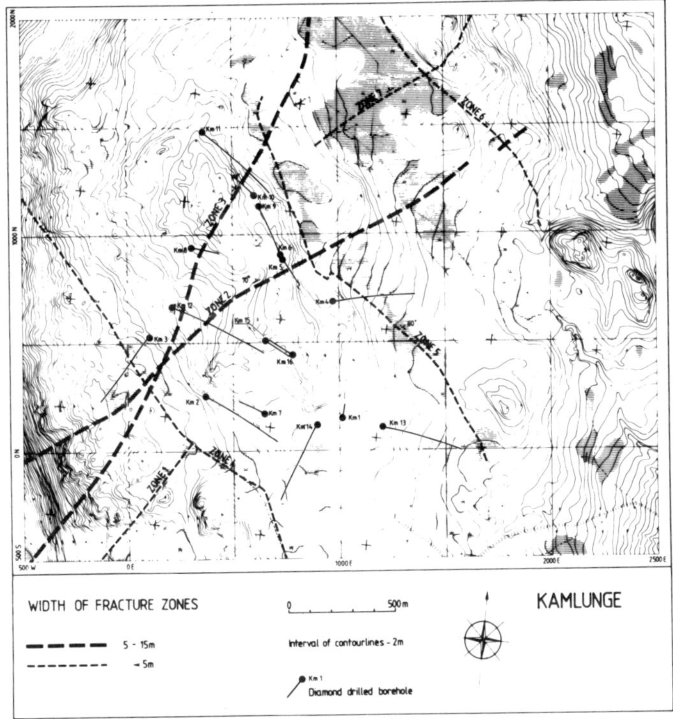

Figure 2. Fracture zones at surface within the study site at Kamlunge.

2.3. Flow in Fractures and Fracture Zones

Water flow in fractured cystalline rock takes place in the fractures or in those parts of the fractures which are open to water flow. Several recent investigations indicate that only a small portion of the fractures carry water. The majority of the fractures carry little or no water.

In the site investigations performed in connection with the Swedish nuclear fuel safety study, KBS-3, the hydraulic conductivities were found to lie in the range ~10^{-11} m/s (the lower measurement limit) and up to 10^{-7} m/s. Zones with conductivities of up to 10^{-8} m/s have been found at even the larger depths. Although they are not common they seem to occur at intervals on the order of hundred meters. Comparisons of the hydraulically conducting zones and the fracture maps from the core logs, have shown that far from all fractures carry water. Figure 5 shows a comparison of the number of visible fractures on the cores. In both sites the frequency

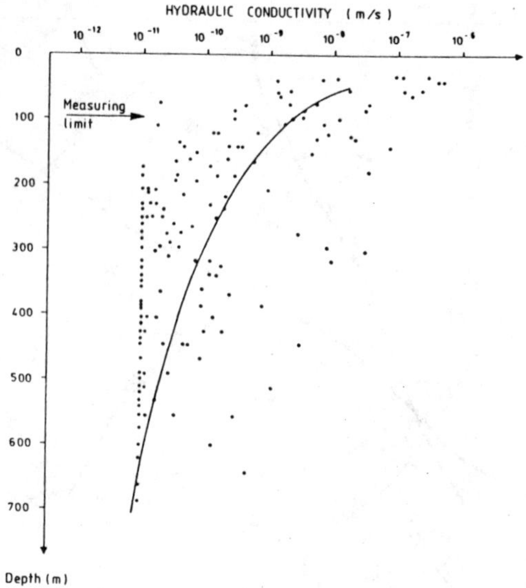

Figure 3. Correlation between hydraulic conductivity and depth for the rock mass at Kamlunge. Values from 4 holes.

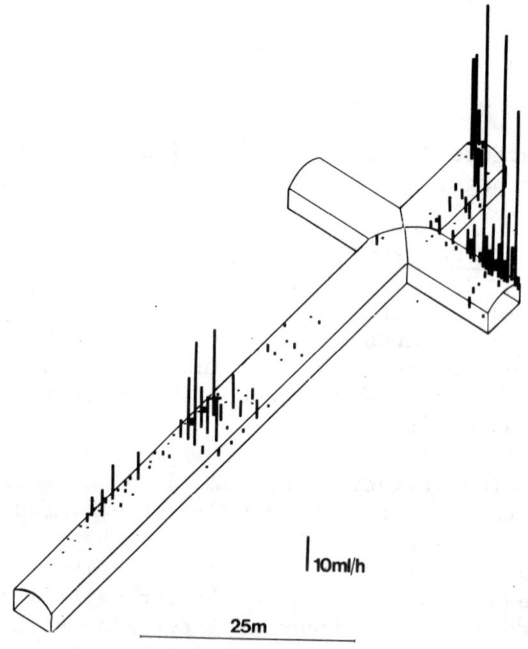

Figure 4. Water flow rate into 3-D drift at Stripa before drilling the injection holes.

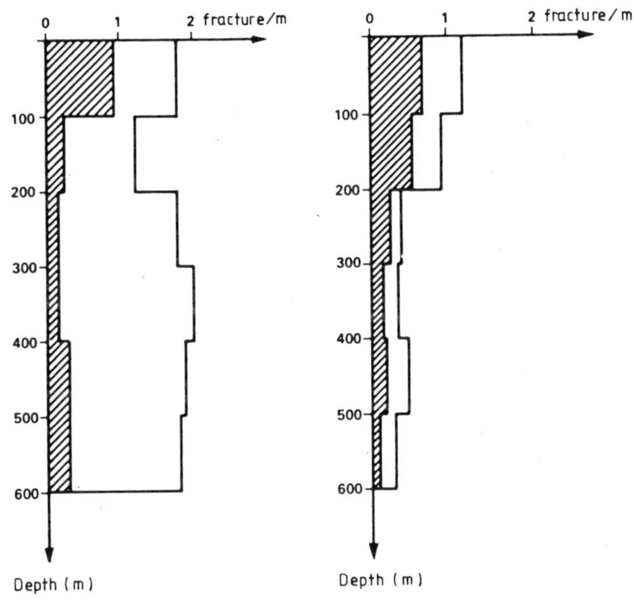

Figure 5. Total fracture frequency and hydraulic fracture frequency in the rock mass within the sites at Finnsjön (left-hand figure) and Sternö (right-hand figure).

of water conducting fractures is on the order of one in every 5 to 10 meters. The number of visible fractures is 2 to 5 times larger. The water conducting fractures include all fractures where measurable conductivity was obtained. This means that even the smallest conductivities which are 4 to 5 orders of magnitude smaller than the largest are counted. The smallest fractures will of course carry a very small fraction of the water flow in the rock.

In the lineaments of fissure zones, the fissuring is very frequent and the rock may even be broken down to particles of cm size in places. Finer material containing altered rock and clays can also be found.

The flow in the lineaments is very little studied. There are indications that the water moves in preferential paths in these zones, but it is also conceivable that the water flows more or less evenly in the lineament. The hydraulic conductivity of the lineaments were found to be higher than that in good rock by about one order of magnitude in the Swedish investigations (KBS-3). In these zones the water will encounter blocks or rock particles of very different sizes and properties.

2.4. Channeling

Investigations in the Stripa mine in mid Sweden have shown that the flow in individual fractures takes place in channels (2). Figure 6 shows how the natural water flow emerges from two fractures as they intersect the face of the drift. The white arrows are proportional to the flow and the figures beside the arrows indicate the magnitude of the flow. Subsequent injection of small amounts of water with nonsorbing tracers (so as not to disturb the natural water flow) showed that the tracers arrived at two of the collection points in the fracture used for the tracer experiment. The collectio was made by drilling 1 m long holes in the plane of the fracture with a spacing varying between .5 and .7 m. This means that every hole collected water from about this length of fracture. These results indicate that about 5-20% of the fracture plane carries more than 90% of water. The actual breadth of the channels is less than 1 m and could be considerably smaller.

Similar observations have been made in a fracture in a granitic body in Wales (9). Over a 2 m long borehole in the plane of the fracture only about 20% of the fracture was found to have any appreciable hydraulic conductivity. This was limited to sections of a few tenths of cm in breadth.

In the program in Switzerland several deep boreholes have been drilled into granitic rock (NGB-5 1985). In one hole in Böttstein in north-eastern Switzerland, very detailed core-mapping and pressure pulse testing in the hole has been performed. It was found that at intervals of less than 100 m there were shear zones which were strongly altered (kakeritic). The porosity is 3-5% compared to .5%

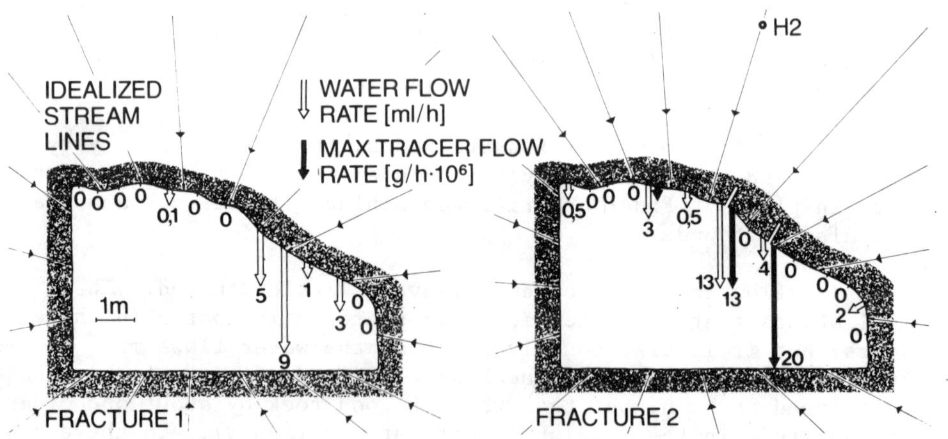

Figure 6. Average water flow rate from fracture 1 and fracture 2 in Stripa.

for unaltered granite, and some of the quartz has been dissolved. The zones are up to a meter wide. In these zones, holes were found with openings up to a cm in size. Some were circular, others were of slitlike shape.

2.5. Dispersion

By dispersion in a broad sense is meant the speading of a species transported by a fluid as the fluid moves in a medium. There are many causes for such spreading: velocity variations in the fluid in a channel, velocity variations between channels in a porous medium, physical interactions with the solid material, and molecular diffusion in the liquid. Also chemical interaction will cause spreading.

Molecular diffusion does not contribute much to the spreading of the front for long transport distances. Diffusion will in fact diminish dispersion by velocity variations within a channel because the concentration difference over the channel width will be decreased by diffusion. For flow in low-permeability fissured rock, the concentrations over the fissure width usually can be considered constant.

Velocity variation between channels is a very important dispersion mechanism. Bear (5) gives a comprehensive treatment on hydrodynamic dispersion theories.

The most detailed models treat the spreading process by modeling more or less randomly oriented conduits combined with some assumptions on how velocities in the channels vary as well as how distribution at channel divisions and mixing at channel intersections occur. An early description is found in the paper by de Josselin de Jong (13). The common basis for practically all these treatments is that the spreading is described by one parameter; the variance σ_z^2 of a pulse as it spreads with distance. The variance increases with traveled distance. Gelhar and Axness (14) show that for certain media there is a theoretical basis for this observation.

For a random process such as molecular diffusion, a dispersion coefficient D_L analogous to the diffusion coefficient could be determined from $D_L = \sigma_z^2/2t$. For porous media with fairly uniform particle size this has been verified experimentally by many different investigators, in the laboratory.

The dispersion coefficient is proportional to the velocity and the particle size: $D_L \propto vd_p$. This also implies that the variance is proportional to the distance:

$$\sigma_z^2 \propto 2tvd_p = 2d_p z$$

In some investigations (Neretnieks(34)), the dispersion coefficient increases considerably with distance. The explanation for this is usually summarized in the word 'channeling' or 'uneven distribution'. Schwartz (46), by computer simulation, has shown that the uneven distribution of resistances in a porous medium may not lead to a variance which increases proportionally to the distance traveled. Mercado (23) and Neretnieks (31) by a different method showed that in a medium where stratification occurs - parallel unconnected strata with transmissivity differences between channels $\sigma_Z \propto z$ instead of $\sigma_Z^2 \propto z$ as in the diffusion-dispersion case above.

Matheron and de Marsily (22) arrived at the same conclusion and also concluded that the usual 'convection diffusion equation' cannot in general be applied even for large distances.

Neretnieks (31) discussed several dispersion mechanisms including stratification and derived a model which includes these effects as well as the effects of physical interaction by diffusion into stagnant zones of water in the matrix of the rock. It was shown that matrix diffusion effects can have a dominating influence on the pulse spreading when the accessed stagnant water volume is large in comparison to the mobile water.

2.6. Diffusion into the Porous Matrix

Crystalline rocks, such as granites and gneisses, have microscopically small fissures between the crystals. These fissures comprise an interconnected pore system containing water. The radionuclides are much smaller than the microfissures and can diffuse into this pore system. The inner surfaces in the rock matrix are many times (thousands and more) larger than the surfaces in the fractures in which the water flows. Penetration into and sorption on the inner crystal surfaces retard the radionuclides far beyond the retardation caused by sorption on the fracture faces. Even nonsorbing nuclides will move slower than the flowing water, since they diffuse into the stagnant water in the pores.

Gneisses and granites in Swedish Precambrian rock have been found to have a continuous pore system consisting of the microfissures between the crystals in the rock matrix. The porosity in this pore system varies between 0.06% and 1% (52) for the rock matrix. Similar results have been obtained by other investigators (10,11). Fracture minerals and rock in crushed zones have higher porosities. Values between 1 and 9% have been measured. Substances dissolved in the water can diffuse into this pore system and sorb on the inner surfaces. Figure 7 shows penetration and sorption in the rock matrix.

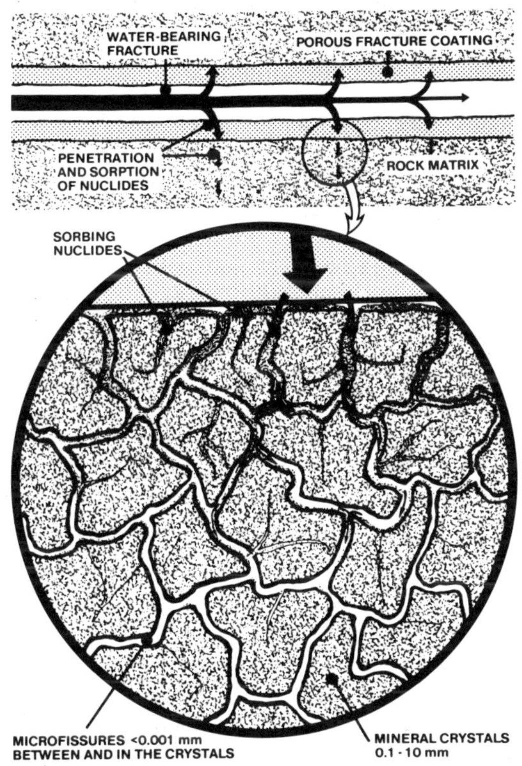

Figure 7. The figure shows the penetration and sorption of the nuclides in the microfissures in the rock matrix.

The penetration depth increases with time. Non-sorbing species penetrate far into the matrix, while sorbing nuclides are retarded since they also have to fill up the sorption sites before they migrate further.

The penetration depth can be calculated with the aid of Fick's laws for diffusion, if diffusivity is known, and by taking retardation due to sorption into account.

2.7. Sorption and Other Interactions

The minerals of crystalline rocks have a considerable cation exchange capacity. They also exhibit a capacity to form surface complexes with anions as well as cations. The capacity to bind different ions to the surface depends on the pH, the concentration of many of the other ions and on the form the species have in the solution. Many cations form complexes with the anions and other complexing agents in the water. The complexed metal ions which would have a positive charge if not complexed, may be cations, neutral species of even negatively charged species, depending on the composition of the water.

The component may be bound (adsorbed, ion exchanged) to the surface of the minerals of the rock, or it may form its own minerals by precipitation. It may also react irreversibly by mineralization into some very stable solids.

The different complexes will of course have different affinity to the mineral surfaces. Because of the complexity of the situation it is often convenient to summarize all the effects into a term "sorption coefficient" of K_d. This expresses how much of a component e.g. Ca^{2+} is bound to the surfaces of the rock and how much is in the water at equilibrium. One neglects which form (complexed or noncomplexed) the various species containing the component have in the solution as well as in the bound state. K_d is the ratio of concentrations of the component in the two phases.

$$K_d = q/C$$

q denotes the concentration on (in) the minerals and C is the concentration in the water.

The component may simultaneously be a part of several species (complexes) which are sorbed in different ratios. K_d often varies very much with changes in the composition of the groundwater (pH, Eh, concentration of complexing agents etc.). It is therefore a very inprecise entity but because of its simplicity it has been found to be very useful in mathematical modeling and in transferring information from chemists to modelers.

Adsorption and ion exchange are often summarily called sorption and then denote fast reversible reactions of the dissolved components with the surface of the rocks. These processes are treated by the K_d-concept. Processes such as precipitation and irreversible mineralization are usually best treated as such.

3. MATHEMATICAL MODELING

The description of the mathematical models will mostly be limited to one dimension. The concepts and models may be extended to 2 or 3 dimensions in a straightforward way. There are, however, very few analytical solutions available for more than one dimension and usually numerical solutions are used. The analytical solutions have been found very valuable when checking complex numeric codes. They also often facilitate the understanding of the influence of the different parameters on the results in a better way than numerical codes or solutions do.

The simplest models are based on the concept of advective flow with a known velocity. A tracer pulse transported by the flow will be dispersed around the mean due to random velocity components. The dispersion is assumed to be of the same nature as molecular diffusion.

3.1. Advection - Dispersion Description

The advection-dispersion equation for a nonreactive tracer for linear flow can be written

$$\frac{\partial C}{\partial t} + v \frac{\partial C}{\partial z} = D_L \frac{\partial^2 C}{\partial z^2} \tag{3.1}$$

A very common case of interest is where the medium is of infinite length and a tracer is suddenly injected at a point z = 0. The injection is kept up forever and is made in such a way that the concentration always is C_o at injection point. This is not a very realistic assumption but the case has been extensively treated in the literature. Other more realistic cases are treated in a comprehensive way by Maloszewski and Zuber (1984). The initial and boundary conditions can be written:

IC	$C = 0$	$z \quad 0$	$t = 0$	
BC1	$C = C_o$	$z = 0$	$t \quad 0$	(3.2a-c)
BC2	$C = 0$	z	$t \quad 0$	

The classical solution to this is given by Lapidus and Amundson (20). It is given below Eq. (3.5) but can be written for short as:

$$C/C_o = f(t,\{t_w,Pec\}) \tag{3.3}$$

indicating that it is a function of time t and that two parameters t_w (water residence time) and Pec (a measure of dispersivity) suffice to fully define the solution. The parameters include velocity, distance, and dispersion coefficient and could have been combined in other ways. However, two such groups suffice to define the solution.

As the use of radioactive tracers is common and the transport of radionuclides from final repositories for nuclear waste is of high interest at present, the formulation of the equation is extended to include radionuclides already at this stage.

Equation (3.1) is directly extended to apply to a decaying species by adding a decay term $- C\lambda$ on the right hand side. If the boundary condition (3.2b) also is $C = C_o \cdot e^{-\lambda t}$, then the solution

(3.3) can be used if it is multiplied by $e^{-\lambda t}$. Extension to the case where instantaneous equilibration of the tracer with the solid material takes place and where the equilibrium is linear is also straightforward. With

$$R_d = 1 + K_d \rho_p \frac{1 - \varepsilon_f}{\varepsilon_f} \tag{3.4a}$$

for instantaneous volume reaction and

$$R_a = 1 + K_a a \tag{3.4b}$$

for instantaneous surface reaction equation (3.3) is changed only by exchanging t_w for $t_o = t_w R$ (where R = either R_a or R_d). The solution to Eq. (3.1 - 3.2a-c) can be written:

$$\frac{C}{C_o \cdot e^{-\lambda t}} = \tfrac{1}{2}\text{erfc}\ \{\tfrac{1}{2}(\text{Pec}\ \frac{t_o}{t})^{\frac{1}{2}}(1 - \frac{t}{t_o})\} +$$

$$\tfrac{1}{2}\ e^{\text{Pec}}\text{erfc}\ \{\tfrac{1}{2}(\text{Pec}\ \frac{t_o}{t})^{\frac{1}{2}}(1 + \frac{t}{t_o})\} \tag{3.5}$$

Equation (3.1) may easily be made to include reaction rates between liquid and solid and to accomodate other initial and boundary conditions. A common boundary condition is that the inlet concentration varies with time

$$C = C(z = 0,\ t) \tag{3.6}$$

The convolution integral may be used to handle this

$$C(t) = \int_0^{\infty} C(0,t-t')\ f'(t',t_w R,\text{Pec})dt' \tag{3.7}$$

f' is the response to an instantaneous pulse.

In the chemical engineering literature and hydrologic literature many other inlet and outlet boundary conditions have been discussed. Very often the simple solutions (3.5) or (3.7) are not sufficient to explain the curve forms of an experiment. Recently Landström et al. (19) in an experiment with non-reacting tracers over 11.8 m in fractured crystalline rock in Studsvik obtained a very long tail which could not be fitted with equation (3.5). A similar result was obtained in another field experiment over 30 m in Finnsjön by Gustafsson and Klockars (16). In both cases it was concluded that this could be caused by the presence of at least three different independent channels through which the water flows. The slower channels will contribute to the tailing. Other mechanisms may also cause such tailing. Several such mechanisms will be treated later.

3.2. Channeling

Fractured rock especially at larger depths may have considerable distances between water bearing fractures. Data from deep holes in the Swedish rock (down to 800 m) indicate that at depths below a few 100 m the spacing of conducting fractures is on the order of 1 fracture per 5 to 10 m (12). Other observations in individual fractures (1,2,3,30) indicate that even in well defined fractures in crystalline rock there is considerable channeling. Abelin et al. (2) found that in 3 well visible fissures in the Stripa deep (360 m) rock laboratory, less than 20% of the fracture breadths carried more than 70% of the flow.

A model was recently tested where all "dispersion" is assumed to be caused by channeling (31). It was tested on some laboratory experiments using a natural fissure (30) and several field experiments (3,4,24,25).

The model is based on the assumption that all channels conduct the flow from inlet to outlet without mixing between channels underway. At the outlet, however, the fluid from all channels is instantaneously mixed. This would simulate a very common way of sampling for tracers. Assume that the channels can be uniquely described by their openings as regards the flow rate and concentration response.

For the case with a discrete distribution $F(\delta_i)$ of channel openings δ_i the effluent concentration in each channel is denoted by $C(t,\delta_i)$. The flow in each channel is $Q(\delta_i)$. Both $C(t,\delta_i)$ are assumed to be functions totally defined for every class of channel opening δ_i. The effluent concentration $C(t,\delta_i)$ may be obtained from the advection-dispersion equation (3.5) when this is applicable or if other mechanisms are active the appropriate function must be used. The underlying assumption is that the time and some characteristic of the channel e.g. δ_i is sufficient to characterize the effluent. Other effects such as dispersion in the channel or reactions with the walls of the channel must be known functions of δ_i. The mixed effluent from all channels has the concentration:

$$C(t) = \frac{\sum_{i=1}^{N} F(\delta_i)Q(\delta_i)C(\delta_i,t)}{\sum_{i=1}^{N} F(\delta_i)Q(\delta_i)} \qquad (3.8a)$$

For a continuous distribution of channel openings $f(\delta)$ we have:

$$C(t) = \frac{\int_0^\infty f(\delta)Q(\delta)C(\delta,t)d\delta}{\int_0^\infty f(\delta)Q(\delta)d\delta} \tag{3.8b}$$

A flow system with channeling will spread a tracer pulse along its pathways and the pulse will also be spread at the observation (mixing point), even if there is no spreading in the individual channels.

Figure 8 shows the response of a stratified system to a Dirac pulse at the inlet. The channel widths in this case are taken to be log normally distributed and the cubic law for flowrate

$$Q = \text{const}_1 \cdot \delta^3 \tag{3.9}$$

is assumed to apply. The velocity in such a system is

$$v = \text{const}_2 \cdot \delta^2 \tag{3.10}$$

For a distribution $F(\delta_i)$ or $f(\delta)$ which is described entirely by a mean μ_1 and variance σ_ℓ^2 the mean transit time $\bar{t}$ can be determined from the first moment. For a Dirac pulse at the inlet

$$\bar{t} = \int_0^\infty C(t)tdt/\int_0^\infty C_o dt \tag{3.11}$$

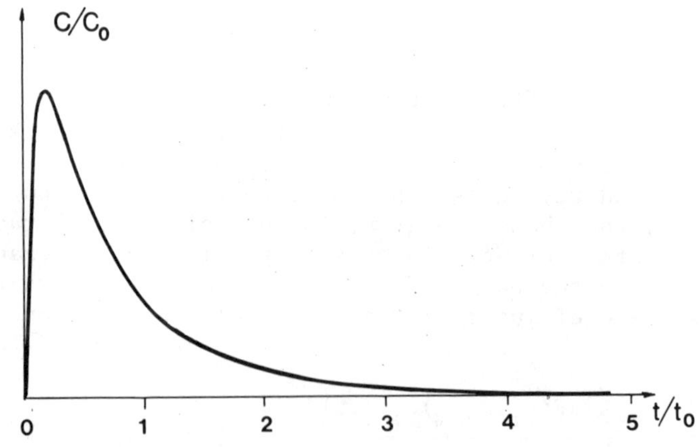

Figure 8. Concentration at the outlet of a medium with parallel fissures which has been injected with a tracer pulse.

From the second moment the variance σ_t^2 can be determined

$$\sigma_t^2 = \int_0^\infty C(t)(t - \bar{t})^2 dt / \int_0^\infty C_o dt \tag{3.12}$$

From the latter the equivalent of the dispersion coefficient for this experiment can be determined

$$D_L = \frac{\sigma_t^2}{\bar{t}^2} \cdot \tfrac{1}{2}\bar{v}z \tag{3.13}$$

For a log normal distribution, from Eqs. (3.11) and (3.12) the following simple expression for the variance is obtained

$$\left(\frac{\sigma_t}{\bar{t}}\right)^2 = e^{4(\ell n 10 \sigma_\ell)^2} - 1 \tag{3.14}$$

where σ_ℓ is the standard deviation in the log normal distribution. From equation (3.14) it is seen that $(\sigma_t/\bar{t})^2$ is a constant for a given σ_ℓ. In fact it has been shown that for an arbitrary distribution $-f(\delta)$ - this applies (31). From equation (3.13) it is found that

$$D_L = \text{const} \cdot \bar{v}z \tag{3.15}$$

It may be concluded from this, that if there is pure channeling in a flow region then an apparent hydrodynamic dispersion coefficient will increase with distance between injection and observation (mixing) points. Equations (3.13) and (3.14) give a direct and simple relation between D_L and σ_ℓ.

The consequences of using the wrong mechanism in predicting the tracer behaviour over longer distances are shown in Figure 9. This figure shows a reference case breakthrough curve for a step injection (the "experimental" results obtained from the channeling model with $\sigma_\ell = 0.208$). From this a Pec = 10 was obtained by fitting with the advection-dispersion model, (Eq. (3.5)). The predicted curves are for a 10 times longer distance. It can be seen that the two models predict very different breakthrough curves. Figure 10 shows some results on dispersion coefficients obtained from experiments in fissured crystalline rock (34). The few data available indicate an increase of D_L with distance.

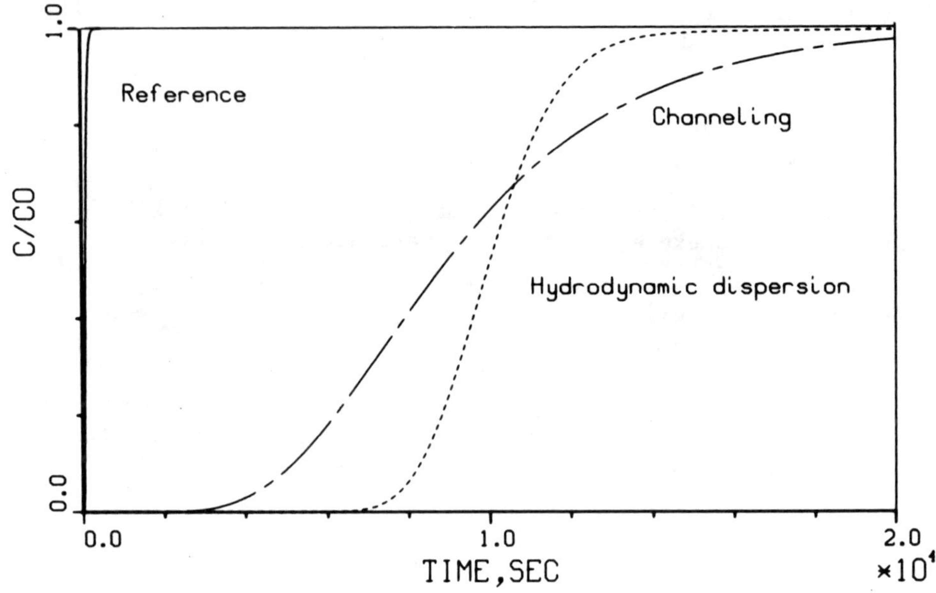

Figure 9. Predicted curves using the hydrodynamic dispersion model and the channeling dispersion model. For nonsorbing tracer for a longer distance (10 times) and a lower flow (Pe = 100, σ = 0.208).

Hydrodynamic dispersion will take place in every channel in addition to the channeling. Equations (3.8a, 3.8b) are quite general and can be used also for such cases. It can be seen from Figure 9 that the influence of hydrodynamic dispersion will have a diminishing influence over longer distances if channeling is in effect.

If the frequency of mixing between the various channels between inlet and outlet is known, it is possible to modify the model to account for this (43).

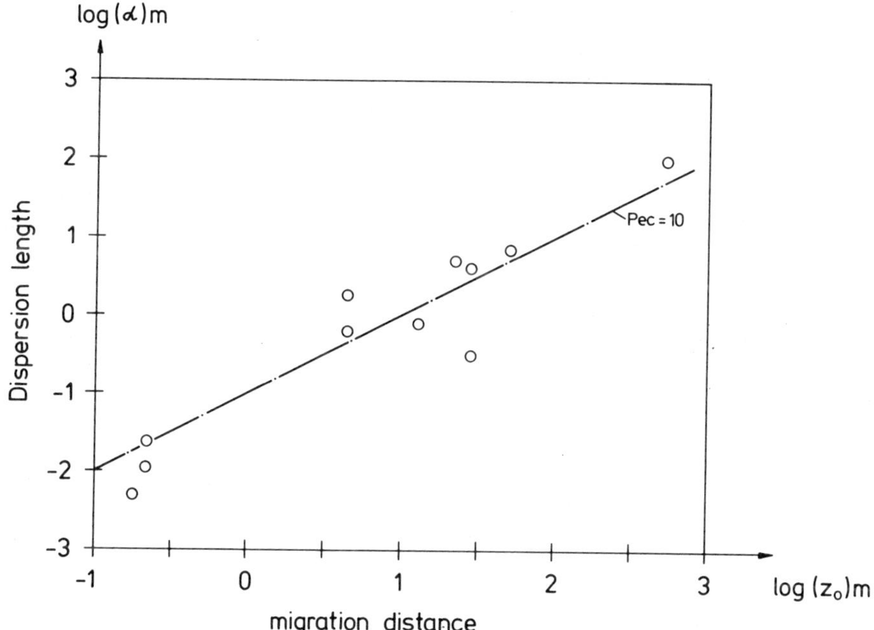

Figure 10. Experimentally determined dispersion lengths in fissured crystalline rock.

3.3. Matrix Diffusion

The mechanism may be active in addition to those previously described. There is considerable experimental evidence on crystalline rock porosities and diffusivities from the laboratory (11,30,31,33, 34,49), and from the field (7), in undisturbed rock. Porosities in unaltered rock range from 0.06 to over 1% and effective diffusivities $D_e = D_p \varepsilon_p$ for small ions and molecules range from 1.10^{-14} - 70.10^{-14} m^2/s.

Dissolved species in the water in a fracture will diffuse into the porous rock matrix. The process can be described by the diffusion equation (including here the effects of sorption with linear equilibrium

$$K \frac{\partial C_p}{\partial t} = D_e\left(\frac{\partial^2 C_p}{\partial x^2} + \frac{\beta}{x}\frac{\partial C_p}{\partial x}\right) - \lambda C_p K \qquad (3.16)$$

if there is no flow in the matrix.

$$K = \varepsilon_p + K_d\rho_p(1 - \varepsilon_p) \tag{3.17}$$

With the low hydraulic conductivities of crystalline rock $<10^{-1[illegible]}$ m/s (10) and low natural hydraulic gradients, transport by advection is negligible compared to that by diffusion.

For short contact times the penetration depths are short and may not deplete the water in the fissure to any large extent. For long contact time the influence can be considerable. Figure 11 shows the penetration depth from a plane surface ($\beta = 0$) versus time for nonsorbing species as well as for species with different sorption coefficients. No radioactive decay ($\lambda = 0$) takes place in this case. The penetration depth $\eta_{0.01}$ is here defined as the distance from the surface at which the concentration in the pore water is 1% of that at the surface. It is obtained by solving equation (3.16) for $\beta = 0$ and using the appropriate boundary and initial conditions. These describe that a body initially with $c_p = 0$ is at time 0 exposed to a concentration at the surface which is then kept constant. The result is $\eta_{0.01} \cong 4\sqrt{(D_e t)/K}$.

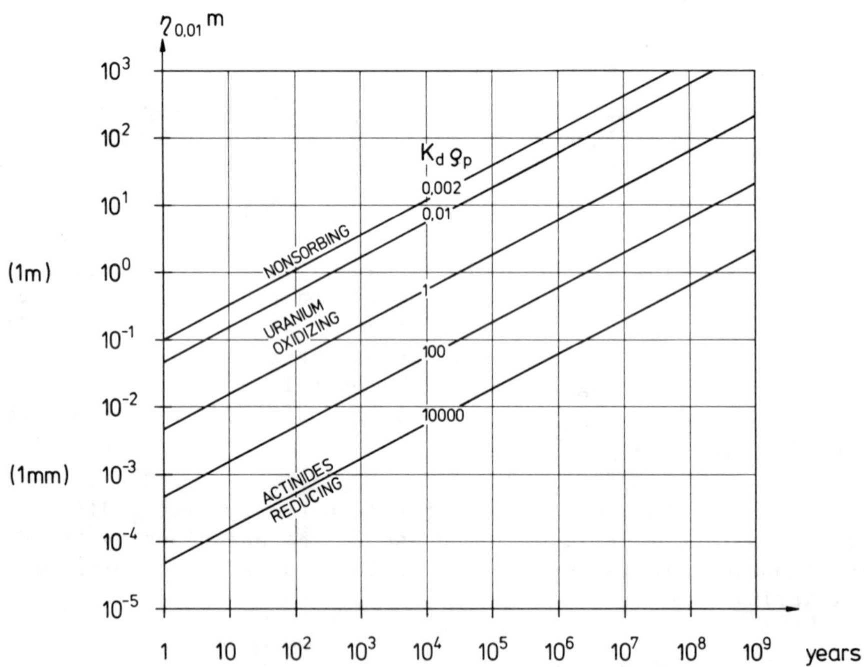

Figure 11. The penetration depth $\eta_{0.01}$ versus time for nonsorbing $K = \varepsilon_p$ and sorbing $K > \varepsilon_p$ species.

To describe the coupled processes of advection-dispersion and matrix diffusion an additional term is added to equation (3.1) which becomes

$$R_a \frac{\partial C}{\partial t} + v \frac{\partial C}{\partial z} = D_L \frac{\partial^2 C}{\partial z^2} + aD_e \left.\frac{\partial C_p}{\partial x}\right|_{x=0} - \lambda C R_a \tag{3.18}$$

The equations (3.16) and (3.18) are coupled by $C = C_p$ at the fissure surface $x = 0$. There are some analytical solutions to equations (3.16) and (3.18) available for different boundary conditions. References to some recent solutions are given in Table 1. The solutions all apply to cases with regular bodies (infinite slabs, cylinders or spheres).

Two inlet boundary conditions are of special interest. In the first $C = C_o.e^{-\lambda t}$ at $z = 0$ for $0 < t < t_o$. This describes a case where the concentration at the inlet decays during the injection time. This is typical for cases where a radioactive tracer solution is prepared at a given time and then used for injecting. In this case the solution for the decaying tracer is obtained as before from the nondecaying case by multiplication with $e^{-\lambda t}$. The second case is when naturally occurring radioactive tracers such as C-14 or H-3 are used. There $C = C_o$ = constant or $C = C(t)$ at $z = 0$ and the solution is not directly obtained from the nondecaying solutions as in the previous case. Solutions with both boundary conditions are given in Table 1.

For the case when there is no hydrodynamic dispersion and when $C = C_o.e^{-\lambda t}$ at $z = 0$ and $t > 0$ the solution is (28)

$$C/C_o = e^{-\lambda t}.\text{erfc}\left\{ \frac{t_w}{\delta} \sqrt{(D_e K)/(t-R_a t_w)} \right\} \tag{3.19}$$

for $t > R_a t_w$, else $C/C_o = 0$.

A criterion for when the matrix diffusion becomes important may be constructed in the following manner. When the time for the breakthrough given by equation (3.19) to reach a value of $C/C_o.e^{\lambda t} = 0.5$ is twice as long as for plug flow i.e. $t = 2t_w.R_a$, matrix diffusion is taken to become important. From equation (3.19) we obtain $0.5 = \text{erfc}(\text{arg})$ giving arg = 0.477. Then

$$\text{arg} = \frac{t_w}{\delta} \sqrt{(D_e K)/(t-R_a t_w)} = 0.477 \tag{3.20}$$

and with $t = 2R_a t_w$ we obtain that for

Table 1. Some recent analytical solutions to equations (3.16) and (3.18) for different geometries and boundary conditions. Initial condition $C_p = C_f = 0$. Boundary condition outlet $C = 0$ at $z \to \infty$.

Reference	Dispersion	Block model	BC inlet	Interface water/block	Comment
Neretnieks 1980a Grisak and Pickens 1981	$D_L = 0$	Infinite thickness	$C = C_o . e^{-\lambda t}$	Equilibrium	-
Tang et al. 1981	$D_L \neq 0$	Infinite thickness	$C = C_o . e^{-\lambda t}$	Equilibrium	-
Sudicky and Frind 1982	$D_L \neq 0$	Finite slabs	$C = C_o$ $C = C_o . e^{-\lambda t}$	Equilibrium	-
Rasmuson and Neretnieks 1980	$D_L \neq 0$	Spheres	$C = C_o . e^{-\lambda t}$	Film resistance	-
Rasmuson 1981	$D_L \neq 0$ $D_R \neq 0$	Spheres	$C = C_o . e^{-\lambda t}$	Film resistance	Radial dispersion 2-D problem
Rasmuson 1984	$D_L \neq 0$	Spheres, slabs	$C = C_o$, $C = C_o . e^{-\lambda t}$	Film resistance	
Rasmuson 1985a	$D_L \neq 0$	Spheres, cylinders, slabs	$C = C_o . e^{-\lambda t}$	Reaction rates + Film resistance	
Rasmuson 1985b	$D_L \neq 0$	Arbitrary mixtures of spheres, cylinders and slabs of different sizes and size dependent properties	$C = C_o . e^{-\lambda t}$	Reaction rates + Film resistance	
Pigford and Chambre 1982	$D_L = 0$	Infinite thickness	$C = C_o . e^{-\lambda t}$	Equilibrium	Chain decay

$$t_w > 0.23 \frac{\delta^2 R_a}{D_e K} \tag{3.21}$$

the matrix diffusion considerably influences the transport of a dissolved species. Note the square dependence on fracture opening. For a nonsorbing species $K = \varepsilon_p$ and $R_a = 1$, and we obtain as an example for $\varepsilon_p = 0.002$ and $D_e = 5.10^{-14}\ m^2/s$, $t_w > 0.23.10^{16}\ \delta^2$ s. For a 100 μm fracture t_w must be larger than 0.7 year to feel the influence of matrix diffusion, whereas for $\delta = 10$ μm, $t_w > 2.7$ days. With a strong matrix sorption $K = 10^4$, even fairly large fractures would influence the process already for short transport times of the water.

The criterion could of course be chosen differently. If the time is chosen where the 50% concentration point has travelled a time $1.1R_a t_w$ instead of $2R_a t_w$, the criterion is sharpened by a factor of 100. This would give $t_w = 2.7$ days in the 100 μm fissure and $t_w = 40$ minutes in the 10 μm fissure.

Even the second criterion gives a clearly noticeable influence on the shape of a breakthrough curve especially in the tail. The values of D_e and ε_p were taken for from data on fresh, unaltered crystalline rocks. For fissures where the surfaces are altered the few available data (51) indicate that the values can be considerably higher.

It may thus be expected that the effect of matrix diffusion will be noticeable in many field experiments and even may become the dominant effect in small fractures and for long water transit time. Maloszewski and Zuber (21) recently discussed the impact of matrix diffusion in field experiments and also concluded that it may have a considerable impact. In short the solutions to the matrix diffusion case equations (3.16) and (3.18) for a stable species can be written

$$C/C_o = f(t, \{Pec, t_w R_a, \frac{D_e K}{\delta^2}, S\}) \tag{3.22}$$

If channeling also must be accounted for, at least an additional parameter describing the channeling distribution e.g. σ_ℓ must be introduced (Moreno et al., 1983, have chosen to use either σ_ℓ or Pec in their models). If the penetration depth $\eta_{0.01}$ is comparable to the block sizes S, the solution is also dependent on this as shown in equation 3.22.

3.4. Transport in a Varying Flow Field and in Media with Varying Block Sizes

In many real field situations the boundary conditions may be more complex than discusses above, the flow may not have constant velocity or the rock blocks may have considerably different sizes and shapes. Very general boundary and initial conditions may be handled by standard numerical methods used for solving equations (3.16) and (3.18) and this is in practice seldom a problem. This group uses standard finite difference methods and the integrated finite difference method as standard tools for numerical calculations.

Varying velocity fields may also be handled with standard numerical techniques. It was shown by Sauty (45) that in a cylindrically radial flow field very small errors are introduced by assuming a constant flow velocity, provided dispersion is reasonably small Pec > 3. Neretnieks and Rasmuson (32) showed that even smaller Peclet numbers can be handled well in an arbitrary flow field by using a weighting technique to determine an "average" Peclet number. It is based on a method of continuously adding the variances of each section of the flow tube.

Blocks of various sizes and shapes and diffusional properties can be handled rigorously by exchanging the $aD_e(\partial C_p/\partial x)\big|_{x=0}$ term in equation (3.18) for

$$\frac{1}{mV_s} \cdot \int_A D_e \frac{\partial C_p}{\partial n}\, dA = \frac{1}{m} \int_0^\infty \frac{\beta+1}{b} \cdot f(b)\, D_e \left.\frac{\partial C_p(b)}{\partial x}\right|_{x=0} db \qquad (3.23)$$

which means that all the solute transported over all the surfaces is accounted for over all the various blocks of different sizes. The block size distribution is f(b) and b is the block radius. This technique, however, makes it necessary to solve equation (3.16) for every block size "i" (and shape and property).

$$K^i \frac{\partial C_p^i}{\partial t} = \frac{1}{r^\beta} \frac{\partial}{\partial r} \left(r^\beta D_e^i \frac{\partial C_p^i}{\partial r}\right) \qquad i = 1.2...N(\infty) \qquad (3.24)$$

where again β = 0,1 or 2 for slabs, cylinders and spheres recpectively.

Some analytical solutions have been obtained by Rasmuson (38) for such cases. The application of numerical techniques to the N equations of (3.24) and one equation of (3.18) increases the work load proportional to N. Neretnieks and Rasmuson (32) and Rasmuson and Neretnieks (42) developed and tested an approximate technique whereby the N equations (3.24) are reduced to one equation. This is based on what they call the PSEUDOBODY approach. The basic idea was originally put forward by Pruess and Narasimhan (35) under the name "MINC". The idea is that all shells of the rock blocks at the same distance from the block surface behave in the same manner. They can thus be lumped together forming PSEUDOBODIES with the same total outer surface and the same total volume as the real blocks. The blocks have a cross-sectional area A(x) as a function of distance from the surface x, which is the same as the average for all the real blocks. The concept is shown in figures 12 and 13. Equation (3.24) simplify to

$$K \frac{\partial C_p}{\partial t} = \frac{1}{A(x)} \frac{\partial}{\partial x} \left(D_e A(x) \frac{\partial C_p}{\partial x}\right) \qquad (3.25)$$

which is equivalent to (3.16) for slabs,cylinders and spheres of equal size if the appropriate A(x) function for these bodies is used. The PSEUDOBODY approach considerably decreases the computational effort, and the cases tested, give surprisingly small differences compared to the exact analytical solutions. Figure 14 shows some comparisons between the exact analytical solution and the results obtained for the PSEUDOBODY method for two cases. The block size distributions are given in Table 2 below:

	Block diameter m	Volume fractions	Other surface m^2
	0.02	0.0909	22.5
Case "small"	0.5	0.9091	9
	∞	-	0.5
	0.1	0.0909	4.5
Case "large"	0.5	0.9091	9
	∞	-	0.5

Table 2. Block size distributions used in the comparisons

The "∞" block size indicates that only a surface area of 0.5 m^2 of the large block is exposed to the flowing water in the representative elementary volume.

Figure 14 also gives curves where the previously mentioned averaging technique for Pec has been used in a flow field which has a velocity ratio of 8000.

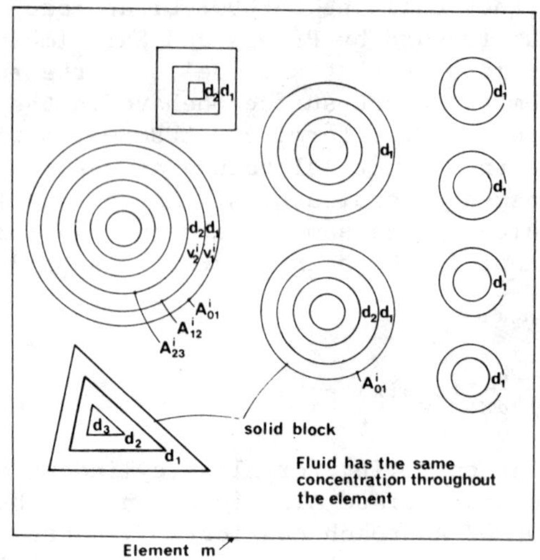

Figure 12. The principle of gathering all shells of equal distance from the surface into one volume.

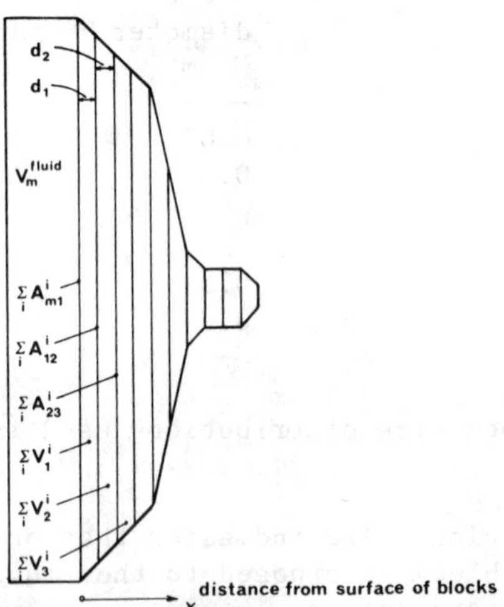

Figure 13. The sum of the areas and volumes of the blocks for various distances from the surface for the pseudobody.

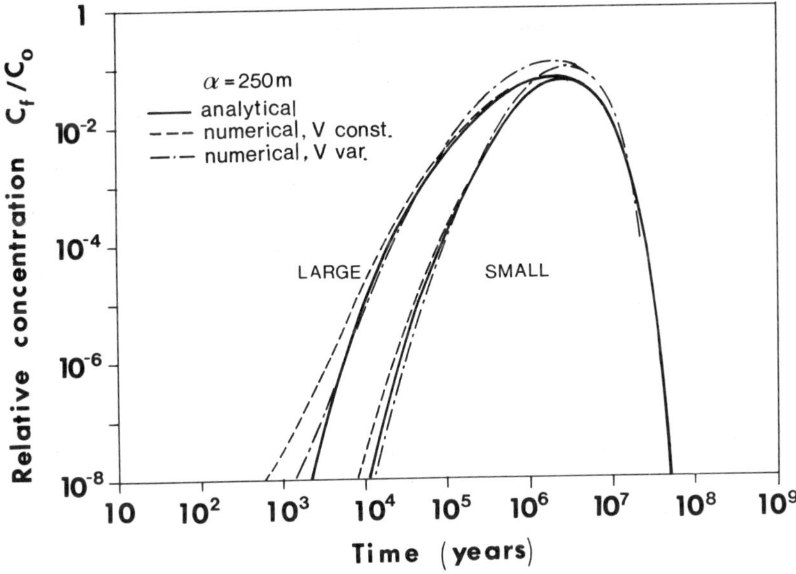

Figure 14. Computed effluent curves for Np-237 from a crushed zone with varying block sizes
1) exact solution -
2) PSEUDOBODY APPROXIMATION -
1) and 2) constant velocity and Pec = 0.875
3) PSEUDOBODY approximation and variable velocity (factor 8000) "exact" solution using dispersion proportional to velocity.

These two approximate techniques give errors which are small compared to the natural variations in data on dispersion, diffusion, porosity, blocksize distribution, velocity and equilibrium data.

The previous models are based on the concept that the fluid moves around bodies of various shapes. There may be cases when it is more appropriate to model flow in a channel in a large body of rock and diffusion out from the channel and into the rock in a diverging manner. Recently in a deep borehole in Switzerland small tube like channels were found (NAGRA NGB 5) which are deemed to carry an important part of the flow. Rasmuson and Neretnieks (44) modeled advection-dispersion and matrix diffusion for various channels geometries.

4. EXPERIMENTAL EVIDENCE

The concepts and models described previously need experimental data. The most important data are: Fracture widths and frequencies, data on dispersion, matrix diffusivity and porosity and sorption data. Some of these data may be obtained in the laboratory but some must be gathered by field experiments.

4.1. Fracture Widths and Frequencies

Fracture widths and the frequency of fractures determine the flow porosity of the rock and thus the velocity for a given flowrate. The fracture frequency determines the available wetted surface with which the dissolved species may interact.

The fracture widths have often been estimated from pressure drop measurements and by the use of the assumption that the fracture behaves like a smooth slit. The further assumption of laminar flow gives a direct relation between the flowrate and the opening. This is the so called cubic law (47, 48).

If fractures are filled with porous material of if there are closed and open parts in the plane of the fracture, the cubic law relation is not expected to hold. Tsang (56) showed theoretically that the cubic law breaks down in the latter case. There are very few experiments where both pressure drop and tracer (water) residence time has been measured simultaneously. In an experiment in Stripa which is discussed in detail later it was found that the "cubic" law fracture width was 1-2 orders of magnitude smaller than the fracture width which is obtained from the residence time.

The fracture frequency has been discussed in section 2.2.

4.2. Dispersion and Channeling

4.2.1. Injection pumping tests

Three, two well tests, have been performed in crystalline rocks within the KBS program. At the Studsvik Site Landström et al. (18) used nonsorbing tracers in a two well test in granitic rock at about 100 m depth. The travelling distances from injection hole to one observation hole was 20 m and the distance to the pumping hole which maintained the drawdown and the gradient was 51 m. The water flow was clearly localized to one section in all three holes in what appears to be one or a few fractures at close distance. Later another experiment was performed in other holes in the same area (19) but only two holes were used and at a closer spacing - 11.8 m. The third experiment was performed at the Finnsjön site (Gustavsson and Klockars) over a distance of 30 m between injection and withdrawal holes. The breakthrough curves of the nonsorbing tracers allow dispersivities to be evaluated using the advection-dispersion equation. An attempt has also been made to use the advection-dispersion-matrix diffusion model where a third parameter is determined. The experimental results have also been interpreted using an advection-channeling model with or without matrix diffusion (25). For nonsorbing tracers and the rather short residence times the impact of matrix diffusion is difficult to separate from other dispersive effects. The obtained dispersion lengths α from the above experiments are shown in figure 10 together with other data which are described below.

4.2.2. Single fracture experiments in laboratory

A series of laboratory experiments have been performed using natural fractures. In the first set of experiments (30) the rock used was a 30-cm-long granitic drill core 20 cm diameter taken from the Stripa mine at a depth of 360 m below ground level. The core has a natural fissure which runs parallel to the axis. The cylindrical surface of the drill core was sealed with a coat of urethane lacquer to prevent any water leaving the rock except through the outlet end of the fissure. The granite cylinder was mounted between two end plates containing inlet and outlet channels (Figure 15).

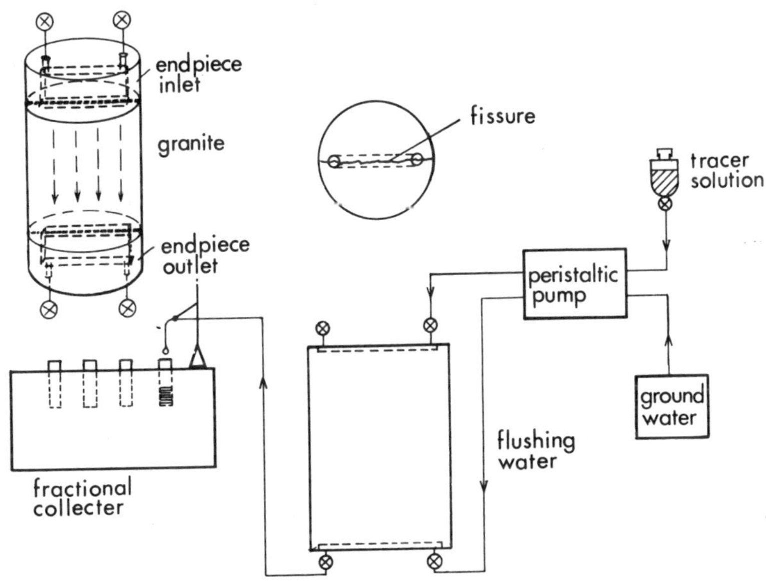

Figure 15. Experimental setup.

Artificial groundwater with a tracer was fed to the upper channe by means of a four-channel peristaltic pump ensuring a steady downward flow through the fissure. At low flow rates, flushing water was simultaneously fed through the lower outlet channel to flush the emerging tracer and so reduce the time delay due to the channel volume of the end piece. The effluent was continuously fed to a fractional collector for analysis of the tracer concentrations. The tracers were introduced, either as a step up or as a step up followed by a step down, after a suitable amount of tracer had been introduced

Figure 16 shows the breakthrough curves for tritiated water for different flow rates. The figure shows plateaus at $C/C_o = 0.7-0.8$ indicating that there are at least two channels. Similar curves were obtained using another nonsorbing tracer - a negatively charged lignosulphonate ion. Neglecting the presence of two channels the best fit to the advection-dispersion equation gives Peclet numbers ranging between 8 and 27 with an average of 14.2. The dispersion length $\alpha \cong 25$ mm. The average fracture width calculated from the residence time was 0.18 mm.

The same data were also analyzed with the advection-channeling model (Eq. 3.8b) assuming that the breakthrough curve is caused by different velocities in a multitude of independent channels. With a log normal distribution the logarithmic standard deviation was found to be 0.094 on the average.

Strontium and cesium which are sorbing tracers were also run in the same core. A typical breakthrough-curve for strontium is shown in Figure 17.

Whereas the tritiated water and the lignosulphonate ion were predicted and found to show negligible effects of penetration of the matrix, the strontium and cesium were predicted to be strongly influenced by sorption within the rock matrix and by sorption on the surface of the fracture. The predicted and the experimental results agree surprisingly well in all 13 runs with sorbing tracers. Figure 17 gives one example.

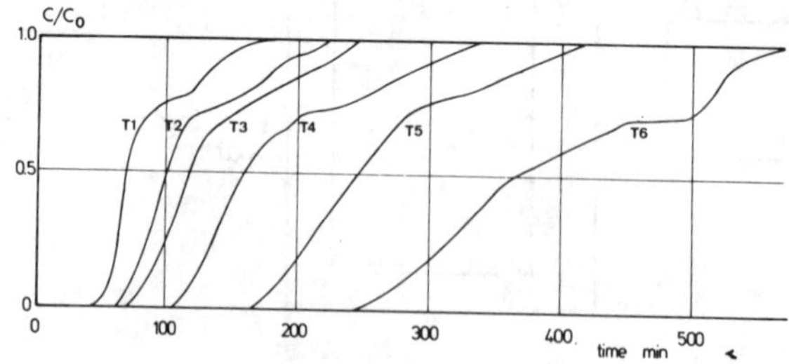

Figure 16. Experimental breakthrough curves for THO.

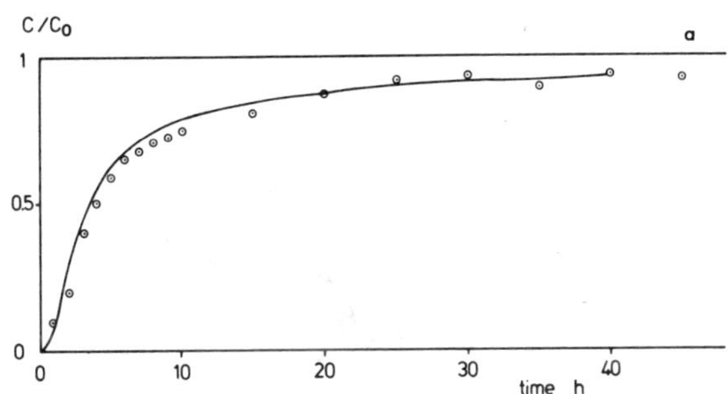

Figure 17. Experimental strontium breakthrough curve and curve predicted by advection-channeling-matrix diffusion model

In a later set of experiments (25) two more cores were used 18.5 and 27 cm long. More refined models were used which also account for the dispersion effects of the inlet and outlet end pieces. Peclet numbers of 20 and 15 and fracture widths of 0.14 and 0.13 mm were found. Predicted and experimental results on diffusivities and sorption equilibria for strontium differed much more for these cores than for the first core. The obtained results are, however, within the large range of diffusivities and sorption data found for crystalline rocks and coating and alteration materials (52).

4.2.3. Single fracture experiments in Stripa

Tracer experiments have been performed in two natural fractures in the Stripa granite 360 m below the ground (3,4). The experiments have utilized the natural inflow to the drifts which are located in water saturated rock.

The design of the experiment was based on the idea that reasonably well defined individual fractures can be located and tracers can be introduced into the natural water flow within a single fracture without a large disturbance of the flow field.

The experiment was run in naturally fractured granitic rock at similar depth as a future underground repository. Conservative (nonsorbing) tracers have been used to characterize the water flow within the fracture. The results from the runs with the nonsorbing tracers and data on sorption and porosity obtained in the laboratory have been used to predict the breakthrough curve for one of the sorbing tracers. The predicted breakthrough curve would later be compared with the experimentally obtained breakthrough curve.

Figure 18 shows the layout of the test site with two fractures and five injection holes intersecting the fracture planes. Only

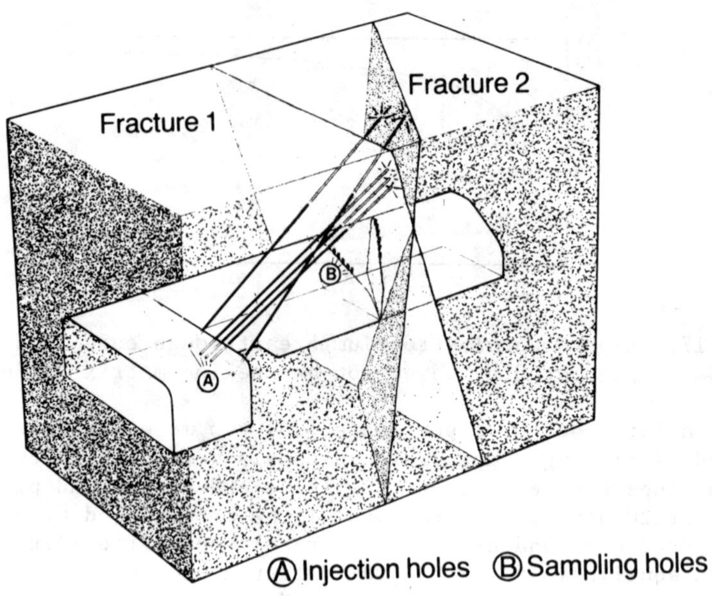

Figure 18. Schematic view of the test site.

fracture number two was utilized when injecting the tracers. To locate the connections between injection holes and sampling holes pressure pulse test were performed.

The following conservative (nonsorbing) tracers were used Brom Thymol Blue, Elbenyl, Eosin, Iodide, Uranine. The following sorbing tracers were used Cs, Sr, Eu, Nd, Th, U.

The tracers were injected into a single fracture which has a "natural" water flow towards the drift, where the water coming out of the fracture is collected. The sorbing tracers were continuously injected at one of the injection points for several months. At the injection points where only nonsorbing tracers were injected, ground water was continuously injected between tracer pulses to maintain the same flow rates everywhere throughout the experiment.

As only Sr was predicted to reach the sampling holes within the time of the experiment and the other sorbing tracers would be sorbed in the vicinity of the injection point, part of the fracture around the injection point was excavated.

Three different model concepts have been tested against experimental data, namely

1) The advection-dispersion surface sorption model equation (3.18).
2) The advection-dispersion matrix diffusion model equations (3.18 and 3.16)
3) The channeling model with no intermixing between channels equations (3.8b + 3.18 (with D_L = 0) + 3.16).

Figures 19 and 20 show the fit between the three different models and the experimental data. As can be seen in the figures it is possible to get a good fit with all the models. It is not possible to select between the models with the results from this experiment only. Some of the mechanisms and their parameters have to be determined independently.

The term fracture width or openings implies that it is a geometric property and that the fracture has a fairly constant opening. Our observations indicate that fractures are closed in some parts and are open in other parts. The opening thus may vary considerably over the "plane" of the fracture. These "channels" have at present unknown extension in the breadth direction as well as in the length direction, i.e. before they loose their identity by connecting with other channels. Any calculations of the fracture openings will thus include some assumptions on other properties.

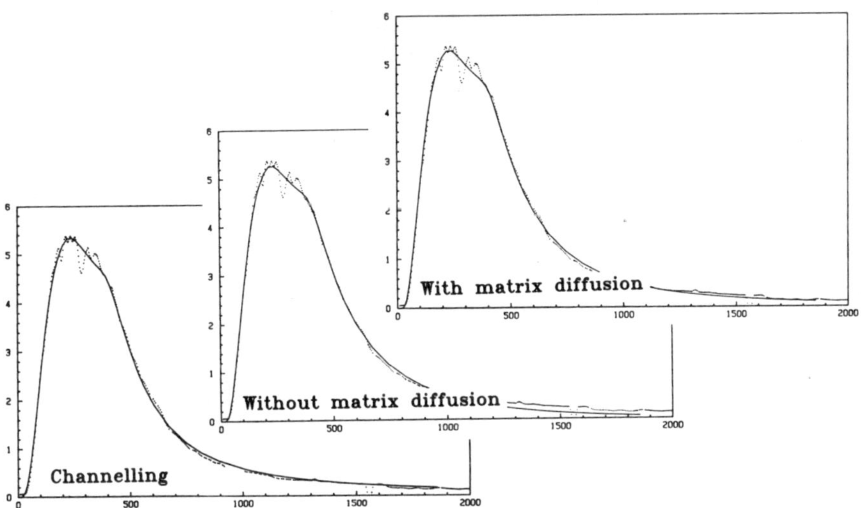

Figure 19. Fit between models and experimental data at sampling hole 2-6.

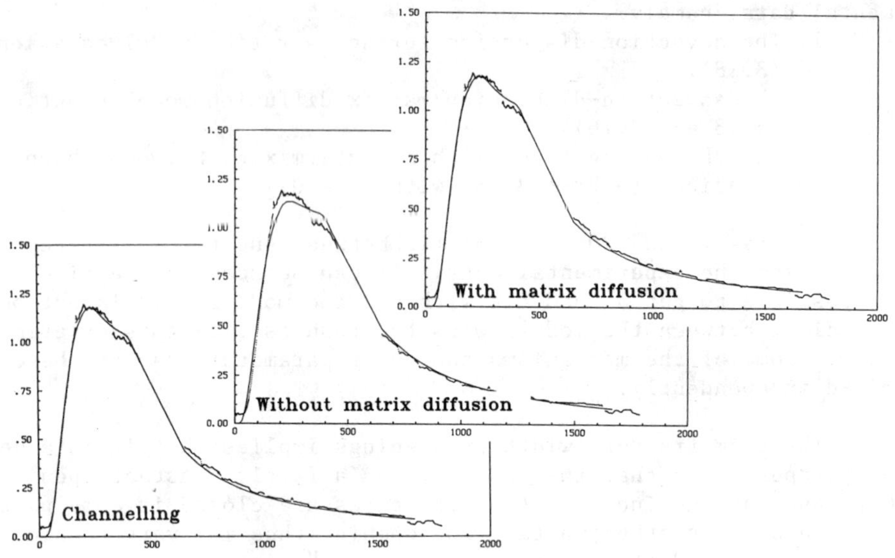

Figure 20. Fit between models and experimental data at sampling hole 2-8.

The average width of the fracture could be determined from measuring the flowrate over a breadth of the fracture and the residence time of the water. We call this *the mass balance fracture width:*

$$\delta_f = f(Q, t_w, L, B)$$

Another property of interest is that equivalent fracture width which would permit a certain flowrate at a given pressure drop. This we call *the cubic law fracture width:*

$$\delta_c = f(Q, \Delta h, L, B)$$

A third equivalent fracture width is that which would give a certain water velocity for a given pressure drop. This we call *the frictional loss fracture width:*

$$\delta_l = f(\Delta h, t_w, L)$$

The results from the calculations are given in Table 3.

Table 3. Equivalent fracture widths (μm) and Peclet numbers

Sampling hole	linear δ_f	δ_1	δ_c	radial δ_f	δ_1	δ_c	Pe
2-6	240	1.1	6.7	130	1.2	5.5	2-2.5
2-8	280	2.0	10	150	2.0	8.5	0.9-37
4*	2200	1.3	16	1200	1.4	13	
5*	1500	0.9	11	820	0.96	9.1	

* Fracture used in the preparatory investigation.

Table 3 shows the large difference between the different calculated equivalent fracture widths. It should be noted that δ_c underestimates the mean residence time for the water and δ_1 corresponds to a fracture that can not carry all the water actually measured, there would have to be several of these fractures within the fracture "plane". When calculating δ_f one assumes that all water collected over the breadth B passes the injection point and that the injected tracer directly enters this flow. If only part of the total flow passes the injection point, the fracture volume will be overestimated and thereby give a too large value of the fracture widths δ_f and δ_c.

The results from the water flow monitoring are presented in a summarized form in Figure 6. As can be seen the water flow is very unevenly distributed within the fracture planes. This type of pattern was also seen in the fracture used in a preparatory investigation (1).

The large span in Peclet number for flowpath 2-8 is due to the way the fitting is done. The use of independent information on matrix diffusivity gives higher dispersion numbers than if the diffusivity is determined in the fitting procedure together with the other parameters. This means that in the field experiment there is either a higher diffusivity in the matrix than laboratory samples indicate or that there are other causes for "dispersion" e.g. diffusion into stagnant volumes of water.

The flow parameters obtained in the runs with the conservative tracers and laboratory data on sorption have been used to predict the breakthrough curve for Sr. Sr was predicted to reach the sampling holes well within the time of the experiment. Sr did not arrive in detectable concentrations at the sampling holes. The rest of the sorbing tracers were predicted to sorb in the vicinity of the injection point. To make it possible to determine how far from the injection points they had traveled and how deep into the rock they had penetrated, part of the fracture around the injection point was excavated. Figure 21a shows the excavated fracture and the

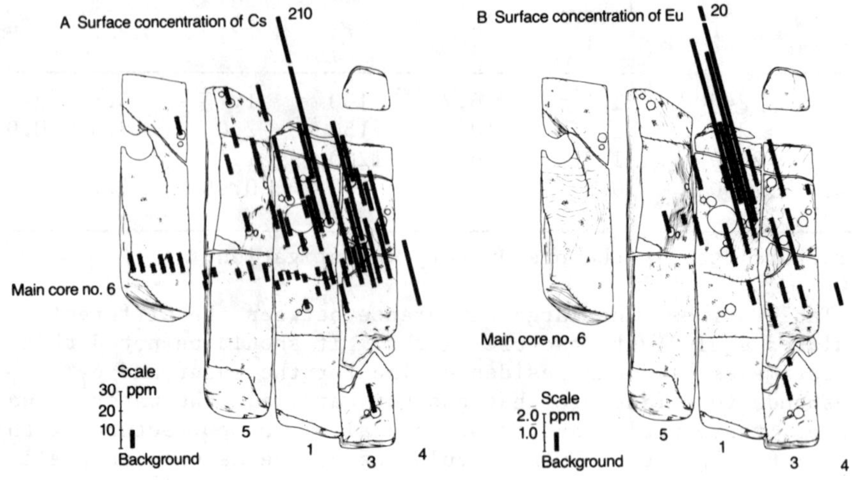

Figure 21. Excavated fracture with concentration of Cs (a) and Eu (b).

concentration of Cs. Figure 21b shows the concentration of Eu.

It can be seen in Figure 21 that there are elevated concentrations of Cs and Eu around the injection point, but there seems to be an area to the right of the injection hole which has the highest concentrations. This could indicate that the injected water was not evenly distributed around the injection point but has a preferred direction of flow.

Figure 22 shows concentration profiles into the rock obtained in one of the samples taken close to the injection hole. It can be seen in Figure 22 that elevated concentrations of Cs,Eu and Nd, at this sampling point, can be found down to a depth of 1.4 mm.

There are some inherent difficulties to interpret tests with sorbing tracers. Among these one could mention the variation in the natural content of the tracer over the fracture surface, variation in mineral composition giving different K_d values at different sites in the fracture and variations of the thickness of the fracture filling and coating materials.

As can be seen in Figures 19 and 20 it is possible to obtain a good fit with all of the models. They model different mechanisms.

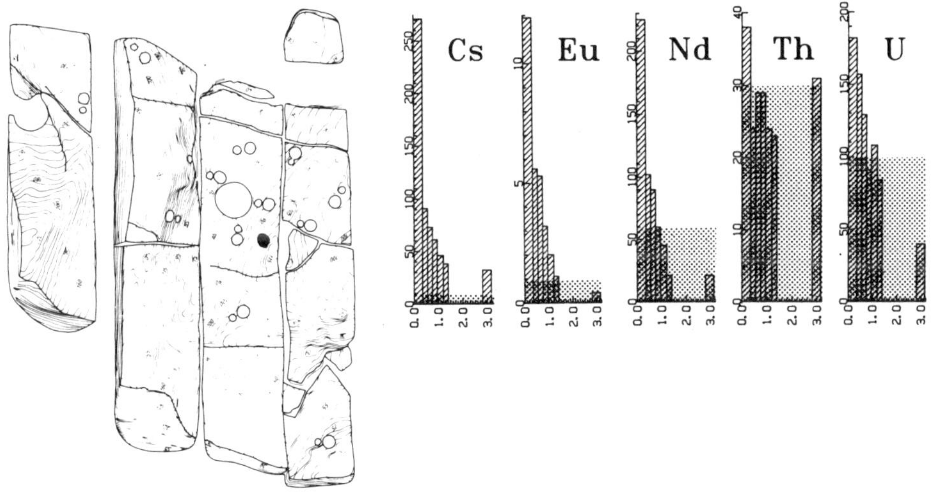

Figure 22. Concentration profiles into the rock.

The fitting thus cannot differentiate between the mechanisms. The correct mechanisms must be selected by some other independent process.

The calculated surface concentrations and the concentration profiles within the rock matrix could partly be confirmed by experimental data. They show sorbing tracers concentration profiles which can be explained by matrix diffusion.

4.2.4. Water flowrates and tracer transport to the 3D drift in Stripa

The natural flowrate to a 75 m long drift in the Stripa mine is monitored in detail (8). Small amounts of conservative (non-sorbing) tracers are injected from 9 separate injection zones located between 10 and 55 m above the drift (test site) at the 360 m level. All water emerging into the upper part of the test site is collected. The total area covered is more than 700 m^2. In order to study the spacial distribution of water flow pathways, this 700 m^2 large area is divided into more than 350 sampling areas.

The experimental drift is separated from the other tunnels and drifts by a 135 m long access drift.

The layout of the test site is shown in Figures 23 and 24. The sampling arrangement is shown in Figure 25.

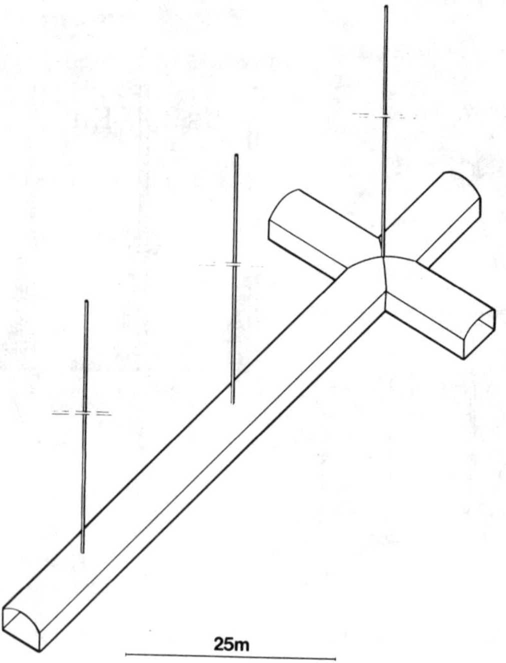

Figure 23. Layout of the test site with the three vertical injection holes

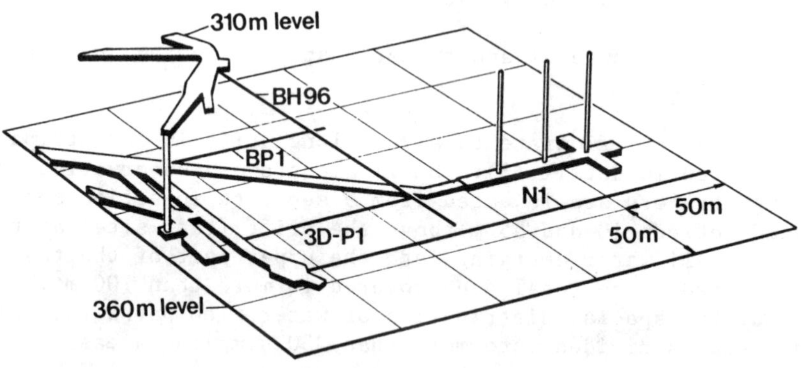

Figure 24. Map over the test site area

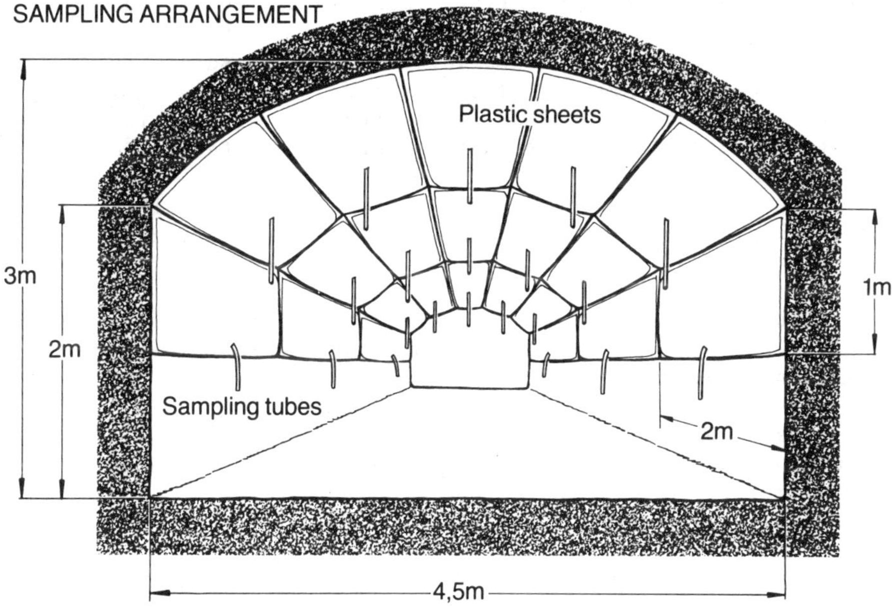

Figure 25. Sampling arrangement

Figure 4 shows the distribution of flowrates over the ceiling and walls of the drift.

When selecting where to inject tracers, information from both the test site and the injection holes were available. The most important sources of information were: water collection, stereo-photographs and logging, water inflow measurements (2 m zones) and radar measurement.

The most valuable information was the data from the water inflow measurement, since the selected zones must have a fairly high hydraulic conductivity (in this case measured as water inflow rate) if it should be possible to inject such large amounts of tracers that the tracer concentrations down in the test site would be detectable, considering that the concentration decreases during the migration due to dilution, dispersion, diffusion into the rock matrix etc.

Because of lack of information about hydraulic connection between sections of the injection holes and the test site, out of the possible zones, those zones closest to the test site were selected for the tracer injection.

With the radar reflection measurement, three fracture zones were found within the measuring limit, about 100 m. Each of the zones intersects at least one of the injection holes. Out of these three fracture zones, one was interpreted to possibly intersect the edge of test site. Most of the injection zones were chosen in such a way that they are located below any fracture zone or intersected by the fracture zone interpreted to intersect the test site.

Figure 26 shows the selected injection zones. The injections into all nine injection zones are carried out simultaneously. Therefore, nine different tracers were selected.

Out of approximately 100 dyes that were tested in a laboratory experiment, only 7 were found to be stable with time and non-sorbing. The remaining two tracers are salts.

The tracers will be injected continuously for one year. The injections are carried out with a "constant" overpressure, approximately 10-15 % above the natural pressure. The different injection flow rates vary between 1 and 20 ml/h.

Approximately 100 samples/day are taken for analysis. Depending on the water flow rates, water is collected using fractional collectors or bottles at places with low flow rates. The time interval between samples from the 65 most conductive sampling areas is about 16 hours. Samples are also taken from another 80 places, such as sampling areas with low water flow rates, wet spots at the floor, from the pilot hole, the access drift etc. The time intervals between samples from these places are 1-5 weeks, depending on the water flow rates.

Hole	Level [m]
I	31 - 33
	17 - 19
II	55 - 57
	33 - 35
	9 - 11
III	36 - 38
	28 - 30
	18 - 20
	12 - 14

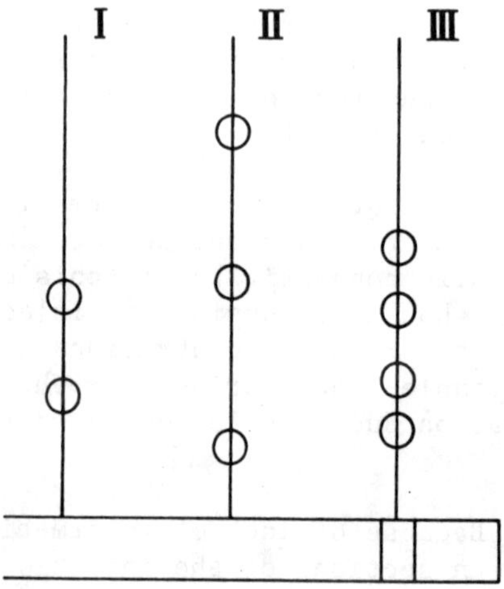

Figure 26. Location of the injection zones

Results up to April 1985:

After six months of injection (April 1985), tracers from five injection zones are seen in about 35 plastic sheets, among which 20 are connected to fractional collectors (see Figures 27 and 28).

Looking at Figures 27 and 28, it is remarkable that Br^-, injected rather close to the ceiling in hole III, has not been found in the sampling areas close to hole III, but in many of the sampling areas around hole II. Figure 28 shows those sampling areas that have "high" water flow rates (i.e. are connected to fractional collectors) and in which tracers have been found.

Present models for describing the water flow in crystalline rock assume that the rock can be described as a homogeneous porous media, at least for large volumes.

The results from the water collection (see Figure 4) clearly shows that water does not flow uniformly in the rock over the scale considered (700 m^2), but seems to be localized to wet areas with large dry areas in between.

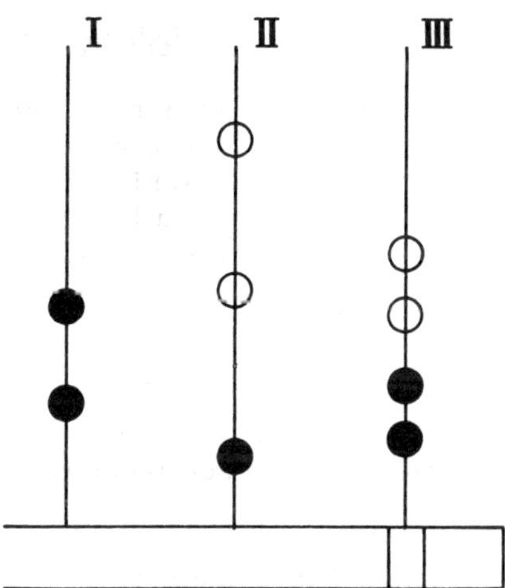

Figure 27. Tracers emerged into the test site (April 1985)

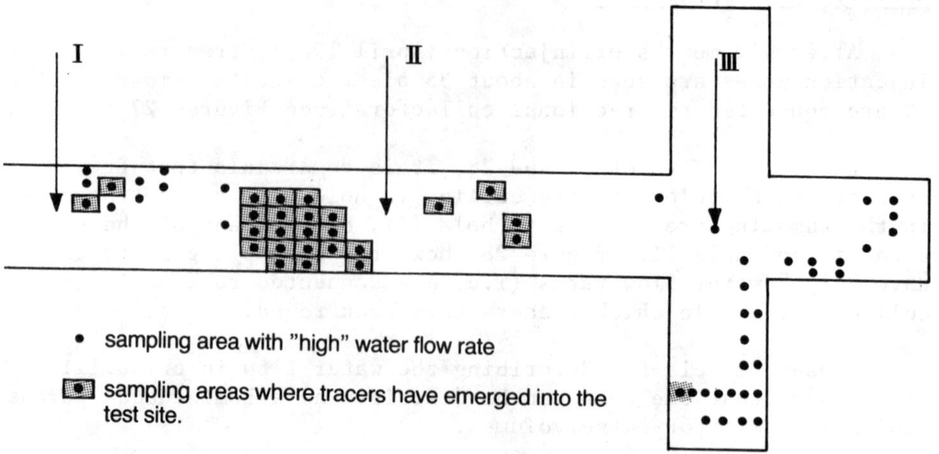

Figure 28. Tracer distribution (April 1985)

Looking at the first preliminary results from the tracer runs which shows where and when tracers have emerged into the test site, it is obvious that the water flow can not be described as a homogeneous porous media flow in this site at this scale. Channeling seems to play an important role.

4.3. Diffusion Experiments in the Laboratory

Crystalline rock has been found to be porous with connected porosities in the range of 0.06 - 1 % or even higher in some samples. Dissolved species can access the matrix porosity by molecular diffusion but not by flow when hydraulic gradients are small as under most natural conditions. Sorbing species may access and sorb on the inner surfaces of the matrix. The capacity for sorption is very much larger there than on the surfaces of the sparse water conducting fractures.

Porosities and diffusivities of granites and gneisses from different areas in Sweden have been measured by Skagius and Neretnieks (50,51,52,53).

Two different methods were used to determine the porosity of the rock pieces studied in the diffusion experiments. The first method was a "water saturation" method (method 1) and was made on the pieces before the diffusion experiment was started. The pieces were dried and weighed. Thereafter the pieces were resaturated with water and weighed. The second method was a leaching method (method 2). The pieces were saturated with a solution of iodide, Uranine or Cr-EDTA of known concentration. The amount of component each piece contained in the saturated condition was determined by leaching out the component from the piece with distilled water. From this information the pore volume and the porosity of the piece were determined. The porosity measurements by the second method were carried out after the diffusion experiments. The method used in the diffusion experiments is straightforward. A hole with the same dimension as the piece of the rock, was made in a 10 mm thick PVC-plate. The piece of the rock was fixed in the hole with silicon glue. The plate with the rock sample was then heated in a vacuum chamber and saturated with distilled water by the same method used in making the porosity measurements, the water saturation method. After saturation two chambers made of transparent PVC were fastened on to the PVC-plate, one on each side (see Figure 29). In the experiments with Uranine and Cr-EDTA one of the chambers was filled with distilled water and the other was filled with a solution containing Uranine (~ 10 g/l) or Cr-EDTA (~ 8 g/l). In the first experiments with iodide one chamber was filled with distilled water and the other with a solution containing 1 mol/l of sodium iodide. Later it was shown that there had been some erosion of the rock samples that had been in contact with the 1 mol/l sodium iodide solution. In all the following experiments a solution containing 0.1 mol/l sodium iodide was used instead. To avoid any osmotic effects the other chamber was filled with 0.1 mol/l sodium nitrate solution instead of distilled water. This gives equal ionic strength on either side of the rock piece.

Samples (10 ml) were taken from the chamber which at the outset contained distilled water or sodium nitrate solution. The concentration of the diffusing component was measured. The iodide concentration was measured using an ion selective electrode, the concentration of Uranine using UV-spectrophotometry and the concentration of Cr-EDTA using atomic absorption spectrometry. Each time a sample was taken out, 10 ml of distilled water or sodium nitrate was added to the chamber to keep the volume in the chamber constant.

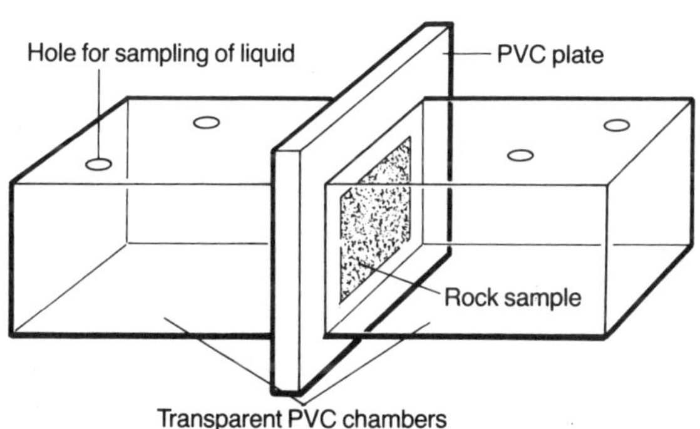

Figure 29. The diffusion cell

The diffusivity and the porosity (or inner capacity) may be determined by comparing the experimental data with the solution to Fick's law. The rate of change of concentration at a point in a one-dimensional system is given by Fick's second law

$$\frac{\partial C}{\partial t} = D \frac{\partial^2 C}{\partial x^2} \tag{4.1}$$

D is the diffusion coefficient. In this case with diffusion in a porous material the apparent diffusion coefficient must be used to account for porosity, tortuosity and sorption effects. The total porosity of the material is here looked upon as the sum of the "transport" porosity and the "storage" porosity. The storage porosity just influences the accumulation in the system. Eq. (4.1) can then be written

$$(\varepsilon_{tot} + K_d \cdot \rho) \frac{\partial C}{\partial t} = D_p \cdot \varepsilon^{+} \frac{\partial^2 C}{\partial x^2} \tag{4.2}$$

Comparing Eq. (4.1) and (4.2) gives

$$D = \frac{D_p \cdot \varepsilon^{+}}{\varepsilon_{tot} + K_d \cdot \rho} = \frac{D_e}{\alpha} \tag{4.3}$$

where $D_e = D_p \cdot \varepsilon^{+}$ is the effective diffusion coefficient and $\alpha = \varepsilon_{tot} + K_d \cdot \rho$ is a rock capacity factor.

The solution of Eq. (4.2) for the case of diffusion through a porous slab initially at zero concentration, with constant inlet concentration C_1 at $x = 0$ and outlet concentration C_2 ($C_2 << C_1$) at $x = \ell$ is

$$\frac{C(x,t)}{C_1} = 1 - \frac{x}{\ell} - \frac{2}{\pi} \sum_{n=1}^{\infty} \frac{1}{n} \sin \frac{n\pi x}{\ell} \exp\left(- \frac{D_e \cdot n^2 \pi^2 \cdot t}{\ell^2 \cdot \alpha}\right) \tag{4.4}$$

The rate at which the diffusing substance emerges from a unit area of the face $x = \ell$ of the slab is given by differentiating Eq. 4.4 and putting it into Fick's first law

$$N = - D_e \left.\frac{\partial C}{\partial x}\right|_{x=\ell} \tag{4.5}$$

By integrating Eq. (4.5) with respect to the time t, the total amount of diffusing substance M which had passed through the slab in time t is obtained

$$\frac{M}{\ell C_1} = \frac{D_e \cdot t}{\ell^2} - \frac{\alpha}{6} \frac{2\alpha}{\pi^2} \sum_{n=1}^{\infty} \frac{(-1)^n}{n^2} \exp\left(- \frac{D_e \cdot n^2 \cdot \pi^2 \cdot t}{\ell^2 \cdot \alpha}\right) \tag{4.6}$$

As $t \to \infty$ Eq. (4.6) approaches the linear relation

$$M = \frac{C_1 \cdot D_e}{\ell} \cdot t - \frac{C_1 \cdot \ell \cdot \alpha}{6} \tag{4.7}$$

with the slope $C_1 \cdot D_e/\ell$ and an intercept on the time axis $t = \ell^2 \cdot \alpha/6 \cdot D_e$.

If the diffusing component is not being sorbed on the material then $\alpha = \varepsilon_{tot}$, which means that the intercept on the time axis gives the total porosity of the material.

If the transport only takes place in the pore water then the relation between the effective diffusivity D_e and the bulk phase diffusivity D_v for a component can formally be written

$$D_e = D_p \cdot \varepsilon^+ = \frac{\varepsilon^+ \cdot \delta_D}{\tau^2} \cdot D_v \tag{4.8}$$

where δ_D is the constrictivity and τ the tortuosity of the porous material. Providing no size factors, pore sizes or/and sizes of the diffusing component, influence the diffusion, the formation factor or the diffusivity will only depend on the properties of the porous material.

The formation factor $\frac{\varepsilon^+ \delta_D}{\tau^2}$ in equation (4.8) may also be obtained by measuring the electric resistivity of the rock sample R_s and its pores are filled with a solution with known resistivity R_o, provided the conduction takes place only in the liquid phase. Then

$$\frac{D_e}{D_v} = \frac{\varepsilon^+ \delta_D}{\tau^2} = \frac{R_o}{R_s} \tag{4.9}$$

In Figure 30, the logarithmic value of the effective diffusivities of iodide in the granites, gneisses and fissure coating materials are plotted versus the logarithmic value of the experimental porosities determined by the leaching method, and in Figure 31 versus the logarithmic value of α from Eq. (4.7). Results obtained by Bradbury et al. (11) for iodide diffusion in different granites from the United Kingdom are also presented in the figures for comparison. For the granites and the gneisses a linear regression has been made

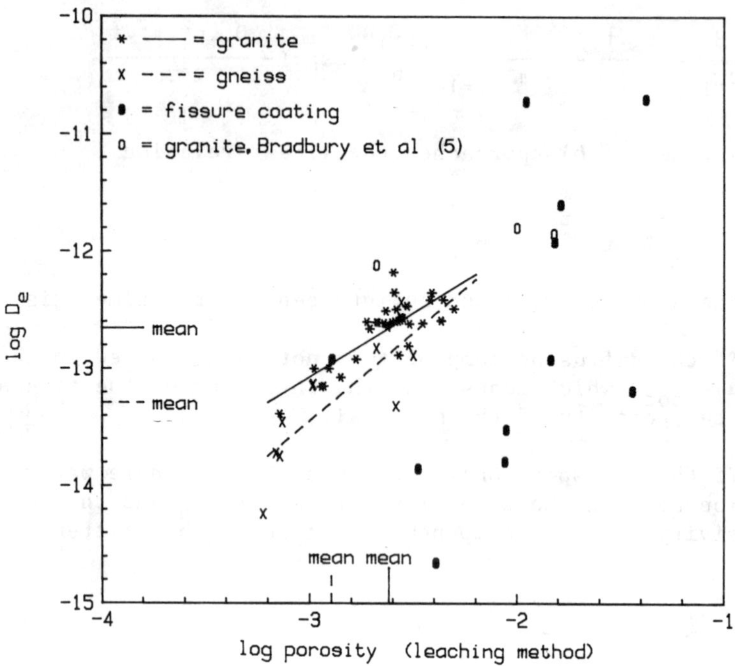

Figure 30. The logarithmic value of the effective diffusivity of iodide in the rock materials versus the logarithmic value of the porosity from the leaching method (P2).

(the lines in the figures) and the mean values, both logarithmic and arithmetic, of the effective diffusivities, porosities and α-values have been calculated. The logarithmic mean values are marked in the figures. The effective diffusivity in the granites, logarithmic mean value = $22.0.10^{-14} m^2/s$ and arithmetic mean value = $25.2.10^{-14} m^2/s$, is higher than in the gneisses, logarithmic mean value = $5.1.10^{-14} m^2/s$ and arithmetic mean value = $9.2.10^{-14} m^2/s$. The mean values of the porosity determined by the leaching method are also higher for the granites, logarithmic = 0.24 % and arithmetic = 0.26 %, than for the gneisses, logarithmic = 0.13 % and arithmetic = 0.15 %. The same holds for the α-value where the granites have a logarithmic mean = 0.29 % and an arithmetic mean = 0.33 % and the gneisses have a logarithmic mean = 0.15 % and an arithmetic mean = 0.29 %. For both granites and gneisses the mean α-values are higher than the mean experimental porosity values.

Figures 30 and 31 show that the effective diffusivity of iodide in the fissure coating materials is of the same order of magnitude

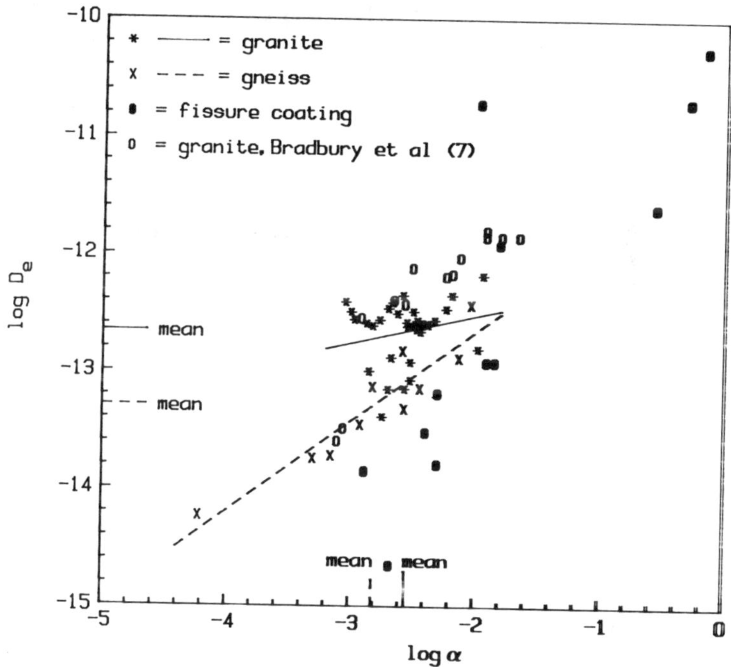

Figure 31. The logarithmic value of the effective diffusivity of iodide in the rock materials versus the logarithmic value of α from Eq. 4.7.

or higher as in the granites and the gneisses. Those samples that have an effective diffusivity that is of the same order of magnitude as the granites and the gneisses have, however, higher α-values and much higher porosity values. This could be due to a higher "storage" porosity in the fissure coating material, or to a lower pore diffusivity in the fissure coating material compared with the granites and the gneisses.

At set of samples from Finnsjön, Fi 88 and Fi 89, are taken at different distances from fissure surfaces. The porosity and diffusivity of iodide were measured to find out if there was any variation with distance from the fissure. Figures 32 and 33 show the experimental porosity versus distance from the fissure. Fi 88 shows no obvious variation in porosity with distance from the fissure. For Fi 89 the porosity decreases with distance up to about 80 mm from the fissure, and then remains rather constant or increases slightly. In Figure 34 and 35, the diffusivities of iodide are plotted versus distance from the fissure. In the samples from Fi 88 and from Fi 89 the diffusivities do not show any obvious dependence on the distance from the fissure. In the samples from Fi 89 which

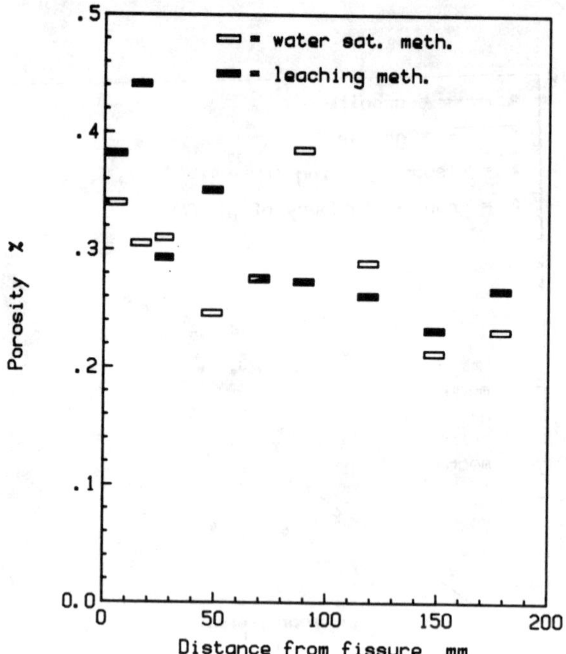

Figure 32. Porosity of samples from Finnsjön, Fi 88, versus distance from a fissure surface.

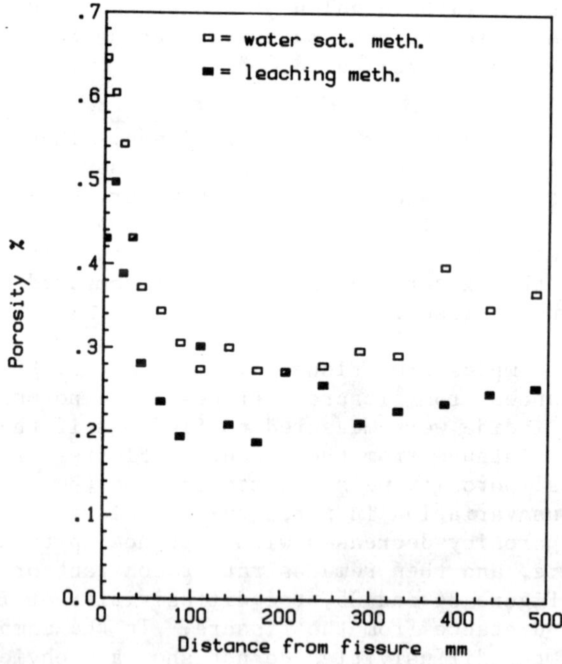

Figure 33. Porosity of samples from Finnsjön, Fi 89, versus distance from a fissure surface.

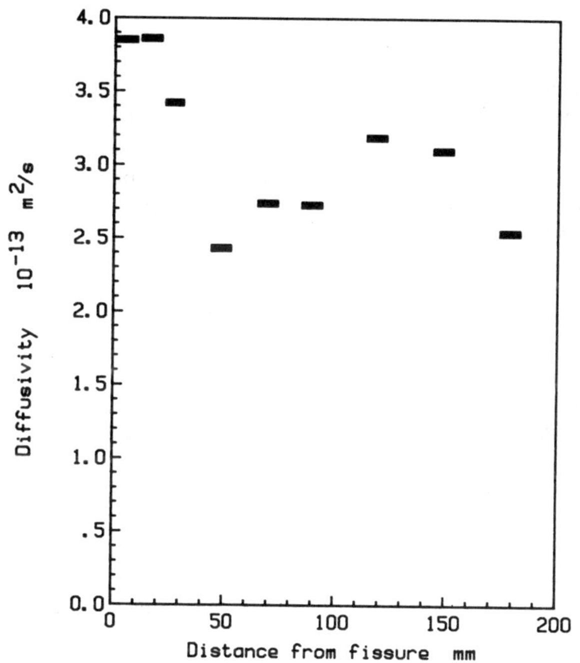

Figure 34. Effective diffusivity in samples from Finnsjön, Fi 88, versus distance from a fissure surface.

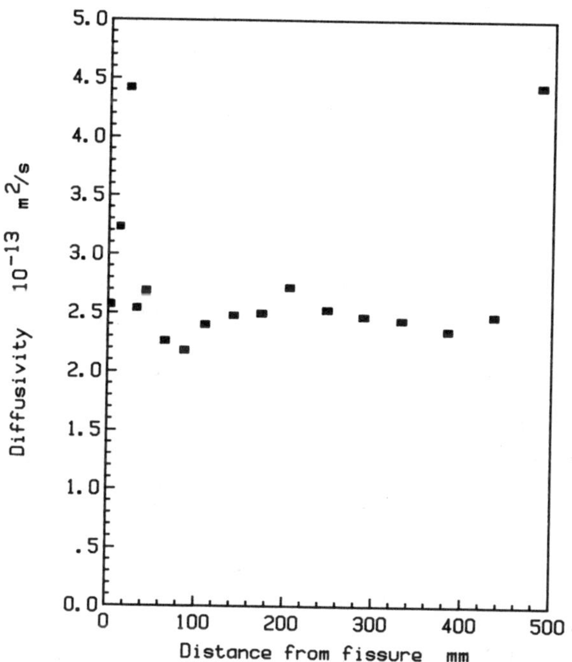

Figure 35. Effective diffusivity in samples from Finnsjön, Fi 89, versus distance from a fissure surface.

were taken near the fissure where the porosity was higher, one would expect a higher diffusivity. The experimental porosity is, however, a total porosity value. The diffusivity is dependent only on the "transport" porosity. Thus a higher total porosity does not have to give a higher diffusivity. The "transport" porosity could have about the same value even if the total porosity increases.

At expected radioactive waste repository depths in the ground the rock is exposed to rather high stresses caused by the large overburden of rock. When drillcores are taken up from the ground this overburden no longer exists. As a result of this there might be an increase in the porosity of the rock samples. The effective diffusivity measured in rock samples under atmospheric pressure in the laboratory would then be higher than the effective diffusivity in the rock "in situ".

To simulate the stress that may exist in the bedrock at large depths, diffusion experiments with iodide and electrical resistivity measurements in rock materials under mechanical stress were performed (53).

The apparatus used in the diffusion experiment is shown in Figure 36. On each side of a water saturated rock sample (ϕ 42 mm, ~ 10 mm thick), a plate of stainless steel with circular channels was mounted. The channels in each plate connect to a circular system.

At the high concentration side a 0.1 mol/l sodium iodide solution, and at the low concentration side a 0.1 mol/l sodium nitrate solution was circulated. At different times small samples were taken out from the storage bottle at the low concentration side.

The apparatus used in the electrical resistivity measurements was similar to the apparatus used in the diffusion experiments. The difference is that insulated electrical wires connect to the end plates instead of the tubes of the circulation system.

In the electric conductivity experiments the rock cores were saturated with 1 mol/l NaCl solution. The high concentration of the salt-water solution ensures that the conductivity of the water in the pore volume dominates over surface conductivity on the mineral surfaces.

Figure 37 shows a plot of the concentration at the low concentration side versus time for iodide diffusion through a piece from Svartboberget. The experiment was started with the piece under atmospheric pressure. After 35 days the pressure was increased to 330 bars.

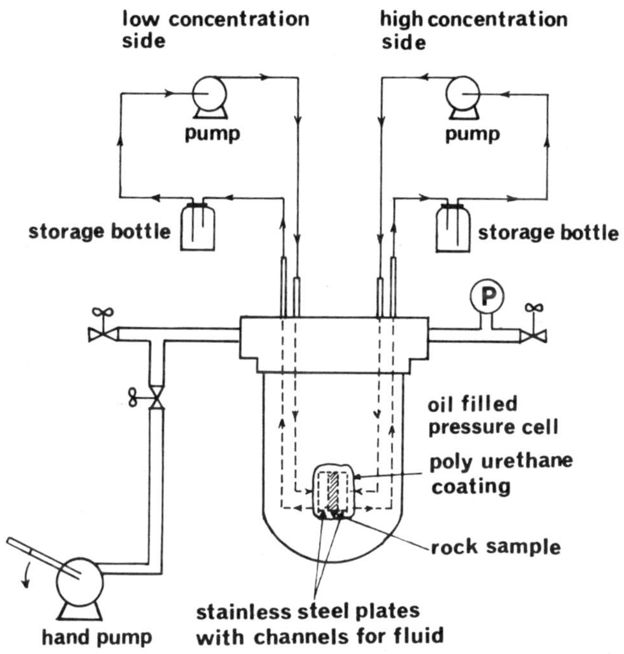

Figure 36. The apparatus used in the diffusion experiments with rock materials under mechanical stress.

It is seen that there is a marked decrease in diffusivity when the pressure is increased.

In the resistivity measurements the pressure was raised in steps, and the resistance was measured at each level. From the resistance the resistivity was calculated, and then the formation factor $\delta^{+}.\delta_{D}/\tau^{2}$, was determined. The resistance was also measured as the pressure was lowered from the maximum value down to atmospheric pressure.

Figure 38 shows the formation factor versus pressure for a granite sample from Finnsjön, where the procedure with increasing and decreasing the pressure have been made two times on the same rock sample. The formation factor decreases with increasing pressure, and then increases again when the pressure is lowered down to atmospheric pressure, however, not to the same values. The second time the procedure with increasing and decreasing the pressure was performed, about the same values as the first time was obtained for pressures higher than 100 bars. The decrease in the formation factor with pressure indicates that the crosssectional area of the

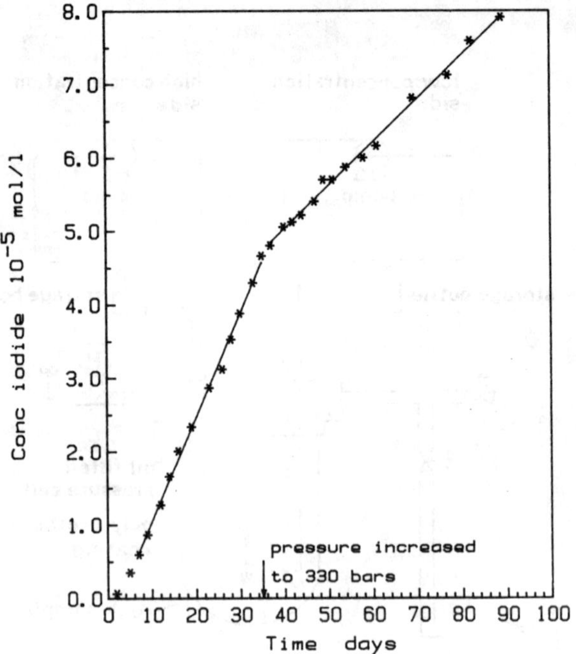

Figure 37. Concentration of iodide versus time, diffusion through biotite gneiss from Svartboberget.

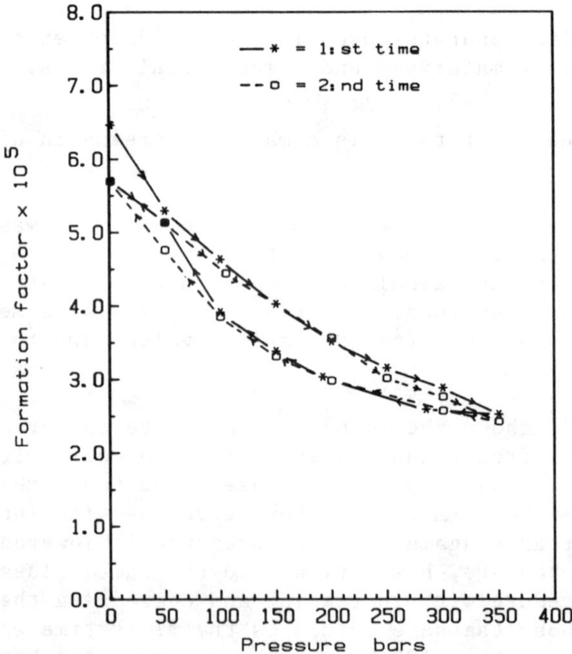

Figure 38. Formation factor versus pressure for a granite sample from Finnsjön.

pores in the sample is decreased when the sample is under mechanical stress.

The formation factors calculated from the diffusion experiments under stress are in fair agreement with the formation factors from the resistivity measurements at higher pressures.

In Figures 39 and 40 the results from the various methods are compared. Figure 39 shows the diffusivities of iodide determined in granites and Figure 40 the diffusivities of iodide determined in gneisses. The open bars represents the diffusivities in samples under stressed conditions and the filled bars the diffusivity in samples under atmospheric pressure. The diffusivity of iodide from the electrical resistivity measurements have been calculated by Eq. 4.9 using the formation factor for the unstressed samples and the formation factor at maximum stress for the stressed samples.

The formation factor in samples at 300-350 bars stress relative to the formation factor in the samples at atmospheric pressure obtained in the diffusion experiments are in fair agreement with those determined from the electrical resistivity measurements for the same rock materials. Electrical resistivity measurements can be used to give approximate values of the diffusivity. The advantage with electrical resistivity measurements is that the experimental time is much shorter than in the diffusion experiments.

The diffusivity or the formation factor in samples at 300-350 bars were in no case lower than about 20% of the value in unstressed samples.

4.4. Diffusion in the Field

Two field migration experiments in the rock matrix have been made (7). The experiments have been carried out in "undisturbed" rock, that is in rock under its natural stress environment. Since the experiments were performed at the 360 m-level (in the Stripa mine), the rock was subject to nearly the same conditions as the rock surrounding a nuclear waste repository as proposed in the Swedish concept (KBS).

The experiments had to be designed so that the influence of the stress field caused by the drift and the drilling itself was eliminated.

Near drillholes and drifts, the rock stresses will be changed compared to "undisturbed" rock. A general rule in these cases is that the rock stresses are changed about 2 hole diameters out from and below the hole. That is, outside these 2 hole diameters essentially "undisturbed" rock exists.

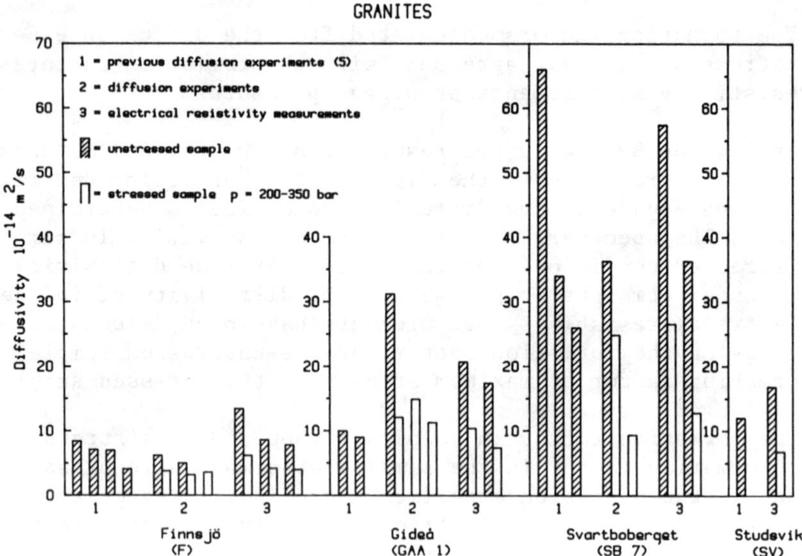

Figure 39. Effective diffusivities in granites.

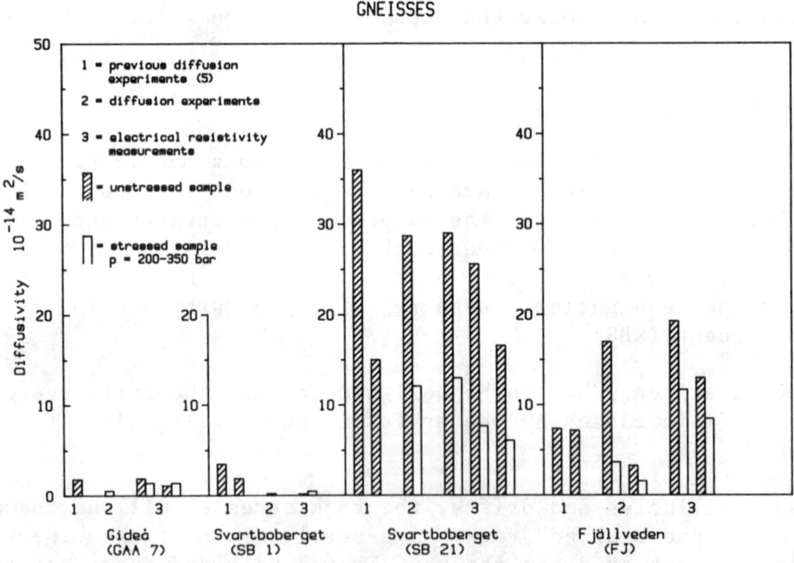

Figure 40. Effective diffusivities in gneisses.

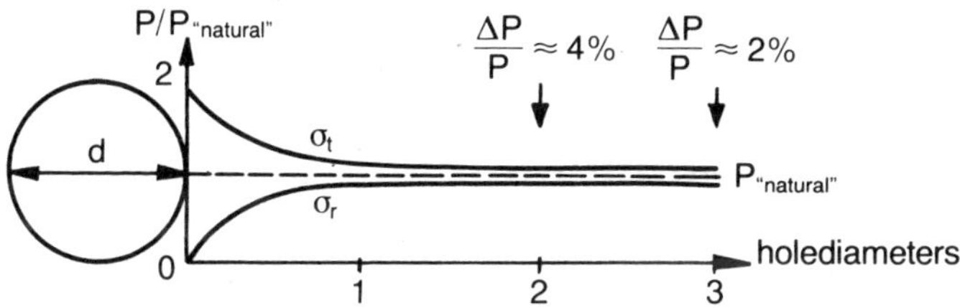

Figure 41. $\frac{\Delta P}{P}$ vs. distance from a hole.

Since the objective with this experiment is to do a migration experiment in "undisturbed" rock, the experiment had to take place more than 2 drift diameters below the drift.

In the second of the experiments which is described here a 15 m deep 146 mm hole was drilled. At this distance from the drift the changes in rock stresses due to the drift can be neglected, i.e. essentially "undisturbed" rock is reached. At the depth of 15.5 - 17.5 m a rock stress measurement was performed by the Swedish State Power Board, which confirmed that "undisturbed" rock was reached.

However, even if the changes due to the drift can be neglected, the existence of the 146 mm hole will cause a further change in the rock stresses approximately 0.3 m (2 hole diameters) outward and below the bottom of the hole. Thus, in the bottom of the 146 mm hole a 20 mm hole (approximately 3 m long) was drilled. This 20 mm hole will cause a change in the rock stresses approximately 4 cm outward, but outside this disturbed zone and 0.3 m below the larger hole essentially "undisturbed" rock is reached.

With the 146 mm packer positioned just above the little hole (see Figure 42), the little hole serves as injection hole in this experiment.

If tracers can migrate from the little hole (injection hole) past the disturbed zone and into "undisturbed" rock, this experiment will indicate the existence of a connected pore system in "undisturbed" rock.

After drilling the holes, one small packer was placed in the little hole and one big packer was placed near the bottom of the big hole (see Figure 42). The small as well as the big packer were mechanically operated. The function with the big packer was just to close off the injection compartment from the rest of the hole. The small packer was used to get a nylon tube down to the bottom of the small hole, in order to get a good circulation when

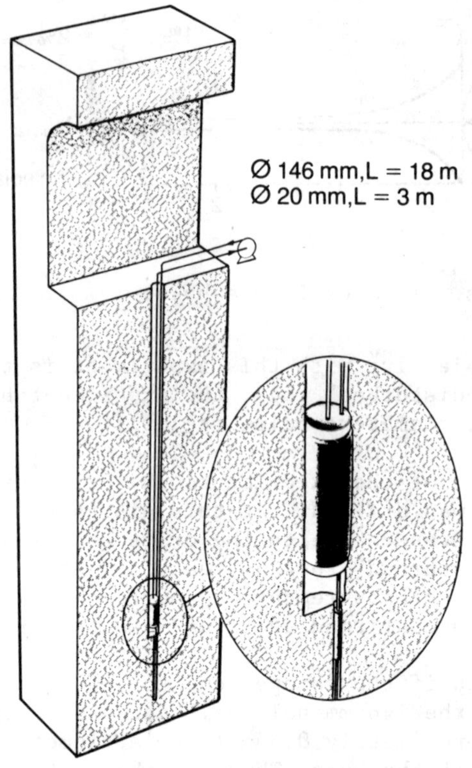

Figure 42. Drilling dimensions and packer positions.

the tracers were injected.

After the installation of the packers, the water flow into the little hole and the water pressure was monitored. Since there was no measurable inflow of water into the injection compartment, no reliable value on the water pressure could be found.

According to other measurements in the Stripa mine at the same level and only ≈ 100 m from the drift where this experiment has been performed the water pressure 18 - 21 m below the drift is expected to be between 1.0 and 1.4 MPa.

A pressure of 1.5 MPa (i.e. 0.1 - 0.5 MPa overpressure) was used during the whole injection time. This small overpressure ensured that the tracers would migrate by flow and diffusion out from the injection hole and into the pore system of the rock matrix. The overpressure was obtained by using compressed nitrogen gas.

The tracers were iodide, Chrome-EDTA and Uranine. All tracers were previously found not to react with the rock.

After about 6 months the injection was terminated. The packers were retrieved and the little hole was overcored. The core from the overcoring had a diameter of 132 mm and was ≈ 3.5 m long, with injection hole (ϕ 20 mm) at the side. The core was cut into ≈ 5 cm long cylinders (see Figure 43)

From these cylinders, a number of sampling cores (ϕ 10 mm) were drilled at different distances from the injection hole. These sampling cores were leached in distilled water (see Figure 44).

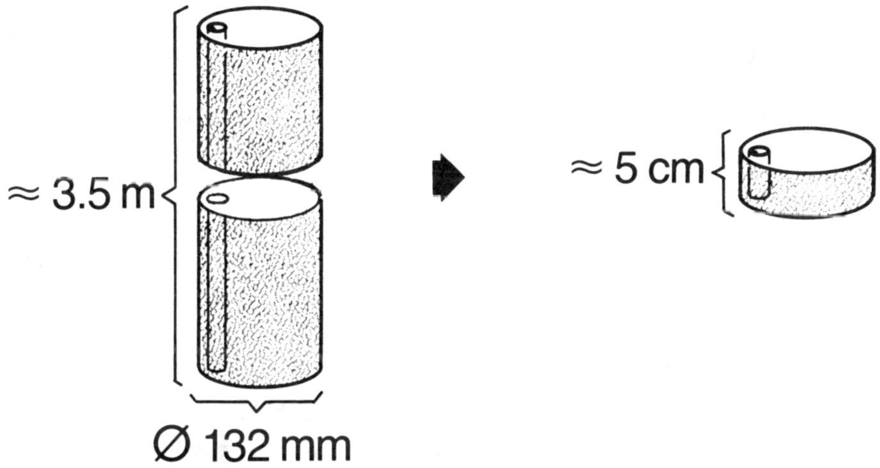

Figure 43. Sampling, step 1.

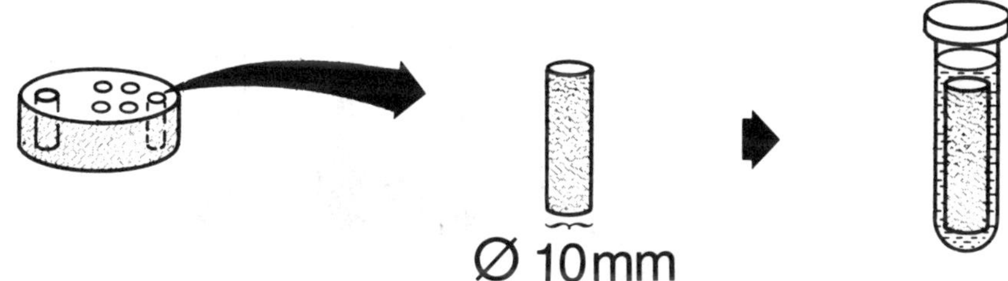

Figure 44. Sampling, step 2.

The tracer concentration in the distilled water was determined and recalculated for the concentration in the pore water. The recalculated concentration is based on the porosity that was obtained for every individual sampling core. Porosity is obtained from the weight difference betweeen wet and dry core. After this overcoring, which made it possible to study the concentration profile approximately 11 cm outward, another hole was drilled (see Figure 45).

The distance between the cores was ≈ 19 mm (i.e. ≈ 5 mm distance between holes) at the depth of interest (18 - 21 m).

With this "extra core", the concentration profile could be studied approximately 25 cm outward from the injection hole.

The sampling procedure for core 2 was the same as for core 1. A total number of ≈ 650 sampling cores were drilled (≈ 400 from core 1 and ≈ 250 from core 2).

From core 1 samples were taken at 22 different depths, which made it possible to study the variation in migration distance in the matrix versus depth.

Because of core 2, the concentration profile could in some cases be followed 25 cm outward from the injection hole at approximately the same depth.

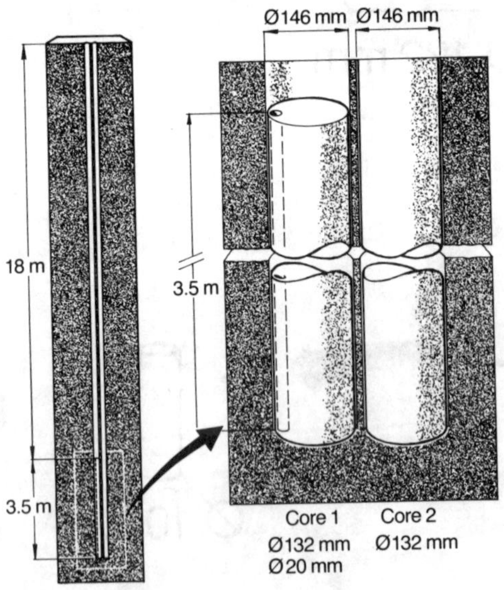

Figure 45. Overcoring arrangements.

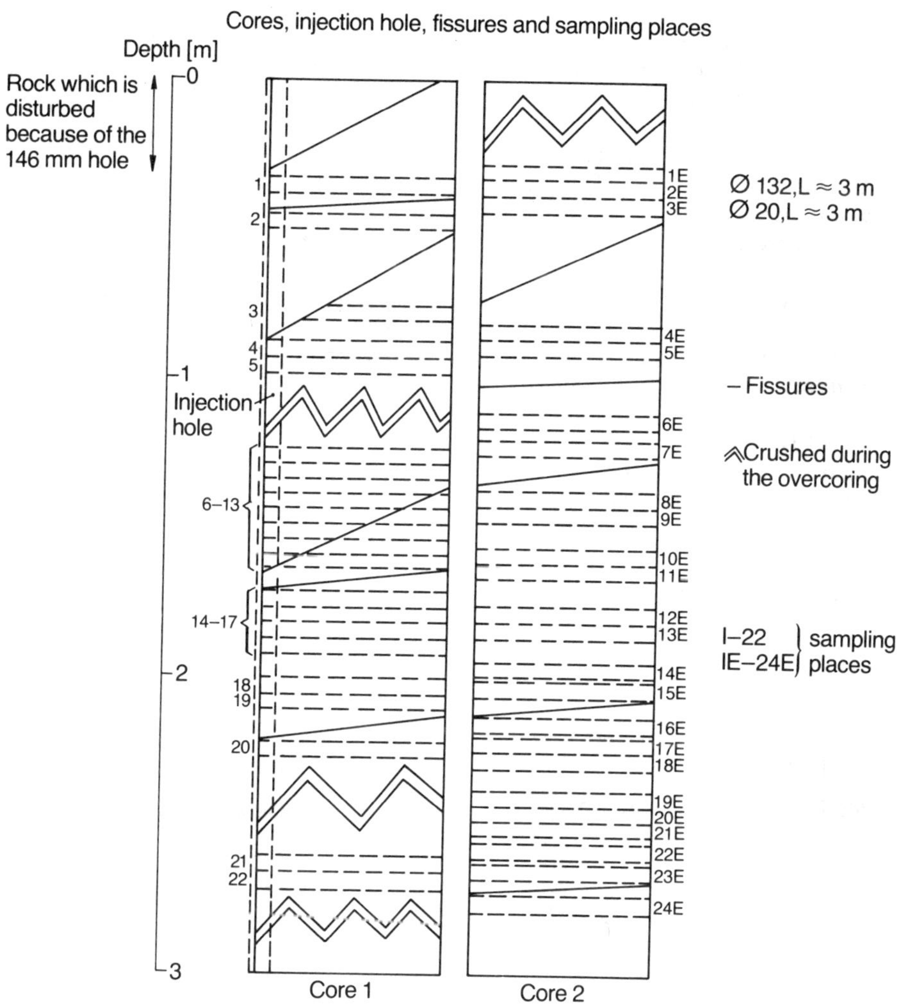

Figure 46. Core description.

The experimentally obtained concentration profiles were compared with theoretically calculated profiles. These were obtained by solving the advection dispersion equation for radial flow. The equations which predict the concentration profile of a non-sorbing component when radial diffusion and flow (convection) occur simultaneously are

Diffusion equation: $$\frac{\partial C}{\partial t} + V_r \frac{\partial C}{\partial r} = D_p \frac{1}{r} \frac{\partial}{\partial r} \left(r \frac{\partial C}{\partial r}\right) \qquad (4.10)$$

Radial flow equation: $V_r = \frac{\text{const.}}{r}$ (4.11)

The initial and boundary conditions used imply that there is no tracer in the rock at the start and constant concentration in the injection hole at all times thereafter and that there is steady flow.

The results of the experiments show a considerable variation in migration distance with depth. The penetration depth could in some cases vary with a factor 3 or more in sampling places that were separated by just a few tens of centimeters in depth.

The pervading trend is that all three tracers have migrated a long distance into the rock matrix at the top and the bottom of the injection hole, while the migration distance is rather short in the middle section of the hole.

Figure 47 shows the concentration profile obtained at a point about 0.4 m below the bottom of the largest hole. Figure 48 shows the same for a depth of 2.6 m.

Cr-EDTA and Uranine usually migrated a somewhat shorter distance (see Figure 49).

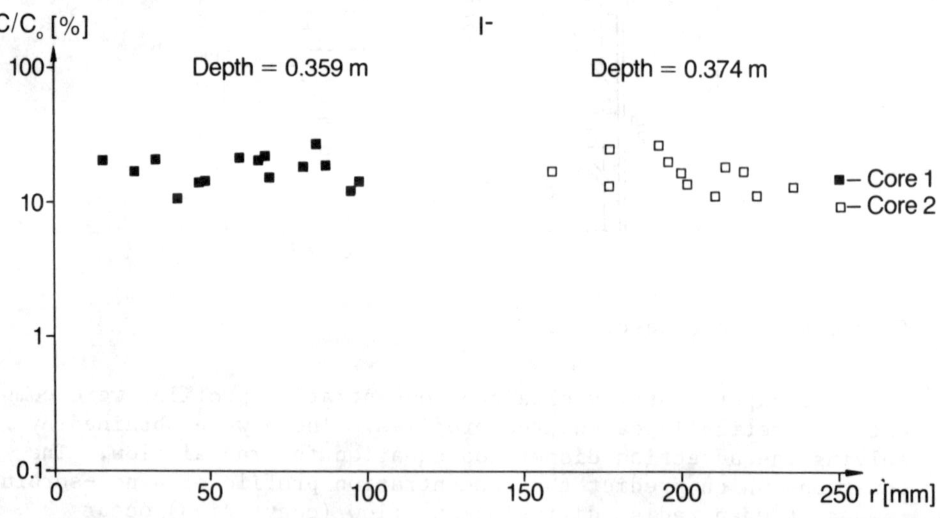

Figure 47. Tracer concentration vs. distance from injection hole for sampling places 1 and 2E.

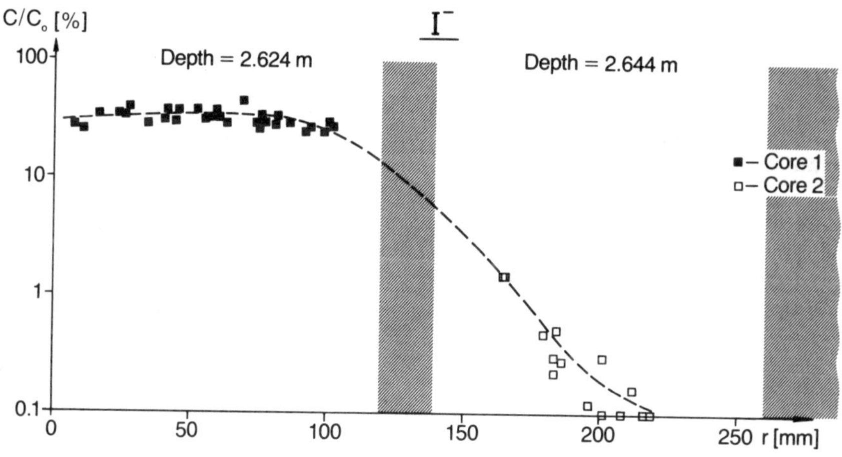

Figure 48. I- concentration vs. distance from injection hole for sampling places 21 and 23E.

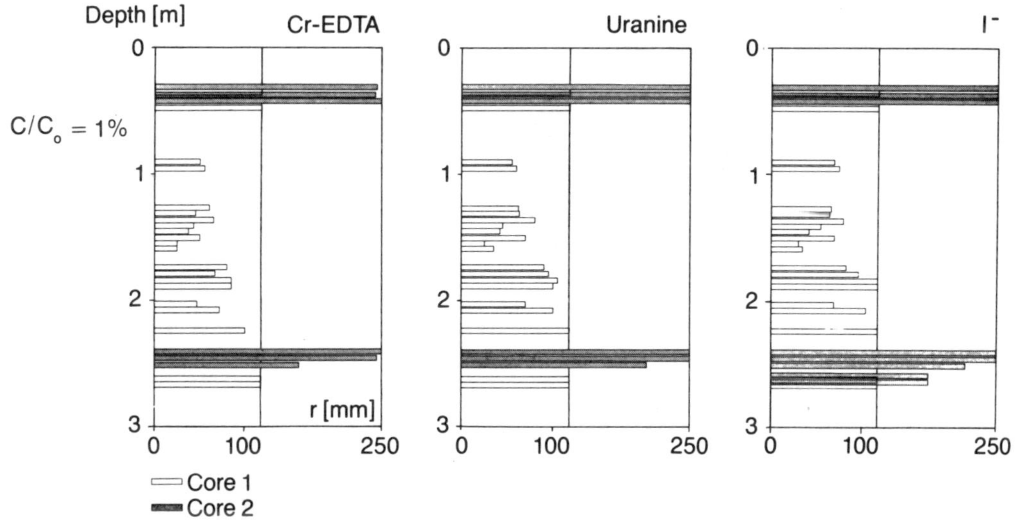

Figure 49. Core 1 and Core 2. Penetration distance for C/C_0 = 1 % for Cr-EDTA, Uranine and I- vs. depth in the cores.

Figure 49 shows that the concentration profiles for all three tracers could be followed at the same depth in core 1 and core 2 in the top of the cores. This was also possible in the bottom of the cores for I-.

The fact that the migration distance is different at different depths can be caused by differences in porosity (ε_p) and differences in the migration parameters (K_p and D_p).

The porosity has been measured for every individual sampling core by comparing the weight difference between wet and dry core. Figure 50 illustrates the mean value of the porosity (+/- the standard deviation) for each sampling place for core 1 and core 2.

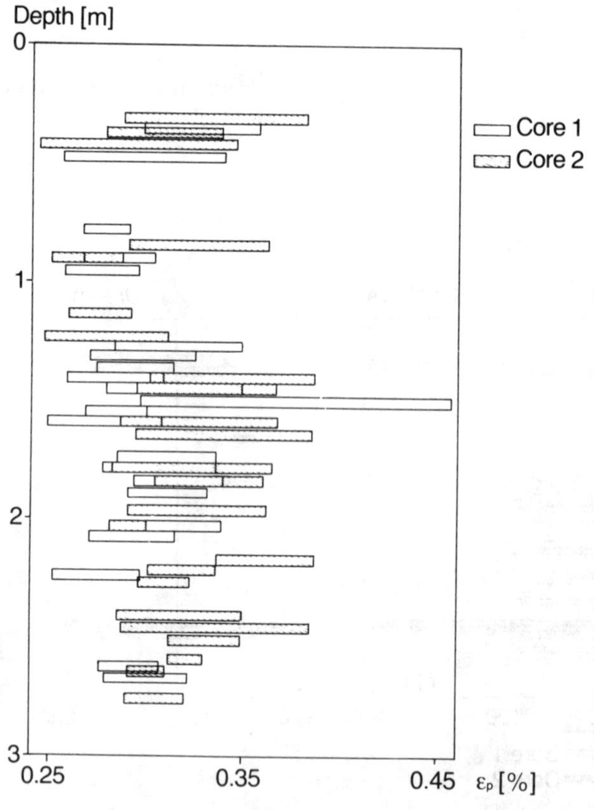

Figure 50. Core 1 and Core 2. The porosity (+/- the standard deviation) vs. depth.

Since the porosity is almost the same for all sampling places, the difference in migration distance with depth cannot be explained by the porosity.

Variation in the migration parameters (K_p and D_p) with location due to inhomogeneities, seem to be the reason for the considerable differences in migration distance for sampling places that were separated by just a few tens of centimeters. By comparing theoretical and experimental curves D_p and K_p were estimated.

These approximate values of D_p and K_p are summarized in table 4 to indicate what the difference in migration distance with depth means in terms of diffusivity and hydraulic conductivity.

Table 4. Approximate values on D_p and K_p for different depths in core 1.

Sampling place	Depth (m)	D_p (m^2/s)	K_p (m/s)
1-2	0.36-0.48	$> 1.10^{-10}$	$> 2\text{-}5.10^{-13}$
3-9	0.78-1.41	$\approx 0.5.10^{-10}$	$\approx 0.1.10^{-13}$
10-13	1.46-1.59	$\approx 0.05.10^{-10}$	$< 0.1.10^{-13}$
14-20	1.74-2.24	$\approx 1.10^{-10}$	$\approx 1.10^{-13}$
21-22	2.62-2.67	$> 1.10^{-10}$	$> 2.5.10^{-13}$

The results indicate that it is possible for tracers (and therefore radionuclides) to migrate a distance into a rock matrix under natural stress conditions and that the in situ diffusivities are comparable to these obtained in the laboratory.

4.5. Experiments with Sorbing Substances

To study the diffusion into and sorption in granite of strontium and cesium a set of batch experiments were performed (50).

Rock pieces taken from the Stripa Mine and from the Finnsjön area were crushed. Each material was graded into the following six fractions: 0.100 to 0.120 mm, 0.200 to 0.250 mm, 0.375 to 0.43 mm, 1.0 to 1.5 mm, 2.0 to 3.0 mm, and 4.0 to 5.0 mm. Before starting the adsorption experiments, all the fractions were washed and dried. The fractions were then contacted with a "synthetic" groundwater to obtain equilibrium between the solids and the water.

After this preparation, part of each solid fraction was mixed in separate glass bottles with synthetic groundwater solutions containing strontium, and in the same way another part of each fraction was mixed with solutions containing cesium. The initial concentration of strontium was 10 ppm and of cesium 15 ppm.

All the bottles were kept in a shaking bath and kept at 25^oC. At different times, small fractions (1 ml) of the solutions were taken out, and the concentration of cesium and strontium was measured by atomic absorption spectrometry. When the concentrations no longer decreased, it was assumed that equilibrium between the solids and the liquids had been obtained. The solids were then separated from the solutions and were contacted with fresh synthetic groundwater in order to desorb the sorbed ions from the solid out into the groundwater. The concentrations of cesium and strontium in the liquid in these desorption experiments were measured in the same way as in the adsorption experiments. Equilibrium was usually reached within 10,000 hours.

The sorption coefficient (K_d) was obtained to be the same for adsorption and desorption of strontium, but not for cesium. This indicates that the sorption coefficient for cesium is dependent on the liquid concentration (i.e. the isotherm is non-linear) or that the sorption process for cesium is partly irreversible.

The experimental data indicate that the amount of sorption is dependent not only on the mass of granite particles, but also to some extent on the size of the particles. These experimental results led us to test the following hypothesis: The granite particles act as porous bodies with active inner surfaces where reversible sorption can take place. In addition, a certain amount of nuclides are being reversibly sorbed on the external surfaces of the particles. The amount of sorption on the inner surfaces per unit mass of granite particles is described by the common Freundlich isotherm.

$$q_V = k_V C_p^{\beta} \tag{4.12}$$

where k_v and β are the Freundlich parameters and C_p is the concentration in pore fluid. The external surfaces are assumed to exhibit a different sorption capacity that is related to the external area of the particles by another type of Freundlich equation

$$q_A = k_A C_p^{\beta} \tag{4.13}$$

The Freundlich parameters are k_A and β, and we assume that β has the same value in both Eqs. (1) and (2).

The total sorption capacity q of the rock particles is obtained by adding the capacities of the inner surfaces and the external surfaces. This yields

$$q = q_V + a_e q_A \quad (4.14)$$

and $a_e = 6/\rho_s d_p$, where d_p is the particle size and ρ_s is the density of the rock particles. The outer surface area of the particles is obtained by assuming the particles to be spherical. Spheres have a surface-to-volume ratio of $6/d_p$.

Equation (4.14) may with Eqs. (4.12) and (4.13) be transformed to

$$q = kC_p^{\beta} \quad (4.15)$$

with $$k = k_V + \frac{6}{\rho_s d_p} k_A \quad (4.16)$$

β, k_V and k_A may be determined from the experiments, β is obtained by plotting total amount adsorbed for a given particle size versus the equilibrium concentration in the liquid. β was obtained to be near 1 for strontium and about 0.6 for cesium for both Finnsjön and Stripa granites.

k_V and k_A are obtained from plots of k versus $1/d_p$ (equation (4.16)). This is shown in Figure 51.

The fraction of the amount sorbed on the outside of the particles decreases from about 15 % to 1 % for both strontium and cesium on Finnsjön granite when particle size increases from 0.1 to 1.2 mm and from about 40 % to 5 % for Stripa granite. For small particles the sorption on the large outer surface created by the crushing thus can account for a considerable fraction of the sorption.

The diffusivity of the species into the particles is obtained from the concentration time curves showing how the concentration in the water in the batch experiment changes with time. Figure 52 shows the results for strontium sorption on Finnsjön granite and the theoretical curves obtained by solving the diffusion equation with initial and boundary equations appropriate for the batch experiment.

The fit to the smaller particles is not good and other processes than diffusion may have played a role. Table 5 compiles the obtained diffusivity data.

The obtained diffusivities are larger than can be explained by diffusion in the water of the micropores. There is some process by which the species migrate in the sorbed state. This has also been found for clays (33).

Similar experiments have been performed with sawed slabs 5 mm thick. These pieces have not been subject to the stresses induced by crushing and not unexpectedly the diffusivities were found to be smaller (on the order of a factor 3)(51).

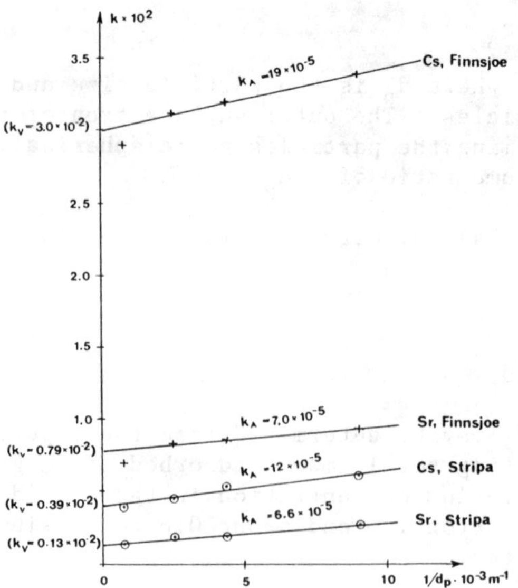

Figure 51. Determination of the constants k_A and k_V from experimental results.

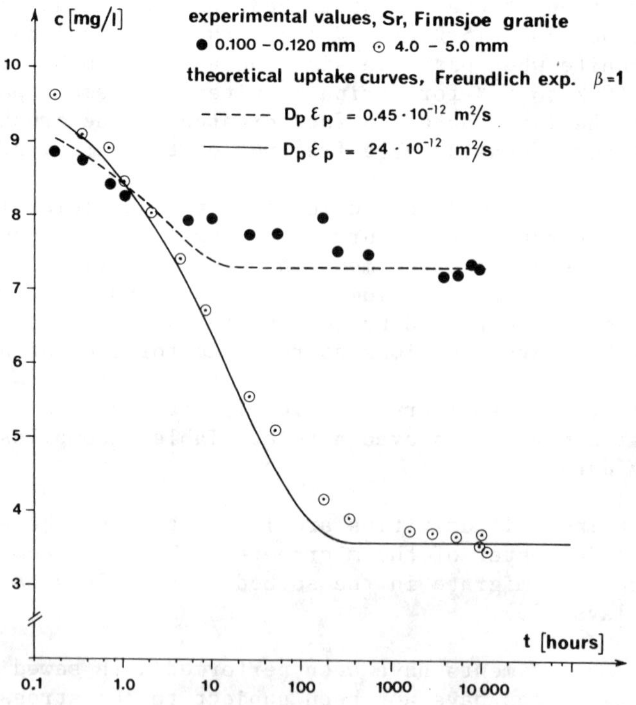

Figure 52. The fit of the diffusion model to the experimental uptake curve for strontium on Finnsjön granite.

Table 5. Effective pore diffusivities $D_p\varepsilon_p$ for cesium and strontium in Finnsjön granite and Stripa granite.

Particle Size Fraction (mm)	Finnsjön granite Cesium $D_p\varepsilon_p$ (m^2/s)	Finnsjön granite Strontium $D_p\varepsilon_p$ (m^2/s)	Stripa granite Cesium $D_p\varepsilon_p$ (m^2/s)	Stripa granite Strontium $D_p\varepsilon_p$ (m^2/s)
0.10 to 0.12	0.85×10^{-12}	0.45×10^{-12}	0.29×10^{-12}	0.011×10^{-12}
0.20 to 0.25	2.8×10^{-12}	1.6×10^{-12}		
0.375 to 0.43	3.8×10^{-12}	3.5×10^{-12}		
1.0 to 1.5	37×10^{-12}	4.4×10^{-12}		
2.0 to 3.0	12×10^{-12}	11×10^{-12}		
4.0 to 5.0	16×10^{-12}	24×10^{-12}	0.98×10^{-12}	1.0×10^{-12}

5. DISCUSSION AND CONCLUSIONS

In sparsely fractured crystalline rock the water moves in fractures which are only partly open. The variation in permeability within fractures and between fractures is so large that it seems reasonable to assume that a few channels in a few fractures carry most of the water. Fracture zones in the rock are not infrequent and have been found in all the areas investigated. Smaller zones are found with frequencies of one or a few per kilometer, whereas larger zones are found at frequencies of one or a few per ten kilometers. The fracture zones have higher hydraulic conductivities than the rock mass itself. The water flowing in the fractures in the rock may or may not mix with such frequency that the concentration of a dissolved species is dispersed in the manner of regular hydrodynamic dispersion - Fickian dispersion - before the waters enter features of the next scale - the fracture zones. The very sparse information on dispersion in fractured rock is not sufficient to clear this question at present. The longest travel distances where dispersion has been observed in deep lying crystalline rock is on the order of 50 m with a possible exception of a 500 m distance. The interpretation of tracer tests is not a straight forward matter because normally there is not sufficient information on the flowpath and many other mechanisms which also cause a spreading. Available information indicates that distances of 50 to 100 m are not sufficient to arrive at a constant dispersion coefficient. On the contrary it seems that the dispersivity increases with the observation distance. This would indicate that there is a very pronounced channeling on this scale.

Most models are based on the advection-dispersion equation and may thus not give good results when used for extrapolation to larger distances than those for which the data were obtained.

The question if the "dispersion" is caused by Fickian dispersion or if it mainly due to channeling becomes especially important for radionuclide migration in low permeability fractured rock. The assumption of Fickian dispersion would lead to an underestimation of the amount of a species which arrives "early" if the extrapolation is large. For decaying radionuclides this may in some cases mean the difference between an extremely large decay and an insufficient decay.

There have only been a few attempts to model channeling. Possibly the fracture network models which are being developed will be of help to model tracer movement over larger distances with better confidence.

One of the difficulties in making measurements over large distances is that experiments will take a very long time. Another difficulty is that for long contact times the tracers will have time to diffuse into the stagnant water of the rock matrix. This effect is so pronounced that it will totally swamp all other mechanisms including the advective transport. This difficulty could be circumvented if enough were known of the flowpaths. Information on how many channels and with what exposed surface are encountered along the path as well the detailed distribution of flow between the channels is needed to account for the uptake of tracers into the matrix. In addition matrix porosity and diffusivity would be needed. At present this information is not available in sufficient detail. It is also by no means sure that the information needed ever can be obtained for this purpose.

The use of natural tracers such as Tritium and Carbon-14 to assess travel times and dispersion has not been successful for low permeability fractured rock by the same reasons (29).

The effects of matrix diffusion, although they cause severe difficulties in evaluating tracer experiments, are very beneficial when there is a need to have long travel times for contaminants. Then the presence of stagnant water volumes into which the species may travel by molecular diffusion causes the species to have a much longer travel time than that of the actually moving water mass. The stagnant water volume is often several orders of magnitude larger than the mobile water.

Species which interact with the inner surfaces of the rock matrix are further retarded. These effects are so large that most of the important radionuclides are expected to have travel times considerably longer than their halflives in typical repository environments in crystalline rock.

6. REFERENCES

1. Abelin, H., Gidlund, J., and Neretnieks, I., "Migration in a Single Fissure," *Scientific Basis for Nuclear Waste Management V*, Elsevier Publishing Co., Berlin, 1982, pp. 529-538.
2. Abelin, H., Gidlund, J., Moreno, L., and Neretnieks, I., "Migration in a Single Fissure in Granitic Rock," *Scientific Basis for Nuclear Waste Management VII*, Elsevier Publishing Co., Boston, 1984, p. 239.
3. Abelin, H., Moreno, L., Tunbrant, S., Gidlund, J., and Neretnieks, I., "Flow and Tracer Movement in Some Natural Fractures in the Stripa Granite," presented at the June 1985, NEA/OECD Stripa Workshop (proceedings in preparation).
4. Abelin, H., Neretnieks, I., Tunbrant, S., and Moreno, L., "Migration in a Single Fracture," *KBS Technical Report* (in print).
5. Bear, J., "Hydrodynamic Dispersion," *Flow Through Porous Media*, R. J. M. de Wiest, ed., Chapter 4, Academic Press, New York, 1969, p. 109.
6. Birgersson, L., and Neretnieks, I., "Diffusion in the Matrix of Granitic Rock, Field Test in the Stripa Mine," *Scientific Basis for Nuclear Waste Management V*, Elsevier Publishing Co., Berlin, June 1982, p. 519.
7. Birgersson, L., and Neretnieks, I., "Diffusion in the Matrix of Granitic Rock. Field Test in the Stripa Mine," *Scientific Basis for Nuclear Waste Management VII*, Elsevier Publishing Co., Boston, Nov. 1983, p. 247.
8. Birgersson, L., Abelin, H., Gidlund, J., and Neretnieks, I., "Water Flowrates and Tracer Transport to the 3D Drift in Stripa," presented at the June 1985, NEA Information Symposium on In Situ Experiments in Granite Associated With the Disposal of Radioactive Waste (Stripa Symposium), held at Stockholm, Sweden (proceedings in print).
9. Bourke, P.J., Durrance, E.M., Heath, M.J., and Hodgkinson, D.P., "Fracture Hydrology Relevant to Radionuclide Transport," *AERE-R 11414*, Atomic Energy Research Establishment, Harwell, England, January 1985.
10. Brace, W.F., Walsh, J.B., and Frangos, W.I., "Permeability of Granite Under High Pressure," *Journal of Geophysical Research*, Vol. 73, 1968, p. 2225.
11. Bradbury, M.H., Lever, D., and Kinsey, D., "Aqueous Phase Diffusion in Crystalline Rock," *Scientific Basis for Nuclear Waste Management V*, Elsevier Publishing Co., Berlin, 1982, pp. 569-578.
12. Carlsson, L., Winberg, A., and Grundfelt, B., "Model Calculations of the Groundwater Flow at Finnsjön, Fjällveden, Gidea and Kamlunge," *KBS TR 83-45*, May 1983.
13. de Josselin de Jong,G., "Longitudinal and Transverse Diffusion in Granular Deposits," *EOS Transactions, AGU*, Vol. 39, 1958, p. 68.

14. Gelhar, L.W., and Axness, C.L., "Three-dimensional Stochastic Analysis of Macro Dispersion in Aquifers," *Water Resources Research*, Vol. 19, 1983, pp. 161-180.
15. Grisak, G.E., and Pickens, J.F., "Solute Transport Through Fractured Media. 1. The Effect of Matrix Diffusion," *Water Resources Research*, Vol. 16, 1980, pp. 719-730.
16. Gustavsson, E., and Klockars, C.E., "Studies on Groundwater Transport in Fractured Crystalline Rock Under Controlled Conditions Using Non-Radioactive Tracers," *KBS TR 81-07*, 1981.
17. "Final Storage of Spent Nuclear Fuel," *KBS-3*, Swedish Nuclear Fuel Supply Co., SKBF, Stockholm, Sweden, May 1983, Parts III, IV.
18. Landström, O., Klockars, C-E, Holmberg, K-E., and Westerberg, S., "In Situ Experiments on Nuclide Migration in Fractured Crystalline Rocks," *KBS TR 110*, 1978.
19. Landström, O., Klockars, C-E., Persson, O., Andersson, K., Torstenfelt, B., Allard, B., Larsson, S.A., and Tullborg, E.L., "A Comparison of in Situ Radionuclide Migration Studies in the Studsvik Area and Laboratory Measurements," *Scientific Basis for Nuclear Waste Management V.*, Elsevier Publishing Co., Berlin, June 1982, pp. 697-706.
20. Lapidus, L., and Amundson, N.R., "Mathematics of Adsorption in Beds," *Journal of Physical Chemistry*, Vol. 56, 1952, p. 984.
21. Maloszewski, P. and Zuber, A., "On the Theory of Tracer Experiments in Fissured Rock with a Porous Matrix," *Rapport 1244/17P*, Institute of Nuclear Physics, Cracow, 1984.
22. Matheron, G., and de Marsily, G., "Is Transport in Porous Media Always Diffusive? A Counterexample," *Water Resources Research*, Vol. 16, 1980, p. 901.
23. Mercado, A., "The Spreading Pattern of Injected Water in a Permeability Stratified Aquifer," *Symposium of Haifa:Artificial Recharge and Management of Aquifers*, IAHS-AISH Publ. Vol. 72, March 19-20, 1967, p. 23.
24. Moreno, L., Neretnieks, I., and Klockars, C-E., "Evaluation of Some Tracer Tests in the Granitic Rock at Finnsjön," *KBS TR 83-38*, 1983.
25. Moreno, L., Neretnieks, I., and Eriksen, T., "Analysis of Some Laboratory Tracer Runs in Natural Fissures," *Water Resources Research*, Vol. 21, 1985, pp. 951-958.
26. Moreno, L., Neretnieks, I., and Landström, O., "Evaluation of Some Tracer Tests at Studsvik," 1984 (Paper for publication)
27. "Endlager für Hochaktive Abfälle: Das System der Sicherheits-barrieren," NAGRA projekt Gewähr, Jan., 1985.
28. Neretnieks, I., "Diffusion in the Rock Matrix: An Important Factor in Radionuclide Retardation?" *Journal of Geophysical Research*, Vol. 85, 1980, p. 4379.
29. Neretnieks, I., "Age Dating of Groundwater in Fissured Rock. Influence of Water in Micropores," *Water Resources Research*, Vol. 17, 1981, p. 421.

30. Neretnieks, I., Eriksen, T., and Tähtinen, P., "Tracer movement in a Single Fissure in Granitic Rock: Some Experimental Results and Their Interpretation," *Water Resources Research*, Vol. 18, 1982, p. 849.
31. Neretnieks, I., "A Note on Fracture Flow Mechanisms in the Ground," *Water Resources Research*, Vol. 19, 1983, p. 364-370.
32. Neretnieks, I., and Rasmuson, A., "An Approach to Modeling Radionuclide Migration in a Medium with Strongly Varying Velocity and Block Sizes Along the Flow Path," *Water Resources Research*, Vol. 20, 1984, pp. 1823-1836.
33. Neretnieks, I., "Diffusivities of Some Dissolved Constituents in Compacted Wet Bentonite Clay and the Impact on Radionuclide Migration in the Buffer," *Nuclear Technology*, Vol. 71, 1985, p. 458.
34. Neretnieks, I., "Transport in Fractured Rock," presented at the Jan., 1985, IAH 17th International Congress on the Hydrology of Rocks of Low Permeability, held at Tucson, Arizona (in print).
35. Pruess, K., and Narasimhan, T.N., "A Practical Method for Modeling Fluid and Heat Flow in Fractured Porous Media," presented at the Jan., 1982, Society of Petroleum Engineers of AIME 6th Symposium, held at New Orleans, La.,.
36. Rasmuson, A., "Diffusion and Sorption in Particles and Two Dimensional Dispersion in a Porous Medium," *Water Resources Research*, Vol. 17, 1981, p. 312.
37. Rasmuson, A., "Migration of Radionuclides in Fissured Rock: Analytical Solutions for the Case of Constant Source Strength," *Water Resources Research*, Vol. 20, 1984, pp. 1435-1442.
38. Rasmuson, A., "The Effect of Particles of Variable Size, Shape and Properties on the Dynamics of Fixed Beds," *Chemical Engineering Science.*, Vol. 40, 1985, pp. 621-629.
39. Rasmuson, A., "The Influence of Particle Shape on the Dynamics of Fixed Beds," *Chemical Engineering Science*, 1985, Vol. 40, 1985, p. 1115.
40. Rasmuson, A., and Neretnieks, I., "Exact Solution of a Model for Diffusion in Particles and Longitudinal Dispersion in Packed Beds," *AIChE Journal*, Vol. 26, 1980, p. 686.
41. Rasmuson, A., Narasimhan, T.N., and Neretnieks, I., "Chemical Transport in a Fissured Rock: Verification of a Numerical Model," *Water Resources Research*, Vol. 18, 1982, p. 1479.
42. Rasmuson, A., and Neretnieks, I., "Radionuclide Migration in Strongly Fissured Zones - The Sensitivity to Some Assumptions and Parameters," *Water Resources Research*, Vol.22, 1986,pp.559.
43. Rasmuson, A., "Analysis of Hydrodynamic Dispersion in Discrete Fracture Networks Using the Method of Moments," *Water Resources Research*, Vol. 21, 1985, pp. 1677-1683.
44. Rasmuson, A., and Neretnieks, I., "Radionuclide Transport in Fast Channels in Crystalline Rock," *Water Resources Research*, Vol. 22, 1986, pp. 1247-1256.
45. Sauty, J.P., "An Analysis of Hydrodispersive Transfer in Aquifers," *Water Resources Research*, Vol. 16, 1980, p. 145.

46. Schwartz, F.W., "Macroscopic Dispersion in Porous Media: The Controlling Factors," *Water Resources Research*, Vol. 13, 1977, p. 743.
47. Snow, D.T., "Rock Fracture Spacings, Openings and Porosities," ASCE Journal of Soil Mechanics and Foundations Division, Vol. 94, 1968, p. 73.
48. Snow, D.T., "The Frequency of Apertures of Fractures in Rock," *International Journal of Rock Mechanics and Mineral Science*, Vol. 7, 1970, p. 23.
49. Skagius, K., and Neretnieks, I., "Diffusion in Crystalline Rocks," *Basis for Nuclear Waste Management V*, Elsevier Publishing Company, 1982, pp. 509-518.
50. Skagius, K., Svedberg, G., and Neretnieks, I., "A Study of Strontium and Cesium Sorption on Granite," *Nuclear Technology*, Vol. 59, 1982, pp. 302-313.
51. Skagius, K., Neretnieks, I., "Porosities of and Diffusivities in Crystalline Rock and Fissure Coating Materials," *Scientific Basis for Nuclear Waste Management VII*, Elsevier Publishing Company, Boston, Nov., 1983, p. 835.
52. Skagius, K., and Neretnieks, I., "Porosities and Diffusivities of Some Nonsorbing Species in Crystalline Rocks," *Water Resources Research*, Vol. 22, 1986, pp. 389-398.
53. Skagius, K., and Neretnieks, I., "Diffusivity Measurements and Electrical Resistivity Measurements in Rock Samples Under Mechanical Stress," IBID, p. 570.
54. Tang, G.H., Frind, E.O., and Sudicky, E.A., "Contaminant Transpor in Fractured Porous Media. An Analytical Solution for a Single Fracture," *Water Resources Research*, Vol. 17, 1981, p. 555.
55. Tsang, Y.W., and Witherspoon, P.A., "Hydromechanical Behaviour of a Deformable Fracture Subject to Normal Stress," *Journal of Geophysical Research*, Vol. 86, 1981, pp. 9287-9298.
56. Tsang, Y.W., and Witherspoon, P.A., "The Dependence of Fracture Mechanical and Fluid Properties on Fracture Roughness and Sample Size," *Journal of Geophysical Research*, Vol. 88, 1983, pp. 2359 2366.

7. LIST OF SYMBOLS

A	Surface area (m^2)
a	Specific surface (m^{-1})
C	Concentration in mobile fluid (mol/m^3)
C_p	Concentration in pore fluid (mol/m^3)
C_o	Concentration at inlet boundary (mol/m^3)
d_p	Particle diameter (m)
D_e	Effective diffusivity (m^2/s)
D_L	Dispersion coefficient (m^2/s)
D_p	Pore diffusivity (m^2/s)

D_v	Diffusivity in bulk liquid (m^2/s)
K	Volume distribution coefficient (m^3/m^3)
K_A	Distribution coefficient for surface sorption (m)
K_d	Distribution coefficient for volume sorption (m^3/kg)
k	Constant
k_A	Constant for surface sorption
k_v	Constant for volume sorption
ℓ	Slab thickness (m)
M	Amount of diffusing component (mol)
m	$\varepsilon_f/(1-\varepsilon_f)$
N	Flowrate of diffusing component (mol/s)
P	Pressure (N/m^2)
Pec	Peclet number vz/D_L
Q	Flowrate of water (m^3/s)
q	Concentration in solid phase (mol/kg)
R	Retardation factor in general
R_a	Retardation factor due to surface sorption
R_d	Retardation due to volume sorption
R_o	Resistivity of salt solution (Ωm)
R_s	Resistivity of salt water saturated rock (Ωm)
r	Radial direction (m)
S	Fracture spacing (m)
t	Time (s)
t_o	Residence time for tracer (s)
t_w	Residence time for water (s)
V_r	Radial water velocity (m/s)
V_s	Volume of rock (m^3)
v	Water velocity (m/s)
x	Distance into rock (m)
z	Distance in flow direction (m)
α	Capacity factor
β	Geometric factor
δ	Fracture width (m)
δ_c	Fracture width determined by "cubic law" (m)
δ_D	Constrictivity (m)
δ_f	Fracture width from residence time (m)
δ_ℓ	Fracture width from laminar velocity (m)
ε_f	Porosity of mobile fluid
ε_p	Porosity of rock matrix
ε_{tot}	Total porosity of matrix
ε^+	"transport porosity" of matrix
$\varepsilon^+\delta_D/\tau^2$	Formation factor
λ	Decay constant (s^{-1})

ρ_p	Density of particle (kg/m^3)
ρ_s	Density of minerals (kg/m^3)
σ_ℓ	Logarithmic standard deviation of fracture widths
σ_z	Standard deviation of water travel distance (m)
σ_t	Standard deviation of water residence time (s)

KBS - technical reports can be obtained from: INIS CLEARING HOUSE, International Atomic Energy Agency, P.O.Box 100, A-1400 VIENNA, Austria.

MULTIPHASE FLOW IN FRACTURED RESERVOIRS

Ole Torsaeter, Jon Kleppe, and Teodor van Golf-Racht

MULTIPHASE FLOW IN FRACTURED RESERVOIRS

Ole Torsaeter, Jon Kleppe, and Teodor van Golf-Racht

Division of Petroleum Engineering and Applied Geophysics
Norwegian Institute of Technology,
University of Trondheim
N-7034 Trondheim-NTH, Norway

ABSTRACT

Naturally fractured reservoirs represent a complex class of reservoirs. Multiphase flow in such reservoirs adds to the complexity and has been studied extensively over the last few years.

This chapter presents a review of the current state of technology regarding certain aspects of such reservoirs. Physical properties of fractures and fractured systems are defined and discussed. Various flow processes and models describing these processes are reviewed. Particular emphasis has been placed on the imbibition flow process.

Experimental results on imbibition in carbonate chalk are presented. A large number of samples have been tested in the laboratory, and the results are discussed in regard to effects of wetting properties, sample shapes, and boundary conditions. Simulation of one of the experiments shows that the prediction of imbibition flow rate by simulation models is critically dependent on a well defined capillary pressure curve.

An improved dual porosity model for simulation of multiphase flow in fractured reservoirs is discussed.

1. INTRODUCTION

Fractured reservoirs exist throughout the world. Table 1 lists fractured carbonate reservoirs in various countries. These reservoirs represent large reserves, and the oil recovery from such reservoirs is clearly an important area of study. At present, interest centers primarily on two aspects of the problem. The first is a thorough understanding of the whole physical process for the development of more realistic mathematical models. The second important aspect is the measurement of multiphase fractured reservoir flow parameters in the laboratory, and the use of these data in mathematical models. In this chapter, some aspects of both experimental studies and mathematical modeling are considered.

To introduce the rock medium and its characteristics, a section on physical properties of fractured rocks is presented. Porosity, permeability, compressibility, and in particular parameters describing fluid-fluid-rock interactions, such as capillary pressure and relative permeability, are discussed. Section 3 includes an extensive introductory section devoted to the fundamental literature on both theoretical and experimental aspects of fluid flow in fractured reservoirs. The rest of the section presents experimental results on studies of capillary imbibition, namely:

1) Imbibition in chalk plugs from the Ekofisk Field in the North Sea;
2) Imbibition in rock samples of various sizes and shapes;
3) Imbibition in core plugs with various boundary conditions; and
4) Imbibition in matrix blocks containing microfractures.

A subsection on numerical simulation of laboratory experiments is also presented in section 3.

Section 4 gives a review of the state of the art of simulation of multiphase flow in naturally fractured reservoirs. Flow equations are presented and the matrix/fracture fluid exchange term has been emphasized. Improvements in model formulations are discussed.

2. PHYSICAL PROPERTIES OF FRACTURED ROCKS

2.1. Introduction

The physical properties of rocks and fluids used as basic data for reservoir engineering studies of conventional reservoirs have been extensively studied in the last three to four decades. Since the objective of this chapter is to examine rock characteristics of fractured reservoirs, the rock properties will be discussed as properties of either fractures or of the fracture-matrix system. Specific matrix properties will not be discussed since these represent classic properties of a conventional reservoir.

Table 1. Fractured carbonate reservoirs.

LOCATION	MATRIX		FRACTURES			REMARKS
	Porosity %	Permeability md	Width mm	Density m^{-1}	Porosity %	
CANADA -Beaver River Field	2-6	2-200			<0.2	
FRANCE -Escau Field	12	1-5	0.1		<0.02	
IRAN -Haft Kel Field (Asmari Form.)	16	<1	a) 50 b) 0.5 c) 0.1	a) 1.10^{-3} b) 1.10^{-2} c) 0.1-8	0.2	a) Normal faults b) Large fractures c) Small fractures
IRAQ -Ain Zalah Field	0-11	<0.1	a) 0.1-0.2 b) 0.1-18	18-35		a) Open fractures b) Fractures partially filled with calcite
-Kirkuk Field	0-30	0-1000	0.1-0.2	1-2		
ITALY -Gela Field	3-5	0.01-0.1				Network of fine fractures
MEXICO -Sitio Grande (Reforma area)	6-11					Fractures and caverns present

Table 1. Continued.

LOCATION	MATRIX		FRACTURES			REMARKS
	Porosity %	Permeability md	Width mm	Density m^{-1}	Porosity %	
NORWAY -Ekofisk -Valhall	 20–40 35–45	 0.1–10 1–10	 $< 5.10^{-2}$ $< 5.10^{-2}$	 1–30 0–200	 <0.01 <0.01	
QATAR -Dukhan Field	 15–20	 15–75	 0.1	 1–3		A clearly defined fracture system is observed
UNITED STATES -West Edmond Field, OK	<10	<1		1		
-Darst Creek Salt Flat Field, TX (Austin Chalk)	12	0.1	<0.2	15		
VENEZUELA -Mara and La Paz Fields	<3	<0.1	<6	0.5		

In addition to porosity and permeability, a review of the geological aspects of fracturing and the compressibility of the fracture-matrix system is presented.

Particular attention has been given to parameters describing fluid-fluid-rock interaction, such as capillary pressure and relative permeability. A system consisting of a matrix block saturated with fluid A surrounded by fractures saturated by a different fluid B is the basis of fluid displacement mechanisms in fractured reservoirs, and will be discussed in detail.

2.2. Fractures

Various definitions of a fracture may be given. From a geomechanical standpoint, a fracture is a surface in which a loss of cohesion has taken place, i.e., a fracture is the result of a rupture. In general, a fracture in which relative displacement has occurred can be defined as a fault, while a fracture in which no noticeable displacement has occurred can be defined as a joint point. Basically, whether a fracture is considered a joint or a fault depends on the scale of investigation, and in this text, the term fracture will correspond to a joint. Discontinuities breaking up the rock beds into blocks is a fairly general definition of fractures relevant to this work.

Fractured reservoirs most likely occur in brittle reservoir rocks where tectonic events have developed. Depending on the rock properties and the causes of fracturing, the extent and openings will vary widely. Table 1 indicates fracture distribution and openings in some fractured carbonate reservoirs. The main causes of fracturing are (45):

1) Diastrophism in the case of folding and faulting;
2) Deep erosion of the overburden that permits the upper parts to expand, uplift, and fracture through planes of weakness; and
3) Volume shrinkage as a result of loss of water in shales and cooling of igneous rocks.

The best quantitative information about fracture parameters such as opening, extension, distribution, nature, and orientation is obtained by direct measurements on outcrops and on cores. Indirect sources of information such as log analysis, well testing, and production history may also give valuable fracture data (1).

Studies of fracture characteristics usually follow a special pattern, beginning with the examination of single fractures, continuing with groups of fractures, and finally, relationships among the various groups are established. Single fracture parameters refer to such characteristics as opening (width), size, nature, and orientation. The multi-fracture parameters refer to the fracture arrangement (geometry) which further generates the bulk unit, called the matrix block. The number of fractures and their orientation are directly related to fracture distribution and density.

Fracture opening is represented by the distance between the fracture walls, and may depend on nature of stresses, reservoir environment, and type of rock. The fracture openings vary considerably as shown in Table 1. *Fracture size* refers to the relationship between fracture length and layer thickness. The terms minor, average and major fractures are often used, where a minor fracture has a length less than a single layer pay, average fractures traverse more than one layer, and major fractures have a large extension (hundreds of meters). *The nature of fractures* refers to filling material and wall characteristics. *Fracture orientation* can be defined by two angles, azimuth and dip angles. Parallel fractures belong to a fracture system, and intercommunicating fracture systems form a fracture network.

Multi-fracture parameters of importance in this context are the parameters necessary to evaluate matrix block size and geometry. The matrix blocks are defined by shape, volume, and height, in relation to the fracture system's dip, strike, and distribution. The shape of the matrix block is irregular, but for practical work, the block units are frequently reduced to simplified geometrical volumes, such as cubes or as elongated or flat parallelepipeds. The term fracture density is used to describe both real and idealized fracture systems. Fracture density expresses the degree of fracturing through various relative ratios. The reference may be bulk volume (volumetric fracture density), area or length (areal or linear fracture density); the analytical expressions are as follows:

- Volumetric fracture density, f_v, is the ratio between fracture bulk surface, s_f, and matrix bulk volume, V_B:

$$f_v = \frac{s_f}{V_B} \tag{2.1}$$

- Areal fracture density, f_s, is the ratio between the cumulative length of the fractures, l_t, and matrix bulk area s in cross flow section:

$$f_s = \frac{l_t}{s} \tag{2.2}$$

- Linear fracture density is the ratio between the number (n) of fractures intersecting a straight line (normal on flowing direction) and the length of the straight line (L):

$$f_w = \frac{n}{L} \tag{2.3}$$

If various matrix block units are grouped into certain geometrical distributions, a number of idealized fractured networks can be obtained. If the models are as shown in Figure 1, the fracture density is modified according to the flow direction and to the impermeability of the block faces (39).

A representative description of the shape and dimensions of the average matrix block unit is very important when considering dynamic aspects of a fractured reservoir. In most studies, the following assumptions are usually made: 1) The fractures are continuous planes; 2) The fractured reservoir is represented by a certain number of networks; and 3) The distance between the fractures of the same plane is reported by a frequency law. Idealized blocks will result from the intersection of fracturing planes.

2.3. Porosity

Fractured reservoirs are made up of two porosity systems; one intergranular formed by void spaces between the grains of the rock (primary porosity), and a second formed by void spaces of fractures and vugs (secondary porosity). The two porosities are expressed by the conventional definitions:

ϕ_1 = matrix void volume/total bulk volume

ϕ_2 = fracture void volume/total bulk volume

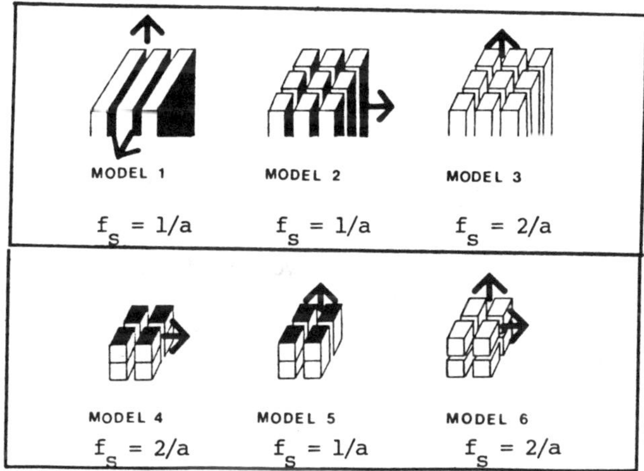

Figure 1. Idealized models and fracture density expressed as a function of flowing direction [after Reiss (39)].

In dealing with fracture reservoirs, we usually refer the matrix porosity to matrix bulk volume i.e.,

$$\phi_m = \text{(volume voids of the matrix)/(matrix bulk volume)}$$

while the definition of fracture porosity is unchanged, i.e.,

$$\phi_f = \phi_2 \tag{2.4}$$

The primary porosity may be expressed in terms of the matrix porosity, and the fracture porosity as:

$$\phi_1 = (1 - \phi_f)\phi_m \tag{2.5}$$

Double porosity is more important for the dynamic evaluation of the reservoir than for the reservoir storage capacity.

2.4. Permeability

In the presence of both matrix and fractures, permeability is redefined as matrix permeability, fracture permeability, and system permeability. In most fractured reservoirs, the distance between fractures is much longer than the core sample length, and in this case, the measured permeability represents the matrix permeability. The fracture permeability may create some confusion because it can be referred to as single fracture permeability, fracture network permeability, or fracture permeability of fracture-bulk volume.

Single fracture permeability can be derived from Navier-Stokes equation for laminar flow between two parallel plates:

$$u = - \frac{w^2 \Delta P}{12 \mu \Delta x} \tag{2.6}$$

where, u is the flow velocity, μ is dynamic viscosity, $\frac{\Delta P}{\Delta x}$ is the pressure gradient in the flow direction and w is the fracture opening.

By analogy to Darcy's law, we obtain the following expression for fracture permeability:

$$k_{ff} = \frac{w^2}{12} \tag{2.7}$$

If the fracture forms an angle (α) to the flow direction, the permeability will be:

$$k_{ff} = \frac{w^2}{12} \cos^2\alpha \tag{2.8}$$

For a fracture network formed by fracture systems α, β, each with its own orientation, Parsons (37) calculates the permeability to be:

$$k_{ff} = \frac{1}{12} [\cos^2\alpha \sum_1^{n_\alpha} w_{\alpha i}^2 + \cos^2\beta \sum_1^{n_\beta} w_{\beta i}^2 + \ldots \qquad (2.9)$$

In conventional fractured reservoirs, the fracture and the associated rock bulk form a hydrodynamic unit. The cross sectional area exposed to flow is therefore expressed by not only the fracture cross sectional area, but by the rock cross sectional area. For a system with parallel fractures in the flow direction (Figure 2), the permeability will be

$$k_f = k_{ff} \frac{w}{a} = \frac{w^3}{12a} \qquad (2.10)$$

where a is the distance between fractures and a >> w.

The permeability of a fracture-matrix system may be represented by the simple addition of the permeabilities of matrix k_m and fracture k_f,

$$k_t = k_m + k_f \qquad (2.11)$$

This equation is, of course, flow direction dependent.

A summary of permeability and porosity relationships for the various block geometries (fracture reservoir models) shown in Figure 1 is given in Table 2.

2.5. Compressibility

The rock compressibility used when dealing with conventional reservoirs reflects only the deformation of the pores, since the matrix volume reduction is negligible in comparison. In fractured reservoirs, the matrix compressibility is normally less than for

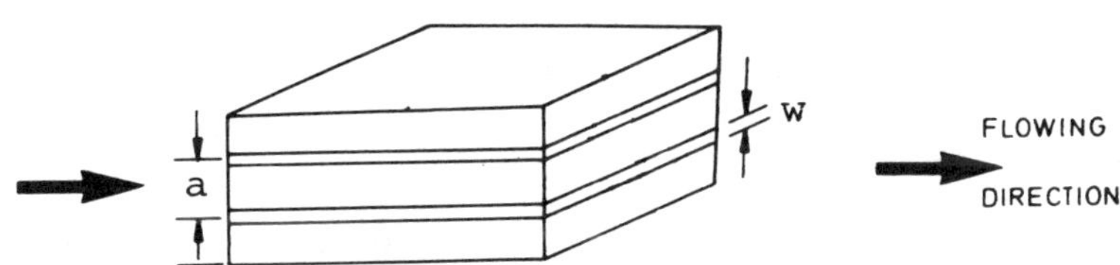

Figure 2. Parallel fractures in the flow direction.

Table 2. Parameters for simplified models (refer to Figure 1) [after Reiss (39)].

Model		Formulas for computing fracture properties			
No.	type	f_s	ϕ_f	$k_f(\phi_f,a)$	$k_f(\phi_f,w)$
1	slides	1/a	w/a	$1/12\ a^2\phi_f^3$	$1/12\ w^2\phi_f$
2	matches	1/a	2w/a	$1/96\ a^2\phi_f^3$	$1/24\ w^2\phi_f$
3	matches	2/a	2w/a	$1/48\ a^2\phi_f^3$	$1/12\ w^2\phi_f$
4	cubes	1/a	2w/a	$1/96\ a^2\phi_f^3$	$1/12\ w^2\phi_f$
5	cubes	2/a	2w/a	$1/48\ a^2\phi_f^3$	$1/12\ w^2\phi_f$
6	cubes	2/a	3w/a	$1/162\ a^2\phi_f^3$	$1/18\ w^2\phi_f$

conventional reservoirs. This is supported by the fact that the fracturing reflects the rigidity of the rock, and the rock has broken rather than deformed elastically. However, the presence of fractures introduces an additional compressibility, which can be defined in two ways:

$$c_{ef} = -\frac{1}{V_t}\left(\frac{\partial V_f}{\partial p}\right)_T \tag{2.12}$$

or

$$c_{pf} = -\frac{1}{V_f}\left(\frac{\partial V_f}{\partial p}\right)_T \tag{2.13}$$

where V_t is bulk volume and V_f is fracture volume. The relationship between these two definitions is:

$$c_{ef} = \phi_f \cdot c_{pf} \tag{2.14}$$

The effective oil compressibility of the total fractured system with oil and water can be calculated as the sum of the individual contributions, given the saturations in fractures and matrix (45):

$$S_{of} \cong 1$$
$$S_{wf} = 0$$
$$S_{om} = 1 - S_{wm}$$
$$c_{eo} = c_o + c_w \frac{\phi_m \cdot S_{wm}}{\phi_m(1-S_{wm}) + \phi_f} + c_{pm} \frac{\phi_m}{\phi_m(1-S_{wm}) + \phi_f} + c_{pf} \frac{\phi_f}{\phi_m(1-S_{wm}) + \phi_f} \tag{2.15}$$

where; S_{wm} is the matrix water saturation; c_o is the oil compressibility; c_w is the water compressibility; c_{pm} is the matrix pore compressibility; and c_{pf} is the fracture pore compressibility.

For most fractured reservoirs, $\phi_f << \phi_m$ and c_{pm} has the same magnitude as c_{pf} giving:

$$C_{eo} = c_o + c_w \frac{S_{wm}}{1 - S_{wm}} + c_{pm} \frac{1}{1 - S_{wm}} \tag{2.16}$$

Since the fracture compressibility term is ignored, Eq. (2.16) is identical to the effective compressibility for oil in a conventional reservoir.

2.6. Relative Permeability

In a fractured reservoir, the evaluation of relative permeability curves is complicated because of the nature of the double porosity system. In the literature, the relative permeability of a given fractured reservoir is seldom examined, while the influence of heterogeneity within a porous medium on relative permeability has been studied in detail.

The heterogeneity problem in rock was for a long time limited to reservoirs having a directional variation of permeability and its consequences on field behavior. Later, the reservoir heterogeneity was examined in relation to relative permeability. Among the various heterogeneities of a reservoir, the most significant are layering and fracturing. Figure 3 shows anomalous shaped relative permeability curves due to fracturing. The following discussion will be concerned with the relative permeability concept in fractured rock.

Fractured reservoirs are often simulated with ordinary single-porosity models. In this case, the relative permeability is adjusted to account for the effect of the fracturation. These relative permeability curves are called pseudo relative permeabilities and are derived in different ways, such as from pilot test

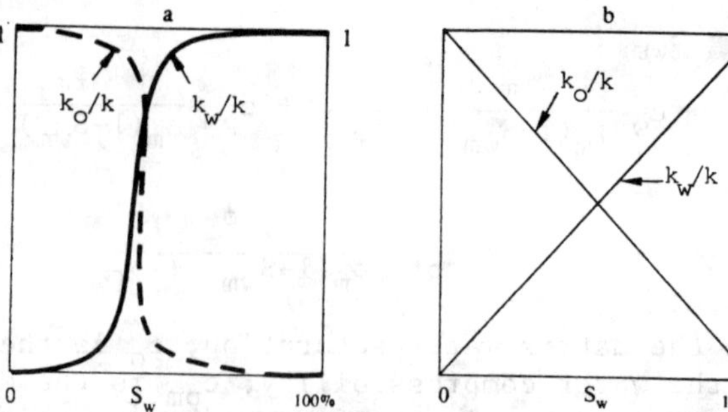

Figure 3. Relative permeabilities for a) cores with fractures not parallel to flow b) cores with fractures parallel to flow.

behavior, fractured cross section models, core permeabilities and fracturation data, or from laboratory experiments of a fractured model. Pseudo relative permeabilities will not be discussed further in this text.

Fractured reservoir models need both matrix and fracture relative permeability relationships. Usually the matrix relative permeability is assumed to be a function of the saturation in the matrix only, and the fracture relative permeability is a function of the fracture saturation. The matrix relative permeability is obtained from laboratory experiments on core plugs. The capillary pressure in the fractures is in most cases assumed to be zero, and the relative permeability in the fractures is approximated to a linear function of saturation in the fractures.

An interesting theoretical study of relative permeability in a fractured reservoir with non-negligible matrix permeability is presented by Braester (11). The basic principles of this model are:

- In a fractured reservoir with dynamic pressure gradients not negligible compared to those occurring during imbibition, a certain fracture-matrix fluid exchange will take place besides the imbibition.
- The wetting and non-wetting phases are circulating from the fractures into the matrix and back into the fractures in the zones saturated with water and oil phases, respectively (Figure 4).
- Through such a model, the relative permeability becomes a function of saturation in the fractures as well as in the matrix.
- The flow is considered continuous flow in both matrix blocks and fractures.

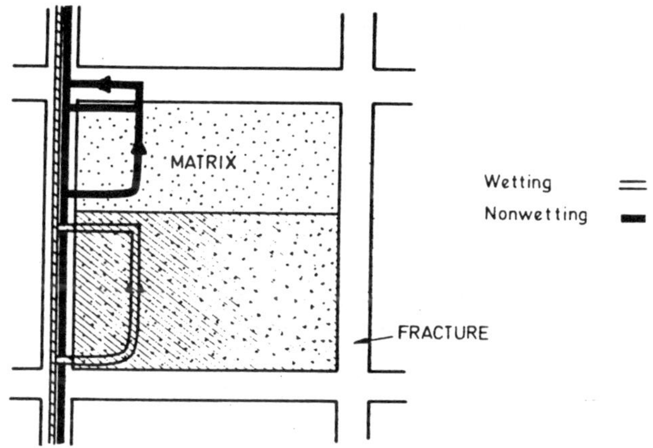

Figure 4. Schematic flow model [after Braester (11)].

Based on this model, the shape of the relative permeability curves will be as indicated in Figure 5. These curves are calculated from relative permeability relationships motivated by similar equations suggested by Brooks and Corey (12):

$$k_{ro} = [\frac{k_2}{k} + (1 - \frac{k_2}{k})(1 - S_{w1}^2)(1 - S_{w1})^2](1 - S_{w2})^2(1 - S_{w2}^2)$$

$$k_{rw} = [\frac{k_2}{k} + (1 - \frac{k_2}{k}) S_{w1}^4] S_{w2}^4 \qquad (2.17)$$

where; k_2 is the permeability of fracture network, k is the total permeability of the porous fractured system, S_{w1} is the water saturation in the matrix, and S_{w2} is the water saturation in the fractures.

The interpretation of the curves in Figure 5 is as follows:

- In point A, the matrix block water saturation and the fracture water saturation are equal to 1 and $k_{rw} = 1$.
- In point B, the water saturation in the blocks is zero and the water saturation in the fractures is 1, giving a low k_{rw}.

This low relative permeability at point B, k_{rwB}, can be derived for an idealized fractured matrix system of cubic blocks as shown in Figure 6.

$$k_{rw} = \frac{k_2}{k} \qquad \text{for } S_{w1} = 0, \quad S_{w2} = 1 \qquad (2.18a)$$

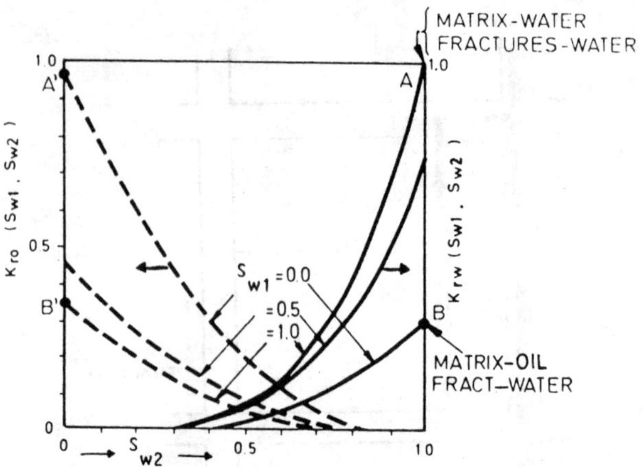

Figure 5. Relative permeability of a matrix-fracture reservoir unit [after Braester (11)].

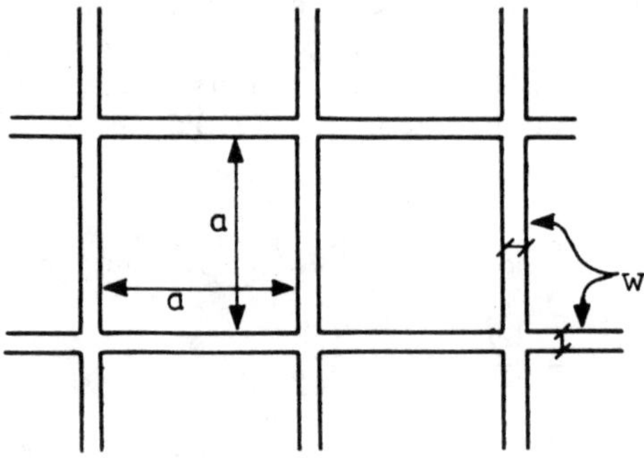

Figure 6. Idealized matrix/fracture system.

Table 3. Displacement processes in a fractured reservoir [after van Golf-Racht (45)].

Case	Fluid saturation: Matrix Block	Fluid saturation: Fracture network	Wetting phase in matrix	Oil Displacement Process
1	oil	water	water	imbibition
2	oil	water	oil	drainage
3	oil	gas	oil	drainage
4	gas	water	water	imbibition
5	water	oil	water	drainage
6	water	gas	water	drainage

where

$$k_2 = \frac{w^3}{6a} \tag{2.18b}$$

$$k = k_1 + \frac{w^3}{6a} \tag{2.18c}$$

giving

$$k_{rw} = \frac{w^3/6a}{k_1 + w^3/6a} \tag{2.19}$$

Figure 7 shows the variation of k_{rwB} for different block and fracture parameters. For given values of matrix permeability and block dimensions, any reduction in fracture width considerably reduces the values of k_{rwB}.

The application of the model of Braester (11) will require a method for deriving relative permeability relationships (as suggested in Eq. (2.17)) for a given fractured reservoir.

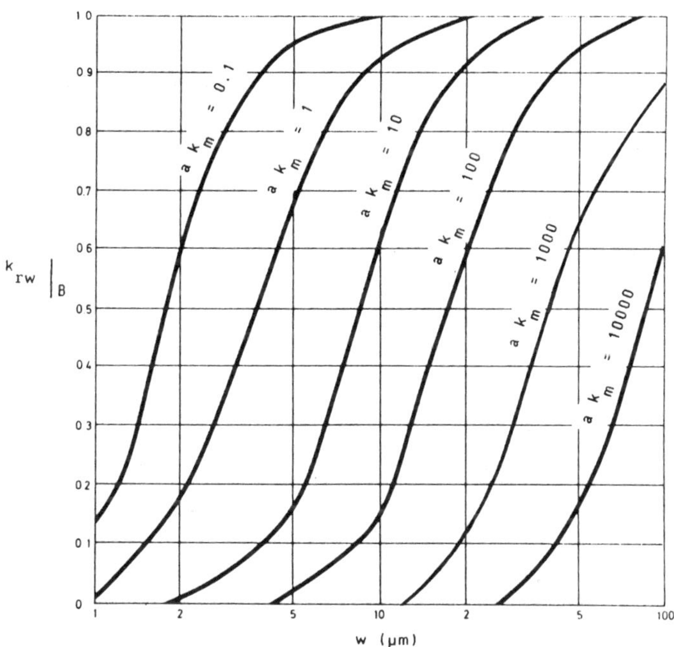

Figure 7. Variation of k_{rw}/B vs. width w(µm) for various block and fracture parameters (a in cm and k_m in mD).

2.7. Capillary Pressure

The definition of capillary pressure is

$$p_c = p_{nw} - p_w \tag{2.20}$$

where subscripts nw and w represent non-wetting phase and wetting phase respectively.

In a fractured reservoir, the capillary pressure curve plays a very important role since capillary forces contribute to the imbibition displacement process and oppose the drainage displacement process. The relationship between the fluid saturating the matrix block and the fluid saturating the fracture will determine the type of process (drainage or imbibition) taking place during production (Table 3).

An idealized element of a fractured rock is shown in Figure 8 with the matrix saturated with oil and partially or completely submerged in water or gas. In this case, two sets of forces will play a role in the displacement process:

- Gravity forces due to the difference in densities between oil and water (or gas), and
- Capillary forces due to the interaction of surface forces within the pores.

The difference in pressure which leads to this displacement is (39):

$$\Delta p = (X - x)(\rho_w - \rho_o) \cdot g + p_c \tag{2.21}$$

where; X and x are distances defined in Figure 8. ρ_w, and ρ_o are the water and oil specific gravities.

The flow rate per unit cross section is given by Darcy's law:

$$u = \frac{k_w \Delta p_w}{\mu_w x} = \frac{k_o \Delta p_o}{\mu_o (a - x)} \tag{2.22}$$

where

Δp_w, Δp_o = Pressure drops in water and oil bearing zones

k_w, k_o, μ_w, μ_o = Permeabilities and viscosities of the two phases

a = Typical vertical dimension of the matrix element

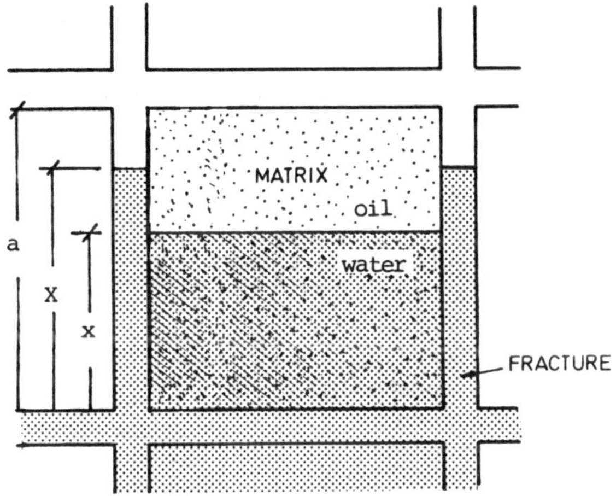

Figure 8. Illustration of fluid distribution in schematic fractured system [after Braester (11)].

It follows that:

$$\Delta p = \Delta p_w + \Delta p_o = u \left[\frac{\mu_w}{k_w} x + \frac{\mu_o}{k_o} (a - x) \right] \tag{2.23}$$

The velocity of the oil/water contact in the matrix is related to the flow rate by

$$u_w = (1 - S_{iw} - S_{or})\, \phi_m \frac{dx}{dt} \tag{2.24}$$

where; S_{iw} is the irreducible water in the matrix and S_{or} is the residual oil in the matrix.

Combining these results leads to a differential equation for x (assuming X = a).

$$\frac{dx}{dt} \left[\left(\frac{\mu_w}{k_w} - \frac{\mu_o}{k_o} \right) x + \frac{\mu_o}{k_o} \cdot a \right] + \frac{g(\rho_w - \rho_o)}{(1 - S_{iw} - S_{or}) \cdot \phi_m} \cdot x$$

$$= \frac{(\rho_w - \rho_o) a \cdot g + p_c}{(1 - S_{iw} - S_{or}) \phi_m} \tag{2.25}$$

For an element of matrix suddenly completely immersed in water with the same mobility as oil ($k_w/\mu_w = k_o/\mu_o$), we get the following expression for oil production per unit cross section:

$$u_o = \frac{k_o}{\mu_o} \frac{(\rho_w - \rho_o)a \cdot g + p_c}{a} \tag{2.26}$$

The condition for oil expulsion in the case expressed by Eq. (2.25) is:

$$a(\rho_w - \rho_o) \cdot g + p_c > 0 \tag{2.27}$$

As observed, this inequality is dependent on block size, capillary pressure, and matrix wettability.

For the case of oil and water in a water wet matrix, both the gravity and imbibition terms are positive. In an intensely fractured water wet rock (as for some of the chalk reservoirs in the North Sea), the gravity term becomes negligible, and the process becomes, for practical purposes, capillary imbibition. For an oil-wet matrix, the displacement is only possible if the gravity force overcomes the threshold capillary pressure p_d:

$$a(\rho_w - \rho_o) \cdot g > p_d \tag{2.28}$$

Note that oil cannot be displaced from the matrix by water in an intensely fractured oil-wet reservoir with discontinuous matrix blocks. The equation for oil production from a matrix block saturated with oil and suddenly surrounded by gas will be:

$$q_o = \frac{k_o}{\mu_o} \frac{(\rho_o - \rho_g)a \cdot g - p_c}{a} \tag{2.29}$$

where ρ_g is the gas density. The requirement for oil production will be analogous to the oil-wet oil-water case:

$$a(\rho_o - \rho_g)g > p_d \tag{2.30}$$

As observed from these conditions for oil production, the choice of which fluid to inject depends on the matrix dimension a. A good geologic description is therefore important when planning fractured reservoir development.

Oil production from matrix blocks is frequently described by transfer functions or imbibition curves. The term imbibition curves is of course strictly correct only in the case of negligible gravity effects. These curves represent the produced oil as a function of time, and can be derived either mathematically or experimentally. The next chapter will discuss these recovery curves more in detail.

3. THE IMBIBITION PROCESS

This section is devoted to two items:

1) A brief summary of proposed theoretical models for description of fractured reservoir behavior, and
2) Experimental studies on capillary imbibition.

The purpose of the presentation of fractured reservoir models is to describe the importance of experimental imbibition data of the type presented in this work.

3.1. Introduction

In the last two decades, advancements in the understanding of fractured reservoirs have been reported in various areas such as geology, reservoir description, flow toward the well, reservoir mechanisms etc. The text book of van Golf-Racht (45) on the subject presents a comprehensive discussion on most aspects of fractured reservoir behavior. This literature survey is limited to presenting the main concepts of the models and will specifically discuss the imbibition part of the models.

Barenblatt, Zheltov and Kochina (4) laid a general foundation for studies of flow behavior in fractured systems with the following principles:

1) The interconnected system of fractures and blocks are overlapping continua.
2) Two sets of parameters are established at each mathematical point.
3) Flow takes place through each medium. The exchange of fluids between fractures and blocks is accounted for by a source/sink function in the conservation of mass equation.

The horizontal flow and conservation of mass equations of an incompressible fluid are:

$$\vec{u}_j = -\frac{k_j}{\mu}\nabla p_j \qquad (3.1)$$

$$\frac{\partial \phi_j}{\partial t} + \nabla . \vec{u}_j + (-1)^{(1+j)} . u^* = 0 \qquad (3.2)$$

where $\vec{u}$ is flux, k is permeability, μ is dynamic viscosity, ϕ is porosity, p is pressure, and u^* represents the transfer of fluid between blocks and fractures. Subscript $j = 2$ denotes the medium of fractures and $j = 1$ denotes the matrix blocks. The source function u^* derived by Barenblatt, Zheltov and Kochina (4) from dimensional analysis considerations is:

$$u^* = \frac{\alpha}{\mu}(p_1 - p_2) \tag{3.3}$$

where α is a dimensionless characteristic parameter of the fractured rock. The source function is defined positive for transfer of liquid from blocks to fractures.

A number of models developed to describe flow towards a well in a fractured rock are based on Barenblatt's concept on single phase flow. These models, like Warren and Root's (47) are widely used to interpret transient pressure data in fractured reservoirs. The transient models also became the starting point in the investigation of multiphase flow in fractured reservoirs.

This section is limited to the water-oil displacement process by capillary imbibition. The imbibition process will be emphasized, but in order to demonstrate how complex a general mathematical model for two-phase flow is, two such models will be discussed. These models involve several parameters that cannot be determined experimentally. However, it is worthwhile studying this theory in order to grasp the important flow properties.

Bokserman, Zheltov and Kocheshkov (10) generalized the continuum approach for one-phase, incompressible flow to flow of oil and water. The conservation of mass and the Darcy equations are:

$$\phi_2 \frac{\partial S_{w2}}{\partial t} + \nabla.\vec{u}_{w2} + u^*_w = 0 \tag{3.4}$$

$$-\phi_2 \frac{\partial S_{o2}}{\partial t} + \nabla.\vec{u}_{o2} + u^*_o = 0 \tag{3.5}$$

$$\vec{u}_{w2} = -\frac{k_2 k_{rw2}(S_{w2})}{\mu_w} \nabla \Phi \tag{3.6}$$

$$\vec{u}_{o2} = \frac{k_2 k_{rO2}(S_{o2})}{\mu_o} \nabla \Phi \tag{3.7}$$

In these equations, the subscripts o and w represent oil and water, respectively. S is saturation, k_r is relative permeability, t is time, and Φ is fluid flow potential (datum pressure).

This model is based on a medium in which the permeability of the blocks can be neglected in comparison to that of the fracture network. Since the volume of the fracture network normally is very small compared to the matrix volume, it is assumed that the total amount of water entering the fractures is consumed in the imbibition of the blocks. In incompressible flow, the volume of water entering the blocks equals the volume of oil produced in the fractures, and therefore $u_o^* = - u_w^* = |u^*|$. From the results of imbibition laboratory experiments performed by Mattax and Kyte (34), Bokserman, Zheltov and Kocheshkov (10) obtained the following source function:

$$u^* = C\phi_1 S_{w1} \frac{s_v^2 \ \sigma \cos\theta (k_1/\phi_1)^{\frac{1}{2}}}{\mu_o} \left(\frac{s_v^2 \ \sigma \cos\theta (k_1/\phi_1)^{\frac{1}{2}}}{\mu_o} \cdot t \right)^{-\frac{1}{2}} \qquad (3.8)$$

where, s_v is volume based specific surface area, σ is interfacial tension, and θ is the contact angle. The expression inside the parenthesis is dimensionless time. Bokserman used this model to describe the flow behavior in the case of a moving imbibition zone, as shown in Figure 9.

Other theoretical works on multiphase fracture flow include works on Barenblatt (5) and Braester (11). Barenblatt formulates the immiscible gas-oil flow for each continuum, and flow between the matrix blocks and the fractures is accounted for by a source function based on Darcy's law. In contrast with Barenblatt's model, Braester considered the flow to be governed by a single macroscopic pressure gradient. Braester's source function is defined in terms of the potential gradient in the fractures, the capillary-pressure difference between the liquid in the fractures and the matrix blocks, and the density difference between liquid phases:

$$u^* = \frac{k_1 s_v}{\mu_w} \left(- \frac{\partial p_w}{\partial x} + \frac{p_c}{L} - S_{w1} \ g \ \Delta\rho \right) F_1(S_{w1}) \ F_2 \ (S_{w2}) \qquad (3.9)$$

where, L is the characteristic length of the block, $\Delta\rho$ is the density difference between oil and water and $F_2(S_{w2})$ and $F_1(S_{w1})$ are functions of saturation in fractures and blocks, respectively.

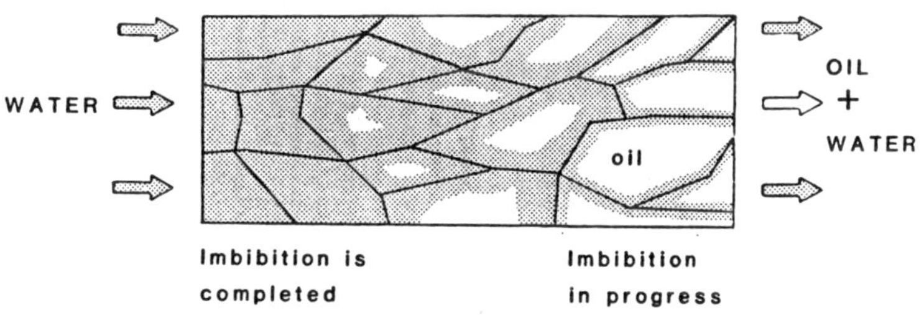

Figure 9. Water imbibition in fractured rock.

This source function was developed by Bear and Braester (7) by using a conceptual model of matrix made up of a bundle of randomly oriented capillary tubes.

3.1.1. Theoretical studies

This subsection is concerned primarily with imbibition aspects of the flow mechanism in the matrix rather than with the total flow problem in the fracture-matrix system. The literature discussed is therefore concentrated on simplified dynamic models describing the flow in a matrix block surrounded by water. The main attention has been given to the forces involved in the process and the oil recovery obtained by spontaneous imbibition.

Bear (8) defines the imbibition process as the spontaneous displacement of the non-wetting fluid by the wetting fluid due to surface tension. As increasingly better lithological descriptions have been made of the reservoirs, it has become obvious that imbibition phenomena which were once considered laboratory curiosities are of practical importance. Serious studies of the imbibition process in reservoir engineering were initiated by the rapid decline of oil productivity of wells in the Spraberry Field in West Texas (13,18). Field data alone proved to be inadequate for the investigation of imbibition, so a number of theoretical studies have been undertaken to provide the understanding of the parameters involved.

Only a few studies have been dedicated to three dimensional exchange of oil and water in a single matrix block. Most of the theoretical approaches and mathematical models describe simplified flow patterns in partially sealed matrix blocks. In addition, numerous studies have been performed on fundamental capillary behavior both in petroleum engineering and other disciplines, but these publications will not be discussed here.

Blair (9) solved the system of equations used by Douglas, Peaceman and Rachford (17) for a cylindrical block which was sealed at the top and bottom. He investigated the influence of the viscosity, the initial saturation of water, and the length in the one-dimensional case. A study of recovery behavior by using this model requires data from capillary pressure and relative permeability tests, which are more time consuming to perform than the imbibition test itself. Blair showed that the time required to imbibe a fixed volume of water is proportional to the square root of the oil viscosity whenever $\mu_o > \mu_w$.

Aronofsky, Masse and Natanson (3) suggested an abstract model based on the variation of recovery with time during the imbibition process. They assumed that the oil production from a small volume is a continuous monotonic function of time, and that it converges to a finite limit. The second basic assumption is that none of the properties which determine the rate of convergence change sufficiently during the process to affect this rate or the limit. Aronofsky, Masse and Natanson (3) suggest a function of the form

$$R = R_{\infty}(1 - e^{-\beta t}) \tag{3.10}$$

where R is recovery of oil in % of original oil in place, β is a constant giving the rate of convergence, and R_{∞} is the limit towards which R converges.

As mentioned previously, Bokserman, Zheltov and Kocheshkov (10) introduced a matrix block imbibition recovery function,

$$R = c(t_D)^{\frac{1}{2}} \tag{3.11}$$

where t_D is dimensionless time and c is a constant (refer to Eq. (3.8)). This expression is similar to experimental results obtained at very low mobility ratios. Bokserman, Zheltov and Kocheshkov (10) assumed that the oil released by the matrix blocks was transferred instantly to the water-oil interphase. In this way, they suggested a total flow model where the oil production is an additive function of individual block contributions. This may be a sound description in large vertical fractures were oil segregates easily, but probably not as good in a system of microfractures where varying saturations in the fractures may affect the imbibition rate.

Braester (11) introduced a method that accounted for the effect of varying saturation in the fractures on imbibition. This was done by using relative permeability relationships which were functions of both matrix and fracture saturations. His relative permeability relationships are discussed in section 2.6. The source function used in Braester's model is Eq. (3.9). The model may be useful for predictions after finding parameters to match observed oil and water productions.

Kleppe and Morse (29) studied the flow behavior in a matrix-fracture system both experimentally and numerically. Their numerical solution was based on flow and continuity equations for the oil and water, and capillary pressure and relative permeabilities were computed at every grid block. They found that for K_f higher than k_m, the recovery was rate sensitive. This was coherent with what has been found previously, but for fracture capacity approximately one tenth the matrix flow capacity, the effect of the rate was negligible. Some results on water-oil ratio vs. recovery for different k_f/k_m, as well as the effect of length in the vertical direction, were also discussed in this chapter.

de Swaan (16) used the experimental and numerical results of Kleppe and Morse to test an analytical model with a new feature. Based on the rate of imbibition function given in Eq. (3.10), de Swaan obtained a model that accounted for varying saturation around the blocks. The blocks downstream in a reservoir subjected to waterflood are necessarily exposed to a varying water saturation resulting from the water imbibition of the upstream blocks. De Swaan transformed Eq. (3.10) to the following form:

$$u_b{}^* = \frac{R_\infty}{\tau} e^{-t/\tau} \tag{3.12}$$

where $u_b{}^*$ is the transfer of fluid between block and fractures, and τ is the time necessary to produce (1-1/e) = 0.63 of the maximum recoverable oil (R_∞). When this matrix block function is known for 100% water environment, the matrix block behavior for varying water saturation at the block surface is given by a convolution. The rate of imbibition per unitary fracture length is:

$$u_u{}^* = \frac{R_{\infty u}}{\tau} \int_o^t e^{-(t-\zeta)/\tau} \frac{\partial S_w}{\partial \zeta} d\zeta \tag{3.13}$$

de Swaan concluded that the decaying exponential function is an adequate description of the water imbibition in matrix blocks, based on the good agreement with both experimental and numerical results found in the literature.

3.1.2. Experimental studies

A number of experimental studies have been performed on the imbibition process in a single matrix block. Since the various authors use very different experimental models and procedures, a few characteristics of the various investigations are presented in Table 4.

One of the earliest experimental work on imbibition was the study of Brownscombe and Dyes (13) related to the Spraberry field. They noticed that the oil recovery was dependent on the boundary conditions. The rate of imbibition is suggested to be proportional to the square root of permeability, since permeability is proportional with the square of the pore size. Their experimental results do not confirm this statement.

Graham and Richardson (24) performed a laboratory investigation of imbibition using a triangular block of fused quartz. They studied the scaling laws and the relation between the rate, the surface tension, and the fracture permeability. The experiments demonstrated that the lower the injection rate into the fractures, the greater is

Table 4. Basic characteristics of experimental models.

Author	Ref. No.	Type of porous medium	Geometry of block	Forces*	Partly sealed (s) All faces open(o)	Type of experiment
Brownscombe	13	Sandstone siltstone	Cylinders Plates	1 + 2	s + o	Conventional
Graham	24	Fused quartz	Triangular	1 + 3	s	"
Mattax	34	Alundum	Semicylindrical tube	1 + 2	s + o	"
Parsons	36	Dolomite	Parallelepiped cube	1 + 2	-	"
Iffly	26	Siltstone	Cylindrical	1 + 2	s + o	"
Du Prey	32	Sandstone	Cylindrical Parallelepiped	1+2+3	s + o	Centrifuge
Kyte	31	Sandstone	Cylindrical	1 + 2	-	"
Kleppe	29	Sandstone	Cylindrical	1+2+3	o	Conventional
Kazemi	28	Sandstone	Cylindrical Parallelepiped	1+2+3	s	"
Mannon	33	Sandstone	Parallelepiped	1 + 3	s	"
Torsaeter	44	Chalk	Cylindrical	1	o	"

* 1. Capillarity, 2. Gravity, 3. Injection pressure

the oil recovery for a given amount of water injected. By solving the basic flow equations, they recognized that the rate of imbibition is proportional to σ, $(k)^{\frac{1}{2}}$, $f(\theta)$ (a function of wettability) and depends on the fluid viscosities and the characteristics of the rocks.

Mattax and Kyte (34) discussed the experimental simulation of the imbibition phenomenon, and verified the scaling laws of Rapoport (38) in this case. They used the dimensionless parameter

$$t_D = \frac{\sigma f(\theta)\ (k/\phi)^{\frac{1}{2}}}{\mu_w L^2} \cdot t \tag{3.14}$$

Their experimental results showed that a single correlation is obtained in each case, regardless of the viscosity level of the fluids and the dimensions and permeability of the porous medium.

The scaling laws were also verified by Parsons and Chaney (36), who simulated in the laboratory a rising water table. They found that the imbibition behavior of an oil saturated rock sample subjected to a slowly rising water table can be synthesized from total immersion imbibition tests on small samples. In most cases, imbibition behavior is adequately described by the values of the maximum recoverable oil for the given rock type.

Iffly, Rousselet and Vermeulen (26) investigated the effect of composition of the rock and fluids on the water imbibition. They concluded that experiments carried out with fluids and/or cores different from those of the investigated field are meaningless, except in very particular cases. The imbibition experiments on siltstone also indicated that oil recovery decreases as the carbonate and organic matter content increases. In the experiments, a naphtenic oil was used and this oil has a great affinity to carbonate material and kerogenic clays. They also checked the experimental results with the recovery relationships given in Eqs. (3.10) and (3.11); Eq. (3.10) gave the best correlation.

The role of capillary-gravity ratio, CGR, was discussed by Iffly, Rousselet and Vermeulen (26) based on experimental results from imbibition tests on cores with different size and properties. The CGR was expressed as

$$CGR = \frac{\sigma . f(\theta)\ (\phi/k)^{\frac{1}{2}}}{\Delta\rho . g . h} \tag{3.15}$$

where h is height of core. The basic conclusion is that recovery time decreases with decreasing CGR.Lefebvre du Prey (32) confirmed this conclusion in experiments similar to those performed by Iffly,

Rousselet and Vermeulen. du Prey also compared centrifugal and conventional imbibition results for various CGR values. His results indicate that centrifuge tests on small samples are not reliable for reproducing the recovery curve of a big matrix block. This is explained by du Prey by the fact that above a certain gravity level, the local properties of two-phase flow change. Kyte (31), who introduced the centrifuge test for predicting matrix block recovery, did not notice this divergence between theory and experiment at high centrifugation speed. The problem of scaling laboratory imbibition data to field conditions is very complex, and several research programs are now devoted to this subject.

As discussed earlier in this literature review, the flow conditions in the fractures surrounding the blocks may affect the imbibition behavior. Graham and Richardson (24) performed the first study on this problem, and Kleppe and Morse (29) performed both experimental and numerical investigation of the effect of the boundary conditions around the blocks. The parameter conductivity ratio defined as,

$$CR = \frac{k_f}{k_m} \tag{3.16}$$

was used in the analysis, and the following conclusions were obtained:

1) Ultimate oil recovery is greatly affected by production rate at $CR > 1$, and
2) For $CR < 0.1$ the effect of production rate on oil recovery is negligible.

Kazemi and Merill (28) performed experimental work on artificially fractured sandstone cores. They injected water into the fracture only, and observed oil production and breakthrough time in relation to injection velocity and capillary pressure. The data indicated that at low water velocities, imbibition caused the water to advance through the matrix faster than it did through the fracture. At high water velocities or at low capillary pressures, water breakthrough in the fractured rock occurred much sooner than in unfractured rock. Kazemi also observed that the final oil recovery by pure imbibition and forced displacement is nearly the same in Berea sandstone.

Mannon and Chilingar (33) studied the effect of water injection rate on imbibition rate using a laboratory model consisting of a horizontal slab of Berea sandstone sealed on all sides but one. A fracture was created by placing an aluminum plate across the open sand face, and as water moved in the fracture, linear counter-current imbibition occurred. They found that the higher the rate of water injection, the greater the imbibition rate and ultimate oil recovery. The laboratory model used by Mannon and Chilingar is very similar to Graham and Richardson's (24) model, but the results are not quite in agreement. Graham and Richardson found that the higher the injection rate, a greater amount of water was required to be injected to produce a given amount of oil. Mannon and Chilingar also indicate that viscous forces may be operative in Graham and Richardson's model under certain conditions, and therefore less suited for pure imbibition studies.

3.1.3. Scaling of experimental data to field conditions

Laboratory imbibition experiments must be performed according to certain scaling requirements to obtain results directly applicable in the reservoir. As discussed previously in the literature review, various dimensionless parameters have been tested with respect to scaling of imbibition experiments. The following is a general discussion of the flow equations and the various forces involved. Since the relative importance of the various forces affects the displacement process considerably, the experimental conditions must be designed to obtain the correct ratios between forces.

The continuity equations and the flow equations for water and oil can be written as follows:

$$\nabla(\rho_o \vec{u}_o) = -\frac{\partial(\rho_o \phi S_o)}{\partial t} \tag{3.17}$$

$$\nabla(\rho_w \vec{u}_w) = -\frac{\partial(\rho_w \phi S_w)}{\partial t} \tag{3.18}$$

$$\vec{u}_o = -\frac{k_o}{\mu_o}(\text{grad } p_o - \rho_o g) \tag{3.19}$$

$$\vec{u}_w = -\frac{k_w}{\mu_w}(\text{grad } p_w - \rho_w g) \tag{3.20}$$

In the case of water displacing oil from a matrix block of height a (as shown in Figure 10), Eqs. (3.19) and (3.20) will become

$$u_{oz} = \frac{k_o}{\mu_o}\left(\frac{\partial p_o}{\partial z} - \rho_o g\right) \tag{3.21}$$

$$u_{wz} = \frac{k_w}{\mu_w}\left(\frac{\partial p_w}{\partial z} - \rho_w g\right) \tag{3.22}$$

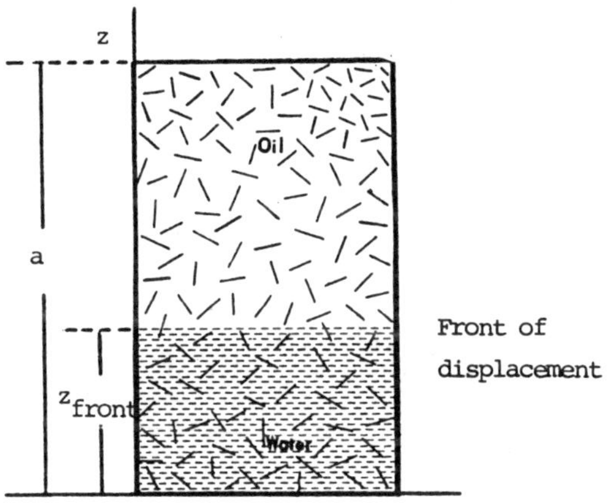

Figure 10. Water advancement in matrix block.

In the following, the Leverett J function is assumed to be the correlating function for capillary pressure. In addition, by introducing an intermediate pressure between oil and water (p) and an arbitrary constant, b, the following relationships for oil and water pressure are obtained:

$$p_o = p + b\sigma\cos\theta(\phi/k)^{\frac{1}{2}}J(S_w) \tag{3.23}$$

$$p_w = p + (b-1)\sigma\cos\theta(\phi/k)^{\frac{1}{2}}J(S_w) \tag{3.24}$$

The velocities, u_{oz} and u_{wz}, and the front height, z, can be transformed to dimensionless terms as follows:

$$u_{ozD} = u_{oz} \cdot \frac{t_o}{a} = u_{oD} \tag{3.25}$$

$$u_{wzD} = u_{wz} \cdot \frac{t_o}{a} = u_{wD} \tag{3.26}$$

$$z_D = \frac{z}{a} \tag{3.27}$$

where the time for frontal advancement from z = 0 to z = a is denoted t_o.

The combination of Eqs. (3.21) - (3.24) gives:

$$u_{oz} = -\frac{k_o}{\mu_o}\left[\frac{\partial p}{\partial z} + b\sigma\cos\theta(\phi/k)^{\frac{1}{2}}\frac{\partial J}{\partial S_o}\frac{dS_o}{dz} - \rho_o g\right] \tag{3.28}$$

$$u_{wz} = -\frac{k_w}{\mu_w}\left[\frac{\partial p}{\partial z} + (b-1)\sigma\cos\theta(\phi/k)^{\frac{1}{2}}\frac{\partial J}{\partial S_o}\frac{dS_o}{dz} - \rho_w g\right] \tag{3.29}$$

By introducing the dimensionless terms u_{ozD}, u_{wzD} and z_D in Eqs. (3.28) and (3.29), the following dimensionless expressions are obtained:

$$u_{ozD} = -\frac{\mu_w}{\mu_o}\frac{\partial}{\partial z_D}\left(\frac{k_o p t_o}{\mu_w a^2}\right) - \frac{\mu_w}{\mu_o} b \left(\frac{k_o t_o \sigma\cos\theta(\phi/k)^{\frac{1}{2}}}{\mu_w a^2}\right)\frac{\partial J}{\partial S_o}\frac{dS_o}{dz_D}$$

$$+ \frac{\mu_w}{\mu_o}\frac{\rho_o}{\rho_w}\left(\frac{k_o t_o \rho_w g}{\mu_w a}\right) \tag{3.30}$$

$$u_{wzD} = -\frac{\partial}{\partial z_D}\left(\frac{k_w p t_o}{\mu_w a^2}\right) - (b-1)\left(\frac{k_w t_o \sigma\cos\theta(\phi/k)^{\frac{1}{2}}}{\mu_w a^2}\right)\frac{\partial J}{\partial S_o}\frac{dS_o}{dz_D}$$

$$+ \left(\frac{k_w t_o \rho_w g}{\mu_w a}\right) \tag{3.31}$$

From these equations, the following important dimensionless groups for scaling the displacement processes are observed:

1) $\dfrac{k_w \Delta p_{inj} t_o}{\mu_w a^2}$ associated with injection pressure

2) $\dfrac{k_w t_o \sigma\cos\theta(\phi/k)^{\frac{1}{2}}}{\mu_w a^2}$ associated with capillary forces

3) $\dfrac{k_w t_o \Delta\rho g}{\mu_w a}$ associated with gravity forces.

To express the dimensionless parameters as velocity ratios, the following velocities are defined:

1) Front velocity: $$u_F = \frac{a}{t_o} \quad (3.32)$$

2) Injection velocity: $$u_{inj} = \frac{k_w \Delta p_{inj}}{\mu_w a} \quad (3.33)$$

3) Capillary velocity: $$u_c = \frac{k_w \sigma \cos\theta (\phi/k)^{\frac{1}{2}}}{\mu_w a} \quad (3.34)$$

4) Gravity velocity: $$u_G = \frac{k_w \Delta\rho g}{\mu_w} \quad (3.35)$$

Table 5 presents dimensionless parameters for each of the forces acting in a displacement process of the type discussed here. The relative importance of the forces is a key factor in scaling problems. From the equations derived in this subsection, Table 6 was obtained. This table shows two and two forces combined as dimensionless ratios.

3.1.4. Conclusions

The important ideas found in the different studies can be summarized as follows:

1. All authors agree that imbibition is the most important mechanism in the displacement of oil by water in fractured reservoirs.

2. The gravity forces and viscous forces can be very important depending on the characteristics of the matrix blocks and the fluids.

3. The boundary conditions existing at the surface of the matrix blocks are important.

4. Recovery versus dimensionless time has shown interesting similarity between theoretical analysis and experimental results. But due to rock heterogeneity and unknown effects of lithology and fluid characteristics, laboratory experiments on the reservoir system are the only valid method for predicting imbibition recovery.

5. The present theory on imbibition states that the imbibition rate is in addition to being time dependent proportional to $k^{\frac{1}{2}}$, $\phi^{-\frac{1}{2}}$, exposed matrix area, fluid characteristics, and rock-fluid interaction relationships.

6. The major shortcoming of the numerical models for simulation of the imbibition production mechanism is the relative

Table 5. Single forces expressed as dimensionless groups.

SINGLE FORCES EXPRESSED AS DIMENSIONLESS GROUPS			
Forces	Expressed as in equations	Expressed as function of pressure	Expressed as velocity ratio
Pressure	$\frac{k_w \Delta p t_o}{\mu_w a^2}$	$\frac{k_w t_o}{\mu_w a^2} \Delta p$	$\frac{u_{inj}}{u_F}$
Capillary	$\frac{k_w t_o \sigma \cos\theta (\phi/k)^{\frac{1}{2}}}{\mu_w a^2}$	$\frac{k_w t_o}{\mu_w a^2} p_c$	$\frac{u_c}{u_F}$
Gravity	$\frac{k_w \Delta\rho g t_o}{\mu_w a}$	-	$\frac{u_G}{u_F}$

Table 6. Combined forces as dimensionless groups.

COMBINED FORCES AS DIMENSIONLESS GROUPS		
	Expressed as in equations	Expressed as function of pressure
Capillary/ gravity ratio	$\frac{(k_w t_o \sigma \cos\theta (\phi/k)^{\frac{1}{2}})/\mu_w a^2}{(k_w \Delta\rho g t_o)/\mu_w a}$	$\frac{p_c/a}{\Delta\rho g}$
Gravity/ pressure ratio	$\frac{(k_w \Delta\rho g t_o)/\mu_w a}{(k_w \Delta p_{inj} t_o)/\mu_w a^2}$	$\frac{\Delta\rho g}{\Delta p_{inj}/a}$
Pressure/ capillary ratio	$\frac{(k_w \Delta p_{inj} t_o)/\mu_w a^2}{(k_w t_o \sigma \cos\theta (\phi/k)^{\frac{1}{2}})/\mu_w a^2}$	$\frac{\Delta p_{inj}}{\sigma \cos\theta} (\frac{\phi}{k})^{\frac{1}{2}}$

permeability concept.

7. The development of a general relationship between fluid-rock-, rock-fluid interacting parameters and imbibition is considered nearly impossible. This task will require a correct micro-description of the rock and the nature of the physico-chemical relations between rock and fluids.

8. Both experimental and numerical techniques should be further developed through comprehensive studies on various reservoir systems combined with basic research on capillary behaviour, wettability, pore structures, and multiphase flow.

3.2. Experimental Studies

Four different experimental studies will be discussed in this section:

1) Imbibition in chalk plugs from the Ekofisk Field in the North Sea;
2) Imbibition in rock samples of various sizes and shapes;
3) Imbibition in core plugs with various boundary conditions; and
4) Imbibition in matrix blocks containing microfractures.

Each study will be presented in the following format with respect to procedures, data, experimental program, and results. A subsection on numerical simulation of laboratory experiments is also presented. An overall discussion of the experimental studies will conclude this section.

3.2.1. Ekofisk imbibition study

The objective of this work was to investigate the capillary imbibition in cleaned Ekofisk Field chalk samples. Wettability tests, imbibition tests, and rock property measurements were performed. The plugs were extracted by alternating toluene and methanol until no more contamination (colour) was observed in the solvents. The plugs were then dried, evacuated, and saturated with simulated formation water. Irreducible water saturation was obtained by centrifuging the plugs in oil. This water saturation will be denoted initial water saturation in the following, since this was the water present at the start of the imbibition experiments. The majority of the plugs was also investigated with respect to the Amott (2) type wettability test. Details about the procedures are given by Torsaeter (43,44).

The reservoir rock in the Ekofisk Field is confined to the Ekofisk and Tor Formations, both chalk formations with similar characteristics. However, the wettability and imbibition studies revealed major differences between the formations. Water imbibition results for the Tor formation is shown in Figure 11, 12, and 13. All the core plugs included in these plots are completely water wet. Figure 11 shows the initial water saturation, S_{wi}, as a function of porosity, where S_{wi} was determined after centrifuging in oil. The water saturation change due to imbibition, ΔS_w, versus porosity is shown in Figure 12. The final water saturation after imbibition, S_{wf}, is the sum of ΔS_w and S_{wi}, and the plot is shown in Figure 13. This diagram shows that the residual oil saturation $(1 - S_{wf})$ after water imbibition is essentially constant independent of porosity, averaging 37.3% of pore volume for the 40 plugs.

The wettability of the plugs from the Ekofisk Formation varied from neutral to water wet, based on the Amott test. The imbibition of core plugs from Ekofisk Formation is therefore not only governed by the pore structure of the rock, but also by the wetting properties. A partial confirmation of this statement is seen in Figure 14, 15, and 16. S_{wi}, ΔS_w and S_{wf} do not correlate with porosity the way the cores from the Tor Formation do. Extraction experiments indicate the difference in imbibition behavior of Tor and Ekofisk Formation is due to surface chemistry differences (43).

3.2.2. Sample shape and boundary conditions

The boundary conditions will affect the imbibition in core plugs. The experiments developed by various authors consider various geometrical and physical elements according to the objective of their study. The sample shapes are generally regular, i.e., cylinders, parallelepipeds, or cubes, and the samples may be sealed in order to examine the effects of flow in each direction (Figure 17).

The present experimental work can be divided into:

1) Capillary imbibition in rock samples of various sizes and shapes, and
2) Capillary imbibition in core plugs with various boundary conditions.

The following procedure was common to all the experiments:

1) The samples were cleaned in a Soxhlet extractor using methanol and toluene and dried.
2) Each sample was weighed dry, then saturated with oil in an evacuated container, and weighed in air and oil. Pore volume and bulk volume then easily could be calculated. All experiments were performed without initial water saturation.
3) The imbibition experiments were performed using imbibition

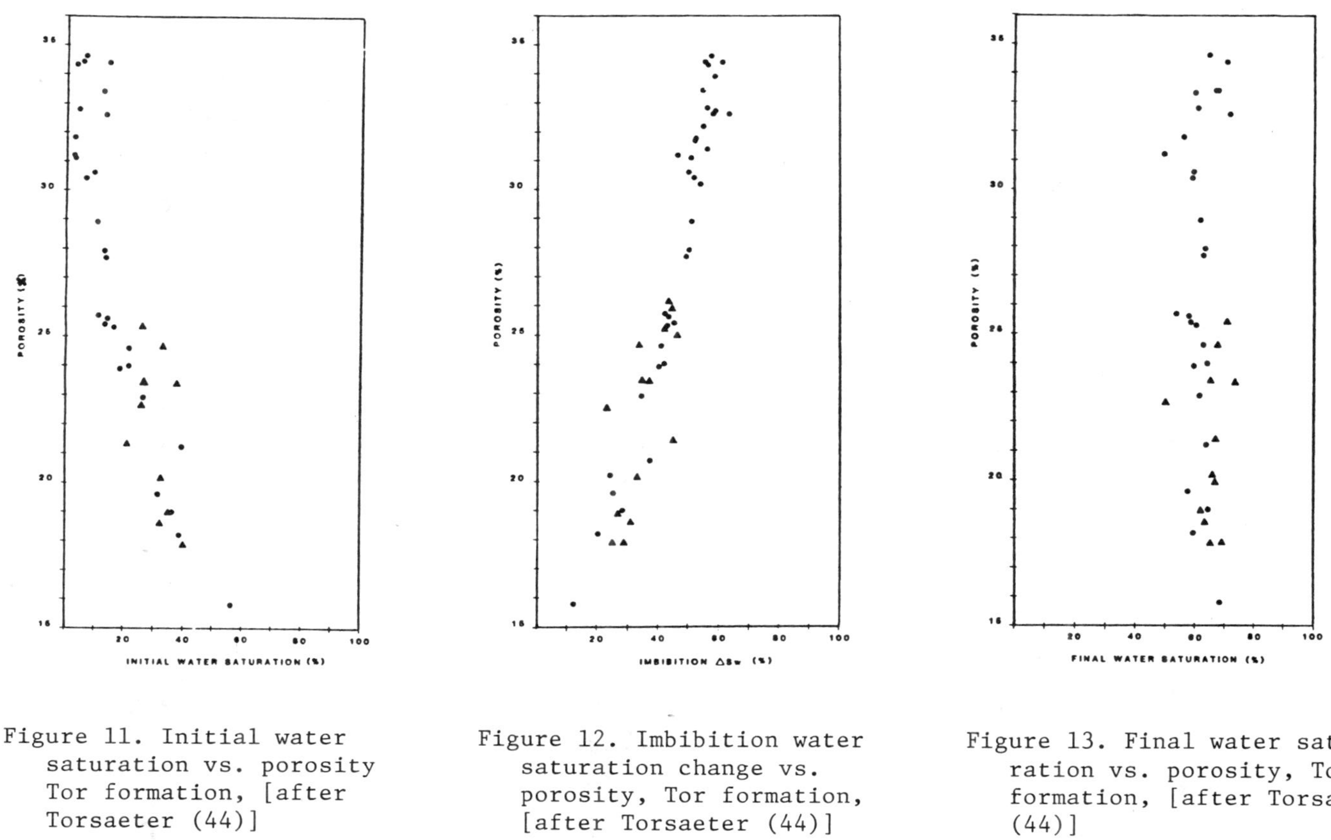

Figure 11. Initial water saturation vs. porosity Tor formation, [after Torsaeter (44)]

Figure 12. Imbibition water saturation change vs. porosity, Tor formation, [after Torsaeter (44)]

Figure 13. Final water saturation vs. porosity, Tor formation, [after Torsaeter (44)]

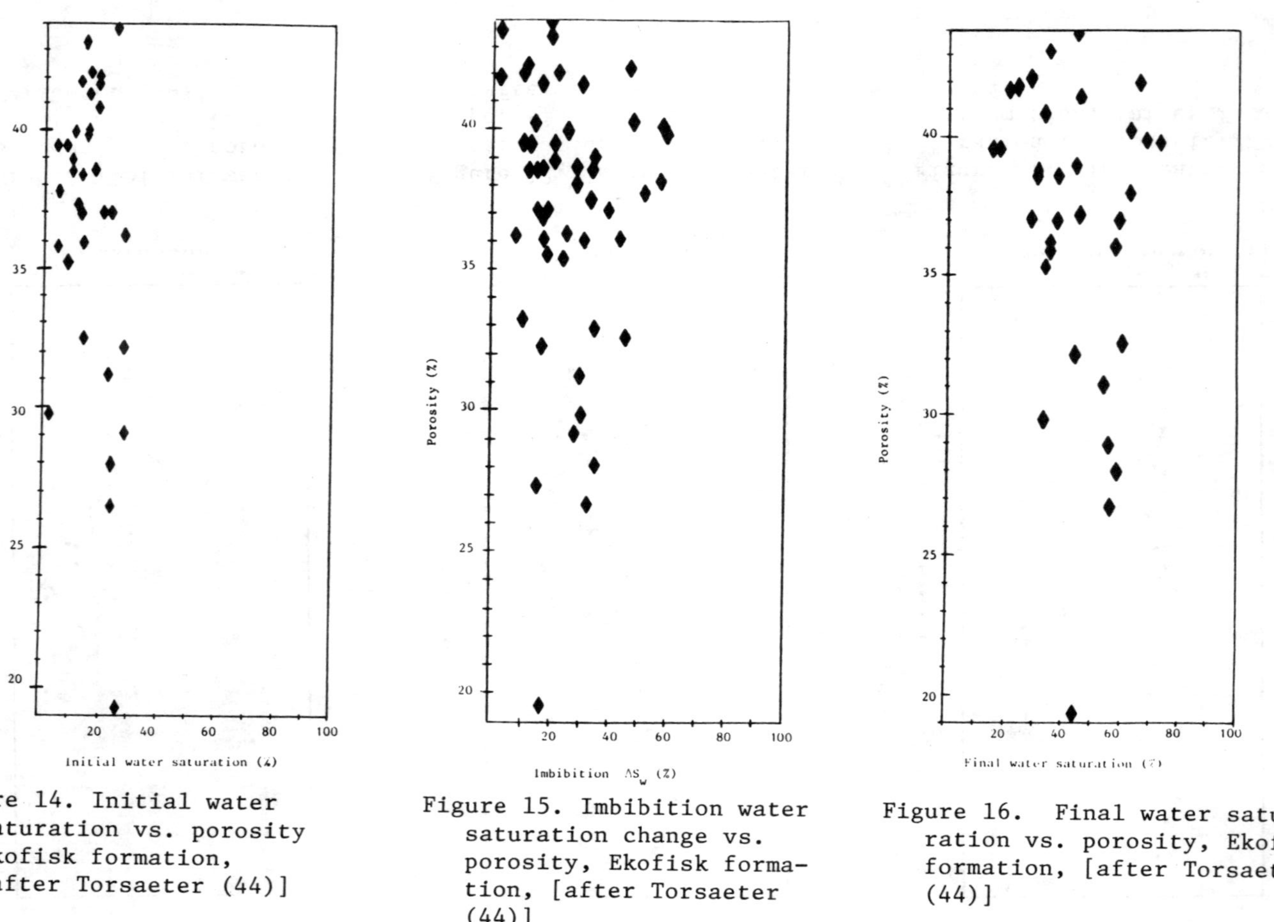

Figure 14. Initial water saturation vs. porosity Ekofisk formation, [after Torsaeter (44)]

Figure 15. Imbibition water saturation change vs. porosity, Ekofisk formation, [after Torsaeter (44)]

Figure 16. Final water saturation vs. porosity, Ekofisk formation, [after Torsaeter (44)]

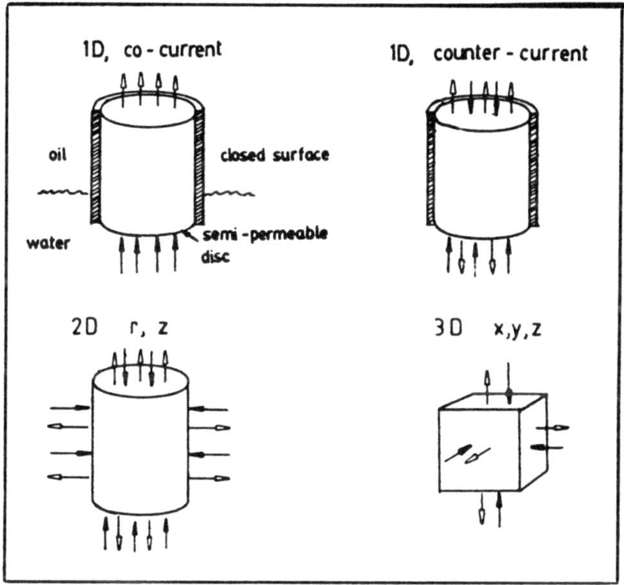

Figure 17. Dimensionality of imbibition experiments.

cells as shown in Figure 18. At a fixed time, the sample was plunged into the water and at certain time intervals, the amount of oil produced was read.

Imbibition in samples of various shapes: Three sets of samples were used: two sets of Berea sandstone, and one set of chalk from a Danish outcrop. Both regular and irregular geometrical shapes were used. Regular shapes facilitated simple measurements of surface area of the samples. The surface area of irregular samples were qualitatively estimated.

All Berea samples as well as all chalk samples were assumed to have the same rock properties (porosity, permeability, capillary pressure curve, wetting characteristics, etc.), respectively. This was believed to be a valid assumption, because for both rock types, all the samples were taken from a small chunk of rock. Any variations in such small pieces of rock are normally negligible, and would not appreciably affect results in this study. The liquid permeability and porosity of the Berea were 0.3 μm^2 and 21.1% respectively. The chalk had a liquid permeability of 0.005 μm^2 and a porosity of 46.9%.

In the experiments, the four regular geometric shapes of cylinder, triangular prism, cube, and parallelepiped were used. In addition, two irregularly shaped Berea samples were used. Figure 19 and Table 7 show the shapes, bulk volumes, surface areas, and

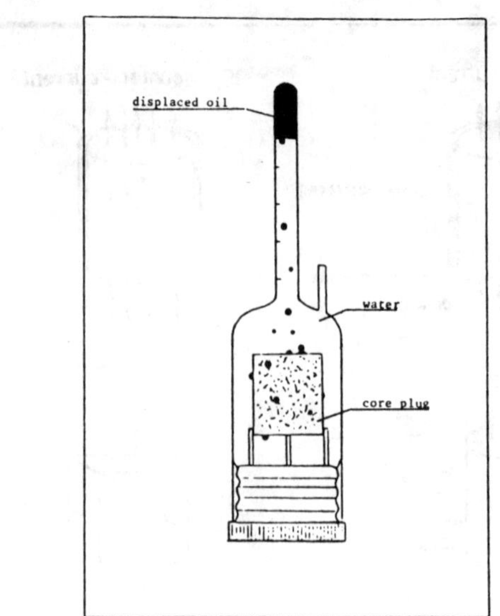

Figure 18. Imbibition cell with oil-saturated core plug.

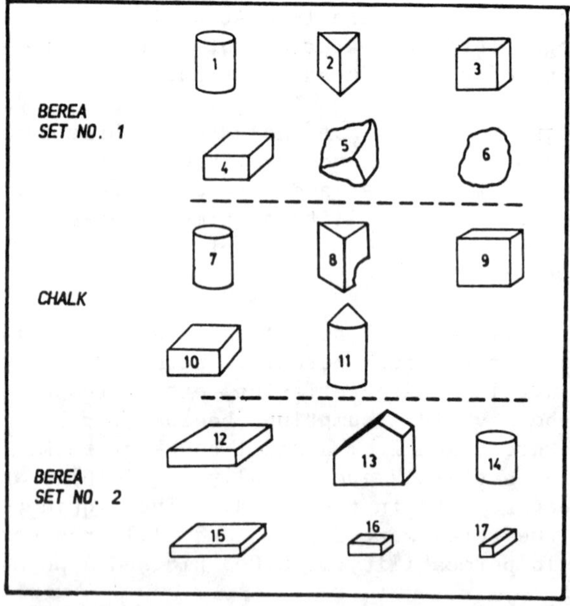

Figure 19. Cores used for studying the influence of shape on imbibition

Table 7. Core sample parameters

Sample	Rock type	Bulk volume [cm^3]	Surface area [cm^2]	Surface to volume ratio [1/cm]
1	Berea, set 1	16,389	37.00	2.27
2	"	19,841	49.00	2.48
3	"	18,161	41.86	2.32
4	"	18,811	46.49	2.45
5	"	18,595	-	-
6	"	19,393	-	-
7	Chalk	47,022	73.01	1.55
8	"	54,129	100.00	1.85
9	"	46,517	78.40	1.69
10	"	47,041	85.88	1.83
11	"	44,238	82.00	1.86
12	Berea, set 2	33,438	78.87	2.36
13	"	33,783	63,88	1.89
14	"	25,833	50.25	1.95
15	"	27,355	74.56	2.73
16	"	11,172	31.22	2.79
17	"	12,670	38.22	3.02

surface-to-volume ratios. A detailed description of the experiments is given by Fossberg (20).

Figure 20 shows the imbibition recovery curves for the first set of Berea samples. The reproducibility of the results were tested, and this is shown in Figure 21. The results of the experiments on the second set of Berea samples are shown in Figures 22, 23, and 24, clearly comparing the imbibition versus shape. The chalk imbibition curves are shown in Figure 25.

Imbibition with various boundary conditions: This part of the experimental work was performed to determine the effects of boundary conditions on capillary imbibition. Cylindrical core plugs of Berea sandstone were used. Three different boundary conditions for one-dimensional imbibition, as shown in Figure 26, were investigated. Eleven plugs were analyzed; all plugs were taken from a small block of Berea sandstone. The absolute permeability of these samples was 0.13 μm^2 and the porosity was 21.5%.

To obtain one-dimensional flow, the sidefaces were coated with epoxy before saturating with paraffin. Imbibition experiments of type A_1 (Figure 26) were performed. The same plugs were then cleaned, the epoxy layers were removed, and new volume measurements were obtained. In this second set of experiments, both the sideface and one endface were coated with epoxy to obtain countercurrent imbibition of type A_2.

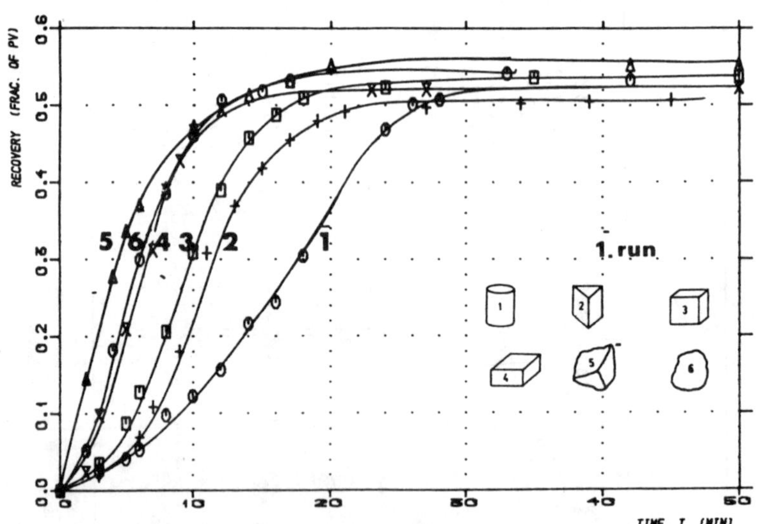

Figure 20. Effect of shape on recovery.
Recovery vs. time for the first set of Berea cores.

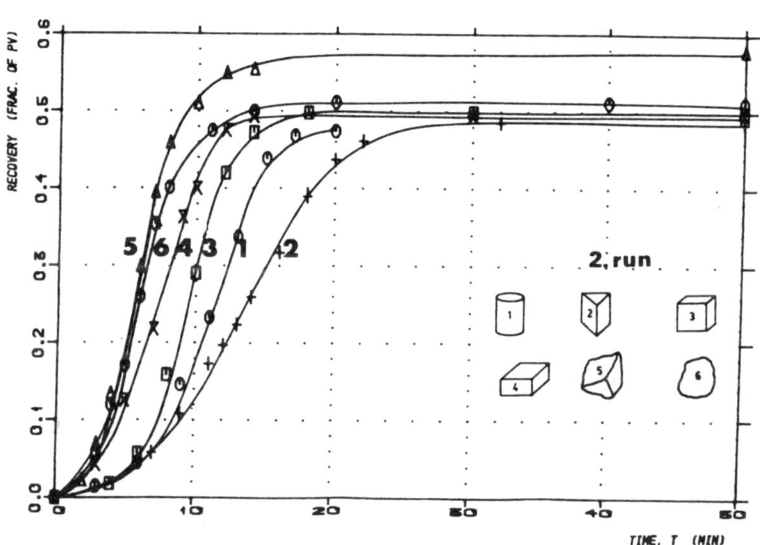

Figure 21. Effect of shape on recovery.
Recovery vs. time for the first set of Berea cores.

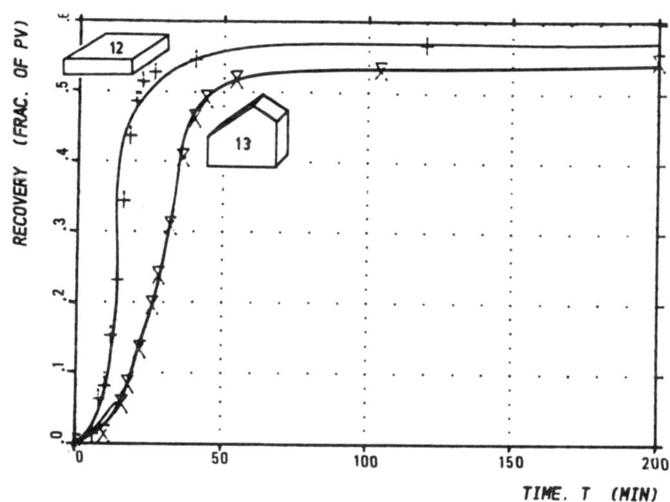

Figure 22. Effect of shape on recovery.
Recovery vs. time for the second set of Berea cores.
Comparison between sample 12 and 13.

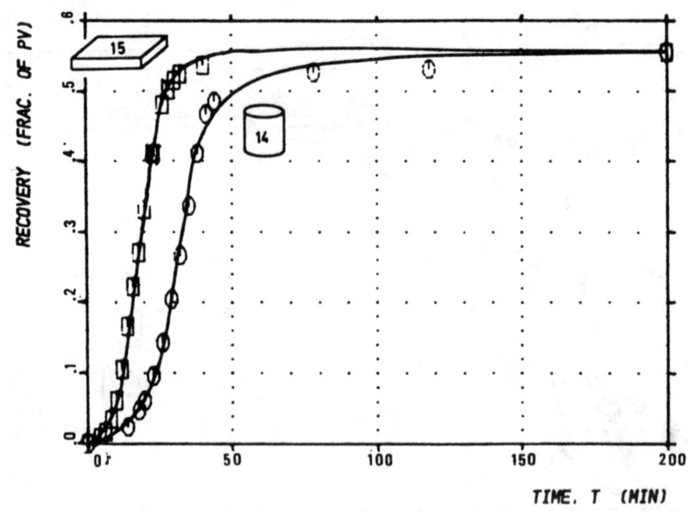

Figure 23. Effect of shape on recovery.
Recovery vs. time for the second set of Berea cores.
Comparison between samples 14 and 15.

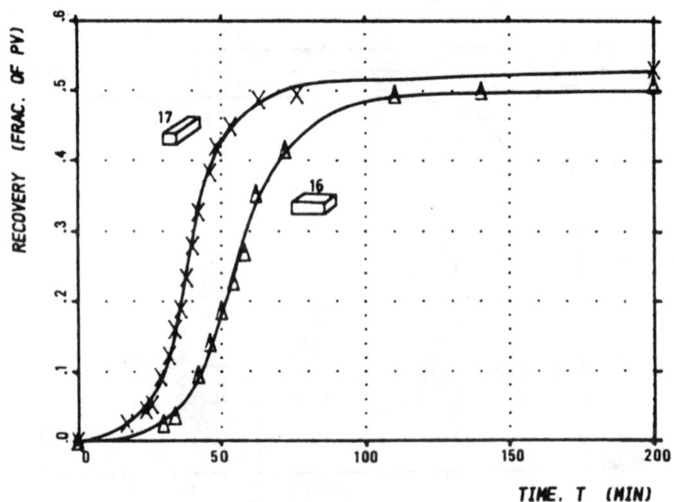

Figure 24. Effect of shape on recovery.
Recovery vs. time for the second set of Berea cores.
Comparison between samples 16 and 17.

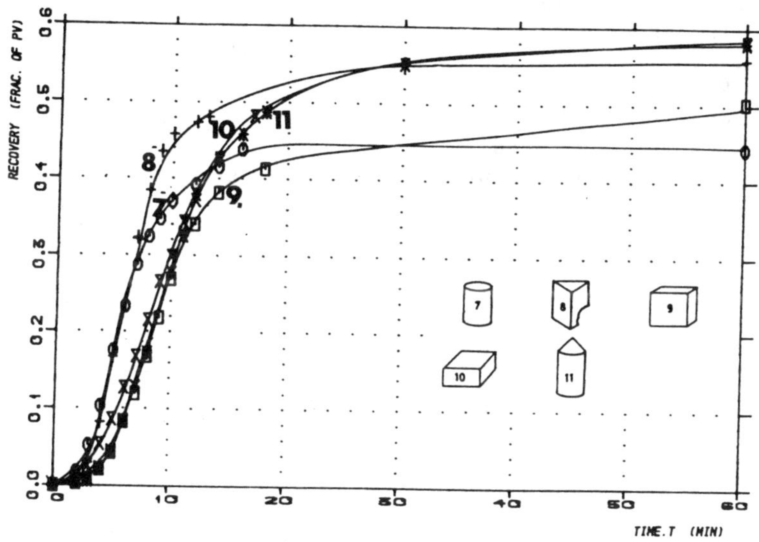

Figure 25. Effect of shape on recovery.
Recovery vs. time for the chalk cores.

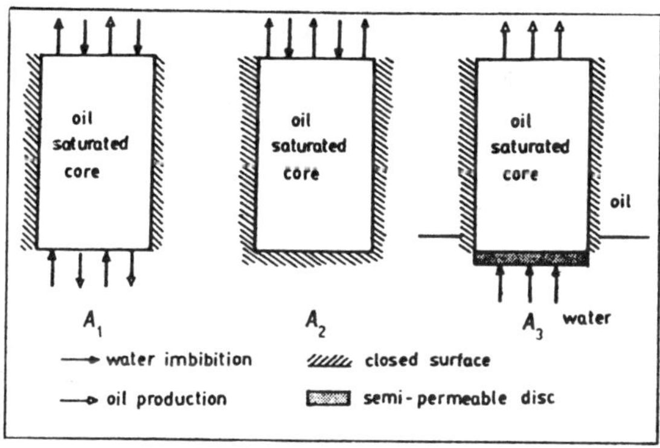

Figure 26. Boundary conditions for one-dimensional imbibition.

The third set of experiments are designed to obtain one-dimensional co-current imbibition (type A_3). In this case, water imbibes in one endface and oil is displaced through the opposite endface. This type of imbibition was obtained by using a 100% water wet chalk disc at the imbibing endface preventing oil to be produced at this endface. The oil saturated core plug with the water saturated chalk plate was submerged in water and oil as shown in Figure 27. The weight was recorded during imbibition, and imbibition recovery was calculated.

A detailed description of the boundary condition tests is given by Johnsen (27).

Imbibition recovery curves for eleven plugs are shown in Figures 28 to 31. Type A_1 and A_2 boundary conditions have been applied on all samples, while type A_3 only has been used on plugs no. 2 and 9.

3.2.3. Imbibition in matrix blocks containing microfractures

It is known that fractured reservoirs are formed by a series of matrix blocks as a result of intersection of "macrofractures network". But it may be expected that additional "microfractures" are contained in the single matrix blocks, and their presence could modify the behavior of imbibition process. In fractured reservoirs in general, there are two types of fractures which could be defined as "macrofractures" and "microfractures".

- Macrofractures are considered to be those fractures of large extension, significant width, and sufficient continuity to assure a complex network, which through their intersection is forming the matrix blocks.

- Microfractures are, on the other hand, considered to be those small, discontinuous fractures whose extension terminates inside the single matrix block. Often those microfractures may form inside the matrix blocks an internal network of limited extension [Figure 32].

The imbibition experiments presented in the following were carried out for samples without microfractures and with microfractures in order to evaluate the contribution of these in the case of under-critical rate of water table advancement (34).

Figure 33 defines some simple models of matrix block containing microfractures. The effect of gravity was eliminated by keeping the water-rock contact at lower open face level. Some comments on each of these simple models are given below.

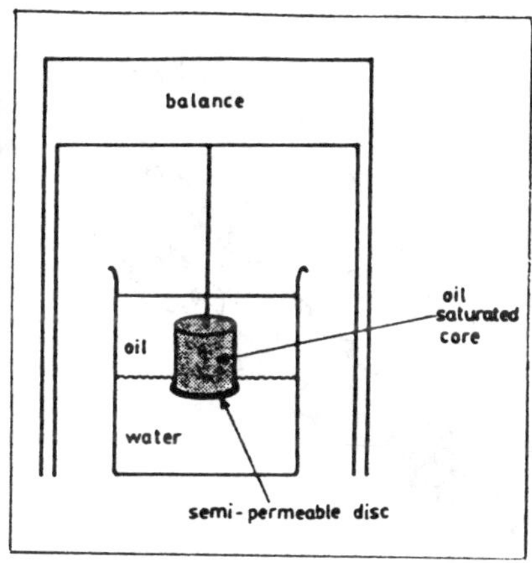

Figure 27. Schematic experimental set-up for type A3 boundary condition.

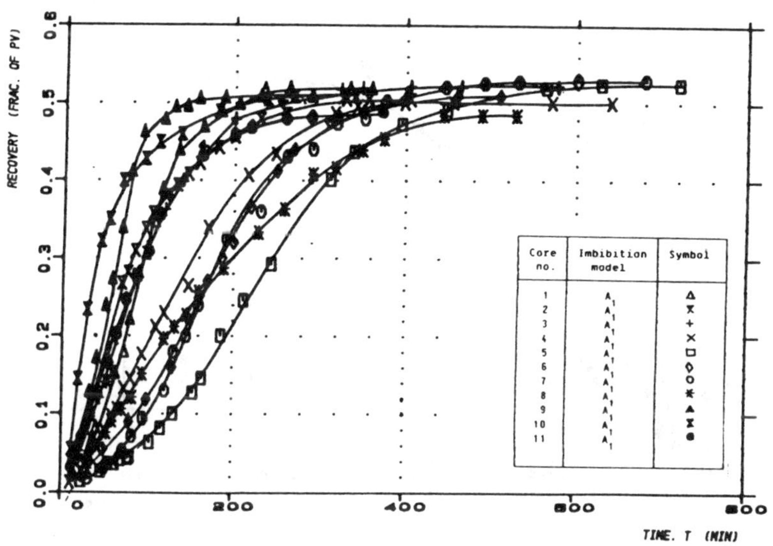

Figure 28. Effect of boundary conditions on recovery. Recovery vs. time for A_1 type boundary conditions.

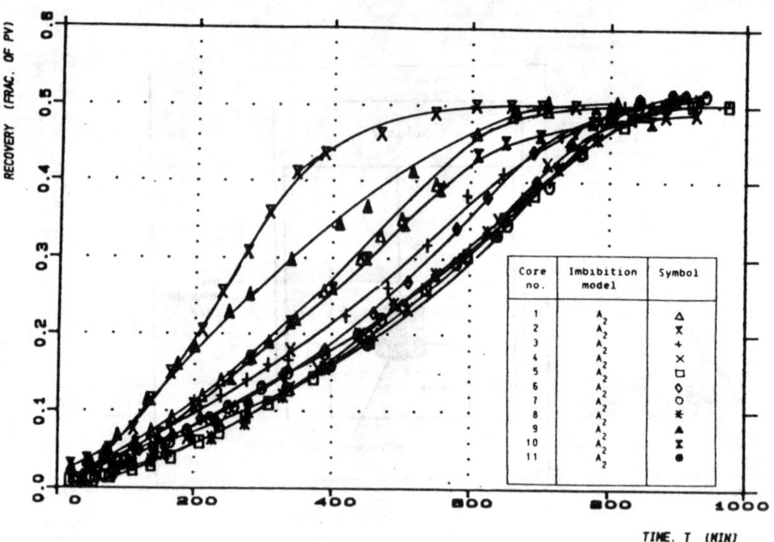

Figure 29. Effect of boundary conditions on recovery. Recovery vs. time for A_2 type boundary conditions.

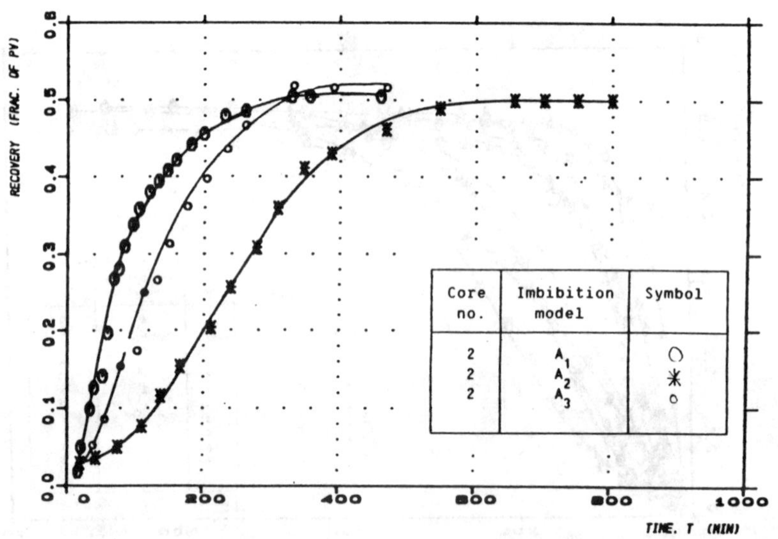

Figure 30. Effect of boundary conditions on recovery. Recovery vs. time for three different types of boundary conditions for core no. 2.

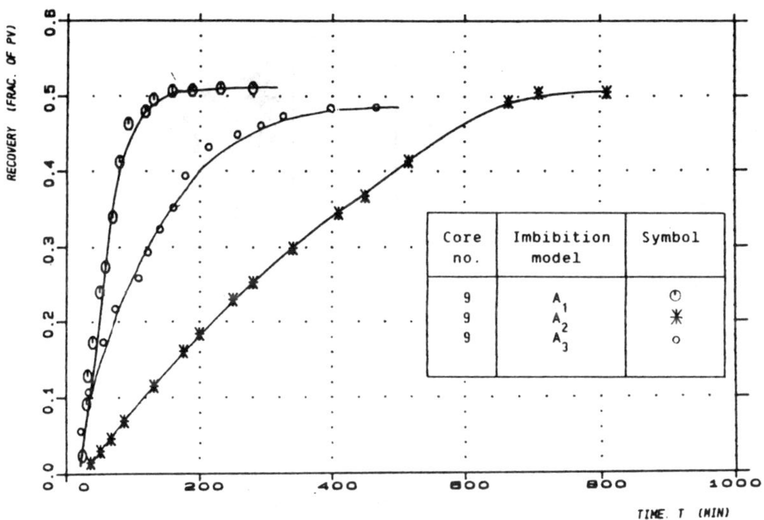

Figure 31. Effect of boundary conditions on recovery. Recovery vs. time for three different types of boundary conditions for core no. 9.

Model A - Without microfractures. Since the block height is around 20 cm, the recovery-time relationship is expected to be strongly influenced by the fact that the capillary height is substantially higher than the block height.

Model B - The horizontal microfracture could contribute to additional spontaneous imbibition starting as soon as the water reaches that level. However, its effect is limited by the supply of water.

Model C - Horizontal and vertical microfractures are here combined as a network. The effect of capillary imbibition developed by horizontal fractures could be accelerated by water supplied through vertical fractures of larger permeability than matrix. A requirement is that sufficient capillary pressure exists in the vertical fracture.

Model D - More vertical fractures are present in the core and again provided that the capillary pressure in the vertical fractures is sufficient, an improved supply of water to the horizontal fracture is obtained, reducing the recovery time. However, an increased counterflow would take place which could reduce the effect.

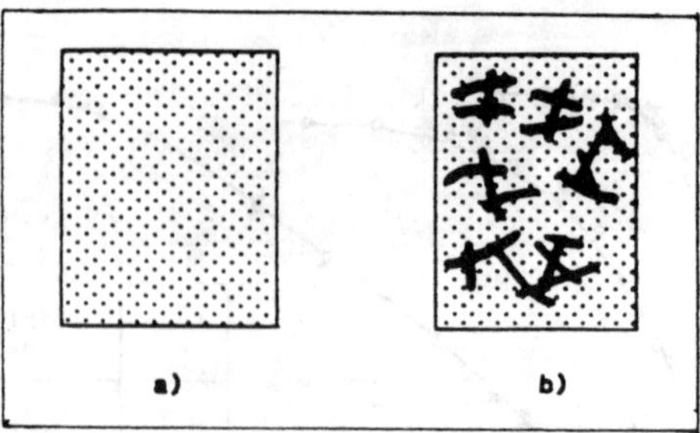

Figure 32. Non-fractured and micro-fractured matrix blocks.

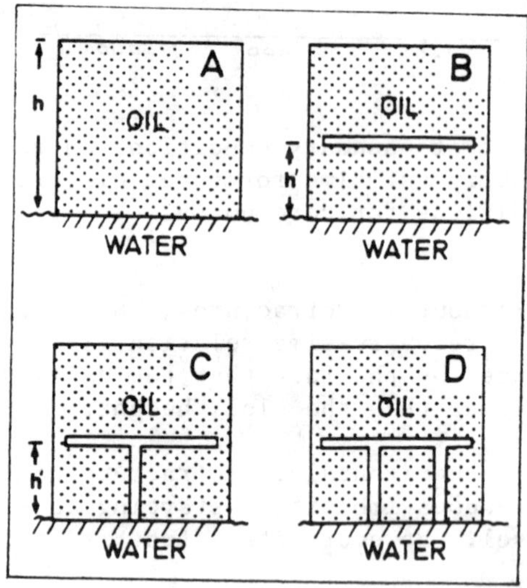

Figure 33. Various configurations of simple micro-fractures in matrix blocks.

The experimental procedure for the imbibition tests was as follows (6): The core samples were cleaned with solvents, dried, and saturated with brine. Absolute permeability to brine was determined. To obtain S_{wi}, viscous oil was injected which was swept by the oil used in the experimental program.

The fracturing of the sample was performed by keeping pore pressure at 0 bar gauge and the lateral pressure at 50 bar. Axial pressure was increased from lateral pressure value to a value resulting in about 0.3% axial compaction.

The imbibition apparatus is shown in Figure 34. Note that the bottom face is swept by a continuous horizontal flow of water.

The core data and imbibition results are summarized in Table 8. The imbibition curves for core number 11, 13, 15 and 18 are shown in Figures 35, 36, 37 and 38.

3.2.4. Simulation of laboratory experiments

Simulation of imbibition in single matrix block systems may be conducted by means of conventional simulation models as reported by Kleppe and Morse (29) and Kazemi and Merril (28). Kvalheim (30) employed the model developed by Kleppe and Morse (29), and simulated imbibition experiments. The experimental procedure is described in section 3.2.1, and the experimental results of the core studied are reported by Torsaeter (43).

The model grid system is shown in Figure 39, and the basic relative permeability and capillary pressure relationships may be found in Figures 40 and 41.

Figure 42 reports oil recovery vs. time for the particular core sample under investigation. As can be seen, there is a wide discrepancy between the model recovery and the experimental results. One explanation of this discrepancy is that the basic relative permeability and/or the capillary pressure curves are in error. Therefore, attempts were made to simulate the system using modified curves.

Modification of the relative permeability curves by multiplying the measured values by a factor of 2, resulted in modest changes in the recovery curve, as is shown in Figure 42. Larger multiplication factors were subsequently employed (not shown) without significant improvements in the model recovery curve.

Next, the capillary pressure curve was modified in a similar fashion. The results of simulations using multiplication factor of 4 and 8, respectively, are shown in Figure 43. As can be seen, the higher capillary pressures have significant effects on the recovery curve. No attempt were made to further match the experimental data. However, it is obvious that by modification of the shape and magnitude of the capillary pressure curve, a match may be obtained.

The purpose of this section is not to discuss simulation of experimental systems in any detail, but to briefly discuss the effects of saturation dependent parameters on the model results. Based on the simulation results presented above, one may include that inaccuracies in relative permeability curves will not have significant effects on the simulation results of imbibition in rock

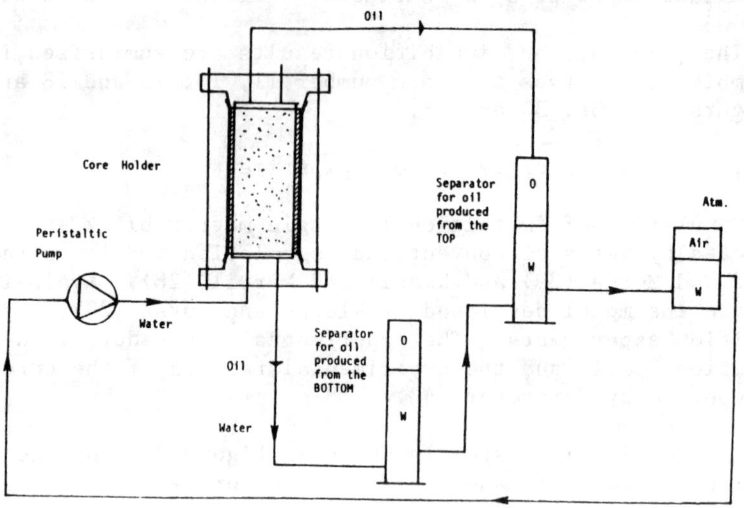

Figure 34. Schematic imbibition apparatus [after Barroux, van Golf-Racht and Mikkelsen (6)].

samples. However, capillary pressures are of outmost importance. Before application of simulated matrix/fracture imbibition flow data in dual porosity models, as discussed in the next chapter, one should therefore carefully examine the measured capillary pressure curves.

3.2.5. Discussion and conclusions

Ekofisk imbibition study: The experiments have brought light upon several aspects of the imbibition behavior of chalk from the North Sea. The porosity correlations in the Tor Formation and the wettability results of the Ekofisk Formation are especially interesting. Two questions arise from the porosity correlations: Why is porosity the correlating parameter, and what causes the residual oil saturation to be constant in the Tor Formation?

1) Permeability or permeability-porosity relationships did not give satisfactory correlations with residual oil saturation. This could be caused by errors in measuring the extremely low permeabilities in chalk and/or the effects of microheterogeneities in the core plugs. Porosity alone was by far the best correlating parameter in the Tor Formation. However, a linear relationship between porosity and average pore throat size has been observed, so combined with results of Figure 11, the initial water saturation increases

Table 8. Core data and imbibition results [after Barroux, van Golf-Racht and Mikkelsen (6)]

Sample no.	Diameter [cm]	Length [cm]	Porosity [%]	Permeability [md]	Fractured	Maximum recovery [% OIP]	Total hours
11	6.5	19.5	44.0	4.4	No	65.6	1200
11	6.5	19.5	44.0	4.4	Yes	64.0	180
13	6.5	21.0	42.6	4.44	No	72.0	950
13	6.5	21.0	42.8	4.44	Yes	*	450
15	6.5	23.0	42.6	2.9	No	73.9	650
15	6.5	23.0	42.6	2.9	Yes	76.9	750
18	6.5	21.1	40.4	2.2	No	54.5	520
18	6.5	21.1	40.4	2.2	Yes	61.5	260

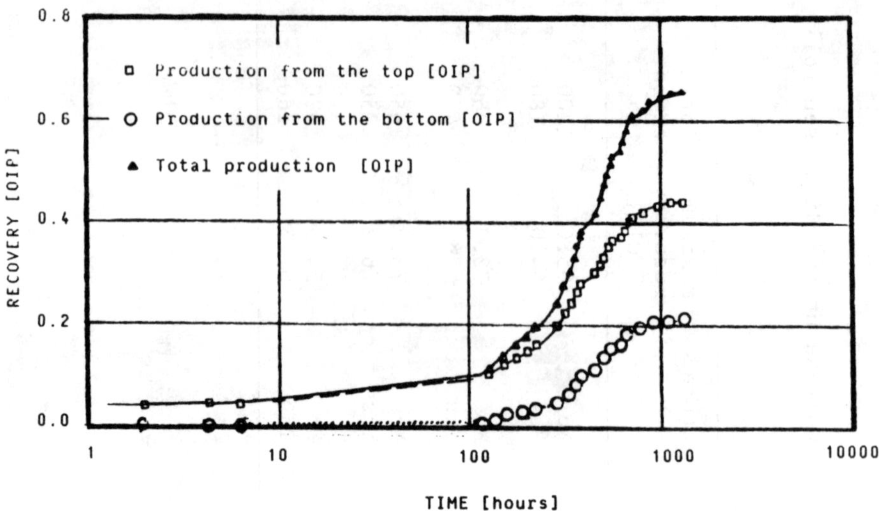

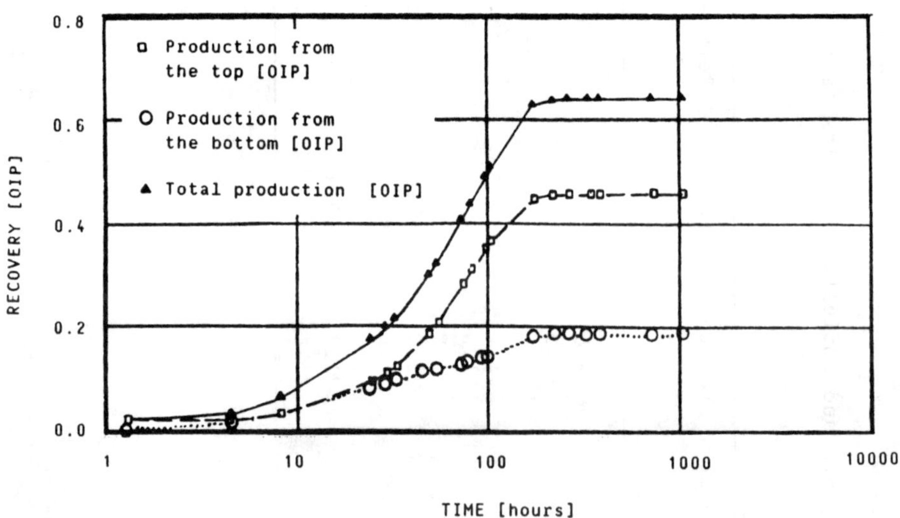

Figure 35. Imbibition recovery curves (sample 11);
top: non-fractured
bottom: fractured
[after Barroux, van Golf-Racht and Mikkelsen (6)].

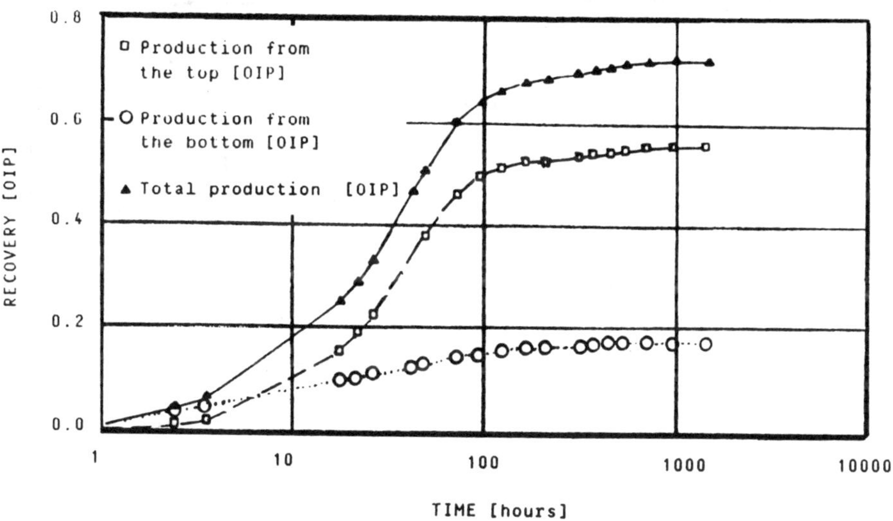

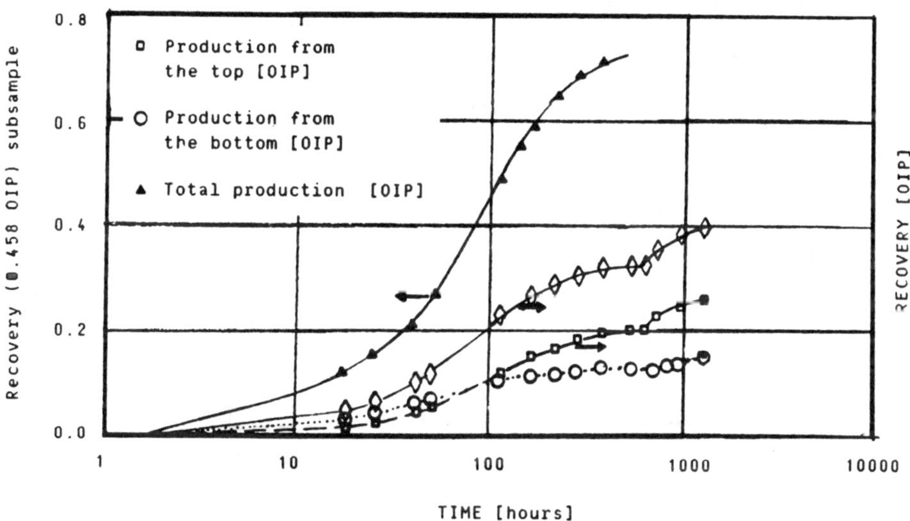

Figure 36. Imbibition recovery curves (sample 13);
top: non-fractured
bottom: fractured
[after Barroux, van Golf-Racht and Mikkelsen (6)]

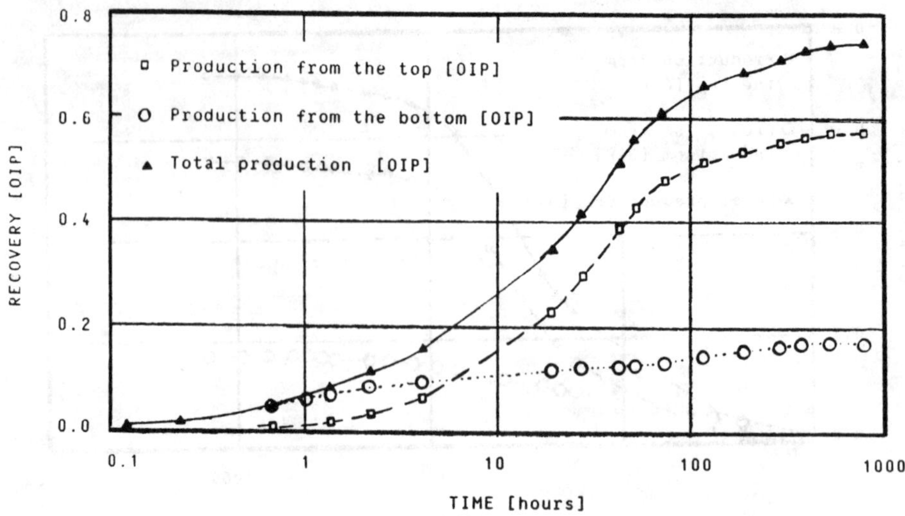

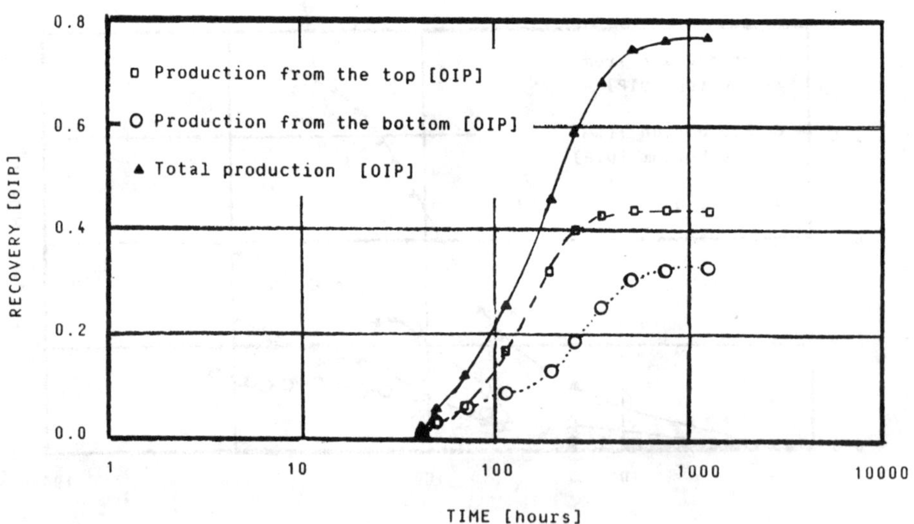

Figure 37. Imbibition recovery curves (sample 15);
top: non-fractured
bottom: fractured
[after Barroux, van Golf-Racht and Mikkelsen (6)]

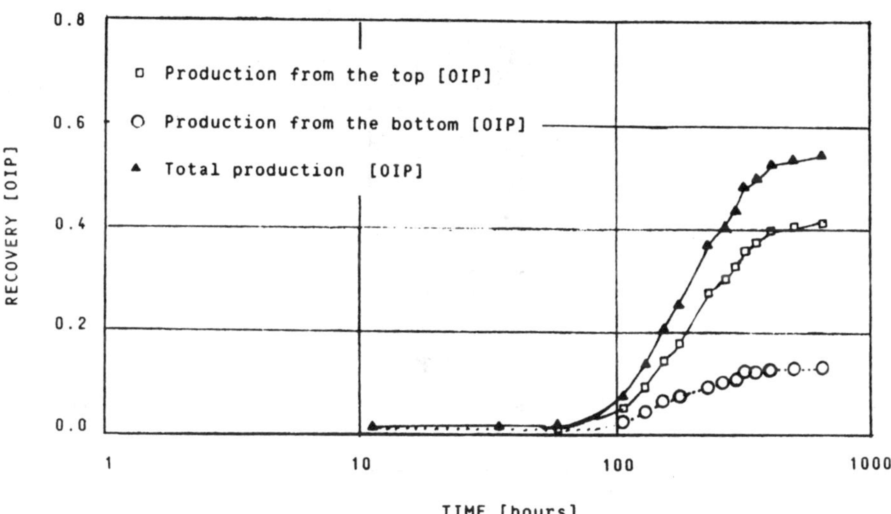

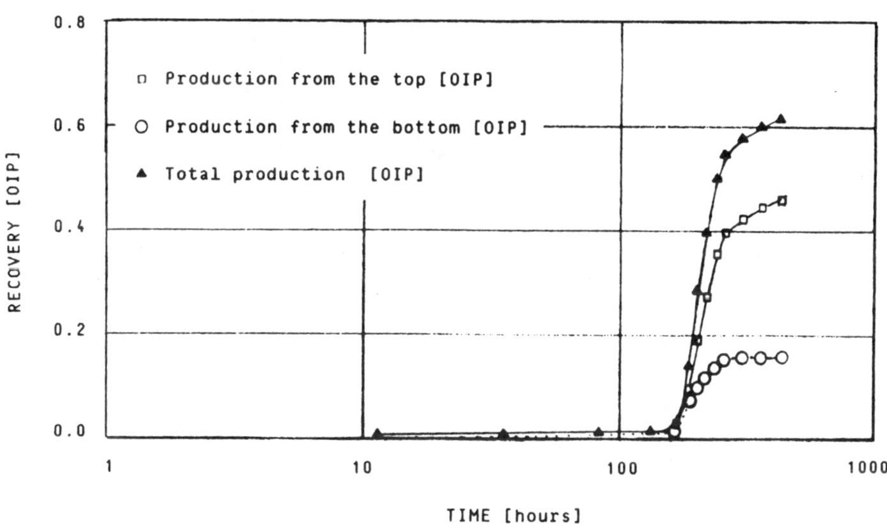

Figure 38. Imbibition recovery curves (sample 18);
top: non-fractured
bottom: fractured
[after Barroux, van Golf-Racht and Mikkelsen (6)]

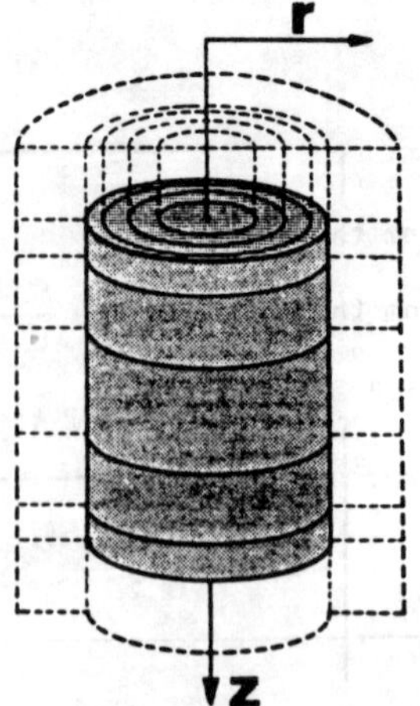

Figure 39. Model grid for simulation of imbibition in cylindrical core.

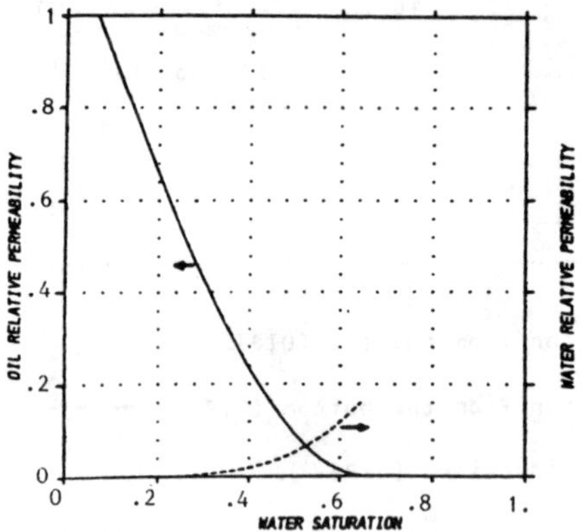

Figure 40. Relative permeability curves.

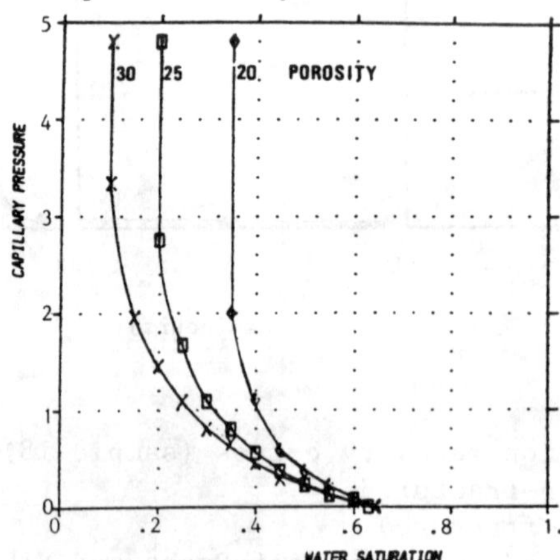

Figure 41. Capillary pressure curves.

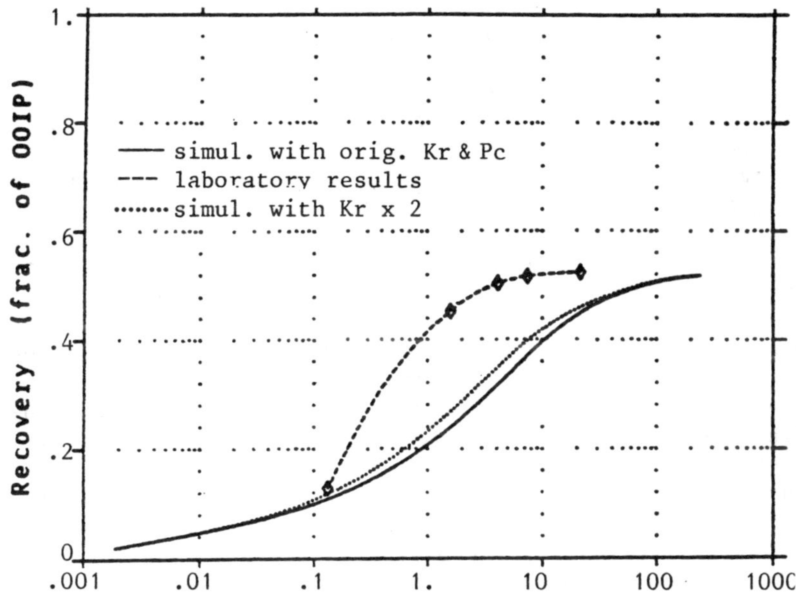

Figure 42. Simulation results vs. lab. (core 582) - effect of relative permeability.

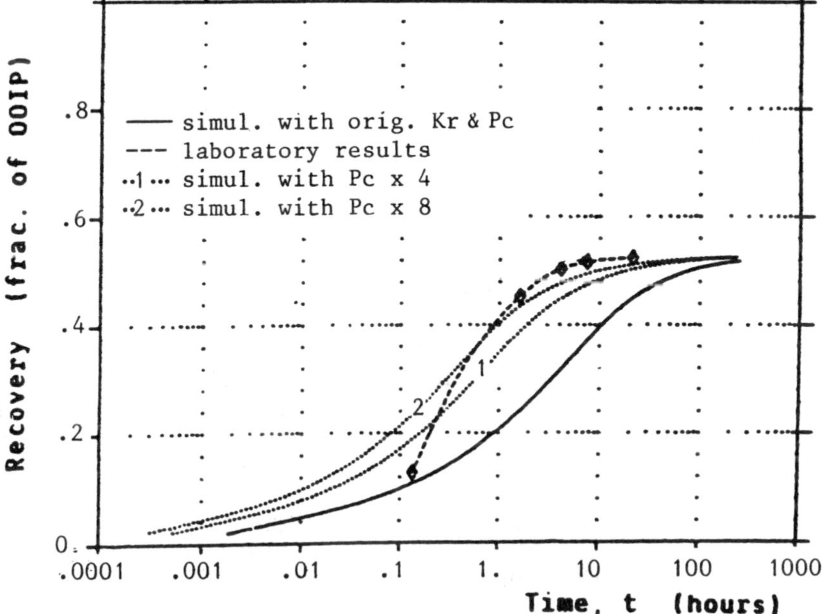

Figure 43. Simulation recovery vs. lab. (core 582) - effect of capillary pressure.

with decreasing pore throat size.

2) Pore parameters are the key to the answer of the second question, since fluid parameters, wettability, and boundary conditions were unchanged in the experiments. The main pore parameters affecting trapping of oil are: pore to throat size ratio, the degree of heterogeneity in the arrangement of pores, throat to pore coordination number, and the properties of the pore surfaces which include composition and degree of roughness. Heterogeneities in pore structure are not pronounced in chalk of the Ekofisk type, but microfractures are observed. These microfractures have a significant effect on permeability, but their influence on the residual saturation after imbibition is not known. The coordination number, i.e., the number of channels connecting each pores, is affecting the trapping of oil. Wardlaw (46) suggests that a low coordination number results in high residual non-wetting phase saturation. But due to the difficulties in quantifying coordination number, the relative importance of the coordination number as one of several variables affecting recovery is not presently known. The surface roughness of pores in reservoir rock varies greatly from the smooth crystal surfaces of some dolomites, to the pitted or clay-coated surfaces of many sandstones. However, in the Ekofisk Field chalks, the variation in surface properties is minor. This is also confirmed by the relatively constant specific surface area measured in this rock (1-2 m^2/g). This discussion has eliminated most pore parameters, except the pore size to pore throat size ratio, as an important factor in trapping mechanism in chalk.

Pore throat size distributions were determined on several core plugs using mercury porosimetry. The average throat size (D_{50}) is defined as the throat size at which 50% of the actual pore volume is saturated with mercury. From scanning electron microscopy (SEM), pore size (D_p) distribution curves were determined, and the pore size to pore throat ratio was calculated and found to be nearly constant for all plugs analyzed.

The fact that the residual saturation after imbibition was close to constant indicates that the chalk samples have a common property, independent of porosity. A combination of several factors may be the reason, but there seem to be strong indications of the D_p/D_{50} ratio being an essential parameter.

The experiments on the Ekofisk Formation core plugs revealed that the imbibition behavior in this formation is governed by wettability variations more than by pore characteristics. The wetting properties of this rock seems to be very complex, and the imbibition recovery is unpredictable. A discussion of analytical investigations for characterization of chalk surfaces and wettability is outside the scope of this chapter.

The conclusions of the Ekofisk imbibition study are:

1) In a water wet chalk, like the Tor Formation, the imbibition oil recovery can be correlated with porosity, and
2) In a chalk formation of complex wetting properties, like Ekofisk Formation, imbibition oil recovery can not be correlated with rock properties.

Sample shape and boundary conditions: The preceding description of the laboratory work shows that the experimental procedures are relatively simple. The errors in volume and weight measurements should not affect the main conclusions of the experiments. However, a more quantitative analysis of these phenomena will require larger samples, and the samples must have exactly the same properties. In the present work, small inhomogeneities in the rock material can be the reason for some of the unexpected effects.

The study of the effect of geometric shape on imbibition indicates that larger surface areas (given a constant volume) yield faster recoveries. The second run on the Berea samples was much like the first one, indicating good reproducibility. Sample No. 2 behaves unexpectedly , but this may very well be due to rock inhomogeneities or wetting alterations in this particular piece of rock.

A quick glance at the results of the final three Berea sample pairs (Figures 22, 23, and 24) will confirm the trend obtained on the first Berea samples. An interesting point, though, is that the largest sample pair imbibed the quickest and the smallest pair the slowest. This is opposite of what one should expect. One explanation may be countercurrent imbibition. It could also be due to minute air bubbles on the surface blocking imbibition. This would have a larger effect on the small samples, relatively speaking.

The chalk samples imbibed very fast. The difference in time to reach a given recovery between the samples is small. Therefore, no conclusions regarding these results are offered.

Systematic studies of imbibition behavior for various sample geometries may be a contribution to solve the problem of scaling laboratory experiments to field scale. However, in practice, due to lack of information, matrix blocks in fractured reservoirs will most likely be approximated by a simple shape. Laboratory experiments will therefore be performed on simple shapes, and the scaling of laboratory tests is considerably reduced. Most fractured reservoirs are modeled with parallelepipedic blocks. An interesting study would therefore be to do laboratory experiments on imbibition in parallelepipedic shapes to determine the application of the Warren and Root (47) type geometric factor to the scaling of laboratory experiments to reservoir conditions.

The study of boundary conditions and imbibition recovery rate show distinct trends, even though a considerable spreading in imbibition rate is observed for samples with the same boundary condition. This spreading is probably due to variations of permeability and porosity of the core and to slight variations in wettability. Trivial laboratory procedures, such as how the samples are shaken before the produced volumes are recorded, may also affect early time imbibition data.

The results also show that the total oil recovery after imbibition is nearly constant, averaging 50% for all three boundary conditions. Comparing type A_1, A_2 and A_3 boundary conditions the following is observed:

- The rate of imbibition is higher for samples with two open endfaces (A_1) than for samples with only one open endface (A_2); and
- Type A_3 boundary conditions will give co-current imbibition, since no oil is produced at the water imbibing endface. Type A_1, A_2 and A_3 are compared in Figures 30 and 31. As seen, the imbibition rate of the sample with co-current flow is slower than the imbibition of type A_1. This is probably mostly due to larger imbibition area in type A_1 samples than in A_3 samples.

The conclusions from this study can be stated as follows:

1) The imbibition recovery versus time relationship is a complex function of shape and volume. The imbibition recovery rate increases with increasing surface area given a constant volume; and
2) Imbibition recovery rate of core plugs is dependent on the type of boundary condition. Core plugs with two endfaces open for imbibition have a higher recovery rate than core plugs with one endface open for imbibition. The rate of imbibition for co-current flow is similar to the rate of imbibition for cores with two endfaces open for imbibition.

Imbibition in matrix blocks containing microfractures: The imbibition recovery versus time relationships for non-fractured and fractured samples presented in Figures 34-37 are commented on in the following.

Sample 11 on Figure 35 shows an extraordinary improvement of imbibition after fracturing. Imbibition time necessary to reach 50% of total recovery ($t_{0,5R}$) was reduced from 340 hours to 51 hours. The result seems to correspond to model C case. Sample 13 (Figure 36) was broken in two samples, and a different recovery-time relationship is observed. The longer imbibition process time after fracturing could be explained by difficult "direct flow" due to the horizontal break.

Sample 15 had abundant vertical fractures and required more time to reach final recovery in the presence of microfractures (760 hours) compared with non-fractured core (650 hours) (Figure 37). This behavior might be explained by a blockage of imbibition as a result of increasing counterflow production. The counterflow production may reduce imbibition in two ways:

- Counterflow production is not swept away from lower open face, and a reduced water-sample contact is obtained; and
- The counterflow production reduces the available cross-section for direct flow.

Sample 18 (Figure 38) had horizontal and vertical fractures similar to model C, and as observed, the microfractures are improving the time of recovery.

Conclusions:

1) If the microfracturing develops a network similar to model C (Figure 33), a reduction in time of capillary imbibition process may be obtained;

2) The lack of vertical fractures, or the development of large horizontal fractures which create distinct subsamples, will result in increased imbibition time after fracturing; and

3) When counterflow rate increases, as a result of abundant vertical microfractures, the imbibition time may increase. This is due to reduced water-sample contact and/or reduced direct flow rate.

4. SIMULATION OF MULTIPHASE FLOW IN NATURALLY FRACTURED RESERVOIRS

4.1. Introduction

Simulation of naturally fractured reservoirs is difficult because of the discontinuous nature of the matrix-fracture system. Ideally, one would simulate the system using a conventional reservoir simulation model fully accounting for the actual geometry and flow processes of the matrix and the fractures. Such studies of small multiphase systems, have been reported by Yamamoto et al. (48), Kleppe and Morse (29), Gilman and Kazemi (22), and Thomas, Dixon and Pierson (42).

Intensely fractured reservoirs, such as in some North Sea carbonate fields, contain a very large number of randomly distributed fractures. The distance between the fractures is normally much smaller than the grid block size employed in reservoir simulation studies. A detailed treatment of the matrix/fracture system of such reservoirs is therefore not practical. As a matter of fact, a given model grid block may, on a small scale, contain a complete fractured system consisting of matrix blocks and fractures. This is illustrated in Figure 44. The fractures may form a continuous network of high permeability channels through which most fluid transport takes place. However, in most cases, a large portion of the fractures may not be interconnected. Such isolated, or partly interconnected fractures, will still have an important influence on the flow behavior of the system.

The representation of the flow mechanisms inside a model grid block of this type is the key problem in the construction of a fractured reservoir simulation model. Fluid transfer will take place between matrix and fractures by viscous displacement, and through the mechanisms of capillary imbibition and drainage by gravity drainage and by fluid expansion. Since it would not be practically possible to use sufficiently small grid blocks to accurately simulate the fluid exchange in field model, some form of simplification must be employed in the models.

Normally, simulation of naturally fractured reservoirs involves a dual porosity model. Here the fractures provide the main path for fluid transport, while the matrix blocks act as sources, feeding fluid to the fractures. Most models consider the matrix/fracture system inside a model grid block to consist of a regular network of fractures and matrix blocks, as shown in Figure 45. The fluid transfer between matrix and fractures in the model is described by means of a mathematical expression or by some type of transfer function.

The usage of mathematical expressions has been reported recently by Hill and Thomas (25), Gilman (23), Gilman and Kazemi (22), Thomas, Dixon and Pierson ((42), and Evans (19). Mass balances are used to predict flow between the matrix blocks and the fractures. To account for various block shapes and sizes, a geometric factor is introduced. This factor was originally proposed for single phase systems (see for instance Warren and Root (47)), but has been employed successfully for multiphase flow. However, as reported by Thomas et al. (41), the results are quite sensitive to the value of the geometric factor. In a reservoir of highly variable matrix block sizes and geometries, which is the case for most naturally fractured reservoirs, the geometric factor represents a major uncertainty. In addition, the interaction fractors should be complex functions of time depending on the type of flow taking place. Gilman (23), by introducing the

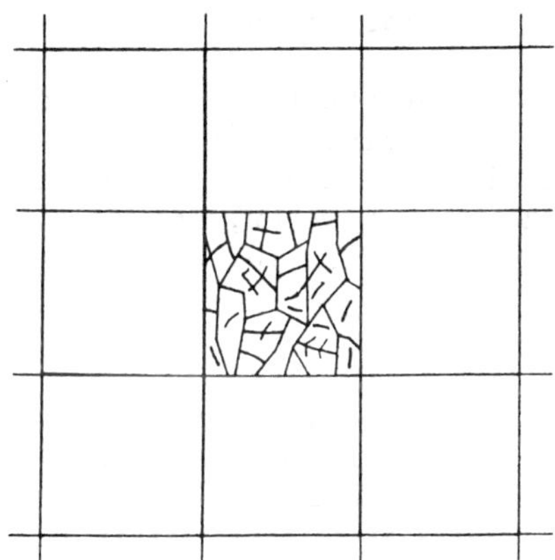

Figure 44. Actual fracture/matrix system inside model grid block.

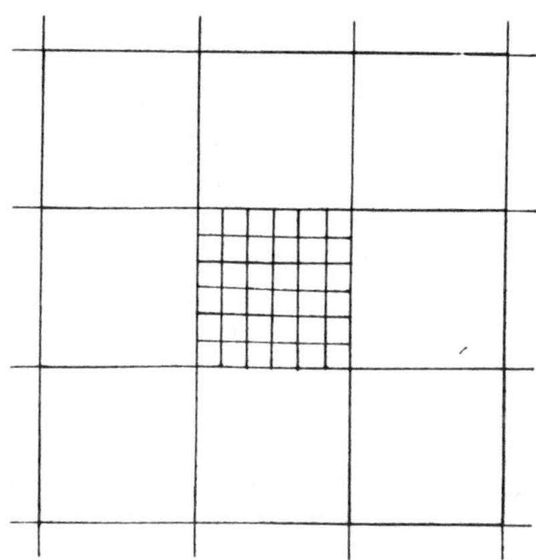

Figure 45. Idealized fracture/ matrix system inside model grid block.

use of smaller matrix subdomains, was able to include gravity effects inside model grid blocks.

Transfer functions are normally determined outside the field simulation model. Various methods may be used in deriving the functions. Rossen (40) used a conventional simulation model to generate recovery curves for single matrix block systems which were included by a semi-implicit formulation in his field model. As reported by de Swaan (16), laboratory experiments may be conducted to define the recovery curves. A third alternative is the derivation of recovery functions by analytical means. de Swaan (15) and Najurieta (35) developed such interaction functions for single-phase systems. A major difficulty in the application of transfer functions is that it is not possible to foresee all the processes to which a given matrix block will be subjected during the production of a fractured reservoir. One particular transfer function may be correct to use for one period of time, while another should be used for the next. de Swaan (14), however, presents an approach of integrating several interaction functions, depending on flow conditions, by using the convolution theorem.

Most papers on dual porosity models consider the fracture network to be the only medium of transport of fluids. The contribution of the matrix blocks to the overall permeability of the reservoir might be significant. Braester (11) in 1972 suggested a formulation that would include the transport capacity of the matrix blocks in the flow equations. Part of his development was the generation of relative permeability curves that were functions of matrix saturation as well as fracture saturation (see Figure 5).

Evans (19) improved the description of the fracture system inside a model grid block by applying a statistical fracture distribution function. An averaging of the fracture system was performed before applying the flow equations. However, statistical methods were not applied to the matrix/fracture flow terms.

4.2. Flow equations

Most dual porosity models reported in the literature consider the fracture network as the continuum while the matrix blocks act as sources or sinks to the fractures. Three dimensional three-phase equations describing flow in such a system are presented in finite difference form in the following.

4.2.1. Fracture flow equations

Oil: $$\Delta[T_o(\Delta p_o - \gamma_o \Delta D)]_f + q_{omf} - q_o = \frac{V_b}{\Delta t} \Delta_t(\phi S_o/B_o)_f \quad (4.1)$$

Gas: $$\Delta[T_g(\Delta p_g - \gamma_g \Delta D) + R_s T_o(\Delta p_o - \gamma_o \Delta D)]_f + q_{gmf} + R_s q_{omf} - q_g - R_s q_o = \frac{V_b}{\Delta_t} \Delta_t(\phi S_g/B_g + \phi S_o R_s/B_o)_f \quad (4.2)$$

Water: $$\Delta[T_w(\Delta p_w - \gamma_w \Delta D)]_f + q_{wmf} - q_w = \frac{V_b}{\Delta t} \Delta_t(\phi S_w/B_w)_f \quad (4.3)$$

where, for fluid α and flow in x-direction:

$$T_\alpha = \left(\frac{k k_{r\alpha} \phi}{\mu_\alpha B_\alpha}\right)_f \left(\frac{\Delta y \Delta z}{\Delta x}\right) \quad (4.4)$$

and $q_{\alpha mf}$ represents the matrix/fracture exchange of fluid α as a source or sink to the fracture system.

4.2.2. Matrix flow equations

In order to relate matrix block saturations to the fluid transfer, mass balances are written for each phase as follows:

Oil: $$- q_{omf} = \frac{V_b}{\Delta t} \Delta_t(\phi S_o/B_o)_m \quad (4.5)$$

Gas: $$- q_{gmf} - R_s q_{omf} = \frac{V_b}{\Delta t} \Delta_t(\phi S_g/B_g + \phi S_o R_s/B_o)_m \quad (4.6)$$

$$\text{Water:} \quad - q_{wmf} = \frac{V_b}{\Delta t} \Delta_t (\phi S_w/B_w)_m \tag{4.7}$$

The matrix/fracture exchange terms, $q_{\alpha mf}$, may be either mathematical expressions or transfer functions, and are described below.

4.2.3. Matrix/fracture transfer

Mathematical expressions for the fluid transfer terms may be written as, for fluid α:

$$q_{\alpha mf} = \lambda_\alpha (p_{\alpha m} - p_{\alpha f}) \tag{4.8}$$

where

$$\lambda_\alpha = \sigma k_{r\alpha} \left(\frac{kV_b}{B_\alpha \mu_\alpha}\right)_m \tag{4.9}$$

The parameter σ is a geometric factor accounting for the size and shape of the matrix block. It is calculated by Gilman and Kazemi (22) as

$$\sigma = 4 \left(\frac{1}{L_x^2} + \frac{1}{L_y^2} + \frac{1}{L_z^2}\right) \tag{4.10}$$

Since the L's are characteristic dimensions of the matrix blocks, the σ becomes a property of the system being simulated and may vary across the reservoir. Inside a given model grid block, however, the matrix blocks are normally assumed to be uniform. In actual reservoirs, it is not possible to determine accurately the size and shape of the matrix blocks, and the determination of σ is therefore highly uncertain.

In order to include effects of flow in the system due to pressure gradients across a matrix block, and effects of diffusion, Thomas, Dixon, and Pierson (42) modified Eq. (4.8) as follows

$$q_{\alpha mf} = \lambda_\alpha (p_{\alpha m} - p_{\alpha f}) + \lambda_\alpha \frac{L_c}{L_B} \Delta p_{\alpha f} \tag{4.11}$$

where L_c is a characteristic length, L_B is the distance across which $\Delta p_{\alpha f}$ acts, and $\Delta p_{\alpha f}$ is the pressure drop across the matrix block.

Transfer functions were generated by Rossen (40) using a conventional simulator on a single matrix block surrounded by gas and water, respectively. The resulting curves of saturation vs. time are shown in Figures 46 and 47. By employing the matrix flow

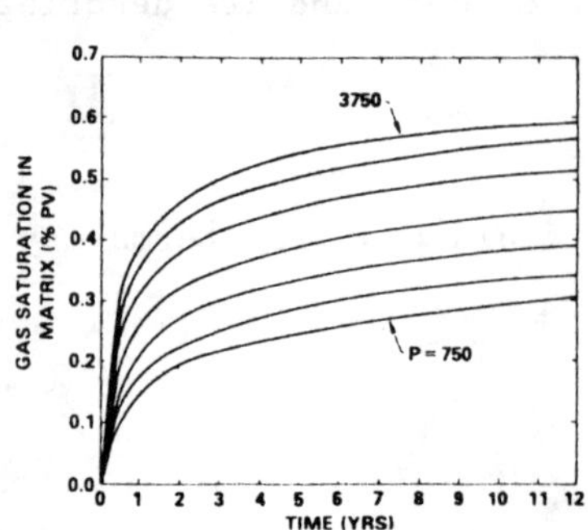

Figure 46. Gas-oil recovery curves [after Rossen (40)].

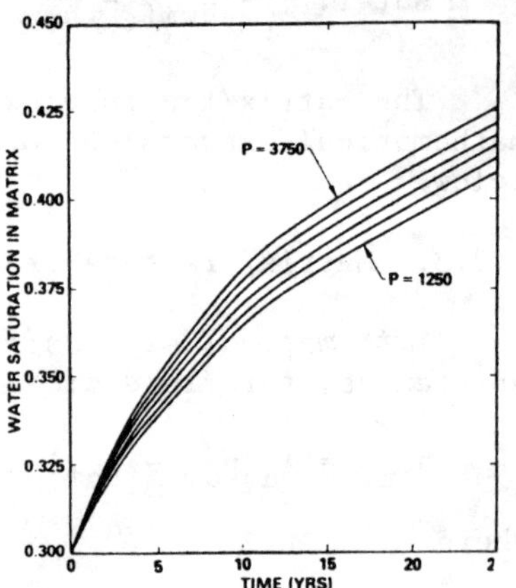

Figure 47. Water-oil recovery curves [after Rossen (40)].

equations, Eqs. (4.5) - (4.7), flow terms are determined that can be put into the fracture flow equations, Eqs. (4.1) - (4.3).

de Swaan (16) fitted the following function to experimental data:

$$q_{mf} = \frac{N_{imb}}{\tau_1} \int_0^t e^{-(t-\theta)/\tau_1} (\partial S_{wf}/\partial\theta) d\theta \tag{4.12}$$

where N_{imb} is the total oil recoverable by imbibition and τ_1 is the time required to reach 63% of N_{imb}.

4.3. An improved model formulation

As discussed in the previous two sections, there are certain limitations to most existing models that might severely influence the results. While the concept of the dual porosity model is an attractive one, the evaluation of the matrix/fracture transfer is very difficult. Although a simple mathematical expression describing the transfer has been used successfully applied (41), its limitations are obvious. The determination of the geometric factor, σ, is difficult and the model results are very sensitive to this factor. Additionally, the interaction parameters involved would, depending on the flow processes taking place, be complex functions of time. Simple expressions, like Eqs. (4.8) - (4.10), would not, in most cases, be representative of these processes. The approach of de Swaan (14) where interaction functions based on analytical derivations, laboratory experiments, or numerical simulations of simple matrix/fracture systems are constructed by convolution is therefore favoured. As the interaction functions would depend on the boundary conditions of the matrix, some technique of keeping track of flood fronts should be included in the formulation.

The fracture network is transporting all fluids between model grid blocks in most formulations. However, the contribution of the matrix permeability might be important. Braester (11) defines flow permeabilities as well as relative permeabilities as functions of both matrix as well as fracture properties. Further improvements would include the solution of the complete matrix flow equation in addition to the fracture flow equation in order to obtain pressures and saturations in fractures as well as matrix. Furthermore, the model should degrade to a conventional reservoir simulation in areas of no fractures.

Normally all matrix blocks inside a model grid block are assumed to be identical (25). Evans (19), however, presented a formulation whereby a statistical fracture distribution function is included and an averaging of the distribution of the fractures inside a model grid block is performed before applying the fracture flow equations. Further improvements would include statistical treatment of the matrix/fracture interaction terms.

In summary, we would pose the following requirements to an improved model:

1) The model should be based on the dual porosity - dual permeability concept;

2) In non-fractured areas, the model should revert to a conventional (non fractured) model;

3) In highly fractured areas, the model should revert to a simple dual porosity model;

4) Matrix/fracture flow terms should be interaction functions, based on all available data such as analytical derivations, laboratory experiments, or numerical simulations;

5) Statistical data on fracture orientation, intensity, and length should be used to generate random fracture flow networks as well as random matrix block geometries. A representative range of generated matrix blocks should then be analyzed in order to generate the required interaction functions for each; and

6) A front tracking technique should be included in the model in order to determine matrix block boundary conditions, and also within individual model grid blocks.

4.3.1. Flow equations

For a fluid, α, the flow equations may be written in finite difference form as

$$\Delta[T_\alpha(\Delta p_\alpha - \gamma_\alpha \Delta D)]_m - q_{\alpha mf} - q_{\alpha m} = \frac{V_b}{\Delta t} \Delta_t(\phi S_\alpha/B_\alpha)_m \tag{4.13}$$

$$\Delta[T_\alpha(\Delta p_\alpha - \gamma_\alpha \Delta D)]_f - q_{\alpha mf} - q_{\alpha f} = \frac{V_b}{\Delta} \Delta_t(\phi S_\alpha/B_\alpha)_f \tag{4.14}$$

where

$$\sum_\alpha S_\alpha = 1 \tag{4.15}$$

These equations may then be solved to yield matrix and fracture pressures and saturations.

4.3.2. Transmissibilities

The transmissibility term in the x-direction are defined as

$$(T_\alpha)_m = \left(\frac{kk_{r\alpha}\phi}{\mu_\alpha B_\alpha}\right)_m \left(\frac{\Delta y \Delta z}{\Delta x}\right) \tag{4.16}$$

$$(T_\alpha)_f = \left(\frac{kk_{r\alpha}\phi}{\mu_\alpha B_\alpha}\right)_f \left(\frac{\Delta y \Delta z}{\Delta x}\right) \tag{4.17}$$

4.3.3. Interaction terms

Explicit knowledge of the term $q_{\alpha mf}$, or the matrix saturation vs. time, as shown in Figures 46 and 47, would enable us to directly apply these in the flow equations. However, the fluid transfer would depend on the type of flow process taking place in the grid block, and cannot be predicted a priori.

However, the approach of de Swaan (14) includes block responses which are functions of several matrix/fracture system properties. He defines

$$q_{mfH} = q_{mfH}(t, x_1, x_2, \ldots\ldots, x_n) \tag{4.18}$$

where q_{mfH} is the matrix/fracture rate response to a unit step disturbance and x_j, j=1, n are the n properties of the system defining the flow process. Then, by convolution

$$q_{mf} = \int_{o}^{f} \frac{\partial y_f(\tau)}{\partial \tau} q_{mfH} (t-\tau, x_1, x_2, ..., x_n) d\tau \qquad (4.19)$$

where y_f is the variable on the matrix surface subjected to a unit change. The best available data on the matrix/fracture flow may be included in the above formulation. No flow process is a priori assumed, and any combination of flow processes may be intersubstituted in the model triggered by corresponding criteria in the matrix and at the matrix surface.

Typically, the matrix/fracture flow term would be a function of matrix block saturation and pressure differential between fracture and matrix.

$$q_{mf} = q_{mf}(S_m, P_m - P_f) \qquad (4.20)$$

By including semi-implicit treatment of this term, we would have

$$q_{mf}^{t+\Delta t} = q_{mf}^{t} + \frac{\partial q_{mf}}{\partial S_m} \Delta_t S_m + \frac{\partial q_{mf}}{\partial (\Delta P)} \Delta_t (P_m - P_f) \qquad (4.21)$$

which would be similar to the mathematical expression case of Eq. (4.8) and Eq. (4.21), however, would more accurately describe the flow process taking place.

In the case of fluid injection or advancement of gas-oil contact/ water-oil contact, some method locating flood fronts should be included in the formulation in order to be able to trigger the correct matrix/fracture flow processes at the arrival of the gas front or the water front respectively.

A simplified case would be the advancement of water-oil contact and gas-oil contact under segregated conditions, as outlined by de Swaan (14). In general, a front tracking method should be incorporated to track fronts of any positions in the system.

4.3.4. Statistical description of matrix blocks fractures

Fractures are generally randomly distributed in naturally fractured reservoirs. The geometry of size of a matrix block would therefore be highly variable. Most models assume that all matrix blocks within a model grid block are uniform. We propose the use of a technique of applying well data on fracture frequency and orientation to generate probable geometries and shapes. Ghez and

Janot (21) presented a method which may serve as an example for this purpose. Figure 48 shows a set of frequency curves of fracture angles and fracture distances that are applied to generate matrix blocks (illustrated at the bottom of the figure). Based on the matrix blocks generated, a limited number of representative matrix block geometries should be associated with an appropriate frequency distribution curve, as shown in Figure 49. A reservoir may be divided into a number of areas of different distribution curves.

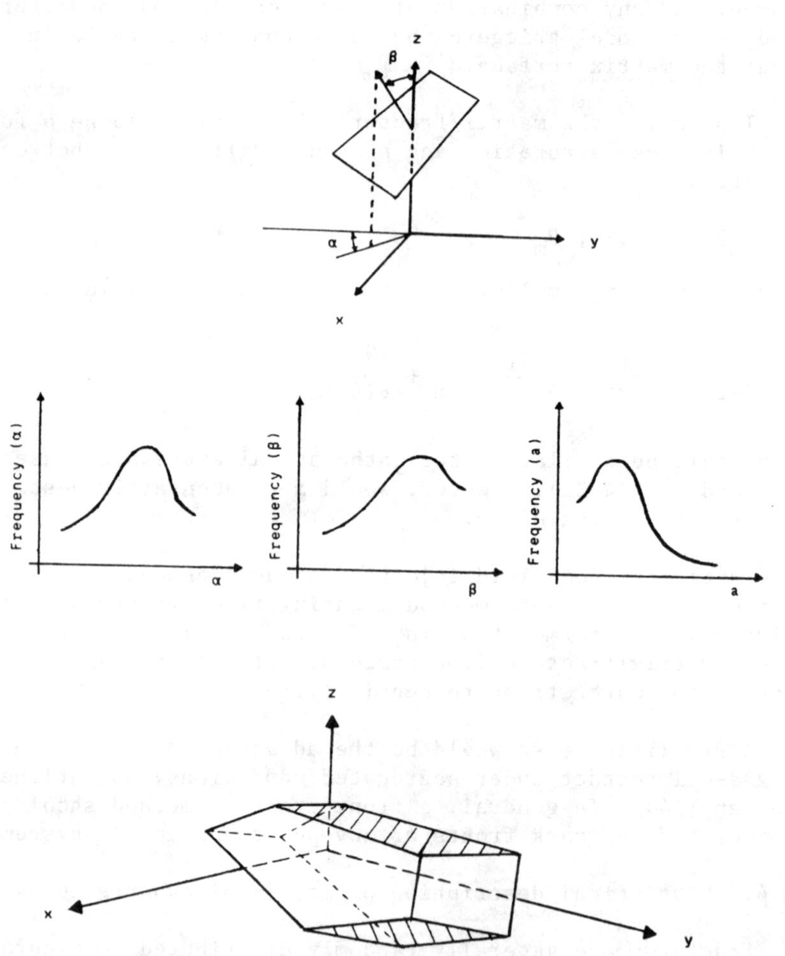

Figure 48. Generation of random matrix blocks [after Ghez and Janot (21)].

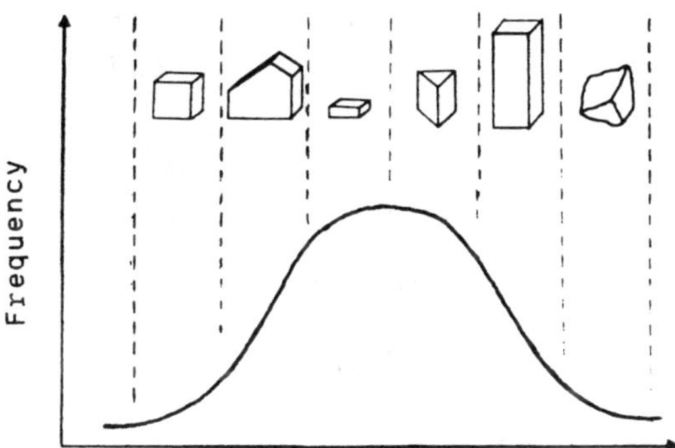

Figure 49. Frequency distribution of representative matrix blocks.

The geometrical figures generated should then be subjected to investigation in the laboratory, by numerical simulation or any other method available in order to determine the interaction functions describing the flow processes that are anticipated. Where required, the systems investigated should be scaled to field conditions by means of the scaling laws discussed previously in this chapter.

5. REFERENCES

1. Aguilera, R., *Naturally Fractured Reservoirs*, PennWell Publishing Company, Tulsa, Oklahoma, 1980.
2. Amott, E., "Observations Relating to the Wettability of Porous Rock," *Transactions*, American Institute of Mining, Metallurgical, and Petroleum Engineers, Vol. 216, pp. 156-162.
3. Aronofsky, J. S., Masse, L., and Natanson, S. C., "A Model for the Mechanism of Oil Recovery from the Porous Matrix Due to Water Invasion in Fractured Reservoirs," *Transactions*, American Institute of Mining, Metallurgical, and Petroleum Engineers, Vol. 213, 1958, pp. 17-19.
4. Barenblatt, G.I., Zheltove, Y.P., and Kochina, I.N., "Basic Concepts in the Theory of Seepage of Homogeneous Liquids in Fissured Rocks," *PMM, (Soviet Applied Mathematics and Mechanics)*, Vol. 24, No. 5, 1960, pp. 852-864 (in Russian).
5. Barenblatt, G. I., "On the Motion of a Gas-Liquid Mixture in a Porous Fissured Medium," *Izvestia Akademii Nauk SSSR, Mekhanika i Mashinostroienie*, No. 3, pp. 47-50 (in Russian).
6. Barroux, C., van Golf-Racht, T.D., and Mikkelsen, M., "Imbibition Oil Recovery From Fractured Reservoir (Matrix Blocks Containing Microfractures), presented at the May 21-22, 1985, Chalk Research Program Symposium of the Norwegian Petroleum

Directorate, held at Stavanger, Norway.

7. Bear, J., Braester, C., "On the Flow of Two Immiscible Fluids in Fractured Porous Media," *Proceedings of the 1st International Symposium on the Fundamentals of Transport Phenomena in Porous Media, IAHR*, Vol. 1, 1969, pp. 177-202.
8. Bear, J., *Dynamics of Fluids in Porous Media*, 1st ed., American Elsevier Environmental Science Series, New York, 1972.
9. Blair, P. M., "Calculation of Oil Displacement by Countercurrent Water Imbibition," presented at the May, 1960, Fourth Biennial Secondary Recovery Symposium of Society of Petroleum Engineers, held at Wichita Falls, Tex., (Preprint 1475 G).
10. Bokserman, A. A., Zheltov, Y.P., Kocheshkov, A.A., "Motion of Immiscible Liquids in a Cracked Porous Medium," *Soviet Physics Doklade*, Vol. 9, No. 4, Oct. 1964, pp. 285-287.
11. Braester, C., "Simultaneous Flow of Immiscible Liquids Through Porous Media," *Society of Petroleum Engineers Journal*, Vol. 12, No. 3, Aug., 1972, pp. 297-305.
12. Brooks, R. H., and Corey, A. T., "Properties of Porous Media Affecting Fluid Flow," *Journal of Irrigation and Drainage Division*, Vol. 92, No. 1R2, June 1966, pp. 61-68.
13. Brownscombe, E. R., and Dyes, A. B., "Water-Imbibition Displacement: A Possibility for the Spraberry," *Transactions*, Drilling and Production Practice, American Petroleum Institute, Vol. 19, 1952, pp. 383-390.
14. de Swaan, A., "A Three Phase Model for Fractured Reservoirs Presenting Fluid Segregation," presented at the Jan. 31-Feb. 3, 1982, 6th Symposium of Reservoir Simulation of Society of Petroleum Engineers, held at New Orleans, La., (preprint SPE 10510).
15. de Swaan, A., "Analytic Solutions for Determining Naturally Fractured Reservoir Properties by Well Testing," *Society of Petroleum Engineers Journal*, Vol. 16, No. 2, June 1976, pp. 117-122.
16. de Swaan, A., "Theory of Waterflooding in Fractured Reservoirs," *Society of Petroleum Engineers Journal*, Vol. 18, No. 1, April 1978, pp. 117-122.
17. Douglas, J., Peaceman, D. W., and Rachford, H. H., "A Method for Calculating Multidimensional Immiscible Displacement," *Transactions*, American Institute of Mining, Metallurgical, and Petroleum Engineer, Vol. 216, 1959, pp. 297-308.
18. Elkins, L. F., and Skov, A. M., "Cyclic Waterflooding the Spraberry Utilizes "End Effects" to Increase Oil Production Rate," *Journal of Petroleum Technology*, Vol. 15, No. 4, April 1963, pp. 877-887.
19. Evans, R.D., "A Proposed Model for Multiphase Flow Through Naturally Fractured Reservoirs," *Society of Petroleum Engineers Journal*, Vol. 22, No. 3, Oct. 1982, pp. 669-680.
20. Fossberg, P.E., "A Study of the Effects of Matrix Block Geometry on the Evaluation of the Imbibition Process in Naturally

Fractured Reservoirs," thesis presented to the University of Trondheim at Trondheim, Norway, in 1983, in partial fulfillment of the requirements for the degree of Sivilingeniør, (M.Sc.).

21. Ghez, F., and Janot, P., "Statistical Calculation of Elementary Matrix Blocks in Fractured Reservoirs," presented at the Oct. 8-11, 1972, 47th Annual Fall Meeting of the Society of the Petroleum Engineers of AIME, held at San Antonio, Tex., (Preprint SPE 4047).

22. Gilman, J. R., and Kazemi, H., "Improvements in Simulation of Naturally Fractured Reservoirs," *Society of Petroleum Engineers Journal*, Vol. 23, No. 3, Aug. 1983, pp. 695-707.

23. Gilman, J. R., "Numerical Simulation of Phase Segregation in Primary Porosity (Matrix Blocks) in Two-Porosity Porous Media," presented at the Nov. 15-18, 1983, Reservoir Simulation Symposium of the Society of Petroleum Engineers of AIME, San Francisco, CA. (Preprint 12271).

24. Graham, J. W., and Richardson, T. G., "Theory and Application of Imbibition Phenomena in Recovery of Oil," *Transactions*, American Institute of Mining, Metallurgical, and Petroleum Engineers, Vol. 216, 1959, pp. 377-385.

25. Hill, A. C., and Thomas, G. W., "A New Approach for Simulating Complex Fractured Reservoirs," presented at the March 11-14, 1985, Middle East Oil Technical Conference and Exhibition of the Society of Petroleum Engineers, held at Bahrein, (preprint SPE 13537).

26. Iffly, R., Rousselet, D. C., and Vermeulen, J.L., "Fundamental Study of Imbibition in Fissured Oil Fields," presented at the Oct. 8-11, 1972, 47th Annual Fall Meeting of the Society of Petroelum Engineers, held at San Antonio, Tex., (preprint SPE 4102).

27. Johnsen, K., "En eksperimentell undersøkelse av en-dimensjonal imbibering i kjerneprøver (An Experimental Study of One Dimensional Imbibition in Core Plugs)", thesis presented to the University of Trondheim, at Trondheim, Norway, in 1984, in partial fulfillment of the requirements for the degree of Sivilingeniør, (M.Sc.).

28. Kazemi, H., and Merill, L.S., "Numerical Simulation of Water Imbibition in Fractured Cores," *Society of Petroleum Engineers Journal*, Vol. 19, No. 3, June 1979, pp. 175-182.

29. Kleppe,J., and Morse, R.A., "Oil Production from Fractured Reservoirs by Water Displacement," presented at the Oct. 6-9, 1974, 49th Annual Fall Meeting of the Society of Petroleum Engineers, held at Houston, Tex., (preprint SPE 5084).

30. Kvalheim, B., "Kapillaer oppsuging i kalkstein: Numerisk simulering av laboratorieforsøk (Capillary Imbibition in Chalk. Numerical Simulation of Laboratory Experiments)", thesis presented to the University of Trondheim, at Trondheim, Norway, in 1984, in partial fulfillment of the requirements for the degree of Sivilingeniør, (M.Sc.).

31. Kyte, J.R., "A Centrifuge Method to Predict Matrix-Block Recovery in Fractured Reservoirs," *Society of Petroleum Engineers Journal*. Vol. 10, No. 2, June 1970, pp. 164-170.
32. Lefebvre du Prey, E., "Gravity and Capillarity Effects on Imbibition in Porous Media," *Society of Petroleum Engineers Journal*, Vol. 18, No. 2, June 1978, pp. 195-205.
33. Mannon, R.W., Chilingar, G.V., "Experiments on Effect of Water Injection Rate on Imbibition Rate in Fractured Reservoirs," presented at the Oct. 8-11, 1972, 47th Annual Fall Meeting of the Society of Petroleum Engineers, held at San Antonio, Tex., (preprint SPE 4101).
34. Mattax, C. C., Kyte, J. R., "Imbibition Oil Recovery from Fractured, Water Drive Reservoir," *Society of Petroleum Engineers Journal*, Vol. 2, No. 2, June 1962, pp. 177-184.
35. Najurieta, H.L., "A Theory for the Pressure Transient Analysis in Naturally Fractured Reservoirs," *Journal of Petroleum Technology*, Vol. 32, No. 7, July 1980, pp. 1241-1250.
36. Parsons, R.W., and Chaney, P.R., "Imbibition Model Studies on Water-Wet Carbonate Rocks," *Society of Petroleum Engineers Journal*, Vol. 6, No. 1, March 1966, pp. 34-36.
37. Parsons, R. W., "Permeability of Idealized Fractured Rock," *Society of Petroleum Engineers Journal*, Vol. 6, No. 2, June 1966, pp. 126-136.
38. Rapoport, L. A., "Scaling Laws for Use in Design and Operation of Water Oil Flow Models," *Transactions*, American Institute of Mining, Metallurgical, and Petroleum Engineers, Vol. 204, 1955, pp. 143-150.
39. Reiss, L. H., *The Reservoir Engineering Aspects of Fractured Formations*, Editions Technip, Paris, France, 1980.
40. Rossen, R. H., "Simulation of Naturally Fractured Reservoirs with Semi-Implicit Source Terms," *Society of Petroleum Engineers Journal*, Vol. 17, No. 2, June 1977, pp. 201-210.
41. Thomas, L. K., Dixon, T. D., Evans, C. E., and Vienot, M. E., "Ekofisk Waterflood Pilot," presented at the Sept. 16-19, 1984, 59th Annual Technical Conference and Exhibition of Society of Petroleum Engineers, held at Houston, Tex., (preprint SPE 13120).
42. Thomas, L. K., Dixon, T. N., and Pierson, R. G., "Fractured Reservoir Simulation," *Society of Petroleum Engineers Journal*, Vol. 23, No. 1, Febr. 1983, pp. 42-54.
43. Torsaeter, O., "An Experimental Study of Water Imbibition in North Sea Chalk," thesis presented to the University of Trondheim, at Trondheim, Norway, in 1983, in partial fulfillment of the requirements for the degree of Doktor Ingeniør.
44. Torsaeter, O., "An Experimental Study of Water Imbibition in Chalk From the Ekofisk Field," presented at the April 15-18, 1984, Fourth Symposium on Enhanced Oil Recovery of the Society of Petroleum Engineers and Department of Energy, held at Tulsa, Ok., (preprint SPE/DOE 12688).

45. van Golf-Racht, T. D., *Fundamentals of Fractured Reservoir Engineering*, Developments in Petroleum Science, 12, Elsevier Scientific Publishing Company, Amsterdam - Oxford - New York, 1982.
46. Wardlaw, N. C., "The Effects of Pore Structure on Displacement Efficiency in Reservoir Rocks and in Glass Micromodels," presented at the April 20-23, 1980, First Joint Symposium on Enhanced Oil Recovery of Society of Petroleum Engineers and Department of Energy, Tulsa, Ok., (preprint SPE/DOE 8843).
47. Warren, J. E., and Root, P. J., "The Behaviour of Naturally Fractured Reservoirs," *Society of Petroleum Engineers Journal*, Vol. 3, No. 3, Sept. 1963, pp. 245-255.
48. Yamamoto, R. H., Padgett, J. B., Ford, W. T., and Boubeguire, A., "Compositional Reservoir Simulator for Fissured Systems - The Single Block Model," *Society of Petroleum Engineers Journal*, Vol. 23, No. 3, Aug. 1983, pp. 695-707.

6. LIST OF SYMBOLS

a	Distance between fractures (m)
B	Formation volume factor
b	Constant
C	Constant [Eq. (3.8)]
CGR	Capillarity-gravity ratio
CR	Conductivity ratio
c	Constant [Eq. (3.11)]
c	Compressibility (Pa^{-1})
c_{ef}	Effective fracture compressibility (Pa^{-1})
c_{pf}	Fracture pore compressibility related to fracture volume (Pa^{-1})
c_{pm}	Matrix pore compressibility (Pa^{-1})
D	Depth (m)
D_p	Mean bulge size of pore (m)
D_{50}	Average pore throat size (m)
$F(s)$	Function of saturation
$f(\theta)$	Dimensionless function of wettability
f_s	Areal fracture density (m^{-1})
f_v	Volumetric fracture density (m^{-1})
f_w	Linear fracture density (m^{-1})
$J(S_w)$	J-function: $[J(S_w) = (p_c/\sigma\cos\theta)(k/\phi)^{\frac{1}{2}}]$
g	Acceleration of gravity (m/s^2)
h	Height of core (m)
k	Absolute permeability (m^2)
k_f	Fracture permeability associated with the rock bulk (m^2)

k_{ff}	Single fracture permeability (m^2)
k_m	Matrix permeability (m^2)
k_o	Effective permeability to oil (m^2)
k_r	Relative permeability
k_w	Effective permeability to water (m^2)
L	Length (m)
l_t	Cumulative length of the fractures (m)
N_{imb}	Total oil recovery (m^3)
n	Number of fractures intersecting a straight line
OIP	Oil in place (m^3)
PV	Pore volume (m^3)
p	Pressure (Pa)
p_d	Treshold capillary pressure (Pa)
q	Flow rate (m^3/s)
R	Recovery in fractions of original oil in place
R_s	Solution gas-oil ratio (m^3/m^3)
R_∞	Total oil recovery from matrix in fractions of original oil in place
S	Saturation
S_{iw}	Irreducible water (wetting phase) saturation
S_{or}	Residual oil saturation
S_{wf}	Final water saturation after imbibition
S_{wi}	Initial water saturation
ΔS_w	Water saturation change due to imbibition
s_B	Matrix bulk area (m^2)
s_f	Fracture bulk surface (m^2)
s_v	Specific surface area (area per unit volume of sample) (m^{-1})
T	Temperature (K)
T_α	Transmissibility (m^3/s . Pa)
t	Time (s)
u	Flow velocity (m/s)
u^*	Matrix-fracture source function (s^{-1})
u^*_b	Transfer of fluid between block and fractures [Eq. (3.12)] (s^{-1})
u^*_u	Rate of imbibition per unitary fracture length (s^{-1})
V	Volume (m^3)
V_B	Matrix bulk volume (m^3)
V_f	Fracture volume (m^3)
V_t	Total rock volume (m^3)

X	Length (m)
x,y,z	Coordinate distances (m)
α	Angle
α	Dimensionless rock parameter [Eq. (3.3)]
β	Angle
β	Constant [Eq. (3.10)]
γ	Fluid gradient (Pa/m)
θ	Contact angle
λ	Interaction parameter (m^3/s . Pa)
μ	Viscosity (Pa . s)
ρ	Density (kg/m^3)
σ	Interfacial tension (N/m)
σ	Geometric factor (m^{-2})
τ	Time necessary to produce 63% of maximum recoverable oil
ϕ	Porosity
ϕ_1	Matrix void volume/total bulk volume
ϕ_2	Fracture void volume/total bulk volume
ϕ_m	Matrix void volume/matrix bulk volume
ϕ_f	Fracture void volume/total bulk volume
Φ	Fluid flow potential (datum pressure) (Pa)
ζ	Integration parameter

<u>Subscripts</u>

c	Capillary
D	Dimensionless
e	Effective
F	Front
f	Fractures
f	Final (section 3.2.1)
G	Gravity
g	Gas
m	Matrix
nw	Non wetting
O	Oil
p	Pore
u	Unitary
w	Water
w	Wetting [Eq. (2.20)]
1	Matrix
2	Fractures

PART 4

UNCERTAINTY AND THE STOCHASTIC APPROACH TO TRANSPORT IN POROUS MEDIA

NON STATIONARY GEOSTATISTICS

G. de Marsily

NON STATIONARY GEOSTATISTICS

Ghislain de Marsily

Paris School of Mines
Fontainebleau, France

ABSTRACT

In geostatistics, variables are often classified as "stationary" or "non stationary". The latter refers to variables which show a definite trend in space, such as the direction of the hydraulic gradient, for hydraulic heads in an aquifer. In (10), the method of simple kriging for stationary hydrologic variables is presented. In this chapter, universal kriging and the use of generalized covariances of order k are summarized, for the estimation of non stationary hydrologic variables, with one example of application.

1. INTRODUCTION

In a previous NATO Advanced Study Institute, we have tried to summarize the application of simple geostatistics (simple kriging) to hydrological problems (10). Only stationary or intrinsic phenomena were considered, i.e., those where the variable of interest does not show a systematic "trend" or "drift" in space. This is most often acceptable for transmissivities, or thickness of a formation, however it only rarely can be applied to hydraulic head, or else rainfall data, which most often have a definite trend in space, that of the general hydraulic gradient, for heads, or of the orographic effects, for rainfall.

Non stationary geostatistics must therefore be used and will be summarized below, illustrated with an example. All of this comes from the work of G. Matheron and co-workers in Fontainebleau, e.g. Matheron (15), Delfiner (12), Delhomme (3,4,5,6), Chiles (1), and also from Kitanidis (8). A more elaborate presentation is also given in Marsily (11).

Let us start by a recollection of the probabilistic approach to the definition of spatial parameters in a domain.

2. PROBABILISTIC DEFINITION OF A PARAMETER IN SPACE

Many of the magnitudes of interest in hydrology (e.g., transmissivity, storage coefficient, hydraulic head in an aquifer, thickness of a layer, rainfall, etc...) are a function of space, but often highly variable. This spatial variability is however not purely random: if measurements are made at two different locations, the closer the measurements, the closer the measured values. In other words, there is some kind of correlation in the spatial distribution of these magnitudes. Matheron (12,13,14) has given the name of "regionalized variables" to these types of quantities: they are variables typical of a phenomenon developing in space (and/or time) and possessing a certain structure. Here, the term "structure" refers to this spatial correlation which, of course, is very different from one magnitude to the other or from one aquifer to the next.

Regionalized variables can be divided into two main categories: stationary and non-stationary. In the latter, the variable has a definite trend in space: for instance, the variable decreases systematically in one direction. This is generally the case of the hydraulic head. On the contrary, there is no systematic trend in space for the stationary variables. This is in general the case of transmissivity.

Here we will address the problem of how to estimate a regionalized variable, which is the most common problem facing the hydrogeologist in the field. Having measured a variable at a set of points (e.g., heads at several piezometers, transmissivities at several wells, rainfall at several rain gauges), how do we estimate the value of the variable at all other locations in order to produce a contour map of the variables, or a discretized map serving as input for a model ? Kriging is an optimal estimation method and its use will be described for the non stationary case.

To make this estimation we use the concept of *random functions* (R.F) which can be presented as follows. Let Z be the magnitude of interest which we want to study in a given aquifer. We will assume that the magnitude Z is a realization of a random process, i.e., that we can conceptually define a R.F. $Z(x,\xi)$, where x is the coordinates (in 1, 2 or 3 dimensions) in space, and ξ is the "state variable" of the process. For a given value ξ_1 of the "state variable", $Z(x,\xi_1)$ is just an ordinary function of space, which is called a "realization" of the R.F. Z. But the state variable ξ is assumed to be able to take an infinite number of values meaning that we can have conceptually an infinite number of realizations of the function Z. At a given location x_o, $Z(x_o,\xi)$ is a random variable, and thus $Z(x_o,\xi_1)$ is just a number, i.e. the value of Z at location x_o in the realization ξ_1. Now we can characterize the random variable $Z(x_o,\xi)$ by its entire probability distribution function (pdf), or only by its first two moments, the expected value and variance, which we will denote $E[Z(x_o,\xi)] = m_o$ and $\mathrm{Var}\,[Z(x_o,\xi)] = E[(Z(x_o,\xi) - m_o)^2] = \sigma^2_{Z_o}$.

We will also use the covariance between two points x_1 and x_2, which is defined by: $C(x_1,x_2) = E[(Z(x_1,\xi) - m_1)(Z(x_2,\xi) - m_2)]$. This covariance will tell us about the correlation between the values of Z taken at two locations, x_1 and x_2.

Note that the expected value sign E is taken here over the state variable ξ, i.e., over all possible realizations of Z, and not over the space variable x: one often talks about "ensemble averages" instead of "space averages".

If we want to give an example, let us consider a sand dune in a large desert, and let its crescent shape be our domain of interest (7). If Z is some property of the dune sand, e.g. its permeability, Z(x) within the dune will be spatially varying and will be our regionalized variable. But now, there are many such dunes in the same desert, and since their formation is the result of the same aeolian process, starting from the same sandy material, we can consider that they are "similar", i.e., that each of them is a "realization" of the same sedimentation process. If the desert is very large, we are very close to having an almost infinite number of realizations of our dune. Of course each dune will be different, having here and there local variations, heterogeneities, etc... So if we measure the permeability of the sand in a given location, e.g., the center of the dune, or one of its tips, we will find a different value for each of them. Now the ensemble average of that permeability will just be the average of all these permeabilities of the tips of all dunes, *and not* the spatial average of the permeability over *one single dune*. The state variable ξ will then just be the number given to each dune on a map of the desert, if we care to number them.

In practice, of course, we have only one aquifer, and thus our R.F. $Z(x,\xi)$ is only conceptual and all the information on Z is contained in the unique realization $Z(x,\xi_1)$ that we can observe. But this concept will prove to be very useful to perform the estimation of $Z(x,\xi_1)$ in space at locations where it has not been measured, and also to understand and quantify the error of this estimation.

For this concept to be of use, we will need to make two working hypotheses on the process characterized by the R.F. Z:

i) It verifies some kind of stationarity: the most stringent one is true stationarity, which requires that the pdF and all moments of Z (at one point or jointly at several points) is invariant by translation, i.e., does not depend on x. Second order stationarity only requires that the first two moments of Z do not depend on x:

$E(Z(x,\xi)) = m$ not a function of x,
$Var\ (Z(x,\xi)) = \sigma^2$ not a function of x,
$C(x_1,x_2) = E[(Z(x_1,\xi) - m)(Z(x_2,\xi) - m)] = C(\underline{h})$
not a function of x_1, x_2, but only of the vector $\underline{h} = x_2 - x_1$

An even weaker hypothesis, which is more useful in practice than 2nd order stationarity, is called the "intrinsic hypothesis", and requires that the first increments of Z be 2nd order stationary:

$E[Z(x + h) - Z(x)] = f(\underline{h})$ not a function of x, only of $\underline{h}$
$Var\ [Z(x + h) - Z(x)] = 2\gamma(\underline{h})$ not a function of x, only of $\underline{h}$

In practice, $f(\underline{h})$ is taken to be zero, i.e., $E(Z(x,\xi))$ is a constant, m, as in the previous case, but the variance of the increments defines a new function γ, the variogram, which can also be written as:

$$\gamma(\underline{h}) = \frac{1}{2} E[(Z(x + h) - Z(x))^2] \qquad (2.1)$$

This intrinsic hypothesis is necessary because many geologic variables do not have a finite variance, and a covariance cannot be defined. This shows up when estimating the experimental variance σ^2 from the sampling points, as a function of the size of the domain or number of samples: one often finds that σ^2 increases when this size or number increases. When both the covariance and variogram exist, i.e., with 2nd order stationarity, it is easy to show that:

$$\gamma(h) = C(o) - C(h)$$

In this chapter, we will focus on even weaker hypotheses of stationarity which will allow the expected value m to be function of x.

ii) The process characterizing the R.F. Z must be ergodic, which means that Z displays in space the same behavior as in the domain of the realizations: in other words, it is possible to infer from one single realization the pdF (or its first moments) of the R.F. Z. A simple example of a non ergodic process could be given by a R.F. (in one dimension): $Z(x,\xi) = a\cos(\omega x) + b$, where a and b are constants, but where ω is a random variable: for each ξ,ω will be different. The pdF of Z at a given location very much depends on the pdF of ξ. But although Z is stationary, looking at one realization in space will never be enough to infer the pdF of either Z or ω.

Note that already in the previous paragraph (i), we used the notion of ergodicity, when it was said that the experimental variance σ^2 from the sampling points could sometimes be a function of the size of the domain: we assumed that the variance obtained by sampling in space could be taken as the variance of the R.F. Z: this is precisely the ergodic hypothesis!

The stationarity hypothesis can be verified in paractice, whereas the ergodic hypothesis is almost impossible to check: since there is only one realization, it is by definition impossible to verify it! But this is not to be considered important: the probabilistic language is only used to build a tool (or model) to perform an estimation; as long as the assumptions made are not in contradiction with the data, they are acceptable working hypotheses if they help to develop the tool. In other words, nature does not have to be ergodic: only the conceptual R.F. Z, which we have invented and which does not exist in nature, has to be ergodic.

In the remainder of this chapter, for the sake of simplicity, the state variable ξ will be omitted and the R.F. Z will simply be noted $Z(x)$. But one must remember that E (expected value) is always taken over the ensemble of the realizations, i.e., for all possible values of ξ.

3. SUMMARY OF THE KRIGING EQUATIONS IN THE INTRINSIC CASE

In the previous publication (10), we presented the kriging equations in the intrinsic case and we will summarize it here. But let us first define what our problem is:

Let $Z(x_i)$, $i = 1, \ldots, n$ be n measurements of Z in a given domain. Our problem is to estimate the value of Z at a location x_o which has not been measured. Kriging is an algorithm which does precisely that. It works in two steps:

3.1. Step 1 - Statistical Inference

One must first determine the variogram of Z. Its definition is given in (2.1) but using the ergodic hypothesis, we will estimate it from the data. From the n measurements $Z_i = Z(x_i)$, one creates $n(n+1)/2$ pairs (Z_i, Z_j). These pairs are grouped into classes of distance, i.e., separations between the points x_i and x_j. The *experimental* variogram is then given by:

$$\gamma(h) = \frac{1}{2n_h} \sum_{1}^{n_h} (Z_i - Z_j)^2$$

where h is the average distance in a given class, n_h is the number of pairs in that class, i and j are two indices of location where $x_i - x_j$ belongs to the class of distance h.

Note that we assume here the variogram to depend only on the distance h, not on the direction of the separation vector $\underline{h}$ between x_1 and x_2. The variogram is then said to be isotropic. If this does not hold, then the variogram becomes also a function of the direction of the separation vector.

Once an experimental variogram is obtained, one must adjust a variogram *model* on it. Only a limited number of functions are used as variograms: linear, exponential, gaussian, cubic, spherical, as a variogram must satisfy two conditions:

(i) $\lim\limits_{h \to \infty} \dfrac{\gamma(h)}{h^2} = 0$

(ii) $-\gamma(h)$ must be conditionally positive definite.

This can be done graphically by trial and error, or automatically (see for example, Kitanidis (9)).

3.2. Step 2 - Estimation

The kriging estimator is a linear combination of all the measurements:

$$Z_o^* = Z^*(x_o) = \sum_{i=1}^{n} \lambda_o^i Z_i \qquad (3.1)$$

Here the star(*) denotes the estimate of the (unknown) exact value Z_o. The λ's are the kriging weights, i.e., the unknowns of the problem; there are as many λ's as measurement points, but these λ's are different for each point x_o to be estimated.

The values of the λ's are given by imposing two conditions on the estimator:

(i) not to be biased: this writes:

$$E(Z_o^*) = E(Z_o) = m$$

(ii) to be optimal, i.e., that the error of estimation be minimal in quadratic form:

$$A = E[(Z_o^* - Z_o)^2] = \mathrm{Var}[(Z_o^* - Z_o)] \qquad \text{minimum}$$

The first condition leads to:

$$\sum_{i=1}^{n} \lambda_o^i = 1 \tag{3.2}$$

The second condition is met by setting to zero all the derivatives of A with respect to the unknowns λ and to a Lagrange multiplier μ introduced in the minimization of A to impose the condition (Eq. (3.2)). After some simple algebra, one finds that the λ's and μ are the solution of a linear system, which writes:

$$\left.\begin{aligned} &\sum_{j=1}^{n} \lambda_o^j \, \gamma(x_i - x_j) + \mu = \gamma(x_i - x_o) , \quad i = 1,\ldots.,n \\ &\sum_{i=1}^{n} \lambda_o^i = 1 \end{aligned}\right\} \tag{3.3}$$

At last the variance of the error of estimation is given by:

$$\text{Var}\,[(Z_o^* - Z_o)] = \sum_{i=1}^{n} \lambda_o^i \, \gamma(x_i - x_o) + \mu \tag{3.4}$$

Kriging is said to be done with a unique neighborhood if all the n measurements are used systematically for the estimation of any Z_o^*. In that case, the kriging matrix of Eq. (3.3) needs only γ be inverted once, as only the right han ide of Eq. (3.3) depends on x_o. If only a subset of the total number of measurements is used (e.g., the n closest measurement points x_i to the point x_o to be estimated), then kriging is made with a "moving neighborhood". Then the kriging matrix changes each time a different subset of n points is selected.

Note that our estimate Z_o^* has been obtained by imposing conditions on the expected value of Z_o^* or $(Z_o^* - Z_o)^2$; i.e., over an ensemble average. This does not mean at all that Z_o^* is an estimator of some kind of an average of the R.F. Z: such an estimator would just be the expected value $E(Z_o) = m$, i.e., a constant over the domain. Instead, Z_o^* is the best estimate of Z_o for the particular realization of interest. But we define "best" by saying that if we were to perform the estimation for a large number of realizations, we want the error of estimation to be zero in the average, and of minimum variance.

Using the ergodic argument, this is equivalent to saying that if we do the estimation for a large number of points in space (which is precisely our goal), we want the error of estimation to be zero in the average, and of minimum variance.

With this summary, we are now ready to go to non stationary geostatistics.

4. NON STATIONARY PROBLEMS

4.1. Definition

In non stationary problems, the mathematical expectation of Z is no longer a constant: $E[Z(x)] = m(x)$ and the variogram cannot be calculated directly from the data since $m(x)$ is unknown:

$$\gamma(h) = \frac{1}{2} \text{Var } [Z(x+h) - Z(x)] = \frac{1}{2} E[(Z(x+h) - Z(x))^2]$$
$$- \frac{1}{2} [m(x+h) - m(x)]^2$$

If we then try to calculate the variogram as shown in the preceding section, i.e., directly from the data, by:

$$\gamma'(h) = \frac{1}{2n_h} \Sigma (Z_i - Z_j)^2$$

where n_h is the number of pairs $(Z_i - Z_j)$ separated by distance h, we find that the variogram $\gamma'(h)$ is anisotropic because the mathematical expectation m is anisotropic and Z has a main direction of drift, e.g., the direction of flow, for hydraulic heads.

In this direction, if m is a linear spatial function, a parabolic function is added to the true variogram γ. Thus the calculation of the variogram in several directions makes it possible to detect the importance of the nonstationarity.

There are several procedures for solving non stationary problems. We start with three solutions of special cases before turning to the general one in Section 4.3.

4.2. Special Cases

In certain cases, it is possible to:

a) Assume that Z is "locally stationary", that is to say that the variogram stays isotropic for a certain neighborhood, and we can krige with the intrinsic hypothesis in that area, with a moving neighborhood.

b) Assume that the mathematical expectation m(x) is known. It may, for instance, be deduced from other types of measurements. Its mathematical expression might also be known for physical reasons (e.g., the general shape of the drawdown in the vicinity of a borehole for the hydraulic heads) and then the constants of this expression may be fitted on the model. We then verify that the residues $Z(x) - m(x)$ are stationary and can be kriged under the assumptions of the intrinsic hypothesis. It is, however, incorrect to fit a polynomial expression (by least squares) arbitrarily on the data, assimilate it to $m(x)$ and work on the residues. Indeed, the fitting by simple least squares assumes that the residues are independent and therefore that no spatial structure exists. It is usual in statistics to test the independence of these residues with the Durbin-Watson test. It is thus contrary to the hypothesis to try to find a variogram for them. It is nevertheless possible to use generalized least squares if we take this spatial correlation into account iteratively, (See Neuman (16)).

c) Assume that the variogram γ is stationary and known. This is an extension of simple kriging which is called "universal kriging" but which is usually difficult to apply because the variogram must be known. However, assume that we know it and that it is stationary:

$$\gamma(h) = \frac{1}{2} \text{Var}\,[Z(x+h) - Z(x)] \quad \text{not a function of } x \qquad (4.1)$$

We have seen that we cannot compute this variogram directly from the data because the average $m(x)$ is not known. As usual the kriging estimation is:

$$Z_o^* = \sum_i \lambda_o^i Z_i \qquad (4.2)$$

but the condition for having an unbiased estimator is different:

$$E(Z_o^*) = E(Z_o)$$

$$E[\sum_i \lambda_o^i Z_i] = E(Z_o) \ ; \quad \text{but } E(Z) = m(x)$$

whence:

$$\sum_i \lambda_o^i m(x_i) = m(x_o) \qquad (4.3)$$

The average $m(x)$ is not known, but we make the assumption that it is regular and that it may be represented *locally* by a known set of basis functions. Polynomial expressions are commonly used for this purpose. For example, in two dimensions, we write:

$$m(x) = a_o + a_1X + a_2Y + a_3X^2 + a_4XY + a_5Y^2 + \ldots$$

where X and Y are the coordinates of point x in two dimensions.

Or: $m(x) = \sum_{k=1}^{\ell} a_k p^k(x)$

where $p^k(x)$ are polynomials in X and Y.

In order to ascertain that the estimator is unbiased, we impose that Eq. (4.3) is satisfied by any value of a_k:

$$\sum_i \lambda_o^i \left(\sum_k a_k p^k(x_i)\right) = \sum_k a_k p^k(x_o)$$

or: $$\sum_k a_k \left(\sum_i \lambda_o^i\, p^k(x_i)\right) = \sum_k a_k p^k(x_o)$$

which is satisfied if:

$$\sum_i \lambda_o^i p^k(x_i) = p^k(x_o) \qquad k = 1, \ldots, \ell \tag{4.4}$$

These conditions are the equivalent of the single condition $\sum_i \lambda_o^i = 1$ which was imposed in the stationary case. We then minimize the estimation variance $\mathrm{Var}(Z_o^* - Z_o)$, subject to the ℓ conditions [Eq.(4.4)], in the same way. The estimation variance is again a function only of the variogram γ because of Eq. (4.4) and the equations in the kriging system write:

$$\left.\begin{aligned} &\sum_j \lambda_o^j \gamma(x_i - x_j) + \sum_k \mu_k p^k(x_i) = \gamma(x_i - x_o) && i = 1, \ldots, n \\ &\sum_i \lambda_o^i p^k(x_i) = p^k(x_o) && k = 1, \ldots, \ell \end{aligned}\right\} \tag{4.5}$$

where the μ_k are Lagrange multipliers. The solution of Eq. (4.5) gives the λ_o^i for calculating Z_o^* with Eq. (4.2) and the μ_k for calculating the estimation variance with:

$$\mathrm{Var}\,[Z_o^* - Z_o] = \sum_i \lambda_o^i \gamma(x_i - x_o) + \sum_k \mu_k p^k(x_o) \tag{4.6}$$

Note that the drift m is fitted only locally (with a moving neighborhood) and that it does not appear directly in the estimation of Z but only in the calculation of the variance. Therefore it is not the same thing to fit one polynomial expression on the system as a whole and to krige the residues. Each new point x_o with a different neighborhood has a new fit for the drift $m(x_o)$, the coefficients of which (the a_k) are never calculated. For this reason, we generally use only low-degree polynomial expressions (linear in X and Y, or quadratic).

However, the serious problem with universal kriging is that the "true" variogram $\gamma(h)$ must be known and cannot be estimated directly from the data. Although efforts have been made to calculate γ iteratively (assume that γ is known, krige, verify γ once m is known), this is not practical. This is why universal kriging is used only if:

a) there is a drift in part of the system, e.g., towards the boundaries. The variogram is fitted in the center, where the phenomenon is stationary, and is then used to krige the entire domain.

b) If there is no drift at all in a given direction in the field as a whole. Then the variogram is determined from the data in this direction only and we use it in all the other directions while assuming that the "true" variogram is stationary and isotropic. However, it is very difficult to verify the validity of such an assumption.

4.3. General Solution: Intrinsic Random Functions of Order k (IRF-k)

4.3.1. Redefinition of the intrinsic hypothesis

Kriging with the intrinsic hypothesis, which we have summarized above, may be described as follows

a) define the weights λ_o^i such as $Z_o^* = \sum_{i=1}^{n} \lambda_o^i Z_i$ (4.7)

b) write the condition $\sum_{i=1}^{n} \lambda_o^i = 1$ (4.8)

c) then the (minimal) estimation error given by the kriging system is, taking Eq. (4.8) into account :

$$Z_o^* - Z_o = \sum_{i=1}^{n} \lambda_o^i Z_i - Z_o = \sum_{i=1}^{n} \lambda_o^i (Z_i - Z_o) \tag{4.9}$$

d) we then assume that the difference $(Z_i - Z_o)$ or $(Z(x+h)-Z(x))$, called first increment of Z, is stationary. It can then be shown that the variance of the estimation error with kriging depends only on the variogram:

$$\text{Var}\,(Z_o^* - Z_o) = E(Z_o^* - Z_o)^2 = - \sum_{i=1}^{n} \sum_{j=1}^{n} \lambda_o^i \lambda_o^j \gamma(x_i - x_j) + 2 \sum_{i=1}^{n} \gamma(x_i - x_o) \tag{4.10}$$

We can formulate all these equations again by arranging them slightly differently. We define $\lambda_o^o = -1$ (i.e., the value of λ_o^i for i = o) and associate the point x_o with the value i = o. Then the Eqs. (4.8) to (4.10) can be written:

$$\sum_{i=o}^{n} \lambda_o^i = 0 \qquad (4.8\text{ bis})$$

$$Z_o^* - Z_o = \sum_{i=o}^{n} \lambda_o^i Z_i \qquad (4.9\text{ bis})$$

$$E[(Z_o^* - Z_o)^2] = - \sum_{i=o}^{n} \sum_{j=o}^{n} \lambda_o^i \lambda_o^j \gamma(x_i - x_j) \qquad (4.10\text{ bis})$$

Eq. (4.9 bis) subject to the condition (4.8 bis) is called an increment of order zero of the random function Z. The intrinsic hypothesis assumes that this increment is stationary. The variance of the estimation error is then a linear function of the variogram. Finally, it is possible to determine the variogram directly from the data, as shown in Section 3.

This method can be called the procedure for the intrinsic random functions of order zero, which will be generalized below.

4.3.2. Intrinsic Random Functions of Order 1(IRF-1) and of Order 2 (IRF-2)

We treat orders 1 and 2 simultaneously and the estimation runs as follows:

a) we define the weights λ_o^i such as $Z_o^* = \sum_{i=1}^{n} \lambda_o^i Z_i$ (4.11)
(in fact, we are looking for the optimal weights λ_o^i).
We define likewise $\lambda_o^o = -1$.

b) we impose three conditions (1st order) or six conditions (2nd order):

$$\text{1st order} \quad \left\{ \begin{array}{l} \sum_{i=o}^{n} \lambda_o^i = 0 \\ \sum_{i=o}^{n} \lambda_o^i X_i = 0 \\ \sum_{i=o}^{n} \lambda_o^i Y_i = 0 \end{array} \right. \qquad (4.12)$$

2nd order
$$\left\{\begin{array}{ll} \sum_{i=o}^{n} \lambda_o^i = 0 & \sum_{i=o}^{n} \lambda_o^i X_i^2 = 0 \\ \sum_{i=o}^{n} \lambda_o^i X_i = 0 & \sum_{i=o}^{n} \lambda_o^i Y_i^2 = 0 \\ \sum_{i=o}^{n} \lambda_o^i Y_i = 0 & \sum_{i=o}^{n} \lambda_o^i X_i Y_i = 0 \end{array}\right\} \quad (4.13)$$

where $x_i = (X_i, Y_i)$ are the two coordinates (in two dimensions) of the point x_i.

c) Then the error of estimation is given by:

$$Z_o^* - Z_o = \sum_{i=o}^{n} \lambda_o^i Z_i \quad (4.14)$$

The quantity $\sum_{i=o}^{n} \lambda_o^i Z_i$, subject to the condition (4.12) or (4.13), is called a "generalized increment of order 1 (or order 2)" because it filters a polynomial expression of order 1 (or order 2). Assume that we define:

$$Z_i' = Z_i + a_o + a_1 X_i + b_1 Y_i, \quad i = o, \ldots, n$$

Then,

$$\sum_{i=o}^{n} \lambda_o^i Z_i' = \sum_{i=o}^{n} \lambda_o^i Z_i + a_o(\sum_{i=o}^{n} \lambda_o^i) + a_1(\sum_{i=o}^{n} \lambda_o^i X_i) + b_1(\sum_{i=o}^{n} \lambda_o^i Y_i)$$

$$= \sum_{i=o}^{n} \lambda_o^i Z_i$$

if the conditions (4.12) are satisfied. The generalized increment of Z ± any polynomial expression of the 1st order is unchanged. The same would be true for the 2nd order.

In one dimension, at the first order, if the measurement points are equally spaced, one can take a set of 3λ's as for instance: $\lambda^1=1$, $\lambda^2=-2$, $\lambda^3=1$. They satisfy the constraints

$$\sum_1^3 \lambda^i = 0 \quad \text{and} \quad \sum_1^3 \lambda^i x_i = 0 \quad \text{if } x_1=a,\ x_2=2a,\ x_3=3a.$$

Then, $\sum_1^3 \lambda^i Z_i = Z_1 - 2Z_2 + Z_3$. This is by definition a 2nd order difference. Generalized increments of order k are therefore just a generalization, in two or more dimensions, of this simple concept.

d) We make the assumption that the generalized increments of the 1st or the 2nd order of Z are stationary (intrinsic hypothesis of order 1 or 2). It is then possible to show, exactly as in the hypothesis of order zero, that the variance of the estimation error may be expressed in the following form:

$$\mathrm{Var}[Z_o^* - Z_o] = \mathrm{Var}\left(\sum_{i=o}^{n} \lambda_o^i Z_i\right) = \sum_{i=o}^{n} \sum_{j=o}^{n} \lambda_o^i \lambda_o^j K(x_i - x_j)$$

where K is a new function, called the "generalized covariance" of the 1st or 2nd order of the IRF Z. K is stationary, i.e., K is only a function of $h = x_i - x_j$.

e) If we assume that K(h) is known, the equations in the kriging system write (when we replace γ by $-K$ in the preceding expressions):

1st order

$$\left\{\begin{array}{l} \sum_{j=1}^{n} \lambda_o^j K(x_i - x_j) - \mu_1 - \mu_2 X_i - \mu_3 Y_i = K(x_i - x_o), \\ \qquad\qquad i = 1,\ldots,n \\ \sum_{i=1}^{n} \lambda_o^i = 1 \;;\; \sum_{i=1}^{n} \lambda_o^i X_i = X_o \;;\; \sum_{i=1}^{n} \lambda_o^i Y_i = Y_o \end{array}\right. \qquad (4.15)$$

2nd order

$$\left\{\begin{array}{ll} \sum_{j=1}^{n} \lambda_o^j K(x_i - x_j) - \sum_{k=o}^{5} \mu_k p^k(x_i) = K(x_i - x_o), & \\ & i = 1,\ldots,n \\ \sum_{i=1}^{n} \lambda_o^i p^k(x_i) = p^k(x_o), & k = 0,\ldots,5 \end{array}\right. \qquad (4.16)$$

$p^o(x),\ldots,p^5(x)$ designate the 6 polynomials in X_i, Y_i from Eq. (4.13) and the estimation variance is given by:

1st order:
$$\mathrm{Var}(Z_o^* - Z_o) = E[(Z_o^* - Z_o)^2] = K(o) + \mu_1 + \mu_2 X_o + \mu_3 Y_o - \sum_{i=1}^{n} \lambda_o^i K(x_i - x_o) \qquad (4.17)$$

2nd order:
$$\mathrm{Var}(Z_o^* - Z_o) = K(o) + \sum_k \mu_k p^k(x_o) - \sum_{i=1}^{n} \lambda_o^i K(x_i - x_o) \qquad (4.18)$$

When these new equations are compared with those of universal kriging equations [(4.4) to (4.6)], they prove to be identical. This is not surprising if we bear in mind the filtering properties of the generalized covariance. The I.R.F. assumes that the drift is locally linear (or quadratic) and we can krige as soon as we know the generalized covariance K(h). Observe that here K(o) is usually zero unless we use integrated values, in which case we can show that this term is given by:

$$K(o) = \frac{1}{S_o^2} \int_{S_o} \int_{S_o} K(x-y)\, dxdy$$

where S_o is the area of integration

Therefore, IRF-k of a higher order can also be defined but practical experience shows that it is enough to use IRF-1 and IRF-2 in most cases.

4.3.3. Statistical inference of the generalized covariance

In order to identify the variogram in the case of an I.R.F.0, it was only necessary to calculate $\gamma(h) = \frac{1}{2} E[(Z(x+h) - Z(x))^2]$, since the first increment (of order zero) was stationary. To do it, we only used the measurement points 2 by 2. Then an analytical expression was fitted on the experimental variogram (see Section 3). For K(h) we actually proceed in the same way but the fitting becomes automatic. The statistical inference runs as follows:

a) Choice of a point x_o where, naturally, $Z_o = Z(x_o)$ is known. We then choose n points close to x_o with $Z_i = Z(x_i)$ known, $i = 1,\dots,n$. A moving neighborhood is generally used just as in kriging.

b) A generalized increment of order k is built, i.e., we compute:

$$G(x_o) = \sum_{i=o}^{n} \lambda_o^i Z_i \qquad (4.19)$$

where the weights λ_o^i satisfy the conditions for being generalized increments of kth order, i.e., for example (4.13) at the order 2. To calculate a set of weights λ which fulfill the conditions (4.13) several methods can be used. We can, for instance, calculate the λ which minimize:

$$\left(\sum_{i=o}^{n} \lambda_o^i Z_i \right)^2 \qquad (4.20)$$

subject to the conditions (4.13). Just as in kriging, these weights are obtained by cancelling the partial derivatives Eq. (4.20) with respect to λ while taking Eq. (4.13) into account through the Lagrange multipliers. We can also calculate the λ as solutions to a problem of universal kriging (see Section 4.2.3) by using any variogram γ and taking $\lambda_o^o = -1$. Another solution is to take a small number of points and solve Eq. (4.13) directly.

c) We assume that the increments $G(x_o)$ are stationary, of zero average (for all x_o and for all sets x_i of neighboring points). Then the variance of these increments $G(x_o)$ may be expressed as a function of the (still unknown) generalized covariance K(h) by:

$$\mathrm{Var}[G(x_o)] = [G(x_o)]^2 = \sum_{i=o}^{n} \sum_{j=o}^{n} \lambda_o^i \lambda_o^j K(x_i - x_j) \tag{4.21}$$

d) We assume that K(h) can be expressed as a prescribed function of h, which only depends linearly on unknown coefficients A_i; a usual form is:

$$K(h) = A_o[1 - \delta(h)] + A_1|h| + A_s h^2 \ell n|h| + A_3|h|^3 \tag{4.22}$$

where δ is the Kronecker symbol, and A_o is the nugget effect. The second, third and fourth terms are called the linear, spline and cubic terms, respectively. In order for K(h) to be a generalized covariance, the A_i must satisfy:

$$A_o \geq 0 \; ; \quad A_1 \leq 0 \; ; \quad A_3 \geq 0 \; ; \quad A_s \geq -\sqrt{-24A_1A_3/\pi^2}$$

in 1 dimension

or: $$A_o \geq 0 \; ; \quad A_1 \leq 0 \; ; \quad A_3 \geq 0 \; ; \quad A_s \geq -\frac{3}{2}\sqrt{-A_1A_3}$$

in two dimensions

In practice, experience has shown that this limited class of generalized covariance functions K(h) is quite sufficient for the study of most problems. Sometimes it is not even necessary to use all the terms: A_o, A_1 or A_o, A_1, A_s, or simply A_s may suffice. We can then write Eq. (4.21) as follows:

$$[G(x_o)]^2 = \sum_{i,j} \lambda_o^i \lambda_o^j \Big(A_o[1-\delta(x_i - x_j)] + A_1|x_i - x_j| + A_s|x_i - x_j|^2 \cdot \ell n|x_i - x_j| + A_3|x_i - x_j|^3 \Big) \tag{4.23}$$

e) The calculation of $G(x_o)$ by Eq. (4.19) is repeated a great number of times (several hundreds or thousands) while varying:
- the point x_o
- as well as the neighboring points x_i, i=1,...,n of each

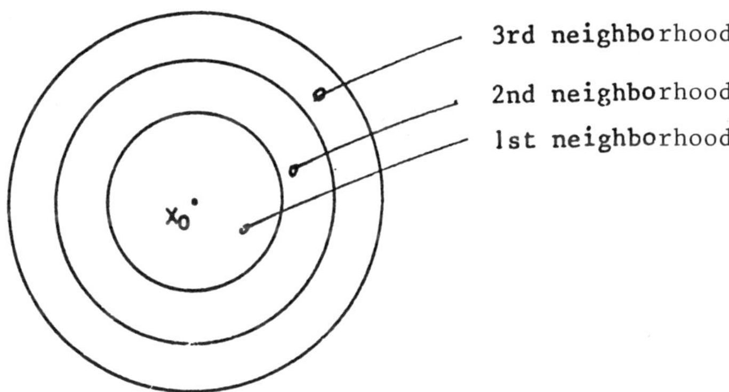

Figure 1.* Neighborhood to calculate the generalized covariance.

point x_o. Often the x_i are chosen in increasingly large circles surrounding the point x_o (Figure 1). The linear combinations λ must be as little correlated as possible (not have many points in common between two of them).

f) The coefficients A_i are determined by simple regression:

$$\underset{A_i}{\text{Min}} \sum_{x_o} \{G(x_o)^2 - \sum_{i,j=o}^{n} \lambda_o^i \lambda_o^j [A_o(1-\delta(h)) + A_1 h + A_3 h^3$$

$$+ A_s h^2 \ell n h] \qquad \text{where } h = |x_i - x_j|$$

The A_i are calculated by cancelling the first derivative of the preceding expression with respect to the A_i (linear system of 4 equations with 4 unknowns). We can also, if we so desire, weight this sum in order not to give too much weight to the large $G(x_o)$ values, which would present too great a variance.

g) Once K(h) is known, the whole kriging procedure is verified by recalculating, one by one, all the measurement points, as follows (this procedure is called "Thomas", by the name of the apostle who would only believe what he could see):

i) one value of Z, say Z_i at location x_i, is taken out of the set used in kriging;
ii) we compute the estimated value Z_i at this point, by kriging

*Originally published (1986) in "Quantitative Hydrogeology: Groundwater Hydrology for Engineers"; Reprinted with kind permission from Academic Press, New York, N.Y.

with the other data;

iii) we can then determine exactly the kriging error at this point, since we know Z_i and Z_i^*. We will compare it with the standard deviation σ_{Z_i} of the estimation error, given by the kriging system at this same point.

iv) by doing so systematically for all data points one by one, it is possible to check that there is no systematic bias:

$$\frac{1}{n} \sum_{i=1}^{n} (Z_i - Z_i^*) \cong 0$$

and that the kriging errors are coherent with their predicted standard deviation:

$$\frac{1}{n} \sum_{n} \left(\frac{Z_i - Z_i^*}{\sigma_{Z_i}}\right)^2 \cong 1$$

If these two conditions are met satisfactorily, one can say that both the assumptions behind the model and the estimated covariance K(h), are coherent with the data and one can then with some confidence use the model for a real estimation. Consequently, to krige with the IRF k, we must:

- choose the order k (0, 1 or 2);
- choose the degree of K(h) (0, 1, 3, spline);
- calculate K(h) and verify its validity as for the variogram.

It is possible to make several test runs in order to choose the best order k or degree of K(h), but experience also allows us to select the most suitable values simply by studying the data. A computer program [BLUEPACK 3-D (17)] has recently been developed to apply non stationary geo-statistics to 2 or 3 dimensional problems, according to these principles. See also Delfiner (2), and Kitanidis (9).

4.4. Example

The example concerns an unconfined aquifer in chalk, at Origny-Sainte-Benoite(Aisne, France). The aquifer is drained by three rivers to the north, east and south. A piezometric survey was made on December 31, 1976, in 88 piezometers. An additional 64 measurement points were introduced into the kriging by taking water levels in the rivers surrounding the aquifer at regular intervals, since these rivers acted as prescribed head boundary conditions for the aquifer.

Since the heads are typically non stationary, generalized covariances were used at order 1 (locally linear drift). The generalized covariance was found to be in h^3: $K(h) = a|h|^3$. It was

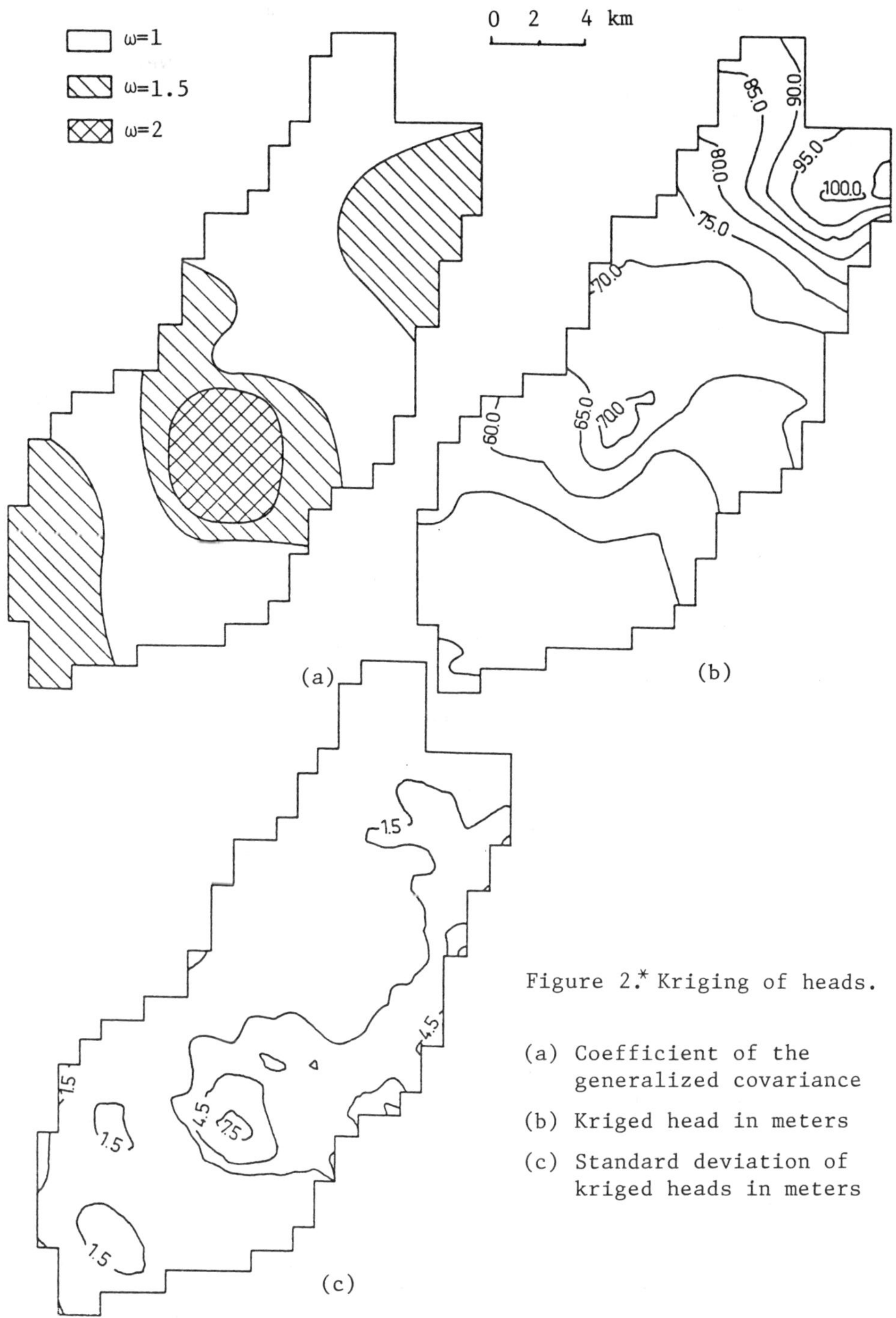

Figure 2.* Kriging of heads.

(a) Coefficient of the generalized covariance

(b) Kriged head in meters

(c) Standard deviation of kriged heads in meters

* Originally published (1986) in "Quantitative Hydrogeology: Groundwater Hydrology for Engineers"; Reprinted with kind permission from Academic Press, New York, N.Y.

however necessary to use a different coefficient "a" for different zones of the aquifer, as the spatial variability was greater under the plateau than under the plains. The values of "a" were adjusted by fitting the estimation error of kriging on the true estimation error when the validity of the model was verified as shown in the "Thomas" procedure.

The map of "a", the kriged map of the heads and the standard deviation of the estimation error are given in Figure 2.

5. REFERENCES

1. Chiles, J.P., "Géostatistique des phénomènes non stationnaires," thesis presented to the University of Nancy I, France, in 1977.
2. Delfiner, P., "Linear Estimation of Non Stationary Spatial Phenomena," *Advanced Geostatistics in the Mining Industry*, M. Guarascio, M. David and C. Huijbregts, eds., NATO-ASI, Ser. C 24, Reidel, Dordrecht, The Netherlands, 1976.
3. Delhomme, J.P., "Application de la Théorie des Variables Régionalisées dans les Sciences de l'Eau," thesis presented to the University of Paris VI, France, in 1976.
4. Delhomme, J.P., "Kriging in Hydroscience," *Advances in Water Resources*, Vol. 1, No. 5, pp. 251-266.
5. Delhomme, J.P., "Application de la Théorie des Variables Régionalisées dans les Sciences de l'Eau," *BRGM*, Ser. 2, Sect. III, No. 4, 1978, pp. 341-375.
6. Delhomme, J.P., "Spatial Variability and Uncertainty in Groundwater Flow Parameters: A Geostatistical Approach," *Water Resources Research*, Vol. 15, No. 2, pp. 269-280.
7. Grolier, J., Personal Communication, 1986.
8. Kitanidis, P.K., "Statistical Estimation of Polynomial Generalized Covariance Functions and Hydrologic Applications," *Water Resources Research*, Vol. 19, No. 4, 1983, pp. 909-921.
9. Kitanidis, P.K., "Minimum Variance Unbiased Quadratic Estimation of Covariance of Regionalized Variables," *Journal of International Association of Mathematical Geology*, Vol. 17, No. 2, pp. 195-208.
10. Marsily, G. de., "Spatial Variability of Properties in Porous Media: A Stochastic Approach," *Fundamentals of Transport Phenomena in Porous Media*, J. Bear and M.Y. Corapcioglu, eds., Martinus Nijhoff Publishers, Dordrecht, The Netherlands, 1984, pp. 719-769.
11. Marsily, G. de., *Quantitative Hydrogeology: Groundwater Hydrology for Engineers*, Academic Press, New York, N.Y., 1986.
12. Matheron, G., *Les Variables Régionalisées et Leur Estimation*, Masson, Paris, 1965.

13. Matheron, G., "La Théorie des Variables Régionalisées et ses Applications," Paris School of Mines, Cahiers du Centre de Morphologie Mathématique, 5, Fontainebleau, France,1970.

14. Matheron, G., "The Theory of Regionalized Variables and its Applications," Paris School of Mines, Cahiers du Centre de Morphologie Mathématique, 5, Fontainebleau, France,1971.

15. Matheron, G., "The Intrinsic Random Functions and Their Applications," *Advances in Applied Probability*, Vol. 5, pp. 438-468.

16. Neuman, S.P., "Role of Geostatistics in Subsurface Hydrology," *Geostatistics for Nature Resources Characterization*, G. Verly, M. David, A.G. Journel, and A. Maréchal, eds., Part 1, Reidel, Dordrecht, The Netherlands, 1984, pp. 787-816.

17. Renard, D., Geoffroy, F., Touffait, Y., and Seguret, S., "Présentation de BLUEPACK 3D, release 4," Paris School of Mines, Centre de Géostatistique et de Morphologie Mathématique, Service Informatique, Fontainebleau, France, 1985.

6. LIST OF SYMBOLS

A_i	Coefficients
a_i	Coefficients
$C(h)$	Covariance function
E	Expected value
$G(x)$	Generalized increment of Z
h	Separation distance between two points
$\underline{h}$	Separation vector between two points
i,j,ℓ,m	Indices
$K(h)$	Generalized covariance function
m	Mean of Z
n	Number of points
$p^k(x)$	Polynomial function of x
S_o	Area of integration (for the estimation of averages over blocks)
Var	Variance
x	Coordinates in space, in 1, 2 or 3 dimensions
X,Y	Coordinates in two dimensions
Z	Regionalized variable, or random function
Z_o^*	Estimation of Z at location x_o
$\gamma(h)$	Variogram
λ	Kriging weight
ℓn	Neperian logarithm
μ	Lagrange multiplier
σ,σ^2	Standard deviation, variance
ξ	State variable of a realization

14. Matheron, G., "La Théorie des Variables Régionalisées et ses Applications," Paris School of Mines, Cahiers du Centre de Morphologie Mathématique, Fontainebleau, France, 1970.

15. Matheron, G., "The Theory of Regionalized Variables and its Applications," Paris School of Mines, Cahiers du Centre de Morphologie Mathématique, 5, Fontainebleau, France, 1971.

16. Matheron, G., "The Intrinsic Random Functions and their Applications," [illegible]

17. [illegible]

18. [illegible] A.G. Journel [illegible]

19. Renard, D., [illegible] "Présentation de BLUEPACK 3D," [illegible] Centre de Géostatistique et de Morphologie Mathématique, [illegible]

LIST OF SYMBOLS

[illegible]

Exp [illegible]
Covariance function [illegible]
[illegible] Euclidean distance between two points
separation vector between two points

[illegible]

Regionalized variable [illegible] function

Variogram
Kriging weight
[illegible]
Standard deviation [illegible]

STOCHASTIC ANALYSIS OF SOLUTE TRANSPORT IN SATURATED AND UNSATURATED POROUS MEDIA

Lynn W. Gelhar

STOCHASTIC ANALYSIS OF SOLUTE TRANSPORT IN SATURATED AND UNSATURATED POROUS MEDIA

Lynn W. Gelhar

Massachusetts Institute of Technology
Department of Civil Engineering
Cambridge, MA 02139, U.S.A.

ABSTRACT

The application of stochastic methods to the analysis and prediction of the large-scale behavior of heterogeneous natural porous earth materials is reviewed emphasizing recent results for solute transport in saturated and unsaturated media. A stochastic methodology based on the representation of natural heterogeneity as a three dimensional statistically anisotropic homogeneous random field is developed. This approach makes use of spectral representation and a pertubation approximation to solve the stochastic partial differential equations governing flow and solute transport under the assumption of local statistical homogeneity. This approach leads to solutions for the large-scale mean behavior in terms of effective parameters such as hydraulic conductivities and macrodispersivities, and for the variance of the dependent variables, head and concentration. Results for the case of saturated solute transport demonstrate especially the near source developing dispersion characteristics that are reflected by the mean behavior, as well as the concentration variance as a measure of the reliability of predictions from the classical transport equation. It is found that the concentration variance is very large near sources of contamination, indicating the large uncertainty that is to be anticipated in classical transport models under those conditions. Field observations of large-scale solute transport in aquifers are also discussed. In the case of unsaturated flow the occurrence of large-scale tension-dependent hydraulic anisotropy and large-scale hysteresis is illustrated, and predictions of macrodispersivity under unsaturated conditions are also presented. Discussions focus on research needs and future directions, emphasizing the need for carefully designed numerical experimentation and large-scale controlled field experiments.

1. INTRODUCTION

The flow properties of natural porous earth materials are observed to be highly variable in space. This feature is illustrated qualitatively by the photograph in Figure 1 which shows an exposure of the sand and gravel deposits located in Switzerland. From such qualitative observations it is evident that factors such as grain size vary enormously even in individual formations. This kind of intraformational heterogeneity can be characterized more quantitatively in terms of parameters such as the saturated hydraulic conductivity. Data, such as that shown in Figure 2 based on small laboratory cores taken from a single borehole, shows that the standard deviation of the natural logarithm of hydraulic conductivity can vary from 0.2 to as much as 5 in various types of natural deposits. An equally important characteristic of this natural intraformational heterogeneity is its spatial structure. Because of the natural variability of the depositional processes which affect permeability, the resulting heterogeneity often has layered or lenticular structure and reflects directional dependence or anisotropy.

Fig. 1. Photograph showing the natural heterogeneity of a sand and gravel deposit in Switzerland (photo by E. Trüeb).

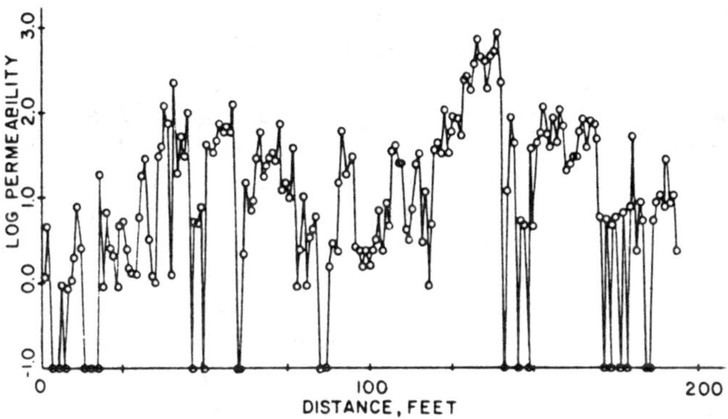

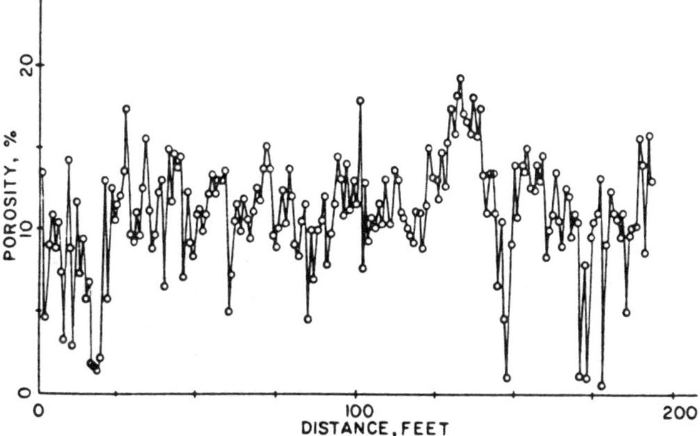

Fig. 2. Permeability (millidarcy) and porosity data from laboratory cores from a borehole in the Mt. Simon aquifer in Illinois.

Many important practical problems of subsurface flow or solute transport require predictions over relatively large time and space scales, up to thousands of years and over kilometers or more. In most cases direct measurements over these time and space scales are simply not feasible so that we require methods of extrapolating relatively small scale laboratory or field observations to these large time or space scales. The key question is how we can realistically incorporate the effects of natural heterogeneity into models which extrapolate to predict behavior at this large time and space scale.

There are several possible approaches to prediction of this large-scale behavior. A first category of approaches involves the use of deterministic numerical models. Within that framework, the classical approach is to treat formations, layers or zones within the aquifer as being homogeneous and to construct a model based on discretization which reflects this zonation or layering. An important weakness to this classical approach is that it ignores the always present intraformational spatial variability. Consequently, classical numerical models do not provide a quantitative measure of the error introduced by this assumption. A second deterministic approach might be to represent the actual detailed heterogeneity of the aquifer in a complex three-dimensional numerical model. The key drawback to this kind of approach is that it would require extremely detailed measurements of the three-dimensional spatial distribution of hydraulic properties. This is clearly an impractical task and, furthermore, sampling an aquifer in such detail may actually alter the hydraulic properties of the medium. A further limitation of this detailed modeling approach is its computational feasibility. One can easily recognize that possibly on the order of 10^9 nodes could be required to describe even a modest sized three-dimensional realistically heterogeneous natural material; numerical solutions of systems of equations of this dimension are extremely challenging even with the largest supercomputers.

Another possible approach to this large-scale prediction problem is to treat the natural heterogeneity in a stochastic sense. The intraformational heterogeneity of flow properties can be represented in that framework by spatial random fields which are characterized by a relatively small number of statistical parameters. Within this framework we seek solutions of flow or transport problems expressed in terms of probability distributions or, more simply, in terms of certain moments of those distributions. For example, the mean solution can be used to determine effective large-scale parameters and the form of the partial differential equation that will describe the large-scale flow or transport process. Similarly, the second moment or variance can be used to characterize the model error or reliability of the mean model. It turns out that the mean equation described in this sense will, in many cases, be equivalent to the classical deterministic equation, and treatment of the variance will give us a quantitative measure of the error to be anticipated in applying the classical model. As will be seen in later discussions, the stochastic approach provides new understanding of large-scale controlling processes.

The primary purpose of this paper is to illustrate, through a summary of several recent results on solute transport in saturated and unsaturated media, the kind of new physical insight which has developed from the stochastic approach. The paper first gives a brief outline of stochastic methodology and then develops some important results on the stochastic behavior of flow in saturated porous media. These flow results are then incorporated into a solute transport analysis which emphasizes especially the near source characteristics of developing macrodispersion as reflected in the mean solute concentration equation, and illustrates the use of the concentration variance to interpret this near source behavior. Several field observations are reviewed in relation to the stochastic results and, finally, some recent results on flow and transport in unsaturated porous media are summarized. The discussion emphasizes new insights which have evolved through this stochastic approach and important problems which remain to be resolved.

2. STOCHASTIC METHODOLOGY

The purpose here is to summarize the general ideas behind the stochastic approach. Detailed development of the techniques is available in Gelhar and Axness (13) and Gelhar (12). The spatial variability of porous medium flow properties such as hydraulic conductivity is represented by three-dimensional spatial random fields which are characterized by their second moment properties, that is, the covariance function or spectral density function. The natural logarithm of hydraulic conductivity is represented as a mean plus a perturbation around that mean as follows:

$$\ln K = F + f; \quad E(\ln K) = F, \quad E(f) = 0 \tag{2.1}$$

and the covariance function of lnK is given by

$$R_{ff}(x_1,x_2) = \mathrm{cov}[f(x_1),f(x_2)] = E[f(x_1)f(x_2)] \tag{2.2}$$

$$= R_{ff}(\xi) \; ; \quad \xi = x_1-x_2$$

where x_1 and x_2 indicate two different spatial locations. When the covariance function depends only on the difference $x_1 - x_2$, the process is said to be stationary or statistically homogeneous. Figure 3 illustrates graphically a one-dimensional stationary covariance function. The separation distance at which the correlation decreases to the e^{-1} level is conveniently designated as the correlation scale λ. In the general case x_1 and x_2 are vectors and, if the process is stationary or statistically homogeneous, the covariance function can always be expressed in terms of the spectral density function or spectrum as follows:

$$R_{ff}(\vec{\xi}) = \iiint_{-\infty}^{\infty} e^{i\vec{k}\cdot\vec{\xi}} \, S_{ff}(\vec{k})d\vec{k} \tag{2.3}$$

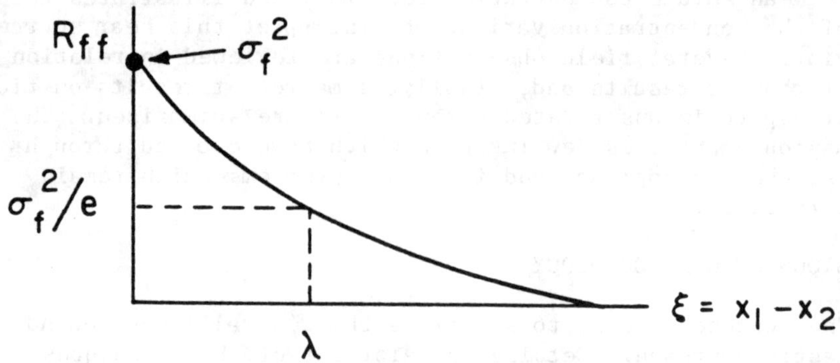

Fig. 3. Schematic graph of a stationary one-dimensional covariance function of lnK showing the correlation scale λ.

where the spectral density function is expressed as

$$S_{ff}(\vec{k}) = \frac{1}{(2\pi)^3} \iiint_{-\infty}^{\infty} e^{-i\vec{k}\cdot\vec{\xi}}\, R_{ff}(\vec{\xi})d\xi \tag{2.4}$$

which depends on the three-dimensional wave number vector k. If a process is statistically homogeneous, it will always have a spectrum, and then it is always possible to represent the function via the spectral representation theorem

$$f(\vec{x}) = \iiint_{-\infty}^{\infty} e^{i\vec{k}\cdot\vec{x}}\, dZ_f(\vec{k}) \tag{2.5}$$

$$E(dZ_f dZ_f^*) = S_{ff}(\vec{k})\, d\vec{k}$$

A simple example of a three-dimensional covariance function is

$$R_{ff}(\vec{\xi}) = \sigma_f^2 \exp\left[-(\xi_1^2/\lambda_1^2 + \xi_2^2/\lambda_2^2 + \xi_3^2/\lambda_3^2)^{1/2}\right] \tag{2.6}$$

where $\sigma_f^2 = E(f^2)$, the variance of lnK. The exponential form in Eq. (2.6) involves three distinct correlation scales λ_1, λ_2, λ_3; this is a statistically anisotropic model which can be used to represent physical anisotropy of the heterogeneity as reflected in layering and lenticular structures. The spectral density function corresponding to Eq. (2.6) is

$$S_{ff}(\vec{k}) = \sigma_f^2 \lambda_1\lambda_2\lambda_3 / \left[\pi^2(1 + \lambda_1^2 k_1^2 + \lambda_2^2 k_2^2 + \lambda_3^2 k_3^2)^2\right] \qquad (2.7)$$

Figure 4 is a sketch of the three-dimensional correlation surface associated with the covariance function of Eq. (2.6). For this model, the parameters which characterize the spatial variability of a hydraulic conductivity are the variance σ_f^2 and the three correlation scales.

The hydraulic conductivity as characterized above is now introduced in the flow equation, and the result is an equation in which the parameters or coefficients are random, that is, a stochastic partial differential equation. The resulting stochastic equation is solved using a perturbation approximation and a spectral representation technique which assumes local statistical homogeneity. The notion of local statistical homogeneity is illustrated in Figure 5. The basic idea is that the mean F(x) and the variance $\sigma_f^2(x)$ must vary slowly relative to the correlation scale of the process in order for the assumption of local statistical homogeneity to be plausible and useful. For example, a length scale associated with the mean trend F(x) is

$$L_F = \left(\frac{1}{F} \frac{dF}{dx} \right)^{-1}$$

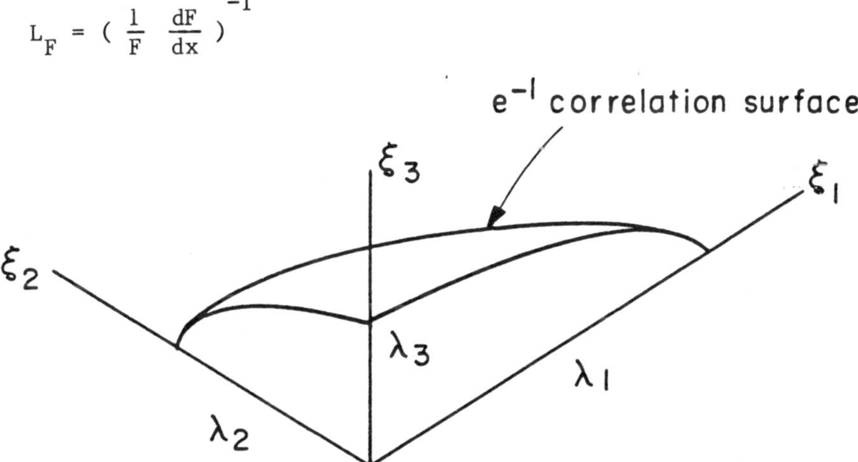

Fig. 4. Three-dimensional statistically anisotropic covariance function illustrated by the e^{-1} correlation surface with correlation scales λ_1, λ_2, and λ_3.

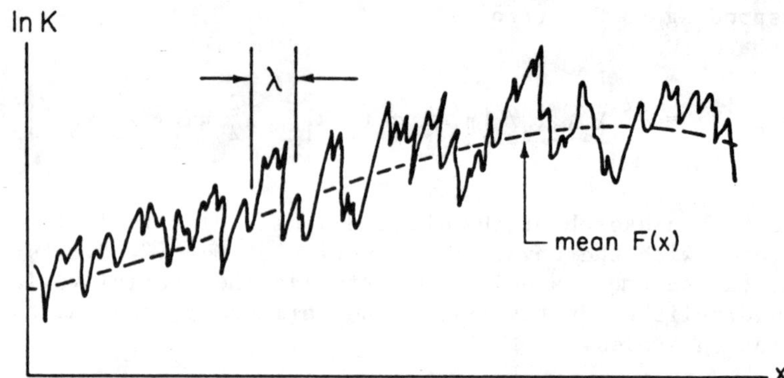

Fig. 5. Hypothetical lnK versus distance graph illustrating a trend in the mean and the notion of local stationarity.

For local statistical homogeneity we require a disparity in scale with $\lambda \ll L_F$. One could also think of a similar length scale associated with the spatial dependence of the variance, and the correlation scale would also have to be much smaller than the scale. In addition, for the notion of statistical homogeneity to be meaningful, the overall scale of the problem must be much greater than the correlation scale.

Solutions for the random head and flow fields are then used to develop generic analytical results for useful moment properties of the solution. These include first the effective large-scale properties, that is, the mean behavior as reflected in an effective hydraulic conductivity for the flow problem and a macrodispersion coefficient for the solute transport problem. Results are also obtained for the variances of the dependent variables, for example the head variance and the concentration variance. These variance results clearly demonstrate, in a generic form, the factors which determine the reliability of the mean model prediction.

3. SATURATED FLOW ANALYSIS

3.1 Head Equations

Steady flow of a constant density fluid in a locally isotropic porous medium is described by

$$\frac{\partial}{\partial x_i} (K \frac{\partial \phi}{\partial x_i}) = 0 \quad , \quad \phi = \frac{p}{\rho g} + z = H + h \tag{3.1}$$

which is expressed in terms of a piezometric head defined in terms of the fluid pressure p, the fluid density ρ, the gravitational constant g and the vertical position z. The piezometric head is decomposed into a mean H and a zero mean perturbation h as shown. This head decomposition and the hydraulic conductivity representation from Eq. (2.1) are introduced into Eq. (3.1) and, after subtracting the mean of the resulting equation from Eq. (3.1), the result is the following equation describing the head perturbation

$$\frac{\partial^2 h}{\partial x_i^2} - J_i \frac{\partial f}{\partial x_i} = \frac{\partial f}{\partial x_i} \frac{\partial h}{\partial x_i} - E\left(\frac{\partial f}{\partial x_i} \frac{\partial h}{\partial x_i} \right) \simeq 0 \tag{3.2}$$

where $J_i = - \partial H / \partial x_i$ is the mean hydraulic gradient. The zero mean term on the right-hand side of Eq. (3.2) is composed of second order terms which involves products of perturbations, whereas the terms on the left-hand side are first order terms in which perturbations appear to the first power. Consequently, if the perturbations are small enough, the right-hand side will be much smaller than the left-hand side; in the perturbation approximation used here, the right-hand side of Eq. (3.2) is neglected. Equation (3.2) is then solved using spectral representation for the f and h process, under the assumption of local stationarity as discussed above. The result is a spectral relationship between the input hydraulic conductivity variations and the resulting head variations. The head variance is found by integrating the spectrum of head over wave number space. For a three-dimensional statistically anisotropic hydraulic conductivity field the head variance is found to be in the form

$$\sigma_h^2 = \frac{\pi}{8} \sigma_f^2 J^2 \lambda_1 \lambda_3 \left(1 + \mathcal{O}(\lambda_3 / \lambda_1)^2\right) \tag{3.3}$$

for the case $\lambda_1 \gg \lambda_3$, with $\lambda_1 = \lambda_2$ and $J_1 = J$, $J_2 = J_3 = 0$.

Equation (3.3) is a local head variance relationship illustrating that the head variability is determined by the input lnK variance, the mean hydraulic gradient and the correlation scales of the input lnK process. Results of this type can be used directly to evaluate the model error in a deterministic model. Given estimates of the variance and correlation scale of the lnK field and values of the mean hydraulic gradient from the deterministic simulation, the likely variation about the mean simulation can be calculated from Eq. (3.3). Similarly, calculations of head variance from Eq. (3.3) can be used in this sense as the calibration target to judge the significance of the difference between a numerical model simulation and observations of head in the field. A generic local head variance relationship such as Eq. (3.3) might also be used in an inverse sense; that is, if the head variance is estimated from variation of the head around a smooth mean solution, this expression could be used to estimate the parameters of the input lnK process.

3.2 Mean Flow

The mean flow characteristics of a heterogeneous porous medium can be evaluated by taking the expected value of the Darcy equation that follows

$$E\ (q_i) = -E\ (K\frac{\partial \phi}{\partial x_1})\ ;\ K = K_\ell e^f,\ E(\ln K) = \ln K_\ell \tag{3.4}$$

$$= -K_\ell E\ [(1 + f + \frac{f^2}{2} + \ldots)\ (\frac{\partial H}{\partial x_i} + \frac{\partial h}{\partial x_i})]$$

$$\simeq K_\ell J_i\ (1 + \frac{\sigma_f^2}{2}) - K_\ell E\ (f\frac{\partial h}{\partial x_i})$$

where third order terms have been neglected in the final expression. The second order cross correlation term which reflects the effect of hydraulic conductivity variations as they alter the hydraulic gradient is evaluated using the spectral solution for the head perturbation. The result is a simple linear relationship between the mean specific discharge and the mean hydraulic gradient in the following form

$$E(q_i) = \bar{K}_{ij} J_j \tag{3.5}$$

where $\bar{K}_{ij}$ is the effective hydraulic conductivity tensor which will be a function of the lnK variance and the ratios λ_1/λ_2 and λ_1/λ_3. A specific example of this dependence, is shown in Figure 6 for the special case where λ_1 and λ_2 are equal. Gelhar and Axness (13) give complete evaluations of the effective conductivity tensor in the general case.

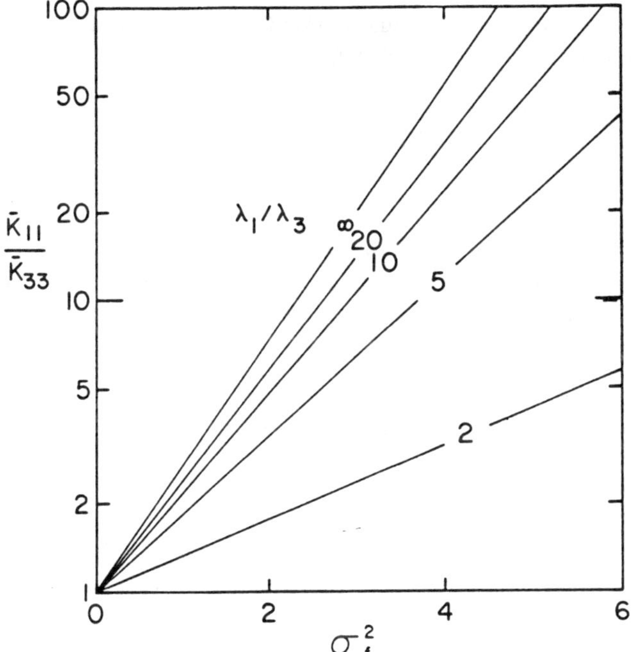

Fig. 6. Hydraulic anisotropy of effective (mean) conductivity as a function of the variance of lnK and the aspect ratio λ_1/λ_3; for $\lambda_1 = \lambda_2$.

There are several possible applications of effective conductivity results such as those in Figure 6. If the parameters of the hydraulic conductivity variation are known, they can obviously be used for direct calculation of hydraulic anisotropy. However, it is more likely that hydraulic anisotropy is determined directly from aquifer tests, in which case it may be possible to evaluate the ratios of correlation scales if the variance of lnK is known from other observations. A very important special case of Eq. (3.5) is that of two-dimensional statistically isotropic flow. This case can be obtained from a three-dimensional case by considering λ_3 to be indefinitely large with $\lambda_1 = \lambda_2$. In this case it is found that

$$\bar{K}_{ij} = K_\ell \delta_{ij} \tag{3.6}$$

$$K_\ell = \exp[E(\ln K)]$$

that is, the effect of hydraulic conductivity or transmissivity for the two-dimensional aquifer situation is essentially the geometric mean. This result forms the basis for the simple rule of thumb that the large-scale transmissivity can be found simply by averaging the logarithms of local transmissivity measurements.

3.3 Flow Perturbation

The variations of the specific discharge about the mean can be found simply by subtracting the mean expression of Eq. (3.4) from the original Darcy equation. The result, retaining only first order terms in the perturbations, is

$$q_i' = q_i - E(q_i) \simeq K_\ell \left(J_i f - \frac{\partial h}{\partial x_i}\right) \tag{3.7}$$

Using this expression and the spectral solution for the head perturbations, the spectrum of the specific discharge is

$$S_{q_i q_j}(\vec{k}) = K_\ell^2 J_m J_n \left(\delta_{im} - \frac{k_i k_m}{k^2}\right)\left(\delta_{jn} - \frac{k_j k_n}{k^2}\right) S_{ff}(\vec{k}) \tag{3.8}$$

which shows how the spectrum of specific discharge fluctuations is related to the spectrum of hydraulic conductivity variations. The expression in Eq. (3.8) can be used to evaluate the specific discharge variation in the different directions. For the statistically isotropic case with $\lambda_i = \lambda$ in Eq. (2.7) and with $J_1 = J$, $J_2 = J_3 = 0$, integration of Eq. (3.8) over the wave number domain yields

$$\sigma_{q_1}^2 = \iiint_{-\infty}^{\infty} S_{q_1 q_1}(\vec{k})\, d\vec{k} = \frac{8}{15} \frac{\sigma_f^2 q^2}{\gamma^2} \tag{3.9}$$

$$\sigma_{q_2}^2 = \sigma_{q_3}^2 = \iiint_{-\infty}^{\infty} S_{q_2 q_2}(\vec{k})\, d\vec{k} = \frac{1}{15} \frac{\sigma_f^2 q^2}{\gamma^2} \tag{3.10}$$

where $\gamma = q/K_\ell J$ with q the mean specific discharge (in the x_1-direction). The ratio of the transverse to longitudinal specific discharge variances is then

$$\sigma^2_{q_2} / \sigma^2_{q_1} = 1/8 \tag{3.11}$$

The variation in longitudinal flow is much stronger than that of the transverse flow. Using a composite media approach, Dagan (4) found a value for this same variance ratio which is 1/4 (see his equation 71).

The characteristics of the specific discharge variation about the mean are especially important in the analysis of solute transport as discussed in the following chapter.

4. SOLUTE TRANSPORT ANALYSIS

4.1 Mean Solute Transport

The purpose of this chapter is to summarize some of the results of Gelhar and Axness (13), and amplify on those results by considering some additional features of the unsteady mean concentration field and developing dispersion. The mean concentration field of an ideal (passive non-reactive) solute in a saturated porous medium is described by [see Gelhar and Axness 13)]

$$n \frac{\partial \bar{c}}{\partial t}\Big|_{\xi_i} + \frac{\partial}{\partial \xi_i} E(c' q_i') = \alpha q \frac{\partial^2 \bar{c}}{\partial \xi_i^2} \tag{4.1}$$

where

n = porosity (assumed constant)

$\xi_1 = x_1 - qt/n$

$\xi_2 = x_2$

$\xi_3 = x_3$

c' = concentration perturbation

α = local dispersivity

q = mean specific discharge (in x_1-direction)

Note that Eq. (4.1) is written in terms of a moving coordinate system translating with the mean velocity q/n. The expected value expression in the middle term in Eq. (4.1) is the key term which must be evaluated. It reflects the additional mass transport due to the correlation between the concentration and specific discharge fluctuations. This term is evaluated by considering the equation describing perturbations in the concentration field,

$$n \left.\frac{\partial c'}{\partial t}\right|_{\xi_i} + q_i' \frac{\partial \bar{c}}{\partial \xi_i} = \alpha q \frac{\partial^2 c'}{\partial \xi_i^2} \tag{4.2}$$

Note that Eq. (4.2) is a first order approximation; terms involving products of perturbations have been neglected in this equation. Detailed development of this equation is presented in Gelhar and Axness, (13); the result is their Equation 75. Equation (4.2) is solved using spectral representation. The resulting differential equation for the Fourier amplitudes is

$$\left(\frac{\partial}{\partial t} + \beta\right) dZ_c = G_j dZq_j/n \tag{4.3}$$

$$G_j = -\frac{\partial \bar{c}}{\partial \xi_j} \quad \text{(assumed locally constant in space)}$$

$$\beta = (ik_1 + \alpha k^2)\, q/n$$

Note that Eq. (4.3) is a differential equation in time only. The spatial coordinates appear only parametrically through the mean concentration gradient G_j. The cross correlation term in Eq. (4.1) is then found by multiplying both sides of Eq. (4.3) by the complex conjugate of dZq_i and solving the resulting first order linear constant coefficient differential equation for the cross-spectrum of c and q_i. Integration of the cross-spectrum over the wave number domain yields

$$E(c' q_i') = \iiint_{-\infty}^{\infty} S_{q_j q_i}(\vec{k}) \left(\int_{\tau=0}^{t} e^{\beta(\tau-t)} n^{-1} G_j(\tau,\xi_i)\, d\tau\right) d\vec{k} \tag{4.4}$$

where a homogeneous initial condition has been assumed for c'. Eq. (4.4) shows that the macrodispersive flux is not in general proportional to the local mean concentration gradient; rather it is influenced by exponentially weighted contributions from earlier concentration gradients. The main contribution to the time integral in Eq. (4.4) will be near the upper limit $\tau = t$. Therefore G_j is expanded as follows

$$G_j = -\frac{\partial \bar{c}}{\partial \xi_i} = G_j(\tau)\Big|_t + \frac{\partial G_j}{\partial \tau}\Big|_t (\tau - t) + \ldots \qquad (4.5)$$

Retaining only the first term in this expansion, Eq. (4.5) can be written

$$E(c'q_i') \simeq G_j(\xi_i, t) \iiint_{-\infty}^{\infty} S_{q_j q_i}(\vec{k}) \int_{\tau=0}^{t} e^{-\beta(t-\tau)} d\tau d\vec{k} \qquad (4.6)$$

Note that the omitted higher order terms in Eq. (4.5) do not contribute to the macrodispersion coefficient in the mean equation; i.e., the coefficient of the second derivative of mean concentration. These higher order terms will introduce higher derivative terms in the mean equation and, as discussed in detail in Gelhar et al. (14), will introduce asymmetry and nonGaussian effects in a mean concentration distribution. However, here the main concern is the behavior of the macrodispersion coefficient as a measure of the behavior of the spatial second moment of the mean concentration distribution. Finally, evaluating the time integral in Eq. (4.6) yields

$$E(c'q_i') \simeq G_j q \iiint_{-\infty}^{\infty} S_{q_j q_i}(\vec{k}) \frac{1-e^{-\beta t}}{q\, n\, \beta} d\vec{k} \qquad (4.7)$$

$$= G_j\, q\, A_{ij}$$

where the macrodispersivity tensor A_{ij} is given by the wave number integral in Eq. (4.7).

From Eq (4.7) it is seen that as $t \to \infty$, the macrodispersivity A_{ij} approaches a constant, and we recover the usual Fickian transport relationship given by

$$E(c'q_i') = - q\, A_{ij} \frac{\partial \bar{c}}{\partial x_j} \tag{4.8}$$

where the macrodispersivity is given by the integral

$$A_{ij} = \iiint_{-\infty}^{\infty} \frac{S_{q_j q_i}(\vec{k})d\vec{k}}{[ik_1 + \alpha_L k_1^2 + \alpha_T(k_2^2 + k_3^2)]\; q^2} \tag{4.9}$$

Note that Eq. (4.9) has been written for the slightly more general case of an isotropic local dispersion with local longitudinal dispersivity α_L and transverse dispersivity α_T. The large time result of Eq. (4.9) is identical to the result found by Gelhar and Axness (13) under the assumption of a strictly steady concentration field (see their Eq. 22'). It is easily shown that Eq. (4.9) also applies to steady unsaturated flow fields provided that the local dispersion coefficient is constant. Note also that Eq. 4.9 is equivalent to the result of Winter et al. (32), developed using an abstract mathematical approach based on semigroup theory.

The result of Eq. (4.9) is not directly usable because the specific discharge spectrum is not generally known or easily measured. However the results of the previous chapter for the flow perturbation provide a relationship between the log hydraulic conductivity spectrum and the specific discharge spectrum [Eq. (3.8)]. Although natural media will not usually be statistically isotropic, it is initially informative to consider the case of a statistically isotropic medium in which $\lambda_1 = \lambda_2 = \lambda_3$ in Eq. (2.7). In this case the wave number integral of Eq. (4.9) can be evaluated analytically as discussed in Gelhar and Axness (13). When the local dispersivity is much smaller than the correlation scale, the longitudinal macrodispersivity is found to be of the form

$$A_{11} = \sigma_f^2\, \lambda / \gamma^2 \tag{4.10}$$

where $\gamma = q/K_\ell J$ and q is the mean specific discharge in the x_1 direction and J is the mean hydraulic gradient in that same direction. The constant γ is simply the ratio of the effective conductivity to the geometric mean conductivity. Under these same conditions the transverse macrodispersivity becomes

$$A_{22} = A_{33} = \sigma_f^2 \, \alpha / \, 3 \, \gamma^2 \tag{4.11}$$

Gelhar and Axness have also evaluated the macrodispersivity tensor for the case when α/λ is not small. Their Figure 1 shows that the inclusion of local dispersion slightly reduces both the longitudinal and transverse dispersion coefficients. This kind of result is physically plausible because one can expect that a local dispersion process will smooth the concentration field and consequently result in a smaller correlation between the concentration and flow fluctuation. Equation (4.11) indicates that the transverse dispersivity is extremely small compared to the longitudinal dispersivity of Eq. (4.10), but field observations show that this ratio is likely to be on the order of 10^{-1}. This unrealistic result is not surprising since the model which assumes statistical isotropy of the hydraulic conductivity heterogeneity is obviously not a realistic one for natural materials.

The analysis leading to Eq. (4.9) was restricted to relatively small hydraulic conductivity variations so that second order terms could be neglected in the flow and solute transport perturbation equations. It is very difficult to assess the affects of that linearization analytically, but comparisons with Monte Carlo simulations give some indication of the range of applicability of these results. Figure 7 shows a comparison of the theoretical result for the longitudinal dispersivity with numerical results from three different Monte Carlo simulations. The results of Heller (17) for a two-dimensional flow system show excellent agreement with the theory [Eq. (4.10) with $\gamma = 1$ for two-dimensional flow]. His results show that the longitudinal dispersivity increases very precisely as the first power of the variance of log hydraulic conductivity. Results of Smith and Schwartz (25) for a two-dimensional flow system are more limited but agree in terms of the general order of magnitude of the simulated dispersion coefficient. The three-dimensional results of Warren and Skiba (31) are also consistent with the theory; the correlation scale in that case would be one-half of the block size, i.e., 1/24.

For the more realistic case of statistically anisotropic log hydraulic conductivity fields, Gelhar and Axness (13) have evaluated Eq. (4.9) for several cases. A specific result for the case with horizontal and vertical anisotropy is shown in Figure 8. In this case the medium is viewed as being imperfectly stratified with a vertical correlation scale λ_3 which is relatively small compared to the horizontal scales λ_1 and λ_2. Horizontal anisotropy is introduced by using a λ_1, which is somewhat larger than λ_2. The figure illustrates, in a horizontal plane, the shape of the assumed statistical anisotropy, i.e., a constant correlation line as the

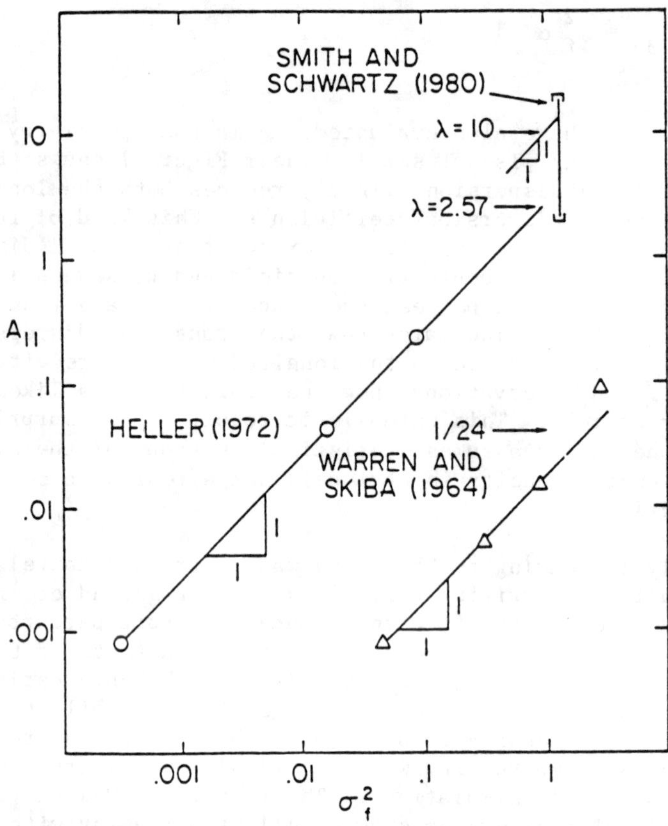

Fig. 7. Comparisons of the longitudinal macrodispersivity from the stochastic theory, Eq. (4.10), and Monte Carlo simulations [from Gelhar and Axness (13)].

dashed line in the figure. The solid line corresponds to an iso-concentration line of the mean concentration field produced as a result of an instantaneous point injection of mass in this uniform mean flow field. The parameters which were used in this calculation are based on estimates for the Mt. Simon aquifer shown in Figure 2. The resulting macrodispersivity tensor shows several important features of this kind of anisotropic medium. The ratio of horizontal transverse to longitudinal dispersivity is about 0.04 in this case, a reasonable value in terms of field observations. The vertical transverse dispersivity is found to be at least an order of magnitude smaller than the horizontal transverse dispersivity. In addition, the off diagonal term A_{12} is numerically significant and is negative. This negative component leads to the rotation of the plume in the horizontal plane in a direction opposite to that of the principal axes of the heterogeneity. This strong anisotropy of transverse dispersion is consistent with field observations.

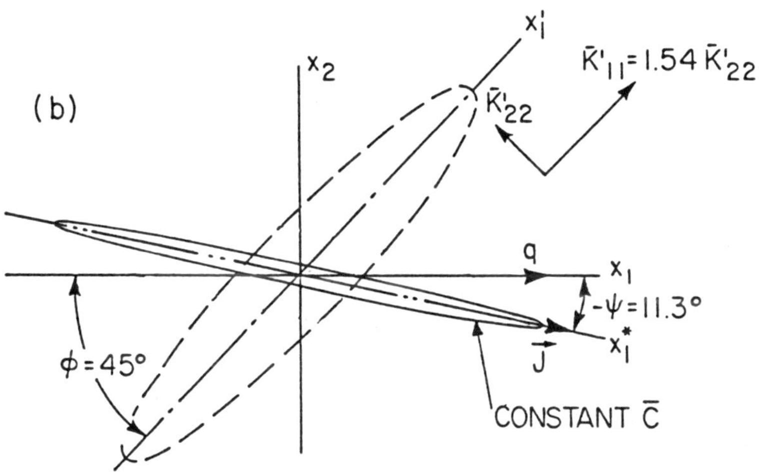

Fig. 8. Configuration of a dispersing mean plume (solid ellipse) and the aquifer heterogeneity (dashed ellipse) for the case with horizontal and vertical statistical anisotropy.
Given: $\sigma_f^2 = 2.5$, $\lambda_1 = 10\text{m}$, $\lambda_1/\lambda_2 = 5$, $\lambda_1/\lambda_3 = 20$
Find: $A_{11} = 1.0\text{m}$, $A_{22}/A_{11} = .041$, $A_{33}/A_{11} = .0014$, $A_{12}/A_{11} = -.20$

Some features of developing dispersion can be recognized by examining Eqs. (4.6) and (4.7) with finite t. Consider first the behavior for very small time; in this case using

$$\frac{1-e^{-\beta t}}{q\,n\,\beta} \to \frac{t}{qn} \quad \text{as } t \to 0$$

Eq. (4.7) becomes

$$A_{ij}(t) = \iiint_{-\infty}^{\infty} \frac{t}{qn}\, S_{q_j q_i}\, d\vec{k} = \frac{E\,(q_j' q_i')}{q^2}\,\bar{x} \qquad (4.12)$$

where $\bar{x} = qt/n$, the mean displacement. For a statistically isotropic medium, with the specific discharge variance from Eqs. (3.9) and (3.10), Eq. (4.12) produces

$$A_{11} = \frac{\sigma_{q_1}^2}{q^2}\,\bar{x} = \frac{8}{15}\,\frac{\sigma_f^2}{\gamma^2}\,\bar{x} \tag{4.13}$$

$$A_{22} = \frac{\sigma_{q_2}^2}{q^2}\,\bar{x} = \frac{1}{15}\,\frac{\sigma_f^2}{\gamma^2}\,\bar{x} \tag{4.14}$$

and consequently the ratio of transverse to longitudinal dispersivity is

$$A_{22}/\,A_{11} = 1/8 \tag{4.15}$$

Both longitudinal and transverse dispersion coefficients increase linearly with mean displacement. The results of Eqs. (4.13) and (4.14) are precisely equivalent to the small time limits for the spatial moments found by Dagan (6) using a Lagrangian analysis (see his equation 4.11).

For intermediate times, the behavior of the macrodispersivity can be evaluated very simply for the case with α, the local dispersivity, very small compared to λ. In this case we can formally set $\alpha = 0$ in Eq. (4.6) and, after interchanging the order of time and wave number space integration, the result is

$$A_{ij}(t) = (qn)^{-1} \int_{\tau=0}^{t} \left(\iiint_{-\infty}^{\infty} e^{-ik_1 u\tau}\, S_{q_j q_i}(\vec{k})\, d\vec{k} \right) d\tau \tag{4.16}$$

From Eq. (2.3) the wave number integral can now be written in terms of the covariance function as

$$A_{ij}(t) = (qn)^{-1} \int_{\tau=0}^{t} R_{q_j q_i}(-q\tau/n,\ 0,\ 0)\, d\tau \tag{4.17}$$

and noting that $R_{q_j q_i}(\vec{v}) = R_{q_i q_j}(-\vec{v})$,

$$A_{ij}(t) = \int_{\xi=0}^{\bar{x}} R_{q_i q_j}(\xi,0,0)/q^2 d\xi \tag{4.18}$$

For large mean displacement ($\bar{x} \to \infty$),

$$A_{ij}(\infty) = \int_0^{\infty} q^{-2} R_{q_i q_j}(\xi,0,0)\, d\xi$$

$$= \int_{\xi=0}^{\infty} [\iiint_{-\infty}^{\infty} e^{ik_1\xi} q^{-2} S_{q_i q_j}(\vec{k})\, d\vec{k}]\, d\xi$$

$$= \iiint_{-\infty}^{\infty} q^{-2} S_{q_i q_j}(\vec{k}) \left(\int_{\xi=0}^{\infty} e^{ik_1\xi}\, d\xi \right) d\vec{k}$$

$$A_{ij}(\infty) = \pi \iint_{-\infty}^{\infty} q^{-2} S_{q_i q_j}(0,k_2,k_3)\, dk_2 dk_3 \tag{4.19}$$

Equation (4.19) is an alternative expression for the asymptotic macrodispersivity in the case when the local dispersivity $\alpha = 0$. This form is equivalent to that of Eq. (64) in Gelhar and Axness (13).

Dagan (6) has evaluated moment expressions which are equivalent to the integral expressed by Eq. (4.18) for the case of a statistically isotropic hydraulic conductivity field. Expressing these results [his Eqs. (4.9) and (4.10)] in terms of dispersivities, i.e., derivatives of the second moment,

$$A_{11}(y)/\ A_{11}(\infty) = 1 - \frac{4}{y^2} + \frac{24}{y^4} - 8\left(\frac{1}{y^2} + \frac{3}{y^3} + \frac{3}{y^4}\right) e^{-y} \tag{4.20}$$

$$A_{22}(y)/\ A_{11}(\infty) = \frac{1}{y^2} - \frac{12}{y^4} + \left(\frac{12}{y^4} + \frac{12}{y^3} + \frac{5}{y^2} + \frac{1}{y}\right) e^{-y} \tag{4.21}$$

where $y = \bar{x}/\lambda$. Note that for large $\bar{x}/\lambda$, $A_{22} \to 0$. This result is consistent with Eq. (4.11) if $\alpha = 0$. The results of Eqs. (4.20) and (4.21) are presented graphically in Figure 9, which shows that A_{11}, the longitudinal dispersivity, increases monotonically to an asymptotic value, whereas A_{22} increases to a maximum and then decreases to zero. Consequently, the ratio of transverse to longitudinal dispersivity decreases with displacement as shown. After a mean displacement of roughly 10 correlation scales, the longitudinal dispersivity has practically attained its asymptotic value. The developing dispersion results have also been evaluated from approximate evaluations of the spectral integral in Eq. (4.7) with $\alpha/\lambda \ll 1$; those results are practically equivalent to Eqs. (4.20) and (4.21).

Also shown in Figure 9 is the asymptotic value of the ratio of transverse to longitudinal dispersivities based on Eqs. (4.10) and

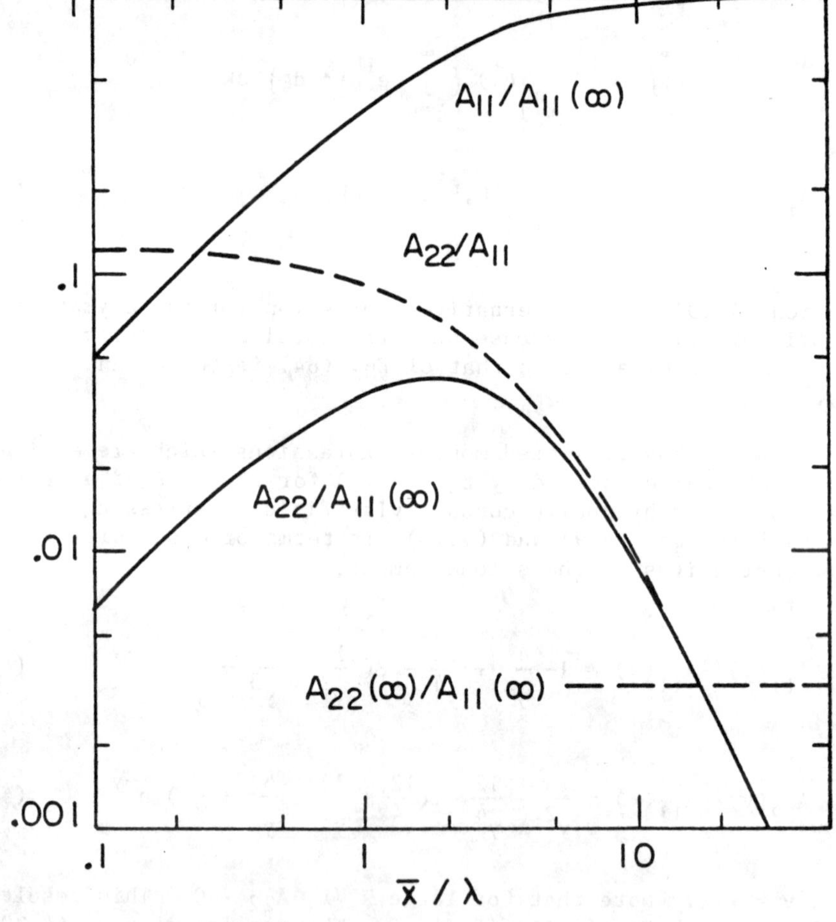

Fig. 9. Developing longitudinal (A_{11}) and transverse (A_{22}) macrodispersivities for a statistically isotropic medium with $\lambda/\alpha = 100$; $\bar{x}$ is the mean displacement.

(4.11) with λ/α = 100. Thus, it is evident that Eq. (4.21), which assumes α = 0, is not valid for indefinitely large $\bar{x}/\lambda$. Local dispersion will ultimately affect the transverse dispersivity.

It is of interest to compare the above results, which are based on the Eulerian form of the transport equation and a solution by a spectral representation, with other theoretical treatments of dispersion in random porous medium. Several studies, beginning with the classical work of Taylor (29) have considered the dispersion effect of a random velocity field with no local dispersion or diffusion. Eq. (4.17) is equivalent to Taylor's original result if the Lagrangian covariance function of the velocity field is approximated by the Eulerian covariance function evaluated at the mean displacement. In the context of flow in porous media, Dieulin et al. (10) obtained the result equivalent to Taylor's and Simmons (24), and also presented results in a one-dimensional case which are similar to Taylor's. As noted above, the results of Dagan (6), and Gelhar and Axness (13), with the extensions discussed above, are precisely equivalent, and have the advantage that they explicitly incorporate the effect of the random hydraulic conductivity field in determining the resulting specific discharge field. As noted by Dagan (5), the analysis by Tang et al. (28) is very difficult to follow in terms of mathematical development and does not seem to be consistent with the classical result or other recent theoretical developments. When the effects of local dispersion or diffusion are included, the result of Gelhar and Axness (13) for the asymptotic dispersivity, i.e., Eq. (4.9), is equivalent to the corresponding semigroup theory development by Winter et al. (32). Dieulin (9) has developed detailed comparisons of the results and approaches in the recent theoretical work on dispersion in heterogeneous porous media.

The Eulerian based spectral approach which has been outlined in this chapter, has the advantage that it is mathematically straightforward and provides direct physical insight about assumptions. The spectral approach leads directly to the form of the transport partial differential equation which governs the mean concentration, whereas other methods such as Dagan (6) require distributional assumptions in order to arrive at a governing mean equation. In fact the nonGaussian effects that are expected near the source could easily be included by considering higher order terms to the expansion of Eq. (4.5). However, it must be recognized that such refinement of the near source analysis may not be justified because the resulting mean concentration represents ensemble behavior. It is questionable whether the ensemble mean behavior is useful to quantify near source transport characteristics. This question can be addressed more quantitatively by considering the concentration variance.

4.2 Concentration Variance

The variation of the concentration around the ensemble means can be characterized by the concentration variance which is easily evaluated within the spectral framework as follows

$$\sigma_c^{\,2} = \iiint_{-\infty}^{\infty} E(dZ_c dZ_c^*) = \iiint_{-\infty}^{\infty} S_{cc}(\vec{k})d\vec{k} \tag{4.22}$$

The spectral integral is evaluated using dZ_c from the steady state version of Eq. (4.3). In this case, a special form of the spectrum for the hydraulic conductivity variation was used as follows

$$S_{ff} = \frac{4}{3}\frac{\sigma_f^2}{\pi^2}\frac{\ell^5 k^2}{(1+k^2\ell^2)^3} \tag{4.23}$$

where ℓ is a correlation scale. This spectral density function will produce a finite variance for the steady state case. The corresponding covariance function for lnK is

$$R_{ff} = \sigma_f^2\,(1-s/3)\,e^{-s} \quad , \quad s = \left|\vec{\xi}\right|/\ell \tag{4.24}$$

Ongoing research at MIT is dealing with the concentration variance under a variety of conditions [see Vomvoris and Gelhar (30)]. An analytical solution has been found by E. Vomvoris, a doctoral student at MIT, for the steady state case. That result is of the form

$$\sigma_c^{\,2} = \frac{\sigma_f^2}{\gamma^2}\left[\frac{2}{3}\frac{\ell^3}{\alpha}\left(\frac{\partial\bar{c}}{\partial x_1}\right)^2 + \frac{1}{9}\ell^2\left(\frac{\partial\bar{c}}{\partial x_2}\right)^2 + \frac{1}{9}\ell^2\left(\frac{\partial\bar{c}}{\partial x_3}\right)^2\right] \tag{4.25}$$

Equation (4.25) was developed for the case with the local dispersivity α much less than the correlation scale λ. An important feature of this result is the dependence of the concentration variance on the local dispersivity as reflected in the first term involving the longitudinal mean concentration gradient. This result indicates that the concentration variance would become very large as the local dispersivity becomes arbitrarily small. This behavior is in contrast to that of the macrodispersivity where we find that the longitudinal macrodispersivity becomes independent of the local dispersivity for very small local dispersivity. Eq. (4.25) can be viewed as a general local gradient relationship which determines the concentration variance in terms of the mean solution as reflected in the components of the mean concentration gradient. Consequently, the concentration variance will be large in areas where the mean concentration gradient is large, such as near sources of contamination. This feature can be illustrated more explicitly by considering a continuous constant strength point source which produces a steady state mean plume described by Gaussian transverse distribution of mean concentration multiplied by x_1^{-1} where x_1 is the longitudinal distance from the point source. The coefficient of variation for concentration can be found from Eq. (4.25) by dividing both sides of the equation by $\bar{c}^2$, in which case it is seen that the coefficient of variation is determined by the spatial derivatives of the logarithm of mean concentration. Thus along the center line of the plume, the coefficient of variation can be written as

$$\frac{\sigma_c}{\bar{c}} = \frac{\sigma_f \ell}{\gamma x_1} \left(\frac{2}{3} \frac{\ell}{\alpha} \right)^{1/2} \tag{4.26}$$

and for the case $\sigma_f = 1$, $\ell = 1\text{m}$, $\alpha = 1$ cm with the isotropic relationship $\gamma = \exp(\sigma_f^2/6,)$

$$\frac{\sigma_c}{\bar{c}} = \frac{7}{x_1}, \quad x_1 \text{ in m} \tag{4.27}$$

Thus for distances less than 10 meters or 10 correlation scales from the source, the coefficient of variation will be near 1 and large variability of concentration is implied near the source. However, at distances of 100 meters or more, the coefficient of variation will be a few percent. It is easily seen from Eq. (4.25) that the coefficient of variation will increase away from the center line of the plume so that the results in Eqs. (4.26) and (4.27) can be viewed as the minimum concentration variability at a given longitudinal distance from the source. Of course, the implied near source behavior with very large coefficient of variation must be viewed as only a qualitative feature since the perturbation approach used in developing Eq. (4.25) was based on the assumption that the concentration variations were relatively small.

These concentration variance results show that there can be large uncertainty in solute transport predictions based on the classical convective dispersion equation which is equivalent to the mean transport equation developed here. The complications with the classical transport equation occur not only because the dispersion coefficients are displacement dependent, as illustrated in Figure 9, but also because of the large variability of the concentration that is implied in the stochastic approach. Essentially, the ergodic hypothesis is not applicable in this near source region. The concentration variance, as discussed here, is best thought of as the variation among different realizations of statistically identical aquifers. Physically then the situation is that the detailed distribution of permeability of the actual aquifer, i.e., a given single realization, has a very strong effect on the local direction of migration of contamination from a localized source. If one is attempting to make predictions within a few correlation scales of the localized source of contamination, it is obviously necessary to measure the local variations in hydraulic conductivity and include these in a deterministic model of the situation. The most direct way to obtain such observation is to carry out a small-scale tracer test in the region of interest, thus directly observing the migration of a solute. However, the most challenging types of contamination problems involve predictions over much larger scales where such direct measurements are not possible. It is under these conditions that a stochastic approach has the greatest utility. From the theoretical results for the concentration variance, it can be expected that a classical transport equation corresponding to the mean stochastic solution will be useful, but there will be variations in concentration around the mean as described by a relationship such as Eq. (4.25).

The possiblity of such concentration variations needs to be recognized when applying the classical transport equation to field situations. First of all, a result such as Eq. (4.25) could serve as a calibration target which could be used to judge the significance of differences between the results of a numerical simulation and field observations of the plume. Also for regulatory purposes, it may be more reasonable to consider, say, the mean concentration plus two standard deviations rather than to draw conclusions from a classical transport model which considers only the mean concentration. Also, we can expect the kind of variability that is identified here to play an important role in the design of systems to monitor contamination plumes. In that case, the spatial correlation structure or the covariance function of the concentration will be the key characteristic.

Dagan (6) has also considered concentration variance, but his analysis is restricted to the near source region where the effects of local dispersivity can be neglected. He also focuses on the behavior of two-dimensional depth averaged plumes which, as will be seen in the next section, are frequently not physically realistic. Dagan contends that the ergodic hypothesis will be applicable and that the ensemble mean behavior will be meaningful near the source if the initial lateral dimensions of the contamination source are much larger than the correlation scale of the aquifer. However, this observation appears to correspond to the trivial case in which the concentration does not change from that of the initial injection concentration.

4.3 Field Observations

Several field studies of solute movement in aquifers have produced results which relate to some of the theoretical features discussed above. The natural gradient tracer test at the Borden site in Canada [Sudicky et al. (27)] clearly shows that near the source the longitudinal dispersion coefficient increases with displacement distance and, in general, it is found that vertical mixing is very limited in these layered outwash sands. The second experiment at this same site [Freyberg et al. (11), Sudicky, (26)] also shows very limited vertical mixing but, in this case, the experiment has been carried out to a distance where it now appears that the dispersivities have attained their asymptotic values.

Several contamination plumes also show features which are consistent with the stochastic theory although, in such cases, the interpretation is not as clear-cut because contamination history is usually not well-defined. The classical chromium plume in outwash sand and gravel on Long Island in New York [Perlmutter and Lieber, (23)] involved detailed three-dimensional monitoring and clearly showed that mixing was not extending over the vertical thickness of the aquifer even though the plume had migrated up to a kilometer horizontally. More recent detailed monitoring of the 40 year old sewage plume in a sand and gravel aquifer on Cape Cod in Massachusetts [LeBlanc (20)] also shows strong vertical stratification of the plume even though contamination has migrated over a mile from the source. Many other recent unpublished investigations have shown similar limited vertical mixing in contamination plumes from hazardous waste sites.

The field observations discussed above emphasize the importance of the three-dimensional structure of contamination plumes in real aquifers. The very limited vertical mixing is consistent in a qualitative sense with calculations such as those summarized in Figure 8 based on the stochastic theory. It is important to consider this three-dimensionality of plumes when designing monitoring systems to locate groundwater contamination. Frequently three-dimensional numerical modeling will be necessary in order to realistically model such field situations. It is also important to recognize this three-dimensionality when designing schemes to control existing contamination in aquifers or restore the quality of contaminated aquifers. My experience is that the three-dimensionality of plumes will be important, except when contaminants are artificially introduced over the entire thickness of an aquifer, for example, through an injection well. Although two-dimensional flow descriptions seem to be useful for many practical groundwater flow problems, it is unlikely that a two-dimensional approach will be adequate for most contamination problems.

A recent detailed compilation of field observations of dispersion in aquifers has been developed by Gelhar et al. (15). Results of that study are summarized in Figure 10 which presents the longitudinal dispersivity determined from each test as a function of the scale of the experiment, that is, the mean displacement distance involved in the experiment. The scale of observation ranges from one meter to 100 kilometers, and dispersivities from one centimeter

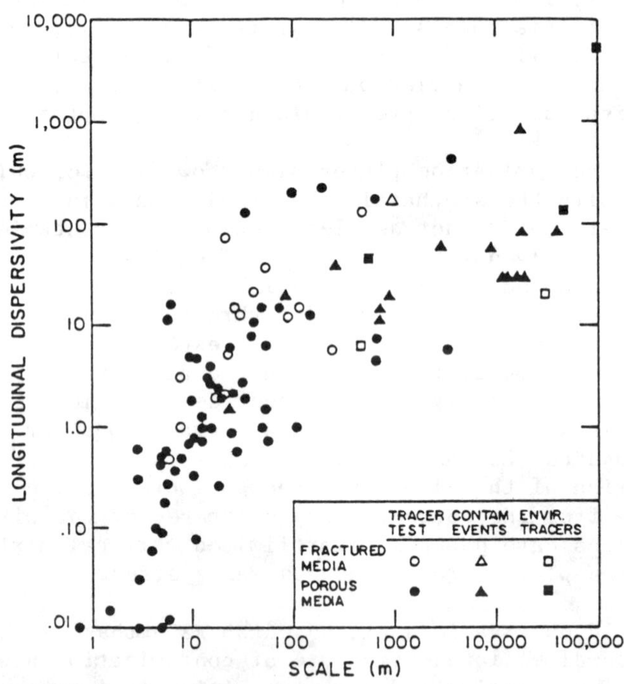

Fig. 10 Longitudinal dispersivity data plotted versus overall displacement scale for different types of observations and media [after Gelhar et al. (15)].

up to 10 kilometers are reported. The data were also segregated according to whether the observations were a tracer test with a known controlled input of solute, contamination events where the nature of the source may be ill-defined, or environmental tracers where natural processes control the nature of the input. Most of the tracer tests have been carried out on a relatively small scale, say, less than a kilometer, whereas the largest scale observations are those based on environmental tracers. The various sites were also classified according to whether the aquifer was a more classical granular material that would be viewed as a porous medium or a fractured medium. Data in Figure 10 do not seem to indicate that there is a distinctive difference between the dispersion characteristics of porous and fractured media.

Part of the review also involved a detailed quality assessment of all of the dispersivity data. Considered in this assessment were the configuration or type of test (e.g., natural gradient test, radial flow test, etc.), type of observations (three-dimensional point sampling, fully penetrating wells, etc.) and the method of interpretation of the data (e.g., spatial moments, breakthrough curve analysis or numerical model calibration). Figure 11 summarizes the results of this quality evaluation of the

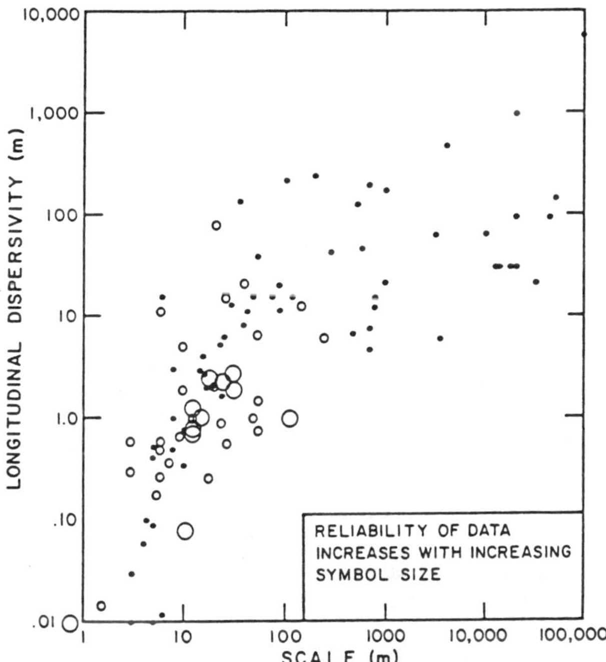

Fig. 11. Field longitudinal dispersivity data classified according to reliability [after Gelhar et al. (15)].

longitudinal dispersivity data. The largest symbols in the figure identify those observations of relatively high reliability whereas the smallest points represent observations in which minimal confidence can be placed. It can be seen that most of the reliable information falls within scale less than about 100 m. Most of the data at scales over a kilometer are of low reliability, so that it is not possible to conclude that the longitudinal dispersivity increases indefinitely with increasing scale as might be inferred from Figure 10. Clearly, there is a need for carefully designed large-scale experiments to resolve questions regarding the behavior of dispersion at these larger scales.

5. UNSATURATED FLOW

5.1 Spatial Variability of Parameters

It can be expected that unsaturated flow processes will be influenced in more complicated ways by spatial variability because of the additional flow parameters and the fundamental nonlinearity of the flow processes. The purpose here is to simply outline the general approach that has been used to analyze some characteristics of unsaturated flow and illustrate through a few summary results some of the new insights to large scale behavior of the unsaturated zone which are developing through the use of stochastic methods. The details of the development are available in Yeh et al. (33, 34, 35) and Mantoglou and Gelhar (22).

A commonly used model of hydraulic conductivity in unsaturated flow is

$$K = K_s e^{-\alpha\psi} \tag{5.1}$$

where K_s is the saturated hydraulic conductivity, α is a soil parameter and ψ is the soil water suction (tension). Eq. (5.1) indicates a linear relationship between lnK and ψ, the tension,

$$\ln K = \ln K_s - \alpha\psi \tag{5.2}$$

This form is convenient because the logarithmic term occurs naturally in the flux portion of the flow equation. Figures 12 and 13 illustrate lnK versus ψ relationship for two different field soils; each individual curve represents a different location in the field.

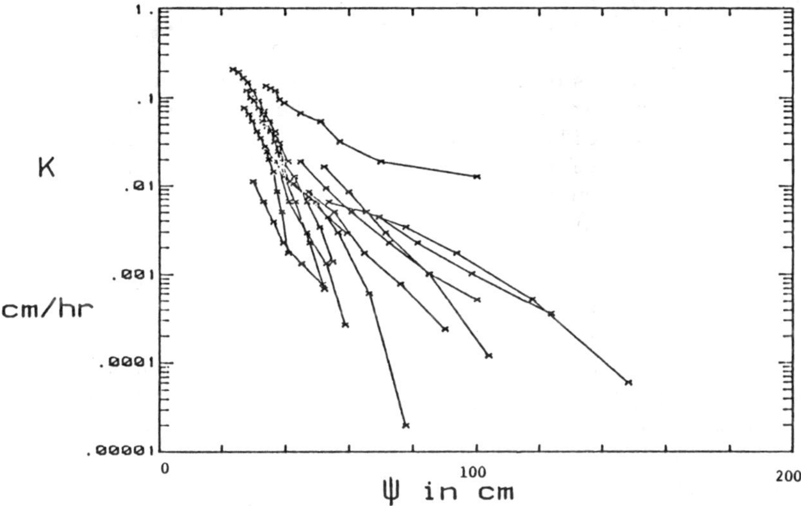

Fig. 12. Unsaturated hydraulic conductivity as a function of tension for Maddock sandy loam [from Yeh et al. (35)].

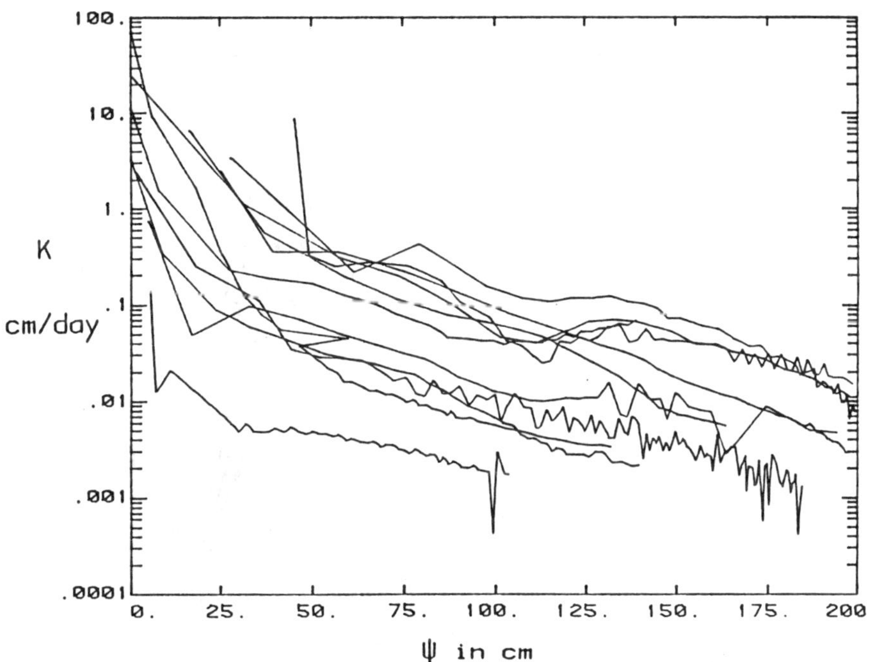

Fig. 13. Unsaturated hydraulic conductivity of Panoche silty clay loam [from Yeh et al. (35)].

The Maddock sandy loam shows a much larger variation of the parameter α than does the Panoche silty clay loam.

When unsteady unsaturated flows are analyzed, it is also necessary to consider the variation of the moisture retention properties of the soil. Figure 14 shows several moisture retention curves from a given field indicating that there is significant variation of the slopes of these curves, i.e., the specific moisture capacity of the soil.

In order to analyze the effects of the variability of these hydraulic properties, each of the parameters is represented as a three-dimensional random field with presumed covariance structure. The general approach is very similar to that in saturated flow except that the governing equation is different, and there are a greater number of random parameters that must be considered. The spectral representation approach is used to solve the resulting approximate perturbation equation describing the tension field, and from that result a tension variance is determined. Effective properties are found by taking the expected value of the original flow equation and using the perturbation solution to evaluate the

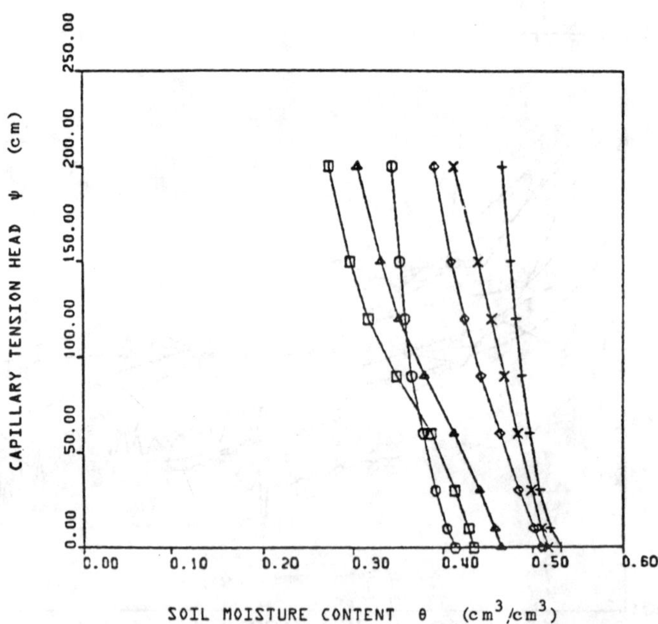

Fig. 14. Capillary tension head ψ versus soil moisture content θ for the Panoche silty clay loam. Each curve corresponds to a different spatial location.

necessary cross-correlation terms which occur. Some of the results are summarized briefly in the following section.

5.2 Results

The variance of capillary pressure head (tension) has been evaluated for steady flow in statistically anisotropic media in a series of papers by Yeh et al. (33, 34). For steady vertical mean infiltration with the horizontal correlation scales $\lambda_2 = \lambda_3 = \lambda$, much greater than λ_1, the vertical correlation scale, they found the following form for the head variance

$$\sigma_h^2 = (\sigma_f^2 + \sigma_a^2 H^2)\, \lambda_1^2\, g(A\, \lambda_1, \lambda/\lambda_1) \tag{5.3}$$

where $\sigma_f^2 = \mathrm{var}\,(\ln K_s)$, $\sigma_a^2 = \mathrm{var}\,(\alpha)$, $H = E(\psi)$, $A = E(\alpha)$, and the function g has the dependence shown in Figure 1 of Yeh et al. (34). Eq. (5.3) shows that the head variance depends on the square of the correlation scale as in the saturated case. No dependence on the mean hydraulic gradient is evident because, in this case, the mean hydraulic gradient is taken to be 1. Note also that it has been assumed that saturated conductivity and α are statistically independent to arrive at this form.

An important new feature of Eq. (5.3) is the strong dependence of the head variance on the mean pressure or tension. Essentially, drier soils with higher mean tension are expected to produce much higher pressure variability. Qualitative indications of this behavior have been reported by Yeh et al. (36), but quantitative confirmation of this behavior is not presently available.

The tensorial behavior of effective hydraulic conductivity for steady unsaturated flow has also been analyzed by Yeh et al. (35). Figure 15 shows the ratio of horizontal to vertical effective hydraulic conductivities for the two soils illustrated in Figures 12 and 13 as a function of the mean tension. This result shows that the Maddock sandy loam, with its much larger variance of α, produces a strong dependence of the anisotropy ratio on the mean tension. On the other hand, the Panoche silty clay loam shows only a very mild dependence of the anisotropy ratio on tension. The behavior of the Maddock soil indicates the possibility of substantial preferential horizontal movement of moisture from localized sources, especially in stratified soils under relatively dry conditions. Yeh et al. (35) discuss the importance of this horizontal spreading phenomena in the context of waste disposal applications.

In the case of unsteady unsaturated flow, the behavior of the effective hydraulic conductivities is much more complicated as has

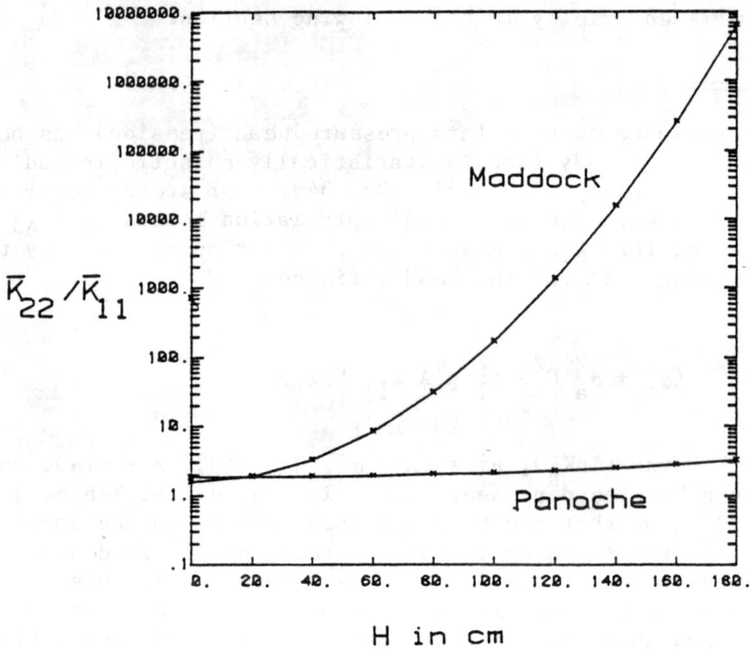

Fig. 15. Anisotropy of effective conductivity of the Maddock and Panoche soils as a function of mean tension; $\bar{K}_{11}$ is the vertical conductivity and $\bar{K}_{22}$ the horizontal.

been demonstrated by the analysis of Mantoglou and Gelhar (22). Again for the case of vertical mean infiltration through a stratified soil as discussed above, it has been found that the effective conductivities exhibit both anisotropy and hysteresis. This behavior is illustrated for the case of the Maddock soil in Figure 16. Hysteresis is reflected in the fact that the effective hydraulic conductivities are dependent on the time rate of change of the mean tension. Figure 16 shows that, under wetting conditions, the horizontal hydraulic conductivity is much lower than the vertical conductivity illustrating the kind of anisotropy that is indicated by Figure 15 for the steady state case. However, for drying conditions, the two effective hydraulic conductivities are more nearly equal and isotropic behavior is indicated. Mantoglou and Gelhar have also shown that the effective moisture retention curve is hysteretic under these conditions.

Dispersive solute transport in heterogeneous unsaturated soils can be analyzed using the approach of Chapter 4. In the case of steady unsaturated flow, Eq. (4.9) for the macrodispersivity tensor

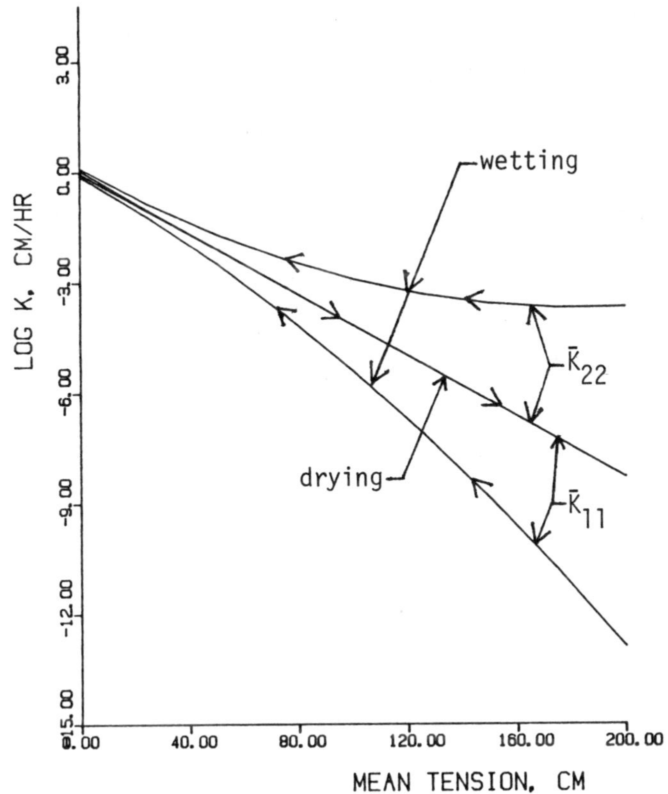

Fig. 16. Effective hydraulic conductivities of Maddock sandy loam for unsteady flow showing predicted hysteresis; $\bar{K}_{11}$ is the vertical conductivity, and $\bar{K}_{22}$ the horizontal.

is applicable, but the specific discharge spectrum must be determined from the solution of the unsaturated flow equation. Some initial analysis of this steady flow problem has been done by Mantoglou and Gelhar (22). They find a longitudinal macrodispersivity for steady vertical mean infiltration in a stratified unsaturated soil in the form

$$A_{11} = \frac{(\sigma_f^2 + \sigma_a^2 H^2)\lambda_1}{\gamma^2 (1 + A\alpha_T)} \tag{5.4}$$

where $\gamma = \bar{K}_{11} / K_m$ with $\bar{K}_{11}$ the vertical effective unsaturated conductivity and $K_m = K_\ell \exp(-AH)$, $\ln K_\ell = E(\ln K_s)$. Of course, the mean vertical hydraulic gradient is again unity. Note that this result reduces to the saturated case, Eq. (4.10), when the expected value of α, $A = 0$, and the variance of $\sigma_a^2 = 0$.

Eq. (5.4) shows that for unsaturated conditions, the longitudinal dispersivity can be strongly dependent on the mean tension and consequently the moisture content. The factor γ in that equation also will depend on the mean tension, usually in such a way as to cause the dispersivity to increase with increasing mean tension.

Field data on dispersion under large-scale unsaturated conditions are very limited, and it is not possible to make quantitative comparisons with the theoretical results described above. Gelhar et al. (15) have developed a compilation of dispersivity data for the unsaturated zone. A summary of their results is shown in Figure 17, a graph showing the dependence of longitudinal dispersivity on the scale of the experiment. Although these data are much more limited than in the saturated case (Figure 10), there seems to be a trend of increasing dispersivity with increasing scale of the experiment. There is a need for carefully controlled large-scale

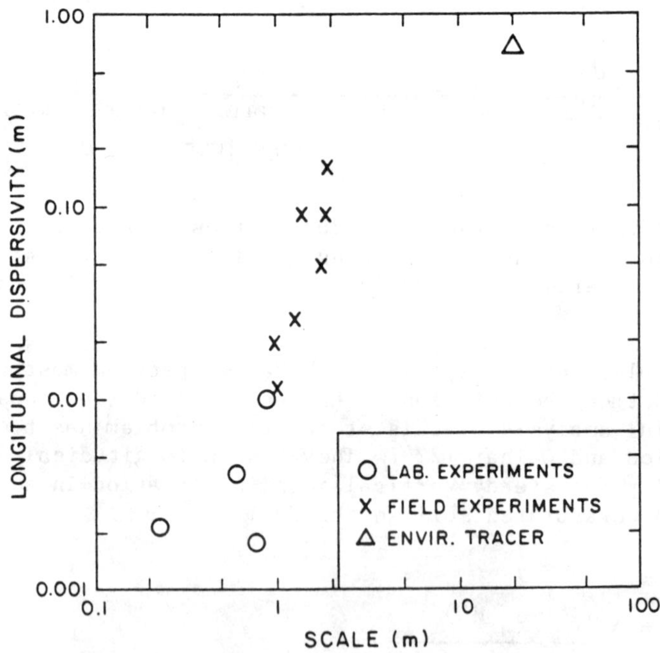

Fig. 17. Observed longitudinal dispersivities for unsaturated flow as a function of overall displacement distance.

solute transport experiments in the unsaturated zone in order to clarify the nature of the dispersive transport process.

The theoretical understanding of large scale dispersion in unsaturated systems is far from complete. There is a need to consider unsteady flow conditions as well as the developing dispersion characteristics relatively near the source of contamination. Other stochastic analyses of solute transport under unsaturated conditions have assumed purely one-dimensional vertical flow [Dagan and Bresler (8), Bresler and Dagan (2)], or a zero-dimensional black box kind of approach [Jury and Stolzy (19)]. Such models can be fit to field observations at relatively small scale, but it is difficult to see how these models can be extrapolated to larger scales, including the effects of real three-dimensional heterogeneity and multidimensional mean flow conditions.

6. COMMENTS AND DIRECTIONS

From the developments in this paper, a number of general features of the stochastic approach are evident. This approach provides not only an analytical methodology but also a basic framework for understanding and quantifying large-scale transport processes. The treatment of macrodispersion in statistically anisotropic media predicts the complete macrodispersivity tensor and provides a theoretical basis for the very limited vertical mixing which has been observed in stratified sediments. Also, the discovery of the anisotropy and hysteresis of effective hydraulic conductivity under unsteady, unsaturated flow conditions is stimulating critical reassessments of the approach to modeling waste disposal activities in the unsaturated zone.

Another important feature of the stochastic approach presented here is its essentially predictive character. The approach, of course, introduces plausible assumptions about the statistical structure of the heterogeneity but does not require direct fitting to observations of the dependent variable (e.g., head or concentration). In this sense the stochastic approach provides a rational framework for extrapolating to time or space scales at which direct observations of such dependent variables are not feasible. Stochastic theory also provides simple, local variance relationships for these dependent variables, and thereby provides a basis for judging reliability of large-scale mean models.

In view of this predictive character, it is possible to design field experiments which can be used to evaluate this theoretical approach. Essentially one needs to measure the spatial variability of hydraulic conductivity and other pertinent hydraulic parameters in enough detail to estimate the input covariance properties required for the stochastic theory. Then it is possible to make independent predictions of the field scale behavior which can be compared with subsequent observations of the dependent variables. Such experiments are under way for the case of saturated zone solute transport [Freyberg et al. (11), LeBlanc (21), Betson et al. (1)], and experiments are currently being designed for the unsaturated zone.

An important current limitation of the overall stochastic framework outlined here is the difficulty of determining the required statistical parameters from a reasonable amount of field data. There is a need to develop more effective statistical methods of estimating these parameters from the limited field data that is available. A recent study of aquifer transmissivity data by Hoeksema and Kitanidis (18) is an example of efforts in this direction. In addition, methods of using indirect information, such as geologic or geomorphic data or observations, or geophysical data, need to be developed systematically to apply to the problem of determining the statistical parameter.

All of the analyses described here, as well as most other stochastic methods, rely on at least an implicit assumption of relatively small perturbations due to the heterogeneity. There is a need for a theoretical analysis and a numerical experimentation in order to evaluate the effects of the small amplitude assumption and develop techniques to accurately treat large amplitude effects. There have been some recent efforts in this direction for the saturated flow situation [Gutjhar (16), Dagan (7)], and these results seem to indicate that the linearized approach is surprisingly robust. In spite of these encouraging results, it can not be assumed that a small perturbation linearization will always be adequate. Especially in the case of extremely heterogeneous media, it may be necessary to develop entirely new approaches which rely on the random characterization of the geometry of heterogeneities. In any case, carefully designed Monte Carlo experiments are required to definitively evaluate the large amplitude effects.

From the discussion of developing dispersion characteristics near a source of contamination, it is evident that the ensemble theory is of limited value in treating that situation. Techniques need to be developed to characterize dilution and mixing in such near source regions where an expanding plume is affected by heterogeneities of larger and larger size as it migrates away from the source. It seems that an approach which incorporates the effects of conditioning due to head or hydraulic conductivity measurements would be useful. Dagan (6) has considered such effects briefly, but for the situations that he considered there seemed to be a very small effect due to conditioning.

The general stochastic framework outlined in this paper has potential applications in many other kinds of problems. Problems such as flow through heterogeneously reactive media, the flow of variable density or viscosity flow, and multiphase flows in porous media can be addressed with this kind of approach. This may also be a useful framework for considering some problems of flow in fractured media; Brown (3) has done some initial work in this area considering flow and transport in a variable aperture single fracture. Of course, the approach could also be used to describe the large-scale behavior of other heterogeneous continua.

7. ACKNOWLEDGMENTS

The research described in this paper has been supported in part by the National Science Foundation (Grant Nos. ECE-8311786 and INT-8212574), the U.S. Environmental Protection Agency (Cooperative Agreement No. CR-811135), the U.S. Nuclear Regulatory Commission (Contract No. NRC-04-83-174), and the Electric Power Research Institute (through the Tennessee Valley Authority Contract No. TV-61664A). However, the views or opinions expressed herein are those of the author and do not necessarily reflect those of these organizations.

8. REFERENCES

1. Betson, R. P., Boggs, J. M., Young, S. C., Waldrop, W. R., and Gelhar, L. W., "Macrodispersion Experiment (MADE): Design of a Field Experiment to Investigate Transport Processes in a Saturated Groundwater Zone," *Research Project 2485-5, EPRI EA-4082*, prepared for Electric Power Research Institute, Palo Alto, CA, 104 pp., June 1985.
2. Bresler, E. and Dagan, G., "Unsaturated Flow in Spatially Variable Fields 3. Solute Transport Models and Their Application to Two Fields," *Water Resources Research*, Vol. 19, No. 2, April, 1983, pp. 429-435.
3. Brown, D.M., "Stochastic Analysis of Flow and Solute Transport in a Variable-Aperture Rock Fracture," S.M. Thesis, Department of Civil Engineering, Massachusetts Institute of Technology, Cambridge, Massachusetts, 1984.
4. Dagan, G., "Models of Groundwater Flow in Statistically Homogeneous Porous Formations," *Water Resources Research*, Vol. 15, No. 1, Feb., 1979, pp. 47-63.
5. Dagan, G., "Comment on 'Stochastic Modeling of Mass Transport in a Random Velocity Field,'" by D. H. Tang, F. W. Schwartz, and L. Smith, (Paper 3W0723), *Water Resources Research*, Vol. 19, No. 4, Aug., 1983, pp. 1049-1052.
6. Dagan, G., "Solute Transport in Heterogeneous Porous Formations," *Journal of Fluid Mechanics*, Vol. 145, Aug., 1984, pp. 151-177.

7. Dagan, G., "A Note on Higher Order Corrections of the Head Covariances in Steady Aquifer Flow," *Water Resources Research*, Vol. 21, No. 4, April, 1985, pp. 573-578.
8. Dagan, G., and Bresler, E., Unsaturated Flow in Spatially Variable Fields 1. Derivation of Models of Infiltration and Redistribution," *Water Resources Research*, Vol. 19, No. 2, April, 1983, pp. 413-420.
9. Dieulin, A., "L'effet d'Echelle en Milieu Poreux," *Rapport CIG LHM/RD/84/17, CEA, CCE*, Ecole des Mines de Paris, Fontainebleau, France, 1984.
10. Dieulin, A. G., Matheron, G., de Marsily, G., and Beaudoin, B., "Time Dependence of an 'Equivalent Dispersion Coefficient' for Transport in Porous Media," *Flow and Transport in Porous Media*, A. Verruijt and F. B. J. Barends, eds., Balkema, Rotterdam, the Netherlands, 1981, pp. 199-202.
11. Freyberg, D. L., McKay, D. M., and Cherry, J. A., "Advection and Dispersion in an Experimental Groundwater Plume," *Proceedings of the Hydraulics Division Specialty Conference in Frontiers in Hydraulics Engineering*, American Society of Civil Engineers, July, 1983, pp. 36-41.
12. Gelhar, L.W., "Stochastic Analysis of Flow in Heterogeneous Porous Media, *Fundamentals of Transport Phenomena in Porous Media*, J. Bear and M. Y. Corapcioglu, eds., Martinus Nijhoff Publishers, Dordrecht, the Netherlands, 1984, pp. 675-717.
13. Gelhar, L. W. and Axness, C. L., "Three-Dimensional Stochastic Analysis of Macrodispersion in Aquifers," *Water Resources Research*, Vol. 19, No. 1, Feb., 1983, pp. 161-180.
14. Gelhar, L. W., Gutjahr, A. L., and Naff, R. L., Stochastic Analysis of Macrodispersion in a Stratified Aquifer," *Water Resources Research*, Vol. 15, No. 6, Dec., 1979, pp. 1387-1397.
15. Gelhar, L. W., Mantoglou, A., Welty, C., and Rehfeldt, K. R. "A Review of Field Scale Physical Solute Transport Processes in Saturated and Unsaturated Porous Media," *Research Project 2485-5, EPRI EA-4190*, Electric Power Research Institute, Palo Alto, California, Aug., 1985, 116 pp.
16. Gutjahr, A., "Stochastic Models of Subsurface Flow: Log Linearized Gaussian Models are 'Exact' for Covariances," *Water Resources Research*, Vol. 20, No. 12, Dec., 1984, pp. 1909-1912.
17. Heller, J. P., "Observations of Mixing and Diffusion in Porous Media," *Proceedings of the Second Symposium on Fundamentals of Transport Phenomena in Porous Media*, International Association of Hydraulic Research and International Soil Science Society, Vol. 1, 1972, pp. 1-26.
18. Hoeksema, R. J., and Kitanidis, P. K., "Analysis of the Spatial Structure of Properties of Selected Aquifers," *Water Resources Research*, Vol. 21, No. 4, April, 1985, pp. 563-572.

19. Jury, W. A., and Stolzy, L. H., "A Field Test of the Transfer Function Model for Predicting Solute Transport," *Water Resources Research*, Vol. 18, No. 2, April, 1982, pp. 369-375.
20. LeBlanc, D. R., "Sewage Plume in a Sand and Gravel Aquifer, Cape Cod, Massachusetts," *U.S. Geological Survey Water Supply Paper 2018*, 1984.
21. LeBlanc, D. R., ed., "Movement and Fate of Solutes in a Plume of Sewage Contaminated Groundwater, Cape Cod, Massachusetts: U.S. Geological Survey, Toxic Waste Groundwater Contamination Program" *Open File Report 84-475*, paper presented at the Toxic Waste Technical Meeting, Tucson, Arizona, March, 1984, 180 pp.
22. Mantoglou, A., and Gelhar, L. W., "Large Scale Models of Transient Unsaturated Flow and Contaminant Transport Using Stochastic Methods," *Technical Report No. 299*, Ralph M. Parsons Laboratory, Department of Civil Engineering, Massachusetts Institute of Technology, Cambridge, Massachusetts, Jan., 1985.
23. Perlmutter, N. M., and Lieber, M., "Disposal of Plating Wastes and Sewage Contaminants in the Groundwater and Surface Water, South Farmingdale-Massapequa Area, Nassau County, New York," *U.S. Geological Survey Water Supply Paper 1879-G*, 1970.
24. Simmons, C. S., "A Stochastic-Convective Transport Representation of Dispersion in One-Dimensional Porous Media Systems," *Water Resources Research*, Vol. 18, No. 4, Aug., 1982, pp. 1193-1214.
25. Smith, L., and Schwartz, F.W., "Mass Transport 1. A Stochastic Analysis of Macroscopic Dispersion," *Water Resources Research*, Vol. 16, No. 2, April, 1980, pp. 303-313.
26. Sudicky, E., "Spatial Variability of Hydraulic Conductivity at the Borden Tracer Test Site," *Preprints of the Symposium on the Stochastic Approach to Subsurface Flow*, G. de Marsily, ed., Fontainebleau, France, June 1985, pp. 150-169.
27. Sudicky, E.A., Cherry, J.A., and Frind, E.O., "Hydrogeological Studies of a Sandy Aquifer at an Abandoned Landfill 4. A Natural Gradient Dispersion Test," *Journal of Hydrology*, Vol. 63, 1983, pp. 81-108.
28. Tang, D. H., Schwartz, F. W., and Smith, L., "Stochastic Modeling of Mass Transport in a Random Velocity Field," *Water Resources Research*, Vol. 18, No. 2, April, 1982, pp. 231-244.
29. Taylor, G. I., "Diffusion by Continuous Movements," *Proceedings*, London Mathematical Society, Vol. 20, 1921, pp. 196-212.
30. Vomvoris, E. G. and Gelhar, L. W., "Transport in Aquifers: Second Moment Estimation," *EOS, Transactions*, American Geophysical Union, Vol. 65, No. 45, Nov. 6, 1984, p. 892.
31. Warren, J. E., and Skiba, F. F., "Macroscopic Dispersion," *Transactions*, American Institute of Mining, Metallurgy and Petroleum Engineers, Vol. 231, 1964, pp. 215-230.

32. Winter, C. L., Newman, C. M., and Neuman, S. P., "A Perturbation Expansion for Diffusion in a Random Velocity Field," *SIAM Journal of Applied Mathematics*, Vol. 44, No. 2, April, 1984, pp. 411-424.
33. Yeh, T.-C. Jim, Gelhar, L. W., and Gutjahr, A. L., "Stochastic Analysis of Unsaturated Flow in Heterogeneous Soils 1. Statistically Isotropic Media," *Water Resources Research*, Vol. 21, No. 4, April, 1985, pp. 447-456.
34. Yeh, T.-C. Jim, Gelhar, L. W., and Gutjahr, A. L., "Stochastic Analysis of Unsaturated Flow in Heterogeneous Soils, 2. Statistically Anisotropic Media with Variable α," *Water Resources Research*, Vol. 21, No. 4, April, 1985, pp. 457-464.
35. Yeh, T.-C. Jim, Gelhar, L. W., and Gutjahr, A. L., "Stochastic Analysis of Unsaturated Flow in Heterogeneous Soils, 3. Observations and Applications," *Water Resources Research*, Vol. 21, No. 4, April, 1985, pp. 465-471.
36. Yeh, T.-C. Jim, Gelhar, L. W., and Wierenga, P. J, "Observation of Spatial Variability of Soil-Water Pressure in a Field Soil," *Soil Science*, In press, 1986.

9. LIST OF SYMBOLS

A_{ij}	Macroscopic dispersivity tensor
c	Solute concentration
E(y)	Expected value of $y(=\bar{y})$
F	E(lnK)
f	lnK - E(lnK)
H	Mean piezometric head, $E(\phi)$
h	Perturbation in piezometric head
J_i	Mean hydraulic gradient
K	Hydraulic conductivity
$\bar{K}_{ij}$	Effective conductivity tensor
K_ℓ	exp[E(lnK)]
k_i	Wave number
q_i	Specific discharge
R_{ff}	Covariance function of f
S_{ff}	Spectrum of f
x_i	Spatial coordinates
λ_i	Integral correlation scales
σ_y^2	Variance of y
ϕ	Piezometric head
ψ	Capillary pressure head

UNCERTAINTY ASSESSMENT FOR FLUID FLOW AND CONTAMINANT TRANSPORT MODELING IN HETEROGENEOUS GROUNDWATER SYSTEMS

R. William Nelson, Elizabeth A. Jacobson, and William Conbere

UNCERTAINTY ASSESSMENT FOR FLUID FLOW AND CONTAMINANT TRANSPORT MODELING IN HETEROGENEOUS GROUNDWATER SYSTEMS

R. William Nelson, Elizabeth A. Jacobson, and William Conbere

Pacific Northwest Laboratory
Earth Sciences Department
Hydrology Section
Richland, Washington 99352, USA

ABSTRACT

There is a growing awareness of the need to quantify uncertainty in groundwater flow and transport model results. Regulatory organizations are beginning to request the statistical distributions of predicted contaminant arrival to the biosphere, so that realistic confidence intervals can be obtained for the modeling results. To meet these needs, methods are being developed to quantify uncertainty in the subsurface flow and transport analysis sequence. A method for evaluating this uncertainty, described in this paper, considers uncertainty in material properties and was applied to an example field problem.

Our analysis begins by using field measurements of transmissivity and hydraulic head in a regional, parameter estimation method to obtain a calibrated fluid flow model and a covariance matrix of the parameter estimation errors. The calibrated model and the covariance matrix are next used in a conditional simulation mode to generate a large number of 'head realizations'. The specific pore water velocity distribution for each realization is calculated from the effective porosity, the aquifer parameter realization, and the associated head values. Each velocity distribution is used to obtain a transport solution for a contaminant originating from the same source. The results are contaminant outflow arrival times for all realizations from which statistical distributions of the arrival times can be calculated. The confidence intervals for the arrival times of contaminant reaching the biosphere are obtained from these statistical distributions.

1. INTRODUCTION

The uncertainty in groundwater flow and transport modeling should be evaluated for subsurface contamination problems. The specific need for an uncertainty analysis depends on the characteristics of the particular system being analyzed. These system characteristics include the system complexity, the uniformity of the results, the magnitude of expected deviations from present conditions, and the importance or seriousness of possible errors in prediction. For example, when the systems are moderately simple, results change little from one situation to the next, deviations from expected conditions are small, and the consequences are of minor seriousness or of low perceived risk, then estimates of uncertainty may not be needed. However, when the systems involve complicated interacting factors, results are variable and so are the interactions, deviations are large, and the consequences are serious or may involve high perceived risk, then estimates of uncertainty are needed and may play an important part in the decision-making process.

Regulatory organizations are emphasizing the need to quantify uncertainty. They are beginning to request the statistical distributions of contaminant arrival at the biosphere, so that confidence intervals can be provided. For example, a draft of proposed U.S.Federal regulations [Draft 40 CFR 191.16 paragraph A, (5)] requests that the confidence intervals be provided for the scenario arrival results to the biosphere. In other words, the arrival results at the biosphere from the assessment modeling should provide both the expected results as well as the statistical distribution from which confidence intervals can be estimated.

The purpose of this work is to provide a groundwater analysis sequence that considers and quantifies, in as much detail as possible, the uncertainties involved in evaluating subsurface contamination scenarios. A significant part of the work involves combining appropriate stochastic and statistical tools with more traditional, physically based, deterministic models. Combining these methods yields an efficient analysis sequence that provides both the primary results and the corresponding uncertainty estimates. The analysis sequence is described in this paper together with the first field site application considering the uncertainty in aquifer parameters. This first application does not consider boundary and source uncertainties.

The analysis sequence begins by characterizing the system through appropriate measurements of flow and transport parameters and application of geostatistical techniques. The flow analysis uses the field measurements to obtain a statistically calibrated fluid flow model from a parameter estimation (inverse) technique which yields both estimates of the model parameters and the associated estimation errors. The calibrated flow model and the covariance matrix of parameter estimation errors are used to generate a large number of transmissivity and corresponding head realizations through conditional simulation. The specific pore velocity distribution for each realization is next calculated from the effective porosity,the transmissivity realization, and the corresponding head values. Each velocity distribution is used to obtain a transport solution for the contaminant originating from the same waste source for all realizations. The results are contaminant arrival times for all realizations from which statistical distributions for the arrival times are calculated. These statistical distributions are then used with the summary methods (i.e., contaminant arrival curves) to provide confidence intervals for the arrival times of contamination reaching the biosphere.

The analysis sequence is presented in greater detail in the following sections and the method is applied to data from the Avra Valley aquifer in Arizona. The data for this site are used because much of the system characterization and fluid flow analysis has been completed and is already published in the literature (3,4). Accordingly, the subsequent analysis steps can proceed using results of these completed steps and the available data. The authors emphasize that this example transport analysis is presented to illustrate the approach and demonstrate a method for uncertainty assessment and does not imply that an important specific pollution problem has been adequately treated.

2. FLOW MODEL CALIBRATION THROUGH STATISTICAL PARAMETER ESTIMATION

Given known boundary conditions, sources and sinks, and a conceptual model, the traditional approach for calibration of a groundwater flow model has been to modify the estimates of aquifer parameters by a trial and error procedure until the simulated hydraulic heads are reasonably close to the measured hydraulic head data. Although such a trial and error procedure may provide a reasonable representation of the measured head data, the estimates of the aquifer parameters are not unique, and their associated uncertainty cannot be determined. To overcome these problems, a unified approach that includes the geostatistical technique of kriging and a statistical inverse or parameter estimations method are used (6, 14, 15, 20).

Kriging is an interpolation technique that yields estimates at points or estimates averaged over areas where no data are available, the associated estimation errors, and covariance information. The kriging technique is applied to the measured aquifer parameter data to provide estimates of the parameters and the covariance of the estimation errors. The kriged estimates are used as prior information in the statistical parameter estimation method, while either the kriged or the measured hydraulic heads are used as the 'observed' head information. Thus, the kriging technique provides the estimates and covariance information needed for part of the statistical parameter estimation method.

The statistical inverse method developed by Neuman (15) is based on prior information about the aquifer parameters as well as observed hydraulic heads. A new transmissivity distribution is obtained by minimizing a generalized least squares criterion composed of two terms The first or model fit term involves minimizing the differences between the measured and calculated hydraulic heads. The second or parameter plausibility term involves minimizing the difference between the calculated values and prior estimates of log transmissivity. Neuman developed the statistical parameter estimation method in terms of log transmissivity instead of transmissivity for specific reasons. One reason is that using log transmissivity guarantees that the transmissivities will always be positive. An additional reason is based on the practical experience that transmissivities appear to be log normal distributed. The generalized least squares criterion is minimized by first obtaining its derivative by the adjoint method (2) and then using a Fletcher-Reeves conjugate gradient algorithm (7) coupled with Newton's method for determining step changes in the parameters. The new transmissivity distribution calculated by this inverse procedure produces hydraulic heads that are reasonably close to the observed hydraulic heads, while keeping the inverse estimates of the parameters reasonably close to the prior estimates. Neuman's inverse method makes use of all available statistical information about the prior estimates of the parameters and the hydraulic heads when determining new estimates of aquifer parameters and the covariance matrix of their estimation errors.

Our application of the analysis sequence uses the field data and previous studies of the Avra Valley aquifer in Southern Arizona as reported by Clifton (3), Clifton and Neuman (4), Neuman and Jacobson (16) and Jacobson (6). Avra Valley is a deep elongated basin containing an extensive alluvial aquifer (Figure 1). Kriged estimates of the aquifer parameters (i.e., log transmissivities) have been obtained by Clifton (3) and used in the two-dimensional, steady-state, statistical inverse model of Neuman (15). Clifton used hand-contoured hydraulic heads for the model calibration using the parameter estimation procedure. One of the authors of this paper (6) used kriged hydraulic heads as the basis for model calibration. This approach has the advantage that estimates of the variance in hydraulic heads are available for use in the statistical parameter estimation procedure. However, in the transport analysis described here, results of Clifton and Neuman (4) were used because only the effect of transmissivity uncertainty was considered. Later work will consider boundary uncertainties in addition to transmissivity uncertainties. In this stepwise approach the significance of

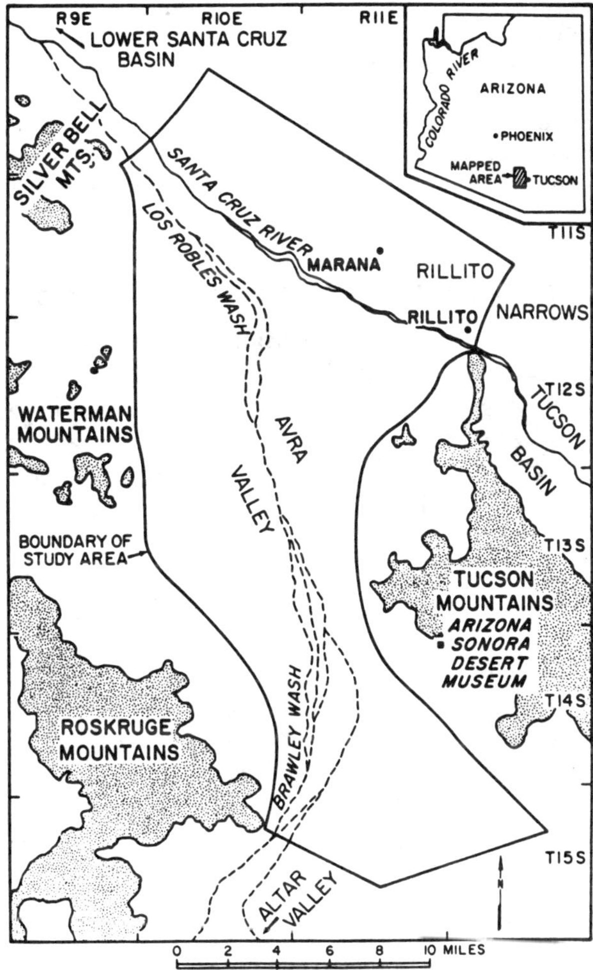

Figure 1. Location and setting of Avra Valley, Arizona, and study area boundary (4)

individual sources of uncertainty on the final assessment results can be observed and studied.

Figure 2, reproduced from Clifton and Neuman (4), is the transmissivity contour map obtained from the statistical parameter estimation model,and the corresponding map of log-transmissivity standard errors is shown in Figure 3. The reader who wishes more detail on the model calibration may consult the paper (4). Figure 4 shows the head contours calculated from the inverse estimates of transmissivity shown in Figure 3. As seen in Figure 4, the heads calculated from the inverse transmissivities are in good agreement with the original contoured heads of Clifton and Neuman (4).

The new distribution of transmissivities obtained from the inverse method together with the known boundary conditions, sources and sinks, and resulting hydraulic head distribution constitute a

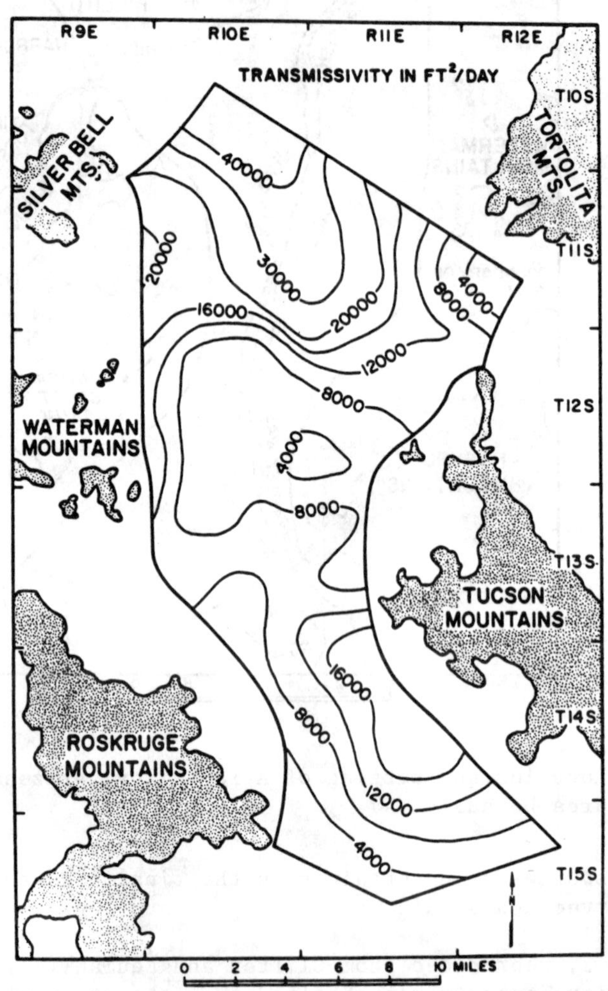

Figure 2. Contour map of the transmissivity field from parameter estimation model (4)

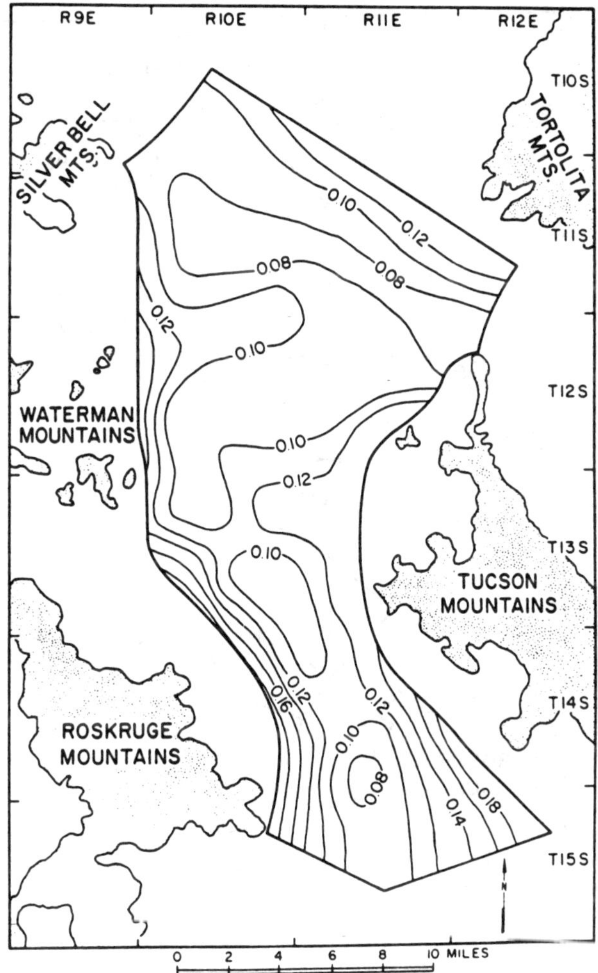

Figure 3. Contour map of the log-transmissivity estimation errors (4).

statistically calibrated groundwater flow model of the Avra Valley aquifer. In addition, an estimate of the parameter uncertainty is obtained from the covariance of the inverse estimation errors. In this way the groundwater flow model is statistically calibrated and the uncertainty associated with the spatially varying aquifer parameters is available. This information is required to continue with the transport uncertainty analysis.

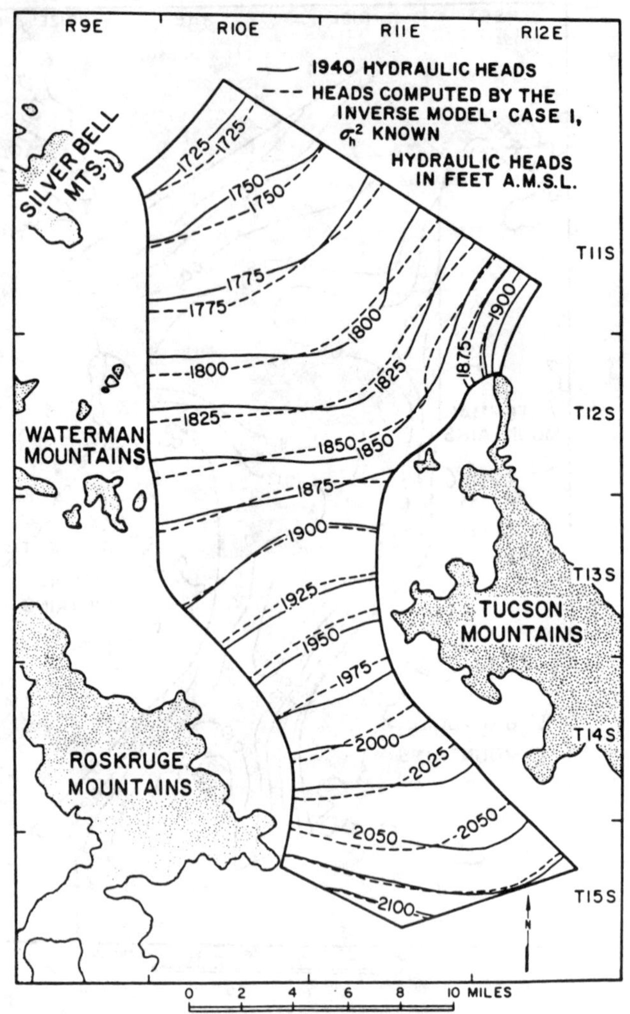

Figure 4. Comparison of the 1940 steady hydraulic heads and the head computed from inverse transmissivities (4)

3. CONDITIONAL FLOW SIMULATIONS USING THE CALIBRATED MODEL

The next step in the analysis uses the calibrated flow model and the covariance information provided by the statistical inverse technique in a conditional simulation mode. Conditional simulations involve generating a large number of 'realizations' of the transmissivities and calculating the associated heads. Each realization is generated in such a manner that the statistical information (mean and covariance) about the log tranmissivities is maintained (4). A realization of the log transmissivity vector $\underline{Y}$, is obtained by setting $\underline{Y} = \hat{\underline{Y}} + \underline{\underline{M}}\,\underline{X}$, where $\hat{\underline{Y}}$ is the vector of the log-transmissivity estimates obtained by the inverse method, $\underline{\underline{M}}$ is a lower triangular matrix defined as $\underline{\underline{M}}\,\underline{\underline{M}}^T = \underline{\underline{V}}$ where $\underline{\underline{V}}$ is the covariance matrix of the estimation errors obtained by the inverse method and $\underline{X}$ is a random vector obtained from a multivariate normal distribution. The transmissivity values for each realization are then calculated from $T_r = 10^Y$, where T_r is the transmissivity. This method of generating the transmissivity realizations (4) appears to be more efficient computationally than using the turning band approach.

The 600 conditional simulations were regenerated for this study based on Clifton's and Neuman's (4) statistically calibrated flow model. The statistics of our 600 realizations and conditional simulations were essentially the same as those shown in Figures 18 and 19 of Clifton and Neuman (4). All 600 realizations, were used in this evaluation because they were inexpensively obtained. Only 7 min of CPU time on the VAX® 11/780 were required to generate the 600 conditional simulation realizations. (VAX® is a registered trademark of the Digital Equipment Corporation, Maynard, Massachusetts.) The subsequent transport solution for the 600 velocity field realizations required 52 CPU min on the same machine.

4. THE ENSEMBLE OF CONTAMINANT PATHLINE REALIZATIONS

The realizations of transmissivities and associated heads from the statistically calibrated fluid flow model are needed to generate velocity fields and the associated pathlines. In generating these velocity fields, we wanted to minimize the problem of discontinuous velocities at the boundaries of adjoining finite elements. Therefore, a new velocity field grid was formed on the original flow model grid and is shown in Figure 5. The nodes of the velocity field grid are at the center of the interior elements and between the exterior nodes on the boundary of the flow model grid. The flow model solution and finite element basis functions were used to calculate the velocity at each of the new nodes of the velocity field grid.

Each conditional simulation realization from the calibrated flow model provides a set of equally likely values of transmissivity over each element and the corresponding head at each node of the flow model. This information, along with the known saturated depth of the aquifer and the known effective porosity, is sufficient to calculate the velocity at each node in the velocity field grid. Unfortunately limited data are available on the saturated depth and the effective porosity for Avra Valley. Accordingly, the

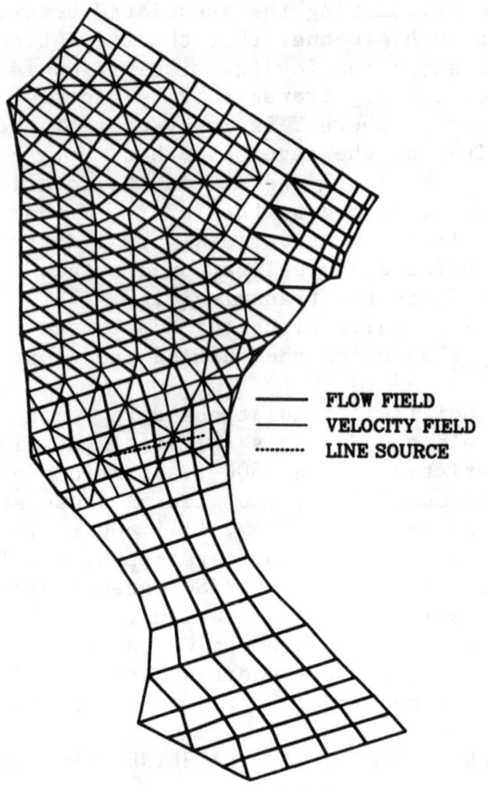

Figure 5. The Avra Valley velocity field grid imposed on the original flow model grid.

saturated depth and effective porosity were assumed constant and assigned values of 500 ft and 0.35, respectively.

The gradient of the head at each node in the velocity field grid was calculated using the flow model head solution and basis functions. Then, for the ith realization and the jth node in the velocity field, we have the transmissivity from the associated flow model element, $(Tr)_{ij}$ and the calculated gradient of the head $(\nabla h)_{ij}$. The pore water velocity, $\overline{v}_{ij}$ is calculated at each node using:

$$\overline{v}_{ij} = - \frac{(Tr)_{ij}}{Pb} (\nabla h)_{ij} \ ; \quad i = 1,2,\ldots,600; \quad j = 1,2,\ldots,J \quad (4.1)$$

where P is the effective porosity, b the aquifer thickness (Pb = 175 ft for this case), and J is the total number of nodes in the pore velocity finite element grid. As indicated by the ith limit in Eq. (4.1), 600 velocity field realizations are available for the pathline analysis.

Because the velocity realizations from Eq. (4.1) are from an asssumed steady-flow condition for Avra Valley, the pathlines and streamlines are identical. Though one could correctly refer to pathlines and streamlines interchangeably in this case, we maintain for clarity the convention of the more general transient systems. The term pathline is used when referrring to transport, fluid particle trajectories, and travel times that involve the time parameterization; whereas, streamline or stream function, whichever is appropriate, is used when fluid flux or the related spatial parameterization is involved.

The assumed distributed source of contamination for this example transport evaluation is a line source approximately 4 miles long that is shown as a dotted line in the lower part of Figure 5. The contaminants are assumed to be deposited dry and mixed instantaneously with the groundwater. In other words, the contaminants enter the system without adding additional fluid to the system. The length of the line source was varied slightly so that the same groundwater flux crossed the line source for all 600 of the conditional simulations. In this way, the same contaminated stream function flux originated at the line source for all realizations and was conveyed to the outflow boundary between the outermost contaminated streamlines.

For each of the 600 velocity field realizations, starting coordinate locations along the line source were generated so that an equal amount of the flux (2%) passed between adjacent streamlines. This spacing of locations gave a total of 51 contaminant starting points along the line source for each realization. Accordingly, at the kth streamline starting location, we define the stream function value by

$$\psi_{ik}\,(x_{oik},\,y_{oik})\;;\qquad k = -25,-24,-23,\ldots,0,1,2,\ldots,25. \qquad (4.2)$$

where ψ_{ik} is the stream function value for the kth streamline at the point along the line source with the coordinates designated inside the parenthesis for the ith realization. x_{oik} is the x-coordinate location of the kth streamline starting location along the line source. y_{oik} is the y-coordinate location of the kth streamline starting location along the line source. The streamline location coordinates along the line source were located such that

$$\psi_{ik}(x_{oik}, y_{oik}) - \psi_{ik-1}(x_{oik-1}, y_{oik-1}) = 0.02; \tag{4.3}$$

for $k = -24,-23,-22,\ldots,0,1,2,\ldots,25$

or in other words the points along the source line, A-B, in Figure 6 are located so that 2% of the total contaminated unit flux flows between any two adjacent streamlines. The total relative contaminated flux is then

$$\psi_{i,25}(x_{o,i,25}, y_{o,i,25}) - \psi_{i,-25}(x_{o,i,-25}, y_{o,i,-25}) = 1 \tag{4.4}$$

Pathlines were next numerically generated from each of the starting points along the line source A-B of known stream function fluxes. Specifically, the kth pathline was generated numerically by solving simultaneously the pair of characteristic differential equations,

$$\left(\frac{dx}{dt}\right)_k = (v_x)_{ik} \quad \text{and} \quad \left(\frac{dy}{dt}\right)_k = (v_y)_{ik} \tag{4.5}$$

for $i = 1,2,3,\ldots,600$; $k = -25,-24,-23,\ldots,0,1,2,3,\ldots,25$.

where $(v_x)_i$ and $(v_y)_i$ are the x and y pore water velocity components for the ith velocity realization. The pathlines described by Eq. (4.5) are generated numerically by stepping in time away from the starting point at zero time using the ith velocity field basis functions to calculate the velocity components v_x and v_y in space before each time step. In Figure 6, pathlines are shown for two realizations: i = 3 and i = 406 with five pathlines shown for each realization. Each of the pathlines reach the outflow boundary denoted as O-N in Figure 6 at the arrival time, T_{ik}, and at outflow location coordinates denoted by

$$(x_{Lik} \quad , \quad y_{Lik}) \tag{4.6}$$

where, T_{ik} is the arrival time to reach the outflow boundary, O-N, along the kth pathline for the ith velocity field realization. The x_{Lik} is the x-coordinate of the kth pathline at the outflow boundary, O-N, for the ith velocity field realization. The y_{Lik} is the y-coordinate of the kth pathline at the outflow boundary, O-N, for the ith velocity field realization. The pathline, however, is also a streamline because the flow is steady. Accordingly, the stream function

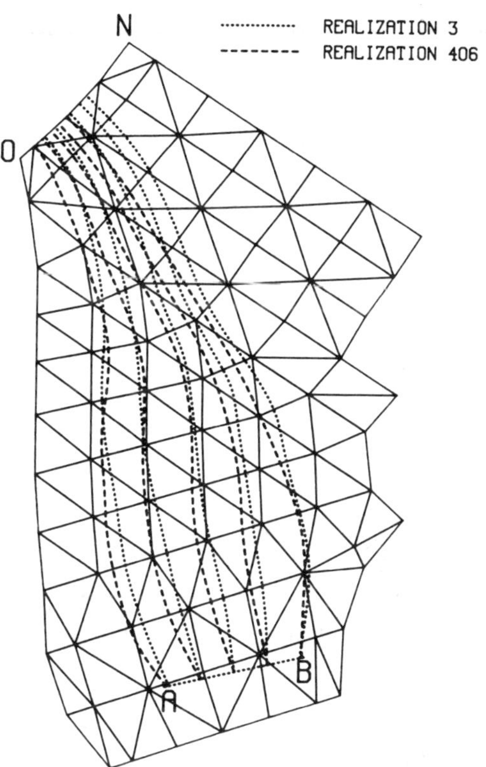

Figure 6. Groundwater pathline from contamination line sources to outflow boundary for realizations 3 and 406.

is constant along the streamline and so

$$\psi_{ik}\ (x_{Lik},\ y_{Lik}) = \psi_{ik}\ (x_{oik},\ y_{oik}) \qquad (4.7)$$

with the right-hand side known from Eq. (4.2) when k is known. Therefore, the arrival time T_{ik} can be plotted as a function of the stream function, ψ_{ik}, for each velocity field realization as shown in Figure 7. A minimum arrival time and two peaks in arrival times are seen in the figure.

The time/stream function flux relationship is more convenient to use if converted to a monotonic increasing function of contaminant arrival time. This is readily accomplished using the steady stream function properties. We begin by noting that the first contaminant arrival at the outflow boundary is at $\psi_{406,-9} = -\ 0.18$ denoting the

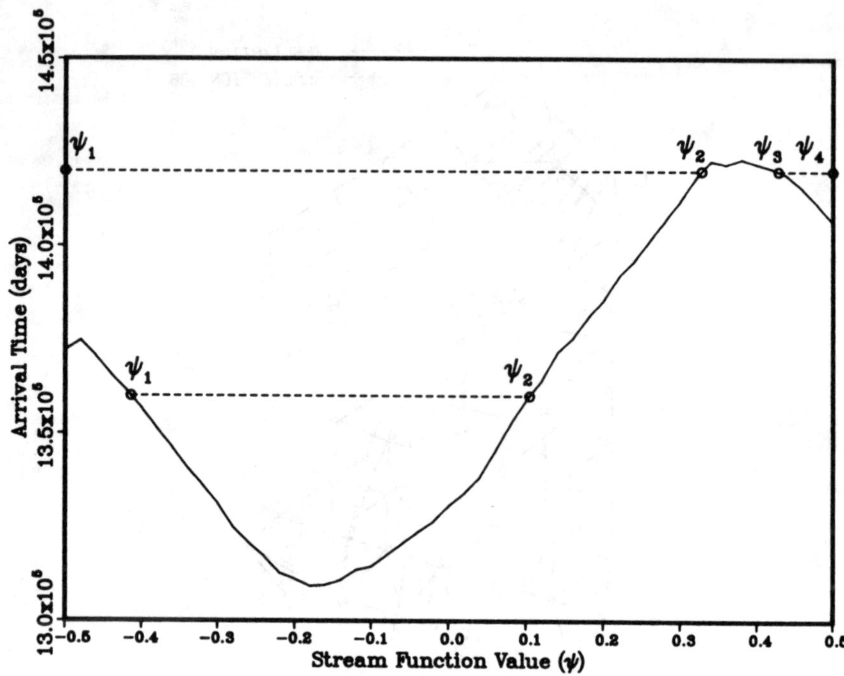

Figure 7. The dependence of contaminant arrival time at the outflow boundary upon the stream function values for velocity realization 406

first streamline reaching the outflow boundary at a travel time (arrival time), $T_{406,-9}$ = 1,309,304 days. Associated with that first arrival time, we define the cumulative outflow of contaminated fluid $(q/Q)_{406} = 0$ (i.e., the cumulative outflow is zero upon first arrival at the outer boundary O-N). At all later arrival times in Figure 7, two or more streamlines as denoted by the stream function values reach the outflow boundary at any given arrival time. Accordingly, at any later arrival time, the cumulative flux that has crossed the outflow boundary, $(q/Q)_i$ is the difference between the stream function values for the bounding streamlines that enter at that time. For example, at some later time, say T = 1,360,000 days, as represented by the lower dashed line in Figure 7, there are two streamlines: $\psi_1 = -0.4138$ and $\psi_2 = 0.1043$. Therefore, the cumulative relative flux, q/Q, that has entered between the two streamlines is the difference in ψ values:

$$\left(\frac{q}{Q}\right)_{406}\Bigg|_{T\,=\,1,360,000\ \text{days}} = \psi_2 - \psi_1 = 0.5181 \tag{4.8}$$

At a still later time (for example at T = 1,420,000 days) the same difference of values approach applies and yields the cumulative relative flux though it may seem slightly more involved. At T = 1,420,000 days, four streamlines are of interest in Figure 7 (i.e., $\psi_1 = -0.5$, $\psi_2 = 0.3279$, $\psi_3 = 0.4288$ and $\psi_4 = 0.5$). Two of the streamlines, ψ_2 and ψ_3 in the figure, represent the two pathlines that arrive at the selected time, T = 1,420,000 days, while the other two stream function values of ψ_1 and ψ_4 represent the outermost bounding streamlines for the contamination plume reaching the outflow boundary. All four of the streamlines are used to provide the cumulative relative flux, $(q/Q)_{406}$, that has outflowed by this time, which is

$$\left(\frac{q}{Q}\right)_{406}\Bigg|_{T\,=\,1,420,000\text{ days}} = (\psi_2 - \psi_1) + (\psi_4 - \psi_3) = 0.8991 \tag{4.9}$$

As additional arrival times are selected and the associated cumulative fluxes, $(q/Q)_S$ are determined from the stream function values, a complete data set for that particular realization is provided. In Figure 8, the subsurface system response function, or the q/Q versus arrival time curve, for realization 406 is given. Also shown in Figure 8 are the contaminant arrival curves for realizations 3, 5, 464, and 531. Though only five realizations of contaminant arrival curves are illustrated in the figure, the cumulative arrival fluxes and associated arrival times at the outflow boundaries were calculated for the full complement of 600 realizations and are available for use in the remaining part of the uncertainty evaluation.

5. THE STATISTICAL CONTAMINANT ARRIVAL DISTRIBUTIONS

The contaminant arrival distribution concepts were developed during the last decade as useful analysis methods in groundwater quality management studies (1,8,9,10,11,12,19). Though the arrival distribution methods have been used previously only with deterministic model results, the concepts now appear just as useful with statistical uncertainty into contaminant arrival concepts, the background deterministic methods are discussed briefly.

The arrival distribution approach can be described either by starting from the theoretical origin in groundwater analysis and logically progressing through the analytical steps to their ultimate use in decision making, or conversely by beginning with the needs of the decision-making process and showing how the contaminant arrival concepts satisfy those needs and can be obtained from system analyses. We focus on the latter or the ultimate use of the arrival concepts by considering those pertinent factors involved in evaluating a subsurface contamination problem.

The major analytical need is to estimate the extent of contamination that will reach the biosphere from the subsurface contamination problem. Specifically, three factors are of major significance to the decision-making process: a) quantity of contaminants reaching the biosphere b) time of contaminant arrival at the biosphere interface c) location of contaminant emergence.

These three factors can be interrelated in two ways. The first and most general approach is to use the outflow location as the predominant variable. This approach provides the arrival distribution summaries, which are described in detail elsewhere (9,10,11,12). The second approach, which is appropriate only for steady systems uses the cumulative quantity of containment outflow as the predominant variable and has become known as the contaminant response functions (13). Response functions, though somewhat less general than the arrival distributions, because they are restricted to analyzing steady-flow systems, are simpler to apply. The response functions using the cumulative contaminated outflow, q/Q, as the coupling variable is used in the continuing uncertainty evaluation because our system involves steady flow and transport. Accordingly, the five individual arrival curves in Figure 8 are recognized as the outflow response functions for velocity field realizations 3, 5, 406, 464, and 531. A separate and unique response function exists for each of the other 595 velocity field realizations. The functional dependence of the response function is expressed as

$$\left(\frac{q}{Q}\right)_i = f_i(T), \quad i = 1,2,3,\ldots,600 \tag{5.1}$$

where $(q/Q)_i$ is the relative cumulative flux for the ith realization and f_i represents the ith functional dependence upon the contaminant arrival time, T at the outflow boundary. The functional dependence for each and every individual response function of Eq. (5.1) is purely deterministic with no stochastic variations between q/Q and T for the same realization, i. The stochastic uncertainties of the response functions enter through considering the arrival times for the 600 realizations at a particular value of q/Q. For example, if one selects q/Q = 0 and then considers all 600 of the first arrival times associated with that zero cumulative outflow, the statistical distribution of first arrival times is obtained and may be shown as a histogram in Figure 9. Although the histogram suggests the arrival times at q/Q = 0 are normally distributed, more careful analysis to confirm this point follows. The later arrival time distributions were examined similarly by analyzing the 600 realizations of arrival time associated with q/Q equal to 0.05 followed by considering 0.10 then 0.15 and so forth increasing each new q/Q value until the last set of arrival times for q/Q = 1 were analyzed. This provided a total of 21 sets of 600 realizations of arrival

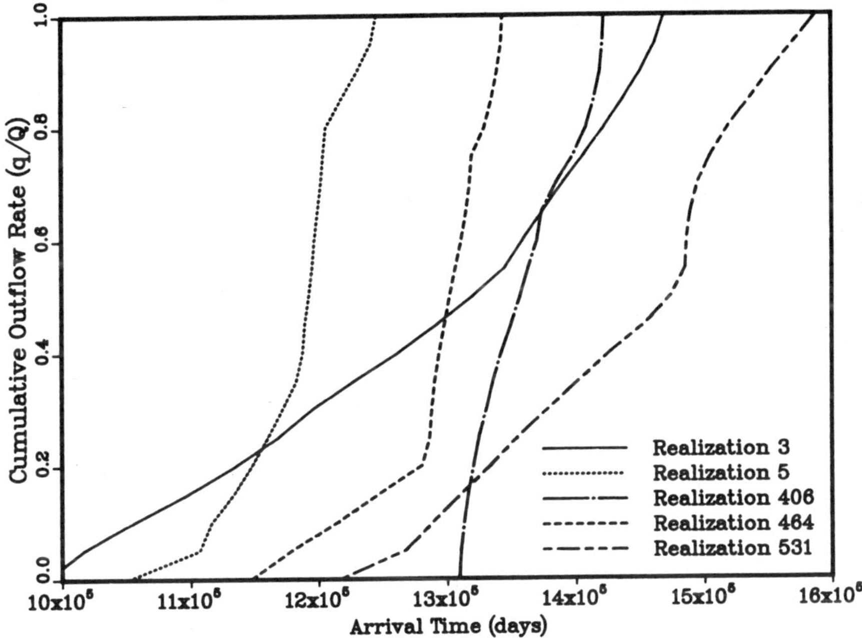

Figure 8. Arrival curves or response functions for velocity field realizations 3, 5, 464, 406, and 531

times with each set of 600 being for a specific value of the cumulative outflow rate q/Q.

A normal probability plot was generated for each set of arrival times corresponding to the different values of q/Q by an approximation method (18). All of the probability plots exhibited straight lines. The 'straightness' of the lines can be measured by the correlation coefficient of the points in the plot. A very powerful test for normality based on this correlation (18) was applied to each set of arrival times, and the results indicate that all the sets of arrival times are normally distributed. Probability plots for q/Q = 0.0, 0.25, 0.50, 0.75, and 1.0 are shown in Figure 10. In the figure, the arrival times are plotted versus multiples of the standard deviation for each of the five cumulative fluxes. Though only five of the data sets are shown in Figure 10, all 21 of the fitted normal distributions are used in constructing subsequent results.

The sample mean and sample standard deviation of the arrival times were calculated for each of the 21 data sets. So for each of the 21 values of q/Q covering the range of the cumulative outflow

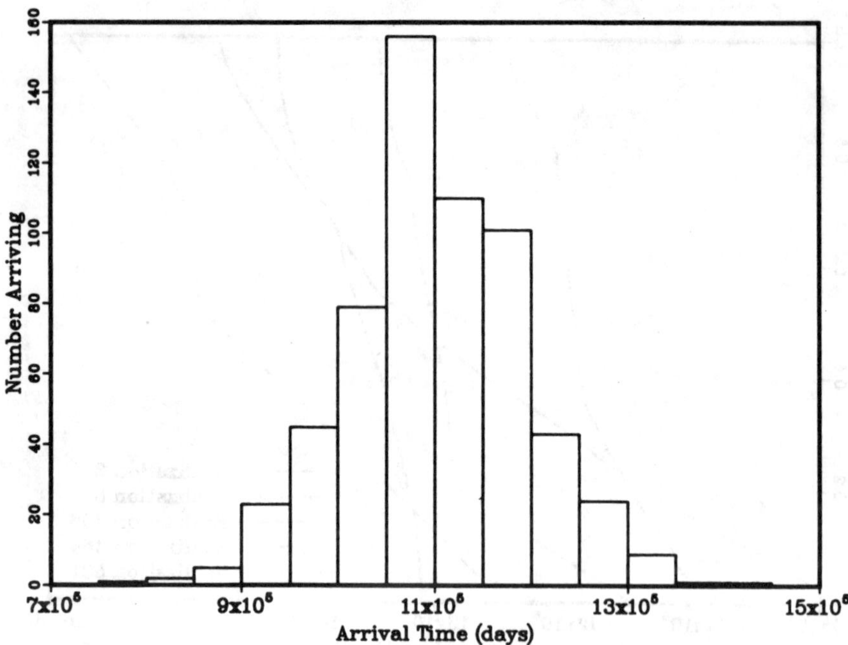

Figure 9. Histogram of the first contaminant arrival times for the relative cumulative outflow flux, q/Q = 0.0

flux, the mean arrival time for each q/Q value is available and can be plotted as the mean response function shown as the solid central curve in Figure 11. Similarly, from the standard deviations for the 21 incremented cumulative flux values, the 95% confidence interval response curves are obtained and are also plotted in Figure 11 as the two dotted curves. For comparison, the dashed curve in Figure 11 is the response curve that was calculated using the spatially varying mean transmissivities determined from the inverse model calibration. This curve probably represents the best, strictly deterministic, flow and transport modeling result for comparison with the mean uncertainty response function and the confidence interval. The agreement between the deterministic and the mean response function is good, although the deterministic response function is steeper and represents, in general, a more uniform velocity field than the stochastic expectation. The results of a trial and error calibration would probably lie somewhere between the best deterministic transmissivity response function (dashed curve) in Figure 11 and some of the more variable stochastic response function realizations (for example, like those shown in Figure 8).

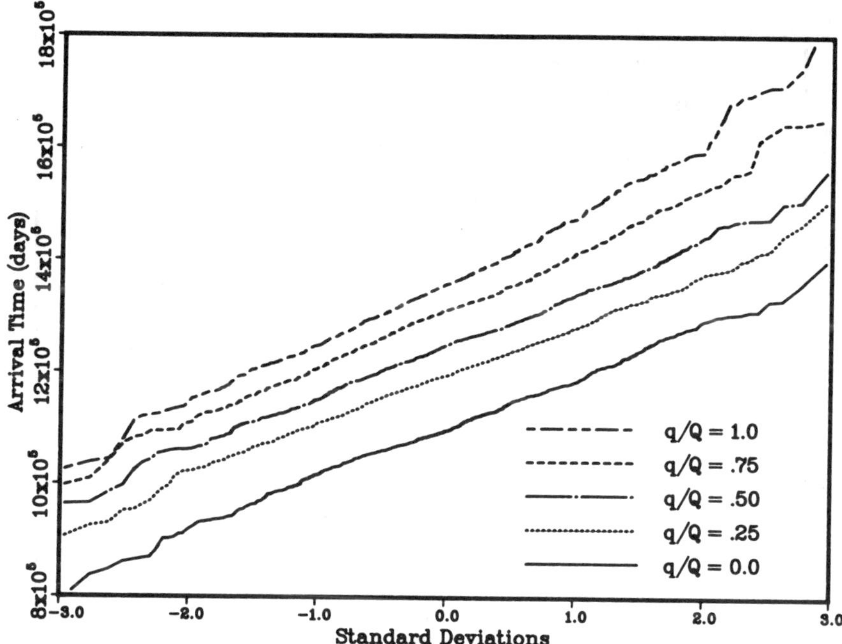

Figure 10. Contaminant arrival time variation with standard deviations for five values of cumulative outflow flux, q/Q

The confidence interval shown in Figure 11 is the more commonly used two-sided confidence interval that includes both the upper and lower limits that bound the mean. Yet as pointed out by both the U.S. Nuclear Regulatory Commission and the Environmental Protection Agency (17), one-sided confidence limits are often more appropriate for water quality controls. Accordingly, convenient one-sided integral confidence intervals were calculated and are shown in Figure 12. The response curve arrival times are lower limits so that the probability is stated as the percentage likelihood of the travel times being equal to or less than shown for the response function arrival times. Specifically, the response functions correspond to arrival times equal to or less than those shown by the curves at probabilities of 1%, 5%, 15%, 25%, and the mean at a probability of 50%.

6. SUMMARY

An analysis sequence is presented and applied to a subsurface flow system. The analysis sequence is used to estimate the uncertainty in the final transport results that arise from our inability

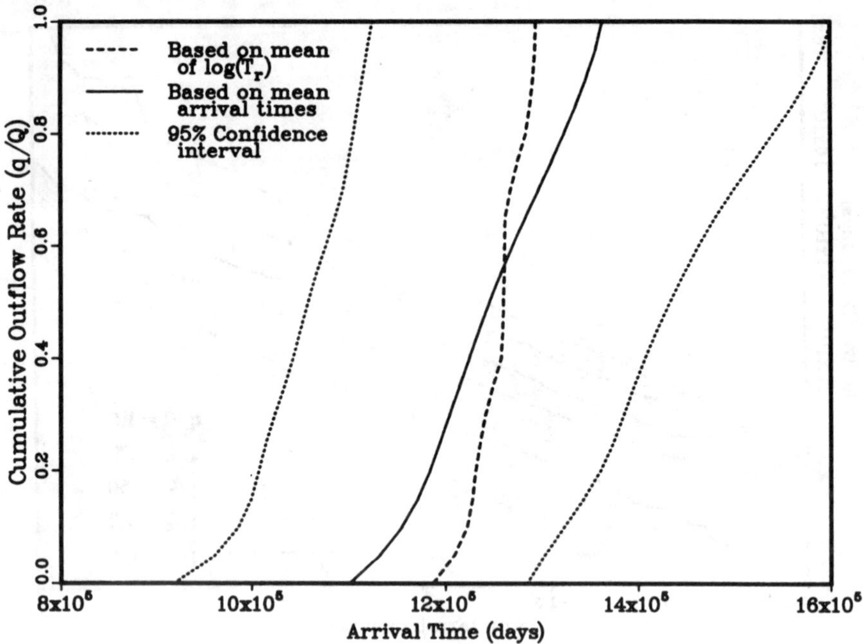

Figure 11. Mean arrival distribution or response function with 95% confidence interval and response function for spatial varying mean transmissivity distribution

to completely characterize subsurface material heterogeneity. The sequence involves combining statistical methods and the classical subsurface deterministic modeling methods to provide the uncertainties in the final analysis results.

Specifically the analysis sequence a) applies geostatistical techniques (kriging) to measured field transmissivity data. b) statistically calibrates a flow model by using the geostatistical results together with the field measured heads in a parameter estimation technique. c) provides the covariance matrix of parameter estimation errors as part of the model calibration. d) uses the covariance matrix and the calibrated flow model to generate a large number of transmissivity and the associated head realizations. e) generates a pore water velocity field and fluid flow paths with the associated travel times from the contaminated source to the outflow boundary for each realization. f) determines the arrival time versus cumulative outflow rate or response function for each realization. g) statistically analyzes the response functions to determine the mean response function and its confidence limits.

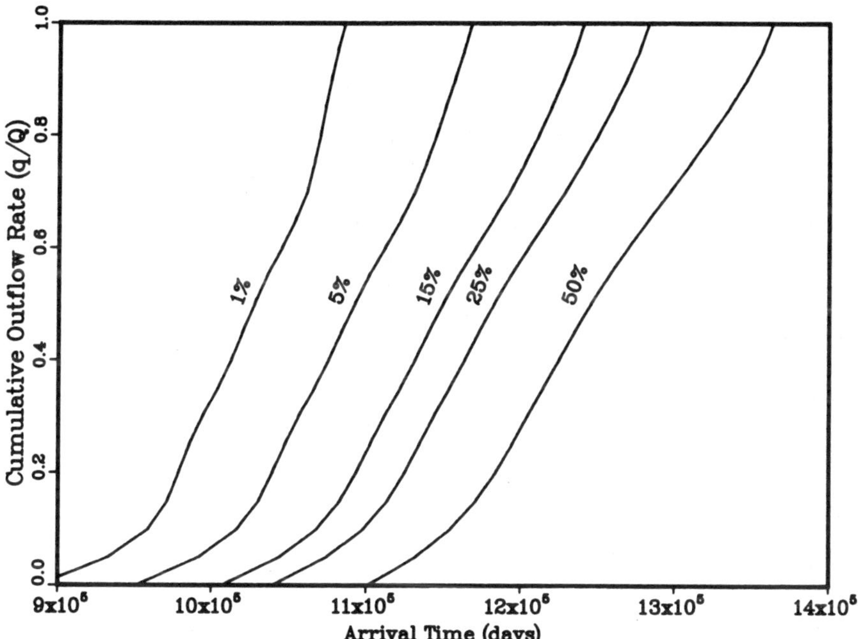

Figure 12. Cumulative arrival curves or response functions where the arrival times are less than or equal to values shown at 1%, 5%, 15%, 25%, and 50% or mean probabilities

In total, we have described a technique or analysis sequence that begins with field data and results in the confidence limits of the response functions at the contaminant outflow boundary. This technique will need to be further developed and expanded in considerable detail in order to satisfy the growing need for quantification of uncertainty in subsurface flow and transport analysis.

7. ACKNOWLEDGMENTS

The authors express special thanks to Peter M. Clifton for his assistance and for making available his previous analysis and published results. The authors also wish to acknowledge word processing support provided by Darla J. Kennedy and special appreciation is expressed to Janelle L. Downs-Berg, technical editor, for help in editing and preparing the final manuscript. This work was supported by the U.S. Department of Energy under Contract DE-AC06-76RLO 1830.

8. REFERENCES

1. Bear, J. *Hydraulics of Groundwater*, McGraw Hill, 1979, pp. 294-299.
2. Chavent, G., Dupuy, M., and Lemonnier, P., "History Matching by Use of Optimal Control Theory," *Society of Petroleum Engineering Journal*, Vol. 15, No. 1, 1975, pp. 74-86.
3. Clifton, P.M., "Statistical Inverse Modeling and Geostatistical Analysis of the Avra Valley Aquifer," thesis presented to the University of Arizona, at Tucson, Arizona, in 1981, in partial fulfillment of the requirements for the degree of Master of Science.
4. Clifton, P.M., and Neuman, S.P., "Effects of Kriging and Inverse Modeling on Conditional Simulation of the Avra Valley Aquifer in Southern Arizona," *Water Resources Research*, Vol. 18, No. 4, pp. 1215-1234.
5. EPA. "U.S. Environmental Protection Agency, 40 CFR 191, Environmental Standards for the Management and Disposal of Spent Nuclear Fuel, High-Level and Transuranic Radioactive Wastes." Subchapter F-Radiation Protection Programs, Working Draft No. 4, 1984, 4/17/84; 5/21/84.
6. Jacobson, E.A., "A Statistical Parameter Estimation Method Using Singular Value Decomposition with Application to Avra Valley Aquifer in Southern Arizona," thesis presented to the University of Arizona at Tucson, Arizona, in 1985, in partial fulfillment of the requirements for the degree of Doctor of Philosophy.
7. Luenberger, D., *Introduction to Linear and Nonlinear Programming*, Addison-Wesley, Reading, Massachusetts, 1973.
8. Mido, K.W., "An Economic Approach to Determining the Extent of Groundwater Contamination and Formulating a Contaminant Removal Plan," *Ground Water*, Vol. 19, No. 1, 1981, pp. 41-47.
9. Nelson, R.W., "Evaluating the Environmental Consequences of Groundwater Contamination. I. An Overview of Contaminant Arrival Distributions as General Evaluation Requirements," *Water Resources Research*, Vol. 14, No. 3, 1978, pp. 408-415.
10. Nelson, R.W., "Evaluating the Environmental Consequences of Groundwater Contamination, II. Obtaining Location/Arrival Time and Location/Outflow Quantity Distribution for Steady Flow Systems," *Water Resources Research*, Vol. 14, No. 3, 1978, p. 416-428.
11. Nelson, R.W., "Evaluating the Environmental Consequences of Groundwater Contamination. III. Obtaining Contaminant Arrival Distributions for Steady Flow in Heterogeneous Systems," *Water Resources Research*, Vol. 14, No. 3, 1978, pp. 429-440.
12. Nelson, R.W., "Evaluating the Environmental Consequences of Groundwater Contamination. IV. Obtaining and Utilizing Contaminant Arrival Distributions in Transient Flow Systems," *Water Resources Research*, Vol. 14, No. 3, 1978, pp. 441-450.

13. Nelson, R.W., "Use of Geohydrologic Response Functions in the Assessment of Deep Nuclear Waste Repositories." *PNL-3817*, Pacific Northwest Laboratory, 1981.
14. Neuman, S.P., and Yakowitz, S., "A Statistical Approach to the Inverse Problem of Aquifer Hydrology: 1. Theory," *Water Resources Research*, Vol. 15, No. 4, 1979, pp. 845-860.
15. Neuman, S.P., "A Statistical Approach to the Inverse Problem of Aquifer Hydrology: 3. Improved Solution Method and Added Perspective," *Water Resources Research*, Vol. 16, No. 2, 1980, pp. 331-346.
16. Neuman, S.P., and Jacobson, E.A., "Analysis of Nonintrinsic Spatial Variability by Residual Kriging with Application to Regional Groundwater Levels," *Mathematical Geology*, Vol. 16, No. 5, 1984, pp. 499-521.
17. Ornstein, P.M., Knapp, M.R and Belote, J.C., *Proceedings of 1983 Civilian Radioactive Waste Management Information Meeting, CONF-831217*, U.S. Department of Energy, Dec. 1983, Washington, D.C., pp. 313-319.
18. Ryan, T.A., Jr., Joiner, B.L., and Ryan, B.F., "Minitab Reference Manual, Minitab Project," Statistics Department, Pennsylvania State University, 1982.
19. Welby, C.W., "A Graphical Approach to Prediction of Groundwater Pollutant Movement," *Bulletin of Association of Engineering Geology*, Vol. 18, No. 3, 1981, pp. 237-243.
20. Wilson, J.L., "Estimation of Aquifer Parameters: Kriging and Its Role in the Unified Approach," presented at the December 3-7, 1979, American Geophysical Union Fall Meeting held at San Francisco, California.

9. LIST OF SYMBOLS

b	Aquifer thickness
J	Total number of nodes in the pore velocity finite element grid
$\underline{\underline{M}}$	Lower triangular matrix
P	Effective porosity
$(q/Q)_i$	Cumulative flux for the ith realization
T_{ik}	The arrival time to reach the outflow boundary 0-N, along the kth pathline for the ith velocity field realization
Tr	Transmissivity
$(Tr)_{ij}$	Transmissivity from the associated flow model element
$\bar{v}_{ij}$	Pore water velocity for the ith realization at the jth element

$\underline{\underline{V}}$	Covariance matrix of the estimation errors obtained by the inverse method
$(v_x)_{ij}$ and $(v_y)_{ij}$	x and y pore water velocity components for the ith velocity realization and at element j.
$\underline{X}$	A random vector obtained from a multivariate normal distribution
x_{Lik}	x-coordinate of the kth pathline at the outflow boundary, O-N, for the ith velocity field realization
x_{oik}	x-coordinate location of the kth streamline starting location along the line source
$\underline{Y}$	Log transmissivity vector
$\hat{\underline{Y}}$	Vector of the log-transmissivity estimate obtained by the inverse method
y_{Lik}	y-coordinate of the kth pathline at the outflow boundary, O-N, for the ith velocity field realization
y_{oik}	y-coordinate location of the kth pathline starting location along the line source
$(\nabla h)_{ij}$	Calculated gradient of the head for the ith realization at the jth element
$\psi_{ik}(,)$	Stream function value for the kth streamline at the point along the line source with the coordinates designated inside the parenthesis for the ith realization

AN OVERVIEW OF THE STOCHASTIC MODELING OF DISPERSION IN FRACTURED MEDIA

Franklin W. Schwartz and Leslie Smith

AN OVERVIEW OF THE STOCHASTIC MODELING OF DISPERSION IN FRACTURED MEDIA

Franklin W. Schwartz

Department of Geology
University of Alberta
Edmonton, Alberta
Canada T6G 2E1

Leslie Smith

Department of Geological Sciences
University of British Columbia
Vancouver, British Columbia
Canada V6T 2B4

ABSTRACT

This chapter reviews the development and application of stochastic modeling techniques for studying mass transport in fractured media. Because of difficulties in developing a conceptual framework with respect to issues concerned, for example, with representative elemental volumes or the connectivity of a network, emphasis in modeling has generally involved discrete rather than continuum approaches. Our work has shown that it is possible to model transport in two and three-dimensional networks of discrete fractures using a particle tracking approach. Characteristics of the internal mass distribution and breakthrough curves can be directly related to features of the network geometry using a Monte Carlo approach. With this technique, realizations of a fracture network are generated probabilistically. Results from these kind of simulations show for example that mass distributions within a fractured medium can be irregular and skewed, and that the mass distribution is sensitive to the mean hydraulic gradient with respect to fracture sets. These characteristics of the internal mass distributions directly determine the shape of the breakthrough curves. More recent studies have suggested that these effects also develop in three-dimensional systems. Computational limitations have restricted the application of the discrete approaches to near-field problems. Work is progressing in developing a new continuum approach for modeling far-field problems, which successfully preserves the unique patterns of transport imparted by the network. Preliminary testing has verified this approach and provided considerable optimism in the ability to deal with large scale networks of fractures.

1. INTRODUCTION

Emphasis in a variety of field and theoretical studies over the past few years has been placed on understanding groundwater flow and mass transport in fractured media. Driving much of this work is the practical interest in developing repositories for high level radioactive waste in rocks that may contain fractures. The purpose of this paper is to review recent developments in stochastic modeling of dispersion in fractured media. Specific objectives are to: i) provide various perspectives on the nature of fractured rocks ii) review some of the approaches that we have developed for modeling flow and mass transport, iii) summarize the results of this work, and iv) describe a new continuum approach for modeling mass transport in fractured systems.

2. CLASSIFICATION OF FRACTURE NETWORKS

A necessary requirement for studying fractured rocks is a conceptual model that incorporates the essential features of the system. The model that has evolved (see Figure 1), although similar in many respects to that often proposed for porous media, is in some ways more complex. Central to this model is the concept of a representative elemental volume (REV), which defines how a parameter of the system such as network permeability varies as a function of volume. An REV is said to exist when there is a negligible change in the value of the parameter accompanying a slight change in the volume over which the parameter is calculated. Shown on Figure 1 are two scales or volumes at which REV's are assumed to exist. Scale 1 is the smallest scale of facturing in the system, while Scale 2 is a larger scale that appears gradually once the volume of interest reaches some minimum size. An REV cannot be defined at the small end of the volume range and within a zone between scales 1 and 2 (Figure 1).

The fundamental relationship depicted in Figure 1 can only be defined when fracture densities are above a so-called critical density. In percolation theory, the critical density is defined as that density above which infinite clusters of fractures appear or, in other words, when connectivity is achieved for the network (see Figure 2 for an example). Robinson (12) considers a cluster of fractures to be percolating within some domain when all sides of a domain are connected. The important point is that if a network is below the critical density its effective hydraulic conductivity will remain zero even if volume is increased.

Assuming that a network is percolating, it still may not be possible to define REV's at some volumes (Figure 1). When the volume of a rock is smaller than that which defines the REV at either scales 1 or 2, the number of fractures in the system is limiting to the extent that changing the volume slightly can change

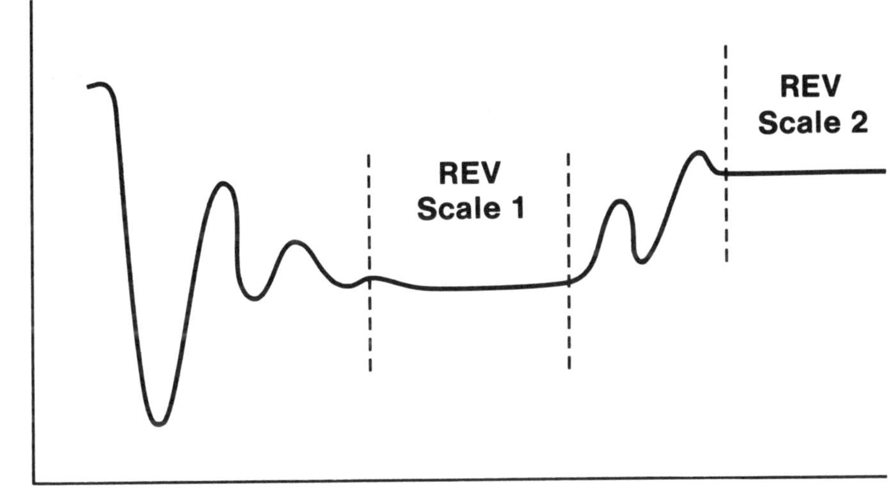

Figure 1. Variability in permeability as a function of volume illustrating REV's at two scales

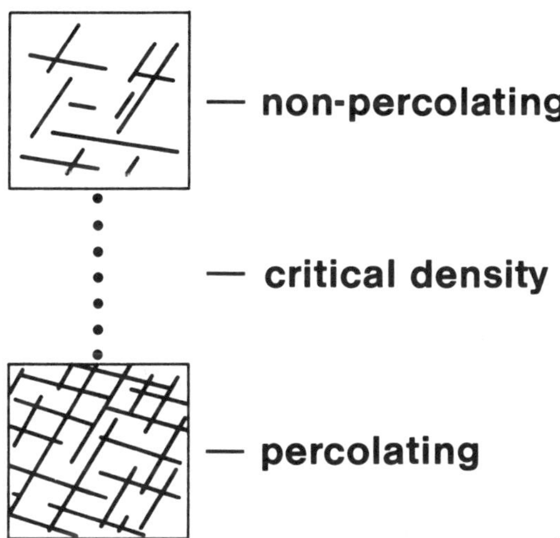

Figure 2. Examples of percolating and non-percolating networks in two dimensions. Critical density is the point where infinite clusters of fractures appear.

fluid flow within the fracture network significantly. As a consequence, there can be considerable variability in a parameter such as effective hydraulic conductivity for the network.

It is in this manner that fractured media may differ considerably from porous media. In porous media, this sub-REV behavior will occur over a small volume range. For example, once volume averaging moves from the volumes of just several pores to hundreds of pores (a relatively small change in volume), we quickly approach the REV (4). In fractured media characterized by relatively low fracture densities, the rock volumes over which this sub-REV behavior may continue could be considerable.

The important question that arises in this respect concerns how flow and mass transport are modeled in fractured rocks. As a generalization, systems for which an REV exists are modeled using continuum concepts, while sub-REV systems are modeled as discrete fracture networks. Although discrete fracture models can in theory be applied to most fracture networks, there are practical limits to the number of fractures that can be considered due to computational constraints. Occasionally, two scales of fracturing can be modeled using equivalent parameters to represent the smaller scale of fracturing and discrete fractures to describe the larger - scale network. The problem in modeling becomes one of deciding whether a particular volume of rock is or is not an REV for various parameters. It is usually assumed in practice that the REV simply exists without providing suitable justification. Many sub-REV systems are probably modeled inappropriately as a result of using continuum approaches.

A case can be made that REV's may not exist at all for fractured rocks. This argument is based on the probable existence of different scales of fracturing within a rock mass. Before any test volume approaches an REV for one scale of fracturing, the next larger scale begins to exert an influence. Thus, hydraulic or transport parameters may vary continuously as a function of volume. This problem could be more pronounced in fractured rather than porous media because the volumes over which sub-REV behavior occurs are in general larger.

What we have tried to show is first that a discrete formulation of the transport problem may be more appropriate than existing continuum methods for problems related to fracture rocks. Second, mass transport in fractured media may be considerably different than in porous media.

3. METHODS FOR MODELING FLOW AND TRANSPORT IN DISCRETE NETWORKS

3.1 Two-Dimensional Systems

The discrete approach to modeling fluid flow and mass transport involves specifying the geometry and hydraulic characteristics of one or more fractures. Although considerable emphasis has been placed on studies of transport in single fractures to evaluate matrix diffusion, fractured networks are less well studied. Besides some early work by Castillo et al. (2,3) and Krizek et al. (7) on simple deterministic networks of infinite fractures, the only other studies are those of Schwartz et al. (16), Endo et al. (5), Robinson (13), and Smith and Schwartz (19). Our work has involved the application of stochastic techniques where individual realizations of a fracture network are generated from a statistical description of the geometric and hydraulic properties of a fracture network. Simulation of transport for a large number of these realizations provides probability distributions on selected output parameters, which is the Monte Carlo procedure for stochastic simulation. All of the studies on fracture networks have considered two-dimensional systems, with emphasis placed on concept validation rather than simulation of field problems.

The basic steps in the stochastic simulation of mass transport include i) the definition of the flow domain and boundary conditions, ii) the probabilistic generation of one realization of the fracture network, iv) the solution of the flow and mass transport problems, and v) the repetition of steps ii to iv and the collection of various output parameters to complete the Monte Carlo simulation. Following is a more complete discussion of this procedure.

The flow domain is chosen to be either square or rectangular with the possibility of specifying known hydraulic heads along all four sides, or along the two ends and no-flow boundaries along the top and bottom. The first set of boundary conditions are more flexible because flow can be defined with a specified orientation to the fracture sets (Figure 3). For these boundary conditions, hydraulic heads are specified at each corner of the domain, and hydraulic head is assumed to vary linearly along the sides. Provision is made to restrict mass transport to the central portion of the network to avoid distortions in flow near the edges caused by the boundary conditions. The second set of boundary conditions are useful because they replicate conditions for what is in effect a large column experiment.

The next step in the procedure is to generate a discrete fracture network. Our initial approach was similar to but yielded a network that is less general than other approaches such as Long et al. (9), Robinson (13), and Rouleau(14). Fractures in two dimensions are represented as sets of linear features. They are

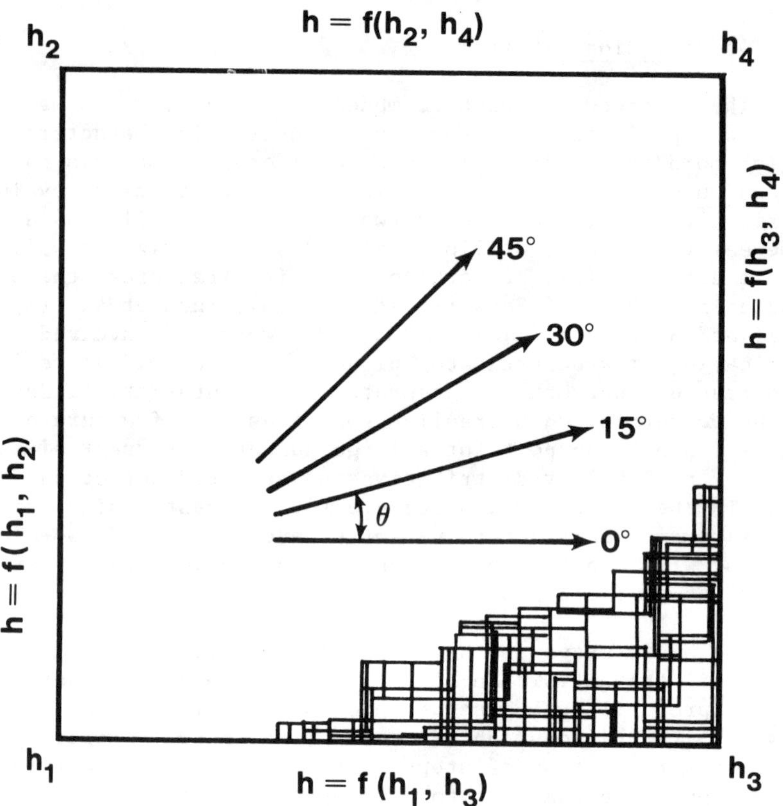

Figure 3. Examples of a discrete fracture network within a square flow domain. Boundary conditions as shown make it possible to orient flow at various angles to the network.

defined in terms of the position of their midpoint, fracture length, attitude and aperture. Some or all of these parameters are considered to be random variables.

In the work we have cited, the midpoints of fractures are located randomly within the domain. This distribution characterizes a homogenous medium where fracturing is not controlled by structural or lithological variability (12). Other distributions, which result in uniform, clustered or mixed distributions are possible but in general have not been considered in work to date. Fractures are assumed to occur in one or more distinct sets. Our work (16, 19) has considered two sets oriented at 90° with respect to one another. Long et al. (9), Robinson (13) and Rouleau (14) are able to consider an arbitrary number of fracture sets with variability in fracture attitudes.

Fracture lengths are treated as random variables with lengths sampled from either a lognormal or negative exponential distribution. Lognormal distributions for fracture apertures are justified on the basis of a limited number of studies such as those of Snow (21).

Fracture density is determined by the number of fractures added to a given domain in relation to its overall size. Most workers out of convenience utilize a constant number of fractures in each set. It is not difficult in a stochastic sense to build variability and spatial correlation into the density of fractures.

The third step in the modeling procedure is to reduce this network to what we term the essential network. The network is simplified by removing dead-end fracture segments and isolated clusters of fractures. Our work and that of Robinson (13) suggests that this procedure, in addition to reducing the computational effort, can improve the stability of the solution of the flow equation. This step has no influence on mass transport because in the absence of diffusion within the system, the dead-end segments in a two-dimensional system are inaccessible to mass.

Step iv in the procedure involves simulating fluid flow and mass transport throughout the network. The description here is a summary of a more detailed one presented by Schwartz et al. (16). Except for details in the way in which the flow equation is solved, Robinson's (13) approach is very similar. We use a finite difference procedure to calculate hydraulic heads at nodes, which are defined as the points where fractures intersect. Implicit in this approach are assumptions of parallel-walled fractures whose apertures can be related to the quantity of flow by a cubic law expression, laminar flow conditions, a rigid fracture network, and no coupling between fluid density and flow. With a sparse matrix solver, we can consider a maximum of 4000 nodes in any realization.

Mass transport is simulated using a particle tracking technique. Moving particles are added to one or more fractures at the start of the simulation, which represents an instantaneous pulse loading. These particles are advected through the network with a velocity that for individual fractures is determined from estimates of hydraulic head, aperture and fixed constants such as fluid density and viscosity. Dispersion of the swarm of particles is generated by the geometry of the network. As individual particles arrive at fracture intersections, transport directions are determined probabilistically with a distribution weighted according to the quantity of flow moving in each direction. This kind of partitioning at a node assumes perfect mixing at an intersection. While there are some experimental data to support this model (7), there are alternative models such as Endo et al. (5) that assumes partial mixing at intersections. The transport model is also constructed with the assumption that mass does not interact chemically with the fractures nor is there diffusion into the intact rock blocks.

Output from the transport simulation for a single realization describes various parameters of the breakthrough curve such as the time required for initial, 25%, 50%, 75% and 90% breakthrough, distributions of mass within the domain, or statistical tests to determine the form of mass distributions within the domain. Over a complete Monte Carlo simulation, it is possible to obtain probability distributions on output parameters of interest. Commonly, results are based on 300 to 500 realizations. At this stage, the Monte Carlo simulation is complete. We will examine in a coming section how such results can be used to learn more about dispersion in fractured media.

3.2 Three-Dimensional Systems

Efforts to study flow and transport in three-dimensional systems have been very limited. Other than work by Long et al. (10) and Long (8), which investigated fluid flow in three-dimensional networks. the only study involving mass transport is that of Smith et al. (20). All of the work in modeling flow and transport is based on deterministic approaches. The added computational effort in moving from two to three dimensions at present only permits trials with a few realizations of a Monte Carlo simulation.

The approach for dealing with three-dimensional networks is in principle the same as that for the two-dimensional problem, progressing from grid generation through the solution of the flow and then the transport problems. Each of these steps however is more difficult in three dimensions than two. Consider first the step of generating the three-dimensional network. There is a problem in defining what the shape of a fracture actually is. As is shown in Figure 4, Smith et al. (20) simulate fractures as rectangular planes, while Long et al. (10) model fractures as disks. The shape of fractures in these studies is defined more for the convenience of the numerical approaches rather than any particular notion about what fracture shapes really exist.

Simulation of fluid flow and mass transport in a three-dimensional network is more difficult than in two dimensions because of the necessity of dealing with the intersection of planes and parts of planes in space. The rough-walled character of individual fracture planes can be modeled by incorporating spatial variability in fracture apertures. For certain correlation structures, it is then possible to simulate channelling of fluid and mass within the network (11). In other respects, the statistical distributions used

● **Smith et al. (1984)**

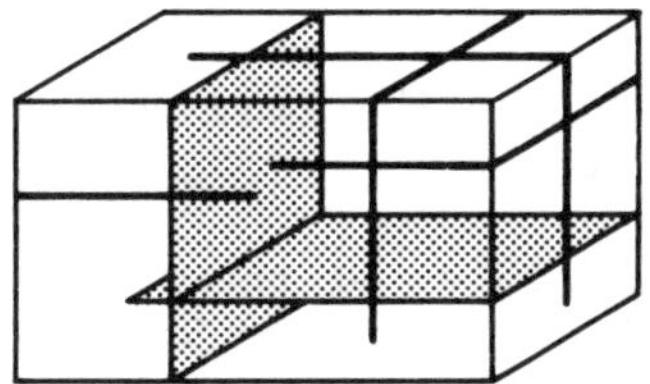

● **Long (1985)**

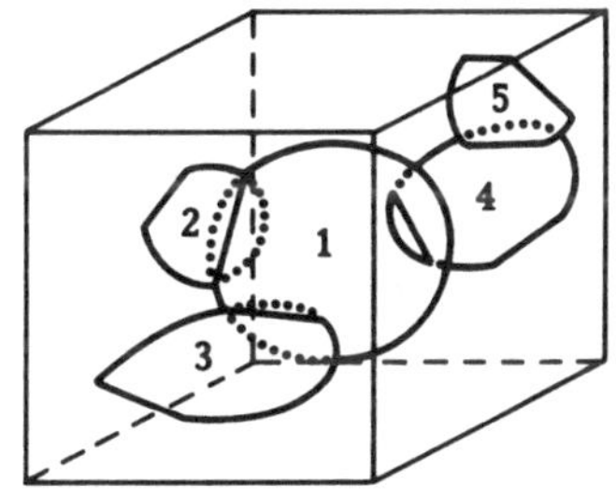

Figure 4. Two different conceptualizations of three-dimensional networks.

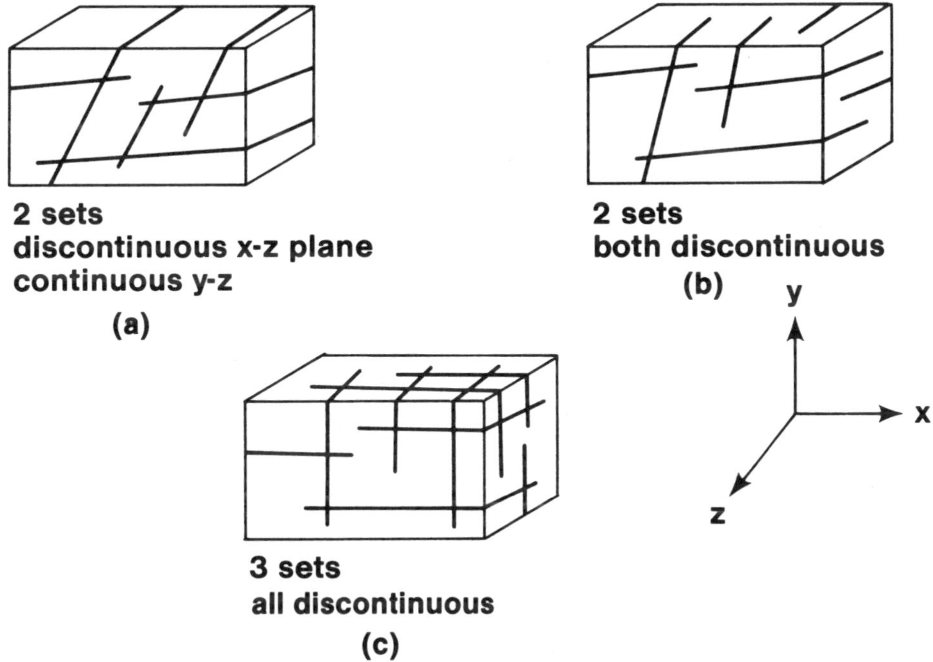

Figure 5. Examples of three different kinds of networks being generated. Moving from a) to c) the geometries become increasingly complex.

to describe the geometry of the network are the same as we discussed earlier.

Examples of the three-dimensional networks we have considered are shown in Figure 5. Moving from Figure 5a to 5c, the networks become more complex. The discussions in this paper will focus on networks of the type shown in Figure 5a. Work is underway to model transport in networks of the type illustrated in Figure 5c.

Because of the network geometry and the spatially-variable aperture, both fluid flow and mass transport is two dimensional within each fracture plane. For this reason, it is necessary to define a row and column grid on each plane and link the planes by common sets of nodes on each. Details of the algorithm used to construct the three-dimensional network of nodes are given in Smith et al. (20). Dead-end fracture segments are retained as an essential part of the network because of the possibility for fluid to circulate within those segments. However, isolated fractures and finite clusters are removed from the system.

A finite element technique is used to solve the steady-state flow problem. Because of common nodes at the fracture intersections, the global matrix equation has a large bandwidth which can approach the size of the entire matrix. Thus direct methods for solving the resulting systems of equations are not appropriate. However, an iterative solution using conjugate gradient acceleration (6) makes it feasible, on a mainframe computer, to solve systems with 50,000 nodes before computing costs become excessive.

As in the two-dimensional case, particle tracking is used to simulate mass transport. Although the implementation of this procedure is more difficult in three dimensions, the basic features of the approach described previously are the same.

4. DISPERSION IN DISCRETE NETWORKS

Discrete network models are particularly useful in studying dispersion in fractured rocks because they provide a realistic analog of the natural process. Macroscopic dispersion occurs because mass is successively partitioned at fracture intersections and spreads to occupy an increasingly larger proportion of the flow domain. The discrete approach provides a very convenient way of linking the character of dispersion to features of the fracture network. Extending this procedure through Monte Carlo simulations makes it possible to evaluate how variability in fracture geometry that may exist from realization to realization generates variability in system behavior.

What we plan to do now is to summarize some of the pertinent results concerning dispersion in two-dimensional networks and touch on results for three-dimensional systems. The two-dimensional simulations all consider relatively sparse networks of orthogonal fractures of the type shown in Figure 6. For this reason, variability in the fracture geometry from realization to realization can result in significant variability in the breakthrough curve or other characteristic parameters of dispersion. We should emphasize that it should be possible to increase fracture densities to the point where the effects on transport of variability in fracture geometries among realizations will be minimized.

For relatively sparse networks, we have observed three notable features in the pattern of dispersion. These are represented schematically in Figure 6. First there is a complex mass distribution within the domain in both the longitudinal (shown in Figure 6) and transverse directions (not shown). These distributions are caused by the relatively limited number of discrete pathways available for mass to pass through the network. A fractured medium can thus differ considerably from a homogeneous porous medium for which a much larger number of pathways can exist. Second there is a consistent negative skew in the mass distributions in a longitudinal direction (Figure 6). This skew is also reflected in the breakthrough curve on Figure 6 as the long tail. The negative skew develops as a consequence of the differing velocities in fracture sets 1 and 2 (horizontal and vertical, respectively). With the mean flow gradient aligned parallel to set 1, vertical gradients for flow can be extremely variable and often very low. The velocity

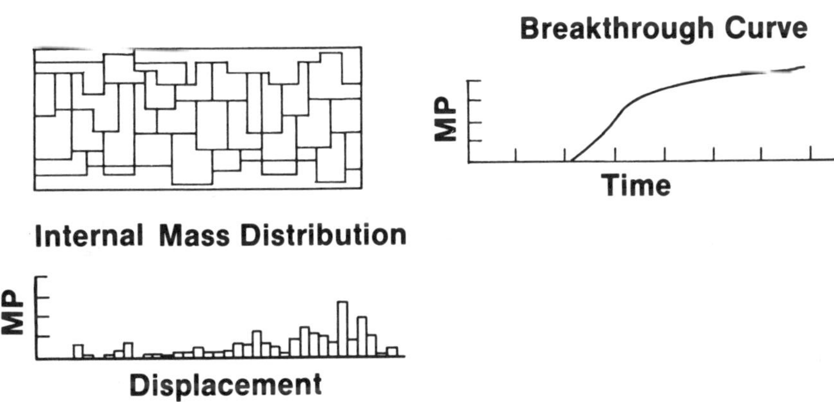

Figure 6. Schematic representation of various distributions in a single realization of a discrete network.

distribution that results as a consequence of these gradients has a mean that is lower and variance that is higher than the velocity distribution in set 1. The negative skew thus reflects the fact that over relatively short distances most of the mass is traveling in the higher velocity pathways dominated by set 1, while small quantities of mass are retarded in the lower velocity segments of set 2. This dissimilarity in the velocity distributions for the various fracture sets is what produces the skew.

The fracture geometry depicted in Figure 6 produces the maximum skew in mass distributions. As the orientation of the fracture sets is changed so that each set is oriented at 45° relative to the mean gradient, the distributions of velocities in each set are identical and no skew develops in the particle distribution.

The third feature of the generalized result depicted in Figure 6 is the very marked longitudinal dispersion that can develop even over a very short distance. Mass is spread from the upstream portion of the system to close to the outflow boundary. As we saw above, this result is produced by the variation in the velocity distribution between sets. However a point yet to be emphasized is that, as the orientation of the sets change relative to the mean hydraulic gradient, the decrease in skew represents a reduction in dispersion. Our simulation results show that for some fractured media dispersion is anisotropic, strongly related to the direction of transport.

The results from these kinds of simulations also have important implications as far as the applicability of classical concepts of dispersion are concerned. Repeated trials, concentrating on the form of internal mass distributions, have shown that the diffusional model of dispersion may not adequately describe the process of dispersion. The Gaussian distributions of mass predicted by the classical diffusional model may exist only for a narrow range of conditions. It appears that a more general model is required for fractured media. Another problem of relating classical theories to transport concerns whether it is possible to describe dispersion by means of a dispersivity value or simple dispersivity function. Our analyses, which have looked at the variance-time derivative of the mass distribution in the longitudinal direction, suggest that the pattern of spreading can be extremely variable. In some cases, the derivative becomes negative indicating that the plume can occasionally contract in size as it moves along. These results are similar to those for porous media (18) that show dispersion to be complex when there is a number of limited pathways available for transport. Mass moving in this manner does not encounter the variety of flow conditions within the network that are necessary to create a constant dispersivity except perhaps over longer transport distances.

This stochastic approach has been extended to examine in detail how fracture geometry influences mass transport (19). Again, the network consists of two orthogonal sets aligned approximately parallel and perpendicular to the mean hydraulic gradient, so that set 1 forms the dominant pathway. The second set of fractures provides connecting pathways for transport between the discontinuous fractures in set 1.

Sensitivity analyses show that mass transport is controlled mainly by the connectivity of the network between the inflow/outflow boundaries. When the connection is more direct, there is an increased probability that mass moves predominantly through set 1. When the connection is less direct, more weight is placed on connecting fractures of set 2 to establish pathways between the discontinuous fractures of set 1. The effects of a less-direct connection through the network include: i) a lower fluid velocity in set 1, ii) a greater variability in velocity within set 1, iii) longer travel times through the network, and iv) a greater variability in arrival times. Features of the fracture geometry that lead to this kind of behavior include: i) a reduction in the number of fractures comprising set 1 and ii) a reduction in the minimum length of fractures forming set 1. The effects of a more direct pathway are just the opposite to those cited above.

The magnitude of dispersive effects is also influenced by the connectivity of the network. An increase in macroscopic dispersion in the longitudinal direction occurs i) if the mean velocity within fracture set 2 decreases relative to set 1, ii) if the variability in fluid velocity, particularly within set 1 increases, and iii) if the average path length through the connecting fractures of set 2 increases.

This concept of connectivity can be used to predict the behavior of mass in a fracture network. In examples where parameters of the geometry were varied to exaggerate indirect connectivity (19), the expected behavior of transport relative to a more-directly connected network was correctly predicted. We are not able to present detailed results here, but experience from this study suggested that close links can be developed between features of the geometry and the pattern of transport. In a later section, we will describe an approach to simulating transport in a continuum that looks at developing this link as a primary objective.

The results from simulations of three-dimensional systems are as yet limited. However, an example for a single realization is presented to illustrate the kind of results we have been able to obtain (20). The fracture network of interest is shown in Figure 7, which depicts the trace of fractures on the xz face of the flow domain. This region of interest is a cube with each face 10 m wide.

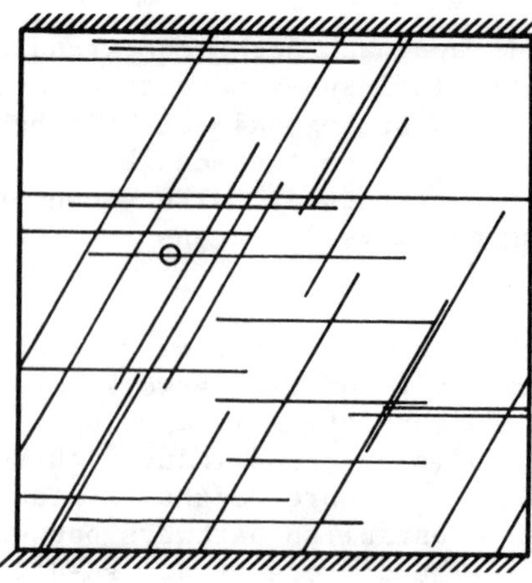

Figure 7. Trace of the network on the xy face of the flow domain. The circle is the projection of an injection well located in the centre of one fracture plane.

Fractures are continuous across the domain in the y-direction with the two sets intersecting at 60°. There are 15 fractures in each of two sets, forming a network with 56 intersections. A total of 5222 nodes are used to define the finite element mesh for this network.

Fluid flow is established between the left and right faces of the domain by assigning constant head values. An injection well adds fluid to a single fracture. The position of the well is depicted on Figure 7. The four other boundaries of the domain are assumed to be no-flow.

Moving particles are added to the system at the injection well. For the sample simulation, 2000 particles are released. Figure 8a-c illustrates the residence time distribution curves (fraction of the total particles released which cross the downstream constant head boundary per day) for this network. A filter has been applied to these breakthrough data to eliminate high frequencies in the residence-time curve introduced by using discrete time steps in the moving particle model. The three curves compare the transport behavior for cases of a constant aperture of 100 microns (Figure 8a) and spatially correlated apertures (Figures 8b and 8c). These latter two examples, which are for two different realizations from the same set of statistical parameters, have a standard deviation in aperture of 0.10 (base 10 logarithms) and integral scales in both coordinate directions of 2.0 m.

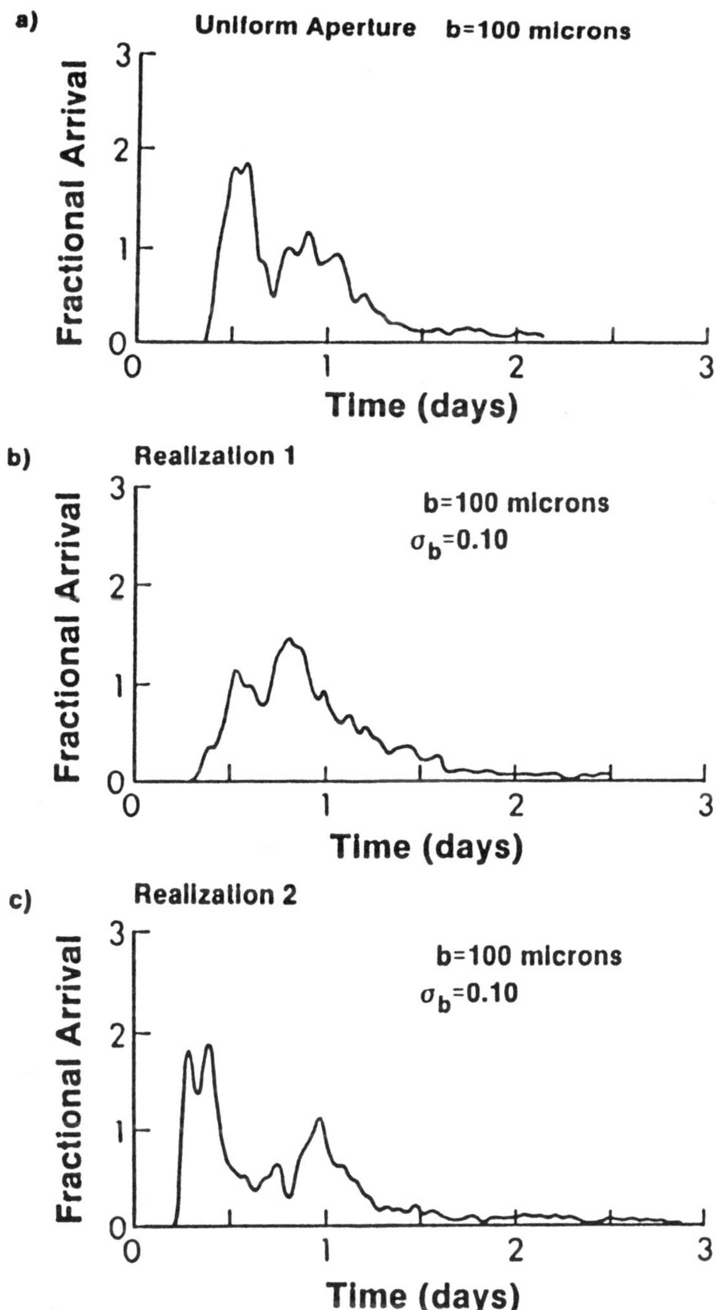

Figure 8. Residence time distribution curves for a) uniform aperture and b) and c) examples of spatially correlated apertures.

The residence-time distribution curves reflect the travel time for particles moving along streamlines established by the interaction of the radial flow field with the regional gradient, together with the effects of mixing that occurs at fracture intersections. The two peaks indicate that there are two preferred pathways for transport from the injection well to the downstream boundary. The long tail occurs because various fractions of the particles either follow circuitous routes to the outflow boundary, enter fractures of set two which have relatively low fluid velocities, or encounter lower fluid velocities near the bounding edges of fractures in set 1. Figures 8b and 8c, for rough-walled fractures, show that although the main features of the parallel plate model are retained, considerable smaller-scale variation can be imposed on the residence time curve by the variable apertures.

At the present time, work is continuing on the development of codes that allow more general fracture geometries and a larger number of fractures. Preliminary results suggest that three-dimensional effects are of considerable importance in determining the character of transport in fractured rocks.

5. NEW CONTINUUM APPROACH

One of the major limitations of the discrete modeling approaches that we have finished discussing is the relatively small number of fractures that can be included in a simulation. What this means is that transport in large networks has to be simulated using some kind of continuum approach. However, such an approach is oversimplified because the diffusional model of dispersion that forms the basis for all existing models may not adequately describe dispersion. We have been working to develop a new continuum approach for dispersion to account more realistically for the influences of fractures and fracture geometry on dispersion.

Although details of the modeling approach have been presented elsewhere (17) a summary will be presented here to help readers understand the method. The essence of the technique is to formulate the transport of mass in a continuum as a random-walk problem (1, 15). In the conventional application of the particle tracking method, advection is accounted for the deterministic particle motion, and dispersion by some probabilistic motion. This latter component of particle motion is modeled using statistics that produce Gaussian mass distributions parallel and perpendicular to the direction of groundwater flow. Where our continuum approach differs from the conventional approach is in the treatment of particle motion. We do not a priori assume a classical model for particle motion. The way in which the reference particles are moved is determined as part of the simulation problem.

To characterize transport, we create a discrete fracture model of the type discussed earlier for one or more sub-domains of the continuum. The sub-domain has a fracture geometry similar to that of the continuum, but the number of fractures involved is typically much smaller. By statistically characterizing how reference particles move through the network, we can generate a unique description of particle motion that can be extrapolated to the continuum. Thus without explicitly including fractures at the continuum scale, mass spreading is simulated in a way that more realistically accounts for the effects of fracturing.

Within the discrete sub-model, a swarm of reference particles, usually 500, is used to compile probability distributions on the negative logarithm of the seepage velocity in the directions of possible transport. For two orthogonal fracture sets, there are four such directions. The velocity distributions are constructed from a relatively large number of observations because each particle experiences approximately 50 to 100 velocity changes in a network consisting of approximately 4000 nodes. In addition, the paths taken by the particles can be interpreted to provide a histogram of fracture length for both sets, and estimates of the probability of a particle moving in different directions at a fracture intersection.

Figure 9 illustrates how these distributions on particle direction, fracture length, and directional velocity are sampled to simulate particle motion through the continuum. Note also that the particle motion within the continuum accounts for the boundary conditions. Particles reflect from no-flow boundaries and pass through flux boundaries. In practice, the swarm of reference particles can either be moved through the domain one after another or moved together through a time step. In the simulation trials to date, the cumulative travel times are used to construct breakthrough curves.

5.1 Verification

The procedure that is used to verify the accuracy of the modeling approach involves a comparison of breakthrough curves for a discrete sub-model with dimensions of 62.25x21.0 m and a continuum of the same size. If the continuum approach successfully models transport in the discrete network, the breakthrough curves for these two examples should be nearly identical. The results of this test are shown in Figure 10 for two different examples. The correspondence between the two approaches is excellent considering that the continuum approach really is only based on rudimentary information about the fracture network.

The success of these preliminary trials provides evidence that the approach is applicable to modeling transport in more complex

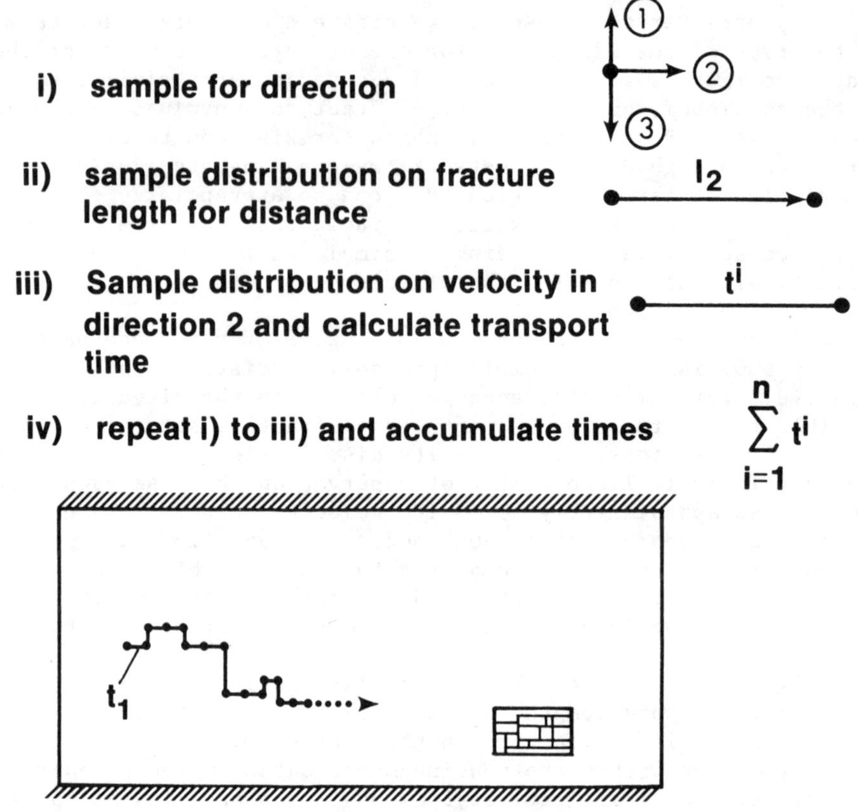

Figure 9. Schematic representation of distributions to be sampled and how an individual particle moves through the continuum.

systems. Work is ongoing to test the method further and to develop realistic applications based on fully two-dimensional systems. More extensive applications are being hindered by conceptual problems that remain to be overcome, the most serious of which is the question of what fracture density is necessary to assure that an ergodic hypothesis holds for the network. When the density of fracturing is relatively low the transport behavior of an individual particle through the network is not the same as that for the particle swarm through time. What gives rise to this non-ergodic behavior is that individual particles moving through the network may not sample the broad range of velocities that can exist, especially in the near-vertical fractures.

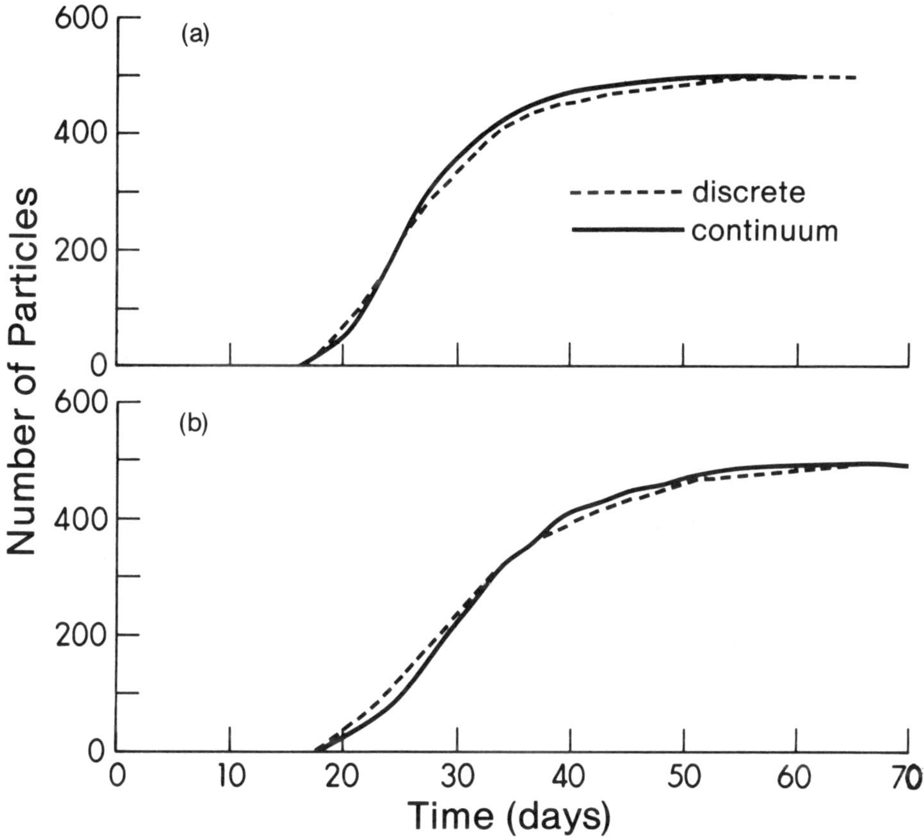

Figure 10. Comparison of breakthrough curves obtained using discrete and continuum approaches.

6. CONCLUDING COMMENTS

Stochastic approaches to modeling mass transport in fractured rock systems have contributed significantly in developing an understanding of this complex subject. However, results are still very modest considering the critical problems of a conceptual-theoretical nature that remain to be studied. The question of how closely parameters of fracture networks conform to the concept of an REV at given scales bears heavily on how well concepts of REV survive spatial scaling. This problem has not been studied in detail and yet it is fundamental to modeling of fractured rocks. There are still major uncertainties in how the hydraulic properties of fractured rocks are idealized in models. A continuing problem concerns how well the parallel plate model describes flow within fractures. Recent experimental evidence suggests that rough-wall models, accounting for both aperture variability and contact area, may be necessary to accurately represent flow.

Contrasting these kinds of problems are those of a practical nature that provide a major impediment to progress. For example, efforts in extending discrete fracture modeling to three dimensions are constrained by computer limitations. Tremendous resources are required in terms of storage and execution times to simulate transport in even sparse networks when, for example, spatial structure is built into apertures. However, given the progress in fractured rock hydrology over the past few years, we remain optimistic that these problems will be overcome.

7. ACKNOWLEDGEMENTS

Our continuing work into mass transport in fractured rocks has been supported primarily through a series of contracts provided by Atomic Energy of Canada Limited as part of the Canadian Nuclear Fuel Waste Management Program. Other funding has been provided to both principal investigators by grants from the Natural Sciences and Engineering Research Council of Canada. Charles Mase is a joint investigator in the three-dimensional model studies.

8. REFERENCES

1. Ahlstrom, S.W., Foote, H.P., Arnette, R.C., Cole, C.R., and Serne, R.J., "Multicomponent mass transport model: Theory and numerical implementation (discrete-parcel-random walk version)", *Rep. BNWL-2127*, Battelle Pacific Northwest Lab., Richland, Wash., 1977.
2. Castillo, E., Karadi, G.M., and Krizek, R.J., "Unconfined flow through jointed rock," *Water Resources Bulletin*, Vol. 8, No. 2, 1972, pp. 266.
3. Castillo, E., Krizek, R.J., and Karadi, G.M. "Comparison of dispersion characteristic in fissured rocks," *Proceedings of the Symposium on Fundamentals of Transport Phenomena in Porous Media*, University of Guelph, Guelph, Ontario, Canada, Vol. 2, 1972, pp. 778 - 797.
4. Dagan, G., "Statistical theory of groundwater flow: pore- to laboratory-, laboratory- to formation- and formation- to regional- scale," *Greco 35 Hydrogeologie*, Ministere De La Recherche et la Technogie Centre National de la Recherche Scientifique, 1985, pp. 1-32.

5. Endo, H.K., Long, J.C.S., Wilson, C.R., and Witherspoon, P.A, "A model for investigating mechanical transport in fracture networks," *Water Resources Research,* Vol 20, No. 10, 1984, pp. 1390-1400.
6. Hageman, L.A., and Young, D.A., *Applied Iterative Methods,* Academic Press, New York, N.Y., 1981.
7. Krizek, H.J., Karadi, R.M., and Socias, E., "Dispersion of a contaminant in fissured rock," presented at the 1972, International Society of Rock Mechanics Symposium on Percolation Through Fissured Rocks, held at Stuttgart, Germany.
8. Long, J.C.S. "Verification and characterization of continuum behaviour of fractured rock at AECL Underground Research Laboratory," *BMI/OCRD-17,* Office of Crystalline Repository Development, Battelle Memorial Institute, 1985, p. 239.
9. Long, J.C.S., Remer, J.S., Wilson, C.R., and Witherspoon, P.A., "Porous media equivalents for networks of discontinous fractures," *Water Resources Research,* Vol. 18, No. 3, 1982, pp. 645-658.
10. Long, J.C.S., Gilmour, P., and Witherspoon, P.A., "A model for steady fluid in random three-dimensional networks of disc-shaped fractures," *Water Resources Research,* Vol. 21, No. 6, 1985, 1985.
11. Moreno, L., Neretnieks, I., and Eriksen, T., "Analysis of some laboratory tracer runs in natural fissures," *Water Resources Research,* Vol. 21, No. 7, 1985, pp. 951-958.
12. Priest, S.D., and Hudson, J.A., "Discontinuity spacing in rock," *International Journal of Rock Mechanics and Mining Sciences,* Vol. 13, 1976, pp. 135-148.
13. Robinson, P.C., "Connectivity flow and transport in network models of fractured media," *TP1072,* Theoretical Physics Division, A.E.R.E., Harwell, U.K. 1984.
14. Rouleau, A., "Statistical Characterization and Numerical Simulation of a Fracture System - Application to Groundwater Flow in the Stripa Granite," thesis presented to the University of Waterloo, at Waterloo, Ontario, Canada, in 1984, in partial fulfillment of the requirements for the degree of Doctor of Philosophy.
15. Schwartz, F.W., "Macroscopic dispersion in a porous media: The controlling factors," *Water Resources Research,* Vol. 13, No. 4, 1977, pp. 743-752.
16. Schwartz, F.W., Smith, L., and Crowe, A.S., "A stochastic analysis of macroscopic dispersion in fractured media," *Water Resources Research,* Vol. 19, No. 5, 1983, pp. 1253-1265.
17. Schwartz, F.W., and Smith, L., "A new continuum approach for modeling dispersion in fractured media," *Proceedings of the 17th International Congress on Hydrogeology of Rocks of Low Permeability,"* International Association of Hydrogeologists, 1985, pp. 538-546.

18. Smith, L., and Schwartz, F.W., "Mass transport, 1, A stochastic analysis of macroscopic dispersion," *Water Resources Research* Vol 16, No. 2, 1980, pp. 303-313.
19. Smith, L., and Schwartz, F.W., "An analysis of the influence of fracture geometry on mass transport in fractured media," *Water Resources Research,* Vol. 20, No. 9, 1984, pp. 1241-1252.
20. Smith, L., Mase, C.W., and Schwartz, F.W., "A stochastic model for transport in networks of planar fractures," *Greco 35 Hydrogeologie,* Ministere De La Recherche et la Technologie Centre National de la Recherche Scientifique, 1985, pp. 523-536.
21. Snow, D.T., "The frequency and apertures of fractures in rock," *International Journal of Rock Mechanics and Mineral Sciences* Vol. 7, 1970, pp. 23-40.

SENSITIVITY ANALYSIS OF GROUND-WATER MODELS

Carl D. McElwee

SENSITIVITY ANALYSIS OF GROUND-WATER MODELS

Carl D. McElwee

Kansas Geological Survey
The University of Kansas
1930 Constant Avenue
Lawrence, Kansas, 66046, U.S.A.

ABSTRACT

One of the most difficult tasks in ground-water modeling is the estimation of aquifer parameters from field measurements of hydraulic head. This paper examines model sensitivity through the use of sensitivity analysis. For each model parameter one can define a sensitivity coefficient. These sensitivity coefficients depend on the choice of model, the spatial coordinates, the time variable, the number and type of model parameters, and the boundary conditions. For good sensitivity to the parameters, all sensitivity coefficients should be independent and as large as possible at the locations and times of interest. Methods for determining sensitivity coefficients are discussed and some typical examples showing certain important characteristics are presented. The sensitivity coefficients can be used to estimate variances and confidence intervals for the aquifer parameters. The model sensitivity can be increased for parameter estimation by applying some general principles from sensitivity analysis. Several examples of improved sensitivity are presented.

1. INTRODUCTION

One of the most difficult tasks in ground-water modeling involves estimation of the aquifer parameters to be used in a predictive ground-water model. Usually some historical hydraulic head data are available along with some field or laboratory estimates of the aquifer parameters. These data are usually sparse and of varying quality. Estimation of the aquifer parameters from hydraulic head data is generally recognized to be difficult and may be unstable or nonunique [Yakowitz and Duckstein (19)]. One goal of this paper is to develop a better understanding of model sensitivity to the aquifer parameters.

Sensitivity analysis is the primary tool used to investigate parameter estimation in this paper. A general first-order sensitivity analysis formalism is presented to calculate the perturbed head caused by a change in an aquifer parameter. A central figure in the formalism is the sensitivity coefficient. When the sensitivity coefficients are known, the parameter variances can be calculated. A large parameter variance means the model is not very sensitive to that parameter. Therefore, some time is devoted to methods for determining sensitivity coefficients. The sensitivity coefficients are affected by the choice of model, the number and type of parameters, and the boundary conditions. In addition, the sensitivity coefficients are functions of space and time. Several examples of sensitivity coefficients for a variety of models and boundary conditions are presented.

The final goal of this paper is to develop some general principles or guidelines for designing models that have the desired sensitivity to model parameters. This is done by looking in detail at how the sensitivity coefficients enter the least squares estimation procedure and how the sensitivity coefficients are affected by the model specification. The model specification includes such things as the model equations, the boundary conditions, and the number of parameters or parameter zones. In general, one might suppose it is desirable to make the sensitivity coefficients as large as possible. Also, it might be that the chosen parameters should be independent. In section 8 these questions are dealt with in more detail and some general conclusions are stated. Several examples of improved sensitivity, obtained by applying the guidelines, are also presented in section 8.

2. GENERAL DEFINITION OF SENSITIVITY COEFFICIENTS

When a model is used to describe the hydraulic head distribution (h) in a ground-water system, the head is assumed to depend uniquely upon the physical parameters input to the model.

$$h = h\left[\underline{x},t;T(\underline{x}),S(\underline{x}),Q(\underline{x},t)\right] \tag{2.1}$$

T, S, and Q are respectively transmissivity, storativity, and flux of water in or out of the system. $\underline{x}$ is a vector containing the appropriate number of coordinates for the model dimensionality (usually one or two). It has been assumed that the transmissivity and storativity do not depend upon time (t). This assumption means that unconfined aquifers and delayed yield will not be explicitly considered. In this work $Q(\underline{x},t)$ will be assumed to be known. Only the variation of the model response to changes in $T(\underline{x})$ and $S(\underline{x})$ will be considered.

In studying the sensitivity of a ground-water-flow system to parameter variations, a mathematical model must be specified:

$$F\left[\underline{x},t,h;T(\underline{x}),S(\underline{x}),Q(\underline{x},t)\right] = 0. \tag{2.2}$$

The model specification represented symbolically by F in Eq. (2.2) includes such things as initial conditions, boundary conditions, and the appropriate differential equations. Eq. (2.2) may be solved analytically or numerically.

2.1 Models with Constant Parameters

For models with constant parameters, the solution to (2.2) can be written as

$$h = h(\underline{x},t;T,S,Q). \tag{2.3}$$

Consider the variation of one of the parameters, T, for example. Eq. (2.2) now becomes

$$F(\underline{x},t,h^*;T+\Delta T,S,Q) = 0 \tag{2.4}$$

where h^* is the perturbed head. Solution of (2.4) gives

$$h^* = h^*(\underline{x},t;T+\Delta T,S,Q). \tag{2.5}$$

The sensitivity coefficient for variations in T is defined as

$$U_T(\underline{x},t;T,S,Q) = \frac{\partial h}{\partial T} = \lim_{\Delta T \to 0} \frac{\Delta h}{\Delta T}, \tag{2.6}$$

where $\Delta h = h^*-h$.

A similar development for the storage coefficient allows the two sensitivity coefficients to be written as

$$U_T(x,t) = \frac{\partial h(x,t;T,S,Q)}{\partial T} \tag{2.7}$$

$$U_S(x,t) = \frac{\partial h(x,t;T,S,Q)}{\partial S} \; . \tag{2.8}$$

The functional dependence on T, S, and Q has been dropped on the left-hand side of Eqs. (2.7) and (2.8) for convenience in writing U_T and U_S. However, the sensitivity coefficients do depend on T, S, Q, the initial conditions, the boundary conditions, and the underlying model Equations.

The solution of the flow equation (2.4) is assumed to depend analytically upon the parameters T and S, and T, S, and Q are independent of each other. Now consider a perturbation of the transmissivity, ΔT . Since it has been assumed that the solutions depend analytically on the parameters, the function $h^*(\underline{x},t;T+\Delta T,S,Q)$ may be expanded into a Taylor series. If ΔT is small, the second- and higher-order terms may be neglected.

$$h^*(\underline{x},t;T+\Delta T,S,Q) \simeq h(\underline{x},t;T,S,Q) + U_T\Delta T \tag{2.9}$$

Thus the new head produced by a perturbation in transmissivity (ΔT) may be calculated from (2.9) if the sensitivity coefficient and the unperturbed head are known. Similarly, if a perturbation in storage coefficient (ΔS) occurs, the perturbed head is given by

$$h^*(x,y,t;T,S+\Delta S,Q) \simeq h(x,y,t;T,S,Q) + U_S\Delta S \tag{2.10}$$

to first order in ΔS .

Eqs. (2.7) and (2.8) show that it would be desirable to calculate U_T and U_S for a given model, if possible. Then the response of the model to various perturbations could be calculated simply from (2.7) or (2.8) without actually evaluating the model equations again. The work of McElwee and Yukler (11) indicates that Eqs. (2.9) and (2.10) should be valid for parameter variations of about 20% or less.

2.2 Models With Spatially Varying Parameters

In the more general case where transmissivity and storativity vary with space, a slightly different procedure is used to define the sensitivity coefficients. h will be the hydraulic head resulting from a transmissivity distribution $T(\underline{x})$. Let h* represent the hydraulic head that results when the transmissivity distribution is changed at one point ($\underline{x}_o$) by a small amount $\Delta T(\underline{x}_o)$.

$$\frac{\Delta h(\underline{x},t;\underline{x}_o)}{\Delta T(\underline{x}_o)} = [h^*(\underline{x},t;T(\underline{x})+\delta(\underline{x}-\underline{x}_o)\Delta T(\underline{x}_o),S(\underline{x}),Q(\underline{x},t))$$

$$-h(\underline{x},t;T(\underline{x}),S(\underline{x}),Q(\underline{x},t))]/\Delta T(\underline{x}_o) \qquad (2.11)$$

The symbol $\delta(\underline{x}-\underline{x}_o)$ represents the Dirac delta function [Lighthill (10)]. x is assumed to be a unitless variable so that $\delta(\underline{x}-\underline{x}_o)$ is also unitless. The sensitivity with respect to variations in transmissivity is defined as

$$U_T(\underline{x},t;\underline{x}_o) = \lim_{\Delta T(\underline{x}_o)\to 0} \frac{\Delta h(\underline{x},t;\underline{x}_o)}{\Delta T(\underline{x}_o)} . \qquad (2.12)$$

This sensitivity coefficient tells how much the head will be changed at point $\underline{x}$ due to a change in transmissivity $\Delta T(\underline{x}_o)$ at point $\underline{x}_o$. Since $\Delta T(\underline{x}_o)$ is assumed to be small, a first-order expansion may be employed to obtain

$$h^* \simeq h + U_T(\underline{x},t;\underline{x}_o)\Delta T(\underline{x}_o). \qquad (2.13)$$

If the transmissivity is changed at more than one point, then the change in head Δh must be found by integrating over the area (or volume), A (or V), where $T(\underline{x})$ is changed

$$h^* - h = \Delta h(\underline{x},t) = \int_A U_T(\underline{x},t;\underline{x}_o)\Delta T(\underline{x}_o)d\underline{x}_o. \qquad (2.14)$$

Outside the region A (or V), $\Delta T(x_o)$ is zero. In the special case when the transmissivity is to be changed by a constant amount ΔT everywhere, Eq. (2.14) becomes

$$h^* - h = \Delta h(\underline{x},t) = \Delta T \cdot U_T(\underline{x},t) \qquad (2.15)$$

where

$$U_T(\underline{x},t) = \int_A U_T(\underline{x},t;\underline{x}_o)d\underline{x}_o . \qquad (2.16)$$

In the preceding development, $\underline{x}$ and $\underline{x}_o$ are assumed to be appropriate dimensionless variables.

A similar development for the sensitivity with respect to storativity (U_S) yields

$$h^* - h = \Delta h(\underline{x},t) = \int_A U_S(\underline{x},t;\underline{x}_o)\Delta S(\underline{x}_o)d\underline{x}_o \quad (2.17)$$

and

$$h^* - h = \Delta h(x,t) = \Delta S \int_A U_S(\underline{x},t;\underline{x}_o)d\underline{x}_o = \Delta S \cdot U_S(\underline{x},t) \quad (2.18)$$

when ΔS is constant over the region of integration. As before,

$$U_S(\underline{x},t;\underline{x}_o) = \lim_{\Delta S(\underline{x}_o)\to 0} \frac{\Delta h(\underline{x},t;\underline{x}_o)}{\Delta S(\underline{x}_o)} \quad (2.19)$$

where

$$\Delta h(\underline{x},t;\underline{x}_o) = h^*(\underline{x},t;T(\underline{x}),S(\underline{x})+\delta(\underline{x}-\underline{x}_o)\Delta S(\underline{x}_o),Q(\underline{x},t)) - h(\underline{x},t;T(\underline{x}),S(\underline{x}),Q(\underline{x},t)). \quad (2.20)$$

The sensitivity coefficients U_T and U_S are seen to be the quantities needed to calculate the response of a model to perturbations in the spatial distribution of transmissivity and storage. Consequently, a discussion of some of the general properties of sensitivity coefficients is in order. In later sections, procedures for determining sensitivity coefficients will be illustrated.

2.3 General Comments About Sensitivity Coefficients

The sensitivity coefficients will depend on the independent variables (space and time), the model parameters (T, S, etc.), and the boundary conditions. Each of these factors may have a dramatic effect on the model sensitivity and thus on any attempt to perform inverse calculations. Much of this paper will be concerned with studying these effects. The general confined flow equation can be written for some region (R, Figure 1) as

$$\frac{\partial}{\partial \underline{x}}\left[\frac{T(\underline{x})}{T_{max}}\frac{\partial h}{\partial \underline{x}}\right] = \left[\frac{S_{max}}{T_{max}}\right]\left[\frac{S(\underline{x})}{S_{max}}\right]\frac{\partial h}{\partial t} - \frac{Q'(\underline{x},t)}{T_{max}} \quad (2.21)$$

where T_{max} and S_{max} are the maximum values of the transmissivity and storativity, respectively. Q' is the specified water flux per unit area of the model (Q/A) and would be caused by pumpage, injection, leakage, etc. In addition to the flow equation, we need boundary and initial conditions for a complete solution to the hydraulic head and the sensitivity coefficients. Typical boundary conditions are head specified

$$h = H \text{ on } \Gamma_1 \tag{2.22}$$

and flux specified

$$Q''(\Gamma_2,t) = -T\left[\frac{\partial h}{\partial n}\right]_{\Gamma_2} \text{ on } \Gamma_2 . \tag{2.23}$$

Γ is the boundary ($\Gamma = \Gamma_1 + \Gamma_2$) of region R (Figure 1). Q" is the specified flux per unit length (Q/ℓ) of boundary Γ_2. The initial condition can be specified as

$$h(\underline{x},0) = f(\underline{x}) \text{ in } R . \tag{2.24}$$

Eqs. (2.21) through (2.24) illustrate how the head and, consequently, the sensitivity coefficients depend upon the transmissivity and storativity. The head can be written symbolically as

$$h = h\left[\underline{x}, \frac{T_{max}}{S_{max}} t; \frac{T(\underline{x})}{T_{max}}, \frac{S(\underline{x})}{S_{max}}, \frac{Q(\underline{x})}{T_{max}}\right] \tag{2.25}$$

provided Q and the boundary conditions do not depend upon time. If Q and the boundary conditions do depend upon time, then additional time dependence besides $[(T_{max}/S_{max})t]$ may be introduced in the head solution. $T(\underline{x})/T_{max}$ and $S(\underline{x})/S_{max}$ represent normalized distributions for the transmissivity and storativity which vary between the limits of zero and one. They specify the shape of the T and S variation, but not the absolute magnitudes.

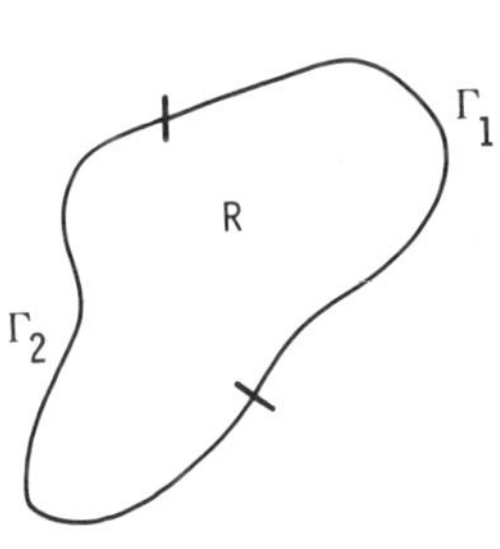

Fig. 1. Region of solution and boundaries for a model.

Next, a few simple observations are made regarding sensitivity coefficients. At steady state ($\partial h/\partial t = 0$), h does not depend on $S(\underline{x})$. Therefore, U_S is zero at steady state and is small as one nears steady state. In the homogeneous case $T(\underline{x}) = T_{max}$ and $S(\underline{x}) = S_{max}$ and Eq. (2.25) reduces to

$$h = h(\underline{x}, \frac{T}{S} t) \tag{2.26}$$

when the specified fluxes, $Q(\underline{x})$, are zero. If all specified fluxes are zero, only barrier or specified head boundaries are used. From Eqs. (2.26), (2.7), and (2.8), the sensitivity coefficients U_T and U_S can be shown to be dependent:

$$U_T = -\frac{S}{T} U_S. \tag{2.27}$$

This means that the inverse problem is nonunique since any value of S and T with the correct ratio will give a good solution.

Even if one allows spatial variation in T and S in Eq. (2.25), the inverse problem is still nonunique since any S_{max} and T_{max} having the same ratio will give an equally good solution provided there are no specified fluxes in the model. Theoretically, a specified T on each streamline will uniquely determine the transmissivity distribution and, consequently, the storativity distribution. Some inverse procedures [Knowles et al. (9)] solve for the fluxes, Q, in addition to the transmissivity and storativity. The foregoing discussion and Eq. (2.25) show that, when both Q and T are adjusted, an additional level of nonuniqueness is introduced since only the ratios of Q and T need to be held constant. In others words, Q and T are not independent parameters.

Some initial condition must be specified for the hydraulic head at the beginning of a model simulation; consequently, this condition determines the initial condition on the sensitivity coefficients. The sensitivity coefficients may or may not start out zero. One commonly used initial condition is a flat head distribution. In this case, the sensitivity coefficients (U_T and U_S) have an initial value of zero. Another commonly used initial condition is a steady-state head distribution. (Additional fluxes, $Q(\underline{x})$, are imposed and future changes are predicted.) For a steady-state initial condition, U_S is zero, but U_T will not in general be zero. The initial values of U_T may be found by solution of equations to be discussed later.

When the sensitivity coefficients are zero or not independent, the inverse process will not work. This result is inherent in the model and does not depend on the details of the inverse process. In actual practice, the inverse process may experience difficulty when the sensitivity coefficients are very small but not zero. In this case, the model is simply not very sensitive to changes in aquifer parameters. By calculating and examining the sensitivity coefficients, one may obtain an indication of the stability of the inverse process. As a rule-of-thumb, the sensitivity coefficients should be as large as possible and be independent for a stable inverse problem.

3. METHODS FOR DETERMINING SENSITIVITY COEFFICIENTS

In this section various methods for determining sensitivity coefficients are considered. Three methods will be considered: analytical expressions, finite difference approximations, and solution of a partial differential equation.

3.1 Analytical Expressions For The Sensitivity Coefficients

Sometimes analytical formulas for the head or drawdown can be found for simple models. In many of these cases, finding convenient analytical expressions for the sensitivity coefficients is also possible. As an example, consider the Theis equation. The Theis equation [Theis (18)] describes radial confined ground-water flow in a uniformly thick, horizontal, homogeneous, isotropic aquifer of infinite areal extent.

$$s = \frac{Q}{4\pi T} \int_{(r^2 S/4Tt)}^{\infty} \frac{e^{-u}}{u}\, du \tag{3.1}$$

In the above equation s is drawdown (L), Q is the discharge (L^3/T), T is the transmissivity (L^2/T) , t is the time (T), S is the dimensionless storage coefficient, and r is the radial observation distance from the pumped well (L).

The sensitivity coefficients may be obtained from Eq. (3.1) by applying the definitions given in equations (2.7) and (2.8). After applying Leibnitz's rule for differentiating an integral [Hildebrand (8)] to Eq. (3.1), one obtains [McElwee and Yukler (11)]

$$U_S = \frac{\partial s}{\partial S} = - \frac{Q}{4\pi TS} \exp(- \frac{r^2 S}{4Tt}) \tag{3.2}$$

and

$$U_T = \frac{\partial s}{\partial T} = - \frac{s}{T} - \frac{S}{T} U_s \tag{3.3}$$

These equations for the sensitivity coefficients may be evaluated quite easily if one can evaluate the drawdown (s). Notice that, if not for the first term on the right of Eq. (3.3), U_T and U_S would be dependent. Eqs. (3.2) and (3.3) will be discussed and plotted in a later section of this paper.

As a further example, consider the leaky confined aquifer. The aquifer system, defined by Hantush and Jacob (7), is composed of a level, isotropic, homogeneous, porous medium of infinite areal extent. The lower aquifer boundary is assumed to be impervious, while the upper boundary is assumed to be a leaky confining bed. Water is derived from the aquifer by elastic expansion of the water and compression of the aquifer matrix as pumping occurs. Leakage through the semiconfining bed is assumed to be proportional to the drawdown in the semiconfined aquifer. It is assumed that no water is removed from storage in the semiconfining unit and that no drawdown occurs in the source bed. The analytical solution for the drawdown is

$$s = \frac{Q}{4\pi T} \int_u^\infty \frac{1}{y} \exp\left(-y - \frac{L^2 r^2}{4y}\right) dy, \tag{3.4}$$

$$u = r^2 S/4Tt, \quad L^2 = K'/Tm'$$

where K' is the permeability of the semiconfining bed, m' is the thickness of the semiconfining bed, and the other quantities are the same as for the Theis equation. By applying Leibnitz's rule for differentiating an integral [Hildebrand (8)], obtaining the sensitivity coefficients with respect to S and T is easy [Cobb, McElwee, and Butt (2)]:

$$U_S = \frac{\partial s}{\partial S} = - \frac{Qr^2}{16\pi T^2 t} \left[\frac{1}{u} \exp\left(-u - \frac{L^2 r^2}{4u}\right)\right] \tag{3.5}$$

$$U_T = \frac{\partial s}{\partial T} = - \frac{s}{T} - \frac{S}{T} U_S. \tag{3.6}$$

As before, these expressions are easy to evaluate if the drawdown has been calculated previously.

These two examples of analytical expressions for the sensitivity coefficients are typical. Many more examples could be presented for simple models.

3.2 Finite Difference Expression For Sensitivity Coefficients

Sometimes it is not convenient or possible to come up with an analytical formula to be evaluated for the sensitivity coefficients. In these cases, one can always evaluate the sensitivity coefficients numerically by a finite difference approximation if the head or drawdown can be calculated. To illustrate this process, we continue with the leaky confined aquifer of the previous section. If we try to evaluate U_L, the sensitivity with respect to leakage, by differentiating Eq. (3.4), we obtain [Cobb, McElwee, and Butt (2)]

$$U_L = \int_u^\infty \{-Lr^2/2y^2\} \exp(-y - L^2r^2/4y)\, dy. \tag{3.7}$$

Note that both U_S and U_T in Eqs. (3.5) and (3.6) can be expressed in such a manner that, after the drawdown (s) is computed, no further numerical integration is required. The sensitivity with respect to leakage, U_L in Eq. (3.7), can be computed only by additional numerical integration that would involve the formulation of a more complex subroutine. Therefore, the decision might be made to generate U_L by a finite difference approximation. The approximation

$$U_L = \partial s/\partial L \simeq \{s(L+\Delta L) - s(L-\Delta L)\} /2\Delta L \tag{3.8}$$

becomes increasingly accurate as ΔL approaches zero. Satisfactory evaluation of U_L occurred for ΔL set equal to .01 L. Plots of U_L will be presented later. This or a similar finite difference scheme could be used to calculate the sensitivity coefficients in many situations.

3.3 A Partial Differential Equation For Sensitivity Coefficients

For the general time-dependent case when transmissivity and storativity can vary spatially, no closed-form expression exists for the head and the sensitivity coefficients. The head is given by the solution of the following partial differential equation [or equivalently Eq. (2.21)].

$$\frac{\partial}{\partial \underline{x}} \left[T(\underline{x}) \frac{\partial h}{\partial \underline{x}}\right] = S(\underline{x}) \frac{\partial h}{\partial t} - Q'(\underline{x},t) \tag{3.9}$$

A partial differential equation for the sensitivity with respect to transmissivity can be developed by applying some of the definitions given earlier. If h* is the new head that results when the transmissivity is changed by $\Delta T(\underline{x}_o)$ at $\underline{x}_o$, then

$$\frac{\partial}{\partial \underline{x}} \left[T(\underline{x}) \frac{\partial h^*}{\partial \underline{x}}\right] + \frac{\partial}{\partial \underline{x}} \left[\delta(\underline{x}-\underline{x}_o)\ \Delta T(\underline{x}_o) \frac{\partial h^*}{\partial \underline{x}}\right] =$$

$$S(\underline{x}) \frac{\partial h^*}{\partial t} - Q'(\underline{x},t). \qquad (3.10)$$

Applying the definition of $U_T(\underline{x},t;\underline{x}_o)$, Eq. (2.12) results in the following expression.

$$\frac{\partial}{\partial \underline{x}} \left[T(\underline{x}) \frac{\partial U_T(\underline{x},t;\underline{x}_o)}{\partial \underline{x}}\right] + \frac{\partial}{\partial \underline{x}} \left[\delta(\underline{x}-\underline{x}_o) \frac{\partial h}{\partial \underline{x}}\right] =$$

$$S(\underline{x}) \frac{\partial U_T(\underline{x},t;\underline{x}_o)}{\partial t} \qquad (3.11)$$

In deriving Eq. (3.11), Eq. (3.9) has been subtracted from Eq. (3.10), the result divided by $\Delta T(\underline{x}_o)$, and the limit taken as $\Delta T(\underline{x}_o) \to 0$.

Eq. (3.11) is a partial differential equation for $U_T(\underline{x},t;\underline{x}_o)$ which looks very much like the original flow equation except for two differences. First, the fluxes $[Q(\underline{x})]$ do not appear in Eq. (3.11). Second, there is an additional term involving the differentiation of a delta function.

Except in very simple cases, numerical methods must be used to solve Eq. (3.9). The question arises as to how equation (3.11) may be used with numerical methods to obtain $U_T(\underline{x},t;\underline{x}_o)$. Only the term involving the differentiation of the delta function will be considered. The other terms in Eq. (3.11) are similar to terms in the flow Eq. (3.9) and may be handled with standard techniques. The elementary central difference formula for the partial derivative in the x direction of an arbitrary function f(x,y) evaluated at point (x_i,y_j) is

$$\left[\frac{\partial f(x,y)}{\partial x}\right]_{\substack{x=x_i\\ y=y_j}} \simeq \frac{f_{i+1/2,j} - f_{i-1/2,j}}{\Delta x} \qquad (3.12)$$

where a uniformly spaced node system is assumed such that $x_i = i\Delta x$ and $y_j = j\Delta y$. At this point let the grid system for specifying T to be arbitrary and simply denote the value of T at some point $\underline{x}_o(x_k y_\ell)$ by $T_{k,\ell}$. The x component of the delta function term in Eq. (3.11) is

$$\frac{\partial}{\partial x}\left[\delta(x-y_k)\delta(y-y_\ell)\,\frac{\partial h}{\partial x}\right]_{\substack{x=x_i\\ y=y_j}} \simeq \frac{\delta_{j,\ell}}{\Delta x^2}\left[\delta_{i+1/2,k}\cdot\right.$$

$$\left.(h_{i+1,j}-h_{i,j}) - \delta_{i-1/2,k}(h_{i,j}-h_{i-1,j})\right] \tag{3.13}$$

where $\delta_{i,j}$ is the Kronecker delta with the following properties

$$\delta_{i,j} = \begin{cases} 1 \text{ if } i=j \\ 0 \text{ if } i\neq j \end{cases} \tag{3.14}$$

Notice that specifying the transmissivity half way between the nodes where h is known is convenient. Eq. (3.13) gives the finite difference numerical approximation of the delta function term in Eq. (3.11). An alternate procedure could be performed for a finite element approximation.

Usually Eq. (3.11) would not be solved for point changes in the transmissivity. Rather, the transmissivity is usually assumed to be constant over a zone which includes several points or nodes. According to Eq. (2.16) we must integrate Eq. (3.11) over that zone to obtain the sensitivity with respect to transmissivity in that zone. This would be equivalent to summing Eq. (3.13) over all k and ℓ values in that zone.

$$\int_{\text{zone}_m} \frac{\partial}{\partial x}\left[\delta(x-x_k)\delta(y-y_\ell)\,\frac{\partial h}{\partial x}\right]_{\substack{x=x_i\\ y=y_j}} dA \simeq \sum_{\substack{k,\ell\\ \text{zone}_m}} \frac{\delta_{j,\ell}}{\Delta x^2}\cdot$$

$$\left[\delta_{i+1/2,k}(h_{i+1,j}-h_{i,j}) - \delta_{i-1/2,k}(h_{i,j}-h_{i-1,j})\right] \tag{3.15}$$

The partial differential equation for the sensitivity with respect to storativity is somewhat easier to obtain. If h* is the new head that results when the storativity is changed by $\Delta S(\underline{x}_o)$ at $\underline{x}_o$ then

$$\frac{\partial}{\partial \underline{x}}\left[T(\underline{x})\,\frac{\partial h^*}{\partial \underline{x}}\right] = \left[S(\underline{x})+\delta(\underline{x}-\underline{x}_o)\Delta S(\underline{x}_o)\right]\frac{\delta h^*}{\delta t} - Q'(\underline{x},t). \tag{3.16}$$

Subtracting Eq. (3.9) from Eq. (3.16), dividing by $\Delta S(\underline{x}_o)$, and taking the limit as $\Delta S(\underline{x}_o)\to 0$ results in the following equation.

$$\frac{\partial}{\partial \underline{x}}\left[T(\underline{x})\,\frac{\partial U_S(\underline{x},t;\underline{x}_o)}{\partial \underline{x}}\right] = S(\underline{x})\,\frac{\partial U_S(\underline{x},t;\underline{x}_o)}{\partial t} + \delta(\underline{x}-\underline{x}_o)\,\frac{\delta h}{\delta t} \qquad (3.17)$$

Recall that the definition of U_S is given by Eq. (2.19).

Once again, Eq. (3.17) looks identical in form to the flow equation except there are no fluxes [Q(x)] and the delta function term has been added. The finite difference numerical solution of Eq. (3.17) is carried out in a manner similar to that used for the flow equation except for the delta function term. If Eq. (3.17) is to be evaluated at $\underline{x} = (x_i, y_j)$ and $\underline{x}_o = (x_k, y_\ell)$ then

$$\left[\delta(x-x_k)\delta(y-y_\ell)\,\frac{\delta h}{\delta t}\right]^{n+1/2}_{\substack{x=x_i\\ y=y_j}} \simeq \frac{(h^{n+1}_{i,j} - h^{n}_{i,j})}{\Delta t}\,\delta_{i,k}\delta_{j,\ell} \qquad (3.18)$$

where Δt is the time step and $h^n_{i,j}$ is the hydraulic head at node point (i,j) after n time steps. As before, to obtain the sensitivity to the storativity in a zone containing several nodes, we must integrate Eq. (3.17) over that zone or sum the right side of Eq. (3.18) over the nodes in that zone.

$$\int_{\substack{zone\\ m}} \left[\delta(x-x_k)\delta(y-y_\ell)\frac{\delta h}{\delta t}\right]^{n+1/2}_{\substack{x=x_i\\ y=y_j}} dA \simeq \sum_{\substack{k,\ell\\ zone\\ m}} \frac{(h^{n+1}_{i,j} - h^{n}_{i,j})}{\Delta t}\,\delta_{i,k}\delta_{j,\ell} \qquad (3.19)$$

If numerical methods have been used to obtain the solution to Eq. (3.9), the flow equation, then the h's appearing in Eq. (3.15) or (3.19) are known and present no obstacle to a numerical solution of the equations for the sensitivity coefficients. The same numerical techniques used to solve the flow equation may be used to solve the sensitivity equations. In fact, the same computer code used for the flow equation can be used for the sensitivity equations by simply replacing the fluxes in Eq. (3.9) by the terms in Eq. (3.15) or (3.19). The system must be solved for each discrete value of x_o or zone. More discussion of the application of numerical methods to the solution of the sensitivity equations is given by McElwee (12). Ultimately, a numerical solution for the following sensitivity coefficients must be obtained.

$$U^n_{Ti,j;k,\ell} = \frac{\partial h^n_{i,j}}{\partial T_{k,\ell}} \tag{3.20}$$

$$U^n_{Si,j;k,\ell} = \frac{\partial h^n_{i,j}}{\partial S_{k,\ell}} \tag{3.21}$$

The superscript n is used to denote the nth time step while the subscripts i, j, k, and ℓ are the usual node indices. If zones are used, Eqs. (3.20) and (3.21) must be summed over all k and ℓ in that zone to obtain the total sensitivity to the parameter in that zone. Sometimes it is convenient to use only one subscript for the spatial variation and one for the transmissivity variation. In that case, the more compact forms for the sensitivity coefficients are $U^n_{Ti;k}$ and $U^n_{Si;k}$.

Along with the flow equation (3.9), initial conditions and boundary conditions must be given for h. An initial condition can be given in the form

$$h(x,y,0) - z = 0 \tag{3.22}$$

for a two-dimensional model. If Z is a constant, we have a flat initial surface. The boundary may be a rather arbitrary function of $\underline{x}$ denoted by

$$B(\underline{x}) = 0. \tag{3.23}$$

The boundary conditions must be specified on this curve. A specified head boundary is given by

$$[h(\underline{x},t)]_{B(\underline{x})} = H(\underline{x}). \tag{3.24}$$

If a specified flow of water is required on the boundary, then

$$\left[T(\underline{x}) \frac{\partial h(\underline{x},t)}{\partial n}\right]_{B(\underline{x})} = -Q''(\underline{x},t) \tag{3.25}$$

where $\partial/\partial n$ denotes the partial derivative in a direction normal to the boundary. If the constant is zero in (3.25), the result is a barrier boundary. In general, the boundary could contain both types of boundary conditions given by Eqs.(3.24) and (3.25).

If the initial condition on h is a steady-state solution $U_S(\underline{x},0;\underline{x}_o)$, the initial condition on the sensitivity with respect to storage is zero. If z is a constant in (3.22), the initial condition on the sensitivity with respect to transmissivity, $U_T(\underline{x},0;\underline{x}_o)$, is also zero. If the initial surface is a steady-state solution, the initial condition $U_T(\underline{x},0;\underline{x}_o)$ is a solution to the steady-state form of Eq. (3.11) with the appropriate boundary conditions. The boundary conditions for U_T and U_S can be found by differentiating (3.24) and (3.25). For a specified head boundary

$$\left[U_T(\underline{x},t;\underline{x}_o)\right]_{B(\underline{x})} = 0 \tag{3.26}$$

$$\left[U_S(\underline{x},t;\underline{x}_o)\right]_{B(\underline{x})} = 0 \ . \tag{3.27}$$

A specified flow boundary condition results in

$$\left[T(\overset{2}{\underline{x}}) \frac{\partial U_T(\underline{x},t;\underline{x}_o)}{\partial n}\right]_{B(\underline{x})} = \delta(\underline{x}-\underline{x}_o)Q''(\underline{x},t) \tag{3.28}$$

$$\left[\frac{\partial U_S(\underline{x},t;\underline{x}_o)}{\partial n}\right]_{B(\underline{x})} = 0. \tag{3.29}$$

The sensitivity equations for the case of constant T and S can be obtained from Eqs. (3.11) and (3.17) by integrating over the whole model region.

$$U_T(\underline{x},t) = \int_R U_T(\underline{x},t;\underline{x}_o)\, dA \tag{3.30}$$

Performing this integration gives [McElwee and Yukler (11)]

$$T\frac{\partial}{\partial \underline{x}}\left[\frac{\partial U_T(\underline{x},t)}{\partial \underline{x}}\right] + \frac{\partial}{\partial \underline{x}}\left[\frac{\partial h(\underline{x},t)}{\partial \underline{x}}\right] = S\, \frac{\partial U_T(\underline{x},t)}{\partial t} \tag{3.31}$$

$$T\frac{\partial}{\partial \underline{x}}\left[\frac{\partial U_S(\underline{x},t)}{\partial \underline{x}}\right] = S\frac{\partial U_S(\underline{x},t)}{\partial t} + \frac{\partial h}{\partial t}(\underline{x},t) \ . \tag{3.32}$$

The boundary conditions for constant T and S are obtained by integrating Eqs. (3.26) to (3.29). For example, Eq. (3.28) becomes

$$\left[\frac{\partial U_T(\underline{x},t)}{\partial n}\right]_{B(\underline{x})} = \frac{Q''(\underline{x},t)}{T^2} . \tag{3.33}$$

4. EXAMPLES OF SENSITIVITY COEFFICIENTS

In the following sections, several examples of sensitivity coefficients will be shown. All the methods of determining sensitivity coefficients discussed in the previous sections will be illustrated.

4.1 The Theis Equation

The sensitivity coefficients for the Theis equation are given by Eqs. (3.2) and (3.3). The sensitivity coefficient for transmissivity, U_T, is shown in Figures 2 and 3 for a well pumping 32,000 ft^3/day with a transmissivity of 3,200 ft^2/day and a storage coefficient of .00095 at a time of .017 days. The radial dependence of U_T is shown in Figure 2. The system is obviously most sensitive to changes in the transmissivity near the well where drawdown is the largest. From Figure 2 the sensitivity coefficient U_T seems to diverge at the well. In fact, it may be shown from Eq. (3.3) that for $r^2S/4Tt >> 1$,

$$U_T \simeq \frac{Q}{4\pi T^2}\left[1.577216 + \ln\left(\frac{r^2 S}{4Tt}\right)\right]. \tag{4.1}$$

Expression (4.1) shows that U_T should diverge logarithmically at the well. The sensitivity function U_T changes sign in the region 300-

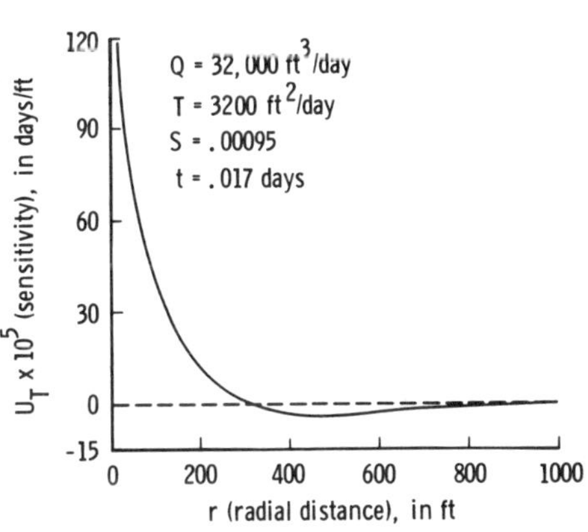

Fig. 2. Radial dependence of U_T for the Theis equation [McElwee and Yukler (11)].

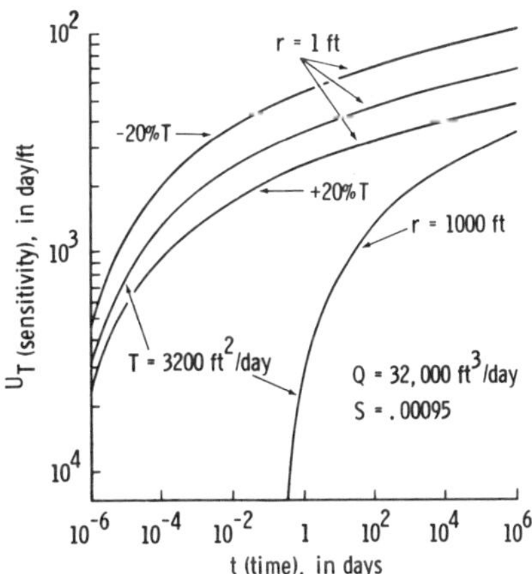

Fig. 3. Time dependence of U_T for the Theis equation [McElwee and Yukler (11)].

320 feet from the well, as it must in order for the cones of depression to have the same volume for differing transmissivities. The magnitude of the sensitivity coefficient U_T is relatively small in the region where U_T is negative.

Figure 3 shows a portion of the time dependence of U_T for two values of radius and two perturbed values of transmissivity. Notice that for large values of t, the dependence of U_T on t is fairly weak, though U_T is not constant. The curves labeled ±20% T show how U_T at a radius of 1 foot changes when the transmissivity is perturbed by ±20%. In this region a larger transmissivity results in decreased sensitivity, and a smaller transmissivity results in increased sensitivity. The curves labeled r = 1 foot and r = 1000 feet, for T = 3200 ft^2/day show the effect of changing r in the evaluation of U_T. These two curves have an identical shape but are displaced from one another along the t axis. The relation of these curves can be seen from Eq. (3.3). The critical ratio is r^2/t; as long as this ratio is the same, U_T will not change. Thus the curve for r = 1 foot at t = 10^{-6} days has the same value as the curve for r = 1000 feet at t = 1 day.

The sensitivity with respect to the storage coefficient, U_S, may be evaluated by using Eq. (3.2). The radial dependence of U_S is shown in Figure 4. U_S does not diverge at the well as does U_T. Eq. (3.2) and Figure 4 show that the radial dependence of U_S is Gaussian. U_S does not change sign because an increase or decrease in S results in a general raising or lowering, respectively, of the

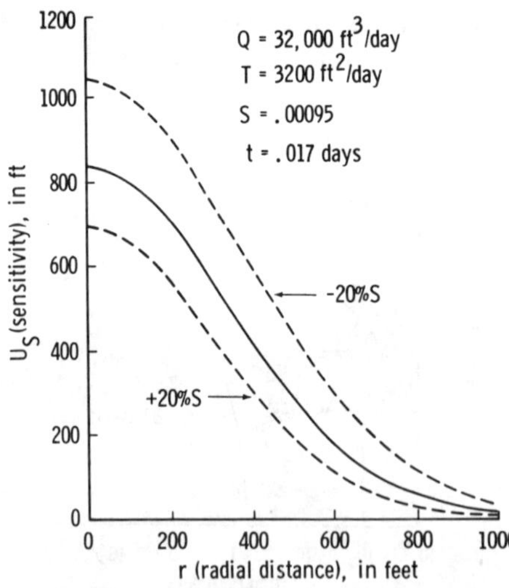

Fig. 4. Radial dependence of U_S for the Theis equation [McElwee and Yukler (11)].

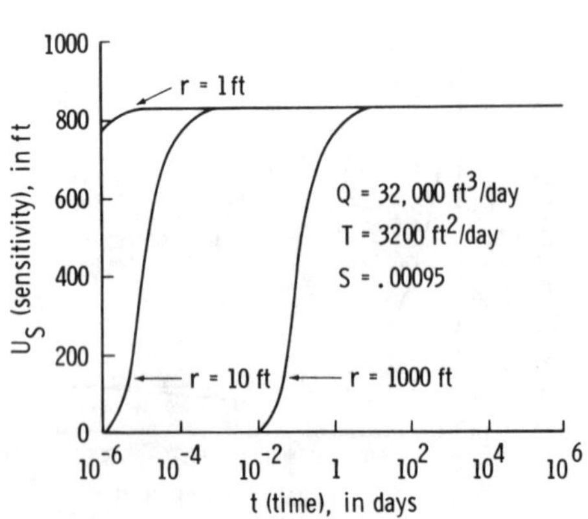

Fig. 5. Time dependence of U_S for the Theis equation [McElwee and Yukler (11)].

cone of depression. The dashed lines in Figure 4 show U_S when S is changed by ±20%. These curves indicate that the system is less sensitive for a larger S and more sensitive for a smaller S. This behavior can also be seen from Eq. (3.2).

The time dependence of U_S is illustrated in Figure 5 for three different r values. As time increases, U_S approaches a constant value. Even for r = 1000 feet, U_S is nearly constant after about 1 day. U_S is practically zero when the drawdown is very small and nearly constant after the drawdown attains 1-2 feet. The three curves shown in Figure 5 are identical except for displacement in time. From Eq. (3.2) it may be seen that U_S has the same value when r^2/t remains constant, provided Q, T, and S are unchanged. Thus the t = 1 day point on the curve for r = 1000 feet is identical to the t = 10^{-4} day point on the curve for r = 10 feet.

4.2 The Hantush Radial Leaky Aquifer

The sensitivity coefficients for the leaky aquifer are given by Eqs. (3.5), (3.6), and (3.8). U_T and U_S are evaluated by analytical expressions. U_L is obtained by a finite difference approximation. Many features of these sensitivity coefficients are similar to those found for the Theis equation. Therefore, description will be brief and new features will be pointed out. The sensitivity coefficients are shown in Figures 6 through 11 for Q = 196,000 ft^3/day, T = 24,300 ft^2/day, S = .002, and L = .0004 ft^{-1}.

The radial dependence of U_T is shown in Figure 6. The function diverges logarithmically near the well. U_T changes sign at some finite value of radius. This demonstrates the fact that when T is changed, the cone of depression deepens in some areas and shallows in others.

Figure 7 depicts a portion of the time dependence of U_T for variations in r and T. Note that U_T is inversely proportional to T. The curves represent a transmissivity of 24,300 ft^2/day and ±20% of that value at a radius of 100 feet and a T of 24,300 ft^2/day at a radius of 1,000 feet. Note that all curves flatten after three to four days. This describes the steady condition caused by deriving the discharge Q totally from leakage. This is a new effect that was not seen in the Theis case. In this case, U_T is constant after some time.

The radial dependence of U_S is shown in Figure 8. This coefficient does not diverge at the well, nor does it change sign. It is inversely proportional to S. The constancy of algebraic sign indicates that as S changes, a general raising or lowering of the cone of depression occurs.

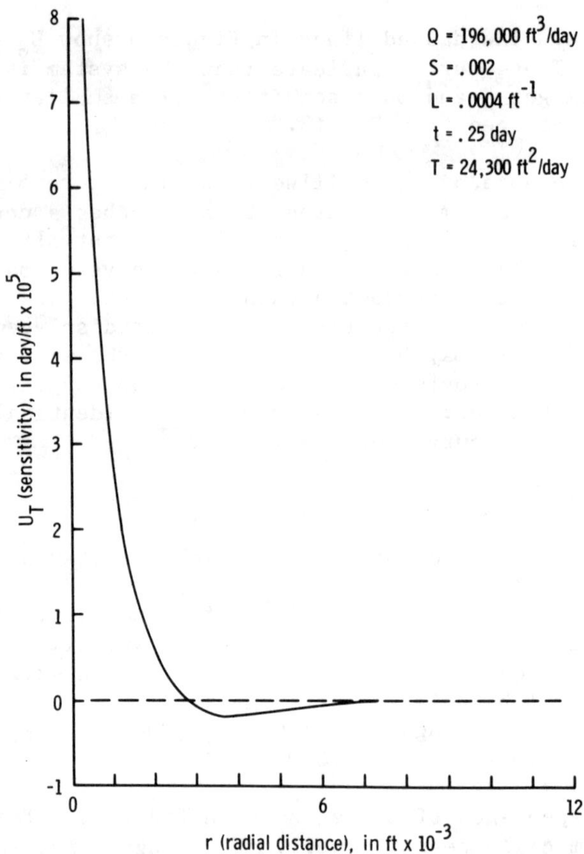

Fig. 6. Radial dependence of U_T for the leaky aquifer [Cobb, McElwee, and Butt (2)].

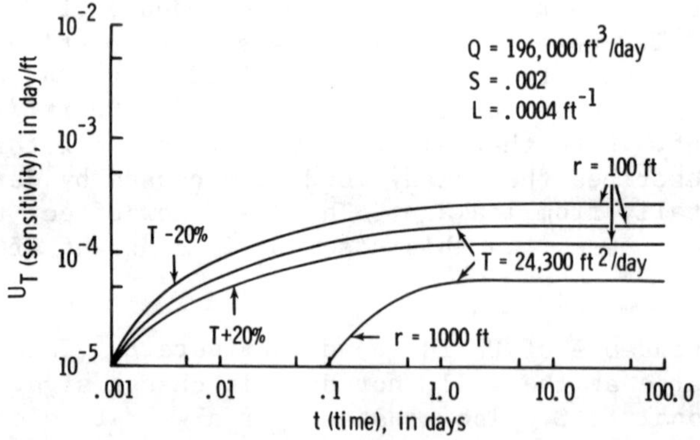

Fig. 7. Time dependence of U_T for the leaky aquifer [Cobb, McElwee, and Butt (2)].

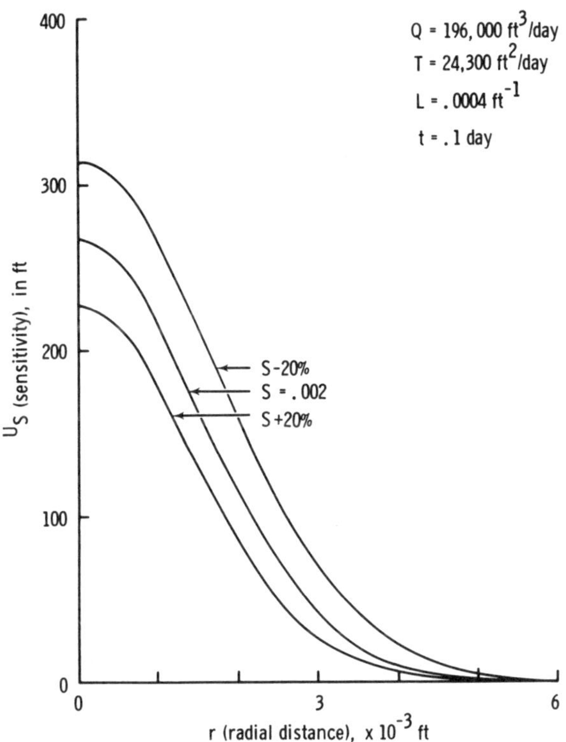

Fig. 8. Radial dependence of U_S for the leaky aquifer [Cobb, McElwee, and Butt (2)].

The time dependence of U_S is presented in Figure 9. Radial variation is indicated by the presence of three curves. Each curve reaches its maximum value for U_S at a time that increases with its radial value. At some finite value of time, each curve approaches zero in value, indicating that a steady state is achieved. Until steady state is attained, a dual source is supplying the pumpage, namely water released from storage and leakage. The curves roll over as leakage starts to dominate the source mechanism. U_S is zero outside the cone of depression and at any time after steady state is attained. Again, this is a new effect. U_S for the Theis model did not go to zero but approached a constant value for large times.

Figure 10 shows the radial dependence of U_L. The sensitivity coefficient U_L does not diverge at the well and approaches zero for large values of r. These are similar to the curves for U_S.

The time dependence of U_L is shown in Figure 11 for two values of r. All curves grow with time until a steady state is achieved where leakage is supplying the entire discharge Q. At that point, U_L is constant in time.

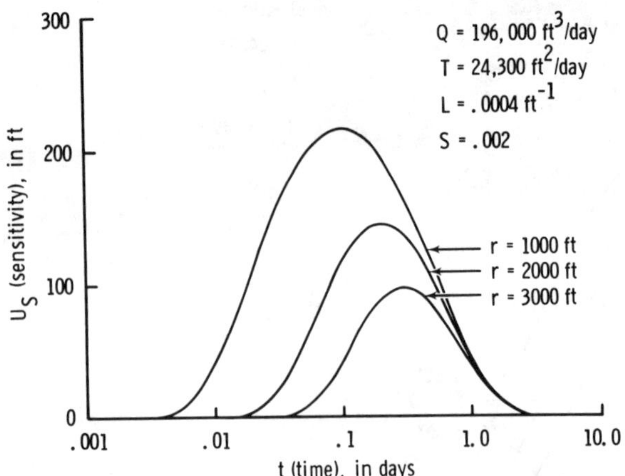

Fig. 9. Time dependence of U_S for the leaky aquifer [Cobb, McElwee, and Butt (2)].

In summary of the Theis and leaky aquifer sensitivity coefficients, a few observations can be made. The radial dependence of all the sensitivity coefficients (U_T, U_S, and U_L) shows that the greatest sensitivity is near the well and that the sensitivity approaches zero as the radial distance increases. The time dependence of all the sensitivity coefficients shows that initially the sensitivity grows with time. In the leaky case, U_S goes to zero as the steady state is approached while U_T and U_L approach constant values. In the Theis case, U_S approaches a constant value.

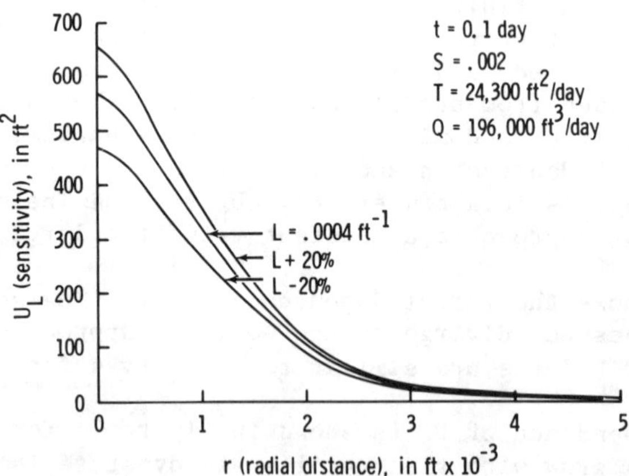

Fig. 10. Radial dependence of U_L for the leaky aquifer [Cobb, McElwee, and Butt (2)].

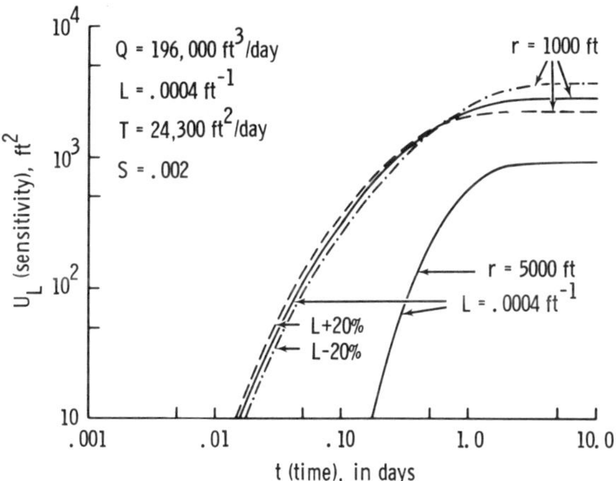

Fig. 11. Time dependence of U_L for the leaky aquifer [Cobb, McElwee, and Butt (2)].

4.3 One-Dimensional Model With Spatially Varying Transmissivity

Consider a steady state one-dimensional model with no interior fluxes. Eq. (3.9) becomes

$$\frac{\partial}{\partial x}\left[T(x)\,\frac{\partial h}{\partial x}\right] = 0. \tag{4.2}$$

The first integration of Eq. (4.2) gives

$$T(x)\,\frac{\partial h}{\partial x} = C = \text{constant}. \tag{4.3}$$

For boundary conditions, assume

$$h = H \quad \text{at } x = R \tag{4.4}$$

and

$$\frac{\partial h}{\partial x} = -\,\frac{Q/\ell}{T(0)} \quad \text{at } x = 0. \tag{4.5}$$

Q/ℓ is the boundary flux per unit length of boundary. Integration of Eq. (4.3) yields the final solution for the hydraulic head

$$h(x) = \frac{QR}{\ell} \int_{x/R}^{1} \frac{dx'}{T(Rx')} + H. \tag{4.6}$$

The normalized variable $x' = x/R$ has been introduced. Eq. (4.6) allows for an arbitrary distribution of the transmissivity.

Consider the case of constant transmissivity in Eq. (4.6). The solution is

$$h(x) = \frac{Q}{\ell T} (R-x) + H. \tag{4.7}$$

Using the definition of U_T from the Eq. (2.7) yields

$$U_T(x) = \frac{\partial h(x)}{\partial T} = \frac{Q}{\ell T^2} (x-R) = -\frac{s}{T}, \tag{4.8}$$

where s is the drawdown with reference to the constant head boundary.

$$s = h(x) - H = \frac{Q}{\ell T} (R-x) \tag{4.9}$$

This form [Eq. (4.8)] of the sensitivity coefficient is rather common [see Eqs. (3.3) and (3.6)] and merely says that the model is more sensitive to transmissivity in areas having larger drawdown. Notice that, as the constant head boundary is approached, the sensitivity coefficient (U_T) goes to zero. The sensitivity with respect to storativity, U_S, is zero since only the steady state is being considered.

The sensitivity coefficients for an arbitrary transmissivity distribution can be found by considering the head solution, Eq. (4.6), and the definition of the sensitivity coefficient, Eq. (2.12). The new head caused by changing the transmissivity at one point (x_o) is

$$h^*(x) = \frac{QR}{\ell} \int_{x/R}^{1} \frac{dx'}{T(Rx') + \delta(Rx'-x_o)\Delta T(x_o)}. \tag{4.10}$$

The sensitivity coefficient developed from (2.12) is as follows

$$U_T(x;x_o) = \begin{cases} -\frac{QR}{\ell}\frac{1}{T^2(x_o)} & \text{if } x \leq x_o \leq R \\ 0 & \text{if } x_o < x \end{cases} \tag{4.11}$$

This sensitivity coefficient is inversely proportional to the square of the transmissivity. Thus, areas of low transmissivity have a larger effect on model results than areas of high transmissivity. Also, notice that the transmissivity of x_o values less than the observation point, x, do not affect model results at the observation point. The sensitivity coefficient resulting from changing the transmissivity, a constant amount ΔT over the whole model area [Eq. (2.16)] is obtained by integrating Eq. (4.11).

$$U_T(x) = \int_0^1 U_T(x;Rx_o') \, dx_o' = -\frac{QR}{\ell}\int_{x/R}^{1}\frac{dx'_o}{T^2(Rx_o')} \tag{4.12}$$

The normalized integration variable $x_o' = x_o/R$ has been introduced. If the transmissivity is constant, Eq. (4.12) becomes identical with Eq. (4.8).

Typically, numerical methods are used to solve the model equations when the transmissivity is allowed to vary in an arbitrary manner. Assume a constant node spacing (Δx) grid system has been set up such that $N\Delta x = R$, where N+1 is the total number of nodes (x = 0 is the first node). The head at point x_i ($x_i = i\Delta x$) is obtained from Eq. (4.6) by the following replacement.

$$R\int_{x_i/R}^{1} dx' \rightarrow \Delta x \sum_{k=i}^{N} \tag{4.13}$$

Assuming that a constant transmissivity exists between points k and k+1 ($T_{k+1/2}$), Eq. (4.6) becomes

$$h_i = \frac{Q\Delta x}{\ell}\sum_{k=i}^{N}\frac{1}{T_{k+1/2}} + H. \tag{4.14}$$

The sensitivity coefficient is obtained by differentiating Eq. (4.14).

$$U_{Ti;k} = \frac{\partial h_i}{\partial T_{k+1/2}} = \begin{cases} -\frac{Q\Delta x}{\ell}\frac{1}{T^2_{k+1/2}} & \text{if } k \geq i \\ 0 & \text{otherwise} \end{cases} \tag{4.15}$$

this result could have been obtained from Eq. (4.11) simply by integrating the effect of a constant transmissivity over one node spacing. The sensitivity coefficient $U_{Ti;k}$ represents the change in hydraulic head at node point i due to a change in the transmissivity at k+ 1/2.

Consider a specific example of Eqs. (4.14) and (4.15). Assume $Q\Delta x/\ell = 10^3$, N = 9, H = 1, and $T_{k+1/2} = (k+1) \times 10^3$. Figure 12 is a plot of the head and Figure 13 is a plot of the absolute value of various sensitivity coefficients. All sensitivity coefficients are zero at node 10 where the head is specified. Since the magnitude of the sensitivity coefficients is inversely proportional to the transmissivity squared, the coefficients decrease dramatically as the transmissivity increases from node 0 to node 9.

4.4 A Simple Two-Dimensional Model With Spatially Varying Transmissivity

For a first look at two-dimensional sensitivity coefficients we use a simple two-dimensional flow model shown in Figure 14. This model has two zones for transmissivity and recharge. There are 25 node points with a node spacing of 500 in the y direction and 1000 in the x direction. Transmissivities of 1000 for zone one and 2000 for zone two are chosen. The recharge is .00625 and .003125 for zones one and two, respectively. No units have been given since any consistent set may be used.

The boundary conditions remain to be specified. Assume that the flux is specified on the x = 0 boundary and that the head is specified on the other three boundaries. Let the flux per unit

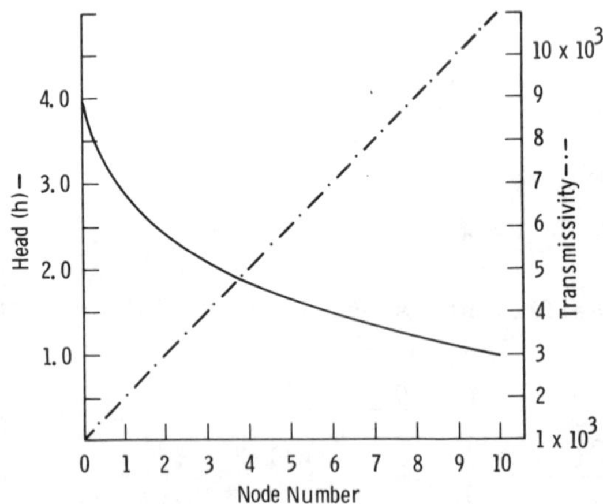

Fig. 12. Head and transmissivity for a simple one-dimensional model.

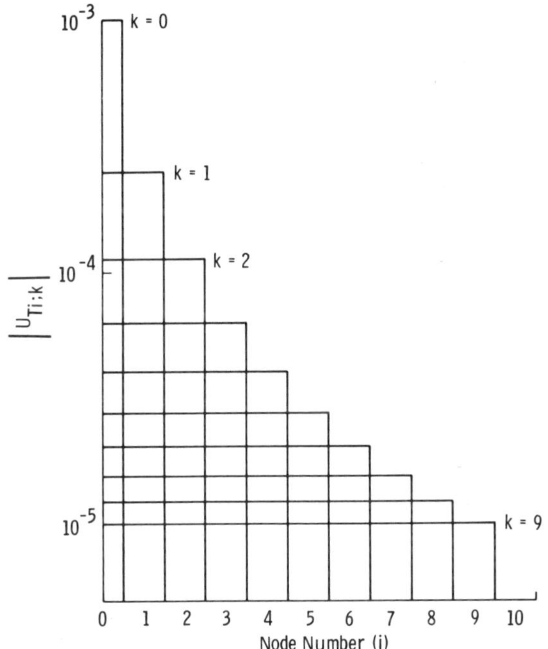

Fig. 13. Sensitivity coefficients for a simple one-dimensional model.

length (Q/ℓ) of the boundary at x = 0 be −50 units (out of the model area). For simplicity, assume the flow is parallel to the x axis and that the head at nodes 1 and 21 is 100. This allows the appropriate head to be specified for all the other boundary nodes. The values are shown in Table 1.

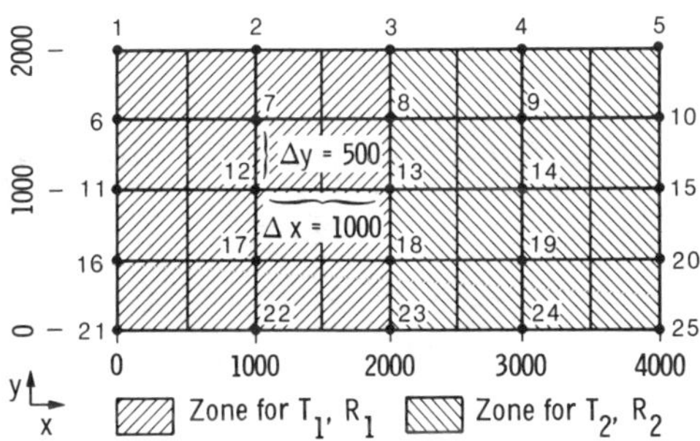

Fig. 14. A simple two-dimensional model.

Table 1. Head values for boundary nodes.

Head	Node Number
100	1, 21
146.9	2, 22
187.5	3, 23
205.5	4, 24
221.9	5, 10, 15, 20, 15

The sensitivity coefficients for the transmissivities in the two zones are shown in Figures 15 and 16. Notice that the coefficients are zero on the head-specified boundaries. Of course, this is required by Eq. (3.26). Also notice that the sensitivity coefficients have their largest magnitudes either in the middle of the flow-specified boundary or in the middle of the model.

5. EFFECT OF BOUNDARY CONDITIONS ON SENSITIVITY COEFFICIENTS

Boundary conditions have an effect on the shape and magnitude of the sensitivity coefficients. This is shown explicitly by Eqs. (3.26) - (3.29). The examples in sections 4.3 and 4.4 were affected by the boundary conditions chosen; however, in those sections, alternate boundary conditions were not shown. The purpose of this section is to compare sensitivity coefficients for various choices of boundary conditions. The examples chosen will be closely related to the examples of previous sections.

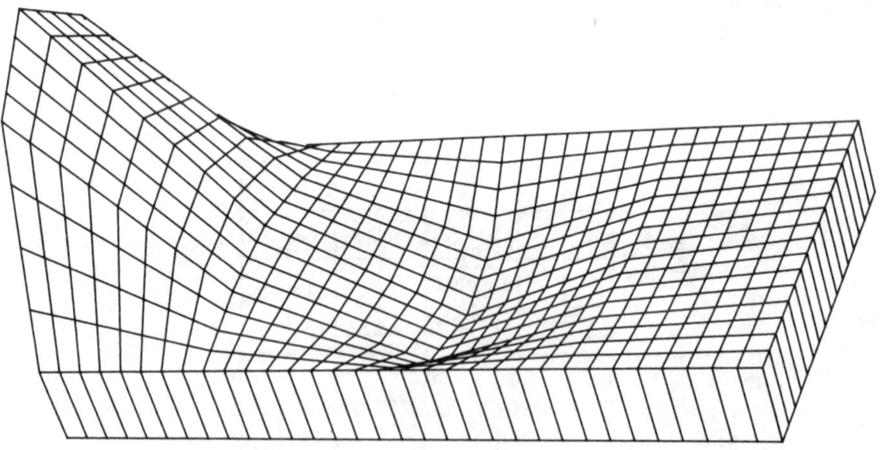

Fig. 15. Sensitivity coefficient for transmissivity in zone one.

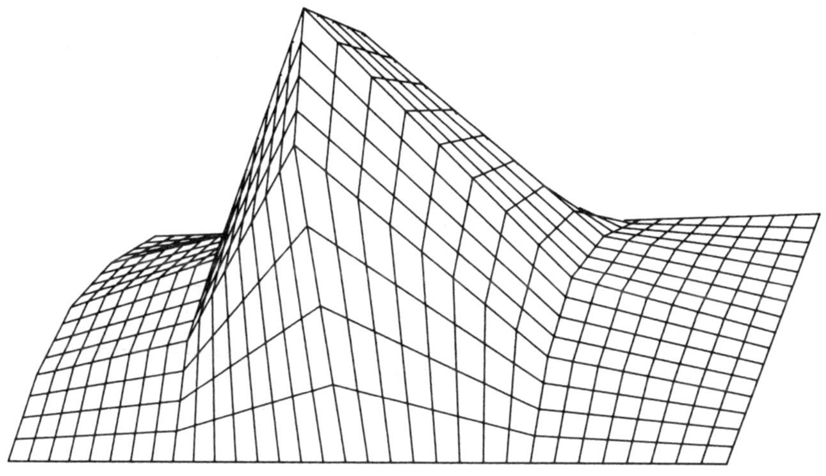

Fig. 16. Sensitivity coefficient for transmissivity in zone two.

5.1 Finite Radial Confined Aquifer

The values of the parameters Q, T, and S for the finite radial numerical model were chosen to be the same as those used in the Theis equation of section 4.1. However, two additional parameters are needed: the radius of the well and the radius of the outer boundary. The radius of the well was taken as 1 foot; and an outer boundary of 10,000 feet was used. The numerical results for U_T, obtained by choosing a constant head boundary or a barrier boundary at 10,000 feet, are shown in Figure 17 along with U_T calculated from the Theis equation. For times less than about 10 days, no difference exists in the three curves for r = 1 foot and for r = 1000 feet.

The constant head boundary at 10,000 feet produces a U_T as shown by the dot-dash curves in Figure 17. The Theis infinite model results are shown as a solid curve. For times greater than about 20 days, the water level is static owing to the constant head boundary, and U_T obtains a constant value in time. The radial dependence of U_T at steady state is a straight line when it is plotted versus log r and goes to zero at the outer boundary where the drawdown is zero.

Note that U_T at steady state is positive for all r. This is to be contrasted with the U_T shown in Figure 2 for the Theis equation. The fact that U_T is positive is a direct consequence of the fact that all the water being pumped is supplied by the constant head boundary; no water is being supplied from storage. Increasing or decreasing T results in a general raising or lowering, respectively, of the hydraulic head at steady state.

The numerical solution for U_T with a barrier boundary condition at 10,000 feet is shown by the dashed curves in Figure 17. After about 10 days, the cone of depression has reached the barrier boundary and U_T becomes constant in time.

In Figure 18 the radial dependence of U_T (for a barrier boundary) is shown for time considerably greater than 10 days, i.e., for time such that U_T is constant in time. Notice that U_T is negative for r greater than about 5500 feet. This is to be contrasted with a positive U_T for a constant head boundary. Because no water can flow into the system with a barrier boundary, all water pumped must come from storage. Thus, for two systems with differing T and the same Q, the cones of depression must contain the same volume at any given instant. The low T system will have a larger drawdown near the well and a smaller drawdown far from the well, because the lower T impedes the flow to the well. This explains why U_T has both negative and positive areas. A change in T will produce greater drawdown in one area and less drawdown in another area.

Figure 19 compares U_S calculated from the Theis equation (solid curve) with U_S calculated numerically for a constant head boundary

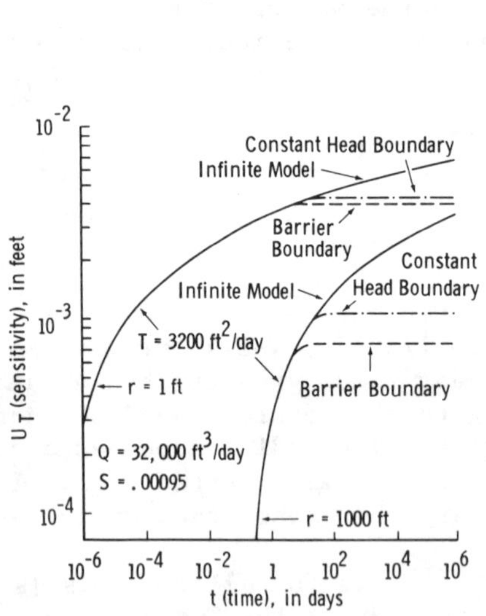

Fig. 17. Effect of the boundary at 10,000 ft on U_T [McElwee and Yukler (11)].

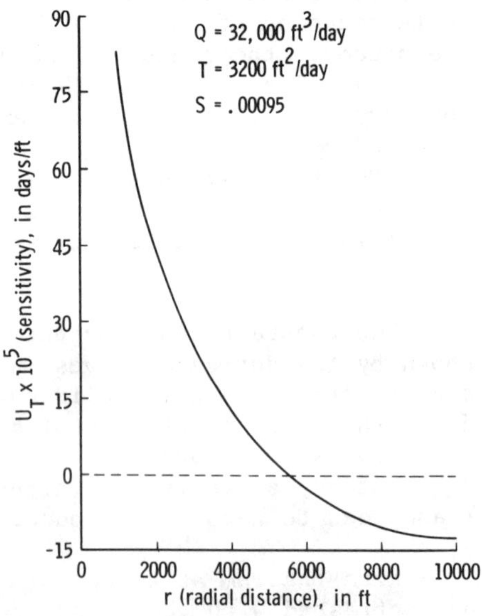

Fig. 18. Radial dependence of U_T at large time for a barrier boundary at 10,000 ft [McElwee and Yukler (11)].

(dot-dash curve) or a barrier boundary (dashed curve) at 10,000 feet. From the earlier discussion, remember that the water level does not change much after 20 days for the constant head boundary condition (approximately steady state). From Figure 19 we see that U_S for the constant head boundary is approximately zero after about 100 days. This behavior of U_S is to be expected, since the solution at steady state is independent of S because no water is coming from storage then.

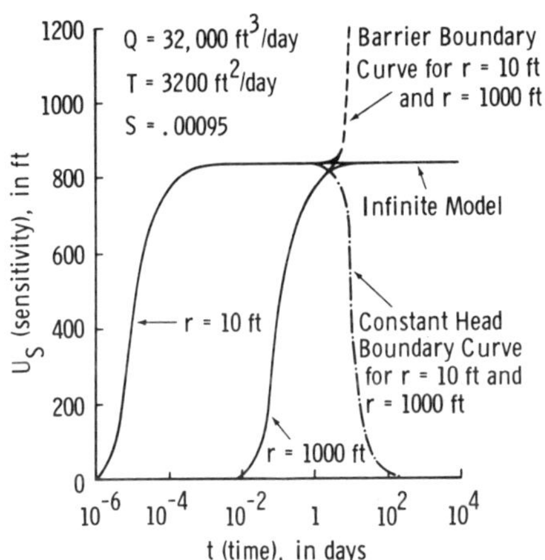

Fig. 19. Effect of the boundary at 10,000 ft on the time dependence of U_S [McElwee and Yukler (11)].

All values of U_S plotted in Figure 19 are positive, indicating that an increase or decrease of S results in a general raising or lowering, respectively, of the hydraulic head. U_S for the barrier boundary condition increases dramatically after a few days time, as is shown in Figure 19. U_s increases linearly with time after about 10 days. Numerical results indicate that U_S is the same for all values of r after about 10 days. The linear increase of U_S with time is due to the fact that the hydraulic head decreases uniformly with time after about 10 days. In short, the system becomes increasingly sensitive to S as the barrier boundary exerts a greater influence on the drawdown.

To summarize, a number of effects have been observed due to boundary conditions. U_T becomes constant after some time and U_S either goes to zero or increases linearly with time as the influence of the boundary is felt. A great deal of similarity exists between the leaky aquifer case of section 4.2 and the constant head boundary results presented here. U_T becomes constant and U_S goes to zero in both cases. On the other hand, U_S increases linearly with time for the barrier boundary at long times. This points out that each system has a characteristic behavior, and sensitivity analysis can help understand that behavior.

5.2 Alternate Boundary Conditions For The One-Dimensional Aquifer

In section 4.3, an analytical expression was derived for the sensitivity with respect to transmissivity for a steady-state one-dimensional model with spatially varying transmissivity and certain boundary conditions. Those sensitivity coefficients were shown in

Figure 13. In this section, the same system will be used except for differing boundary conditions.

Consider a steady-state one-dimensional model with the head specified at both boundaries.

$$h = H_1 \text{ at } x = 0 \tag{5.1}$$

$$h = H_2 \text{ at } x = R \tag{5.2}$$

Eqs. (4.2) and (4.3) are still valid for this model. Integrating Eq. (4.3) yields

$$h(x) = C \int_{x/R}^{1} \frac{dx'}{T(Rx')} + H_2 , \tag{5.3}$$

where C is a constant to be determined from the boundary condition at x = 0. Putting x = 0 in Eq. (5.3) results in the following expression for C.

$$C = (H_1-H_2)/ \int_0^1 \frac{dx'}{T(Rx')} \tag{5.4}$$

If T is constant, Eqs. (5.3) and (5.4) yield

$$h(x) = (\frac{H_2-H_1}{R})\, x + H_1. \tag{5.5}$$

Since Eq. (5.5) does not depend on the transmissivity,

$$U_T(x) = \frac{\partial h(x)}{\partial T} = 0. \tag{5.6}$$

When the transmissivity is not constant, the sensitivity coefficients can be obtained by applying the definition, Eq. (2.12),

$$U_T(x;x_o) = \left[h(x)-H_2 + (H_2-H_1)\Theta(x - x_o)\right]/\left[T^2(x_o) \int_0^1 \frac{dx'}{T(Rx')}\right]. \tag{5.7}$$

$\Theta(x_o - x)$ is the Heaviside unit step function.

$$\Theta(x_o - x) = \begin{cases} 1 & \text{if } x_o \geq x \\ 0 & \text{if } x_o < x \end{cases} \tag{5.8}$$

$h(x)$ is given by Eqs. (5.3) and (5.4). Eq. (5.7) shows that $U_T(x;x_o)$ has both negative and positive areas. If we assume that $H_1 \geq h(x) \geq H_2$, then $U_T(x;x_o)$ negative for $x_o \geq x$ and positive for $x_o < x$. If T is constant $U_T(x)$ is zero, as was already known from Eq. (5.6). Thus, the model becomes less sensitive to the value of T as a constant value of T is approached.

The numerical solution for the above model with constant node spacing (Δx) may be obtained as before by replacing the integrals in Eqs. (5.3) and (5.4) with the appropriate summations.

$$h_i = \frac{(H_1 - H_2)}{\sum_{k=0}^{N} \frac{1}{T_{k+1/2}}} \sum_{k=i}^{N} \frac{1}{T_{k+1/2}} + H_2 \tag{5.9}$$

The sensitivity coefficient $U_{Ti;k}$ can be obtained from Eq. (5.9) by differentiation.

$$U_{Ti;k} = [h_i - H_2 + (H_2 - H_1)\Theta(k-i)] / [T^2_{k+1/2} \sum_{\ell=0}^{N} \frac{1}{T_{k+1/2}}] \tag{5.10}$$

This equation is the discrete equivalent of Eq. (5.7). $U_{Ti;k}$ represents the change in hydraulic head at node point i due to a change in transmissivity at k+1/2. Eq. (5.10) can be written in slightly different form.

$$U_{Ti;k} = 1/[T^2_{k+1/2} \sum_{\ell=0}^{N} \frac{1}{T_{\ell+1/2}}] \begin{cases} (h_i - H_2) & \text{if } k < i \\ (h_i - H_1) & \text{if } k \geq i \end{cases} \tag{5.11}$$

This form shows that the sensitivity coefficients can be determined from just two functions, $T^2_{1/2}\, U_{Ti;0}$ and $T^2_{N+1/2}\, U_{Ti;N}$. All other sensitivity coefficients can be generated from these two. When k = i, $T^2_{k+1/2}\, U_{Ti;k}$ switches from one curve to the other.

As an example of Eq. (5.11), consider the simple model of section 4.3, i.e., N = 9, and $T_{k+1/2} = (k+1) \times 10^3$. This time the head will be specified at both ends. From Figure 12 we can see that H_1 = 3.93 and H_2 = 1.0. Figure 20 shows a plot of $T^2_{1/2}\ U_{Ti;0}$ and $T^2_{9+1/2}\ U_{Ti;9}$; in addition, $T^2_{4+1/2}\ U_{Ti;4}$ is shown to illustrate the crossover between the two curves. Notice that the sensitivity coefficients are zero at both boundaries since head is specified there. $U_{Ti;0}$ is everywhere positive while $U_{Ti;9}$ is everywhere negative. All the other sensitivity coefficients have some negative and some positive areas. These sensitivity coefficients are very different from those shown in Figure 13; yet the only difference in the models is that the head is specified at the left boundary in this case.

5.3 Alternate Boundary Conditions For The Simple Two-Dimensional Model

For the model defined in section 4.4 and Figure 14, it is possible to specify different boundary conditions and observe the effect on the sensitivity coefficients. First, let the boundaries at y = 0 and y = 2000 be barrier boundaries instead of head-specified boundaries. The resulting sensitivity coefficients are shown in Figures 21 and 22. Notice that the sensitivity coefficients no longer show a variation in the y direction. If, in addition to the barrier boundaries at y = 0 and y = 2000, the flow-specified boundary at x = 0 is changed to a head-specified boundary, the resulting sensitivity coefficients are shown in Figures 23 and

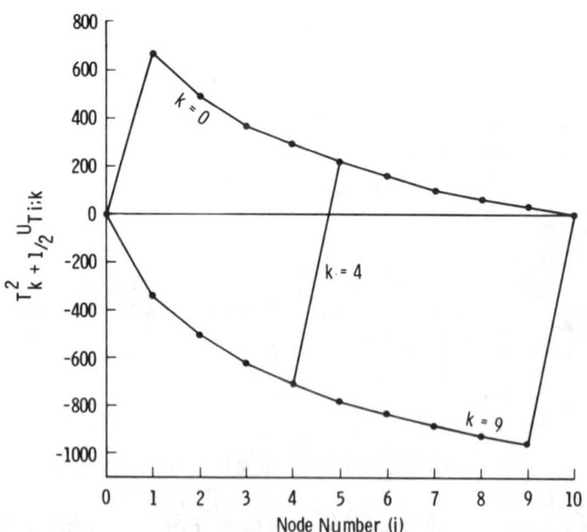

Fig. 20. Sensitivity coefficients for the one-dimensional model with head specified on both ends.

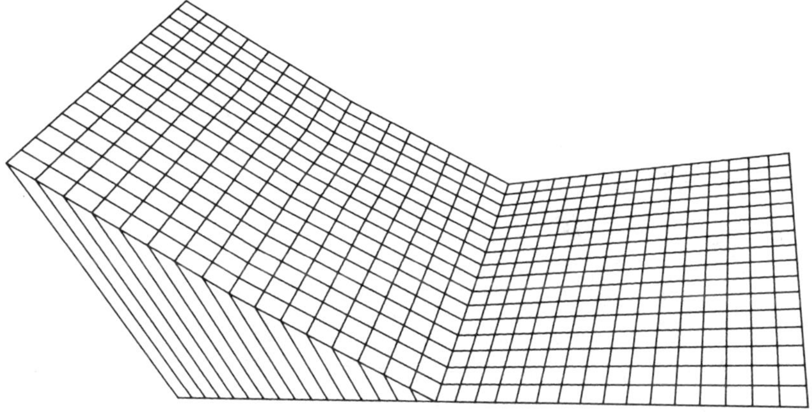

Fig. 21. Sensitivity coefficient for transmissivity in zone one, barrier boundaries at y = 0 and y = 2000.

24. It is very clear from these examples that the boundary conditions exert a large influence on the sensitivity coefficients.

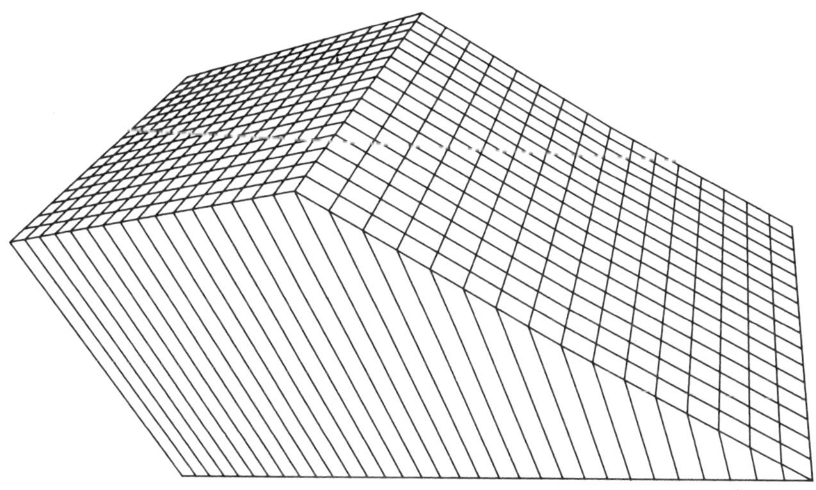

Fig. 22. Sensitivity coefficient for transmissivity in zone two, barrier boundaries at y = 0 and y = 2000.

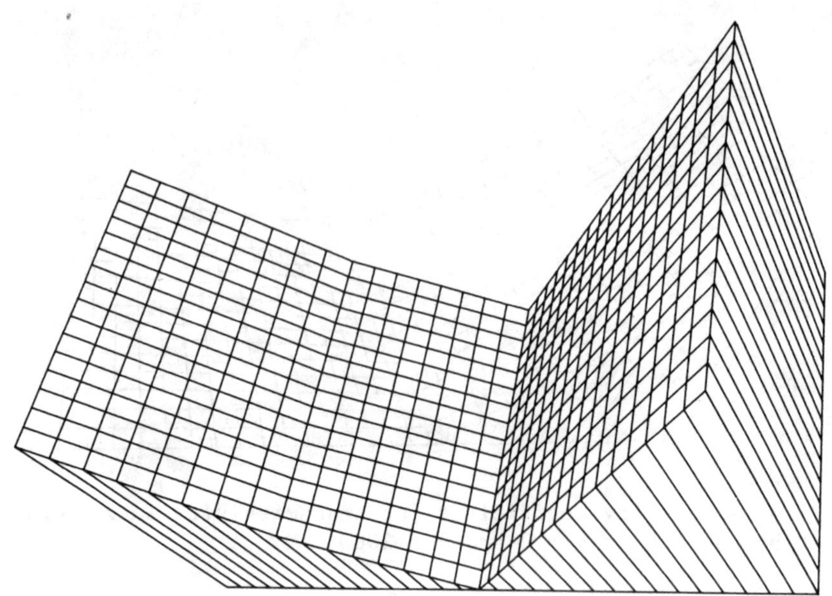

Fig. 23. Sensitivity coefficient for transmissivity in zone one for barrier boundaries at y = 0 and y = 2000 and head specified at x = 0 and x = 4000.

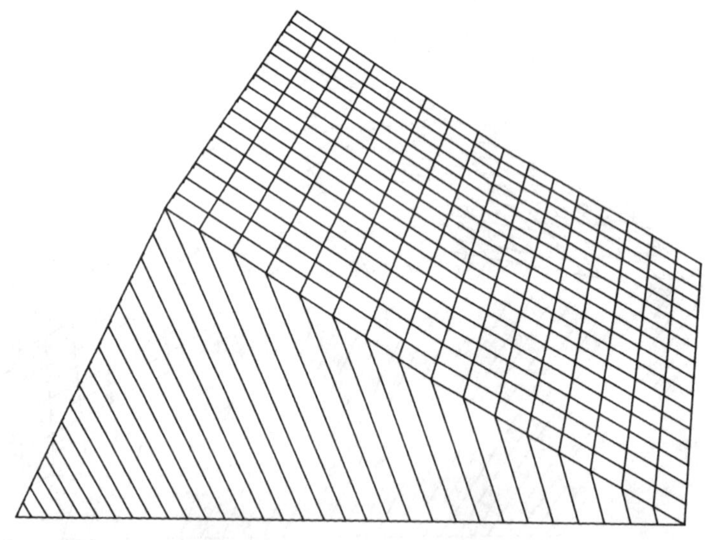

Fig. 24. Sensitivity coefficient for transmissivity in zone two for barrier boundaries at y = 0 and y = 2000 and head specified at x = 0 and x = 4000.

6. THE ROLE OF SENSITIVITY COEFFICIENTS IN PARAMETER ESTIMATION

The so-called "indirect" inverse procedures attempt to calculate the "best" transmissivity and storativity by minimizing some error functional [Cooley (3), Cooley (4), Neuman and Yakowitz (15), and Neuman (16)]. In this case, the aquifer parameters which are obtained by minimization do not exactly satisfy the direct equations. Rather, the best average solution is obtained over the historical period of record. Suppose that initial estimates for T and S can be made, and that h_i^n is the head calculated from the model at node point i and time step n. If all the aquifer parameters are changed by some amount (ΔT_k or ΔS_k, k is the zone index), the new head h_i^{*n} is given by

$$h_i^{*n} = h_i^n + \sum_k^{M_1} U_{Ti;k}^n \Delta T_k + \sum_k^{M_2} U_{Si;k}^n \Delta S_k \tag{6.1}$$

where M_1 is the number of T zones and M_2 is the number of S zones. $U_{Ti;k}^n$ represents the change in head at node i for time step n due to a change in the transmissivity in zone k. A similar definition applies for $U_{Si;k}^n$. These are the sensitivity coefficients that have been discussed at length in earlier sections.

6.1 Ordinary Least Squares

If he_i^n is the experimentally measured head at node point i for time step n, it would be desirable to choose ΔT_k and ΔS_k in such a way as to minimize the difference between h* and he. The error functional chosen to be minimized is the sum of the squared errors over all node points and time steps.

$$E(\Delta T_k, \Delta S_k) = \sum_n \sum_i \left[he_i^n - h_i^{*n}\right]^2 \tag{6.2}$$

This error functional assumes that measurement accuracy and any other source of error is the same for all points and times. (If this is not true, a weighting function could be used. If a weighting function other than one is used, we have generalized least squares, which will be discussed briefly in the next section.)

A necessary condition for minimization of $E(\Delta T_k, \Delta S_k)$ is that the partial derivatives with respect to ΔT_k or ΔS_k be zero. We can drop the subscript T and S on $U_{Ti;k}^n$ and $U_{Si;k}^n$ without ambiguity by letting k go from one to $M_1 + M_2 = p$. In general, the number and locations of the zones for T and S will be different.

$$U^n_{i;k} = \begin{cases} U^n_{Ti;k} & k < M_1 \\ U^n_{Si;k} & M_1 < k > M_2 \end{cases} \tag{6.3}$$

$\sum_n U^n_{i;k}$ can now be considered as one element of the matrix $\underset{\sim}{U}$. Note that, in general, $\underset{\sim}{U}$ is not a square matrix. Similarly define a parameter vector, $\underline{P}$. Note, + indicates the transpose of a vector or matrix.)

$$\underline{P}^+ = [T_1, T_2, \cdots T_{M_1}, S_1, S_2, \cdots S_{M_2}] \tag{6.4}$$

The minimization condition can now be written as [Beck and Arnold (1)]

$$(\underset{\sim}{U}^+\underset{\sim}{U})\underline{\Delta P} = \underline{R} \tag{6.5}$$

or

$$\underline{\Delta P} = (\underset{\sim}{U}^+\underset{\sim}{U})^{-1}\ \underline{R} \tag{6.6}$$

where $\underline{\Delta P}$ is a vector of parameter changes required to minimize the functional. One element of the vector $\underline{R}$ (R_k) is given by

$$R_k = \sum_n \sum_i U^n_{i;k}\ (he^n_i - h^n_i). \tag{6.7}$$

Any standard matrix routine may now be used to solve Eq. (6.5) for the parameter changes which will minimize the error functional. Since it is a nonlinear problem ($\underset{\sim}{U}$ depends on $\underline{P}$), iteration until convergence occurs is necessary; i.e., until $\underline{\Delta P}$ approaches zero.

In order to perform a more detailed statistical analysis on the parameter reliability, it is necessary to find the parameter covariance matrix. For ordinary least squares and certain statistical assumptions (additive, zero mean, uncorrelated and constant variance errors), the parameter covariance is given by [Beck and Arnold (1)]

$$\underset{\sim}{P} = \text{cov}\ (\underline{P}) \approx (\underset{\sim}{U}^+\underset{\sim}{U})^{-1}\sigma^2 \tag{6.8}$$

where σ^2 is the head error variance. The estimated standard error of the parameters is given by the square root of the diagonal elements of the matrix $\underset{\sim}{P}$ in Eq. (6.8).

$$\text{Estimated Standard Error of } P_k = P_{kk}^{1/2} \tag{6.9}$$

If σ^2, the head error variance, is unknown, it can be approximated by

$$\sigma^2 \simeq s^2 = \sum_n \sum_i [he_i^n - h_i^{*n}]^2/(m-p) \tag{6.10}$$

where m is the total number of observations in space and time and p is the number of parameters to be estimated.

The approximate parameter correlation matrix, which shows the degree of dependence among parameters, can be found from Eq. (6.8). An element of the matrix has the form [Beck and Arnold (1)]

$$r_{ik} = P_{ik} \, (P_{ii} \, P_{kk})^{-1/2}. \tag{6.11}$$

The diagonal elements are all unity. The off-diagonal elements are between -1 and 1. As the magnitude of an off-diagonal element approaches one, it indicates a high correlation between parameters. When this occurs, the two parameters are nearly dependent and it may not be possible to estimate both.

6.2 Other Methods

One might possibly want to weight some measurements more than others. Also some prior information may exist on the aquifer parameters. In these cases, the function to be minimized would not be given by Eq. (6.2) but by a more general function such as [Beck and Arnold (1)]

$$E(\Delta T_k, \Delta S_k) = [\underline{he}-\underline{h}^*]^+ \underset{\sim}{W} [\underline{he}-\underline{h}^*] + [\underline{P}-\underline{P}_o]^+ \underset{\sim}{V} [\underline{P}-\underline{P}_o] \tag{6.12}$$

where

$$[\underline{he}- \underline{h}^*] = [(he_1^1- h_1^1), \cdots (he_i^n- h_i^{*n}) \cdots]. \tag{6.13}$$

$\underset{\sim}{W}$ is the weight matrix for head measurements, $\underline{P}_o$ is the vector of prior estimates for the aquifer parameters, and $\underset{\sim}{V}$ is the weight matrix for the prior estimates. $\underset{\sim}{W}$ is a symmetric square matrix whose dimensions are equal to the total number of head measurements. $\underset{\sim}{V}$ is a symmetric square matrix with dimensions equal to the number of prior estimates. By the proper choice of $\underset{\sim}{W}$ and $\underset{\sim}{V}$ one can perform weighted least squares (WLS) estimation, maximum likelihood (ML) estimation, or maximum a posteriori (MAP) estimation.

In this more general case, the extension of Eq. (6.6) becomes

$$\Delta \underline{P} = \left[\underset{\sim}{U}^{+}\underset{\sim}{W}\,\underset{\sim}{U} + \underset{\sim}{V}\right]^{-1}\left[\underset{\sim}{U}^{+}\underset{\sim}{W}(\underline{he} - \underline{h}^{*}) + \underset{\sim}{V}(\underline{P} - \underline{P}_{o})\right]. \qquad (6.14)$$

The parameter covariance matrix Eq. (6.8) must also be extended.

$$\underset{\sim}{P} = \mathrm{cov}(\underline{P}) \simeq \left[\underset{\sim}{U}^{+}\underset{\sim}{W}\,\underset{\sim}{U} + \underset{\sim}{V}\right]^{-1} \qquad (6.15)$$

The work presented here will deal only with the ordinary least squares estimation procedure because the main purpose is to show how sensitivity coefficients can be used to perform a model sensitivity analysis. From this section we can see that sensitivity coefficients clearly play a central role in any common estimation technique. Therefore, the procedures discussed in this work for the least squares case can be generalized for more sophisticated estimation techniques.

7. USING SENSITIVITY COEFFICIENTS TO ESTIMATE CONFIDENCE INTERVALS

In the following sections we shall briefly indicate how confidence intervals can be estimated for both the estimated parameters and the calculated heads.

7.1 Confidence Intervals And Regions For Estimated Parameters

With the statistical assumptions invoked in section 6.1, it can be shown that the quantity $(P_k - \rho_k)/P_{kk}^{1/2}$ is described by a t(m-p) distribution. P_k is the parameter value estimated by least squares, ρ_k is the correct value, P_{kk} (the square of the estimated standard error of P_k) is the diagonal element from Eq. (6.8), and m-p is the number of degrees of freedom used in calculating s^2 in Eq. (6.10). The 100(1-α)% confidence interval is approximated by [Beck and Arnold (1)]

$$\delta P_k = \pm\, P_{kk}^{1/2}\; t_{1-\alpha/2}\,(m-p). \qquad (7.1)$$

When σ^2 is estimated by s^2, as in Eq. (6.10), the approximate boundary of the 100(1-α)% confidence region is an hyperellipsoid given by [Beck and Arnold (1)]

$$(\underline{P} - \underline{\rho})^{+}\underset{\sim}{U}^{+}\underset{\sim}{U}(\underline{P} - \underline{\rho}) = r^{2} \qquad (7.2a)$$

$$\delta \underline{P}^{+} \underset{\sim}{U}^{+} \underset{\sim}{U} \delta \underline{P} = ps^{2} F_{1-\alpha}(p,m-p). \tag{7.2b}$$

r^2 has an F-distribution with the two degrees of freedom p and m-p, if σ is unknown. $r^2 = \sigma^2 \ell^2_{1-\alpha}(p)$ if σ is known [Beck and Arnold (1), page 294].

7.2 Confidence Intervals for Calculated Head

For the least squares case, with the assumptions of section 6.1, the covariance matrix for the calculated heads ($\underline{h}_c$) is [Beck and Arnold (1)]

$$\mathrm{cov}(\underline{h}_c) = \underset{\sim}{U} \underset{\sim}{P} \underset{\sim}{U}^{+} \tag{7.3}$$

where the $\underset{\sim}{U}$ are the usual sensitivity matrices and $\underset{\sim}{P}$ is the parameter covariance matrix from Eq. (6.8). The diagonal elements of Eq. (7.3) give the variance of the calculated head. The square root of these diagonal elements gives the estimated standard error in calculated head.

8. MODEL DESIGN FOR MAXIMUM SENSITIVITY

In this section, a few general observations will be made regarding model sensitivity. Some examples will be given to illustrate these principles.

8.1 Minimize The Estimated Errors Or Confidence Intervals

If the confidence intervals or the estimated standard errors of the parameters can be made small, then the model has good sensitivity. From Eqs. (6.8) and (6.9), it is seen that the diagonal elements of $(\underset{\sim}{U}^{+}\underset{\sim}{U})^{-1}$ are critical in this regard. Therefore, striving to maximize the model sensitivity is equivalent to minimizing the diagonal elements of $(\underset{\sim}{U}^{+}\underset{\sim}{U})^{-1}$. The inverse of the matrix is given by its adjoint divided by its determinant.

$$(\underset{\sim}{U}^{+}\underset{\sim}{U})^{-1} = \mathrm{adj}(\underset{\sim}{U}^{+}\underset{\sim}{U})/|\underset{\sim}{U}^{+}\underset{\sim}{U}| \tag{8.1}$$

One approach to achieve maximum sensitivity is to try to increase the determinant $|\underset{\sim}{U}^{+}\underset{\sim}{U}|$ [Beck and Arnold (1)]. However, the determinant can be increased by a simple scaling of $\underset{\sim}{U}^{+}\underset{\sim}{U}$ which does nothing for the accuracy. So one must be careful to increase $|\underset{\sim}{U}^{+}\underset{\sim}{U}|$ in such a way that the accuracy actually increases.

From Eq. (8.1), $\underset{\sim}{U}^+\underset{\sim}{U}$ clearly should not be singular ($|\underset{\sim}{U}^+\underset{\sim}{U}| \neq 0$). This means that the sensitivity coefficients and the parameters must be independent. For example, consider the case of two parameters, T and S. The least squares matrix is

$$(\underset{\sim}{U}^+\underset{\sim}{U}) = \begin{bmatrix} \sum\limits_{i,n} (U^n_{Ti})^2 & \sum\limits_{i,n} U^n_{Ti} U^n_{Si} \\ \sum\limits_{i,n} U^n_{Ti} U^n_{Si} & \sum\limits_{i,n} (U^n_{Si})^2 \end{bmatrix} ; \tag{8.2}$$

i and n denote spatial and temporal measurement locations. If U_T and U_S happen to be dependent as in Eq. (2.27), showing that $|\underset{\sim}{U}^+\underset{\sim}{U}| = 0$ is not difficult. Clearly, none of the sensitivity coefficients may be dependent. However, problems may arise when two or more of the sensitivity coefficients are nearly dependent. Sometimes this condition can be seen by plotting and comparing sensitivity coefficients. Another way to examine dependence between sensitivity coefficients is to calculate the sensitivity correlation matrix, an element of which has the form

$$(sr)_{ik} = (\underset{\sim}{U}^+\underset{\sim}{U})_{ik} / [(\underset{\sim}{U}^+\underset{\sim}{U})_{ii} (\underset{\sim}{U}^+\underset{\sim}{U})_{kk}]^{1/2}. \tag{8.3}$$

If any off-diagonal elements are in the range of .9 or greater, significant correlation exists between those sensitivity coefficients, which may result in a smaller value for $|\underset{\sim}{U}^+\underset{\sim}{U}|$ and problems in finding an accurate inverse.

After an inverse, $(\underset{\sim}{U}^+\underset{\sim}{U})^{-1}$, has been found, it is possible to examine the dependence between the aquifer parameters by looking at the elements of the parameter correlation matrix given by Eq. (6.11). If any of the off-diagonal elements approach a value of 1, a significant correlation exits between those two parameters. This indicates that the two parameters are nearly dependent and should not both be estimated.

Another indicator of the stability of the inverse [Eq. (6.5)] is the condition number of $\underset{\sim}{U}^+\underset{\sim}{U}$ [Strang (17)]. The condition number indicates how errors in $\underline{R}$ or $\underset{\sim}{U}^+\underset{\sim}{U}$ are amplified in the final solution for $\underline{\Delta P}$. Ideally, if the condition number is one, no amplification occurs. If the condition number is large, small changes in $\underline{R}$ or $\underset{\sim}{U}^+\underset{\sim}{U}$ might produce large changes in the solution $\underline{\Delta P}$. One might suspect that as $|\underset{\sim}{U}^+\underset{\sim}{U}|$ approaches zero, the condition number would grow. Thus, the condition number should be an indicator of the dependence or near dependence of sensitivity coefficients or parameters. In the work that follows, the reciprocal condition number will usually be examined rather than the condition number. If the reciprocal condition number is near one, the matrix is well conditioned and errors are not amplified. On the other hand, if the reciprocal condition number is very small, the matrix is ill-conditioned. If the reciprocal condition number is 10^{-k} and k approaches the number of significant digits used by the computer, the matrix is said to be nearly singular at that working precision [Dongarra et al. (5)].

In many cases where the aquifer parameters vary considerably in magnitude over the model, it is helpful for accuracy and convergence to use normalized sensitivity coefficients (PU_P) and solve for relative changes in the parameters ($\Delta P/P$). Eqs. (6.3) to (6.7) can be written for these changes with minor modifications. First, define the normalized sensitivity coefficients as

$$U'^{n}_{i;k} = P_k U^{n}_{i;k} \tag{8.4}$$

and the relative parameter changes as

$$\Delta P'_k = \Delta P_k / P_k. \tag{8.5}$$

As in section 6.1, let $\sum_n U'^{n}_{i;k}$ be one element of the matrix $\underset{\sim}{U}'$. The normalized least squares equations can now be written as

$$(\underset{\sim}{U}'^{+}\underset{\sim}{U}')\underline{\Delta P}' = \underline{R}' \tag{8.6}$$

where one element of $\underline{R}'$ is

$$R'_k = P_k R_k. \tag{8.7}$$

As will be seen later, it is much easier visually to compare normalized sensitivity coefficients and relative parameter changes.

Earlier sections have dealt with methods for determining sensitivity coefficients. Several examples of sensitivity coefficients have been given for various models. Clearly from that work, the sensitivity functions may vary in magnitude considerably over space and time. Obviously better sensitivity will result if the measurement points in space and time, which go into the matrix $\underset{\sim}{U}$, are chosen to primarily sample the regions where the sensitivity coefficients have their largest values. On the other hand, if the measurement points cannot be adjusted, clearly the estimated parameters will have greater uncertainty when using data from areas of low sensitivity. As a simple example of this principle, consider the sensitivity with respect to storativity for the leaky aquifer shown in Figure 9. If we have a choice, larger sensitivities clearly result when the observation well is closer to the pumped well. For an observation well at 1000 feet, head measurements taken before .01 day or after 1.0 day will contribute little to defining the storativity parameter.

In dealing with the ground-water inverse problem, we work with a set of head measurements. In many cases, more than one model specification is consistent with this head data. Other geohydrologic information may further restrict the suite of possible models. However, usually considerable latitude exists in specifying the model. In this case one should choose the model that has the greatest sensitivity to the aquifer parameters. In particular, consider the effect of boundary conditions. From earlier discussion, Eqs. (3.26) and (3.27), it was concluded that the sensitivity coefficients go to zero at a specified head boundary. However, this is not the case for a specified flow boundary. As an example of this, consider the sensitivity coefficients shown in Figures 15 and 21. Exactly the same head data is appropriate for both of these cases; only the boundary conditions at $y = 0$ and $y = 2,000$ are different. In Figure 15, $y = 0$ and $y = 2,000$ represent specified head boundaries, while in Figure 21 they are barrier boundaries. Clearly the model in Figure 21 has the potential for greater sensitivity. In general, specified head boundaries, with their resultant zero-sensitivity coefficients, result in smaller model sensitivity and larger parameter uncertainty.

There is, of course, a limit on the number of parameters that can be determined, which is the number of measurements made. As discussed earlier, the uncertainty of each parameter is related to the sensitivity of the model and the measurement errors. It makes no sense to try to determine a given parameter if the model sensitivity is very low. In practice, what is done many times is to assume that a number of nodes have a common parameter value. This collection of nodes is called a zone. It seems reasonable that the total sensitivity for the zone would be the sum of the individual nodal sensitivities. Actually, the preceeding statement is just the discrete analog of Eq. (2.16), where the area of integration is just the zone. From this discussion it seems reasonable that, through a proper summation of nodes into zones, it might be possible to obtain an acceptable level of sensitivity in each resulting zone.

If summing into zones occurs, the least squares Eq. (6.5) will be modified. To see how, consider the case of three zones. The least squares matrix is

$$
(\underset{\sim}{U}^{+}\underset{\sim}{U}) = \begin{bmatrix} \sum_i U_{i;1}^2 & \sum_i U_{i;1}\,U_{i;2} & \sum_i U_{i;1}\,U_{i;3} \\ \sum_i U_{i;2}\,U_{i;1} & \sum_i U_{i;2}^2 & \sum_i U_{i;2}\,U_{i;3} \\ \sum_i U_{i;3}\,U_{i;1} & \sum_i U_{i;3}\,U_{i;2} & \sum_i U_{i;3}^2 \end{bmatrix} . \quad (8.8)
$$

Suppose zones 1 and 2 are to be combined into a single zone. The total sensitivity for the combined zone at node i is $(U_{i;1} + U_{i;2})$. If we sum the first and second row and the first and second column in $(\underset{\sim}{U}^{+}\underset{\sim}{U})$ we obtain

$$
(\underset{\sim}{U}^{+}\underset{\sim}{U})_s = \begin{bmatrix} \sum_i (U_{i;1} + U_{i;2})^2 & \sum_i (U_{i;1} + U_{i;2})U_{i;3} \\ \sum_i (U_{i;1} + U_{i;2})U_{i;3} & \sum_i U_{i;3}^2 \end{bmatrix} . \quad (8.9)
$$

Similarly, if all three zones are to be combined into a single zone, the total sensitivity for the combined zone at node i is $(U_{i;1} + U_{i;2} + U_{i;3})$. By summing all the rows and columns in either Eq. (8.8) or (8.9), we obtain

$$
(\underset{\sim}{U}^{+}\underset{\sim}{U})_s = \sum_i (U_{i;1} + U_{i;2} + U_{i;3})^2 . \quad (8.10)
$$

From the above discussion, it is clear that the least squares equation (6.5) may be collapsed to any convenient number of zones by summing the appropriate rows and columns of $\underset{\sim}{U}^{+}\underset{\sim}{U}$ and summing the appropriate elements of $\underline{\Delta P}$ and $\underline{R}$. It seems that it might be possible to collapse the number of parameter zones and to adjust the zone shapes with the above technique such that a minimum parameter sensitivity or maximum parameter error is achieved. Of course, this procedure would have to be tempered by knowledge of the geohydrology. The zone formation would have to satisfy the joint goals of increasing parameter sensitivity and being consistent with the known hydrogeology.

8.2 Examples Of Methods For Maximizing Model Sensitivity

In the following sections various examples will be given which illustrate techniques for increasing model sensitivity. The goal is to increase model sensitivity until an acceptable level of error in the estimated parameters is achieved.

8.2.1 The Theis aquifer

For the Theis equation, the sensitivity coefficients attain their maximum magnitudes at infinite time and zero radius (Figures 2 and 3). Therefore, logically, observation wells should be located very close to the pumping well and observed for very long times [Yeh and Sun (20)]. Practically, there are some problems with this approach. The transmissivity value obtained from the pumping test is an average of the transmissivity in the region of the cone of depression. Locating too close to the pumping well restricts the region sampled. Also, perturbing influences may cause the drawdown near the well to deviate from the assumed model (for example, partial penetration and well construction). It is not possible to continue a pumping test indefinitely; usually a maximum duration is dictated by external influences (cost, manpower, equipment, etc.). Therefore, having a way to terminate the pumping test when the aquifer parameters had been determined accurately enough would be desirable.

As discussed in the previous section, maximizing $|U^+U|$ is one way to try and minimize parameter uncertainty. Figure 25 is a plot of $|U^+U|$ versus time for the Theis equation with the same parameters as used earlier to generate Figures 2-5. The observation well distance (r) is 1,000 feet. Notice that, as expected, $|U^+U|$ continues to increase with time. This was expected since U_T continues to increase with time (Figure 3) even though U_S is approximately constant after about one day (Figure 5). The non-smooth character of $|U^+U|$ at log cycle boundaries in Figure 25 is due to a different time sample rate in each log cycle of time (10 samples per log cycle). The dotted line extension at 1 day is the curve that would result if the same time sample interval (.1 day) had been used on two successive log cycles from .1 day to 10 days.

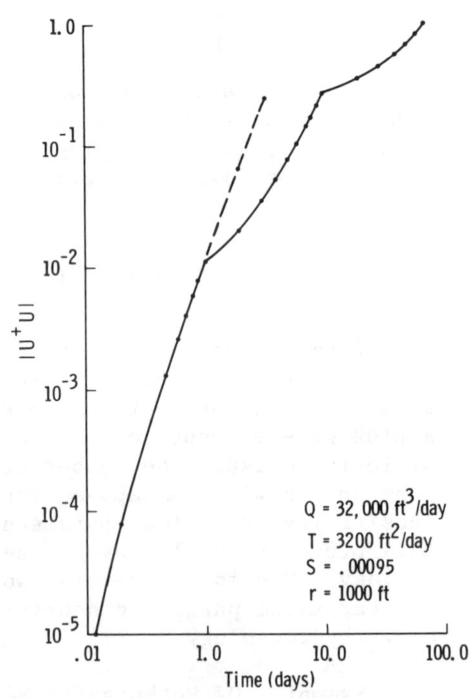

Fig. 25. $|U^+U|$ versus time for the Theis equation.

The 95% confidence intervals for transmissivity (δT) and storage (δS), as defined by Eq. (7.1), are plotted versus time in Figure 26. The head data (he) has been rounded to the nearest .1 foot and Eq. (6.10) has been used to estimate the head variance. The accuracy of T continues to increase with time. However, the curve for δS is fairly flat after about 1 day. (The confidence interval curves in Figure 26 are not monotonically decreasing because of errors in estimating σ, T, and S from the rounded data). These results might have been anticipated from the sensitivity coefficients (Figures 3 and 5). At .3 day the 95% confidence interval expressed as a percent of the parameter is 6.5% for S and 21.5% for T. At 2 days it is 3.9% for S and 3.7% for T. At 20 days it is 3.1% for S and 1.3% for T. From these results the model is seen to be more sensitive to S than to T at early times. However, as time increases, the model becomes more sensitive to T than to S.

These results suggest that sensitivity analysis could determine the duration of a pumping test needed for a given accuracy of the aquifer parameters (assuming that the Theis equation is a reasonable model). This could be done two ways. In the office before the pumping test, if one can estimate the accuracy of the head data (σ) and a range for T and S, then sensitivity analysis could predict the maximum duration needed for the desired accuracy. On the other hand, if a microcomputer can be taken to the field to record and analyse the data in real time, then sensitivity analysis would allow the computer to inform the supervisor of the current estimation accuracy and to stop the test when the desired accuracy had been reached. In this case, σ, T, and S would be estimated from the data as they are collected.

The results presented in this section have been for an observation well at 1,000 feet from the pumped well. However, the results could be used for any r value by simply scaling the time such that r^2/t remains constant. For example, at r = 100 feet, the accuracy of 3.1% for S and 1.3% for T occurs at .2 days.

8.2.2 The leaky aquifer

The leaky sensitivity coefficients are somewhat different from the Theis case (Figures 6-11). However, they still have their maximum value for small r. The time dependence shows some new features. There are three sensitivity coefficients with respect to the three parameters: transmissivity, storage, and leakage. Remember from Figure 9 that U_S has a maximum value at some time and then decreases to zero as the time increases. The other two (U_T and U_L from Figures 7 and 11) reach their maximum values and are constant after some time. For data of a certain accuracy, one might expect the accuracy of S to be constant after some time.

The 95% confidence intervals for the three parameters, as defined by Eq. (7.1), are plotted versus time in Figure 27. However, instead of plotting δP as for the Theis equation, $\delta P/P$ as a percent has been plotted; this procedure gives a much better comparison of relative sensitivities. The head data (he) has been rounded to the nearest .1 foot and Eq. (6.10) has been used to estimate the head variance. After about one day, the 95% confidence interval for S is about 5.5%. The confidence intervals of T and L continue to decrease, but at a fairly slow rate after one day. This is to be expected since $|U^+U|$ continues to increase slowly because U_T and U_L are nearly constant. At the end of 10 days, the confidence intervals for S, T, and L are 5.5%, 12.5%, and 16.7%, respectively. From Figure 27 we can see that the model is most sensitive to S and least sensitive to L, with the sensitivity to T falling between these two.

In the above discussion plotting $\delta P/P$ rather than just δP was more convenient. In the same way, it is easier to compare sensitivity coefficients if PU_P is plotted rather than U_P. The position of the maximum and the areas of zero values or constant values can be determined from either. For example, in Figure 9 the drawdown data for times from .01 to 1.0 days clearly is the most critical for determining the storage coefficient. However, to see relative sensitivity at a glance, it is much better to look at PU_P.

Comparing Figures 7 and 11, one might guess that a high correlation exists between T and L since the sensitivity coefficients look so similar. The correlation matrix, calculated from Eq. (6.11) for the data set of this section, bears this out. The correlation

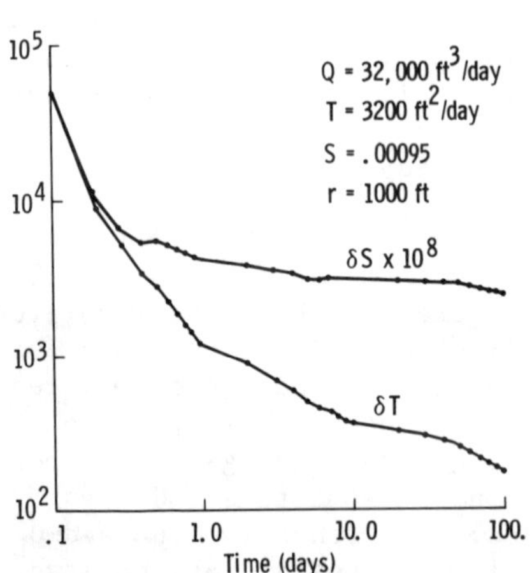

Fig. 26. 95% confidence intervals for the Theis equation, data rounded to nearest .1 ft.

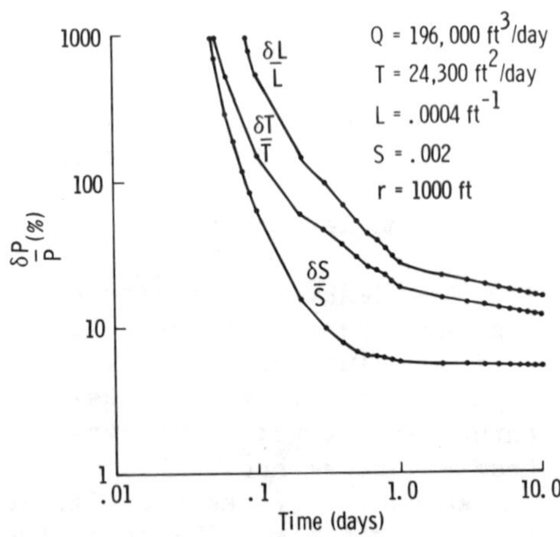

Fig. 27. 95% confidence intervals for the leaky aquifer, data rounded to nearest .1 ft.

matrix element for T and L is -.995. This implies that it will be difficult to obtain good estimates for both T and L with poorer quality data. Figures 6 and 10 indicate that U_T and U_L differ considerably in their radial dependences. This suggests that if more observation well data at various radii is available, the correlation between T and L would be reduced and better estimates could be made. This has been verified by numerical experiment. Adding observation wells at 100 feet and 500 feet reduces the correlation between T and L to -.80. At the same time, the 95% confidence intervals for S, T, and L reduce to 2.1%, .7%, and 2.3%, respectively, at the end of one day of pumping.

As for the Theis case, one could make a decision before the pumping test or in real time as to the duration needed for a given parameter accuracy. Of course, one must be able to give a range of aquifer parameters and data accuracy or estimate them in real time to perform this kind of sensitivity analysis. This example has pointed out that sampling in the spatial domain also affects the sensitivity. Therefore, one can choose the configuration of observation wells and duration of the pumping test based on sensitivity analysis to give the desired accuracy of the aquifer parameters. All of this assumes the leaky model is an adequate representation of the real world aquifer. In many cases, this may not be true.

8.2.3 One-dimensional steady-state model with spatially varying transmissivity

In sections 4.3 and 5.2, analytical expressions for the sensitivity coefficients were derived for two different sets of boundary conditions for the simple model described by Eq. (4.2) and Figure 12. The sensitivity coefficients, for Q specified at one boundary and head at the other, are shown in Figure 13. Some of the coefficients, for head specified at both boundaries, are shown in Figure 20. As remarked earlier, they are very different in character. At this point, we would like to know if one formulation of the problem is better than another for estimating the transmissivity distribution.

To make this judgement, we will look at three things: the condition number of U^+U, the estimated standard error, and the parameter correlation matrix. When Q is specified, the reciprocal condition number is $.362 \times 10^{-4}$ and the estimated standard error in transmissivity (see Figure 29) increases smoothly from about 35 for $T_{1/2}$ to 2,500 for $T_{9+1/2}$ (about 3.5% to 25%). The results for the standard error were obtained from Eq. (6.9), assuming $\sigma = .025$, which roughly corresponds to the head data being accurate to the nearest .1. As expected and predicted by Eq. (4.15), the model is less sensitive in areas having a larger transmissivity. However, the estimated errors seem reasonable and the inverse is well defined. On the other hand, when the head is specified at both ends, the reciprocal condition number is $.320 \times 10^{-8}$. This indicates that the matrix U^+U is nearly singular in single precision arithmetic on most computers. The estimated standard errors range from $.2 \times 10^5$ for $T_{1/2}$ to $.2 \times 10^6$ for $T_{9+1/2}$. In other words, no meaningful estimates can be made. From Eqs. (5.5) and 5.6), we know the model is insensitive to T when T is constant. Eq. (5.10) and Figure 20 indicate that the sensitivity coefficients are not zero if T varies in space. The values of T vary by an order of magnitude across this model; however, the results indicate very low sensitivity. This is a direct result of the boundary conditions chosen.

The parameter correlation matrix calculated from Eq. (6.11) for these two models is revealing. When Q is specified, the transmissivities are only related to their nearest neighbors. The parameter correlation matrix has the structure

$$\underset{\sim}{P} = \begin{bmatrix} 1 & -.5 & & & & \\ -.5 & 1 & -.5 & & & 0 \\ & \cdot & \cdot & \cdot & & \\ & & \cdot & \cdot & \cdot & \\ & 0 & & -.5 & 1 & -.707 \\ & & & & -.707 & 1 \end{bmatrix} . \tag{8.11}$$

On the other hand, when the head is specified at both ends, all the transmissivities are highly correlated. All the entries in the parameter correlation matrix are nearly one (for example .9999). If instead of solving the 10 x 10 matrix of U^+U for all ΔT's, the first nine rows and columns of U^+U are used, the system is much better behaved. In this case, the estimated standard errors and the condition number are larger, but comparable to those found when Q is specified. This amounts to requiring $\Delta T_{9+1/2}$ to be zero, indicating the initial guess is known to be very good. The better results should not be too surprising since a known T value would allow Q to be specified. However, the parameter correlation matrix for the 9 x 9 case still shows significant correlation between several of the T's (at the .8-.9 level). All of this points out the importance of being able to specify Q in the model.

As discussed in section 8.1 resulting in Eq. (8.6), it may be helpful to use normalized sensitivity coefficients (PU_P) and solve for relative changes in the parameters, $\Delta P/P$. For the model with Q specified at one end and head specified at the other, the reciprocal condition number improved from $.362 \times 10^{-4}$ to $.161 \times 10^{-2}$ simply by using the normalized version of the least squares equations. The estimated standard errors and the parameter correlation matrix were not significantly affected since the input head values were given accurately enough already. However, the improved condition number means that the effect of errors in head would not be amplified as much during the inverse process to find the ΔT's. This should be a great advantage in a real-world situation.

At this point we would like to consider the structure of the parameter variance for this model. From Eq. (4.15) and Figure 13, we see that if $T_{k+1/2}$ is a constant for all k, then all the $U_{Ti;k}$ have the same nonzero region as in Figure 13, but they all have the same amplitude. This leads to the following structure for U^+U.

$$\underset{\sim}{U}^+\underset{\sim}{U} = \left[\frac{Q\Delta x}{\ell T^2}\right]^2 \cdot \begin{bmatrix} 1 & 1 & 1 & 1 & 1 \\ 1 & 2 & 2 & 2 & 2 \\ 1 & 2 & 3 & 3 & 3 \\ 1 & 2 & 3 & 4 & 4 \\ 1 & 2 & 3 & 4 & 5 \end{bmatrix} \qquad (8.12)$$

To conserve space, the matrix is shown for only five transmissivities; the extension for any number is obvious. The inverse of this matrix is

$$(\underset{\sim}{U}^+\underset{\sim}{U})^{-1} = \left[\frac{\ell T^2}{Q\Delta x}\right]^2 \cdot \begin{bmatrix} 2 & -1 & 0 & 0 & 0 \\ -1 & 2 & -1 & 0 & 0 \\ 0 & -1 & 2 & -1 & 0 \\ 0 & 0 & -1 & 2 & -1 \\ 0 & 0 & 0 & -1 & 1 \end{bmatrix} . \qquad (8.13)$$

Again, the extension to larger matrices is obvious. We see that the estimated standard error (E.S.E.) is $\sqrt{2}\,\sigma \ell T^2/Q\Delta x$ for all except the last transmissivity and it simply has the 2 replaced by 1. This is the basic structure of the variance due to the model.

The extension of Eq. (8.12) to the case where transmissivity varies spatially is

$$\underset{\sim}{U}^+\underset{\sim}{U} = \left[\frac{Q\Delta x}{\ell}\right]^2 \cdot \begin{bmatrix} T_{1/2}^{-4} & T_{1/2}^{-2}\,T_{3/2}^{-2} & T_{1/2}^{-2}\,T_{5/2}^{-2} \\ T_{3/2}^{-2}\,T_{1/2}^{-2} & 2T_{3/2}^{-4} & 2T_{3/2}^{-2}\,T_{5/2}^{-2} \\ T_{5/2}^{-2}\,T_{1/2}^{-2} & 2T_{5/2}^{-2}\,T_{3/2}^{-2} & 3T_{5/2}^{-4} \end{bmatrix} . \qquad (8.14)$$

For simplicity a 3 x 3 matrix has been shown; the extension to any size is apparent. Likewise, the inverse is the extension of Eq. (8.13).

$$(\underset{\sim}{U}^+\underset{\sim}{U})^{-1} = [\frac{\ell}{Q\Delta x}]^2 \cdot \begin{bmatrix} 2T^4_{1/2} & -T^2_{1/2}\,T^2_{3/2} & 0 \\ -T^2_{3/2}\,T^2_{1/2} & 2T^4_{3/2} & -T^2_{3/2}\,T^2_{5/2} \\ 0 & -T^2_{5/2}\,T^2_{3/2} & T^4_{5/2} \end{bmatrix}. \tag{8.15}$$

We can see from the extension of Eq. (8.15) to an N x N matrix that

$$\text{E.S.E. of } T_{k+1/2} = \frac{\ell T^2_{k+1/2}}{Q\Delta x}\,\sigma \begin{cases} \sqrt{2} & 0 \leq k < N \\ 1 & k = N \end{cases} . \tag{8.16}$$

The factor out front is the contribution to the standard error due to the transmissivity distribution; while the remaining factor is due to the model structure. This shows very clearly that the error will be greater in areas of large T since the model is less sensitive there.

The parameter variance structure and standard error estimates given in the last paragraph are actually much more important than might be supposed. The sensitivity coefficients along a streamline in a two-dimensional model are given by [McElwee (14)]

$$U_{Ti;k} = \begin{cases} -\dfrac{Q_{k+1/2}\;\Delta x_{k+1/2}}{\ell_{k+1/2}\;T^2_{k+1/2}} & k \geq i \\ 0 & \text{otherwise} \end{cases} \tag{8.17}$$

where $Q_{k+1/2}$ is the flux of water in the streamtube between nodes k and k + 1, $\ell_{k+1/2}$ is the streamline width at that point, and $\Delta x_{k+1/2}$ is the node spacing between k and k + 1 along the streamtube. We see that this is just a straightforward extension of Eq. (4.15). Therefore, the estimated standard error of transmissivity determined along a streamline for a two-dimensional model is just the extension of Eq. (8.16) with the subscript k+1/2 attached to Q, Δx, and ℓ.

As an example of how Eq. (7.2) is used to calculate a confidence region, consider the 2 x 2 versions of Eqs. (8.12) and (8.13)

$$\underset{\sim}{U}^+\underset{\sim}{U} = C^2 \begin{bmatrix} 1 & 1 \\ 1 & 2 \end{bmatrix} \tag{8.18}$$

where $C^2 = \left[\frac{Q\Delta x}{\ell T^2}\right]^2$ (8.19)

$$(\underset{\sim}{U}^+\underset{\sim}{U})^{-1} = C^{-2} \begin{bmatrix} 2 & -1 \\ -1 & 1 \end{bmatrix} \tag{8.20}$$

The estimated standard error is $\sqrt{2}\ \sigma/C$ for T_1 and σ/C for T_2. The matrix in Eq. (8.18) may be diagonalized by finding its eigenvalues and eigenvectors. The eigenvectors can be used to form a transformation matrix that will diagonalize Eq. (8.18) and transform to a new set of transmissivities, T_1' and T_2'. The transformed least squares matrix is

$$(U^+U)' = C^2 \begin{bmatrix} \frac{3+\sqrt{5}}{2} & 0 \\ 0 & \frac{3-\sqrt{5}}{2} \end{bmatrix} \tag{8.21}$$

and the new transmissivities are

$$\Delta T_1' = \cos\Theta\ \Delta T_1 + \sin\Theta\ \Delta T_2 \tag{8.22a}$$

$$\Delta T_1' = -\ \sin\Theta\ \Delta T_1 + \cos\Theta\ \Delta T_2, \tag{8.22b}$$

$$\Theta = 58.3^0.$$

The above equations are those for a simple axis rotation of angle Θ. The inverse of Eq. (8.21) is easily found to be

$$(\underset{\sim}{U}^+\underset{\sim}{U})'^{-1} = C^{-2} \begin{bmatrix} \frac{2}{3+\sqrt{5}} & 0 \\ 0 & \frac{2}{3-\sqrt{5}} \end{bmatrix} = C^{-2} \begin{bmatrix} .3820 & 0 \\ 0 & 2.6180 \end{bmatrix}. \tag{8.23}$$

The estimated standard error is $.6180\sigma/C$ for T'_1 and $1.6180\sigma/C$ for T'_2.

The connection between the estimated standard errors in the original and transformed systems is given by Eq. (7.2) with $\ell = 1$. This corresponds to the 39% confidence limit [Beck and Arnold (1), page 294]. Eq. (7.2) defines an ellipse, the interior of which is the confidence region. Figure 28 shows this confidence region and its relationship to the original and transformed transmissivities. In the figure, σ/C has been set to one for convenience. The equation of the ellipse (for $\sigma/C = 1$) is

$$\frac{(\delta T_1')^2}{.3820} + \frac{(\delta T_2')^2}{2.6180} = 1 \ . \tag{8.24}$$

As mentioned in section 8.1, it is possible to sum nodes into zones to achieve an improved or acceptable level of sensitivity for the resulting number of parameters. For example, consider the 4 x 4 version of Eq. (8.12).

$$\underset{\sim}{U}^+\underset{\sim}{U} = \left[\frac{Q\Delta x}{\ell T^2}\right]^2 \cdot \begin{bmatrix} 1 & 1 & 1 & 1 \\ 1 & 2 & 2 & 2 \\ 1 & 2 & 3 & 3 \\ 1 & 2 & 3 & 4 \end{bmatrix} \tag{8.25}$$

Summing the third and fourth rows and columns results in

$$(\underset{\sim}{U}^+\underset{\sim}{U})_s = C^2 \begin{bmatrix} 1 & 1 & 2 \\ 1 & 2 & 4 \\ 2 & 4 & 13 \end{bmatrix} . \tag{8.26}$$

C^2 is defined in Eq. (8.19). The inverse of this matrix is

$$(\underset{\sim}{U}^+\underset{\sim}{U})_s^{-1} = C^{-2} \begin{bmatrix} 2 & -1 & 0 \\ -1 & 9/5 & -2/5 \\ 0 & -2/5 & 1/5 \end{bmatrix} . \tag{8.27}$$

It can be seen from Eq. (8.27) that the sensitivity to the T in the combined node zone has increased. From Eq. (8.27) the estimated standard error for the combined node zone is $\sigma/(\sqrt{5}\ C)$; this is to be compared to the original estimated standard errors of $\sqrt{2}\ \sigma/C$ for T_3 and σ/C for T_4. Assuming, of course, that it is acceptable to sum these two T's, the estimated standard errors are improved by factors of .45 and .32. This assumes σ does not change much.

As an example of the effect of summing nodes into zones, the simple model described by Eq. (4.2) and Figure 12 will be considered. The estimated standard error for ten zones has been discussed earlier and is shown in Figure 29 by the top curve. If pairs of nodes are summed into zones, the resultant estimated standard error for five zones is shown by the middle curve in Figure 29. In

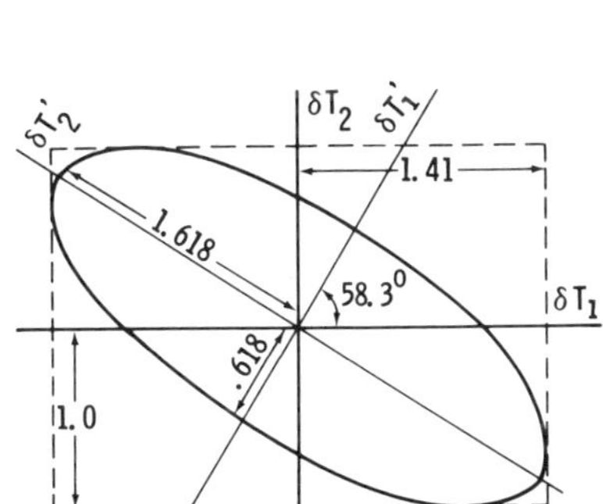

Fig. 28. 39% confidence region for the two-parameter model.

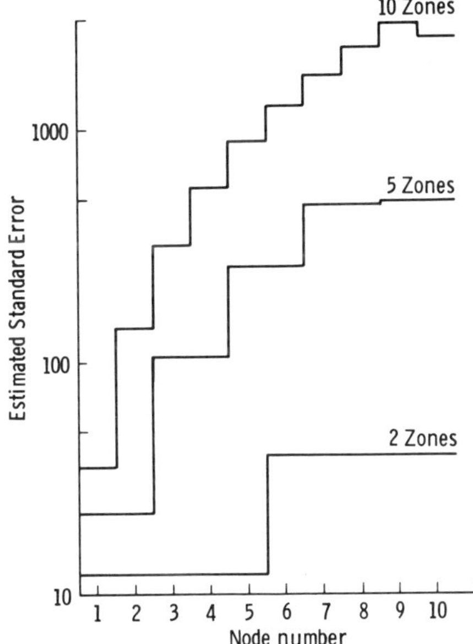

Fig. 29. Estimated standard error in transmissivity for various numbers of zones.

the extreme that five adjacent nodes are summed giving two zones, the lower curve in Figure 29 indicates the estimated standard error. Summing also improves the reciprocal condition number of the resulting least squares matrix which is solved for the parameter changes. In the ten-zone case the reciprocal condition number is .0016, while it is .0054 for five zones and .12 for two zones. Clearly, summing nodes into larger zones can be beneficial mathematically; however, this reduces the resolution of the transmissivity distribution. The maximum acceptable size of the zones would have to be determined from geohydrology considerations.

It should be noted that neither summing nor using the normalized least squares Eq. (8.6) can significantly alter an earlier conclusion: specifying the head as a boundary condition at both ends of the model leads to severe instability in the inverse process. Being able to specify Q or alternately one transmissivity value along the model has a tremendous stabilizing influence on the inverse process.

8.2.4 One-dimensional transient model with spatially varying parameters

Figure 30 shows an idealized one-dimensional transient model chosen to illustrate the use of sensitivity analysis. The model has a constant head boundary on the right (at node 11) and a barrier boundary on the left (between nodes 0 and 1). The model has a node spacing of 1,000 feet. Therefore, it is 10,500 feet between boundaries. There are 10 equations (nodes 1 to 10) to be solved for the 10 unknown head values as a function of time. The transmissivity is specified midway between node points (for example, $T_{3/2}$ occurs between nodes 1 and 2), while storativity is specified at the node points. Therefore, 10 values of transmissivity and 10 values of storativity are needed to describe this model. $T_{3/2}$ is 52,000 gpd/ft (6,952 ft^2/day) and the transmissivity increases by 2,000 gpd/ft (267 ft^2/day) as the node number increases by one. Thus, $T_{10+1/2}$ is 70,000 gpd/ft (9,358 ft^2/day). The storativity at node 1 (S_1) is .0050 and increases by .0005 as the node number increases by one. Thus, S_{10} is .0095.

The initial head distribution is assumed to be flat. The aquifer is being pumped at a rate Q equal to 1,500 gal/day (200 ft^3/day) per unit transverse length. The pumping represents a line sink located at node 7, 4,000 feet away from the constant head boundary. The model has a steady-state solution where all water being pumped comes from the constant head boundary, and the head distribution to the left of the well is flat and somewhat lower than the original level.

The correct values for the transmissivity at the 10 intermediate node points and for the storativity at the 10 node points are shown in the second column of Table 2. These values, along with the other model parameters discussed previously, were used to generate hypothetical "field" data for the hydraulic head as a function of time. These values of hydraulic head (accurate to five decimal places) were then used in an ordinary least squares inverse initial estimate for transmissivity was 61,000 gpd/ft (8,155 ft^2/day) and for storativity was .00725.

The transmissivity and storativity values calculated for early time by the inverse procedure are shown in the third column of Table 2. The early time calculations were made using hydraulic heads for 10 time steps with the total time slightly less than 2 days. For these early times the drawdown is less than 19 feet at the

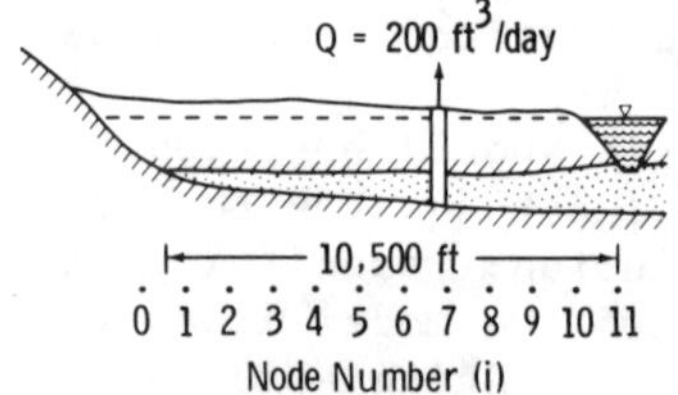

Fig. 30. A simple one-dimensional transient model [McElwee (13)].

line sink and is less than one foot farther than 3 node points away from it. The values of calculated transmissivity and storativity are within 20% of the actual values, but there is no clear evidence that the inverse procedure has been successful in finding the spatial trend of increasing T and S. At early time periods the drawdown is small and the model is fairly insensitive to the transmissivity and storativity. This insensitivity can be seen very easily by examining the sensitivity coefficients or by calculating the estimated standard error. Without a sensitivity analysis one might have expected better answers, since the head is given so accurately.

At middle times (column 4 of Table 2), when the drawdown is substantial and hydraulic heads are changing fairly rapidly with time, the greatest sensitivity and best inverse solution results. The middle time inverse calculations were made using hydraulic heads for 10 time steps between 2 days and 112 days. The system rapidly approaches steady-state for times greater than 112 days. The largest error in transmissivity is less than 7% and most values are very close to the correct values. The storativity calculations have less than 10% error and most are very close to the correct values. The largest errors occur near node 1. The reason for this will be discussed later. These results were obtained using head data accurate to 5 decimal places and a Gauss-Seidel Iterative (GSI) solution to Eq. (6.5). Using this solution routine, it was not possible to lower the error in aquifer parameters below 7-10% near node 1. However, additional work using a high-efficiency matrix package (HEMP) direct solution technique showed that the error in T and S could be made very small by using head data accurate to 5 decimal places. However, the largest of the small errors (.001%) still occurred near node 1. The reciprocal condition number of the least squares matrix for this middle time data is $.33 \times 10^{-6}$. This is much smaller than one would like and indicates that doing the matrix inversion in single precision arithmetic may lead to problems if not done efficiently. Apparently this is why the HEMP solution could achieve better accuracy. A plot of the estimated standard error ($\delta P/P$) calculated by a sensitivity analysis is shown in Figure 31. Notice that the HEMP solution near node 1 is consistent with Figure 31. However, the middle time errors in Table 2 near node 1 are much bigger than predicted by Figure 31, presumably caused by the inefficiency of the GSI solution.

The last column in Table 2 shows the transmissivity calculations as the model approaches steady-state. The storativity calculations have become unstable and cannot be made because of low sensitivity. The late time inverse calculations have been made using hydraulic heads from 5 time steps between 112 days and 850 days. The transmissivity calculations for the last 4 node points are very good. However, the calculated transmissivities for the first 6 node points are fairly bad. This can be explained by looking at the sensitivity coefficients of the estimated standard errors. The sensitivity at the last 4 nodes is about 3 or 4 orders of magnitude greater than for the first 6 nodes. This lack of sensitivity is because $\partial h/\partial x$ is practically zero for the first 6 node points for times greater than 112 days.

Table 2. Inverse Calculations Over Various Time Periods (Head Data Accurate to 5 Decimal Places) [McElwee (13)]

Grid Number	Correct Value	Early Time	Middle Time	Late Time
		TRANSMISSIVITY (gpd/ft)		
1 + 1/2	52,000	59,575	48,415	72,213
2 + 1/2	54,000	62,107	55,064	69,827
3 + 1/2	56,000	64,365	55,980	71,049
4 + 1/2	58,000	66,684	57,955	70,520
5 + 1/2	60,000	69,070	60,000	71,130
6 + 1/2	62,000	71,283	62,002	243,374
7 + 1/2	64,000	55,152	64,001	64,006
8 + 1/2	66,000	56,931	66,000	65,998
9 + 1/2	68,000	58,603	68,000	67,998
10 + 1/2	70,000	60,330	70,000	69,998
		STORATIVITY		
1	.0050	.0057	.004656	Unstable
2	.0055	.0063	.006037	Unstable
3	.0060	.0069	.005801	Unstable
4	.0065	.0075	.006491	Unstable
5	.0070	.0081	.007015	Unstable
6	.0075	.0086	.007501	Unstable
7	.0080	.0080	.007998	Unstable
8	.0085	.0073	.008501	Unstable
9	.0090	.0078	.009000	Unstable
10	.0095	.0080	.009500	Unstable

Notice that for the middle time calculations shown in Table 2, the largest error in transmissivity occurs at node 1 and decreases considerably at higher numbered nodes. This should mean that the model is least sensitivity to T at node 1. A look at the sensitivity coefficients or the estimated standard should verify this. The solid curves in Figure 31 represent the estimated standard errors ($\delta P/P$) for the transmissivity and storativity at each node for $\sigma = .25 \times 10^{-5}$ (values for other σ's may be determined by the appropriate factor). The largest errors in both T and S occur at the lower numbered nodes. This is a little surprising because the sensitivity coefficients for S do not have their lowest value at node 1 but at node 10 [McElwee (13)] near the constant head boundary. Also, notice that both solid curves in Figure 31 have practically the same $\delta P/P$ for node 1. One might suspect that T_1 and S_1 are not both independent parameters in this model. The element in the parameter correlation matrix (6.11) corresponding to T_1 and S_1 is .9917, which verifies the suspicion. Thus, the larger error in S at the lower node numbers is due to a dependence between T and S and not specifically due to low sensitivity values there.

The dashed curves in Figure 31 result when ΔT_1 is dropped from the matrix equations (6.5). This means that T_1 is assumed known and solve for S_1. Notice that the estimated standard error in S_1 drops by almost an order of magnitude. The condition number of the least squares matrix is also improved by a factor of about 6 when ΔT_1 is dropped from consideration. Eq. (2.27) suggests that maybe the model is only sensitive to the ratio of S_1/T_1. Taking values of T_1 and S_1 from Table 2, we can see that $.005/52000 \simeq .004656/48415$. Thus, a basic nonuniqueness exists near node 1. The sensitivity to transmissivity is low near node 1, which causes substantial error in the calculation of T_1. Since S_1 is dependent upon T_1, there is substantial error in S_1 also even though the sensitivity coefficients for S_1 are not particularly low near node 1.

8.2.5 Two-dimensional steady-state models

The simple two-dimensional model of section 4.4 is considered here with various boundary conditions. There are three different cases. Case 1 is defined in section 4.4 and the sensitivity coefficients are shown in Figures 15 and 16. The head is specified on three sides and the sensitivity coefficients go to zero on these boundaries. The flux is specified at $x = 0$. Cases 2 and 3 are defined in section 5.3.

Case 2 changes the boundary condition at $y = 0$ and $y - 2{,}000$ to a barrier boundary. These sensitivity coefficients are shown in Figures 21 and 22. Notice that the coefficients do not go to zero at $y = 0$ and $y = 2{,}000$. Since they are nonzero over a larger area than case 1, one might expect the estimated standard error to be smaller for case 2 than case 1. It cannot be seen from Figures 15, 16, 21, and 22 because the plotting routine normalizes the maximum values; however, the sensitivity coefficients for case 2 are 3 to 4 times larger than those for case 1. This is another reason to expect the estimated standard error for case 2 to be lower.

Case 3 is defined with the same boundary conditions as case 2 except the head is specified at $x = 0$ instead of the flux. These sensitivity coefficients are shown in Figures 23 and 24. They have about the same maximum value as case 1. The sensitivity for zone 1 is negative, while the sensitivity for zone 2 is positive. However,

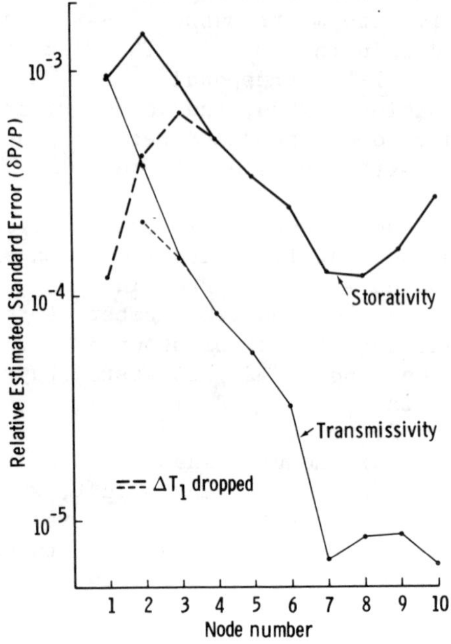

Fig. 31. Estimated standard error for one-dimensional transient model.

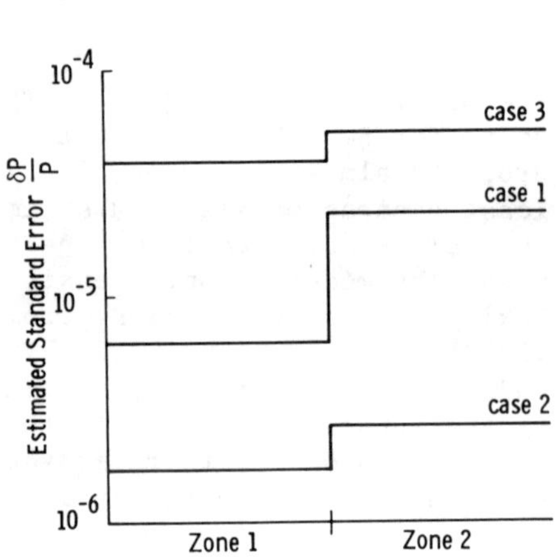

Fig. 32. Estimated standard error for the simple two-dimensional case.

except for the sign, these two sensitivity coefficients look very much alike. This is born out by the parameter correlation matrix whose off-diagonal element is .997. Therefore, the parameters T_1 and T_2 are highly correlated, and we would expect to have trouble estimating both simultaneously.

Figure 32 shows the estimated standard error for the transmissivity in the two zones for σ = .00025 (about three decimal place accuracy) for all three cases discussed above. As we expected, case 2 has the lowest error and case 3 has the highest. Case 3 has lower values of the sensitivity coefficients than case 2, and they appear to be nearly dependent. Therefore, logically case 3 should have the largest estimated standard error. Case 1 is somewhat intermediate, the sensitivity coefficients are considerably lower than case 2, but the parameters have a low correlation. With only two large zones, the zero values of the sensitivity coefficients on three boundaries for case 1 do not play a large role. However, if we had several more zones, clearly any small zone near one of the head specified boundaries would have a fairly large estimated standard error due to the low sensitivity induced by the boundary condition.

The above examples illustrate several important points, but they are not very realistic; therefore, we looked at the data and model presented by Yakowitz and Duckstein (19) for a part of the Tucson basin. The head and transmissivity maps are shown in Figures 33 and 34. The model area is shown by the smaller square in each figure. Yakowitz and Duckstein have concluded that the model is not very sensitive to the transmissivity and that the error for least squares estimation of parameters should be quite high, even for only nine zones. They do not give a detailed discussion of the boundary conditions used in their model. However, it seems clear from one sentence in their paper and the above results that they were using a head-specified boundary condition all the way around the model. From the earlier examples in this section and previous discussion in this paper, we would expect this to be the most unstable case.

We have performed a sensitivity analysis for the data and model presented by Yakowitz and Duckstein for various boundary conditions and number of zones. For a large number of zones (81), we also find the estimated standard error to be quite large for most boundary conditions. However, noticable improvement occurs when some boundary fluxes are specified. When the head is specified as a boundary condition all the way around the model, the estimated standard error is so large that no meaningful estimate can be made at all. If the model is zoned into nine zones and some fluxes specified on the boundaries, we find the estimated standard error for the zones ranges from 9.3% to 100%. Once again, if head is specified on all boundaries, no meaningful estimate can be made and high correlation is observed between parameters. If 4 transmis-

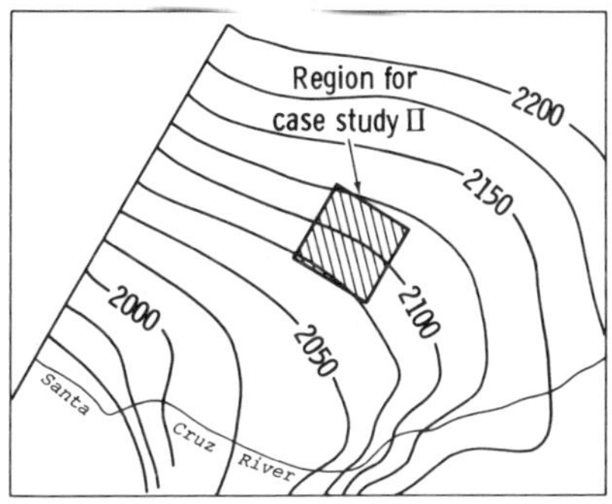

Fig. 33. Head map for the Yakowitz and Duckstein (19) model.

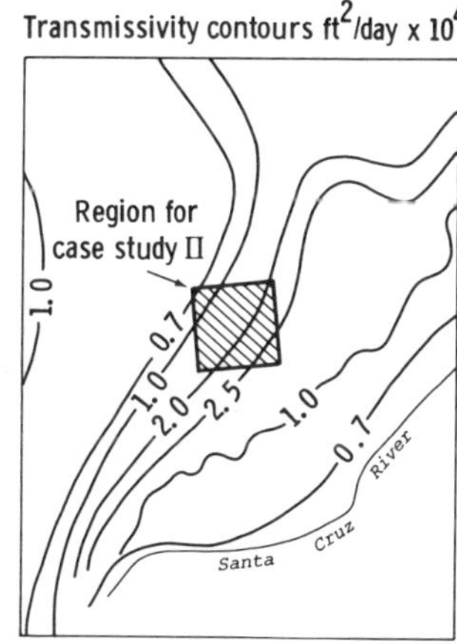

Fig. 34. Transmissivity map for the Yakowitz and Duckstein (19) model.

sivity zones are used and some boundary fluxes are given, we find the estimated standard error ranges from 4.3% to 16.9%. As before, if the head is specified all the way around, the estimated error is too large to make any meaningful estimate and the parameters show a high correlation. In all this work, the value of σ was estimated from the data and was in the range of .8 to .9 feet.

This example from a real-world situation shows clearly the advantages of using some of the techniques discussed in this paper to make the parameter-estimation procedure better conditioned. In paricular, boundary conditions are very important. The sensitivity should be made as large as possible over a large extent of the model. One must insure that the sensitivity coefficients and the parameters are not dependent. Zonation by summing nodes can be used to increase sensitivity.

9. SUMMARY AND CONCLUSIONS

This paper has given the definition of a sensitivity coefficient and shown how they may be used to perform a sensitivity analysis. A first-order sensitivity formalism has been presented to calculate the perturbed head due to a change in an aquifer parameter. We have discussed various methods of determining sensitivity coefficients and examined the effect of boundary conditions. Several examples of sensitivity coefficients have been given in some detail for a variety of models and boundary conditions. We have shown how the sensitivity coefficients enter the least squares parameter-estimation formalism and how confidence intervals and estimated standard errors can be found. We have seen that sensitivity coefficients are an essential part of parameter-estimation and that an intelligent sensitivity analysis can lead to a more stable and better conditioned inverse process.

Some general guidelines can be given for increasing the model sensitivity, leading to more accurate parameter estimation. For maximum sensitivity, the measurements of head should occur at locations and times where the sensitivity coefficients are near their maximum values. Sensitivity coefficients are greatly influenced by boundary conditions; therefore, the boundary conditions giving the largest sensitivity, consistent with the data and known hydrogeology, should be chosen. We have seen the importance of being able to specify some fluxes for the model. Emsellem and de Marsily (6) have commented on the importance of knowing some fluxes or transmissivities in the groundwater inverse problem. This requires fairly accurate knowledge of the slope of the hydraulic head and the transmissivity at some locations. Judging from the improved stability and accuracy that can be obtained, our field methods should be geared toward this goal. The sensitivity coefficients and the parameters should be independent. This can be examined by looking at plots of the sensitivity coefficients and by calculating the sensitivity and parameter correlation matrices. Zoning by summing nodal sensitivities can lead to improved model sensitivity; however, that must be balanced with a loss of transmissivity resolution. Using these guidelines, the goal is to select a model with maximum (or at least a certain minimum) sensitivity to the aquifer parameters, based on a knowledge of sensitivity coefficients and their properties.

10. REFERENCES

1. Beck, J. V., and Arnold, K. J., *Parameter Estimation in Engineering and Science,* John Wiley and Sons, New York, N.Y., 1977.
2. Cobb, P. M., McElwee, C. D., and Butt, M. A., "Analysis of Leaky Aquifer Pumping-Test Data: An automated Numerical Solution Using Sensitivity Analysis," *Ground Water,* Vol. 20, No. 3, 1982, pp. 325-333.
3. Cooley, R. L., "A Method of Estimating Parameters and Assessing Realiability of Models of Steady State Groundwater Flow. 1. Theory and Numerical Properties," *Water Resources Research,* Vol. 13, No. 2, 1977, pp. 318-324.
4. Cooley, R. L., "A Method of Estimating Parameters and Assessing Reliability for Models of Steady State Groundwater Flow. 2. Application of Statistical Analysis," *Water Resources Research,* Vol. 15, No. 3, 1979, pp. 603-617.
5. Dongarra, J. J., Bunch, J. R., Moler, C. B., and Stewart, G. W., *Linpack Users' Guide,* 3rd ed., Society for Industrial and Applied Mathematics, Philadelphia, PA., 1979.
6. Emsellem, Y., and de Marsily, G., "An Automatic Solution for the Inverse Problem," *Water Resources Research,* Vol. 7, No. 5, 1971, pp. 1264-1283.
7. Hantush, M. S., and Jacob, C. E., "Non-Steady Radial Flow in an Infinite Leaky Aquifer," *Transactions of the American Geophysical Union,* Vol. 36, 1955, pp. 95-100.
8. Hildebrand, F. B., *Advanced Calculus for Applications,* Prentice-Hall, Englewood Cliffs, New Jersey, 1962, pp. 360.
9. Knowles, T. R., Claborn, B. J., and Wells, D. M., "A Computerized Procedure to Determine Aquifer Characteristics," *Water Resources Center Pub. No. WRC-72-5,* Texas Technological University, Lubbock, Texas, 1972.
10. Lighthill, M. J., *Introduction to Fourier Analysis and Generalized Functions,* Cambridge University Press, New York, N.Y., 1958.
11. McElwee, C. D., and Yukler, M. A., "Sensitivity of Groundwater Models With Respect to Variations in Transmissivity and Storage," *Water Resources Research,* Vol. 14, No. 3, 1978, pp. 451-459.
12. McElwee, C. D., "A One-Dimensional Study of the Groundwater Inverse Problem Using Sensitivity Analysis," *Open File Report 82-12,* Kansas Geological Survey, Lawrence, Kansas, 1982.

13. McElwee, C. D., "Sensitivity Analysis and the Ground-Water Inverse Problem," *Ground Water,* Vol. 20, No. 6, 1982, pp. 723-735.
14. McElwee, C. D., "The 2-D Inverse Problem as a Suite of 1-D Problems," *Open File Report 82-13,* Kansas Geological Survey, Lawrence, Kansas, 1982.
15. Neuman, S. P., and Yakowitz, S., "A Statistical Approach to the Inverse Problem of Aquifer Hydrology. 1. Theory," *Water Resources Research,* Vol. 15, No. 4, 1979, pp. 845-860.
16. Neuman, S. P., "A Statistical Approach to the Inverse Problem of Aquifer Hydrology. 3. Improved Solution Method and Added Perspective," *Water Resources Research,* Vol. 16, No. 2, 1980, pp. 331-346.
17. Strang, G., *Linear Algebra and Its Applications,* 2nd ed., Academic Press, Inc., New York, N.Y., 1980.
18. Theis, C. V., "The Relation Between the Lowering of the Piezometric Surface and the Rate and Duration of Discharge of a Well Using Groundwater Storage," *Transactions American Geophysical Union,* Vol. 16, 1935, pp. 519-524.
19. Yakowitz, S., and Duckstein, L., "Instability in Aquifer Identification: Theory and Case Studies," *Water Resources Research,* Vol. 16, No. 6, 1980, pp. 1045-1064.
20. Yeh, W. W.-G., and Sun, N.-Z., "An Extended Identifiability in Aquifer Parameter Identification and Optimal Pumping Test Design," *Water Resources Research,* Vol. 20, No. 12, 1984, pp. 1837-1847.

11. LIST OF SYMBOLS

A	Area of integration
B	Model boundary
C	Constant
E	Squared error function
F	Arbitrary function or statistical F distribution
H	Specified hydraulic head
K'	Permeability of semiconfining bed
L	Leakage factor for leaky aquifer
N	Total number of grid spaces for 1-D model
$\underline{P}$	Parameter vector
$\underset{\sim}{P}$	Parameter covariance matrix
Q	Water flux for a model
Q'	Water flux per unit model area
Q"	Water flux per unit length of boundary
R	Maximum x value for 1-D model, region of interest for a model, or right hand side of the least squares equation
S	Storage coefficient
T	Transmissivity

U_T	Sensitivity coefficient for transmissivity
U_S	Sensitivity coefficient for storage coefficient
$\underset{\sim}{U}$	Sensitivity matrix
$\underset{\sim}{U}'$	Normalized sensitivity matrix
$\underset{\sim}{V}$	Weight matrix for prior estimates of parameters
$\underset{\sim}{W}$	Weight matrix for head observations
f	Arbitrary function
h	Hydraulic head
h^*	Perturbed hydraulic head
h_e	Experimentally measured head
ℓ_2	Perpendicular direction in cross-sectional models
$\ell^2_{1-\alpha}$	Statistical ℓ distribution
m'	Thickness of semiconfining bed
p	Number of parameters
r	Radial distance from pumped well, or an F distribution in Eq. (7.2a).
$\underset{\sim}{r}$	Parameter correlation matrix
s	Drawdown
t	Time
$t_{1-\alpha/2}$	Statistical t distribution
u,y	Dummy variables of integration
$x,\underline{x}$	Cartesian coordinate value or vector
α	Related to confidence interval of statistical distributions
Γ	Model boundary
Δ	Change in some quantity
Δh	Change in head
ΔL	Change in leakage factor
$\underline{\Delta P}$	Vector of parameter changes
ΔS	Change in storage coefficient
ΔT	Change in transmissivity
Δx	Grid spacing
δ_{ij}	Kronecker delta
δP_k	Confidence interval for parameter P_k
$\delta(x-x_o)$	Delta function
$\theta(x-x_o)$	Heaviside function
$\rho_k,\underline{\rho}$	Correct value of P_k or parameter vector
σ	Variance of error in head
$\frac{\partial}{\partial n}$	Normal derivative
i,j,k,ℓ	Node indices, used as subscripts
m	Zone index, used as a subscript
n	Time index, used as a superscript
'	Used to denote normalized or transformed quantities
—	Underline indicates a vector quantity
~	Underline indicates a matrix
+	Superscript indicating transpose of a vector or matrix

PART 5

ADVANCES IN NUMERICAL METHODS

MATHEMATICAL MODELING OF THE BEHAVIOR OF HYDROCARBON RESERVOIRS - THE PRESENT AND THE FUTURE

Aziz S. Odeh

MATHEMATICAL MODELING OF THE BEHAVIOR OF HYDROCARBON RESERVOIRS - THE PRESENT AND THE FUTURE

Aziz S. Odeh

Mobil Research and Development Corporation
Research Department, Dallas Research Division
P.O.Box 819047
Dallas, Texas 75381 USA

ABSTRACT

The derivation of the flow equations which describe the behavior of a black oil reservoir is outlined. The functional relations among the variables which result in the nonlinearity of the equations are explained in the context of reservoir engineering fundamentals. The state of the art of simulating black oil reservoirs and enhanced oil recovery processes is discussed. The effect of discretization required by the commonly used numerical methods on the physics of the processes is examined, and the shortcomings are identified. Finally, some of the outstanding problems that have not been adequately solved but need to be solved for belivable answers are addressed.

1. INTRODUCTION

The objective of reservoir engineering is to economically optimize the recovery of hydrocarbon. To do this the reservoir engineer should be able to predict the performance of the reservoir under various exploitation schemes. This requires an adequate physical description of the reservoir, and a method to calculate the hydrocarbon and pressure distributions as a function of time. The physical description is obtained from seismic and geologic data, from core and log analysis, and from single and multiple well testing. Hydrocarbon and pressure distributions are calculated by mathematical models or simulators which are the modern tools of the reservoir engineer.

All reservoirs are three dimensional. Therefore, from a theoretical point of view the engineer should always use a three-dimensional simulator to study their behavior. In some cases, the variation of properties in one dimension such as the vertical may not be significant, or one is interested in a detailed behavior of the reservoir in the vertical plane. In such cases a two-dimensional model is used. If the reservoir is shaped like a sandbar, a one-dimensional model may be adequate. Some fundamental reservoir engineering calculations are sometimes made using average properties or a zero-dimensional (tank) model. One- or two- or three-dimensional models may be viewed as composed of many interacting tank units (21). All these models are used in reservoir engineering. Generally speaking, as the dimensionality increases, so does the cost of the study. A graphical representation of the various models is shown in Figure 1.

The availability of larger and faster computers has given impetus to the extensive use of simulators especially in a three-dimensional mode for managing hydrocarbon reservoirs. Literally, hundreds of millions of dollars of expenditures based on the simulators' results, are being made. Because of this, one needs to fully understand the strength and weakness of these models.

By and large, finite difference has been the principle method to numerically solve the equations of interest. This numerical method has been adequate for simulators of the black oil type. It has not been as successful with enhanced oil recovery (EOR) models where moving fronts exist, though it is still widely used in these models. In this paper, the disadvantage of the finite difference method is explained in the context of EOR simulators.

Most of the effort in the simulation technology has been concerned with the efficient discretization of the appropriate equations, and the solution of the resulting matrices. Hardly any attention is being paid to the effect of discretization on the physics. Hydrocarbon reservoirs are continuous in space and time. Their mathematical models break up this continuity into discrete parts. This paper discusses how this procedure could result in erroneous reservoir behavior.

Finally, the paper identifies some fo the outstanding problems that need to be addressed and solved, to give the EOR simulators an acceptable degree of accurate representation of the physical system.

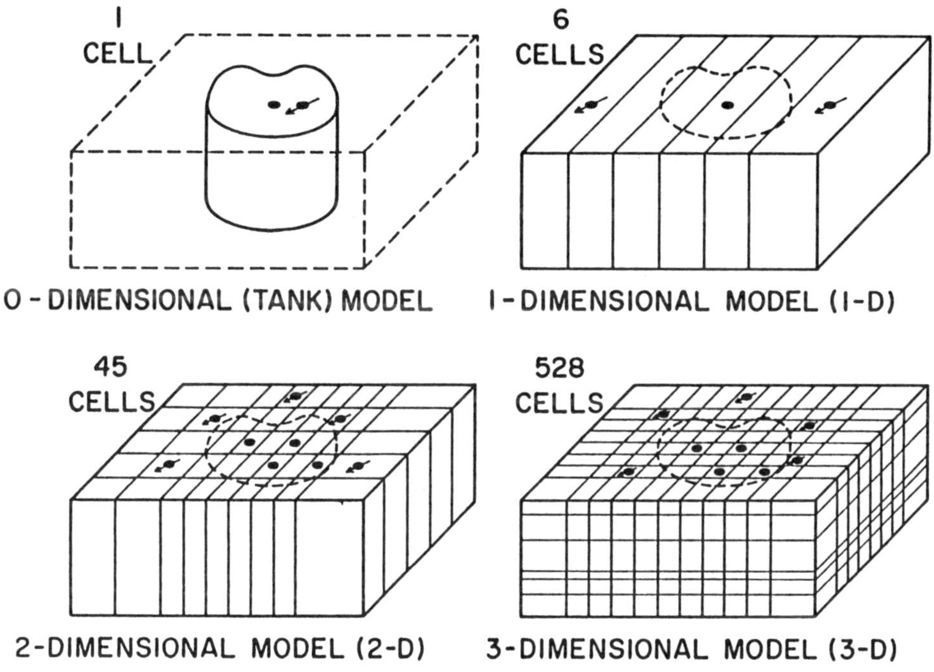

Figure 1. Classification of reservoir simulators by dimensions.

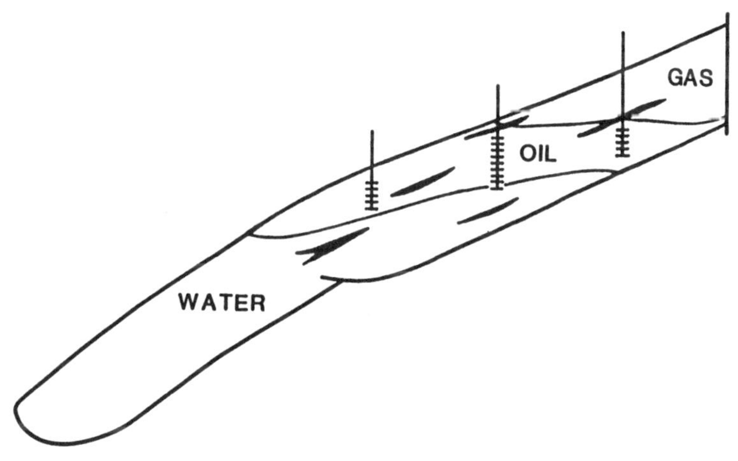

Figure 2. Reservoir cross-section.

2. THE RESERVOIR, ITS FLUIDS, AND PROPERTIES

Figure 2 is a cross section of a typical oil reservoir at discovery. Gravity and geological time cause the water to segregate to the bottom, the gas to the top, and the oil to be in between. When the wells are put on production, the oil-gas and the water-oil interfaces will move in an irregular manner depending on the location of the wells. Each well may undergo a different production history. In Figure 3 a typical history of a well is illustrated. The well is drilled and perforated in the oil leg which always contains water that may or may not be mobile. The oil contains gas in solution. Due to the pressure decrease caused by oil production, a free gas phase may evolve out of solution. Part of the gas could segregate to the top of formation while part of it remains in the oil leg. Thus, the oil leg could have simultaneously mobile oil, gas and water. Due to the pressure gradient a water cone is formed as the well is produced, causing water production. When the water production becomes excesive, the bottom perforations are plugged. As production continues a gas cusp or a cone may be formed which could cause excessive gas production. In many cases a high gas production causes an unnecessary loss of the natural energy of the reservoir. When this happens the well is recompleted or shut-in. The behavior of the well is just an example of what a mathematical model should be capable of simulating.

The system of partial differential equations which describes flow of immiscible fluids (gas, oil, and water), through porous media contains parameters which are functions of the properties of the fluids and the porous system. In Figure 4 the pressure-volume behavior of oil and gas in a black oil reservoir is given. A black oil reservoir is that reservoir where the properties of the oil is a function of pressure only, and where the composition of the oil is assumed constant throughout the life of the reservoir. Most oil reservoirs fall in this category.

In the figure the discovery pressure is shown to be 4000 psi. At this pressure no free gas is present. All the gas is in solution. The oil is undersaturated. As the pressure decreases the oil expands until the bubblepoint or saturation pressure is reached. This is the pressure below which gas comes out of solution and forms a free gas phase. Below this pressure, the oil shrinks as the pressure decreases. However, the total volume of oil and gas increases. Finally, at atmospheric conditions, the maximum shrinkage of oil and the maximum volume of oil and gas occur. The ratio of one volume of oil at reservoir condition to its resulting volume under atmospheric condition is known as the formation volume factor, B. This parameter is a function of pressure; it increases until the bubblepoint pressure is reached, and then decreases. Thus, it has a slope discontinuity at the bubblepoint pressure which could cause

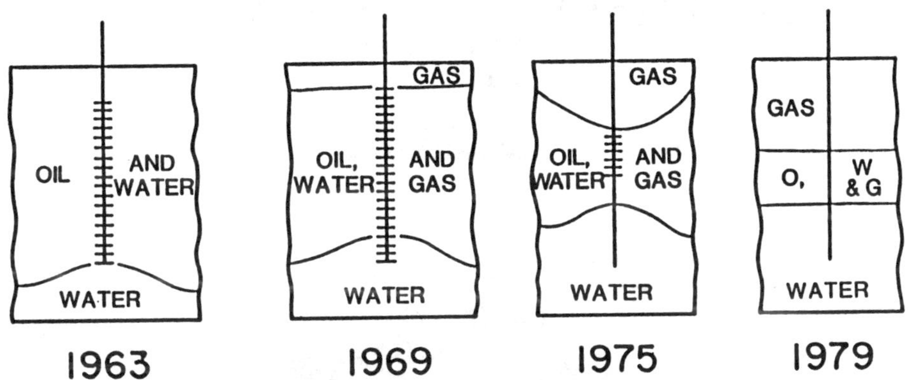

Figure 3. History of a well.

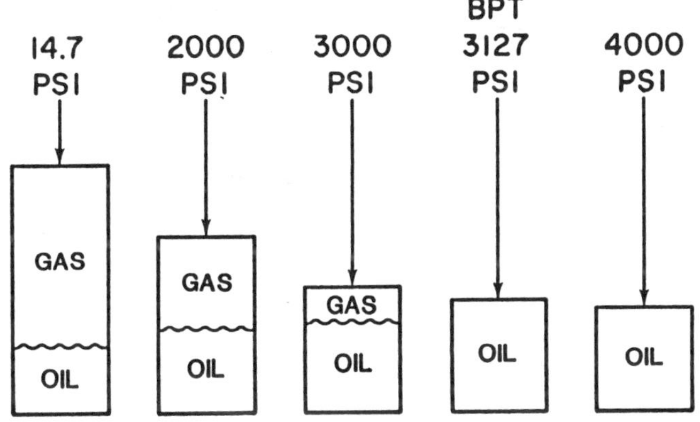

Figure 4. Pressure-volume behavior of oil and gas.

numerical difficulties.

The capacity of the porous medium to store fluids is indicated by the porosity, ϕ, defined in Figure 5. Porosity is the ratio of the volume of the void space to the bulk volume. It is a function of pressure and decreases as the pressure of the reservoir decreases. The ability of the porous medium to transmit fluids is given by the permeability, k. When the porous system contains one fluid only, its permeability termed specific permeability is generally not considered a function of the fluid. There is however some evidence that the permeability, like the porosity, is affected by the pressure. Most simulators do not account for this effect.

POROSITY - VOID VOLUME / TOTAL VOLUME

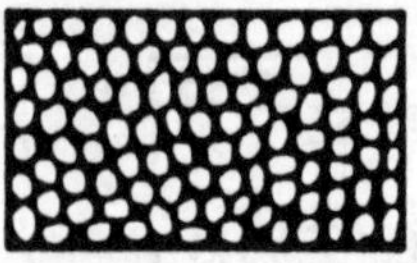

PERMEABILITY - ABILITY TO FLOW

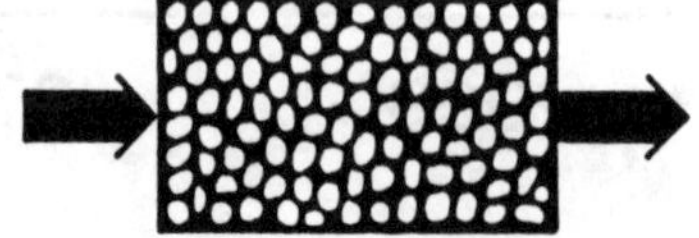

SATURATION - FRACTION OF VOID VOLUME OCCUPIED BY PARTICULAR PHASE

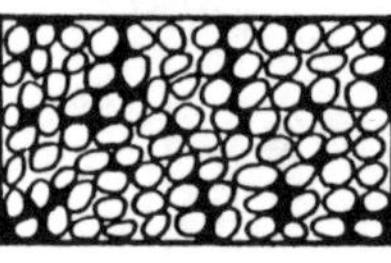

Figure 5. Definitions of porosity, permeability and saturation

When more than one fluid occupies the porous system we indicate the relative volume of each fluid by the saturation, S. This is the ratio of the volume of the fluid to the pore volume. The sum of saturations is equal to one. If oil and water occupy the void space as in Figure 6, for example, the permeability to either fluid is not constant. It is a function of the saturation. It is indicated by k_r, the relative permeability, which is the ratio of the permeability at any saturation to the permeability at 100% saturation. Figure 6 shows typical oil-water relative permeability curves. When gas is present the three-phase relative permeability relations are more complex. The oil relative permeability, k_{ro}, is a function of the gas and water saturations. This complicates the partial differential equations that will be discussed later.

Three forces are responsible for the distribution of fluids in the reservoirs. These forces are shown in Figure 7. The viscous forces generally termed the Darcy forces. The gravitational forces are due to the difference in density. The capillary forces are due

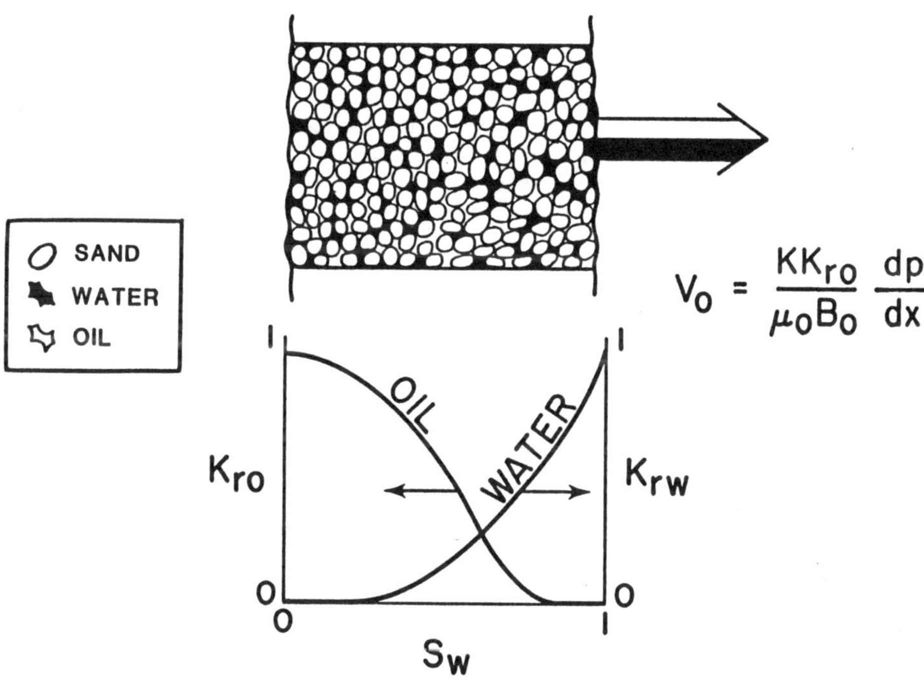

Figure 6. Relative permeability and fractional flow.

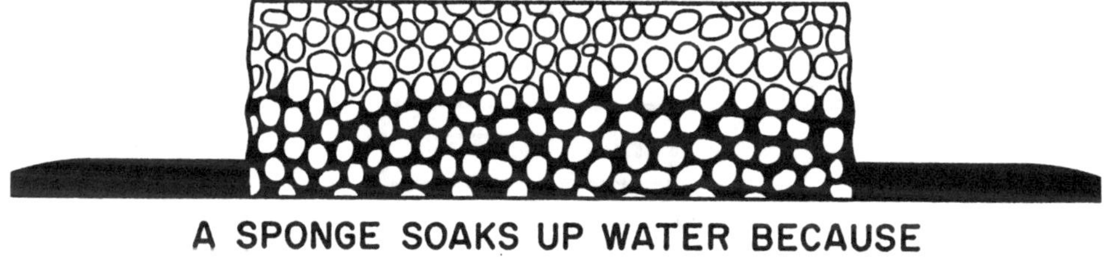

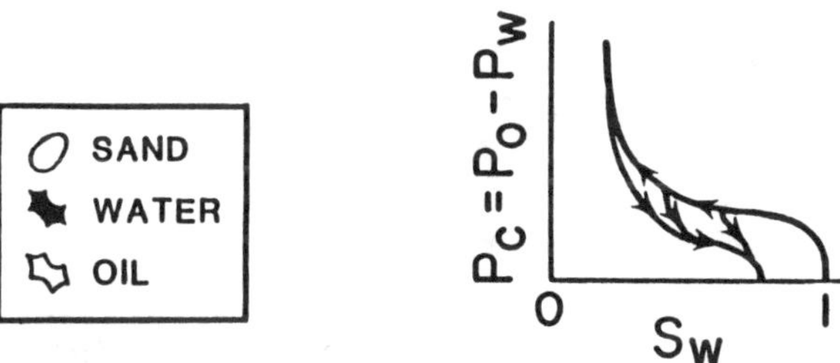

Figure 7. Forces responsible for the distribution of fluids in the reservoirs: a) viscous, b) gravitational, and c) capillary, and the capillary pressure (p_c) as a function of saturation (S_w).

to the affinity of the porous system to one fluid more than to the other. Thus we speak of a water wet or an oil wet reservoir. A water wet rock, for example, imbibes water, resulting in a curved interface between the water and oil. Across this interface a pressure drip exists which is indicated by the capillary pressure. It is a function of saturation as shown in the figure. Furthermore, its value varies depending on whether the water saturation is increasing or decreasing. This is another example of the many relations that a simulator should contain.

3. COMPONENTS OF A RESERVOIR SIMULATOR

Figure 8 gives the template of a reservoir simulator. The physics and chemistry of the process are described by mathematical equations. The solution of these equations is the heart of a simulator. The solution should accurately reproduce the physics and chemistry of the process which the equations describe. The simulator should be able to account for the operational features or constraints practised during the exploitation of the reservoir. This ability is what makes a simulator a useful tool for managing the reservoir.

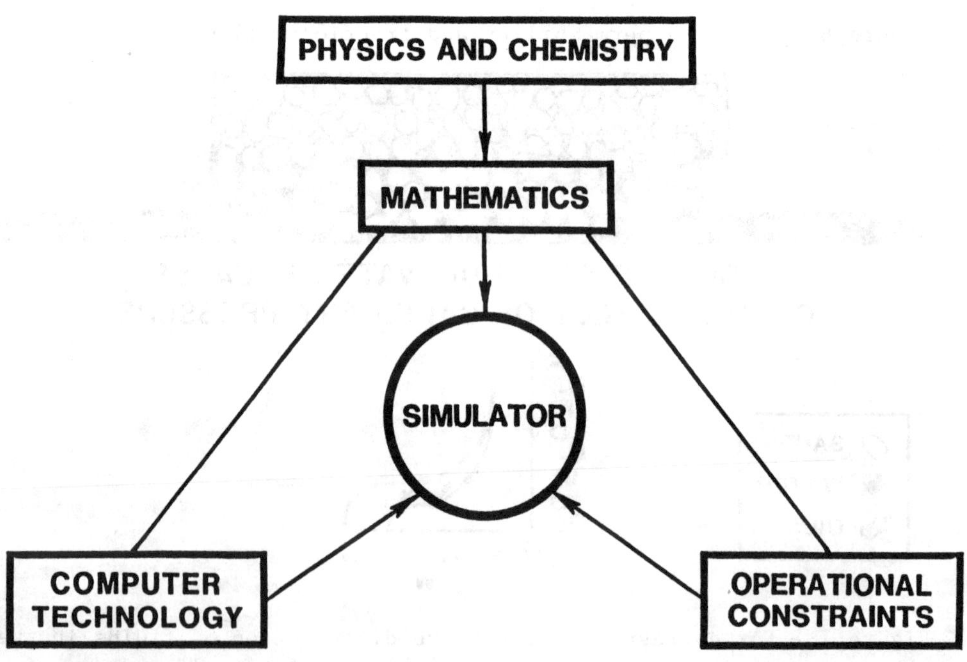

Figure 8. Reservoir simulator template.

The third component is the computer technology. The state of the computer technology strongly impacts how rigorous a solution we can afford, and how much operational constraints we can include. The computer technology will not be discussed in this paper.

These three components when interfaced in a computer program result in what we call a reservoir simulator.

4. CLASSES OF SIMULATORS

There are two classes of simulators: a) Those that model a process where no distinct front exists, or if it exists, lack of accurate determination of the concentration profile at the leading edge of the front does not significantly affect recovery, and b) Those that model processes where a moving front exists. Accurate determination of the concentration profile is essential to correctly model the physical process.

An example of the first class is the black oil simulator. In case of water injection into an oil reservoir a water front exists. Failure to accurately describe the water saturation at the oil-water interface may result in water breaking through into a well sooner or later than what actually happens. However this does not significantly affect the hydrocarbon recovery calculated by the model.

An example of the second class is the Enhanced Oil Recovery (EOR) simulators such as the thermal simulators to model the movement of combustion fronts in the reservoir. Failure to correctly solve for the temperature profile resulting from the combustion front significantly affect the answers. This is so because the kinetics of combustion are an exponential function of temperature.

5. RESERVOIR FLOW EQUATIONS FOR A BLACK OIL SYSTEM

In reservoir engineering we deal with two basic equations : a) the material balance or continuity equation, and b) the equation of motion (Darcy's law).

The continuity equations in differential form for the oil, gas, and water are:

oil $$\frac{\partial}{\partial t}\left(\frac{\phi S_o}{B_o}\right) = -q_o - \frac{\partial v_o}{\partial x} \tag{5.1}$$

water $$\frac{\partial}{\partial t}\left(\frac{\phi S_w}{B_w}\right) = -q_w - \frac{\partial v_w}{\partial x} \tag{5.2}$$

gas $$\frac{\partial}{\partial t}\left(\frac{\phi S_g}{B_g} + \frac{\phi R_{so} S_o}{B_o} + \frac{\phi R_{sw} S_w}{B_w}\right) \tag{5.3}$$

$$= -q_g - \frac{\partial v_g}{\partial x} - \frac{\partial}{\partial x}(R_{so} v_o) - \frac{\partial}{\partial x}(R_{sw} v_w)$$

where t is the time, ϕ is the porosity, S is the saturation, B is the formation volume factor, R is the gas in solution in the oil and water, v is the velocity, q is the production rate, and the subscripts o, g, and w refer to oil, gas, and water. The equations of motion are:

$$v_o = -\frac{kk_{ro}}{B_o \mu_o}\frac{\partial \phi_o}{\partial x} \; ; \qquad \frac{\partial \phi_o}{\partial x} = \frac{\partial p}{\partial x} - g\rho_o \frac{\partial D}{\partial x} \tag{5.4}$$

$$v_w = -\frac{kk_{rw}}{B_w \mu_w}\frac{\partial \phi_w}{\partial x} \; ; \qquad \frac{\partial \phi_w}{\partial x} = \frac{\partial p}{\partial x} - g\rho_w \frac{\partial D}{\partial x} - \frac{\partial p_{cow}}{\partial x} \tag{5.5}$$

$$v_g = -\frac{kk_{rg}}{B_g \mu_g}\frac{\partial \phi_g}{\partial x} \; ; \qquad \frac{\partial \phi_g}{\partial x} = \frac{\partial p}{\partial x} - g\rho_g \frac{\partial D}{\partial x} + \frac{\partial p_{cgo}}{\partial x} \tag{5.6}$$

where p is the pressure, D is the vertical distance, $g\rho\partial D/\partial x$ is the gravity gradient, p_{cgo} is the capillary pressure between gas and oil, p_{cow} is the capillary pressure between oil and water, and ρ is the density.

Combining the continuity equations with the equation of motion results in the partial differential equations for the oil, gas, and water which describe the flow through the porous medium. These are:

$$\nabla \cdot (\lambda_o \nabla \phi_o) - q_o = \frac{\partial}{\partial t}\left(\frac{\phi S_o}{B_o}\right) \tag{5.7}$$

$$\nabla \cdot (\lambda_w \nabla \phi_w) - q_w = \frac{\partial}{\partial t}\left(\frac{\phi S_w}{B_w}\right), \text{ where } \lambda_i = \frac{kk_{ri}}{B_i \mu_i} \tag{5.8}$$

$$\nabla \cdot (\lambda_g \nabla\phi_g) + \nabla \cdot (R_{so}\lambda_o \nabla\phi_o) + \nabla \cdot (R_{sw}\lambda_w \nabla\phi_w) - q_g \qquad (5.9)$$

$$= \frac{\partial}{\partial t} \left\{ \phi \left(\frac{S_g}{B_g} + R_{so} \frac{S_o}{B_o} + R_{sw} \frac{S_w}{B_w} \right) \right\}$$

6. WELL CONSTRAINTS EQUATIONS

In addition to these equations we need equations which describe the constraints imposed on the wells due to the plan of exploiting the reservoir. Examples of such constraints are the specification of the maximum water or gas production rate from the well or the field, the maximum and minimum production oil rate, and maximum produced water-oil ratio. These constraints are a function of the following bottom hole pressure of the well, p_w. Different constraints are expressed by different equations, with p_w being an unknown.

We now have three partial differential equations for the reservoir and a constraint equation for each well for the five unknowns p, S_o, S_g, S_w and p_w. This set is completed by the identity:

$$S_o + S_g + S_w = 1, \qquad (6.1)$$

together with the appropriate initial and boundary conditions. For the boundary conditions it is customary to represent the flow across wells' boundaries by sources and sinks, and by no flow boundary condition on the entire reservoir. The initial condition is represented by a static equilibrium in which velocities of all phases are zero. This reflects the balance of capillary and gravity forces.

Let us examine the first three equations. We note that the coefficients such as λ are functions of the unknowns p and S, while B and ϕ are functions of p. Thus, the equations are nonlinear. In addition, the gas equation contains the oil and water saturations, and the k_{ro} in λ_o is a function of S_w and S_g as was indicated previously. Therefore we are dealing with a set of nonlinear coupled partial differential equations. The heart of the reservoir simulator is the efficient and accurate solution of these equations together with the well constraints. These equations are for black oil resrvoirs (i.e., the oil is nonvolatile). Equations for other systems will be discussed later.

7. DEVELOPMENT OF A BLACK OIL RESERVOIR SIMULATOR

Figure 9 summarizes the development of a reservoir simulator. We will discuss briefly the steps shown in the figure:

7.1 Discretization

Finite difference is the most commonly used. The functions discretized generally have the form $\partial(\lambda\ \partial p/\partial x)/\partial x$. The practice is to use λ upstream, upwind, or donor cell. This generally results in first order correct formulation.

7.2. Solution Procedure

Two nonlinear solution procedures are practiced. These are: a) simultaneous or strongly coupled (2,33) and b) sequential or weakly coupled (13,29).

In the simultaneous or strongly coupled procedure all flow and well equations are solved simultaneously for the unknowns for the entire grid system. The Newton-Raphson method is used to linearize. The advantage of this procedure is that it gives stability. The disadvantage is that it results in a large matrix with each element being a 3x3 submatrix.

In the sequential solution, the three continuity equations are manipulated to give the following pressure equation:

$$\sum_{i=1}^{3} B_i \nabla \cdot (\lambda_i \nabla \phi_i) - q_T = \phi C_T \frac{\partial p}{\partial t}\ , \tag{7.1}$$

where q_T is the total production rate, C_T is the total compressibility of the system.

Two methods are practiced in the sequential procedure: the implicit pressure, explicit saturation, IMPES (13), and the total velocity (29). In both methods the pressure equation is solved first starting with values for the coefficients referred to the beginning of the time step. One may or may not update the pressure dependent coefficients. In the IMPES method, once the pressures are obtained, the saturations are solved for explicitly from the discretized forms of the previously indicated oil, gas, and water partial differential equations.

In the total velocity method it is assumed that flux of oil, gas, and water among the cells remain constant during a time step.

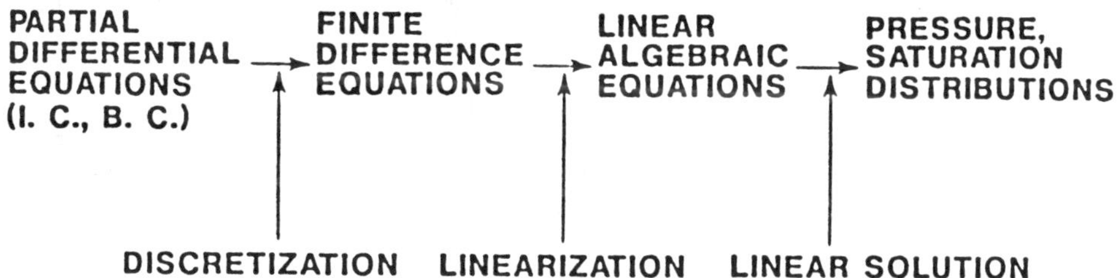

Figure 9. What is a reservoir simulator.

The saturation equation has the following form:

$$U_T \nabla f_i - \nabla \cdot \left[f(\lambda) \nabla p_c \right] - q_i = \frac{\partial}{\partial t} (\phi \frac{S_i}{B_i}) , \tag{7.2}$$

$$U_T = \Sigma U_i , \tag{7.3}$$

$$f_i = \frac{\lambda_i}{\lambda_T} , \qquad \lambda_o = f(S_w, S_g) , \tag{7.4}$$

$$f(\lambda) = \frac{\lambda_o \lambda_w \lambda_g}{\lambda_T} , \tag{7.5}$$

where p_c is the capillary pressure, and U_T is the total flux. The saturation equations are coupled and solved simultaneously.

7.3. Solution of Linear Equations

We must solve 5-diagonal (two-dimensional), or 7-diagonal (three-dimensional) matrices. Direct elimination (27, 33), and iterative methods (9, 28, 34, 38) are used. Both methods have advantages and disadvantages. Reliability is the main advantage of the direct methods. Their main disadvantage is the large storage requirement. While direct methods may be practical for two-dimensional problems, they cannot be used efficiently in most three-dimensional problems, they cannot be used efficiently in most three-dimensional cases because the band widths are too large, resulting in excessive computer storage and time. The methods are very competitive for systems with small band widths, and normally are not used for problems with band widths over a hundred.

The main advantage of the iterative methods is that they require relatively little storage and thus can be applied to very large systems. Their main disadvantage is their sensitivity to the iteration parameters. If the proper parameters are not selected, slow convergence may result. Examples of the iterative methods used by the industry are:

7.3.1. Strongly Implicit Procedure

Strongly Implicit Procedure, SIP (30, 37) is a partial factorization method that requires iterative parameters to be selected by the user.

7.3.2. Successive Over Relaxation Methods

Successive Over Relaxation Method, SOR (34, 38) is user selected one iterative parameter method. It requires diagonal dominance in addition to the proper iterative parameter for a fast rate of convergence.

7.3.3. SOR + Additive Corrections

SOR + Additive Corrections (35) method is based on the idea that by adding a correction to the computed values at an iteration level, the convergence is accelerated. It is an attempt to eliminate the dominant eigen value of the matrix.

7.3.4. Conjugate Gradient Method

Conjugate Gradient method (20, 36) is a self-parameterizing iterative method which works best with symmetric matrices, and non-symmetric ones with the appropriate pre-conditioner.

7.4. Iterative and Non-Iterative Methods (2)

Once pressure and saturation values are obtained they may be accepted and the calculations proceed to the next time step. This is called a non-iterative method. On the other hand, the pressure and saturation dependent coefficient may be updated, and the cycle repeated until a convergence criterion is achieved. This is known as an iterative method. These two methods are practiced in both the simultaneous and sequential procedures.

Compared to the iterative method, the non-iterative method requires less computer time. It is adequate when the gradients are small, otherwise unreasonably small time steps have to be used for a satisfactory solution.

The state of the mathematical formulation, and the nonlinear and linear solution procedures for black oil simulators is good. The availability of vector computers has swung the tide toward the strongly coupled nonlinear solution procedure which has proven to be more stable and more effective in simulating difficult reservoir conditions.

8. MORE COMPLICATED SIMULATORS, EOR MODELS

So far we have concentrated on simulators that model non-volatile oil reservoirs. These are the least complicated models in teh suite of reservoir simulators. Examples of more complicated ones are:

8.1. Compositional Simulators

These models (15, 17) simulate volatile oil reservoirs, miscible displacements, and any process that requires the hydrocarbon to be divided into N number of components - in place of just oil and gas.

For these models a species continuity equation for each of the N hydrocarbon components can be written as :

$$\frac{\partial}{\partial t}\left[\phi(\rho_o S_o x_i + \rho_g S_g y_i)\right] = \nabla \cdot \left[\left(\frac{\rho_o k k_{ro}}{\mu_o} x_i + \frac{\rho_g k k_{rg}}{\mu_g} y_i\right)\nabla\phi\right] + q_i, \quad i = 1.......N \tag{8.1}$$

The water balance is

$$\frac{\partial}{\partial t}\left(\frac{\phi\rho_w S_w}{\partial t}\right) = \nabla \cdot \left(\frac{k k_{rw}\rho_w}{\mu_w} \nabla\phi_w\right) + q_w \tag{8.2}$$

where ρ is the density, x_i is the liquid mole fraction of component i, and y_i is the gas mole fraction of component i. In addition, a thermodynamic constraint, usually in terms of fugacities must be satisfied. The constraint is:

$$f_{io} = f_{ig} \tag{8.3}$$

An equation of state such as the Peng-Robinson (23) can be used to express fugacity in less complex functions (15).

8.2. Thermal Simulator (5, 7, 16)

Heat injection can take the form of steam, or in-situ combustion. In both processes, the temperature is not constant and an energy balance equation is required.

The energy balance equation has the following form:

$$\frac{\partial}{\partial} \sum_i \rho_i S_i U_i = \sum_i \nabla \cdot (\lambda_j H_j \nabla p) + \nabla \cdot (T_C \nabla T) - q_L - q_H , \tag{8.4}$$

where U is the internal energy, H is the entalphy, q_L is the heat loss to the formation, q_H is the heat sink or source through a well, T is the temperature, T_C is the heat conduction transmiscibility.

In addition, chemical reactions such as vaporization, condensation, and cracking need to be accounted for.

8.3. Simulator for Chemical Injections (25)

An additional equation which describes the concentration distribution of the chemical, C_i, such as a surfactant, is needed. It takes the following form:

$$\sum_i \frac{\partial}{\partial t}(\phi S_i \rho_i C_i) + \sum_i \nabla \cdot (\rho_i C_i v_i) - \sum_i \nabla \cdot (D_i \nabla C_i) \tag{8.5}$$

$$+ \frac{\partial}{\partial t} \left[(1 - \phi)\rho_r C_r\right] = 0,$$

where D_i is the diffusion coefficient, and v_i is the velocity in phase i. The subscript r indicates the rock.

In addition to the above equations, the effect of surfactant concentration, for example on the fluid flow behavior and the interaction between the chemicals and the rock and among the chemicals themselves, must be accounted for.

8.4. Naturally Fractured Reservoir Simulators

These simulators may belong to the black oil or to the EOR categories. Their distinguishing characteristic is the presence of two types of porous media, the fractures and the matrix. The fractures are normally a fraction of a centimeter in width and are highly conductive. The distribution of the fractures is very difficult to obtain. Even if it is known it has been intractable to adequately simulate such a physical system.

9. THE PHYSICAL MODEL VERSUS THE MATHEMATICAL MODEL

9.1. The Black-Oil Type Simulators

The principle difference between the two models is continuity versus discretization. The reservoir behavior is based on continuity in space and time while the mathematical model behavior is based on discretization in both space and time. The effect of discretization on the physics will be considered in two parts: a) The effect on the flow behavior away from the wells, and b) The effect on the flow behavior in the immediate vicinity of the wells.

Finite difference method of solution basically deals with averages. In reservoir studies the grid blocks are relatively large, a couple of hundred feet in the x and y directions and several feet in the z direction. To every grid block average physical properties such as permeability, porosity, etc., are assigned. These are obtained from contour maps. We thus end up with somewhat discontinuous properties' distribution in place of smoothly varying ones. Needless to say, this impacts the character of flow. The extent of the impact has not been investigated. In some cases we know that the effect strongly alters the physics because the local gradients are lost and replaced by average gradients. For example, in many naturally fractured reservoirs water is injected, and the oil is produced from the matrix blocks by imbibition. The water is imbibed into the matrix by capillary forces and the oil flows counter current through the fractures into the wells. This mechanism is mainly controlled by the capillary and saturation gradients at the fracture-matrix interfaces. As a result, imbibition for all practical purpose, is lost. The physics of flow is thus not truly simulated.

We know that when gas contacts oil under pressure part of the gas dissolves in the oil. In the reservoir, when gas is injected it partially dissolves in the oil that it contacts, pressure permitting. In simulating this process the gas is dissolved in all the oil in a grid block regardless whether all the oil has been contacted or not. This is done, because every grid block has to be homogeneous in properties. The other extreme is not to let any gas go into solution. The reservoir behavior falls in between.

The effect of discretization in time is best understood by considering Figure 6. One may view Figure 6 as giving the oil and water fractions in an oil-water flowing stream. The flow in the reservoir continuously tracks the curves. In the model this does not occur. Relative permeabilities at the (n+1)-time level are calculated from the n-time level by:

$$k_r^{n+1} = k_r^n + \left(\frac{dk_r}{ds}\right)^n \Delta S \tag{9.1}$$

Since relative permeability is not a straight line function of S, if ΔS is large, during a time step one may end up with a k_r^{n+1} used in the model considerably different from the actual value. This results in significant change in the character of flow. It is unfortunate that customarily one attempts to take as large a time step as possible to reduce computer cost.

The effect of discretization on the well behavior is starting to get some attention (22). The actual well behavior is based on a continuous physical system which considers the total thickness of the productive sand and some average permeability in the drainage area of the well normally calculated from well established flow tests (19), In reservoir engineering calculation , the flow behavior of a well has reached a high degree of sophistication. The calculations can account for drainage area geometry, effect of well's completion, and restricted entry to flow caused by completing in part of the sand. These effects are accounted for by parameters obtained from analyzing flow tests data. The analysis is based on a physical system which includes the well and its drainage area. In the simulator most of the assumptions upon which the calculations of the well's flow behavior are based are violated. The grid block thickness in place of the total sand thickness, the grid block permeability and pressure in place of the average permeability and pressure in the drainage area are used in calculating the well behavior. However, the parameters to account for the effect of well completion, restricted entry, etc., normally obtained from analysis of a flow test are used in the simulator. Until recently, these were used without any scaling (22). More work is needed on this problem.

Another well problem occurs when the perforations do not extend along the entire thickness of the grid block. For example, if the perforations are located at the top, and water is present, it is produced when the saturation exceeds its critical value. In actuality due to difference in density between oil and water, water tends to segregate to the bottom and will not be produced until it reaches a saturation value which, in some cases, is considerably higher than that of the simulator. This difference in behavior is due to the zero dimensionality of the grid block, i.e., the stirred tank model. It is normally corrected by using pseudo transfer functions.

9.2. More Complicated Simulators

Finite difference has been the principle numerical method used by the oil industry in the development of the more complicated models. It results in all the difficulties described above plus: a) less rigor in the mathematical formulation, b) erroneous concentration profile, c) grid orientation effects, and d) masking of the microscopic behavior at the front.

9.2.1. Less rigor in the mathematical formulation

This results from the numerical methods of solution. It is not a product of only finite difference. The number of unknowns in the EOR models, for example, are at least twice or three times that of the black oil model. Therefore, the strongly coupled formulation practised in the black oil model becomes impractical, since present day simulation runs require several thousand mesh points. Therefore, we resort to decoupling of some of the knowns. This results in less rigor, and in stability problems.

9.2.2. Erroneous concentration profile

Finite difference results represent averages. In the case of a moving chemical slug or a heat front, the results represent the average concentration or temperature in the grid block where the slug or front resides. Generally speaking the width of the front is considerably less than the length of the grid block. Thus the correct profile is lost. Since the behavior of the procees depends on the correct concentration profile, averages give erroneous results. Figure 10, for example, the effect of averaging on the kinetics of combustion is shown. In the case of surfactant flooding, the wrong concentration leads to erroneous phase behavior and thus erroneous recovery.

9.2.3. Grid orientation effect (1, 39)

The classical five-star finite difference representation leads to what is known as the grid orientation effect. Shown in Figure 11 are two grid systems. The one on the right is the parallel grid

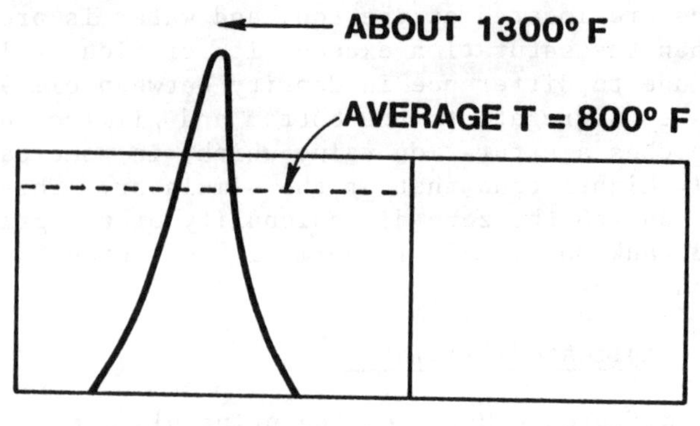

Figure 10. A Moving combustion front.

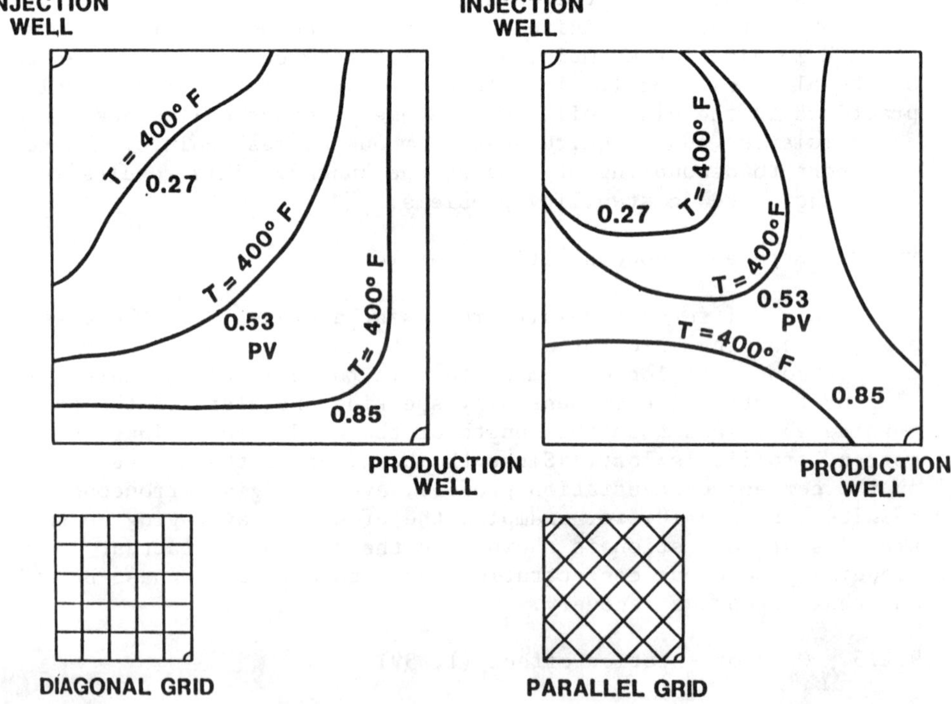

Figure 11. Grid orientation effect - Steam injection - Classical formulation.

while the one on the left is the diagonal grid. Under certain conditions these two grid systems produce two different answers. The difference can be appreciable especially when the mobilities of the fluids differ significantly. Both solutions are incorrect. The correct answer falls in between. Recent use of the nine-point star discretization procedure has resulted in some reduction of the grid orientation effect.

9.2.4. Masking of microscopic behavior (4, 24)

Frontal instability or viscous fingering is an important physical phenomenon which occurs in fluid displacement when the displacing fluid is more mobile than the displaced one. Figure 12 is a schematic representation of viscous fingering. No satisfactory method has been developed to simulate viscous fingering. Two schemes (18, 31) available for miscible displacement leave a lot to be desired.

In case of immiscible displacement, i.e., water displacing oil (Class I simulators),simulator results show that the water breakthrough into a well is delayed because they neglect viscous fingering. However, ultimate recovery is not strongly affected. On the other hand, for miscible displacement (Class II simulators), i.e., solvent displacing oil, results that do not account for viscous fingering show recovery significantly less than actual. Viscous fingering opens up surfaces where chemical exchange occurs. This could result in miscibility and thus higher recovery. Simulation of miscible displacement processes, to be viable, must account for frontal instability.

10. THE FUTURE

We have concentrated our effort on overcoming problems associated with large matrices and very large code (about 1/2 million FORTRAN statements). By and large the effort has been successful.

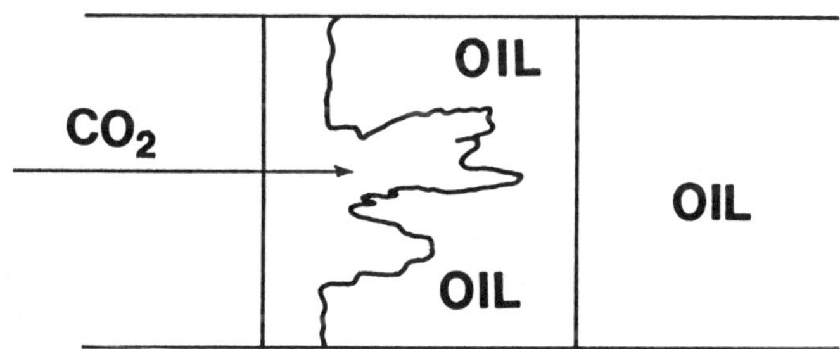

Figure 12. Microscopic behavior - Viscous fingering - Unfavorable mobility ratio.

Now we need to pay more attention to the effect of discretization on the physics, and thus the validity of the results. It is possible that the finite difference method that served us well in the development of black oil models is not adequate for the new generation of models. Some of the new methods that are being looked upon or should be looked upon by the industry are:

10.1. Adaptive Grid Refinement (8)

Here the grid is refined where it is needed according to a set of preassigned criteria. The small mesh system may be adequate to simulate a moving front. However, small grid blocks lead to small time steps. This may render the simulation uneconomical when dealing with field-scale cases. Many difficult problems need to be solved before the viability of this scheme is established.

10.2. High Order Numerical Techniques (3, 32)

Presently used finite difference methods tend to smear the fronts due to numerical dispersion or diffusion. Higher order methods are being investigated. Oscillation seems to be an outcome of these methods.

10.3. Non-Finite Difference Methods

Examples of such methods are: methods of characteristics (12, 14), finite element, Galerkin, and Petrov-Galerkin (10, 11, 6 , 26, 40). The methods of characteristics - moving points have shown great promise in some preliminary studies but still have several technical problems when applied to three-dimensional, field-scale simulation. Finite element methods and their variations have been under investigation for the last ten years. The methods have shown success in control of numerical dispersion and grid orientation effects. These methods are in general, computationally more expensive than finite difference methods. Only recently a relatively small two-dimensional simulator using Petrov-Galerkin was reported (6).

10.4. Combining Analytical with Numerical Methods

The idea here is to solve for the concentration of the slug or front analytically using approximate boundary and initial conditions obtained from the numerical solution. For example, in case of a combustion process one can analytically solve for the temperature profile in the cell where the combustion is occuring. The analytical results are used to arrive at a better value for the effective temperature for the cell than the average temperature calculated by the finite difference method.

11. ACKNOWLEDGEMENT

The author thanks the management of Mobil Research and Development Corporation for permission to publish the paper.

12. REFERENCES

1. Aziz, K., and Settari, A. *Petroleum Reservoir Simulation*, Applied Science Publishers Ltd., London, 1979.
2. Bansal, P. P., Harper, J. L., McDonald, A. E., Moreland, E. E., Odeh, A. S., and Trimble, R. H., "A Strongly Coupled, Fully Implicit, Three-Dimensional, Three-Phase Reservoir Simulator," presented at the September 23-26, 1979, Society of Petroleum Engineers 54th Annual Technical Conference, held at Las Vegas, Nevada (Preprint 8329).
3. Bell, J. B., and Shubin, G. R., "Higher-Order Godunov Methods For Reducing Numerical Dispersion in Reservoir Simulation," presented at the February 10-13, 1985, 8th SPE Symposium on Reservoir Simulation, held at Dallas, Texas (Preprint SPE 13514).
4. Blackwell, R. J., Payne, J.R., and Terry, W. M., "Factors Influencing the Efficiency of Miscible Displacement," *Transactions*, American Institute of Mining Engineers, Vol. 216, 1959, p. 1.
5. Coats, K. H., "A Fully Implicit Steamflood Model," *Society of Petroleum Engineering Journal*, Oct., 1978, pp. 369-383.
6. Cohen, M. F., "Finite Element Methods for Enhanced Oil Recovery Simulation," presented at the February 10-13, 1985, 8th SPE Symposium on Reservoir Simulation, held at Dallas, Texas (Preprint SPE 13512)
7. Crookston, R. B., Culham, W. E., and Chen, W. H., "Numerical Simulation Model for Thermal Recovery Processes," *Society of Petroleum Engineering Journal*, February, 1979, pp. 37-58.
8. Diaz, J. C., Ewing, R. E., Jones, R. W., McDonald, A. E., and Uhler, L. M., "Self-Adaptive Local Grid Refinement Application in Enhanced Oil Recovery," *Proceedings of the 5th International Conference on Finite Elements in Flow Problems*, Austin, Texas, January 23-26, 1984.
9. Douglas, J., Peaceman, D., and Rachford, H., " A Method for Calculating Multi-Dimensional Immiscible Displacement," *Transactions*, American Institute of Mining Engineers, Vol. 216, 1959, p. 297.
10. Douglas, J. Jr., Dupont, T., and Rachford, H. H. Jr., "The Application of Variational Methods to Waterflooding Problems," *Journal of Petroleum Technology*, Vol. 8, July-Sept., 1969, pp. 79-85.

11. Ewing, R. E., and Wheeler, M. F., "Galerkin Methods for Miscible Displacement Problems in Porous Media," *SIAM Journal of Numerical Analysis*, Vol. 17, 1980, pp. 351-365.
12. Ewing, R. E., and Russell, T. F., "Multistep Galerkin Methods Along Characteristics for Convection-Diffusion Problems," *Advances in Computer Methods for Partial Differential Equations-IV IMACS*, 1981.
13. Fagin R. G., and Stewart, Jr., "A New Approach to the Two-Dimensional Multiphase Reservoir Simulation," presented at the Sept., 1973, SPE 48th Annual Fall Meeting, held at Las Vegas, Nevada (Preprint SPE 4542)
14. Glimm, J., Isaacson, E., Marchesin, D., and McBryan O., "Front Tracking For Hyperbolic Systems," *Advances in Applied Mathematics*, Vol. 2, 1981, pp. 91-119.
15. Harper, J. L., Heinemann, R. F., Ray, M.B., and Stephenson, P.E., "A Compositional Simulator for Performing Large Field Studies in a Vector Computing Environment," presented at the 1985, 4th Middle East Oil Show, held at Bahrain (Preprint SPE 13714).
16. Hwang, M. K., Jines, W. R., and Odeh, A. S., "An In-Situ Combustion Process Simulator with a Moving Front Representation," *Transactions*, American Institute of Mining Engineers, Vol. 273, 1982.
17. Kazemi, H., and Vestal, C. R., "An Efficient Multicomponent Numerical Simulator," presented at the October 9-12, 1977, 52nd Annual Fall Meeting, held at Denver, Colorado (Preprint SPE 6890).
18. Koval, E. J., "A Method for Predicting the Performance of Unstable Miscible Displacement in Heterogeneous Media," *Transactions*,American Institute of Mining Engineers, Vol. 228, 1963, p. 145.
19. Matthews, C. S., and Russell, D. G., *Pressure Buildup and Flow Tests in Wells*, Society of Petroleum Engineers Monograph, Vol. 1, 1967.
20. Northrup, E. J., and Woo, P. T., "Application of Preconditioned Conjugate-Gradient-Like Methods in Reservoir Simulation," presented at the February 10-13, 1985, 8th SPE Symposium on reservoir Simulation, held at Dallas, Texas (Preprint SPE 13532).
21. Odeh, A. S., "Reservoir Simulaiton-What Is It?" *Journal of Petroleum Technology*, November 1969, pp. 1383-88.
22. Odeh, A. S., "The Proper Interpretation of Field Determined Pressure and Skin Values for Simulator Use," *Society of Petroleum Engineers Journal*, February 1985.
23. Peng, D. Y., and Robinson, D. B., "A New Two-Constant Equation of State," *Industrial and Engineering Chemistry Fundamentals*, Vol. 15, 1976, pp. 59-64.
24. Perkins, T. K., Johnston, O. C., and Hoffman, R. N., "Mechanics of Viscous Fingering in Miscible Systems," *Transactions*, American Institute of Mining Engineers, Vol. 234, 1965, p. 301.

25. Pope, G. A., and Nelson, R. C., "A Chemical Flooding Compositional Simulator," *Society of Petroleum Engineering Journal*, Oct., 1978, pp. 339-354.
26. Price, H. S., Cavendish, J. C., and Varga, R. A., "Numerical Methods of Higher Order Accuracy for Diffusion-Convection Equations," *Transactions*, American Institute of Mining Engineers, Vol. 253, 1972, pp. 293-303.
27. Price, H. S., and Coats, K. H., "Direct Methods for Reservoir Simulators," *Society of Petroleum Engineering Journal*, June 1974, pp. 295-328.
28. Price, H. S., and Coats, K. H., "Direct Methods in Resrevoir Simulation," *Transactions*, American Institute of Mining Engineers , Vol. 257, 1974, pp. 295-308.
29. Spillette, H. G., Hillestad, J. G., and Stone, H. L., "A High Sequential Solution Approach to Reservoir Simulation," presented at the Sept., 1973, SPE 48th Annual Fall Meeting, held at Las Vegas, Nevada (Preprint SPE 4542).
30. Stone, H. L. J., "Iterative Solution of Implicit Approximations of Multidimensional Partial Differential Equations," *SIAM Journal of Numerical Analysis*, Vol. 5., Sept., 1968, pp. 530-558.
31. Todd, M. R., and Longstaff, W. J., "The Development, Testing, and Application of a Numerical Simulator for Predicting Miscible Flood Performance," *Transactions*, American Institute of Mining Engineers, Vol. 253, 1972, p. 874.
32. Todd, M. R., O'Dell, P. M., and Hirasaki, G. J., "Methods for Increased Accuracy in Numerical Simulation," *Society of Petroleum Engineering Journal*, Dec., 1972, pp. 515-530.
33. Trimble, R. H., and McDonald, A. E., "A Strongly Coupled, Implicit Well Coning Model," *Society of Petroleum Engineering Journal*, August, 1981, pp. 454-458.
34. Wattenbarger, R. A., and Thurnau, D. H., "Application of SSOR to Three-Dimensional Reservoir Problems," presented at the February 19-20, 1976, SPE 4th Symposium on Numerical Simulation of Reservoir Performance, held at Los Angeles, California (Preprint SPE 5728).
35. Watts, J. W., "An Iterative Matrix Solution Method Suitable for Anisotropic Problems," *Transactions*, American Institute of Mining Engineers, Vol. 251, 1971, p. 47.
36. Watts, J. W., "A Conjugate Gradient Truncated Direct Method for the Iterative Solution of the Reservoir Simulated Pressure Equation," presented at the Sept., 23-26, 1979, SPE 54th Annual Technical Conference, held at Las Vegas, Nevada (Preprint SPE 8252).
37. Weinstein, H. G., Stone, H. L., and Kwan, T. V., "Iterative Procedure for Solution of Systems of Parabolic and Elliptic Equations in Three Dimensions," *Industrial and Engineering Chemistry Fundamentals*, Vol. 8, No. 2, May, 1969, pp. 281-287.

38. Woo, P. T., and Emanuel, A. S., "A Block Successive Over-Relaxation Method for Coupled Equations," presented at the Oct., 3-6, 1976, SPE 51st Annual Fall Technical Conference, held at New Orleans, Lousiana (Preprint 6106).
39. Yanosk, J. L., and McCracken, T. A., "A Nine-Point Finite Difference Reservoir Simulator for Realistic Prediction of Adverse Mobility Ratio Displacements," *Transactions*, American Institute of Mining Engineers, Vol. 267, 1979, pp. 253-262.
40. Young, L. C., "A Finite Element Method for Reservoir Simulation," *Society of Petroleum Engineering Journal*, Vol. 2, 1981, pp. 115-128.

13. LIST OF SYMBOLS

B	Formation volume factor
C_T	Total compressibility of the system
D	Vertical distance
D_i	Diffusion coefficient
f_i	λ_i/λ_T, fugacity of phase i
$f(\lambda)$	$\lambda_o\lambda_w\lambda_g/\lambda_T$
g	Gravitational acceleration
H	Entalphy
k	Intrinsic permeability
k_r	Relative permeability
P	Pressure
P_{cgo}	Capillary pressure between gas and oil
P_{cow}	Capillary pressure between oil and water
q	Production rate
q_L	Heat loss to the formation
q_H	Heat sink or source through a well
q_T	Total production rate
R	The gas in solution in the oil and water
S	Saturation
T	Temperature
T_c	Heat conduction transmiscibility
t	Time
U	Internal energy
U_T	Total flux
v	Velocity
v_i	Velocity in phase i
x_i	Liquid mole fraction of component i
y_i	Gas mole fraction of component i
ϕ	Porosity
λ	$kk_r/(B\mu)$
ρ	Density
μ	Viscosity

Subscripts

g,o,T,w	Gas,oil,total,water

NUMERICAL MODELING OF MULTIPHASE FLOW IN POROUS MEDIA

Myron B. Allen III

NUMERICAL MODELING OF MULTIPHASE FLOW IN POROUS MEDIA

Myron B. Allen III

Department of Mathematics
University of Wyoming
Laramie, Wyoming 82071, USA

ABSTRACT

The simultaneous flow of immiscible fluids in porous media occurs in a wide variety of applications. The equations governing these flows are inherently nonlinear, and the geometries and material properties characterizing many problems in petroleum and groundwater engineering can be quite irregular. As a result, numerical simulation offers the only viable approach to the mathematical modeling of multiphase flows. This chapter provides an overview of the types of models that are used in this field and highlights some of the numerical techniques that have appeared recently. The exposition includes discusssions of multiphase, multispecies flows in which chemical transport and interphase mass transfers play important roles. This chapter also examines some of the outstanding physical and mathematical problems in multiphase flow simulation. The scope ot the chapter is limited to isothermal flows in natural porous media; however, many of the special techniques and difficulties discussed also arise in artificial porous media and multiphase flows with thermal effects.

1. INTRODUCTION

1.1. Importance of Multiphase Flow in Porous Media

Multiphase flows in porous media occur in a variety of settings in applied science. The earliest applications involving the simultaneous flow of two fluids through a porous solid appear in the soil science literature, where the flow of water in soils partly occupied by air has fundamental importance (128). This *unsaturated flow* in some ways represents the simplest of multiphase flows. Yet, as we shall see, it exemplifies a fact underlying the continued growth in research in this area: multiphase flows in porous media are inherently nonlinear. Consequently, numerical simulation often furnishes the only effective strategy for understanding their behavior quantitatively.

Although the earliest studies of multiphase flows in porous media concern unsaturated flows, the most concentrated research in this field over the past four decades has focused on flows in underground petroleum reservoirs. Natural oil deposits almost always contain connate water and occasionally contain free natural gas as well. The simultaneous flow of oil, gas, and water in porous media therefore affects practically every aspect of the reservoir engineer's job of optimizing the recovery of hydrocarbons. Here, again, the physics of multiphase fluid flows give rise to nonlinear governing equations. The difficulty imposed by the nonlinearities along with the irregular geometries and transient behavior associated with typical oil reservoirs make numerical simulation an essential tool in petroleum engineering. The advent of various enhanced oil recovery technologies has added to this field further levels of complexity and hence an even greater degree of reliance on numerical methods.

Most recently, multiphase flows have generated serious interest among hydrologists concerned with groundwater quality. There is growing awareness that many contaminants threatening our groundwater resources enter water-bearing rock formations as separate, nonaqueous phases. These oily liquids may come from underground or near-surface storage facilities, landfills at which chemical wastes are dumped, industrial sites such as oil refineries or wood-treatment plants, or illegal waste disposal. Regardless of the source of the contaminants, our ability to understand and predict their flows underground is crucial to the design of sound remedial measures. This is a fairly new frontier in multiphase porous-media flows, and again the inherent complexity of the physics leads to governing equations for which the only practical way to produce solutions may be numerical simulation.

1.2. Scope of the Article

The purpose of this article is to review some of the more salient applications of numerical simulation in multiphase porous-media flows. In light of the history and breadth of these applications, a review of this kind must choose between the impossibly ambitious goal of thoroughness and the risks of narrowness that accompany selective coverage. This article steers toward selective coverage. The aim here is to survey several multiphase flows that have attracted substantial scientific interest and to discuss a few aspects of their numerical simulation that have appeared in the recent technical literature. I confess at the outset that some important multiphase flows receive no attention here at all, and, even for the flows discussed, many potentially far-reaching contributions to numerical simulation get no mention. Perhaps the references given throughout the article can compensate in part for these shortcomings.

In particular, we shall restrict our attention here to underground flows in natural porous media. This restriction excludes many applications in chemical engineering, one notable example being flows in packed beds of catalysts. Also, the article considers only isothermal flows. Therefore we do not discuss steam-water flows in geothermal reservoirs or such thermal methods of enhanced oil recovery as steam injection or fireflooding. Several numerical methods also receive scant or no mention. Among these are integrated finite differences, subdomain finite elements, spectral methods, and boundary-element techniques. Some of these approaches undoubtedly hold promise for future applications in multiphase flows in porous media. For the present, however, we concentrate on developments based on the trinity of more standard discrete approximations: finite differences, Galerkin finite elements, and collocation.

2. BACKGROUND

2.1. Definitions

From a quantitative point of view, one of the most fruitful ways of examining multiphase flows in porous media is through the framework of continuum mixture theory. In contrast to a single continuum, a *mixture* is a set of overlapping continua called *constituents*. Any point in a mixture can in principle be the locus of material from each constituent, and each constituent possesses its own kinematic and kinetic variables such as density, velocity, stress and so forth. How one decomposes a physical mixture into constituents depends largely on one's theoretical aims, but in analyzing porous media we commonly identify the solid matrix as one constituent and each of the fluids occupying its interstices as another.

In discussions of porous-media physics it is important to distinguish between multiphase mixtures and multispecies mixtures. A mixture consists of several phases if, on a microscopic length scale comparable, say, to typical pore apertures, one observes sharp interfaces in material properties. In this sense all porous-media flows involve multiphase mixtures, owing to the distinct boundary between the solid matrix and the interstitial fluids. At this boundary, density, for example, changes abruptly from its value in the solid to that in the fluid. More complicated multiphase mixtures occur, common examples being the simultaneous flows of air and water, oil and water, or oil and gas through porous rock. Here, in addition to rock-fluid interfaces, we observe interfaces between the various immiscible fluids at the microscopic scale. While the detailed structures of these interfaces and the volumes they bound are inaccessible to macroscopic observation, their geometry influences the mechanics of the mixture. This, at least intuitively, is why volume fractions play an important role in multiphase mixture theory. The *volume fraction* ϕ_α of phase α is a dimensionless scalar function of position and time such that $0 \le \phi_\alpha \le 1$, and, for any spatial region R in the mixture, $\int_R \phi_\alpha dx$ gives the fraction of the volume of R occupied by phase α. The sum of the fluid volume fractions in a saturated solid matrix is the porosity α.

On the other hand, there are mixtures in which no microscopic interfaces appear. Saltwater is an example. Here the constituents are ionic or chemical *species*, and spatial segregation of these constituents is not observable except, perhaps, at intermolecular length scales. Air is another multispecies mixture, consisting of N_2, O_2, CO_2, and some trace gases. Multispecies mixtures differ from multiphase mixtures in that volume fractions do not appear in the kinematics of the former.

It is possible to have multiphase, multispecies mixtures. These *compositional flows* occur in porous-media physics when there are several fluid phases, each of which comprises several chemical species. Such mixtures arise in many flows of practical interest, two important examples being multiple-contact miscible displacement in oil reservoirs and the contamination of groundwater by nonaqueous liquids. In these cases the transfer of chemical species between phases is a salient feature of the mixture mechanics. More detailed treatment of compositional flows appears later in this article.

2.2. Review of the Basic Physics

While the theory of mixtures dates at least to Eringen and Ingram (61), its foundations are still the focus of active inquiry, as reviewed by Atkin and Craine (17). Among the applications of mixture theory to multiphase mixtures and porous media are investigations by Prévost (124), Bowen (29,30), Passman, Nunziato, and Walsh (112), and Raats (126). The aims of the present article in this respect are much more limited in scope than those just cited. What follows is a brief review of the basic physics of multiphase flows in porous media, using the language of mixture theory as a vehicle for the development of governing equations (7).

For concreteness, assume that the mixture under investigation has three phases: rock (R) and two fluids (N,W). (The extension of this exposition to mixtures with more fluid phases is straightforward.) Each phase α has its own instrinsic mass density ρ_α, measured in kg/m^3; velocity $\underset{\sim}{v}_\alpha$, measured in m/s; and volume fraction ϕ_α. From their definitions, the volume fractions clearly must obey the constraint $\Sigma_\alpha \phi_\alpha = 1$. In terms of these mechanical variables, the mass balance for any particular phase α is

$$\frac{\partial}{\partial t}(\phi_\alpha \rho_\alpha) + \nabla.(\phi_\alpha \rho_\alpha \underset{\sim}{v}_\alpha) = r_\alpha \tag{2.1}$$

where r_α stands for the rate of mass transfer into phase α from other phases. To guarantee mass conservation in the overall mixture, the reaction rates must obey the constraint $\sum_\alpha r_\alpha = 0$.

We can rewrite Eq. (2.1) in a more common form by noting that the porosity is $\phi = 1 - \phi_R$ and defining the fluid saturations $S_N = \phi_N/\phi$, $S_W = \phi_W/\phi$. Thus

$$\frac{\partial}{\partial t}[(1-\phi)\rho_R] + \nabla.[(1-\phi)\rho_R \underset{\sim}{v}_R] = r_R$$

for the rock phase, and

$$\frac{\partial}{\partial t}(\phi S_\alpha \rho_\alpha) + \nabla.(\phi S_\alpha \rho_\alpha \underset{\sim}{v}_\alpha) = r_\alpha, \quad \alpha = N,W, \tag{2.2}$$

for the fluids.

Each phase also obeys a momentum balance. In its primitive form this equation relates the phase's inertia to its stress $\underset{\approx}{t}_\alpha$, body forces $\underset{\sim}{b}_\alpha$, and rate $\underset{\sim}{m}_\alpha$ of momentum exchange from other phases. Thus,

$$\phi_\alpha \rho_\alpha \left(\frac{\partial \underset{\sim}{v}_\alpha}{\partial t} + v_\alpha . \nabla v_\alpha\right) - \nabla.\underset{\approx}{t}_\alpha - \phi_\alpha \rho_\alpha \underset{\sim}{b}_\alpha = \underset{\sim}{m}_\alpha - \underset{\sim}{v}_\alpha r_\alpha \tag{2.3}$$

If we assume that the rock phase is chemically inert, so $r_R = 0$, and fix a coordinate system in which $\underset{\sim}{v}_R = 0$, then the momentum balance for rock reduces to

$$\nabla.\underset{\approx}{t}_R - \phi_R \rho_R \underset{\sim}{b}_R = \underset{\sim}{m}_R$$

Let us assume that each fluid is Newtonian and that momentum transfer via shear stresses within the fluid is negligible compared with momentum exchange to the rock matrix. In this case $\underset{\approx}{t}_\alpha = -p_\alpha \underset{\approx}{1}$, where p_α is the *mechanical pressure* in fluid α and $\underset{\approx}{1}$ is the unit isotropic tensor. If gravity is the only body force acting on fluid phase α, then $\phi_\alpha \underset{\sim}{b}_\alpha = g\nabla Z$, where g stands for the magnitude of gravitational acceleration and Z represents depth below some datum. For the momentum exchange terms, the assumption common to most theories of porous media is that momentum losses to the solid matrix take the form of possibly anisotropic Stokes drags,

$$\underset{\approx}{\Lambda}_\alpha \underset{\sim}{m}_\alpha = \phi(\underset{\sim}{v}_R - \underset{\sim}{v}_\alpha) = -\phi \underset{\sim}{v}_\alpha$$

where $\underset{\approx}{\Lambda}_\alpha$ is a tensor called the *mobility* of phase α. If we assume further that the inertial effects in the fluid are negligible compared with rock-fluid interactions and that there is no interphase mass transfer, then Eq. (2.3) yields

$$\underset{\sim}{v}_\alpha = - \frac{\underset{\approx}{\Lambda}_\alpha}{\phi S_\alpha} (\nabla p_\alpha - \rho_\alpha g \nabla Z) \tag{2.4}$$

which is familiar as Darcy's law.

Clearly, the mobility $\underset{\approx}{\Lambda}_\alpha$ appearing in Eq. (2.4) accounts for much of the predictive power of Darcy's law in any particular rock-fluid system. Constitutive laws for mobility are largely phenomenological, the most common versions having the form $\underset{\approx}{\Lambda}_\alpha = \underset{\approx}{k} k_{r\alpha}/\mu_\alpha$, where μ_α is the dynamic viscosity of fluid phase α, $\underset{\approx}{k}$ is the *permeability*, and the *relative permeability* $k_{r\alpha}$ is a coefficient describing the effects of other fluids in obstructing the flow of fluid α.

For a two-fluid system with no interphase mass transfer, the relative permeabilities typically vary with saturation, and the curves $k_{rN}(S_W)$, $k_{rW}(S_W)$ look roughly like those drawn in Figure 1 (102). The vanishing-point saturations S_{Nr} and S_{Wr} are called *residual* or *irreducible* saturations, and they account for the fact that, for a particular fluid to flow, it must be present at a sufficient degree of saturation to permit the formation of connected flow channels consisting of that phase. Actually, this picture of relative permeabilities is quite simplistic. In nature relative permeabilities often exhibit significant hysteresis, and the verification of the relative-permeability model in the presence of three or more fluid phases (92,144,101) or compositional effects (22,14) is still not clear.

Eq. (2.4) allows each fluid phase to have its own pressure at any point in the reservoir. These pressure differences indeed occur in nature. At the microscopic scale the effects of interfacial tension and pore geometry on the curvatures of fluid-fluid interfaces lead to capillary effects. Leverett (91) uses the classical thermodynamics of Gibbs (75) to describe these effects, while more recent works such as those of Morrow (103) and Davis and Scriven (51) draw connections with microscopic effects and molecular theories of interfacial tension. These theories imply that, at a macroscopic scale, there will be a pressure difference, or *capillary pressure*, between any two fluid phases in a porous medium. In two-phase systems, for example, there is a single capillary pressure $p_{CNW} = p_N - p_W$. In simple models p_{CNW} is a function of saturation; however, in actual flows the capillary pressure exhibits rather pronounced hysteresis (103,82,134) and dependence on fluid composition (42).

Given velocity field equations such as Eq. (2.4), we can expand the mass balances for the fluid phases to get flow equations for each fluid. Using the customary decomposition of the mobility $\underset{\approx}{\Lambda}_\alpha$ and directly substituting Eq. (2.4) into Eq. (2.2) yields, for a two-phase system,

$$\frac{\partial}{\partial t}(\phi S_N \rho_N) - \nabla \cdot \left(\frac{\rho_N \underset{\approx}{k} k_{rN}}{\mu_N} (\nabla p_W + \nabla p_{CNW} - \rho_N g \nabla Z) \right) = 0$$

$$\frac{\partial}{\partial t}(\phi S_W \rho_W) - \nabla \cdot \left(\frac{\rho_W \underset{\approx}{k} k_{rW}}{\mu_W} (\nabla p_W - \rho_W g \nabla Z) \right) = 0 \qquad (2.5)$$

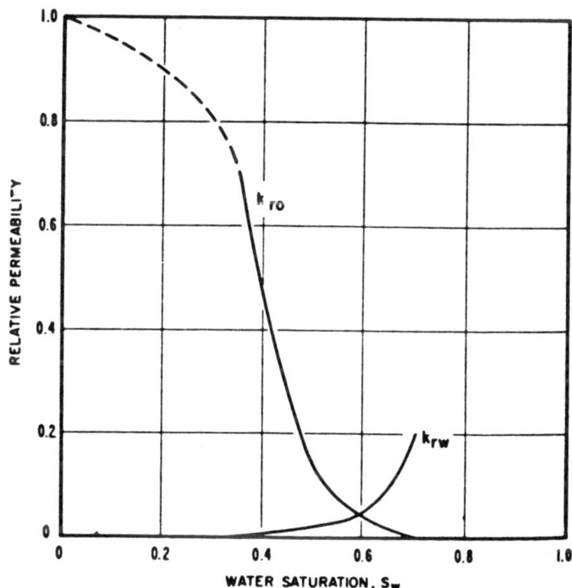

Figure 1. Typical relative permeability curves (102)

Flow equations for systems having more fluid phases will be similar, except that if P phases coexist, then P - 1 independent capillary pressure functions will appear in the system.

2.3. Early Investigations

The picture of multiphase flows in porous media outlined above evolved over several decades beginning in the 1930's. The use of an extended version of the single-phase form of Darcy's law in multiphase flows appears to have begun with Richards (128) in his work on unsaturated flows in the soil physics literature. The explicit use of a separate velocity field equation for each fluid began in the petroleum industry. Here the pioneering work of Muskat et al. (104), Wykoff and Botset (161), Buckley and Leverett (34), Fatt and Dykstra (68), and Welge (159), among others, promoted the widespread acceptance of Darcy's equation altered by the incorporation of relative permeabilities. Today this model is the one most widely used in the prediction of multiphase flows in porous media.

Despite its broad appeal in applications, the multiphase version of Darcy's law has some limitations. Relative permeabilities are not strictly functions of saturation, the most glaring violation being the phenomenon of hysteresis or dependence on saturation history. Such microscopic phenomena as gas slippage at the solid walls, turbulence, and adsorption can also invalidate the Darcy model in certain flows (33). These limitations are worthy of consideration in the application of the multiphase Darcy law to any new rock-fluid system.

3. TWO-PHASE FLOWS

The simplest multiphase flows in porous media are those in which two fluids flow simultaneously but do not exchange mass or react with the solid matrix. While many flows of practical interest exhibit more complex physics, two-phase flows have drawn attention in many applications. Among these are unsaturated groundwater flows, saltwater intrusion in coastal aquifers, and the Buckley-Leverett problem in petroleum engineering.

3.1. Unsaturated Groundwater Flow

In typical soil profiles some distance separates the earth's surface from the water table, which is the upper limit of completely water-saturated soil. In this intervening zone the water saturation varies between 0 and 1, the rest of the pore space normally being occupied by air. Water flow in this *unsaturated zone* is complicated by the fact that the soil's permeability to water depends on its water saturation. Let us derive the common form of the governing equation and examine some of the computational difficulties that arise in its solution.

Most formulations of unsaturated flow rest on the assumption that the motion of air has negligible effect on the motion of water. Therefore one usually neglects the flow equation for air, assuming that the air pressure equals the constant atmospheric pressure at the surface, that is, $p_A = p_{atm}$. Then we can define the *pressure head* in the water by $\psi = (p_W - p_A)/(\rho_w g)$, having the dimensions of length and being negative in the unsaturated zone where $S_W < 1$. Also, instead of saturation, soil physicists typically refer to the soil's *moisture content*, defined by $\Theta = \phi S_W$. In terms of these new variables the capillary pressure relationship for the air-water system becomes $\psi = \psi(\Theta)$ or, provided ψ is an invertible function, $\Theta = \Theta(\psi)$. From Eq. (2.5b), the flow equation for water thus transforms to

$$\frac{\partial}{\partial t}(\Theta\rho_W) = \nabla.[\underset{\approx}{K}.\rho_W(\nabla\psi + \nabla Z)]$$

where $\underset{\approx}{K} = \rho_W g \underset{\approx}{k} k_{rW}/\mu_W$ is the *hydraulic conductivity* of the soil. Notice that $\underset{\approx}{K}$ is a function of ψ, since relative permeability depends on saturation, which varies with ψ according to the capillarity relationship.

In many unsaturated flows the compressibility effects in water are small, so that time derivatives and spatial gradients of ρ_W may be neglected. If this approximation holds, then the flow equation reduces to

$$\frac{\partial\Theta}{\partial t} = \nabla.[\underset{\approx}{K}.(\nabla\psi + \nabla Z)] \qquad (3.1)$$

To get an equation in which ψ is the principal unknown, we simply use the chain rule to expand the time derivative on the left, giving

$$C(\psi)\frac{\partial\psi}{\partial t} = \nabla.[\underset{\approx}{K}(\psi)(\nabla\psi + \nabla Z)]$$

where $C(\psi) = d\Theta/d\psi$ is the *specific moisture capacity*. If the flow is essentially one-dimensional in the vertical direction, then this equation collapses to

$$C(\psi)\frac{\partial\psi}{\partial t} = \frac{\partial}{\partial z}\left[K(\psi)\left(\frac{\partial\psi}{\partial z} + 1\right)\right] \qquad (3.2)$$

which is Richard's equation (128).

Several investigators in hydrology have examined the unsaturated flow equation from analytic viewpoints. Philip (119) gives one of the earliest theoretical treatments of Richard's equation, proposing asymptotic solutions for a nonlinear problem. The equation has also attracted interest in the applied mathematics community, including investigations by Aronson (16), Peletier (116), and Nakano (105). Aronson (16), for example, observes that, while the classical linear heat equation admits solutions in which disturbances propagate with infinite speeds, the nonlinear Eq. (3.2) may propagate disturbances with only finite speed. This implies that a moving interface, or *wetting front*, can form between the downward-moving zone of high moisture content Θ and the zone yet uncontacted by the wave of infiltrating water. Under certain initial conditions this moving boundary can exhibit steep spatial gradients in Θ and consequently in ψ. The resulting sharp fronts pose considerable difficulty in the construction of numerical schemes, since the discrete approximations used typically have lowest-order error terms that increase with the norm of the solution's gradient. We shall discuss this difficulty in more detail in Section 6.

Numerical work by a variety of investigators have corroborated the existence of wetting fronts. Much of this work appeared during the 1970's, and it includes articles by Bresler (33), Neuman (107), Reeves and Duguid (127), Narasimhan and Witherspoon (106), and Segol (136). Van Genuchten (151,152) presents solution schemes for the one- and two-dimensional versions of Richard's equation using both finite differences and finite-element Galerkin methods employing Hermite cubic basis functions. His work furnishes a good comparison of the finite-difference and finite-element approaches to the approximation of wetting fronts.

Van Genuchten's investigation also demonstrates another difficulty in solving Richard's equation numerically. This problem owes its existence to the nonlinear coefficient $C(\psi)$ appearing in the accumulation term of Eq. (3.2). Because the equation itself is nonlinear, implicit time-stepping algorithms must incorporate an iterative procedure for advancing the approximate solution from one time step to the next. There then arises a question regarding the proper time level at which to evaluate $C(\psi)$. Van Genuchten demonstrates that evaluating this coefficient in a fully implicit fashion can lead to material balance errors in certain schemes, among them the Galerkin scheme using two-point Gauss quadrature to evaluate the mass and stiffness matrix elements. Figure 2 shows how this scheme produces a wetting front that lags the true solution. Milly (99) advances an iterative method for evaluating $C(\psi)$ at the correct time level to guarantee good global material balances.

Allen and Murphy (11) propose another approach to the time-stepping problem in unsaturated flows. While the method is used in

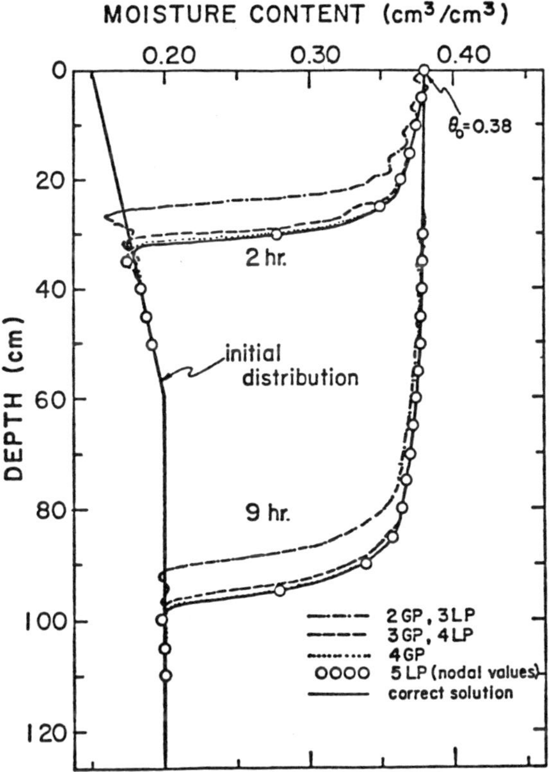

Figure 2. Solutions to the unsaturated flow equation using various finite-element Galerkin schemes (151)

connection with finite-element collocation-a technique closely related to the two-point Gauss quadrature scheme mentioned above-the basic idea should be applicable with most spatial discretizations. If we return to the original form of the accumulation term, Eq. (3.2) becomes

$$\frac{\partial \Theta}{\partial t} = \frac{\partial}{\partial z}\left[K(\psi)\left(\frac{\partial \psi}{\partial z} + 1\right)\right]$$

One can circumvent the difficulties encountered in solving an equation in both Θ and ψ by properly formulating an iterative procedure. Let us approximate the time derivative using implicit finite differences:

$$\frac{\Theta(\psi^{n+1}) - \Theta(\psi^n)}{\Delta t} = \frac{\partial}{\partial z}\left[K(\psi^{n+1})\left(\frac{\partial \psi^{n+1}}{\partial z} + 1\right)\right] + O(\Delta t)$$

We can linearize the flux terms in this approximation by establishing an iterative scheme in which $\psi^{n+1,m}$ represents the value of ψ at the most recent known iteration level and $\psi^{n+1,m+1} = \psi^{n+1,m} + \delta\psi^{n+1,m+1}$ represents the value at the sought iterative level:

$$\frac{\partial}{\partial z}\{K(\psi^{n+1,m})[\frac{\partial}{\partial z}(\psi^{n+1,m} + \delta\psi^{n+1,m+1}) + 1]\}$$

This expression allows the nonlinear coefficient $K(\psi^{n+1})$ to lag by an iteration.

In the accumulation term we also lag $\Theta(\psi^{n+1})$, but in addition we linearly project forward to the next iterative level using the Newton-like extrapolation

$$\frac{1}{\Delta t}[\Theta(\psi^{n+1,m}) + C(\psi^{n+1,m})\delta\psi^{n+1,m+1} - \Theta(\psi^n)]$$

Here, recall that $C(\psi) = d\Theta/d\psi$. The value ψ^n of pressure head at the old time level represents the value furnished by the iterative scheme after convergence, which a computer code can test using either of two criteria. First, one can check whether the iterative increment $\delta\psi^{n+1,m+1}$ is small enough in magnitude or norm to warrant stopping the iteration. Second, one can observe that collecting the terms involving the unknown $\delta\psi^{n+1,m+1}$ on the left and ignoring truncation error leaves the known quantity

$$-\frac{\Theta(\psi^{n+1,m}) - \Theta(\psi^n)}{\Delta t} + \frac{\partial}{\partial z}\left[K(\psi^{n+1,m})\left(\frac{\partial \psi^{n+1,m}}{\partial z} + 1\right)\right] \equiv -R^{n+1,m}$$

acting as a right-hand side in the linearization. This quantity is precisely the residual to the flow equation at the m-th iteration. Whenever $\|R^{n+1,m}\|$ is small in some appropriate norm, the resulting increment $\delta\psi^{n+1,m+1}$ will be small and, more to the point, we shall have solved the time-differenced equation to within a very small error.

It is easy to see why such a scheme conserves mass, at least to within limits imposed by the iterative convergence criteria. If we integrate the residual $R^{n+1,m}(z)$ over the spatial domain Ω of the problem, we find

$$- \int_{\Omega} \frac{\Theta(\psi^{n+1,m}) - \Theta(\psi^{n})}{\Delta t} dz + K(\psi^{n+1,m}) \left(\frac{\partial \psi^{n+1,m}}{\partial z} + 1 \right) \Bigg|_{\partial\Omega} = - \int_{\Omega} R^{n+1,m} dz$$

If the integral on the right were zero, this equation would be precisely the global mass balance for vertical unsaturated flow. Thus by iterating until $\|R^{n+1,m}\|$ is small, we implicitly enforce the global mass balance to a desired level of accuracy.

3.2. Saltwater Intrusion

In coastal aquifers both fresh water and salt water are usually present. Being denser, the salt water underlies the fresh water, the latter forming a lens whose shape and thickness may vary with changes in pumping and recharge. Figure 3 depicts a typical coastal aquifer in cross-section. When the upper portion of the aquifer acts as a source of fresh water, it becomes important to design pumping and recharge strategies that prevent the flow of salt water into production wells.

Strictly speaking, salt water and fresh water are not separate phases. In fact they are completely miscible as fluids, and in a coastal aquifer there exists a zone lying between the two fluids in which salt concentration varies continuously. To be rigorously faithful to the physics of the problem, then one would solve a single-phase flow equation coupled with a transport equation for salt. Indeed, one of the earliest numerical treatments of saltwater intrusion used just this approach (120). Nevertheless, the transition zone between salt and fresh water is often quite narrow in comparison with the overall thickness of the aquifer, and for computational purposes we may consider it to be a sharp interface. Such a sharp-interface approximation serves as justification for treating saltwater intrusion into coastal aquifers as a multiphase flow.

Let us consider the problem of modeling the areal movement of salt and fresh water. To get vertically averaged flow equations, we first write the equations in terms of *hydraulic heads*, defined in the fresh water (F) and salt water (S) as follows:

$$h_\alpha = \frac{1}{g} \int_{p_{ref}}^{p_\alpha} \frac{dp'}{\rho_\alpha(p')} + z, \qquad \alpha = F \text{ or } S,$$

where p_{ref} is some reference value of pressure, and $\sigma_\alpha(p)$ gives the functional dependence of density on pressure. Then, after an application of the chain rule to the accumulation terms, each of equations (2.5) assumes the form

$$\nabla\cdot(K_\alpha \nabla h_\alpha) = S_{s,\alpha} \frac{\partial h_\alpha}{\partial t}, \qquad \alpha = F \text{ or } S \tag{3.3}$$

where $K_\alpha = \rho_\alpha g k k_{r_\alpha}/\mu_\alpha$ is the hydraulic conductivity of fluid α and $S_{s,\alpha} = \rho_\alpha g[d\phi/dp_\alpha+(\phi/\rho_\alpha)d\rho_\alpha/dp_\alpha]$ is the *specific storage* of fluid α. For simplicity, let us assume that the rock matrix is isotropic, so that K_α effectively acts as a scalar coefficient. Next we average the flow equations (3.3) vertically by integrating with respect to z between the lower and upper limits of each zone, using Leibnitz's rule (84). For the freshwater zone, this gives (see Figure 3)

$$\bar\nabla\cdot(T_F\bar\nabla\bar h_F) - (v_F - s_y v_\Phi)\big|_{z=c}\cdot e_z + (v_F - s_y v_\Phi)\big|_{z=c}\cdot\nabla c - (v_F\cdot\nabla b - v_F\cdot e_z)\big|_{z=b} = C_F \frac{\partial \bar h_F}{\partial t}$$

Here $\bar\nabla = (\partial/\partial x, \partial/\partial y)$ in Cartesian coordinates; e_z signifies the unit vector in the z-direction; $T_F = K_F \ell_F$; $C_F = S_{s,F}\ell_F + s_y$; and $\bar h = \ell_F^{-1}\int_b^c h_F dz$ is the vertically averaged freshwater head. The vector v_Σ represents the velocity of the freshwater-saltwater interface; v_Φ is the velocity of the free surface z=c, and s_y is the specific yield, defined as the rate of change in storage with respect to changes in the free surface level.

In the absence of mass transfer between salt water and fresh water, a material point initially on the interface Σ will stay there. Since Σ is the locus of points where z-b=0, this *free surface condition* takes the form

$$\left(\frac{\partial}{\partial t} + v_\Sigma\cdot\nabla\right)(z-b)\big|_\Sigma = \left(-\frac{\partial b}{\partial t} - v_\Sigma\cdot\nabla b + v_\Sigma\cdot e_z\right)\bigg|_{z=b} = 0$$

Multiplying this equation by ϕ and subtracting from the vertically averaged equation above yields

$$\bar\nabla\cdot(T_F\bar\nabla\bar h_F) + q_F\big|_{z=c} - q_F\big|_{z=b} = C_F\frac{\partial \bar h_F}{\partial t} - \phi\frac{\partial b}{\partial t} \tag{3.4}$$

where

$$q_F\big|_{z=c} = -(v_F - s_y v_\Phi)\big|_{z=c}\cdot(e_z - \nabla c)$$

is the effective rate of withdrawal from the freshwater zone, and

$$q_F\big|_{z=b} = -(v_F - \phi v_\Sigma)\big|_{z=b}\cdot(e_z - \nabla b)$$

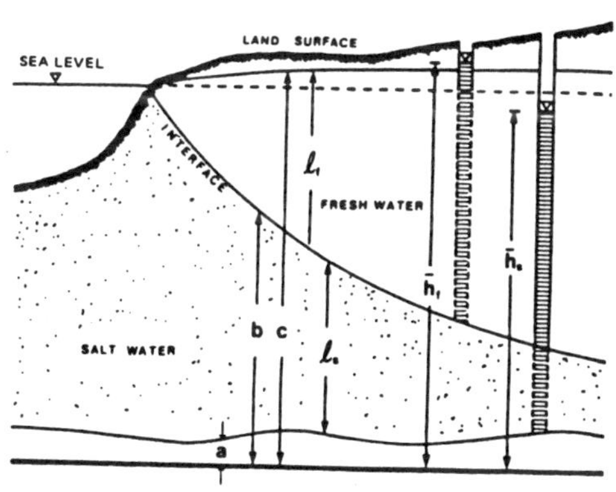

Figure 3. Schematic cross-section of a coastal aquifer (84).

$$K_{ij}^{F} = \int_{\Omega} T_F \nabla N_i . \nabla N_j d\underset{\sim}{x},$$

$$Q_{ij}^{F} = \oint_{\partial\Omega} T_F N_j \nabla N_i . \underset{\sim}{n} dx,$$

$$M_{ij}^{F,1} = - \int_{\Omega} (C_F + \phi\rho_F^{*}) N_i N_j d\underset{\sim}{x},$$

$$M_{ij}^{F,2} = \int_{\Omega} \phi\rho_S^{*} N_i N_j d\underset{\sim}{x},$$

$$B_j^{F} = - \int_{\Omega} T_F \nabla h_{F,\partial} . \nabla N_j d\underset{\sim}{x} + \oint_{\partial\Omega} N_j T_F \nabla h_{F,\partial} . \underset{\sim}{n} d\underset{\sim}{x} + \int_{\Omega} (C_F + \phi\rho_F^{*}) N_j \frac{\partial h_{F,\partial}}{\partial t} d\underset{\sim}{x} - \int_{\Omega} \phi\rho_S^{*} N_j \frac{\partial h_{S,\partial}}{\partial t} d\underset{\sim}{x}$$

$$K_{ij}^{S} = \int_{\Omega} T_F \nabla N_i . \nabla N_j d\underset{\sim}{x},$$

$$Q_{ij}^{S} = \oint_{\partial\Omega} T_F N_j \nabla N_i . \underset{\sim}{n} d\underset{\sim}{x},$$

$$M_{ij}^{S,1} = \int_{\Omega} \phi\rho_F^{*} N_i N_j d\underset{\sim}{x},$$

$$M_{ij}^{S,2} = - \int_{\Omega} (C_S + \phi\rho_S^{*}) N_i N_j d\underset{\sim}{x},$$

$$B_j^{S} = - \int_{\Omega} T_S \nabla h_{S,\partial} . \nabla N_j d\underset{\sim}{x} + \oint_{\partial\Omega} N_j T_S \nabla h_{S,\partial} . \underset{\sim}{n} d\underset{\sim}{x} + \int_{\Omega} (C_S + \phi\rho_S^{*}) N_j \frac{\partial h_{S,\partial}}{\partial t} d\underset{\sim}{x} - \int_{\Omega} \phi\rho_F^{*} N_j \frac{\partial h_{F,\partial}}{\partial t} d\underset{\sim}{x}$$

Table 1. Galerkin integrals appearing in the salt-water intrusion equations.

is the effective rate of exchange of freshwater across the interface Σ, which we have assumed to be zero.

A similar development for salt water leads to the vertically averaged flow equation

$$\overline{\nabla}.(T_S \overline{\nabla} \overline{h}_S) + q_S \Big|_{z=b} - q_S \Big|_{z=a} = C_S \frac{\partial \overline{h}_S}{\partial t} + \phi \frac{\partial b}{\partial t} \tag{3.5}$$

Here $T_S = K_S \ell_S$ and $C = S_{s,S} \ell_S$. The sink terms in this equation are

$$q_S \Big|_{z=b} = - (\underset{\sim}{v}_S - \phi \underset{\sim}{v}_{\Sigma}) \Big|_{z=b} . (\underset{\sim}{e}_z - \nabla b),$$

which represents the effective rate of withdrawal from the saltwater zone, and

$$q_S \Big|_{z=a} = - \underset{\sim}{v}_S \Big|_{z=a} . (\underset{\sim}{e}_z - \nabla a),$$

which gives the effective rate of saltwater leakage into the lower confining layer, whose depth is fixed.

To solve this system we need an equation relating $\bar{h}_F$ and $\bar{h}_S$. In this case, since the two fluids are miscible at the microscopic scale, there will be no head difference between the fluids where they are in contact. Thus the head is continuous across the interface Σ: $h_F = h_S$ at $z = b$. As Huyakorn and Pinder (84) show, this condition allows us to solve for $\partial b/\partial t$ in terms of heads:

$$\frac{\partial b}{\partial t} = \rho_S^* \frac{\partial \bar{h}_S}{\partial t} - \rho_F^* \frac{\partial \bar{h}_F}{\partial t} \tag{3.6}$$

where $\rho_\alpha^* = \rho_\alpha/(\rho_S - \rho_F)$. Combining Eq. (3.6) with Eqs. (3.4) and (3.5) yields the coupled system of flow equations

$$\nabla . \begin{pmatrix} T_F & 0 \\ 0 & T_S \end{pmatrix} \nabla \begin{pmatrix} \bar{h}_F \\ \bar{h}_S \end{pmatrix} + \begin{pmatrix} q_F\big|_{z=c} \\ q_S\big|_{z=b} - q_S\big|_{z=a} \end{pmatrix}$$

$$= \begin{pmatrix} C_F + \phi\rho_F^* & -\phi\rho_S^* \\ -\phi\rho_F^* & C_S + \phi\rho_S^* \end{pmatrix} \frac{\partial}{\partial t} \begin{pmatrix} \bar{h}_F \\ \bar{h}_S \end{pmatrix} \tag{3.7}$$

Let us examine the approximate numerical solution to Eq. (3.7) using finite-element Galerkin methods. In these methods we replace the unknown functions $\bar{h}_F(\underset{\sim}{x},t)$ and $\bar{h}_S(\underset{\sim}{x},t)$ by trial functions

$$\hat{h}_F(\underset{\sim}{x},t) = h_{F,\partial}(\underset{\sim}{x},t) + \sum_{i=1}^{I} h_{F,i}(t)N_i(\underset{\sim}{x})$$

$$\hat{h}_S(\underset{\sim}{x},t) = h_{S,\partial}(\underset{\sim}{x},t) + \sum_{i=1}^{I} h_{S,i}(t)N_i(\underset{\sim}{x})$$

The functions $h_{F,\partial}$ and $h_{S,\partial}$ satisfy the essential boundary conditions for the problem at hand, and each of the sums on the right satisfies homogeneous boundary conditions. For continuous interpolating trial functions, $h_{F,i}(t)$ and $h_{S,i}(t)$ usually stand for the values of head at the i-th spatial node, while the *basis functions* $N_i(\underset{\sim}{x})$ dictate the variation between nodes.

To form the Galerkin equations corresponding to Eq. (3.7), we substitute $\hat{h}_F$ and $\hat{h}_S$ for $\overline{h}_F$ and $\overline{h}_S$ in Eq. (3.7), multiply each equation by each of the basis functions $N_1(\underset{\sim}{x}),\ldots,N_I(\underset{\sim}{x})$, and force the integral of the result over the spatial domain Ω to vanish. Doing this leads to time evolution equations for the unknown nodal values $h_{F,i}(t)$ and $h_{S,i}(t)$. For the freshwater equation, there results

$$\sum_{i=1}^{I} \left(K_{ij}^{F} h_{F,i} - Q_{ij}^{F} h_{F,i} + M_{ij}^{F,1} \frac{dh_{F,i}}{dt} + M_{ij}^{F,2} \frac{dh_{S,i}}{dt} \right)$$

$$+ \int_{\Omega} q_F \Big|_{z=c} N_j d\underset{\sim}{x} = B_j^F , \qquad j = 1,\ldots,I$$

where K_{ij}^F, Q_{ij}^F, $M_{ij}^{F,1}$, $M_{ij}^{F,2}$, and B_j^F have the meanings assigned in Table 1.

A similar collection of evolution equations arises from the saltwater flow equation:

$$\sum_{i=1}^{I} \left(K_{ij}^{S} h_{S,i} - Q_{ij}^{S} h_{S,i} + M_{ij}^{S,1} \frac{dh_{F,i}}{dt} + M_{ij}^{S,2} \frac{dh_{S,i}}{dt} \right)$$

$$+ \int_{\Omega} \left(q_S \Big|_{z=b} - q_S \Big|_{z=a}\right) N_j d\underset{\sim}{x} = b_j^S , \qquad j=1,\ldots,I$$

where the definitions of K_{ij}^S, Q_{ij}^S, $M_{ij}^{S,1}$, $M_{ij}^{S,2}$, and B_j^S again appear In Table 1. Rewriting this set of 2I evolution equations in matrix form gives a system having the structure

$$[K]\{h\} + [M] \frac{\partial}{\partial t} \{h\} = \{r\} \tag{3.8}$$

where $\{h\}$ signifies a vector containing the 2I unknown nodal values of head, [K] and [M] are the stiffness and mass matrices arising from flux and accumulation terms, respectively, and $\{r\}$ is a vector containing known boundary data and withdrawal rates.

The system (3.8) is nonlinear, owing to the dependence of the zonal thicknesses ℓ_F and ℓ_S on the unknown heads. Thus any temporal discretization of these ordinary differential equations will have to be iterative in nature to guarantee consistency between the numerical solution and the flow coefficients at each time level. Pinder and Page (121) advance one such iterative scheme.

The saltwater interface problem exhibits a peculiar computational difficulty associated with the saltwater-freshwater interface Σ. This problem manifests itself as the saltwater wedge retreats or advances. Under these circumstances the intersection of Σ with the lower confining layer, called the *saltwater toe*, moves horizontally. This moving boundary allows for the possibility that the interface may not exist at some areal locations, and at these locations the free surface condition becomes degenerate (97). To accommodate this degeneracy, it becomes necessary to track the moving boundary as the flow calculations proceed.

Shamir and Dagan (139) present a finite-difference algorithm for tracking the saltwater toe in a vertically integrated, immiscible setting. By examining a one-dimensional flow, they develop a scheme for regenerating the spatial grid to guarantee that the toe lies on a computational node. Thus on the ocean side of the separating node they solve the simultaneous flow equations for saltwater and freshwater heads, while on the inland side they solve the equation for freshwater head only. This approach obviously involves a great deal of computational complexity in two or three dimensions, since it requires the construction of multidimensional moving finite-difference grids. However, an analogous idea for finite-element grids in two dimensions has proved promising (55).

In another approach, Sá da Costa and Wilson (131) use a fixed, two-dimensional, quadrilateral finite-element grid to model the immiscible flow equations. They devise a toe-tracking algorithm based on the Gauss points used to compute the integrals contributing to the matrix entries in Eq. (3.8). At Gauss points inland of the toe the model assigns a very small nonzero saltwater transmissibility T_S. Thus, while the saltwater wedge never actually disappears in the numerical scheme, inland of the toe the flow of salt water is negligible.

3.3. The Buckley-Leverett Problem

The Buckley-Leverett problem serves as a fairly simple model of two-phase flow in a porous medium. The problem, introduced by Buckley and Leverett (34), has particular relevance in the petroleum industry, where gas and water injection are two common techniques for displacing oil toward production wells in underground reservoirs. The simplicity of the Buckley-Leverett problem arises from three basic assumptions. First, the total flow rate of oil and displacing fluid (say water) remains constant. Second, the rock matrix and fluids are incompressible. Third, the effects of capillary pressure gradients on the flow field are negligible compared with the pressure gradients applied through pumping. These assumptions are too restrictive to permit widespread application of the Buckley-Leverett model, but, as we shall argue below, the simplified model acts as a paradigm for the numerical difficulties that occur in more complicated models of oil reservoirs.

To derive the Buckley-Leverett model, we begin with Eqs. (2.5), identifying N as oil and W as water and assuming an isotropic porous medium:

$$\frac{\partial}{\partial t}(\phi S_N \rho_N) - \nabla.[\rho_N \Lambda_N(\nabla p_W + \nabla p_{CNW} - \rho_N g \nabla Z)] = 0$$

$$\frac{\partial}{\partial t}(\phi S_W \rho_W) - \nabla.[\rho_W \Lambda_W(\nabla p_W - \rho_W g \nabla Z)] = 0$$

where $\Lambda_\alpha = kk_{r\alpha}/\mu_\alpha$ is the mobility of fluid α. Coupled to these flow equations are the constraint $S_N + S_W = 1$ and a capillarity relationship $p_{CNW} = p_{CNW}(S_W)$. If we restrict our attention to one-dimensional flow in a homogeneous reservoir of uniform cross-section and assume that gravity effects are absent, then the flow equations collapse to

$$\frac{\partial}{\partial t}(\phi S_N \rho_N) - \frac{\partial}{\partial x}\left[\rho_N \Lambda_N \left(\frac{\partial p_W}{\partial x} + \frac{\partial p_{CNW}}{\partial x}\right)\right] = 0$$

$$\frac{\partial}{\partial t}(\phi S_W \rho_W) - \frac{\partial}{\partial x}(\rho_W \Lambda_W \frac{\partial p_W}{\partial x}) = 0$$

Now we invoke the assumption that capillarity has negligible effect on the flow field- wide, so that $\partial p_{CNW}/\partial x \cong 0$. Further, the incompressibility assumption implies that ϕ, ρ_N, and ρ_W are constant in time and that the fluid densities are uniform in space, so that

$$\phi \frac{\partial}{\partial t}(1 - S_W) - \frac{\partial}{\partial x}(\Lambda_N \frac{\partial p_W}{\partial x}) = 0 \tag{3.9a}$$

$$\phi \frac{\partial S_W}{\partial t} - \frac{\partial}{\partial x}(\Lambda_W \frac{\partial p_W}{\partial x}) = 0 \tag{3.9b}$$

Now observe that $- \Lambda_\alpha \partial p_W/\partial x$ is the Darcy flux q_α of phase α. Also by assumption, the total flow rate $q = q_W + q_N$ is a constant. Thus we need only solve one of Eqs. (3.9), using the constant value of q to solve the other equation by subtraction.

Let us solve the water equation (3.9b). Since $-\Lambda_W \partial p_W/\partial x = q_W = \Lambda_W q/(\Lambda_W + \Lambda_N)$, we arrive at the Buckley-Leverett saturation equation

$$\frac{\partial S_W}{\partial t} + \frac{\partial}{\partial x}\left(\frac{qf_W}{\phi}\right) = 0 \qquad (3,10)$$

where $f_W = \Lambda_W/(\Lambda_N + \Lambda_W)$ is the *fractional flow* of water. Eq. (3.10) is clearly nonlinear, since f_W depends on the unknown water saturation S_W through the fluid mobilities. While the functional form of $f_W(S_W)$ depends on the particular rock-fluid system being modeled, fractional flow functions typically have an "S-shaped" profile over their supports ($S_{Wr} = 1 - S_{Nr}$), as shown in Figure 4.

Difficulties in solving Cauchy problems involving Eq. (3.10) arise from two sources. First, the equation itself is a nonlinear, hyperbolic conservation law. Its hyperbolicity owes to our neglect of capillary pressure gradients, inclusion of which would have led to an additional second-order term of the form

$$\frac{\partial}{\partial x}\left[\phi^{-1}\Lambda_W p'_{CNW}(S_W)\frac{\partial S_W}{\partial x}\right]$$

Thus Eq. (3.10) is, in effect, an approximation to a singularly perturbed parabolic problem in which we have neglected the dissipative effects of capillarity.

Second, the flux function qf_W/ϕ appearing in Eq. (3.10) is nonconvex, its S-shaped form implying the existence of an inflection point somewhere in its support. The literature on hyperbolic conservation laws with nonconvex flux functions is quite extensive, including important contributions by Lax (90) and Oleinik (111) and a general discussion by Chorin and Marsden (39). Of special importance in the present context are the following facts. Cauchy problems based on Eq. (3.10) may have no solutions that are classical in the sense of being continuously differentiable over their (x,t)-domains $\Omega \times J$. Instead, such problems may admit only *weak solutions* $S_W(x,t)$. These solutions need only satisfy the integral relation

$$\int_{\Omega \times J}\left(S_W\frac{\partial \Psi}{\partial t} + \frac{q}{\phi}f_W(S_W)\frac{\partial \Psi}{\partial x}\right)dxdt = 0 \qquad (3.11)$$

for all infinitely differentiable functions $\Psi(x,t)$ that vanish on the boundary $\partial(\Omega \times J)$ (126). In contrast to Eq. (3.10), Eq. (3.11) admits functions $S_W(x,t)$ that have discontinuities, or saturation shocks. Unfortunately, weak solutions may not be unique: there may be several different functions $S_W(x,t)$ that satisfy the integral equation (3.11).

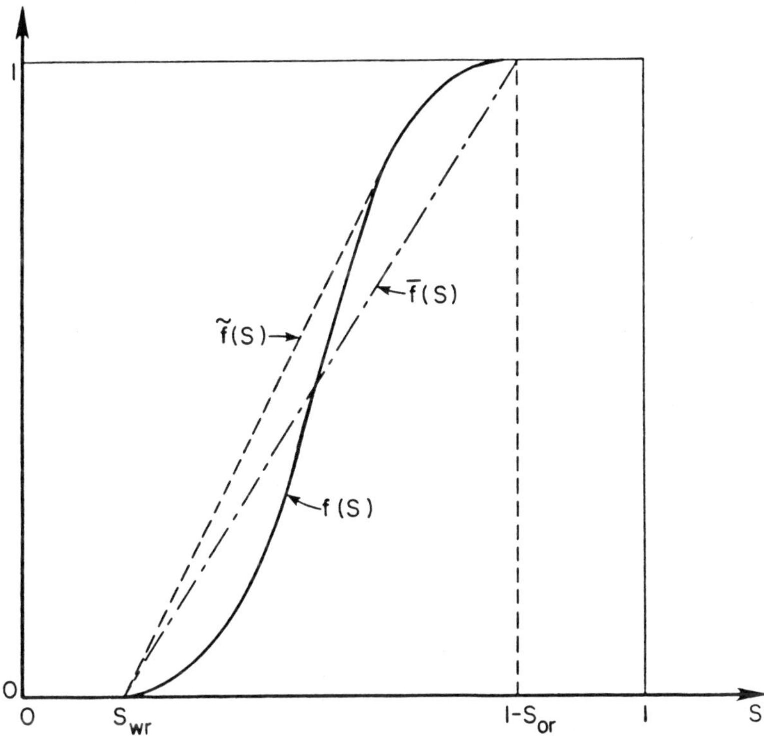

Figure 4. Typical nonconvex fractional flow function f and related convex functions (13).

Nature admits only one solution to the Buckley-Leverett problem. Much of the research into hyperbolic conservation laws has aimed at identifying physically correct weak solutions from among the class of functions obeying Eq. (3.11). To specify the physical solution requires an additional constraint known as the *entropy condition*. There are several equivalent forms of this constraint, including the following (13): i) The solution must depend continuously and stably on the initial data, implying that characteristics on both sides of a discontinuity must intersect the initial curve. ii) The solution must be the same as that obtained using the method of characteristics with $f_W(S_W)$ replaced by its convex hull. iii) The solution must be the limit of solutions, for the same initial data, to a parabolic problem differing from the hyperbolic one by a dissipative second-order term (in this case, capillarity) of vanishing influence.

The tangent construction advanced by Welge (159) explicitly implements condition (ii) while, as Welge shows in his paper, the "equal-area" rule of Buckley and Leverett (34) imposes this same constraint in a slightly different fashion.

Any numerical scheme for solving the Buckley-Leverett problem, or even more complicated models of multiphase flows that are hyperbolic in character, must respect the entropy condition or else risk producing nonphysical results. Douglas et al. (57), for example, propose adding an artificial capillarity to the Buckley-Leverett equation to force convergence to the correct physical solution. An equivalent effect can be achieved by using certain numerical approximations whose lowest-order error terms mimic the desired dissipative phenomena (8). This tactic is perhaps easiest to see in finite-difference approximations. Here, an upstream biased difference analog of the flux term $\partial f/\partial x$ gives

$$\frac{f_i - f_{i-1}}{\Delta x} = \left.\frac{\partial f}{\partial x}\right|_i - \frac{\Delta x}{2}\frac{\partial}{\partial x}\left.\left[f'(S)\frac{\partial S}{\partial x}\right]\right|_i + O(\Delta x^2)$$

Since $f'(S) > 0$ over the support of f, the lowest-order error term acts like the capillarity term neglected in Eq. (3.10) while vanishing linearly as $\Delta x \to 0$. Thus upstream weighting imposes a numerical version of condition (iii) while maintaining consistency in the numerical approximation.

Several investigators have examined upstream-weighted finite-element methods for the Buckley-Leverett problem. Mercer and Faust (96) and Huyakorn and Pinder (83), for example, discuss upstream-weighted Galerkin techniques. Shapiro and Pinder (140) advance a finite-element collocation scheme for the Buckley-Leverett problem using asymmetric basis functions.

Allen and Pinder (12,13) introduce a collocation scheme for the same problem in which upstream biasing of the collocation points leads to the appropriate numerical version of condition (ii). To implement this method, we begin with a continuously differentiable trial function for saturation:

$$\hat{S}(x,t) = \sum_{i=0}^{I} [S_i(t)H_{0,i}(x) + S_i'(t)H_{1,i}(x)],$$

where the basis functions $H_{0,i}(x)$, $H_{1,i}(x)$ are piecewise Hermite cubic polynomials (5). $S_i(t)$, $S_i'(t)$ are the unknown nodal values of S_W and $\partial S_W/\partial x$, respectively. One can similarly represent the nonlinear flux function f_W:

$$\hat{f} = \sum_{i=1}^{I} [f_W(S_i)H_{0,i} + \frac{df_W}{dS_W}(S_i)S_i'H_{1,i}]$$

In the standard collocation we derive ordinary differential equations for the unknown values S_i, S_i', by setting

$$\frac{\partial \hat{S}}{\partial t}(\bar{x}_k,t) + \frac{q}{\phi}\frac{\partial \hat{f}}{\partial x}(x_k,t) = 0$$

at enough points $\bar{x}_k$ in the spatial domain to give one equation for each unknown. Douglas and Dupont (58) show that, on a uniform partition $x_0 < \cdots < x_I = x_0 + I\Delta x$, one can achieve $O(\Delta x^4)$ accuracy in parabolic problems by choosing the Gauss points $x_i+\Delta x/2\pm\Delta x/\sqrt{3}$, $i = 1,\ldots,I-1$, as the collocation points. As Allen and Pinder (13) demonstrate, however, this highly accurate scheme violates the entropy condition in Eq. (3.10). One can force convergence to the correct solution by evaluating the flux term at collocation points upstream of the Gauss points, as in the equation

$$\frac{\partial \hat{S}}{\partial t}(\bar{x}_k,t) + \frac{q}{\phi}\frac{\partial \hat{f}}{\partial x}(\bar{x}_k^*,t) = 0$$

Here, for flow in the positive x direction, $\bar{x}_k^* < \bar{x}_k$. Allen (7) presents an error analysis showing how this scheme introduces artificial capillarity. Figures 5 and 6 compare the results of standard collocation and upstream collocation respectively.

Several investigators have examined the use of upstream weighting in more sophisticated models of multiphase flow. Among the many such studies are those by Peaceman (113), Settari and Aziz (137), and Young (164), each of which offers a good overview of numerical approximations used to model two-phase flows. We shall consider upstream weighting further in Section 4.

One unfortunate aspect of upstream-biased approximations is that their artificially dissipative effects, while guaranteeing convergence, produce unrealistically smeared sharp fronts when the spatial grid mesh is large. What is "large" in this sense depends on the physics of the problem and not the computational resources of the modeler. Therefore, in some problems, unacceptable smearing on uniform grids can occur even when the grid mesh approaches limits in affordable fineness. One approach to resolving this dilemma is

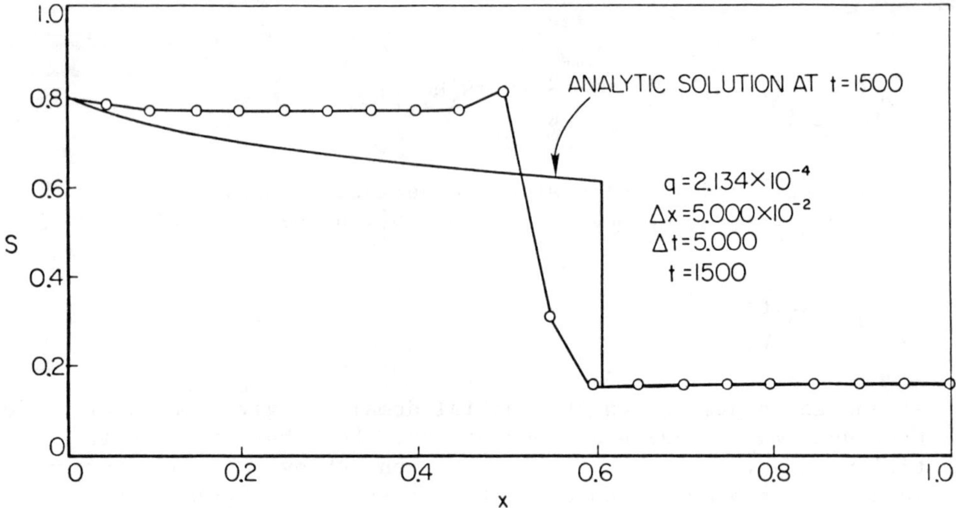

Figure 5. Solution to the Buckley-Leverett problem generated by orthogonal collocation with $\Delta x = 0.1$ (12).

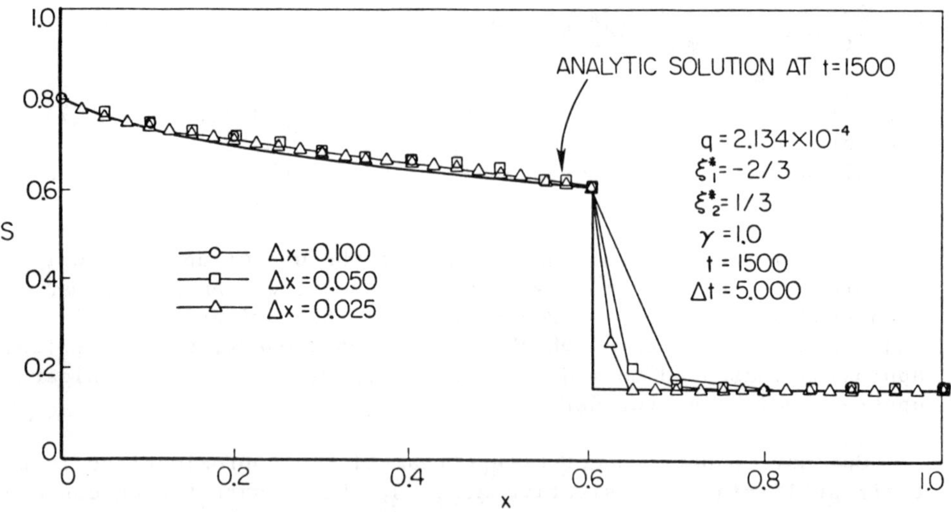

Figure 6. Solutions to the Buckley-Leverett problem generated by upstream collocation with $\Delta x = 0.1, 0.05, 0.025$ (12).

to refine the spatial grid only in the vicinity of the steep front. Since the front itself moves as the flow progresses, such a strategy calls for self-adaptive local grid refinement, a topic discussed in Section 6.

4. FLOWS WITH INTERPHASE MASS TRANSFER

In many multiphase flows of interest in engineering the exchange of chemical species among the fluid phases is crucial to the behavior of the flows. Historically, concern with the compositional aspects of multiphase flows in porous media originated in the petroleum industry, where the effects of gas dissolution, retrograde condensation, and vaporization and condensation of injected gases have substantial implications in oil recovery operations. As the complexities of groundwater contamination by organic wastes become more urgent, however, interest in multiphase flows with mass transfer has spread to the hydrology community. In this section we shall focus on the more established modeling efforts in the petroleum industry, leaving discussion of the newer applications in hydrology to Section 5.

4.1. Compositional Oil Reservoir Flows

In compositional flows there are several fluid phases in which some number of chemical species reside. It is therefore necessary to extend the mixture-theoretic formalism to accommodate two different categories of constituents: phases and species. A more detailed exposition of the development given below appears in Allen (7). For simplicity, let us assume that there are three fluid phases, namely water (W), oil (O), and gas (G) with chemical species indexed by $i = 1,\ldots,N + 1$. As before, let us label the rock phase by the index R. Conceivably, at least, each species can exist in any phase and can transfer between phases via dissolution, evaporation, condensation, and so forth, subject to thermodynamic constraints. We shall assume here that the rock is chemically inert and that there are no intraphase or stoichiometric chemical reactions, although in such applications as enhanced oil recovery by alkaline fluid injection reactions of this kind may be important.

In our new formalism, each pair (i,α), with i chosen from the species indices and α chosen from the phases, is a constituent. Thus, for example, CH_4 in the gas phase is one constituent, CH_4 in oil another, and $n\text{-}C_4H_{10}$ in oil yet another. Each constituent (i,α) has its own *instrinsic mass density* ρ_i^α, measured as mass of i per unit volume of α, and its own velocity $\underset{\sim}{v}_i^\alpha$. To accommodate the familiar kinematics of phases, we shall still associate with each phase α its volume fraction ϕ_α, and if $\phi = 1 - \phi_R$ as before, then we define the saturation of fluid phase α as $S_\alpha = \phi_\alpha/\phi$. Using these basic quantities, we define the following variables:

$$\rho^{\alpha} = \sum_{i=1}^{N} \rho_i^{\alpha} = \text{intrinsic mass density of phase } \alpha,$$

$$w_i^{\alpha} = \rho_i^{\alpha}/\rho^{\alpha} = \text{mass fraction of species i in phase } \alpha,$$

$$\rho = \phi \sum_{\alpha \neq R} S_{\alpha}\rho^{\alpha} = \text{bulk density of fluids,}$$

$$w_i = (\phi/\rho) \sum_{\alpha \neq R} S_{\alpha}\rho^{\alpha} w_i^{\alpha} = \text{total mass fraction of species i in the fluids,}$$

$$\underset{\sim}{v}^{\alpha} = (1/\rho^{\alpha}) \sum_{i=1}^{N} \rho_i^{\alpha} \underset{\sim}{v}_i^{\alpha} = \text{barycentric velocity of phase } \alpha,$$

$$\underset{\sim}{u}_i^{\alpha} = \underset{\sim}{v}_i^{\alpha} - \underset{\sim}{v}^{\alpha} = \text{diffusion velocity of species i in phase } \alpha.$$

If the index N+1 represents the species making up the inert rock phase, then the following constraints hold:

$$\sum_{i=1}^{N} w_i = \sum_{i=1}^{N} w_i^{\alpha} = \sum_{\alpha} \phi_{\alpha} = \sum_{\alpha \neq R} S_{\alpha} = 1$$

where the index α in the second sum can represent any fluid phase, and

$$\sum_{i=1}^{N} \underset{\sim}{u}_i^{\alpha} = 0$$

Each constituent (i,α) has its own mass balance, given by analogy with Eq. (2.1) as

$$\frac{\partial}{\partial t} (\phi_{\alpha}\rho_i^{\alpha}) + \nabla.(\phi_{\alpha}\rho_i^{\alpha}\underset{\sim}{v}_i^{\alpha}) = r_i^{\alpha}$$

where the exchange terms r_i must obey the restriction $\sum_{i=1}^{N} \sum_{\alpha \neq R} r_i^{\alpha} = 0$. If we impose the further constraint that there are no intraphase chemical reactions, then we have in addition $\sum_{\alpha \neq R} r_i^{\alpha} = 0$ for each species $i = 1,\ldots,N$. Since phase velocities are typically more accessible to measurement than species velocities, it is convenient to rewrite the constituent mass balance as

$$\frac{\partial}{\partial t}(\phi S_\alpha \rho^\alpha w_i^\alpha) + \nabla.(\phi S_\alpha \rho^\alpha w_i^\alpha \underset{\sim}{v}^\alpha) + \nabla.\underset{\sim}{j}_i^\alpha = r_i^\alpha \tag{4.1}$$

where $\underset{\sim}{j}_i^\alpha = \phi S_\alpha \rho^\alpha w_i^\alpha \underset{\sim}{u}_i^\alpha$ stands for the *diffusive flux* of constituent (i,α). Summing this equation over all fluid phases α and using the restrictions gives a total mass balance for each species i:

$$\frac{\partial}{\partial t}(\rho w_i) + \nabla.[\phi(S_W \rho^W w_i^W \underset{\sim}{v}^W + S_O \rho^O w_i^O \underset{\sim}{v}^O + S_G \rho^G w_i^G \underset{\sim}{v}^G)] + \nabla.(\underset{\sim}{j}_i^W + \underset{\sim}{j}_i^O + \underset{\sim}{j}_i^G) = 0, \quad i=1,\dots,N$$

To establish flow equations for each species, we need velocity field equations for each fluid phase and some constitutive equations for the diffusive fluxes $\underset{\sim}{j}_i^\alpha$. For the fluid velocities we may postulate Darcy's law, Eq. (2.4), assuming in addition that the porous medium is isotropic. For the diffusive fluxes the appropriate assumption is not so clear. In single-phase flows through porous media, the diffusive flux of a species with respect to the fluid's barycentric velocity is called *hydrodynamic dispersion*. As reviewed in Section 5, theories of hydrodynamic dispersion in multiphase flows remain poorly developed. The most common approach in oil reservoir simulation is to assume that hydrodynamic dispersion is a small enough effect that the diffusive fluxes in the mass balance for each species are negligible. Thus we arrive at the flow equation for species i in the fluids:

$$\frac{\partial}{\partial t}[\phi(S_W \rho^W w_i^W + S_O \rho^O w_i^O + S_G \rho^G w_i^G)]$$

$$- \nabla.\left\{\frac{kk_{rW}\rho^W w_i^W}{\mu_W}(\nabla p_W - \rho^W g \nabla Z) + \frac{kk_{rO}\rho^O w_i^O}{\mu_O}(\nabla p_O - \rho^O g \nabla Z) + \frac{kk_{rG}\rho^G w_i^G}{\mu_G}(\nabla p_G - \rho^G g \nabla Z)\right\} = 0, \quad i = 1,\dots,N$$

To close this set of equations, we need some supplementary constraints giving relationships among the variables. One class of supplementary constraints consists of the thermodynamic relationships giving phase densities and compositions as functions of pressure and overall fluid mixture composition. Conceptually, these relationships take the forms

$$\rho^\alpha = \rho^\alpha(w_1^\alpha,\dots,w_{N-1}^\alpha,p_\alpha), \qquad \alpha = W,O,G$$

$$w_i^\alpha = w_i^\alpha(w_1,\ldots,w_{N-1},p_\alpha), \qquad \alpha = W,O,G; \qquad i = 1,\ldots,N-1$$

$$S_\alpha = S_\alpha(w_1,\ldots,w_{N-1},p_\alpha), \qquad \alpha = W,O,G$$

However, it is important from a computational viewpoint to observe that the actual mathematical statements of these relationships may constitute simultaneous sets of nonlinear algebraic equations giving phase densities, compositions, and saturations implicitly. This occurs, for example, when one uses equal-fugacity constraints in conjunction with an equation of state to solve for local thermodynamic equilibria, as discussed further below.

The other class of supplementary constraints includes constitutive relationships for the particular rock-fluid system being modeled. These relationships may take the following forms:

$$p_{COW} = p_{COW}(S_O,S_G)$$

$$p_{CGO} = p_{CGO}(S_O,S_G)$$

$$k_{r\alpha} = k_{r\alpha}(S_O,S_G), \qquad \alpha = W,O,G$$

Here, as mentioned in Section 2, we have greatly simplified the physics of many compositional flows by omitting possible dependencies on fluid composition through variations in interfacial tension.

4.2. Black-Oil Simulation

Black oil models are special cases of the general compositional equations that allow limited interphase mass transfer, the composition of each phase depending on pressures only. This class of models has become a standard engineering tool in the petroleum industry. As a consequence the literature on the numerics of black-oil simulation, which apparently began in 1948 with a consulting report by John von Neumann (156), has become quite extensive. Indeed, there are now several books in print devoted to black-oil simulation (114, 18). Since any attempt to cover this field in an article of the present scope would be futile, we shall merely review the formulation of the black-oil equations and discuss selected aspects of their numerical solution.

The fundamental premise of the black-oil model is that a highly simplified, three-species system can often serve as an adequate model of the complex mixtures of brine and hydrocarbons found in natural petroleum reservoirs. For practical purposes, petroleum engineers define these three pseudo-species according to what appears at the surface, at stock-tank conditions (STC), after production of the reservoir fluids. Thus, we have the species o, which is stock-tank oil; g, which is stock-tank gas, and w, which is stock-tank water. Underground, at reservoir conditions (RC), these species may partition themselves among the three fluid phases O,G, and W in a distribution depending on the pressures in the formation.

Now we impose a set of thermodynamic constraints on this partitioning of species. First, we assume that there is no exchange of water w into the nonaqueous phases O and G, so that $w_w^W = 1$, and $w_w^O = w_w^G = 0$. Second, we allow no exchange of oil o into the vapor phase G or the aqueous liquid W, so that $w_o^O = 1$, and $w_o^W = w_o^G = 0$. Third, we prohibit the dissolution of gas g into the aqueous liquid W, so that $w_g^W = 0$. However, we allow the gas g to dissolve in the hydrocarbon liquid O according to a pressure-dependent relationship called the *solution gas-oil ratio*, defined by

$$R_S(p_O) = \frac{\text{volume of g in solution at RC}}{\text{volume of o}}$$

where the volumes refer to volumes at STC.

To facilitate further reference to volumes of species at STC, we relate the phase densities ρ^α at RC to the species densities ρ_i^{STC} at STC by defining the *formation volume factors*. For W and G these definitions are fairly simple:

$$B_W(p_W) = \rho_w^{STC}/\rho^W , \qquad B_G(p_W) = \rho_g^{STC}/\rho^G$$

For the hydrocarbon liquid O, however, we must also account for the mass of dissolved gas at RC:

$$B_O(p_O) = (\rho_o^{STC} + R_S\rho_g^{STC})/\rho^O$$

If we substitute these definitions into the flow equations (4.1) for the species o, g, w and divide through by the constants ρ_i^{STC}, we obtain the three black-oil equations

$$\frac{\partial}{\partial t}\left(\frac{\phi S_W}{B_W}\right) - \nabla.[\lambda_W(\nabla p_W - \gamma_W \nabla Z)] = 0 \tag{4.2a}$$

$$\frac{\partial}{\partial t}\left(\frac{\phi S_O}{B_O}\right) - \nabla.[\lambda_O(\nabla p_O - \gamma_O \nabla Z)] = 0 \tag{4.2b}$$

$$\frac{\partial}{\partial t}\left[\phi\left(\frac{S_G}{B_G}+\frac{R_S S_O}{B_O}\right)\right] - \nabla.[\lambda_G(\nabla p_G - \gamma_G \nabla Z)]$$

$$- \nabla.[R_S\lambda_O(\nabla p_O - \gamma_O\nabla Z)] = 0 \qquad (4.2c)$$

where $\lambda_\alpha = \Lambda_\alpha/B_\alpha$ and $\gamma_\alpha = \rho^\alpha g$.

These equations constitute a system of coupled, nonlinear, time-dependent partial differential equations. Each of the equations is formally parabolic in appearance. However, as suggested by the greatly simplified development in Section 3.3, the sytem can exhibit behavior more typical of hyperbolic equations if capillary influences are small. To see this, consider the two-phase version of Eq. (4.2) in which gas is absent, porosity is constant, and fluid compressibilities and gravity forces have no effect. The flow equations in this case reduce to

$$-\phi\frac{\partial S_W}{\partial t} = \nabla.(\lambda_O\nabla p_O)$$

$$\phi\frac{\partial S_W}{\partial t} = \nabla.(\lambda_W\nabla p_W)$$

Adding these equations gives a total flow equation $\nabla.\underset{\sim}{q} = 0$, where $\underset{\sim}{q} = -\lambda_O\nabla p_O - \lambda_W\nabla p_W$. Calling $\lambda = \lambda_O + \lambda_W$ and $p = (p_O + p_W)/2$, we we can rewrite the total flow equation as

$$\nabla.(\lambda\nabla p) - \left(\frac{\lambda_W - \lambda_O}{2}\right)\nabla p_{COW} = 0$$

If we examine the case when $\nabla p_{COW} \cong 0$, the total flow equation reduces to an elliptic *pressure equation*

$$\nabla.(\lambda\nabla p) = 0 \qquad (4.3a)$$

Then, recalling the fractional flow function $f_W = \lambda_W/(\lambda_O + \lambda_W)$, we can rewrite the water flow equation as

$$\phi\frac{\partial S_W}{\partial t} + \underset{\sim}{q}.\nabla f_W(S_W) = 0 \qquad (4.3b)$$

This *saturation equation* is the hyperbolic analog of the one-dimensional Buckley-Leverett problem.

Several approaches to solving the general system (4.2) numerically have appeared in the petroleum engineering literature. We shall review two of the most popular methods: the simultaneous solution (SS) method and the implicit pressure-explicit saturation (IMPES) method.

The SS method, introduced by Douglas, Peaceman, and Rachford (59), and further developed by Coats et al. (45), treats the flow equations (4.2) as simultaneous equations for the fluid pressures p_O, p_G, and p_W. Inverting the capillarity relationships and imposing the restriction on fluid saturations then yields the saturations S_O, S_G, and S_W. For ease of presentation, let us examine the two-phase case, assuming that the vapor phase G does not appear and that the porosity ϕ is constant.

The first step in the formulation is to rewrite the flow equations so that the pressures p_O and p_W appear as explicit unknowns. To do this, we apply the chain rule to the accumulation terms, giving

$$\frac{\partial}{\partial t}\left(\frac{\phi S_W}{B_W}\right) = \phi S_W b_W \frac{\partial p_W}{\partial t} + \frac{\phi S_W'}{B_W}\left(\frac{\partial p_O}{\partial t} - \frac{\partial p_W}{\partial t}\right)$$

$$\frac{\partial}{\partial t}\left(\frac{\phi S_O}{B_O}\right) = S_O b_O \frac{\partial p_O}{\partial t} - \frac{\phi S_W'}{B_O}\left(\frac{\partial p_O}{\partial t} - \frac{\partial p_W}{\partial t}\right)$$

where $b_\alpha = d(1/B_\alpha)/dp_\alpha$ and S_W' signifies the derivative of the inverted capillarity relationship $S_W(p_{COW})$. This device allows us to write the system (4.2) as follows:

$$\phi \begin{pmatrix} (S_W b_W - S_W'/B_W) & (S_W'/B_O) \\ (S_W'/B_O) & (S_O b_O - S_W'/B_O) \end{pmatrix} \frac{\partial}{\partial t}\begin{pmatrix} p_W \\ p_O \end{pmatrix}$$

$$- \nabla . \begin{pmatrix} (\Lambda_W/B_W)\nabla & 0 \\ 0 & (\Lambda_O/B_O)\nabla \end{pmatrix}\begin{pmatrix} p_W \\ p_O \end{pmatrix}$$

$$+ \begin{pmatrix} \rho^W g \nabla Z \\ \rho^O g \nabla Z \end{pmatrix} = \begin{pmatrix} 0 \\ 0 \end{pmatrix} \qquad (4.3)$$

Now we can employ some finite-difference or finite element method to approximate the spatial derivative in Eq. (4.3), getting a system of evolution equations having the form

$$[M] \frac{d}{dt}\{p\} + [K]\{p\} = \{f\}$$

Here [M] is the mass matrix, [K] is the stiffness matrix, {p} represents the vector of unknown nodal values of oil and water pressure, and {f} is a vector containing information from the discretized boundary conditions. Since the entries of [M] and [K] vary with the unknown pressures, this system is nonlinear. Therefore the time-stepping approximation must be iterative. As an example, we might use a Newton-like procedure analogous to that presented in Section 3.1, yielding

$$\left(\frac{1}{\Delta t} [M]^{n+1,m} + [K]^{n+1,m}\right)\{\delta p\}^{n+1,m+1}$$

$$= -\left(\frac{1}{\Delta t} [M]^{n+1,m}(\{p\}^{n+1,m} - \{p\}^{n}) - [K]^{n+1,m}\{p\}^{n+1,m} + \{f\}^{n+1,m}\right)$$

$$= - \{R\}^{n+1,m}$$

In this scheme the notation $\{R\}^{n+1,m}$ suggests that we regard the right side as a residual, iterating at each step until $\|\{R\}^{n+1,m}\|$ is small enough in some norm.

The formulation presented above is not unique. In fact, several variants of the SS method have appeared, including formulations treating different sets of variables as principal unknowns. Aziz and Setari (18) provide a survey of these alternative approaches.

In the IMPES formulation, the basic idea is to combine the flow equations (4.2) to get an equation for one of the fluid pressures (32). Solving this equation implicitly provides the information necessary to update the saturations explicitly at each time step, using an independent set of flow equations and the restriction that saturations sum to unity. Sheldon, Zondek, and Cardwell (141) and Stone and Garder (145) introduced this method.

The development follows a line of reasoning paralleling that leading to Eqs. (4.3). We begin, as in the SS method, by expanding the accumulation terms, this time leaving saturations and pressures as principal unknowns. For the three-phase system, this leads to the following finite-difference approximations

$$\phi\frac{\partial}{\partial t} \left(\frac{S_W}{B}\right) = \frac{1}{\Delta t} (C_1\Delta_t S_W + C_2\Delta_t p_W) + O(\Delta t)$$

$$\phi \frac{\partial}{\partial t}(\frac{S_O}{B_O}) = \frac{1}{\Delta t} (C_3\Delta_t S_O + C_4\Delta_t p_O) + \mathcal{O}(\Delta t)$$

$$\phi \frac{\partial}{\partial t} (\frac{S_G}{B_G} + \frac{R_S S_O}{B_O}) = \frac{1}{\Delta t} (C_5\Delta_t S_G + C_6\Delta_t p_G + C_7\Delta_t S_O + C_8\Delta_t p_O) + \mathcal{O}(\Delta t)$$

The coefficients $C_1,\ldots,C_8$ appearing here stand for the appropriate derivatives extracted using the chain rule, and $\Delta_t u = u^{n+1} - u^n$ defines the time-difference operator.

The next step involves the crucial assumption that the capillary pressures p_{COW}, p_{CGO} change negligibly over a time step. This assumption implies that $\Delta_t p_O = \Delta_t p_W = \Delta_t p_G$ and, furthermore, that we can treat the capillary contributions to the flux terms explicitly. Thus, our implicit, temporally discrete approximations to Eq. (4.2) become

$$C_1\Delta_t S_W + C_2\Delta_t p_O = \Delta t\nabla.[\lambda_W^{n+1}(\nabla p_O^{n+1} - \nabla p_{COW}^{n} - \gamma_W^{n+1}\nabla Z)] \tag{4.4a}$$

$$C_3\Delta_t S_O + C_4\Delta_t p_O = \Delta t\nabla.[\lambda_O^{n+1}(\nabla p_O^{n+1} - \gamma_O^{n+1}\nabla Z)] \tag{4.4b}$$

$$\begin{aligned} C_5\Delta_t S_G + C_7\Delta_t S_O + (C_6 + C_8)\Delta_t p_O \\ = \Delta t\nabla.[\lambda_G^{n+1}(\nabla p_O^{n+1} + \nabla p_{CGO}^{n+1} - \gamma_G^{n+1}\nabla Z) \\ + R_S^{n+1}\lambda_O^{n+1}(\nabla p_O - \gamma_O^{n+1}\nabla Z)] \end{aligned} \tag{4.4c}$$

To get a single pressure equation from this set, we multiply Eq. (4.4c) by the coefficient $B = C_3/(C_7 - C_5)$, multiply Eq. (4.4a) by $A = BC_5/C_1$, add Eqs. (4.4a-c), and observe that the saturation differences in the accumulation terms now sum to an expression proportional to $\Delta_t(S_W + S_O + S_G) = 0$. Therefore our weighted sum of the time-differenced flow equations yields

$$\begin{aligned} C^{n+1}\Delta_t p_O = \Delta t\{A^{n+1}\nabla.(\lambda_W^{n+1}\nabla p_O^{n+1}) + \nabla.(\lambda_O^{n+1}\nabla p_O^{n+1}) \\ + B^{n+1}\nabla.[(\lambda_G^{n+1} + R_S^{n+1}\lambda_O^{n+1})\nabla p_O^{n+1}] - \Gamma^{n+1}\} \end{aligned} \tag{4.5}$$

The new parameter Γ is shorthand for the weighted sum of the gravity terms, and $C = AC_2 + C_4 + B(C_6 + C_8)$. Eq. (4.5) is the *pressure equation*.

Now, provided we have an appropriate technique for producing discrete approximations to the spatial derivatives appearing in these equations, we can implement the following time-stepping procedure. (i) Solve Eq. (4.5) implicitly, using some iterative scheme. (ii) Solve Eq. (4.4a) explicitly for $\Delta_t S_W$ and update the water saturation; solve (4.4b) for $\Delta_t S_O$ and update the oil saturation, setting $S_G^{n+1} = 1 - S_W^{n+1} - S_O^{n+1}$. (iii) Compute p_{COW}^{n+1} and p_{CGO}^{n+1} using the new saturations; then use these to update p_W and p_G. (iv) Begin the next time step. Notice that, in contrast to the SS formulation, the IMPES approach requires the implicit solution of only one flow equation at each time step. As with the SS methods, variants on this development have appeared; see Aziz and Settari (18) for a survey.

The IMPES approach offers the obvious advantage that, with only one implicit equation to solve per time step, the algoritm requires smaller matrix inversions at each iteration. The resulting computational savings can be significant in problems involving large numbers of grid points. On the other hand, because it treats capillary pressures explicitly, the IMPES method suffers instability when the time step Δt exceeds a critical value. This limitation can be inconvenient if the critical value of Δt is unknown or small compared with the life of a field project. The SS method, while requiring more computation per time step, boasts greater stability. This can prove to be a decided advantage when the problem to be solved exhibits strongly nonlinear phenomena, such as coning near wellbores or liquid hydrocarbons passing through bubble points.

The performance of black-oil models is quite sensitive to the treatment of nonlinear coefficients in the discrete flow equations. Consider, for example, the spatial treatment of the flux coefficients λ_α. It is standard practice to use upstream-weighted approximations to these coefficients. To see why, examine the results of Figure 7, showing predictions of a one-dimensional black-oil model using several midpoint and upstream approximations to λ_α. These plots show that upstream-biased analogs of the flux coefficients force the numerical solution to converge to the correct physical solution when capillarity is small. This result corroborates our discussion of the Buckley-Leverett problem in Section 3.3, since, as we have argued, the black-oil system exhibits similar hyperbolic features.

The temporal weighting of the flux coefficients also affects the solution to the black-oil equations. It is a fairly common practice to treat these coefficients explicitly. As Settari and Aziz show, however, this tactic leads to limits on time steps allowable for stable solutions. The limitation is especially severe

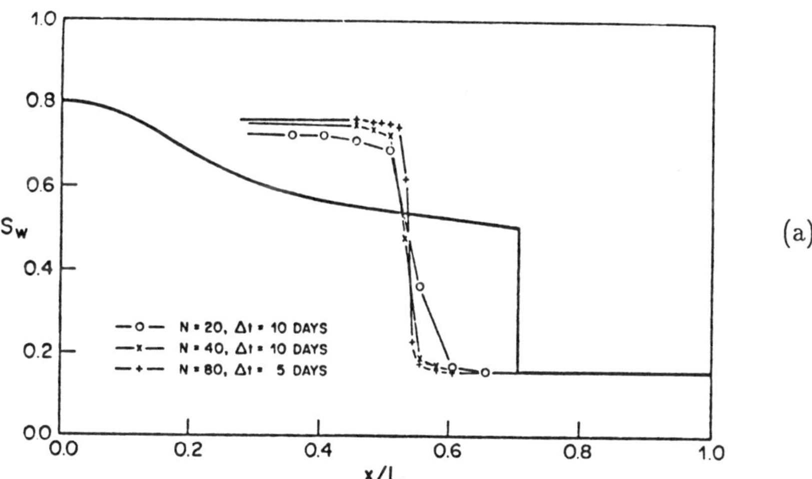

(a)

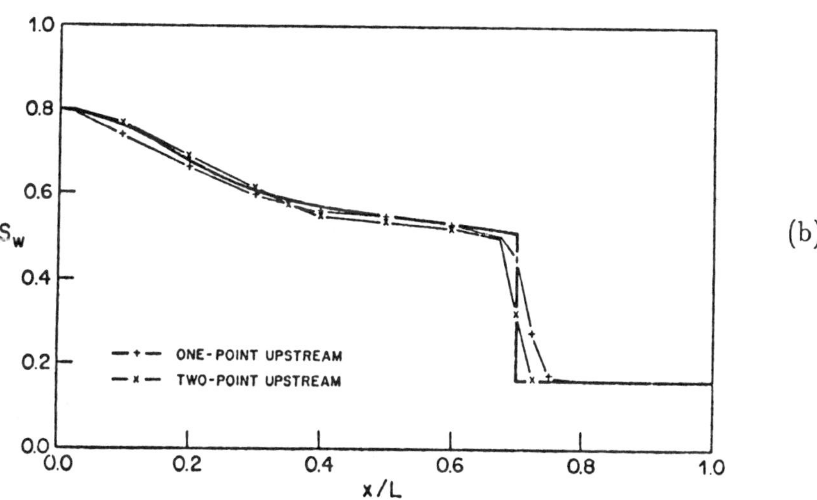

(b)

Figure 7. Black-oil model solutions using (a) midpoint-weighted flux coefficients and (b) upstream-weighted flux coefficients (18).

in problems with gas percolation, which occurs when the fluid mixture pressure drops below the bubble point. Blair and Weinaug (28) introduce the implicit treatment of the flux coefficients that alleviates this stability problem. As Coats (43) reviews, this highly stable method has proved attractive in simulating other, more complex oil-reservoir flows.

One of the most important problems in black-oil simulation, and in fact in reservoir simulation more generally, is the computational inefficiency associated with the solution of large systems of linear algebraic equations. In either the SS or the IMPES approach, the iterative time-stepping scheme calls for the solution of matrix equations at each iteration of each time step. For simulations at practical scales these calculations alone can tax the storage and CPU-time resources of the largest machines currently available. A great deal of recent research has focused on the development of fast iterative techniques for the solution of the large matrix systems arising in applications.

Among the oldest of these iterative techniques are the *block-iterative* methods. These methods use the blocked, sparse structure of the linear systems to solve the equations iteratively, block-by-block (27). Block iterative methods, such as block-successive overrelaxation, tend to be quite sensitive to "tunable" iteration parameters such as overrelaxation coefficients.

Another fairly old class of iterative techniques consists of *alternating direction* methods. These methods, introduced in the context of finite differences by Peaceman and Rachford (115), Douglas and Rachford (60), and Douglas (56), reduce the computational effort in multidimensional problems by implicitly solving over one space dimension at a time. While interest in alternating direction techniques for finite differences has waned in recent years, interest in alternating-direction Galerkin and collocation methods has been growing; see, for example, Ewing (62) and Celia and Pinder (36).

In a different approach, Stone (143) proposes the *strongly implicit procedure* (SIP) for solving matrix equations implicitly. The idea here is to replace a matrix equation having the form $[A]\{p\} = -\{R\}$ by an iterative scheme having the form

$$([A] + [N])\{p\}^{m+1} = ([A] + [N])\{p\}^{m} - ([A]\{p\}^{m} + \{R\})$$

By properly choosing the matrix [N], one can efficiently factor ([A] + [N]) into a product of sparse upper- and lower-triangular matrices. This idea gives rise to an algorithm that gives relatively rapid convergence to the solution {p} of the original equation.

Finally, much recent interest has focused on *conjugate gradient* methods for solving large matrix equations. These methods have their theoretical roots in the equivalence between linear systems and minimization problems for positive self-adjoint matrices (95). However, the methods admit extensions to the nonself-adjoint operators that arise in fluid flow problems, especially in conjunction with such preconditioning methods as incomplete LU factorization and nested factorization (15,110,150,122). The motivation for preconditioning is that, for parabolic flow equations, fine spatial grids can yield iteration equations $[A]\{p\} = -\{R\}$ in which the condition number of [A] is large. By "preconditioning" [A] with another matrix $[A^*]^{-1}$, one can arrive at an equivalent system

$$[A^*]^{-1}[A]\{p\} = -[A^*]^{-1}\{R\}$$

that is better conditioned. Clever choices of $[A^*]^{-1}$ ensure that $[A^*]^{-1}\{R\}$ will be easy to compute at each iteration, thus promoting computational efficiency. It is reasonable to expect that preconditioned conjugate-gradient methods will play a larger role in oil reservoir simulation as the technology continues to advance.

4.3. Compositonal Simulation

The most ambitous applications of the equations for compositional flows arise in the simulation of enhanced oil recovery processes. Many of these processes depend for their success on the effects of interphase mass transfer on fluid flow properties. One noteworthy example of such a process is miscible gas flooding. This technology consists of injecting an originally immiscible gas, such as CO_2, into an oil reservoir with the aim of developing a miscible displacement front in situ. In successful projects, miscibility develops through continuous interphase mass transfers, leading the fluid mixture toward its critical composition and hence reducing the interfacial tension between the resident oil and the displacing fluid. Compositional modeling serves as an important tool in other oil recovery problems, too, including production from gas condensate reservoirs and recovery of volatile oils.

There are several ways to classify compositional simulators. One way is to characterize the models according to their treatment of fluid-phase thermodynamics. There are at least two forms in which the thermodynamic constraints mentioned in Section 4.1 can appear. The oldest form consists of tabular data for the equilibrium ratios w_i^G/w_i^O of species mass (or mole) fractions in the vapor and liquid hydrocarbon phases. Thus, given overall hydrocarbon pressures and compositons at a point in the reservoir, one can compute fluid saturations, densities, and compositions by performing "flash" calculations familiar to chemical engineers (109).

The other form of the thermodynamic constraints is the requirement that vapor and liquid fugacities be equal for each component: $f_i^G = f_i^O$, $i = 1,\ldots,N$. This approach is especially attractive when used in conjunction with an equation of state such as that proposed by Peng and Robinson (117). Equation-of-state methods have the advantage of thermodynamic consistency near fluid critical points, leading to calculations with better convergence properties in models of miscible gas floods. In either the equilibrium-ratio approach or the equation-of-state approach, though, the thermodynamic constraints amount to a system of nonlinear algebraic equations giving fluid saturations, densities, and compositions implicitly.

Another way to classify compositional models is according to the manner in which they solve the flow equations (4.1). Two general schemes have appeared. One of these treats the flow equations sequentially, solving an overall pressure equation and then updating the remaining N-1 composition equations and the thermodynamic constraints at each time step or iteration. This approach parallels the IMPES method in black-oil simulation, and, as one might expect, it offers computational speed at the expense of some stability. The other scheme solves the entire system of flow equations and thermodynamic constraints simultaneously at each time step. This approach, analogous to the SS method of Section 4.2, leads to enormous matrix equations at each iteration. However, it enjoys greater stability than the sequential schemes. Given adequate computers, this *fully implicit* approach is quite attractive, since the compositional equations can exhibit behavior that is too complex to permit a priori estimates of stability constraints.

Among the simulators using sequential methods are those advanced by Roebuck et al. (130); Nolen (109); Van Quy, Corteville, and Simandoux (153); Kazemi, Vestal, and Shank (88); Nghiem, Fong, and Aziz (108); Watts (158), and Allen (6,7). Let us examine the time-stepping structure of one such model (7), restricting attention to an oil-gas system in which gravity has no effect. Summing the flow equations over all N species gives an overall fluid mass balance

$$\frac{\partial \rho}{\partial t} = \nabla\cdot(T_T\nabla p_G - T_O\nabla p_{CGO}) \qquad (4.6)$$

where $T_\alpha = kk_{r\alpha}\rho^\alpha/\mu_\alpha$ for each fluid α and $T_T = T_G + T_O$. This leaves N - 1 independent species balances

$$\frac{\partial(\rho w_i)}{\partial t} = \nabla\cdot(T_i\nabla p_G - T_O w_i^O\nabla p_{CGO}), \qquad i = 1,\ldots,N-1 \qquad (4.7)$$

where $T_i = T_G w_i^G + T_O w_i^O$. We can regard Eq. (4.6) as an equation for the pressure p_G, using Eq. (4.7) to solve for the overall species mass fractions w_i. The thermodynamic constraints then give the saturation, densities, and compositions of the liquid and vapor phases.

To solve these equations sequentially, we first discretize the pressure equation (4.6) in time, using the following Newton-like iterative scheme:

$$\rho^{n+1,m} + \left(\frac{\partial\rho}{\partial p_G}\right)^{n+1,m} \delta p_G^{n+1,m+1} - \rho^n$$

$$= \Delta t\nabla.[T_T^{n+1,m}\nabla(p_G^{n+1,m} + \delta p_G^{n+1,m+1}) - T_O^{n+1,m}\nabla p_{CGO}^{n+1,m}] \qquad (4.8)$$

This scheme is similar to that used in the unsaturated flow equation of Section 3.1. After solving for $\delta p_G^{n+1,m+1}$, we update the pressure iterate by setting $p_G^{n+1,m+1} = p_G^{n+1,m} + \delta p_G^{n+1,m+1}$. Then we can update each mass fraction $w_1,\ldots,w_{N-1}$ using the finite difference approximation

$$\Delta_t w_i^{n+1,m+1} = \frac{1}{\rho^{n+1,m+1}}\{\Delta t\nabla.[T_i^{n+1,m}\nabla p_G^{n+1,m+1}$$

$$- (T_O w_i)^{n+1,m}\nabla p_{CGO}^{n+1,m}] - w_i^n\nabla_t\rho^{n+1,m+1}\} \qquad (4.9)$$

to Eq. (4.7), setting $w_i^{n+1,m+1} = w_i^n + \Delta_t w_i^{n+1,m+1}$. This update calls for values of $\rho^{n+1,m+1}$, which are available from the latest iteration of Eq. (4.8) as

$$\rho^{n+1,m+1} = \Delta t\nabla(T_T^{n+1,m}\nabla p_G^{n+1,m+1} - T_O^{n+1,m}\nabla p_{CGO}^{n+1,m}) + \rho^n$$

This iterative sequence requires the solution of a matrix equation only in the spatially discrete analog of Eq. (4.8), since Eq. (4.9) has an "explicit" form at each iteration. Notice that, while the scheme is not fully implicit, it calls for implicit treatment of the flux coefficients, which lends to the stability of the formulation. Figure 8 shows a flow chart for the time-stepping algorithm, and Figure 9 shows a profile of vapor-liquid interfacial tensions in a simulated vaporizing gas drive (7). The wave of decreasing tensions indicates the development of a zone in which the fluid displacement is very nearly miscible.

With the advent of large, fast digital computers, interest has

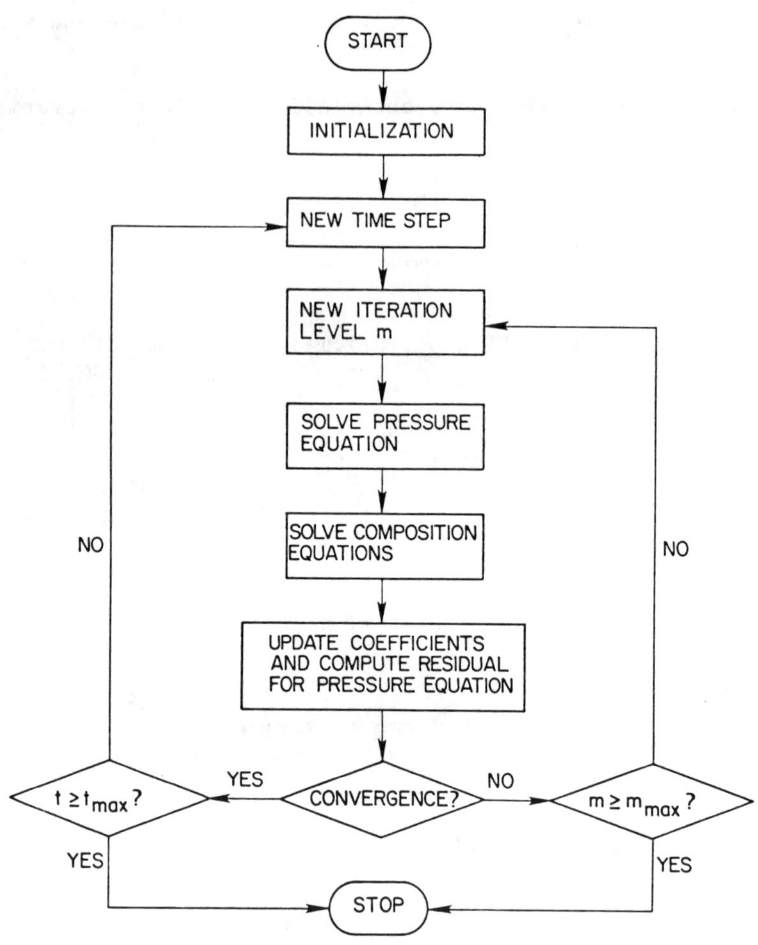

Figure 8. Flow chart of time-stepping procedure for a sequential compositional simulator (7).

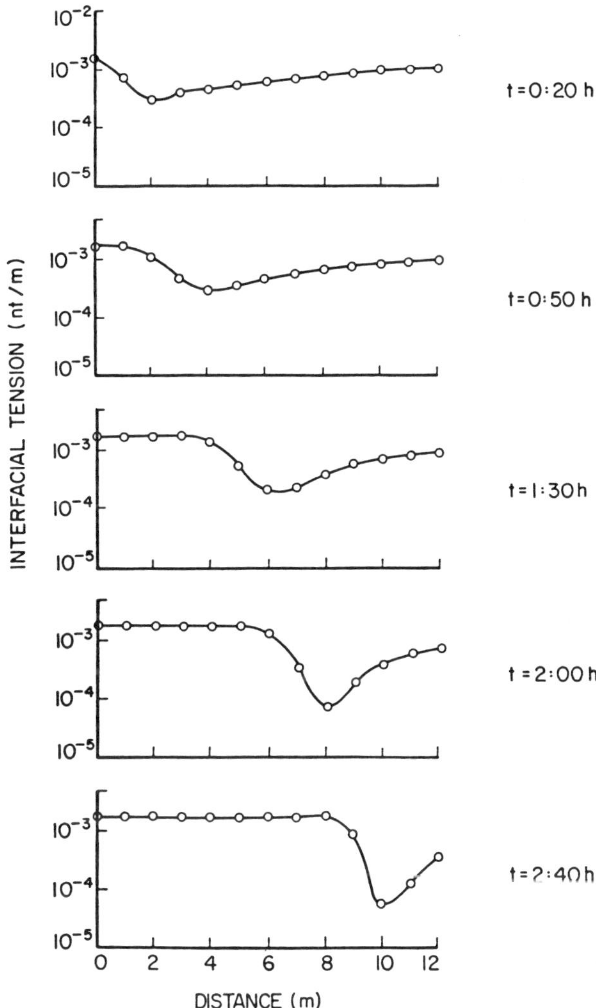

Figure 9. Interfacial tension profile at various times for a compositional simulation of a vaporizing gas drive (7).

grown in the fully implicit approach to compositional simulation. Among the models based on this approach are those reported by Fussell and Fussell (72), Coats (42), Heinemann (80), and Chien, Lee and Chen (38). This class of formulations treats the discretized flow equations and thermodynamic constraints as a set of simultaneous nonlinear algebraic equations, generally using some Newton-like iterative scheme to advance between time steps. The implicit nature of the formulations leads to great stability at the expense of solving large matrix equations of the form $[A]\{y\} = -\{R\}$ at each iteration. Moreover, the iteration matrix $[A]$ typically has less sparseness than the matrices arising from sequential schemes, since simultaneous schemes account for more of the nonlinear coupling between variables. Young and Stephenson (165) present one approach to mitigating this complication by evaluating the flux coefficients explicitly. As should be expected, this scheme reduces the computational effort of the fully implicit approach while sacrificing some of its stability.

There are several areas of difficulty common to practically all compositional simulators. One class of problems concerns the mathematical representation of fluid phase behavior. Most research in compositional simulation now focuses on methods using cubic equations of state coupled with equal-fugacity constraints to represent the fluid thermodynamics. While this approach guarantees thermodynamic consistency and therefore ensures smooth behavior of fluid densities, it requires the solution of highly nonlinear algebraic equations in addition to the discretized flow equations. Furthermore, the numerical solution of these thermodynamic constraints often suffers poor convergence when fluid pressures and compositions approach critical points (129). While the numerical problems associated with fluid phase behavior calculations pose serious challenges to the petroleum industry, an extensive discussion of research in this area would carry us far afield.

Another problem affecting compositional simulation is the numerical smearing introduced by upstream weighting. While this source of error affects other numerical models using upstream weighting, it is particularly problematic in compositional simulation. Because compositional models require so much storage and CPU time per spatial node, field-scale simulations often must use relatively few nodes and correspondingly coarser grids. The artificial diffusion that results can introduce large errors in species mass fractions and thus lead to unreal thermodynamics.

Several investigators have proposed methods for alleviating numerical diffusion in compositional simulators. Chase (37), for example, proposes local grid refinement methods for use with Galerkin finite elements. Section 6.4 discusses local grid refinement in more detail. Wilson, Tan, and Casinader (160) advance a method for selecting upstream-weighted difference approximations that yield reduced artificial smearing. Ewing and Heinemann (64,65) discuss the use of mixed finite-element methods to reduce smearing in compositional models. These authors propose that inaccurate fluid velocities, obtained by numerically differentiating pressure fields, aggravate numerical smearing. By incorporating mixed methods into their numerical scheme, they compute more accurate velocities and thereby help preserve sharp composition fronts in the numerical solution. Section 6 discusses mixed finite-element methods more thoroughly.

Finally, the growing appeal of the fully implicit approach implies that the computational effort associated with the inversion of large linear systems will become an increasingly important concern. The stakes involved in the linear algebra of compositional modeling are much higher than in black-oil simulation, since a typical fully implicit compositional model must solve the discretized flow equations and equal-fugacity constraints for between seven and ten species. This avenue of research should be active for quite some time to come.

5. OUTSTANDING PROBLEMS:PHYSICS

The next two sections review some of the outstanding problems in simulating multiphase flows in porous media. Roughly speaking, these problems fall into two categories:difficulties arising because our knowledge of the physics of multiphase flows is incomplete and difficulties in devising mathematical methods to capture known physics. The two categories are not as distinct as this description suggests. For some phenomena our lack of physical understanding hinders attempts to model them mathematically. Viscous fingering is an example, as discussed below. For other phenomena, the mathematical difficulties are evidence of physical complications that lead to peculiar behavior in the governing equations. The occurrence of sharp fronts in immiscible flows is an example of this coupling. Nevertheless, the distinction between physical and numerical difficulties makes some sense if we interpret it as suggesting strategies for future research. In this section we consider several physical problems.

5.1. Viscous Fingering

Often, in two-phase flows, the bulk of one fluid lies upstream of the other. In this case we say that the "upstream" phase displaces the "downstream" phase, even though there may be large regions where both phases flow simultaneously. The global behavior of such flows depends strongly on whether the mobility of the displacing fluid is greater or less than that of the displaced fluid.

In the latter case, when the *mobility ratio* $\Lambda_{displacing}/\Lambda_{displaced} \equiv M < 1$, the flow proceeds stably. This implies that velocity fields and saturations depend continuously on the boundary and initial conditions and well rates. When $M > 1$, however, channels of high displacing-fluid saturation can bypass zones of displaced fluid in a geometrically irregular pattern. These irregularities in the fluid displacement reflect the instability of immiscible displacements at high mobility ratios. The channeling phenomenon is called *viscous fingering*. While this phenomenon occurs in both single-phase and multiphase flows, we shall restrict our attention to the multiphase case.

Viscous fingering is economically important in oil reservoir engineering, where displacement of oil by some injected fluid is common to almost all recovery processes past primary production. In many cases the injected fluid is water, a gas such as CO_2 or N_2, or a surfactant solution. These fluids tend to be more mobile than common crude oils; therefore viscous fingering can occur. As a result, such a displacement scheme may sweep only a small fraction of the oil-bearing rock between an injection well and a production well. This inefficiency motivates reservoir engineers to add mobility control agents, such as hydrolyzed polymers, to injected fluids to lower their mobility.

Investigations into the physics of viscous fingering in immiscible displacements began in the late 1950's. Saffman and Taylor (132) investigated an analogy between porous-medium flows and Hele-Shaw flows, confirming that $M > 1$ leads to frontal instability. Chuoke, van Meurs, and van der Pol (41) applied perturbation techniques to show the existence of a critical wavelength for unstable fingers. From these early papers through the 1970's the literature on viscous fingering mushroomed. Ewing and George (63) provide a fairly extensive review of this body of work.

Research into viscous fingering has continued in recent years (76,118,163,94,87). While controversy still exists, there seems to be broad agreement that unstable fingers are triggered by heterogeneities in the porous medium observable at the microscopic scale. However, the macroscopic governing equations based on Darcy's law do not explicitly account for microscopic heterogeneities. Mathematical models based on the macroscopic equations and assuming a macroscopically homogeneous porous medium therefore have no mechanism for initiating fingers. Consequently, the homogeneous model will not exhibit instability, even though it is present in nature. One might, as an analogy, consider the mathematical model of an ideal pendulum hung vertically upward with zero velocity. The idealized model predicts that the pendulum is at equilibrium, whereas a natural pendulum in such a configuration is unlikely to stay there.

This failure to capture microscopic physics has unfortunate implications in numerical simulation. The response of a mathematical model to unstable immiscible displacements depends on the degree of heterogeneity in the data of the problem. Discrete models can represent spatial heterogeneity only within the limits imposed by the fineness of the spatial grid. Hence, models of immiscible displacement in media exhibiting heterogeneity at many scales can produce qualitatively different results depending on the spatial discretization used.

Several articles have appeared reporting efforts to produce better numerical representations of viscous fingering, given the inherent limitations of discrete methods. Among these are papers by Glimm, Marchesin, and McBryan (78), who propose the random choice method for solving the flow equations, and Ewing, Russell, and Wheeler (66), who examine a mixed method in conjunction with a modified method of characteristics to give accurate approximations of fluid interfaces. Another set of approaches has been to incorporate the "average" effects of fingering on the mixing of fluids in numerical simulators. This line of research began with Koval (89) and became a common simulation tool with the introduction of a mixing model by Todd and Longstaff (148). This "averaging" approach, while currently lacking in rigor, may offer fertile ground for the interaction of sound physical reasoning with the development of numerical techniques.

Finally, there is a great need for more empirical work on viscous fingering. Among the many sources of uncertainty regarding the nature of fingering is the paucity of field-scale data characterizing its effects. As Settari, Price, and Dupont (138) assert,

> The study of unstable displacements, particularly viscous fingering, is distinguished by the fact that in no other area of reservoir engineering is there less agreement. There is not even complete agreement on the existence of viscous fingering as a real phenomenon for reservoir conditions, let alone agreement as to the magnitude and interaction of the various mechanisms involved.

5.2. Multiphase Hydrodynamic Dispersion

As the derivation of the compositional equations in Section 4.1 demonstrates, individual species within a fluid phase need not move with the barycentric velocity of the phase. In porous-media flows, the deviation of species motions with respect to the mean flow of the fluid is called hydrodynamic dispersion. This diffusion-like phenomenon is familiar in the context of single-phase flows such as miscible displacement in petroleum engineering or soluble contaminant transport in groundwater hydrology. However, the literature on hydrodynamic dispersion in multiphase flows is frustratingly sparse.

One likely reason for this sparseness is the difficulty of understanding the physics of hydrodynamic dispersion even in single-fluid flows. Dispersion in porous media actually comprises a set of phenomena, including the following (71): (i) molecular diffusion, which to macroscopic observers appears retarded owing to the tortuosity of the solid matrix; (ii) Taylor diffusion (146), an effect whereby no-slip boundary conditions at the solid walls cause solutes in small-diameter pore channels to spread with respect to their mean motion; (iii) stream splitting, in which parcels of solute-bearing fluid divide at pore-channel intersections, and (iv) transit-time deviations, in which the dissimilar tortuosities of adjacent flow paths cause nearby fluid parcels to have different net velocities in the mean flow direction. Notice that the descriptions of these phenomena belong to the microscopic level of observation, and hence the use of hydrodynamic dispersion to account for their macroscopic effects imposes an inherent loss of information. To modelers, this smearing of small-scale heterogeneities has undesirable implications. Indeed, in models of solute transport in porous media, hydrodynamic dispersion is often the most poorly quantified of all physical parameters fed into the simulator.

Relatively few investigators have ventured to propose quantitative forms for hydrodynamic dispersion in the multiphase setting. Among the earliest laboratory studies of multiphase hydrodynamic dispersion is that of Thomas, Countryman, and Fatt (147). These authors find that, when two phases flow in a porous medium, each fluid alters the effective pore-size distribution available to the other fluid. Thus the degree of saturation of a given phase has pronounced effects on the observed level of dispersion. More recently, Delshad et al. (53) confirm the dependence of multiphase dispersion on saturations.

As Section 4.1 mentions, most mathematical models of species transport in multiphase systems ignore hydrodynamic dispersion. There are, however, at least three noteworthy exceptions. The first is the compositonal model developed by Young (164), who assumes the second-order tensor form

$$\underset{\approx}{D}_{\alpha} = (D_{\alpha,mol} + \alpha_{\alpha,t}|\underset{\sim}{v}^{\alpha}|)\underset{\approx}{1} + (\alpha_{\alpha,\ell} - \alpha_{\alpha,t})\frac{\underset{\sim}{v}^{\alpha}\underset{\sim}{v}^{\alpha}}{|\underset{\sim}{v}^{\alpha}|}$$

for each fluid phase α. Here $D_{\alpha,mol}$ stands for the molecular diffusion coefficient in phase α, and $\alpha_{\alpha,\ell}$ and $\alpha_{\alpha,t}$ signify the longitudinal and transverse dispersivities, respectively, in phase α. This formulation amounts to a natural extension of the standard hydrodynamic dispersion model to multiphase flows, A model described by Abriola (1) and Abriola and Pinder (2) assumes a related form for dispersion within a phase, namely

$$\underset{\approx}{D}_\alpha = \underset{\approx}{D}_{\alpha,mol} + \underset{\approx}{D}_1 : \underset{\sim}{v}^\alpha \underset{\sim}{v}^\alpha$$

where $\underset{\approx}{D}_{\alpha,mol}$ is a second-order tensor accounting for the effects of molecular diffusion in phase α, modified by the matrix tortuosity, and $\underset{\approx}{D}_1$ is a fourth-order tensor. This form extends the tensor equation proposed by Bear (23) on theoretical grounds. Finally, Baehr and Corapcioglu (20) and Corapcioglu and Baehr (49) derive a set of flow equations for immiscible contaminant transport incorporating a dispersion tensor for each phase; however, they do not postulate a precise tensorial form for dispersion.

Multiphase hydrodynamic dispersion appears to be one area of uncertainty where numerical simulation cannot shed much light. The fundamental questions that plague modelers are the same ones that arise in single-phase flows. What is the mathematical form of dispersion? How can we measure it? Do scale dependencies and asymmetric effects influence dispersion? It seems apparent that these questions address themselves primarily to experimentalists, guided ideally by theoretical studies of continuum mixtures such as that advanced by Bowen (31).

5.3. Multiphase Contaminant Flows

In recent years interest has arisen in multiphase flows involving immiscible groundwater contaminants. It has long been common practice to store or dispose of hazardous chemicals in near-surface or underground sites, and fluids escaping from these sites pose serious threats to groundwater supplies. Many hazardous chemicals and wastes take the form of nonaqueous-phase liquids, or NAPL. Common examples include gasoline, polychlorinated biphenyls (PCB), chlorinated hydrocarbons, coal tars, and creosotes (154). However, many dumpsites harbor a menagerie of chemical wastes, making it difficult to characterize the NAPL chemically. The multiphase flows that lead to contamination of groundwater are physically quite complex, and despite the pressing need for predictive tools, numerical simulation of NAPL flows remains in its infancy.

One type of flow that is important in this context is the simultaneous flow of NAPL and water in the unsaturated zone. This soil layer usually lies between near-surface NAPL sources and the water table and therefore acts as the main pathway for groundwater contamination. As we illustrated in Section 3.1, the flow of a single liquid in the unsaturated zone already poses a difficult nonlinear problem, so one might expect that multiliquid flows will be even harder to simulate numerically.

Current efforts in multiphase unsaturated flows focus mainly on developing physical understanding. Schwille (135), for example, discusses the migration of immiscible organics in the unsaturated zone, reviewing such fundamental processes as capillary action, volatilization of the organic species, and microbial degradation. Allen (9) applies continuum mixture theory to develop a set of flow equations for two liquids in the unsaturated zone. By analyzing a medium containing air (A), NAPL (N), and water (W), he derives a pair of partial differential equations, each resembling Richards' equation in form:

$$\left(C_\alpha + \frac{\Theta_\alpha S_{s,\alpha}}{\phi}\right) \frac{\partial \psi_\alpha}{\partial t} = \nabla . [k_{r\alpha} \underset{\approx}{K} . (\nabla \psi_\alpha + \nabla Z)] \tag{5.1}$$

for α = N or W. Several variables appearing in this equation are analogous to those appearing in the single-liquid case: C_α is the specific moisture capacity of phase α; Θ_α is the moisture content of α; ψ_α is the pressure head in phase α ; $\underset{\approx}{K}$ is the soil's hydraulic conductivity, and Z is depth below some datum. Also appearing are the variables $S_{s,\alpha}$ which is the specific storage associated with phase α, and $k_{r\alpha}$, signifying the relative permeability of the soil matrix to phase α. The pair of flow equations given by Eq. (5.1) constitutes a nonlinear system. Coupling between the equations occurs through the dependence of Θ_α, $S_{s,\alpha}$, and $k_{r\alpha}$ on the pressure heads ψ_α; the capillarity relationships $\psi_\alpha = \psi_\alpha(\Theta_N, \Theta_W)$, and the restriction $\Theta_N + \Theta_W = \phi(1 - S_A)$.

In what appears to be the first effort at numerically simulating multiphase unsaturated flows, Faust (69) develops a two-dimensional finite-difference model for the flow of water and NAPL. This model uses a two-equation formulation similar to that given by Eq. (5.1). To solve the discretized flow equations, Faust devises a fully implicit scheme akin to the SS method used in black-oil simulation. As with other models of multiphase flows, Faust's simulator uses upstream-weighted relative permeabilities to accommodate possible hyperbolic behavior, as explained in Section 4.2.

As a practical matter, the simultaneous flow of NAPL and water is only part of the multiphase contamination problem. Groundwater contamination itself occurs because of mass transfer between NAPL and water. Even though NAPL may be immiscible with water, some of its constituent species may dissolve in water at very small concentrations. While highly dilute, the resulting solution of organics in water is often toxic or carcinogenic. Therefore, a complete mathematical description of multiphase contaminant flows ought to incorporate phase-exchange effects more familiar in the setting of compositional reservoir simulators.

Very little work has been done in this area. Baehr and Corapcioglu (20) propose a model consisting of individual flow equations for each species. Since their model aims principally at predicting pollution from gasoline spills, they include in their formulation such effects as microbial degradation, equilibrium partitioning among fluid phase, and adsorption onto the solid phase. Abriola (1) and Abriola and Pinder (2,3) present a finite-difference model of species transport in an air-water-NAPL system. This simulator accommodates interphase mass transfer through the use of equilibrium ratios analogous to those discussed in Section 4.3. The model solves the nonlinear algebraic equations resulting from the finite-difference approximation using a scheme patterned after the SS method reviewed in Section 4.2. Considering the range of problems solved and the analyses given of the code's performance, this is perhaps the best-documented model of multiphase, multispecies contaminant transport appearing in the literature at this writing.

6. OUTSTANDING PROBLEMS:NUMERICS

Quite a few of the difficulties arising in numerical simulation of multiphase flows concern the limitations of the numerical methods themselves. Here the problem is that the numerical techniques in common use produce approximations that are in some way unrealistic based on our understanding of the flows that they model. In this case the challenge to researchers is to devise new methods or to modify existing approaches to permit more accurate simulations. We shall examine three types of numerical difficulties of topical interest: grid orientation effects, front tracking, and local grid refinement.

6.1. Grid-Orientation Effects

Since the early 1970's, petroleum engineers have recognized that many discrete methods for solving fluid flow equations give qualitatively different results when one changes the orientation of the spatial grid with respect to the geometry of the physical flow. Todd, O'Dell, and Hirasaki (149) first reported this phenomenon in a simulator of immiscible flow. They noted that the effects of grid orientation are especially pronounced at large mobility ratios. A severe example occurs in steamflood simulation (44), where solutions generated using different grid orientations apparently converge to different answers. Since these investigations, a substantial body of research has developed in the effort to overcome or mitigate grid-orientation effects in reservoir simulators.

One of the first effective techniques for reducing grid-orientation effects appeared in 1979, when Yanosik and McCracken (162) presented a nine-point finite-difference scheme that reduces grid-orientation effects for square grids. The nine-point scheme approximates derivatives at a point (x_i, y_i) in two-dimensional domains by using values at all adjacent nodes instead of the corner nodes only. Thus the nine-point analog of the Laplacian on a uniform grid is

$$\nabla^2 u\Big|_{i,j} = \frac{1}{6\Delta x^2}\,[4(u_{i+1,j} + u_{i-1,j} + u_{i,j+1} + u_{i,j-1}) + (u_{i+1,j-1} + u_{i-1,j-1} + u_{i-1,j+1} + u_{i+1,j-1}) - 20u_{i,j}]$$

Coats and Ramesh (46) observe that the nine-point formulation exhibits poor behavior when used on nonuniform spatial grids. Bertiger and Padmanabhan (25) explain this poor performance by demonstrating that the usual nine-point formulation on nonuniform grids yields an inconsistent approximation to ∇^2. These authors then propose a modified nine-point scheme that restores consistency while reducing the grid-orientation effect. In another approach, Potempa (123) advances a finite-element technique that is closely related to the Yanosik-McCracken nine-point difference scheme but again preserves consistency. Several other investigators have devised modified finite-difference schemes yielding solutions that are largely independent of grid-orientation effects; among them are Vinsome and Au (155); Frauenthal, Di Franco, and Towler (70); Shubin and Bell (142), and Preuss and Bodvarsson (125).

Finite-element techniques also admit variants that reduce grid-orientation effects. Among the more promising groups of finite-element schemes in this regard are mixed methods (50,67,10). The motivation behind these techniques is to compute accurate Darcy velocities explicitly rather than incurring the loss of accuracy associated with standard schemes requiring the differentiation of fluid pressures. Thus, for example, we factor the second-order pressure equation

$$\nabla.(k\nabla p) = 0$$

into two first-order equations

$$\underset{\sim}{v} = -k\nabla p$$

$$-\nabla.\underset{\sim}{v} = 0$$

By properly choosing the trial functions for $\underset{\sim}{v}$ and p, we can compute pressures and velocities having the same order of accuracy. In problems involving the effects of species transport the mixed method is especially effective when used in conjunction with time-stepping procedures based on modified methods of characteristics (67). A variety of numerical experiments reported in the references cited above demonstrate the method's ability to give good numerical results even in problems with highly variable material properties.

6.2. Front-Tracking Methods

As we have seen in previous sections, several multiphase flows in porous media exhibit sharp fronts that can be modeled as discontinuous fluid interfaces. The saltwater toe and the Buckley-Leverett saturation shock are two examples of such discontinuities. Discrete approximations using fixed finite elements or finite-difference cells have difficulty in capturing the behavior of these sharp fronts, since the computational procedures tend to smear information over the spatial subregions of the discretizations. Front-tracking methods aim at circumventing this difficulty by assigning computational degrees of freedom to the unknown location of the front. Solving for the frontal locations along with the variables characterizing the smooth parts of the flow allows the modeler to track the front explicitly without introducing numerical diffusion. Since one can concentrate many degrees of freedom at the interface, front tracking methods also hold great promise in the simulation of viscous fingering.

Front-tracking methods have their roots in numerical applications of the method of characteristics in convection-dominated flows. The first applications of this approach in porous-media simulation addressed the miscible transport of solutes in single-phase flows (73, 120). In the method of characteristics, one replaces a partial differential equation by a system of ordinary differential equations valid along curves where the original equation agrees with the chain rule. For example, by comparing the Buckley-Leverett saturation equation

$$\frac{\partial S_W}{\partial t} + \frac{q f_W'(S_W)}{\phi} = 0 \qquad (6.1)$$

with the chain rule

$$\frac{dt}{d\xi}\frac{\partial S_W}{\partial t} + \frac{dx}{d\xi}\frac{\partial S_W}{\partial x} = \frac{dS_W}{d\xi}$$

one can see that $dS_W/d\xi = 0$ along curves $\xi(x,t)$ in the (x,t)-plane where $dx/dt = q f_W'(S_W)/\phi$. Loci of constant S_W therefore travel with speed $q f_W'(S_W)/\phi$. This fact allows us to compute the position of the constant-saturation shock as it moves across a one-dimensional domain.

Perhaps the most extensively applied front-tracking scheme in the current literature is that of Glimm and his coworkers (77,79,93).

This approach uses an IMPES formulation for two-dimensional immiscible displacements in the absence of capillarity. The scheme solves the pressure equation on a finite-element grid whose element boundaries move to align themselves with the saturation shock. To update saturations, the scheme uses standard interior methods in regions where the saturation is smooth and couples to the smooth solution a Riemann problem propagating the interface. This frontal propagation relies on a method of characteristics akin to the one-dimensional version outlined above, taking advantage of a local coordinate system aligned with the shock to advance the discontinuity in its normal direction. Thus the actual computations required to track the front reduce to locally one-dimensional ordinary differential equations.

Jensen and Finlayson (85,86) introduce an alternative scheme for front-tracking that gives good results in convection-dominated species-transport problems. This method defines a set of moving coordinates based on the method of characteristics for the hyperbolic, or purely convective, part of the partial differential equations. Within this moving coordinate system, the convection-dominated transport problem reduces to a problem of the diffusion type. Jensen and Finlayson construct a finite-element grid attached to the moving coordinates, ensuring that the grid in the vicinity of the sharp front is sufficiently fine to avoid the occurrence of nonphysical oscillations in the numerical solution.

In a third approach to front tracking, the Mathematics Group at the Lawrence Berkeley Laboratory applies the theory of Riemann problems for first order hyperbolic systems to solve the immiscible flow equations using the random choice method (47,48,4). The random choice method, developed as a numerical technique by Chorin (40), is an effective procedure for approximating nonlinear hyperbolic conservation laws such as Eq. (6.1). The method replaces the unknown function $S_W(x,t)$ by a piecewise constant approximation $\hat{S}_W(x,t)$ and then solves a sequence of Riemann problems, each advancing the numerical solution by sampling the piecewise constant function $\hat{S}$ to determine initial data. When the solution possesses shocks, the random choice method preserves their sharp fronts, since the sampling at each time step avoids the introduction of spurious intermediate values in the numerical solution. However, the method allows small errors in the shock location since the sampling identifies the frontal position only to within the resolution limits imposed by the spatial grid. Although developed for one-dimensional flows, the random choice method admits extensions to two-dimensional problems. Colella, Concus, and Sethian (47) describe the use of operator splitting techniques to decompose a two-dimensional equation into a sequence of one-dimensional equations.

6.3. Adaptive Local Grid Refinement

Many problems involving multiphase flows in porous media exhibit behavior whose structure is localized in small subregions of the spatial domain. We have already encountered such phenomena in the form of wetting fronts and saturation shocks. Similar localized behavior occurs near wellbores or in the moving concentration fronts found in convection-dominated species transport processes. To capture the essential physics of these features often requires a spatial grid capable of providing high resolution in their vicinity. Grid refinement is especially important in view of the common use of low-order upstream-weighted approximations in near-hyperbolic flows. As frequently applied, these approximations introduce a numerical diffusion error whose magniture is $O(\Delta x)$ for grids of mesh Δx. By refining the spatial grid in the vicinity of the front, one reduces numerical diffusion by shrinking Δx, all the while preserving the desirable effects of upstream weighting.

Generating this extra resolution usually poses few difficulties if the locus of highly structured behavior remains constant in time. However, in the case of moving fronts, for example, the zones where increased resolution is needed move through the spatial domain as time progresses. Under these circumstances the refined portion of the grid must be capable of moving in time to follow the localized structure of the solution. Such schemes fall under the rubric of *adaptive local grid refinement* (ALGR). While ALGR schemes are generally difficult to implement, the technical literature in this area is vast. Therefore the review that follows merely highlights results that appear relevant in multiphase flow simulation.

There are three basic approaches to ALGR. One of these is to increase the polynomial degree of the approximation to the solution in regions needing refinement. Such techniques are called p-methods. Another approach is to add computational degrees of freedom in the regions of refinement, keeping the polynomial degree of the approximation constant. These techniques are perhaps most appropriate when used in conjunction with upstream weighting, since they reduce numerical diffusion by shrinking Δx. Such methods are called *h-methods*. Finally, there are several techniques that allow the location of the spatial nodes in the grid to act as variables in the numerical approximation. By solving for the nodal locations and nodal solution values simultaneously, one effectively forces the grid to move in time to accommodate the structure of the solution. These methods are called *moving finite element (*MFE*)* techniques.

Reports of underground flow simulators using p-methods are not very numerous. Chase (37) describes a chemical flood simulator based on a finite-element Galerkin method that employs hybrid trial functions. These trial functions use C^0 piecewise bilinear Lagrange functions in smooth regions of the flow but insert C^1 piecewise bicubic Hermite functions in the vicinity of steep gradients. Mohsen (100) describes another p-method applied in finite-element collocation solutions of the Buckley-Leverett equation. This approach refines a coarse grid consisting of C^1 piecewise cubic Hermite functions by substituting C^1 piecewise quintic functions near the saturation shock.

The use of h-methods has been more popular. One reason for this fact may be a general aversion to the oscillatory tendencies associated with polynomial approximations of high degree. Another reason is undoubtedly that h-methods fit more naturally into the framework of finite-difference approximations, which do not explicitly use trial functions. Quite a few ALGR schemes for finite differences have appeared; among them are the methods of von Rosenberg (157), Heinemann and van Handelmann (81), and Douglas et al. (57), who present both finite-difference and finite-element schemes. A considerable amount of theoretical work and numerical experimentation has focused on finite-element schemes with ALGR (19,52,24,54). One of the problems that arises in the construction of adaptive refinement codes is the management of the data defining the grid as its structure changes. There are great computational advantages associated with the invention of data structures that can accommodate the dynamic refinement and unrefinement of a grid without destroying the efficiency of matrix solution algorithms (21).

MFE methods adopt a somewhat different approach (35,98,74,55). For an equation of the form $\partial u/\partial t - \mathcal{A}u = 0$, where $\mathcal{A}$ is a spatial differential operator, we begin with a piecewise polynomial trial function u having unknown time-dependent coefficients $u_1(t),\ldots,u_N(t)$. In addition, we allow the coordinates of the spatial nodes $\bar{x}_1,\ldots,\bar{x}_N$ to be variable. By choosing $\{\partial u_i/\partial t, \bar{x}_i\}_{i=1}^N$ to minimize $\|\partial\hat{u}/\partial t - \mathcal{A}\hat{u}\|_2$ in a Galerkin sense, one can devleop a finite-element approximation in which the nodes tend to concentrate around regions where the solution exhibits localized structure. To prevent all of the nodes from accumulating near shocks, however, one must impose certain penalties on the clustering of nodes. A variety of internodal spring functions and viscosity-like devices exist to help preserve good global approximations by maintaining adequate separation between nodes.

ALGR techniques have a wide range of potential applications in general computational mechanics. Fluid flows in particular exhibit highly localized behaviors for which local refinement is an attractive alternative to globally fine grids. Gasdynamic shocks, hydraulic jumps, moving interfaces, and such singularities as sources, sinks, and corners are just a few examples of these features.

7. CONCLUSIONS

Throughout this review we have seen several facets of multiphase flows in porous media reappear in various applications. These physical and computational peculiarities emerge as major themes in the numerical simulation of flows. Let us close by recapitulating these themes.

Every flow we have examined obeys a nonlinear, time-dependent partial differential equation. Nonlinearity is a characteristic feature of multiphase porous-media flows, owing to the fact that the permeability of the rock matrix to one fluid varies with the saturation of any other fluid. Further nonlinearities can arise when storage or compressibility effects imply pronounced dependence on pressure in the accumulation terms or when there is strong coupling within a system of flow equations. The nonlinear governing differential equations generally give rise to nonlinear algebraic equations in the approximating discretizations. These algebraic systems, in turn, demand iterative solution, and therefore one commonly finds Newton-Raphson schemes or related procedures imbedded in implicit time-stepping methods for these problems.

Another common feature in multiphase porous-media flows is the occurrence of sharp fronts or moving boundaries in the fluid system. The Buckley-Leverett saturation shock stands as a classic example. Similar interfaces arise in other contexts: unsaturated flows can give rise to wetting fronts, and the saltwater intrusion problem exhibits a moving boundary in the toe of the salt-water wedge. Sharp fronts pose difficulties to the numerical analyst, since they require high spatial resolution to model and are sometimes associated with uniqueness issues. In certain classes of flows they can also exhibit instability, as when viscous fingering occurs in displacements at adverse mobility ratios. The most natural solutions to these sharp-front difficulties are front-tracking methods and adaptive local grid refinement.

Finally, various numerical aspects of modeling multiphase flows combine to require truly large-scale computations. A typical simulator solves large, sparse matrix equations at every iteration of every time step. When compositional effects are present, the code must solve nonlinear thermodynamic constraints as well. The desirability of local grid enrichment, front-tracking algorithms, or moving grid schemes adds to this scale of calculation both in complexity and in computational effort. Scientists who model multiphase underground flows have every reason to applaud the emerging generation of supercomputers and parallel architectures, since these machines may spell the difference between compromise in the approximation of complex flows and the practical achievement of realistic simulations.

8. ACKNOWLEDGMENTS

The Wyoming Water Research Center supported much of this work through a grant to study multiphase contaminant flows. The National Science Foundation provided support through grant number CEE-8404266.

9. REFERENCES

1. Abriola, L.M., *Multiphase Migration of Organic Compounds in a Porous Medium*, Springer-Verlag, Berlin, 1984.
2. Abriola, L.M., and Pinder, G.F., "A Multiphase Approach to the Modeling of Porous Media Contamination by Organic Compounds 1. Equation Development," *Water Resources Research*, Vol. 21, 1985, pp. 11-18.
3. Abriola, L.M., and Pinder, G.F., "A Multiphase Approach to the Modeling of Porous Media Contamination by Organic Compounds 2. Numerical Simulation," *Water Resources Research*, Vol. 21, 1985, pp. 19-26.
4. Albright, N., and Concus, P., "On Calculating Flows with Sharp Fronts in a Porous Medium," *Fluid Mechanics in Energy Conservation*, J.D. Buckmaster, ed., Society for Industrial and Applied Mathematics, Philadelphia, Pa., 1980, pp. 172-184.
5. Allen, M.B., "How Upstream Collocation Works," *International Journal for Numerical Methods in Engineering*, Vol. 19, 1983, pp. 1753-1763.
6. Allen, M.B., "A Collocation Model of Compositional Oil Reservoir Flows," *Mathematical Methods in Energy Research*, K.I. Gross, ed., Society for Industrial and Applied Mathematics, Philadelphia, Pa., 1984, pp. 157-180.
7. Allen, M.B., *Collocation Techniques for Compositional Flows in Oil Reservoirs*, Springer-Verlag, Berlin, 1984.
8. Allen, M.B., "Why Upwinding is Reasonable," *Finite Elements in Water Resources*, J.P. Laible et al., eds., Springer-Verlag, Berlin, 1984, pp. 13-23.
9. Allen, M.B., "Mechanics of Multiphase Fluid Flows in Variably Saturated Porous Media," *International Journal for Engineering Science*, 1985, to appear.
10. Allen, M.B., Ewing, R.E., and Koebbe, J.V., "Mixed Finite-Element Methods for Computing Groundwater Velocities," *Numerical Methods for Partial Differential Equations*, 1985, to appear.
11. Allen, M.B., and Murphy, C., "A Finite-Element Collocation Method for Variably Saturated Flows in Porous Media," *Numerical Methods for Partial Differential Equations*, to appear.
12. Allen, M.B., and Pinder, G.F., "Collocation Simulation of Multiphase Porous-Medium Flow," *Society of Petroleum Engineers Journal*, 1983, pp. 135-142.

13. Allen, M.B., and Pinder, G.F., "The Convergence of Upstream Collocation in the Buckley-Leverett Problem," *Society of Petroleum Engineers Journal*, 1985, to appear.
14. Amaefule, J.O., and Handy, L.L., "The Effect of Interfacial Tensions on Relative Oil/Water Permeabilities of Consolidated Porous Media," *Society of Petroleum Engineers Journal*, 1982, pp. 371-381.
15. Appleyard, J.R., Cheshire, I.M., and Pollard, R.K., "Special Techniques for Fully-Implicit Simulators," *Third European Symposium on Enhanced Oil Recovery*, F.J. Fayers, ed., Elsevier Scientific Publishing Co., Amsterdam, 1981, pp. 395-408.
16. Aronson, D.G., "Nonlinear Diffusion Problems," *Free Boundary Problems:Theory and Applications*, A. Fasano and M. Primicerio, eds., Vol.1, Pitman Advanced Publishing Program, Boston, Mass., 1983, pp. 135-149.
17. Atkin, R.J., and Craine, R.E., "Continuum Theories of Mixtures: Basic Theory and Historical Development," *Quarterly Journal of Applied Mathematics*, Vol. 29, No.2, 1976, pp. 209-244.
18. Aziz, K., and Settari, A., *Petroleum Reservoir Simulation*, Applied Science Publishers, London, 1979.
19. Babuska, I., and Rheinboldt, W.C., "Error Estimates for Adaptive Finite Element Computations," *Society for Industrial and Applied Mathematics Journal of Numerical Analysis*, Vol. 15, 1978, pp. 736-754.
20. Baehr, A., and Corapcioglu, M.Y., "A Predictive Model for Pollution from Gasoline in Soils and Groundwater," *Proceedings*, Petroleum Hydrocarbons and Organic Chemicals in Groundwater: Prevention,Detection, and Restoration, Houston, Tex., 1984,p.144.
21. Bank, R.E., and Sherman, A.H., "A Refinement Algorithm and Dynamic Data Structure for Finite Element Meshes," *Center for Numerical Analysis Technical Report 166*, University of Texas, Austin, Tex., 1980.
22. Bardon, C., and Longeron, D.G., "Influence of Interfacial Tensions on Relative Permeability," *Society of Petroleum Engineers Journal*, 1980, pp. 391-401.
23. Bear, J., "On the Tensor Form of Dispersion in Porous Media," *Journal of Geophysical Research*, Vol. 66, 1961, pp. 1185-1197.
24. Beiterman, M., "A Posteriori Error Estimation and Adaptive Finite Element Grids for Parabolic Equations," *Adaptive Computational Methods for Partial Differential Equations*, I. Babuska et al., eds., Society for Industrial and Applied Mathematics, Philadelphia, Pa., 1983, pp. 123-141.
25. Bertiger, W.I., and Padmanabhan, L., "Finite Difference Solutions to Grid Orientation Problems Using IMPES," *Proceedings*, Seventh Symposium on Reservoir Simulation, Society of Petroleum Engineers, 1983, pp. 163-172.
26. Birkhoff, G., "Numerical Fluid Dynamics," *Society for Industrial and Applied Mathematics Review*, Vol. 25, 1983, pp. 1-34.

27. Bjordammen, J.B., and Coats, K.H., "Comparison of Alternating Direction and Successive Overrelaxation Techniques in Simulation of Reservoir Fluid Flow," *Society of Petroleum Engineers Journal*, 1969, pp. 47-58.
28. Blair, P.M., and Weinaug, C.F., "Solution of Two-Phase Flow Problems Using Implicit Finite Differences," *Society of Petroleum Engineers Journal*, 1969, pp. 417-424.
29. Bowen, R.M., "Incompressible Porous Media Models by the Use of the Theory of Mixtures," *International Journal of Engineering Science*, Vol.18, 1980, pp. 1129-1148.
30. Bowen, R.M., "Compressible Porous Media Models by the Use of the Theory of Mixtures," *International Journal of Engineering Science*, Vol. 20, 1982, pp. 697-735.
31. Bowen, R.M., "Diffusion Models Implied by the Theory of Mixtures," Appendix 5A to *Rational Thermodynamics*, 2nd ed., by C.A. Truesdell, Springer-Verlag, New York, N.Y., 1984, pp. 237-263.
32. Breitenbach, E.A., Thurnau, D.H., and van Poolen, H.K., "Solution of the Immiscible Fluid Flow Simulation Equation," *Society of Petroleum Engineers Journal*, 1969, pp. 155-169.
33. Bresler, E., "Simultaneous Transport of Solutes and Water Under Transient Unsaturated Flow Conditions," *Water Resources Research*, Vol. 9, 1973, pp. 975-986.
34. Buckley, S.E., and Leverett, M.C., "Mechanism of Fluid Displacement in Sands," *Transactions*, American Institute of Mining, Metallurgical, and Petroleum Engineers, Vol. 146, 1942, pp. 107-116.
35. Carey, G.F., Mueller, A., Sephehrnoori, K., and Thrasher, R.L., "Moving Elements for Reservoir Transport Processes," *Proceedings*, Fifth Symposium on Reservoir Simulation, Society of Petroleum Engineers, 1979, pp. 135-144.
36. Celia, M.A., and Pinder, G.F., "An Analysis of Alternating-Direction Methods for Parabolic Equations," *Numerical Methods for Partial Differential Equations*, Vol. 1, 1985, pp. 57-70.
37. Chase, C.A., "Variational Simulation with Numerical Decoupling and Local Mesh Refinement," *Proceedings*, Fifth Symposium on Reservoir Simulation, Society of Petroleum Engineers, 1979, pp. 73-94.
38. Chien, M.C.H., Lee, S.T., and Chen, W.H., "A New Fully Implicit Compositional Simulator," *Proceedings*, Eight Symposium on Reservoir Simulation, Society of Petroleum Engineers, Feb., 1985, pp. 7-30.
39. Chorin, A.J., and Marsden, J.E., *A Mathematical Introduction to Fluid Mechanics*, Springer-Verlag, New York, N.Y., 1979, Chapter 3.
40. Chorin, A.J., "Random Choice Solution of Hyperbolic Systems," *Journal of Computational Physics*, Vol. 22, 1976, pp. 517-533.
41. Chuoke, R.L., van Meurs, P., and van der Poel, C., "The Instability of Slow, Immiscible, Liquid-Liquid Displacements in Permeable Media," *Society of Petroleum Engineers Journal*, 1959,

pp. 188-194.
42. Coats, K., "An Equation-of-State Compositional Model," *Society of Petroleum Engineers Journal*, 1980, pp. 363-376.
43. Coats, K., "Reservoir Simulation:State of the Art," *Journal of Petroleum Technology*, 1982, pp. 1633-1642.
44. Coats, K.H., George, W.D., Chu, C., and Marcum, B.E., "Three Dimensional Simulation of Steamflooding," presented at the Sept. 30-Oct. 3, 1973, Annual Fall Technical Conference and Exhibition of the Society of Petroleum Engineers.
45. Coats, K.H., Nielsen, R.L., Terhune, M.H., and Weber, A.G., "Simulation of Three-Dimensional, Two-Phase Flow in Oil and Gas Reservoirs," *Society of Petroleum Engineers Journal*, 1967, pp. 377-388.
46. Coats, K.H., and Ramesh, A.B., "Effect of Grid Type and Difference Scheme on Pattern Steamflood Results," presented at the Sept. 26-29, 1982, Fall Technical Conference and Exhibition of the Society of Petroleum Engineers.
47. Colella, P., Concus, P., and Sethian, J., "Some Numerical Methods for Discontinuous Flows in Porous Media," *The Mathematics of Reservoir Simulation*, R.E. Ewing, ed., Society for Industrial and Applied Mathematics, Philadelphia, Pa., 1983, pp. 161-186.
48. Concus, P., "Calculation of Shocks in Oil Reservoir Modeling and Porous Flow," *Numerical Methods for Fluid Dynamics*, K.W. Morton and M.J. Baines, eds., Academic Press, New York, N.Y., 1982, pp. 165-178.
49. Corapcioglu, M.Y., and Baehr, A., "Immiscible Contaminant Transport in Soils and Groundwater with an Emphasis on Petroleum Hydrocarbons: System of Differential Equations vs. Single Cell Model," *Water Sciences and Technology*, Vol. 17, No. 9, Sept., 1985, pp. 23-37.
50. Darlow, B.L., Ewing, R.E., and Wheeler, M.F., "Mixed Finite Element Methods for Miscible Displacement Problems in Porous Media," *Proceedings*, Sixth Symposium on Reservoir Simulation, Society of Petroleum Engineers, 1982, pp. 137-146.
51. Davis, H.T., and Scriven, L.E., "The Origins of Low Interfacial Tensions for Enhanced Oil Recovery," presented at the September 21-24, 1980, Annual Fall Technical Conference and Exhibition of the Society of Petroleum Engineers of AIME.
52. Davis, S., and Flaherty, J., "An Adaptive Finite Element Method for Initial-Boundary Value Problems for Partial Differential Equations," *Society for Industrial and Applied Mathematics Journal for Scientific and Statistical Computing*, Vol. 3, 1983, pp. 6 -27.
53. Delshad, M., MacAllister, D.J., Pope, G.A., and Rouse, B.A., "Multiphase Dispersion and Relative Permeability Experiments," presented at the Sept., 1982, Society of Petroleum Engineers Annual Fall Technical Conference and Exhibition, held in New Orleans, La.

54. Diaz, J.C., Ewing, R.E., Jones, R.W., McDonald, A.E., Uhler, L.M., and von Rosenberg, D.U., "Self Adaptive Local Grid Refinement in Enhanced Oil Recovery," *Proceedings*, Fifth International Symposium on Finite Elements and Flow Problems, Austin, Tex., Jan. 23-26, 1984, pp. 479-484.
55. Djomehri, M.J., and Miller, K., "A Moving Finite Element Code for General Systems of Partial Differential Equations in Two Dimensions," *University of California Center for Pure and Applied Mathematics Report PAM-57*, Berkeley, Calif., Oct., 1981.
56. Douglas J., Jr., "Alternating Direction Methods for Three Space Variables," *Numerische Mathematik*, Vol.2, 1962, pp. 91-98.
57. Douglas, J., Jr., Darlow, B.L., Wheeler, M.F., and Kendall, R.P., "Self-Adaptive Finite Element and Finite Difference Methods for One-Dimensional Two-Phase Immiscible Flow," *Computer Methods in Applied Mechanics and Engineering*, Vol.47, 1984, pp. 119-130.
58. Douglas, J., Jr., and Dupont, T., "A Finite-Element Collocation Method for Quasilinear Parabolic Equations," *Mathematics of Computation*, Vol. 27, pp. 17-28.
59. Douglas, J., Jr., Peaceman, D.W., and Rachford, H.H., Jr., "A Method for Calculating Multidimensional Immiscible Displacement," *Transactions*, American Institute of Mining, Metallurgical, and Petroleum Engineers, Vol. 216, 1959, pp. 297-306.
60. Douglas, J., Jr., and Rachford, H.H., Jr., "On the Numerical Solution of Heat Conduction Problems in Two and Three Space Variables," *Transactions*, American Mathematical Society, Vol. 82, 1956, pp. 421-439.
61. Eringen, A.C., and Ingram, J.D., "A Continuum Theory of Chemically Reacting Media-I," *International Journal of Engineering Science*, Vol. 3, 1965, pp. 197-212.
62. Ewing, R.E., "Alternating-Direction Galerkin Methods for Parabolic, Hyperbolic, and Sobolev Partial Differential Equations," *Proceedings of the Special Year in Numerical Analysis*, University of Maryland, College Park, Md., 1981, pp. 123-150.
63. Ewing, R.E., and George, J.H., "Viscous Fingering in Hydrocarbon Recovery Processes," *Mathematical Methods in Energy Research*, K.I. Gross, ed., Society for Industrial and Applied Mathematics, Philadelphia, Pa., 1984, pp. 194-213.
64. Ewing, R.E., and Heinemann, R.F., "Incorporation of Mixed Finite Element Methods in Compositional Simulation for Reduction of Numerical Dispersion," *Proceedings*, Seventh Symposium on Reservoir Simulation, Society of Petroleum Engineers, Nov., 1983, pp. 341-347.
65. Ewing, R.E., and Heinemann, R.F., "Mixed Finite Element Approximation of Phase Velocities in Compositional Reservoir Simulation," *Computer Methods in Applied Mechanics and Engineering*, Vol. 47, 1984, pp. 161-176.
66. Ewing, R.E., Russell, T.F., and Wheeler, M.F., "Simulation of Miscible Displacement Using Mixed Methods and a Modified Method of Characteristics," *Proceedings*, Seventh Symposium on Reservoir

Simulation, Society of Petroleum Engineers, 1983, pp. 71-82.
67. Ewing, R.E., Russell, T.F., and Wheeler, M.F., "Convergence Analysis of an Approximation of Miscible Displacement in Porous Media by Mixed Finite Elements and a Modified Method of Characteristics," *Computer Methods in Applied Mechanics and Engineering*, Vol. 47, 1984, pp. 73-92.
68. Fatt, I., and Dykstra, H., "Relative Permeability Studies," *Transactions*, American Institute of Mining, Metallurgical, and Petroleum Engineers, Vol. 192, 1951, pp. 249-256.
69. Faust, C.R., "Transport of Immiscible Fluids Within and Below the Unsaturated Zone: A Numerical Model," *Water Resources Research*, Vol. 21, 1985, pp. 587-596.
70. Frauenthal, J.C., di Franco, R.B., and Towler, B.F., "Reduction of Grid Orientation Effects in Reservoir Simulation Using Generalized Upstream Weighting," *Proceedings*, Seventh Symposium on Reservoir Simulation, Society of Petroleum Engineers, 1983, pp. 7-16.
71. Fried, J.J., *Groundwater Pollution*, Elsevier Scientific Publishing Co., Amsterdam, 1975.
72. Fussell, L.T., and Fussell, D.D., "An Iterative Technique for Compositional Reservoir Models," *Society of Petroleum Engineers Journal*, 1979, pp. 211-220.
73. Garder, A.O., Peaceman, D.W., and Pozzi, A.L., "Numerical Calculation of Multidimensional Miscible Displacement by the Method of Characteristics," *Society of Petroleum Engineers Journal*, 1964, pp. 26-36.
74. Gelinas, R.J., Doss, S.K., and Miller, K., "The Moving Finite Element Method: Applications to General Partial Differential Equations with Large Multiple Gradients," *Journal of Computational Physics*, Vol. 40, 1981, pp. 202-221.
75. Gibbs, J.W., "On the Equilibrium of Heterogeneous Substances," *Transactions*, Connecticut Academy, Vol. 3, 1876, pp. 108-248, and 1878, pp. 343-524.
76. Glimm, J., Isaacson, E.L., Lindquist, B., McBryan, O., and Yaniv, S., "Statistical Fluid Dynamics:The Influence of Geometry on Surface Instabilities," *The Mathematics of Reservoir Simulation*, R.E. Ewing, ed., Society for Industrial and Applied Mathematics, Philadelphia, Pa., 1983, pp. 137-160.
77. Glimm, J., Lindquist, B., McBryan, O., and Padmanabhan, L., "A Front Tracking Reservoir Simulator, Five-Spot Validation Studies, and the Water Coning Problem," *The Mathematics of Reservoir Simulation*, R.E. Ewing, ed., Society for Industrial and Applied Mathematics, Philadelphia, Pa., 1983, pp. 107-135.
78. Glimm, J., Marchesin, D., and McBryan, O., "Unstable Fingers in Two-Phase Flow," *Communications in Pure and Applied Mathematics*, Vol. 34, 1981, pp. 53-75.
79. Glimm, J., Marchesin, D., and McBryan, O., "A Numerical Method for Two Phase Flows with an Unstable Interface," *Journal of Computational Physics*, Vol. 39, 1981, pp. 91-119.

80. Heinemann, R.F., "Simulation of Compositional Oil Recovery Processes," *Mathematical Methods in Energy Research,* K.I. Gross, ed., Society for Industrial and Applied Mathematics, Philadelphia, Pa., 1984, pp. 214-235.
81. Heinemann, Z.E., and von Hantelmann, G., "Using Local Grid Refinement in a Multiple Application Reservoir Simulator," *Proceedings,* Seventh Symposium on Reservoir Simulation, Society of Petroleum Engineers, 1983, pp. 205-218.
82. Hoa, N.T., Gaudu, R., and Thirriot, C., "Influence of the Hysteresis Effect on Transient Flows in Saturated-Unsaturated Porous Media," *Water Resources Research,* Vol. 13, 1977, pp. 992-996.
83. Huyakorn, P.S., and Pinder, G.F., "A New Finite Element Technique for the Solution of Two-Phase Flow Through Porous Media," *Advances in Water Resources,* Vol. 1, 1978, pp. 285-298.
84. Huyakorn, P.S., and Pinder, G.F., *Computational Methods in Subsurface Flow,* Academic Press, New York, N.Y., 1983.
85. Jensen, O.K., and Finlayson, B.A., "A Solution of the Transport Equation Using a Moving Coordinate System," *Advances in Water Resources,* Vol. 3, 1980, pp. 9-18.
86. Jensen, O.K., and Finlayson, B.A., "A Numerical Technique for Tracking Sharp Fronts in Studies of Tertiary Oil Recovery Pilots," *Proceedings,* Seventh Symposium on Reservoir Simulation, Society of Petroleum Engineers, 1983, pp. 61-70.
87. Jerrauld, G.R., Davis, H.T., and Scriven, L.E., "Frontal Structure and Stability in Immiscible Displacement," *Proceedings,* Fourth Symposium on Enhanced Oil Recovery, Society of Petroleum Engineers and U.S. Department of Energy, Vol. 2, 1984, pp. 135-144.
88. Kazemi, H., Vestal, C.R., and Shank, G.D., "An Efficient Multicomponent Numerical Simulator," *Society of Petroleum Engineers Journal,* 1978, pp. 355-368.
89. Koval, E.J., "A Method for Predicting the Performance of Unsteady Miscible Displacement," *Society of Petroleum Engineers,* 1963, pp. 145-154.
90. Lax, P.D., "Hyperbolic Systems of Conservation Laws II," *Communications in Pure and Applied Mathematics,* Vol. 10, 1957, pp. 537-566.
91. Leverett, M.C., "Capillary Behavior in Porous Solids," *Transactions,* American Institute of Mining, Metallurgical, and Petroleum Engineers, Vol. 142, 1941, pp. 159-169.
92. Leverett, M.C., and Lewis, W.B., "Steady Flow of Gas-Oil-Water Mixtures Through Unconsolidated Sands," *Transactions,* American Institute of Mining, Metallurgical, and Petroleum Engineers, Vol. 142, 1941, pp. 107-120.
93. McBryan, O., "Shock Tracking Methods in 2-D Flows," *Ninth National Applied Mechanics Congress,* Cornell University, Ithaca, N.Y., 1982.

94. McLean, J.W., "The Fingering Problem in Flow Through Porous Media," thesis presented to the California Institute of Technology, at Pasadena, Cal., in 1980 in partial fulfillment of the requirements for the degree of Doctor of Philosophy.
95. Meijerink, J.A., and Vander Vorst, H.A., "An Iterative Solution Method for Linear Systems of Which the Coefficient Matrix is a Symmetric M-Matrix," *Mathematics of Computation*, Vol. 31, 1977, pp. 148-162.
96. Mercer, J.W., and Faust, C.R., "The Application of Finite Element Techniques to Immiscible Flow in Porous Media," *Finite Elements in Water Resources*, W.G. Gray et al., eds., London, Pentech Press, 1977, pp. 21-43.
97. Mercer, J.W., Larson, S.P., and Faust, C.R., "Simulation of Salt-Water Interface Motion," *Ground Water*, Vol. 18, 1980, pp. 374-385.
98. Miller, K., and Miller, R.N., "Continuously Deforming Finite Elements, I," *Society for Industrial and Applied Mathematics Journal of Numerical Analysis*, Vol. 18, 1981, pp. 1019-1032.
99. Milly, P.C.D., "A Mass-Conservative Procedure for Time-Stepping in Models of Unsaturated Flow," *Finite Elements in Water Resources*, J.P. Laible et al., eds., Springer-Verlag, London, 1984, pp. 103-112.
100. Mohsen, M.F.N., "Numerical Experiments Using 'Adaptive' Finite Elements with Collocation," *Finite Elements in Water Resources*, J.P. Laible et al., eds., Springer-Verlag, Berlin, 1984, pp. 45-61.
101. Molina, N.M., "A Systematic Approach to the Relative Permeability Problem in Reservoir Simulation," presented at the September 21-24, 1980, Annual Fall Technical Conference and Exhibition of the Society of Petroleum Engineers of AIME.
102. Morel-Seytoux, H.J., "Introduction to Flow of Immiscible Liquids in Porous Media," *Flow Through Porous Media*, R.J.M. De Weist, ed., Academic Press, New York, N.Y., 1969, pp. 456-516.
103. Morrow, N.R., "Physics and Thermodynamics of Capillary Action in Porous Media," *Symposium on Flow Through Porous Media*, R. Nunge, ed., American Chemical Society Publications, Washington, D.C., 1969, pp. 83-108.
104. Muskat, M., Wykoff, R.D., Botset, H.G., and Meres, M.W., "Flow of Gas-Liquid Mixtures Through Sands," *Transactions*, American Institute of Mining, Metallurgical, and Petroleum Engineers, Vol. 123, 1937, pp. 69-96.
105. Nakano, Y., "Application of Recent Results in Functional Analysis to the Problem of Wettting Fronts," *Water Resources Research*, Vol. 16, 1980, pp. 314-318.
106. Narasimhan, T.N., and Witherspoon, P.A., "Numerical Model for Saturated-Unsaturated Flow in Deformable Porous Media 3. Applications," *Water Resources Research*, Vol. 14, 1978, pp. 1017-1033.

107. Neumann, S.P., "Saturated-Unsaturated Seepage by Finite Elements," *Journal of the Hydraulics Division of ASCE*, Vol. 99, 1973, pp. 2233-2251.

108. Nghiem, L.X., Fong, D.K., and Aziz, K., "Compositional Modeling with an Equation of State," *Society of Petroleum Engineers Journal*, 1981, pp. 687-698.

109. Nolen, J.S., "Numerical Simulation of Compositional Phenomena in Petroleum Reservoirs," *Proceedings*, Third Symposium on Reservoir Simulation, Society of Petroleum Engineers, Jan., 1973, pp. 269-284.

110. Northrup, E.J., and Woo, P.T., "Application of Preconditioned Conjugate-Gradient-Like Methods in Reservoir Simulation," *Proceedings, Eight Symposium on Reservoir Simulation*, Society of Petroleum Engineers, Feb., 1985, pp. 375-386.

111. Oleinik, O.A., "Uniqueness and Stability of the Generalized Solution to the Cauchy Problem for a Quasi-Linear Equation," *American Mathematical Society Translations, Series 2*, Vol. 33, 1973, pp. 285-290.

112. Passman, S.L., Nunziato, J.W., and Walsh, E.K., "A Theory of Multiphase Mixtures," Appendix 5C to *Rational Thermodynamics*, 2nd ed., by C.A. Truesdell, Springer-Verlag, New York, N.Y., 1984, pp. 286-325.

113. Peaceman D.W., "Numerical Solution of the Nonlinear Equations for Two-Phase Flow Through Porous Media," *Nonlinear Partial Differential Equations: A Symposium on Methods of Solution*, W.F. Ames, ed., Academic Press, New York, N.Y., 1967, pp. 171-191.

114. Peaceman, D.W., *Fundamentals of Numerical Reservoir Simulation*, Elsevier Scientific Publishing Co., Amsterdam, 1977.

115. Peaceman, D.W., and Rachford, H.H., Jr., "The Numerical Solution of Parabolic and Elliptic Differential Equations," *Journal of the Society for Industrial and Applied Mathematics*, Vol. 3, 1955, pp. 28-41.

116. Peletier, L.A., "The Porous Medium Equation," *Applications of Nonlinear Analysis in the Physical Sciences*, H. Amann et al., eds., Pitman Advanced Publishing Program, Boston, Mass., 1981, pp. 229-240.

117. Peng, D.-Y., and Robinson, D.R., "A New Two-Constant Equation of State," *Industrial and Engineering Chemistry:Fundamentals*, Vol. 15, 1976, pp. 53-64.

118. Peters, E.J., and Flock, D.L., "The Onset of Instability During Two-Phase Immiscible Displacement in Porous Media," *Society of Petroleum Engineers Journal*, 1981, pp. 249-258.

119. Philip, J.R., "Numerical Solution of Equations of the Diffusion Type with Diffusivity Concentration Dependent II," *Australian Journal of Physics*, Vol. 10, 1957, pp. 29-40.

120. Pinder, G.F., and Cooper, H.H., Jr., "A Numerical Technique for Calculating the Transient Position of the Saltwater Front," *Water Resources Research*, Vol. 6, 1970, pp. 875-882.

121. Pinder, G.F., and Page, R.F., "Finite Element Simulation of Salt Water Intrusion on the South Fork of Long Island," *Finite Elements in Water Resources*, W.G. Gray et al., eds., Pentech Press, Plymouth, U.K., 1977, pp. 2.51-2.69.
122. Ponting, D.K., Foster, B.A., Naccache, P.F., Nichols, M.O., Pollard, R.K., Rae,J., Banks, D., and Walsh, S.K., "An Efficient Fully Implicit Simulator," *Society of Petroleum Engineers Journal*, 1983, pp. 544-552.
123. Potempa, T., "Finite Element Methods for Convection Dominated Transport Problems," thesis presented to Rice University, at Houston, Tex., in 1982 in partial fulfillment of the requirements for the degree of Doctor of Philosophy.
124. Prévost, J.H., "Mechanics of Continuous Porous Media," *International Journal of Engineering Science*, Vol. 18, pp. 787-800.
125. Preuss, K., and Bödvarsson, G.S., "A Seven-Point Finite Difference Method for Improved Grid Orientation in Pattern Steamfloods," *Proceedings*, Seventh Symposium on Reservoir Simulation, Society of Petroleum Engineers, 1983, pp. 175-184.
126. Raats, P.A.C., "Applications of the Theory of Mixtures in Soil Physics," Springer-Verlag, New York, N.Y., 1984, pp. 326-343.
127. Reeves, M., and Duguid, J.O., "Water Movement Through Saturated-Unsaturated Porous Media:A Finite-Element Galerkin Model," *Oak Ridge National Laboratories Report 4927*, Oak Ridge, Tenn., 1975.
128. Richards, L.A., "Capillary Conduction of Liquids in Porous Media," *Physics*, Vol. 1, 1931, pp. 318-333.
129. Risnes, R., Dalen, V., and Jensen, J.I., "Phase Equilibrium Calculations in the Near-Critical Region," *Proceedings*, Third European Symposium on Enhanced Oil Recovery, F.J. Fayers, ed., Elsevier Scientific Publishing Co., Amsterdam, 1981, pp. 329-350.
130. Roebuck, I.F., Henderson, G.E., Douglas, J., and Ford, W.T., "The Compositional Reservoir Simulator:Case I-The Linear Model," "*Society of Petroleum Engineers Journal*, 1969, pp. 115-130.
131. Sá da Costa, A.A.G., and Wilson, J.L., "Coastal Seawater Intrusion:A Moving Boundary Problem," *Finite Elements in Water Resources*, S.Y. Wang, et al., eds., University of Mississippi, Oxford, Miss., 1980, pp. 2.209-2.218.
132. Saffman, P.G., and Taylor, G.I., "The Penetration of a Fluid into a Porous Medium or Hele-Shaw Cell Containing a More Viscous Liquid," *Proceedings of the Royal Society of London A*, Vol. 245, 1958, pp. 312-329.
133. Scheidegger, A.E., *The Physics of Flow Through Porous Media*, 3rd ed., University of Toronto Press, Toronto, 1974.
134. Schiegg, H.O., "Experimental Contribution to the Dynamic Capillary Fringe," *Symposium on Hydrodynamic Diffusion and Dispersion in Porous Media*, Pavia-Italia, Rome, 1977.
135. Schwille, F., "Migration of Organic Fluids Immiscible with Water in the Unsaturated Zone," *Pollutants in Porous Media*, B. Yaron, G. Dagan, and J. Goldshmid, eds., Springer-Verlag, Berlin, 1984, pp. 27-49.

136. Segol, G., "A Three-Dimensional Galerkin-Finite Element Model for the Analysis of Contaminant Transport in Saturated-Unsaturated Porous Media," *Finite Elements in Water Resources*, W.G. Gray et al., eds., Pentech Press, London, 1977, pp. 2.123-2.144.
137. Settari, A., and Aziz, K., "Treatment of the Nonlinear Terms in the Numerical Solution of Partial Differential Equations for Multiphase Flow in Porous Media," *International Journal of Multiphase Flow*, Vol. 1, 1975, pp. 817-844.
138. Settari, A., Price, H.S., and Dupont, T., "Development and Application of Variational Methods for Simulation of Miscible Displacement in Porous Media," *Society of Petroleum Engineers Journal*, 1977, pp. 228-246.
139. Shamir, V., and Dagan, G., "Motion of the Seawater Interface in a Coastal Aquifer:A Numerical Solution," *Water Resources Research*, Vol. 7, 1971, pp. 644-657.
140. Shapiro, A.M., and Pinder, G.F., "Solution of Immiscible Displacement in Porous Media Using the Collocation Finite Element Method," *Finite Elements in Water Resources*, K.P. Holz et al., eds., Springer-Verlag, Berlin, 1982, pp. 9.61-9.70.
141. Sheldon, J.W., Zondek, B., and Cardwell, W. T., "One-Dimensional, Incompressible, Non-Capillary Two-Phase Flow in a Porous Medium," *Transactions*, American Institute of Mining, Metallurgical, and Petroleum Engineers, Vol. 216, 1959, pp. 290-296.
142. Shubin, G.R., and Bell, J.B., "An Analysis of the Grid Orientation Effect in Numerical Simulation of Miscible Displacement," *Computer Methods in Applied Mechanics and Engineering*, Vol. 47, 1984, pp. 47-72.
143. Stone, H.L., "Iterative Solution of Implicit Approximations of Multidimensional Partial Differential Equations," *Society for Industrial and Applied Mathematics Journal of Numerical Analysis*, Vol. 5, 1968, pp. 530-568.
144. Stone, H.L., "Estimation of Three-Phase Relative Permeability and Residual Oil Data," *Journal of Canadian Petroleum Technology*, Vol. 12, No. 4, 1973, pp. 53-61.
145. Stone, H.L., and Garder, A.O., Jr., "Analysis of Gas-Cap or Dissolved-Gas Reservoirs," *Transactions*, American Institute of Mining, Metallurgical, and Petroleum Engineers, Vol. 222, 1961, pp. 92-104.
146. Taylor, G.I., "Dispersion of Soluble Matter in Solvent Flowing Slowly Through a Tube," *Proceedings*, Royal Society of London, Series A, Vol. 219, 1953, pp. 186-203.
147. Thomas, G.H., Countryman, G.R., and Fatt, I., "Miscible Displacement in a Multiphase System," *Society of Petroleum Engineers Journal*, 1963, pp. 189-196.
148. Todd, M.R., and Longstaff, W.J., "The Development, Testing, and Application of a Numerical Simulator for Predicting Miscible Flood Performance," *Journal of Petroleum Technology*, 1972, pp. 874-882.

149. Todd, M.R., O'Dell, P.M., and Hirasaki, G.J., "Methods for Increased Accuracy in Numerical Reservoir Simulators," *Society of Petroleum Engineers Journal*, 1972, pp. 515-530.
150. Towler, B.F., and Killough, J.E., "Comparison of Preconditioners for the Conjugate Gradient Method in Reservoir Simulation," *Proceedings*, Sixth Symposium on Reservoir Simulation, Society of Petroleum Engineers, Feb., 1982, pp. 7-14.
151. van Genuchten, M. Th., "A Comparison of Numerical Solutions of the One-Dimensional Unsaturated-Saturated Flow and Mass Transport Equations," *Advances in Water Resources*, Vol. 5, 1982, pp. 47-55.
152. van Genuchten, M. Th., "An Hermitian Finite Element Solution of the Two-Dimensional Saturated-Unsaturated Flow Equation," *Advances in Water Resources*, Vol. 6, 1983, pp. 106-111.
153. Van Quy, N., Simandoux, P., and Corteville, J., "A Numerical Study of Diphasic Multicomponent Flow," *Society of Petroleum Engineers Journal*, 1972, pp. 171-184.
154. Villaume, J.F., "Investigations at Sites Contaminated with Dense, Non-Aqueous Phase Liquids (NAPLs)," *Ground Water Monitoring Review*, Spring, 1985, pp. 60-74.
155. Vinsome, P.K., and Au, A., "One Approach to the Grid Orientation Problem in Reservoir Simulation," presented at the Sept. 23-26, 1979, Annual Fall Technical Conference and Exhibition of the Society of Petroleum Engineers.
156. von Neumann, J., "First Report on the Numerical Calculation of Flow Problems," *Collected Works*, A.H. Taub, ed., Vol. 5, The MacMillan Co., New York, N.Y., 1963, pp. 664-712.
157. von Rosenberg, D.U., "Local Mesh Refinement for Finite Difference Methods," presented at the Sept. 26-29, 1982, Fall Technical Conference and Exhibition of the Society of Petroleum Engineers.
158. Watts, J.W., "A Compositional Formulation of the Pressure and Saturation Equations," *Proceedings*, Seventh Symposium on Reservoir Simulation, Society of Petroleum Engineers, Nov., 1983, pp. 113-138.
159. Welge, H.J., "A Simplified Method for Computing Oil Recovery by Gas or Water Drive," *Transactions*, American Institute of Mining, Metallurgical, and Petroleum Engineers, Vol. 195, 1952, pp. 91-98.
160. Wilson, D.C., Tan, T.C., and Casinader, P.C., "Control of Numerical Dispersion in Compositional Simulation," *Third European Symposium on Enhanced Oil Recovery*, F.J. Fayers, ed., Elsevier Scientific Publishing Co., Amsterdam, 1981, pp. 425-440.
161. Wykoff, R.D., and Botset, H.G., "The Flow of Gas-Liquid Mixtures Through Unconsolidated Sands," *Physics*, Vol. 7, 1936, pp. 325-345.
162. Yanosik, J.L., and McCracken, T.A., "A Nine Point Finite Difference Simulator for Realistic Prediction of Adverse Mobility Ratio Displacements," *Society of Petroleum Engineers Journal*, 1979, pp. 253-262.

163. Yortsos, Y.C., and Huang, A.B., "Linear Stability of Immiscible Displacement Including Continuously Changing Mobility and Capillary Effects," *Proceedings,* Fourth Symposium on Enhanced Oil Recovery, Society of Petroleum Engineers and U.S. Department of Energy, Vol. 2, 1984, pp. 145-162.
164. Young, L.C., "A Study of Spatial Approximations for Simulating Fluid Displacements in Petroleum Reservoirs," *Computer Methods in Applied Mechanics and Engineering,* Vol. 47, 1984, pp. 3-46.
165. Young, L.C., and Stephenson, R.E., "A Generalized Compositional Approach for Reservoir Simulation," *Society of Petroleum Engineers Journal,* 1983, pp. 727-742.

10. LIST OF SYMBOLS

A	coefficient
$[A]$	system matrix
$\mathcal{A}$	spatial operator
a	z-coordinate of confining layer
B	formation volume factor;coefficient
B_j	(see Table 1)
b	z-coordinate of interface;derivative of 1/B
$\underset{\sim}{b}$	body force
C	specific moisture capacity;compressibility; coefficient
c	z-coordinate of free surface
D	diffusion coefficient
$\underset{\approx}{D}$	hydrodynamic dispersion tensor
$\underset{\sim}{e}$	unit vector
f	fractional flow;fugacity
$\{f\}$	boundary data vector
g	gravitational acceleration
H_0	Hermite cubic of the first kind
H_1	Hermite cubic of the second kind
h	hydraulic head
$\{h\}$	vector of hydraulic heads
I	number of nodes
J	temporal domain
j	diffusive flux
K	hydraulic conductivity scalar
$\underset{\approx}{K}$	hydraulic conductivity tensor
K_{ij}	(see Table 1)
$[K]$	stiffness matrix
k	permeability scalar
k_r	relative permeability
$\underset{\approx}{k}$	permeability tensor
ℓ	vertical thickness

M	mobility ratio
M_{ij}	(see Table 1)
$[M]$	mass matrix
$\underset{\sim}{m}$	momentum exchange rate
N	basis function;number of species
$[N]$	SIP matrix
$\mathcal{O}$	order symbol
P	number of phases
p	pressure
$\{p\}$	vector of pressures
Q_{ij}	(see Table 1)
q	flux;flow rate
$\underset{\sim}{q}$	flow rate vector
R	residual
R_S	solution gas-oil ratio
$\{R\}$	residual vector
$\mathcal{R}$	spatial region
r	mass exchange rate
$\{r\}$	right-hand side vector
S	saturation
S_s	specific storage
s_y	specific yield
T	transmissibility(defined variously)
t	time
$\underset{\approx}{t}$	stress tensor
u	unknown function
$\underset{\sim}{u}$	diffusion velocity vector
$\underset{\sim}{v}$	velocity vector
x	horizontal space coordinate
$\underset{\sim}{x}$	spatial position vector
$\underset{\sim}{y}$	vector of unknowns
Z	depth below datum
z	vertical space coordinate
α	dispersivity
γ	gravity coefficient
δ	iterative increment operator
Δ_t	time-difference operator
Δt	time increment
Δx	space increment
Θ	moisture content
ξ	curve in (x,t)-plane
λ	flux coefficient
Λ	mobility scalar
$\underset{\approx}{\Lambda}$	mobility tensor

μ	dynamic viscosity
ρ	density
Σ	interface
T	compositional flux coefficients
Φ	free surface
ϕ	volume fraction;porosity
Ψ	test function
ψ	pressure head
Ω	spatial domain
ω	mass fraction
$-$	average;collocation point
$\hat{}$	approximation
$'$	derivative;dummy variable
$*$	dimensionless;upstream;preconditioner

Subscripts and Superscripts

A	air
atm	atmospheric
C	capillary
F	freshwater
G	gas phase
g	gas species
i,j	node indices;species indices
k	collocation point index
ℓ	longitudinal
m	iteration level
mol	molecular
N	nonaqueous liquid phase
n	time level
O	oil phase
o	oil species
R	rock phase
RC	reservoir conditions
ref	reference
S	saltwater
STC	stock-tank conditions
T	total
t	transverse
W	water phase
w	water species
z	z-direction
α	phase index
∂	boundary

ADVECTION-DISPERSION WITH ADAPTIVE EULERIAN-LAGRANGIAN FINITE ELEMENTS

Ralph Cady and Shlomo P. Neuman

ADVECTION-DISPERSION WITH ADAPTIVE EULERIAN-LAGRANGIAN FINITE ELEMENTS

Ralph Cady

Conservation & Survey Division
University of Nebraska-Lincoln
Lincoln, Nebraska, 68588, USA

Shlomo P. Neuman

Dept. of Hydrology & Water Res.
University of Arizona
Tucson, Arizona, 85721, USA

ABSTRACT

Advection-dispersion is generally solved numerically with methods that treat the problem from one of three perspectives. These are described as the Eulerian reference, the Lagrangian reference or a combination of the two that will be referred to as Eulerian-Lagrangian. Methods that use the Eulerian-Lagrangian approach incorporate the computational power of the Lagrangian treatment of advection with the simplicity of the fixed Eulerian grid. A modified version of a relatively new adaptive Eulerian-Lagrangian finite element method is presented for the simulation of advection-dispersion. In the vicinity of steep concentration fronts, moving particles are used to define the concentration field. A modified method of characteristics called single-step reverse particle tracking is used away from steep concentration fronts. An adaptive technique is used to insert and delete moving particles as needed during the simulation. Dispersion is simulated by a finite element formulation that involves only symmetric and diagonal matrices. Based on preliminary tests on problems with analytical solutions, the method seems capable of simulating the entire range of Peclet numbers with Courant numbers well in excess of 1.

1. INTRODUCTION

The advection-dispersion equation can often be characterized by the dimensionless Peclet number

$$Pe = \frac{|\underset{\sim}{v}|\ L}{\|\underset{\approx}{D}\|} \tag{1.1}$$

where $\underset{\sim}{v}$ is the velocity vector, L is a characteristic length, and $\underset{\approx}{D}$ is the dispersion tensor. For the one dimensional spreading of an inert chemical due to molecular diffusion, the governing equation may be written

$$\frac{\partial c}{\partial t} = \frac{\partial^2 c}{\partial x^2} - Pe \frac{\partial c}{\partial x} \tag{1.2}$$

where c is concentration, t is time, x is the spatial coordinate defined relative to L, and Pe is the Peclet number defined by Eq. (1.1) with the dispersion tensor consisting only of molecular diffusion. As Pe approaches zero, the equation becomes parabolic and diffusion dominates. As Pe approaches infinity, the equation becomes hyperbolic and advection dominates. Clearly, the character of the equation may vary from parabolic to hyperbolic over space and time if the velocity or the dispersion tensor vary.

In addition to the Peclet number, the success of many methods for the solution of the advection-dispersion equation may be affected by the Courant number

$$\alpha_c = \frac{v\Delta t}{\Delta x} \tag{1.3}$$

where α_c is the dimensionless Courant number, v is the velocity, Δt is the size of the time step and Δx is the distance between grid points or nodes.

Numerical methods to solve the advection-dispersion equation may be categorized by the emphasis placed on the parabolic or hyperbolic nature of the problem. In Eulerian methods, discretization of the equation is performed relative to a grid that is fixed in space. Lagrangian methods are based on a discretization over a deforming grid or a fixed grid in deforming coordinates. Eulerian-Lagrangian methods combine aspects of both techniques in order to merge the computational power of the Lagrangian reference with the simplicity of a fixed Eulerian grid. Neuman (17) provides a literature survey and offers some general observations relative to the strengths and weaknesses of these methods while Smith and Hutton (25) offer a comparison of a number of methods for the solution of steady two-dimensional advection-dispersion.

2. EULERIAN METHODS

Finite difference or finite element techniques are often used to solve the advection-dispersion equation by discretizing the equation over a fixed or Eulerian grid. Numerical experiments have demonstrated that these techniques perform quite well when dispersion dominates the problem and the distribution of concentration is relatively smooth. Eulerian methods are generally conservative; i.e., they often accurately maintain a mass balance. Neuman (17) concludes that Eulerian methods offer convenient means to address complex problems with spatially nonuniform material properties but often suffer from restrictions on the duration of time steps and the size of spatial grid increments. For example, Daus and Frind (3) propose an alternating-direction finite element method that is constrained to Courant numbers less than 1 and Peclet numbers less than 2. Patel, Markatos and Cross (21) contrast the accuracy of some Eulerian schemes for the simulation of steady advection-dispersion. Gupta, Manohar and Stephenson (7) propose a scheme that appears to relax some of the constraints for Eulerian methods.

3. LAGRANGIAN METHODS

The power of Lagrangian methods stems from the reduction or elimination of the advective terms from the equations discretized over a moving grid. This formulation is often better suited to the solution of simple problems that are dominated by advection. For example, McBride and Rutherford (16) describe a simple Lagrangian method for simulating advection-dispersion in one-dimension for non-uniform flow in rivers. Jensen and Finlayson (13) use a Lagrangian coordinate system with its origin at the center of the moving front to remove much of the hyperbolic nature of the advective-dispersion equation. This later technique is not subject to oscillation and is subject to slight numerical dispersion only at high Peclet numbers. However, techniques that involve a moving reference may be difficult to implement under certain circumstances. Prickett, Naymik and Lonnquist (22) have popularized a Lagrangian advection-dispersion method that advects particles along pathlines and simulates dispersion by applying a random-walk adjustment to the particle locations. Lagrangian methods are often not strictly conservative; i.e., it is frequently difficult to maintain mass balance. Neuman (17) discusses some of the difficulties encountered when applying Lagrangian methods to complex subsurface environments.

4. EULERIAN-LAGRANGIAN METHODS

The combination of the Lagrangian treatment of advection with an Eulerian fixed grid offers some distinct advantages in the solution of complex problems. The review by Neuman (17) refers to a few different methods to treat advection. One is the particle tracking method suggested by Garder, Peaceman and Pozzi (6) which is often referred to as the method of characteristics. This method consists of advecting a set of particles along characteristics and has been applied by Konikow and Bredehoeft (14) and others. Another method, which was suggested by Hinstrup, Kej and Kroszynski (10), is called single-step reverse particle tracking by Neuman (18) or a modified method of characteristics by others including Ewing, Russell and Wheeler (5). This method determines, for the beginning of a time step, the location and concentration of a fictitious particle that will coincide with the node at the end of the time step. This location is determined by tracking backward in time along the characteristic through the node. This technique has been used by a number of investigators including Baptista, Adams and Stolzenbach (1), Cheng, Casulli and Milford (2), Holly and Usseglio-Polatera (11), and Rice and Schnipke (23). The method described by Neuman (17) relies on two grids; one fixed and another that temporarily deforms due to advection. This deforming grid is redefined after each timestep. Similar schemes are proposed by Hasbani, Livne and Bercovier (9) and Le Roux and Quesseveur (15). Doughty, et al. (4) describe another method that requires a fixed grid with a spacing between nodes such that the advected field moves exactly one grid unit per time step. Guven et al. (8) apply this technique to a single-well tracer test in a horizontally stratified aquifer. As in Lagrangian methods, maintaining strict mass balance is often difficult for many Eulerian-Lagrangian methods.

Eulerian techniques to simulate dispersion generally estimate the dispersive change from the concentration field defined by nodes on a fixed grid. For example, Konikow and Bredehoeft (14) apply explicit finite differences on a fixed grid. Cheng, Casulli, and Milford (2) use a novel explicit formulation for dispersion in which the dispersive flux is estimated at the location that will be advected to the node at the end of the time step.

5. ADAPTIVE EULERIAN-LAGRANGIAN METHOD

Neuman (17) presents an Eulerian-Lagrangian method that formally decouples the advection-dispersion problem into two parts, one consisting of pure advection and the other dominated by dispersion. The advection problem is solved using the method of characteristics on a fixed space-time grid coupled with a solution to the dispersion problem on another grid using finite elements. The solution to the dispersion problem uses a Lagrangian finite element formulation consisting of only symmetric and diagonal matrices. This particular method is subject to some artificial dispersion that has been attributed by Neuman and Sorek (20) to interpolation between the grids.

An alternative method that adaptively incorporates moving particles surrounding steep concentration fronts and a modified method of characteristics away from steep fronts was presented by Neuman (18). This method seemed capable of simulating the entire range of Peclet numbers with Courant numbers well in excess of 1.

The method presented in this paper is a modification of Neuman (18).

6. THEORY

Consider the advection-dispersion equation

$$R \frac{\partial c}{\partial t} = \nabla . (\underset{\approx}{D} \nabla c - \underset{\sim}{v} c) - R\lambda c + q/\varepsilon \tag{6.1}$$

where c is concentration, R is the retardation factor, t is time, ∇ is the gradient operator, $\underset{\approx}{D}$ is the dispersion tensor, $\underset{\sim}{v}$ is the seepage velocity vector, λ is the radioactive decay coefficient, q is a source term, and ε is porosity. Initial conditions are given by

$$c(\underset{\sim}{x},0) = C0(\underset{\sim}{x}) \tag{6.2}$$

where $\underset{\sim}{x}$ is the position vector and C0 is a prescribed function. Conditions on inflow and noflow boundaries are given by

$$(- \underset{\approx}{D} \nabla c + \underset{\sim}{v} c) . \underset{\sim}{n} + \alpha (c - C) = Q \qquad \text{on} \quad \Gamma_i \tag{6.3}$$

where $\underset{\sim}{n}$ is the unit outward normal vector along the boundary Γ, C and Q are prescribed functions and α determines the nature of the boundary condition acting on Γ. The condition along the boundary can vary from prescribed flux (α is zero) to prescribed concentration (α is infinite). With intermediate values of α, the condition is mixed.

Outflow boundaries are often represented by the following condition

$$\underset{\approx}{D} \nabla c . \underset{\sim}{n} = 0 \qquad \text{on} \quad \Gamma_0 \tag{6.4}$$

While this is a convenient formulation, it is often not a very realistic representation for certain outflow boundaries. Eq. (6.4) implies that there is no flux across an outflow boundary due to dispersion. We will discuss another condition that may be applied to outflow boundaries later in this chapter.

Using the hydrodynamic derivative modified for retardation

$$\frac{D}{Dt} = \frac{\partial}{\partial t} + \frac{\underset{\sim}{v} . \nabla}{R} \tag{6.5}$$

we can express Eq. (6.1) in Lagrangian form as

$$R \frac{Dc}{Dt} = \nabla.(\underset{\approx}{D}\nabla c) - c\nabla.\underset{\sim}{v} - R\lambda c + q/\varepsilon \qquad (6.6)$$

In this formulation, c represents the concentration of a particle of fluid moving with a velocity equal to $\underset{\sim}{v}/R$.

Neuman (17) introduces an expression of c as the sum of two functions

$$c(\underset{\sim}{x},t) = \bar{c}(\underset{\sim}{x},t) + \dot{c}(\underset{\sim}{x},t) \qquad (6.7)$$

In the present case, we require $\bar{c}$ to satisfy the nonhomogeneous differential equation

$$R \frac{D\bar{c}}{Dt} = - \bar{c}\nabla.v - R\lambda\bar{c} \qquad (6.8)$$

subject to the initial condition

$$\bar{c}(\underset{\sim}{x},t_k) = c(\underset{\sim}{x},t_k) \qquad (6.9)$$

where t_k is time at the beginning of a time step, and the Cauchy condition

$$\underset{\sim}{v}\bar{c}.\underset{\sim}{n} + \alpha(\bar{c} - C) = Q \qquad \text{on } \Gamma_i \qquad (6.10)$$

on inflow and noflow boundaries. Since $\bar{c}$ is the result of pure advection, the outflow boundaries have no influence on $\bar{c}$. Adding Eq. (6.8) to Eq. (6.6) gives

$$R \frac{Dc}{Dt} = \nabla.(\underset{\approx}{D}\nabla c) - (c - \bar{c})\nabla.\underset{\sim}{v} - R\lambda(c - \bar{c}) + q/\varepsilon + R \frac{D\bar{c}}{Dt} \qquad (6.11)$$

7. NUMERICAL APPROACH

Suppose that c is known at time t_k, and we wish the solution at time $t_{k+1} = t_k + \Delta t$. We define $c(\underset{\sim}{x},t_k)$ according to Eq. (6.9). The advection problem is solved by single-step reverse particle tracking combined with an analytic solution to Eq. (6.8) over the time step. Forward particle tracking (method of characteristics) is used in the vicinity of steep concentration fronts to define the concentration field. The residual dispersion problem is solved to obtain the nodal concentrations at the end of the time step by using finite elements on a fixed grid. Adjustments to the particle concentrations are determined by applying a separate finite element analysis to particles within individual elements. A finite element function over the fixed grid, $c^N(\underset{\sim}{x},t)$, is used to approximate $c(\underset{\sim}{x},t)$ and is defined as

$$c(\underset{\sim}{x},t) \cong c^N(\underset{\sim}{x},t) = \sum_n c_n(t)\xi_n(\underset{\sim}{x}) \tag{7.1}$$

where N is the total number of nodes in the finite element grid, c_n is the concentration at node n and ξ_n is the finite element basis function for node n. Basis functions are chosen so that $\sum_n \xi_n(\underset{\sim}{x})=1.0$. A finite element approximation for $\bar{c}(\underset{\sim}{x},t)$ is given by $\bar{c}^N(\underset{\sim}{x},t)$ as

$$\bar{c}(\underset{\sim}{x},t) \cong \bar{c}^N(\underset{\sim}{x},t) = \sum_n \bar{c}_n(t)\xi_n(\underset{\sim}{x}) \tag{7.2}$$

7.1. Single-step Reverse Particle Tracking

Consider a fictitious moving particle that reaches node n at time t_{k+1}. The location of this point at time t_k is

$$^k\underset{\sim}{x}_n = \underset{\sim}{x}_n^{k+1} - \int_{t_k}^{t_{k+1}} (\underset{\sim}{v}/R)\, Dt \tag{7.3}$$

If the duration of the time step does not change and the velocity is constant, $^k\underset{\sim}{x}_n$ remains constant for each n and must be computed only once.

Define ^kc_n as the concentration at time t_k of a fictitious moving particle that reaches node n at time t_{k+1}. In the absence of particles, ^kc_n is determined by evaluating Eq. (7.1) at location $^k\underset{\sim}{x}_n$. The value of $\bar{c}_n^{k+1}$ is calculated from ^kc_n by evaluating the analytic solution to Eq. (6.8) over the timestep from t_k to t_{k+1} as

$$\bar{c}_n^{k+1} = {}^kc_n \exp\left[\int_{t_k}^{t_{k+1}} (\nabla.\underset{\sim}{v} + R\lambda)Dt\right] \tag{7.4}$$

7.2. Continuous Forward Particle Tracking

In the vicinity of steep concentration fronts, particles are introduced and their positions continually tracked along pathlines. The initial concentration for each particle is estimated by applying Eq. (7.1) to the initial location of the particle. Along inflow boundaries, any new particle, r, introduced at $(\underset{\sim}{x}_r,t_k)$ is assigned concentration given by

$$\bar{c}_r^k = \left.\frac{\alpha C + Q}{\underset{\sim}{v}.\underset{\sim}{n} + \alpha}\right|_{\underset{\sim}{x}_r,\, t_k} \tag{7.5}$$

according to Eq. (6.10).

After advection, each particle, p, is located at a new position

$$\underset{\sim}{x}_p^{k+1} = \underset{\sim}{x}_p^k + \int_{t_k}^{t_{k+1}} (\underset{\sim}{v}/R)\, Dt \qquad (7.6)$$

To estimate $\bar{c}_p^{k+1}$, we use the analytic solution to Eq. (6.8)

$$\bar{c}_p^{k+1} = c_p^k \exp\left[\int_{t_k}^{t_{k+1}} (\nabla . \underset{\sim}{v} + R\lambda) Dt\right] \qquad (7.7)$$

7.3. Projection of Particle Concentration upon Nodes

Various methods can be used to estimate the concentration $\bar{c}_n^{k+1}$ from the concentration field defined by the particles. Three have been investigated in this chapter.

The first two are based upon the method used by Neuman (18); weighting the concentration of particles in elements surrounding the node by the reciprocal of the distance between the particle and the node. One of these methods uses all of the particles in the elements adjacent to the node and the other uses only one particle in each of the adjacent elements; the one nearest the node. Nodes that are not surrounded by particles simply assume the concentration due to single-step reverse particle tracking given by Eq. (7.4).

The third method estimates the concentration field defined by the particles by using a locally optimal triangulation of particles within individual elements.

If $^k\underset{\sim}{x}_n$ is in an element containing particles, ^kc_n is estimated by performing a triangulation of the particles and the nodes defining the element by an algorithm presented by Sloan and Houlsby (24). $^k\underset{\sim}{x}_n$ is located in the proper particle (or particle-node) triangle and the value of ^kc_n is estimated from the particle concentrations, c_p^k, the nodal concentrations, c_n^k, and linear particle (or particle-node) basis functions in a fashion similar to Eq. (7.1).

$\bar{c}_n^{k+1}$, the concentration advected to the node, is calculated from the estimate for ^kc_n by applying Eq. (7.4).

7.4. Dispersion by Finite Elements

Applying the Galerkin orthogonalization procedure to Eq. (6.11) yields

$$\int_{Re} [R\ (\frac{Dc}{Dt} - \frac{D\bar{c}}{Dt}) - \nabla.(\underset{\approx}{D}\nabla c) + (\nabla.\underset{\sim}{v} + R\lambda)(c - \bar{c}) - q/\varepsilon]\xi_n dRe = 0$$

$$n = 1,2,\ldots,N \qquad (7.8)$$

where Re is the region bounded by Γ and ξ_n is the basis function for node n. For the time being, we will exclude the finite element approximation for c given by Eq. (7.1) from the time derivatives of Eq. (7.8).

Each node of the finite element grid is treated as a particle that has reached $\underset{\sim}{x}_n$ at t_{k+1}. Following the suggestions of Neuman and Narasimhan (19), the parabolic formulation allows us to apply the lumped-mass approach to the time derivatives in Eq. (7.8)

$$\int_{Re} R\ \frac{Dc}{Dt}\ \xi_n dRe \cong \frac{Dc_n}{Dt} \int_{Re} R\xi_n dRe \qquad (7.9)$$

$$\int_{Re} R\ \frac{D\bar{c}}{Dt}\ \xi_n dRe \cong \frac{D\bar{c}_n}{Dt} \int_{Re} R\xi_n dRe \qquad (7.10)$$

Time derivatives are then approximated by finite differences using a backward difference formulation

$$\frac{Dc_n}{Dt} \cong \frac{c_n^{k+1} - {}^k c_n}{\Delta t} \qquad (7.11)$$

$$\frac{D\bar{c}_n}{Dt} \cong \frac{\bar{c}_n^{k+1} - {}^k c_n}{\Delta t} \qquad (7.12)$$

Applying Green's first identity to the dispersion term gives

$$\int_{Re} -\nabla.(\underset{\approx}{D}\nabla c^N)\xi_n dRe = \int_{Re} \underset{\approx}{D}\nabla c^N.\nabla\xi_n dRe - \int_{\Gamma} \underset{\approx}{D}\nabla c^N.\underset{\sim}{n}\xi_n d\Gamma \qquad (7.13)$$

The boundary integral given in Eq. (7.13) may be calculated from the boundary condition defined by Eq. (6.3) on inflow and noflow boundaries giving

$$\int_{\Gamma} \underset{\approx}{D}\nabla c^N.\underset{\sim}{n}\xi_n d\Gamma = \int_{\Gamma} [\underset{\sim}{v}c^N.\underset{\sim}{n} + \alpha(c^N - C) - Q]\xi_n d\Gamma \qquad (7.14)$$

We define a 'dispersion matrix' that is symmetric, positive semi-definite, and of order N by

$$A_{nm} = \int_{Re} \underset{\approx}{D}\nabla\xi_n . \nabla\xi_m dRe \tag{7.15}$$

$\underset{\approx}{B}$ is a symmetric 'boundary matrix' of order N defined as

$$B_{nm} = \int_{\Gamma_i} -(\underset{\sim}{v}.\underset{\sim}{n} + \alpha)\xi_n\xi_m d\Gamma_i \qquad \alpha < \infty \tag{7.16}$$

if n and m are on an inflow or noflow boundary and zero otherwise. If α is infinity, the concentration at that node is known and an equation for the node is not needed. $\underset{\approx}{F}$ is a symmetric matrix whose terms incorporate radioactive decay and the divergence of velocity as

$$F_{nm} = \int_{Re} (\nabla.\underset{\sim}{v} + R\lambda)\xi_n\xi_m dRe \tag{7.17}$$

$\underset{\approx}{W}$, the 'capacity matrix', is diagonal and of order N, with diagonal terms

$$W_{nn} = \int_{Re} R\xi_n dRe \tag{7.18}$$

$\underset{\sim}{V}$ is a 'source vector' of order N generally defined as

$$V_n = \int_{Re} \frac{q}{\varepsilon}\,\xi_n dRe \tag{7.19}$$

If n is on an inflow or noflow boundary and α is not infinite (the concentration at n is not prescribed), an additional term is incorporated into V_n to account for the boundary condition. In this case

$$V_n = \int_{Re} \frac{q}{\varepsilon}\,\xi_n dRe - \int_{\Gamma_i} (\alpha C + Q)\xi_n d\Gamma_i \tag{7.20}$$

Along outflow boundaries, the condition specified by Eq. (6.4) is easily applied since the value integrated along the boundary segment is always zero. If the dispersive flux across an outflow boundary can not be assumed to be zero, the boundary integral in Eq. (7.13) is non-zero. When expressed as a matrix product with a vector of concentrations, the matrix representing the integral is not symmetric. In order to maintain symmetric coefficient matrices for the unknown concentrations after dispersion, we approximate the integral by substituting the finite element approximation $\bar{c}^N(\underset{\sim}{x},t)$ for $c^N(\underset{\sim}{x},t)$.

$$\int_{\Gamma_o} \underset{\approx}{D}\nabla c^N.\underset{\sim}{n}\xi_n d\Gamma_o \cong \int_{\Gamma_o} \underset{\approx}{D}\nabla\bar{c}^N.\underset{\sim}{n}\xi_n d\Gamma_o = \sum_m c_m \int_{\Gamma_o} \underset{\approx}{D}\nabla\xi_m.\underset{\sim}{n}\xi_n d\Gamma_o \tag{7.21}$$

The suitability of this approximation relies on the gradient of $\bar{c}$ to adequately represent the gradient of c. This approximation is reasonable if the concentration field has dispersed sufficiently prior to advection to the outflow boundary. This approximation is incorporated into the equation for node n by adding the estimate in Eq. (7.21) to V_n.

7.4.1. Implicit dispersion

Incorporating Eq. (7.9)-(7.20) into Eq. (7.8) with the dispersive flux estimated implicitly gives

$$(\underset{\approx}{A} + \underset{\approx}{B} + \underset{\approx}{F} + \underset{\approx}{W}/\Delta t)c^{k+1} = \underset{\sim}{V} + (\underset{\approx}{F} + \underset{\approx}{W}/\Delta t)\bar{c}^{k+1} \tag{7.22}$$

In this form, the dispersive flux and boundary flux terms are estimated implicitly with the concentrations c^{k+1} and a correction is included on both sides by the F matrix to account for the divergence of velocity and radioactive decay. These equations may be solved by many symmetric matrix techniques; we are currently using Cholesky factorization.

7.4.2. Explicit dispersion

An explicit approach estimates the dispersive flux from known concentrations. We may use $\bar{c}^{k+1}$ and omit the corrective F matrix terms from both c and $\bar{c}$ giving a set of simultaneous equations of the form

$$(\underset{\approx}{W}/\Delta t)c^{k+1} = \underset{\sim}{V} + (\underset{\approx}{A} + \underset{\approx}{B} + \underset{\approx}{W}/\Delta t)\bar{c}^{k+1} \tag{7.23}$$

Neuman and Narasimhan (19) discuss stability limitations imposed on the length of the time step of this formulation given by

$$\Delta t \leq \min_n \left[W_{nn}/(A_{nn} + B_{nn})\right] \tag{7.24}$$

Eq. (7.23) is easily solved because it involves only one unknown per equation.

7.4.3. Mixed implicit-explicit dispersion

Following the same rational used by Neuman and Narasimhan (19), we evaluate the stability of nodal concentration by evaluating the length of a stable time step for each node. Those nodes with stable time steps less than Δt are solved implicitly. Eq. (7.22) is applied to all unstable nodes and all stable nodes that are adjacent to unstable nodes. This is slightly less efficient than solving implicitly for only the unstable nodes, but eliminates the need to correct the dispersive flux between nodes when one the flux is solved explicitly for one node and implicitly for the other.

7.5. Adjustment of Particle Concentration for Dispersion

After the dispersion problem has been solved for the nodes, the concentrations of the particles must be adjusted for dispersion.

Neuman (18) adjusts the particle concentrations by an amount determined by the change in the concentration field defined by the finite-element approximations given by Eq. (7.1) and (7.2)

$$c_p^{k+1} = \bar{c}_p^{k+1} + [c^N(\underset{\sim}{x}_p,t_{k+1}) - \bar{c}^N(\underset{\sim}{x}_p,t_{k+1})] \tag{7.25}$$

This adjustment may be sufficient if the gradient of concentration between particles is not sufficiently different from the gradient of concentration defined by the nodal concentrations.

In the event that the particle concentration gradients are significantly different from the nodal concentration gradients, an additional dispersion of the particle concentrations might improve the accuracy of the simulation. Particle concentrations are dispersed by applying a local finite element analysis to the concentration of particles within an element using the results for the dispersion at the nodes as boundary conditions.

Let $\overline{cp}^{k+1}$ be the vector of particle concentrations at the end of the time step but prior to dispersion and cp^{k+1} be the vector of particle concentrations at the end of the time step and after dispersion. Also, let $cp^{N,k+1}$ be the vector of $c^N(\underset{\sim}{x}_p,t_{k+1})$ for all particles, that is, the finite element approximation of the concentration at the particle location after the dispersion problem has been solved for the nodes.

To maintain conservation of mass during the dispersion step, we wish to adjust the particle concentrations such that the change in mass within an element is the same for the particle approximation as for the finite element approximation. To assure consistency, we perform a localized finite element analysis applied to the particle concentration field bounded by an individual finite element. Particles within an element are triangulated according to the algorithm of Sloan and Houlsby (24) and the net change of mass within the element is partitioned to the particles within the element to form the vector $\underset{\sim}{M}$ according to

$$M_p = \int_{Re} [c^N(\underset{\sim}{x}_p,t_{k+1}) - \bar{c}^N(\underset{\sim}{x}_p,t_{k+1})]\xi_p dRe \tag{7.26}$$

where M_p is the net change in mass associated with particle p determined by the triangulation of particles within elements. ξ_p is defined for the particle over the triangulation within the element such that $\sum_p \xi_p = 1.0$. This change in mass represents a cumulative source term in the equation for each particle. A 'dispersion matrix', $\underset{\approx}{A}'$, is assembled for the particles within an element.

$$A'_{pr} = \int_{Re} \underset{\approx}{D} \nabla\xi_p . \nabla\xi_r dRe \qquad (7.27)$$

The p and r refer to particles within the element. A 'capacity matrix', $\underset{\approx}{W}'$ much like that in Eq. (7.18) is formed for particles within the element.

$$W'_{pp} = \int_{Re} \xi_p dRe \qquad (7.28)$$

Because retardation, decay, sources and large-scale dispersion are incorporated in Eq. (7.26), the final equation for each particle requires only the local dispersion, capacity and the net change of mass

$$(\underset{\approx}{A}' + \underset{\approx}{W}'/\Delta t)c_p^{k+1} = (\underset{\approx}{W}'/\Delta t)\bar{c}_p^{k+1} + \underset{\approx}{A}'c_p^{N,k+1} + \underset{\approx}{M}/\Delta t \qquad (7.29)$$

This system of equations is solved and the process repeated for every element containing particles that violate the gradient criterion.

7.6. Adaptive Mechanism

Forward tracked particles are added during simulation along inflow boundaries, at prescribed-concentration nodes and in the vicinity of steep concentration fronts. Particles are deleted once they cross outflow boundaries and as fronts disperse to such a degree that forward tracked particles are no longer necessary to prevent numerical dispersion.

Particle insertion and deletion associated with steep fronts is based upon an evaluation of the variation of the concentration gradient across an element. If the gradient is essentially constant in the vicinity of an element, forward tracked particles are not necessary for the advection of the concentration field without numerical dispersion. On the other hand, if the variation of the concentration gradient is significant, particles are needed to prevent numerical dispersion.

A single criterion based upon a user-supplied value controls the automatic insertion and deletion of particles due to variation of local concentration gradients.

7.7. Mass Balance

Mass-balance errors may be introduced during the advection phase of this technique. Other than the inaccuracies associated with moving along pathlines, errors in mass balance may result as a front disperses beyond the particles that define the front or when the particle concentrations are projected to the nodes and potentially, the method used to reduce the particle concentrations for dispersion. The introduction of new particles to define the front is based upon an evaluation of the change of the concentration gradient across an element that does not contain any particles. If the change of gradient exceeds a user-specified amount, particles are introduced within that element. Adjusting the user-specified value will alter the mass-balance error due to this source.

8. EXAMPLES

The two-dimensional dispersion of a rectangular wave in a steady, uniform velocity field over an infinite two-dimensional region is governed by

$$D_x \frac{\partial^2 c}{\partial x^2} + D_y \frac{\partial^2 c}{\partial y^2} - v_x \frac{\partial c}{\partial x} - v_y \frac{\partial c}{\partial y} = \frac{\partial c}{\partial t} \qquad (8.1)$$

subject to

$$c(x,y,0) = 1.0 \qquad \text{when } x_c - a \le x \le x_c + a, \quad y_c - b \le y \le y_c + b \qquad (8.2)$$

$$c(x,y,0) = 0.0 \qquad \text{otherwise} \qquad (8.3)$$

$$c(-\infty,y,t) = c(\infty,y,t) = c(x,\infty,t) = c(x,-\infty,t) = 0.0 \qquad (8.4)$$

The analytical solution is

$$c(x,y,t) = \frac{1}{4}\{\mathrm{erf}[\frac{a-(x-x_c)+v_x t}{\sqrt{4D_x t}}] + \mathrm{erf}[\frac{a+(x-x_c)-v_x t}{\sqrt{4D_x t}}]\}$$
$$*\{\mathrm{erf}[\frac{b-(y-y_c)+v_y t}{\sqrt{4D_y t}}] + \mathrm{erf}[\frac{b+(y-y_c)-v_y t}{\sqrt{4D_y t}}]\} \qquad (8.5)$$

For this example, $D_x = 1$, $D_y = 0.1$, $v_x = 1000$, $v_y = 33.33$, $a = 0.05$, $x_c = 0.15$, $b = 0.01$ and $y_c = 0.02$. The finite element grid consists of 2000 rectangular elements with an x-distance between nodes of 0.01 and a y-distance between nodes of 0.004. Boundary conditions were simulated as

$$c(0.0,y,t) = 0.0 \tag{8.6}$$

$$c(x,-0.04,t) = 0.0 \tag{8.7}$$

$$\frac{\partial}{\partial x}[c(1.0,y,t)] = 0.0 \tag{8.8}$$

$$\frac{\partial}{\partial y}[c(x,0.088,t)]\ 0.0 \tag{8.9}$$

The directional Peclet numbers are

$$Pe_x = v_x \Delta x / D_x = 10.0 \tag{8.10}$$

$$Pe_y = v_y \Delta y / D_y = 1.33 \tag{8.11}$$

The analytical solution is for a perfectly square wave, that is, a wave with infinite concentration gradients where the concentration drops from 1.0 to zero. This infinite gradient poses a number of serious problems for this method. In order to minimize these problems while maintaining the overall form of the problem,we will approximate the initial concentration field over the fixed Eulerian grid such that the concentration field forms a trapazoidal wave rather than a perfectly square wave. To approximate the square wave both in initial mass and shape, the wave was initialized by setting the initial nodal concentration to

$$c_n(0) = 1.0 \quad \text{when} \quad .11 \le x_n \le .19, \quad .012 \le y_n \le .028 \tag{8.12}$$

$$c_n(0) = 0.5 \quad \text{when } x_n = .10 \text{ or } x_n = .20, \quad .012 \le y_n \le .0.28 \tag{8.13}$$

$$c_n(0) = 0.0 \quad \text{otherwise} \tag{8.14}$$

If particles are introduced during the simulation, their initial concentration was determined by applying either Eq. (7.1) or (7.5).

Figures 1-4 depict results along lines through the peak of the analytical solution at time = 0.0006 using either 32 time steps or 11 time steps. The analytical solution is plotted as a solid line while the simulated nodal values are plotted as crosses. For 32 time steps, the Courant numbers are 1.87 in the x direction and 0.16 in the y direction. For 11 time steps, the Courant numbers are 5.45 in the x direction and 0.45 in the y direction. Because the time step for the simulation with 32 time steps satisfies Eq. (7.24), this problem is stable under the explicit formulation. The simulations with 32 time steps depicted in Figures 1-4 were solved explicitly while the simulations with 11 time steps were solved implicitly.

The left half (parts a and c) of the Figures 1-4 depict results along the line y = 0.04 while the right half (parts b and d) depict results along the line x = 0.75.

Figure 1 represents the results of simulations without particles. Numerical dispersion from the analytical solution is quite apparent in both simulations in Figure 1 with greater dispersion as the number of time steps increase.

Figures 2-4 represent simulations with particles automatically inserted to define the wave as advection and dispersion proceed. About 1500 particles were required to define the wave at the end of simulation with 32 time steps and about 900 particles for the 11 time step simulation. The particle simulations differed in the way that the advected nodal concentration was estimated from the particle concentration. The methods are described in slightly greater detail in section 7.3 of this paper. Results from simulations with advected nodal concentrations estimated by reciprocal distance weighting of all particles in adjacent elements are represented in Figure 2. Figure 3 represents simulations with advection nodal concentrations determined by weighting only the nearest particle in each element adjacent to the node. Results from simulations implementing the triangulation of particles are represented in Figure 4. Table 1 summarizes the departure of the numerical simulations from the analytical solution to the problem by listing the percentage difference between the numerical and analytic concentrations at the peak of the wave (0.543 for the analytical solution) and the difference between the mass of the numerical wave and the mass of the analytical solution wave (0.002 for the analytical solution).

To assess the significance of the explicit solution versus the implicit solution, the problem was simulated with and without particles for 32 time steps using an implicit solution for the residual dispersion problem. The results of these simulations are depicted in Figure 5. There is no significant difference between the explicit simulations depicted in Figures 1 and 4 and the implicit-simulation results in Figure 5.

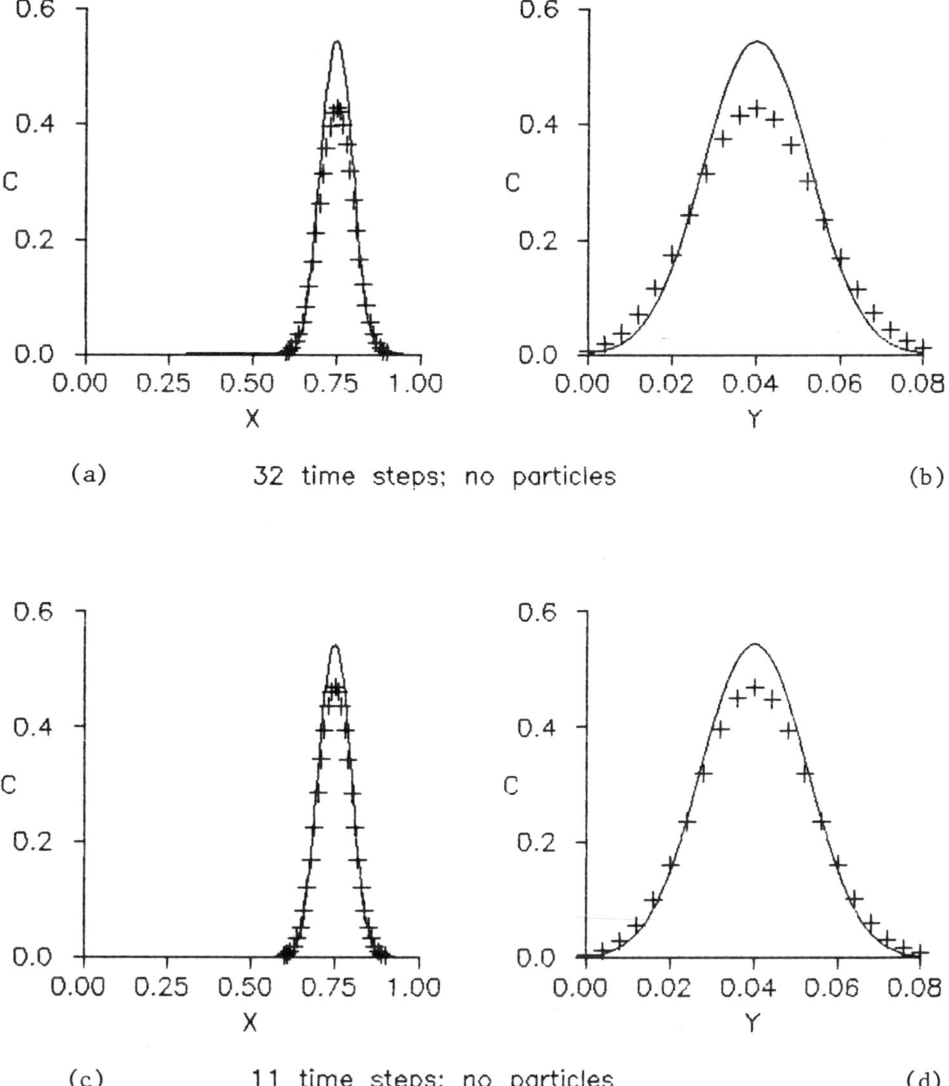

Figure 1. Simulation of problem 1. without forward-particle tracking.

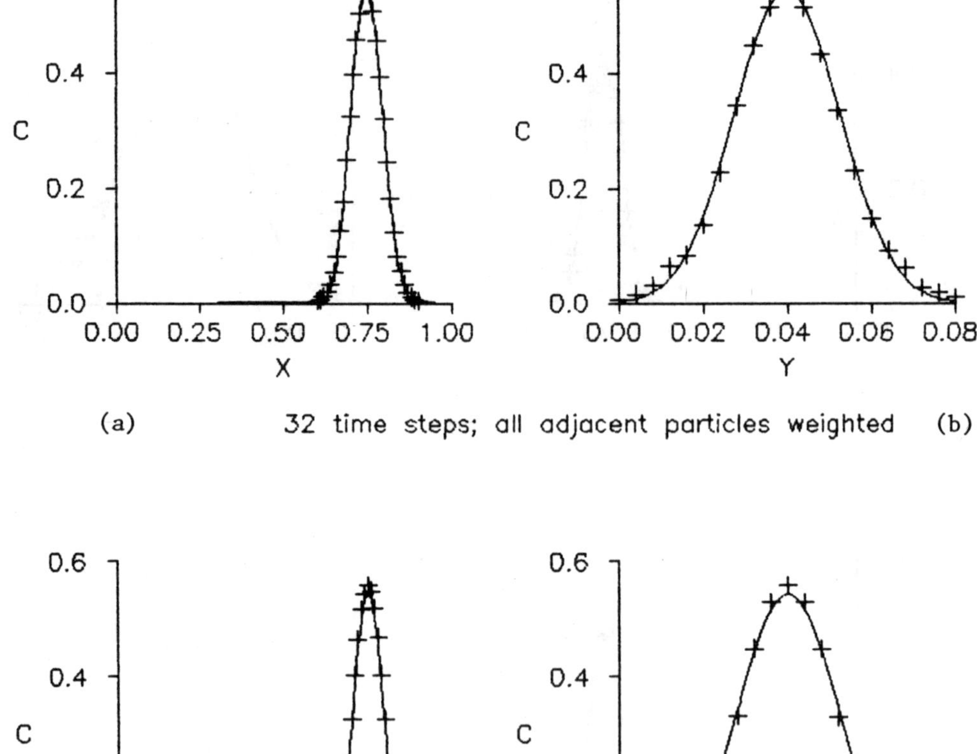

Figure 2. Simulation of problem 1. with all adjacent particles included in weighting of nodal concentrations.

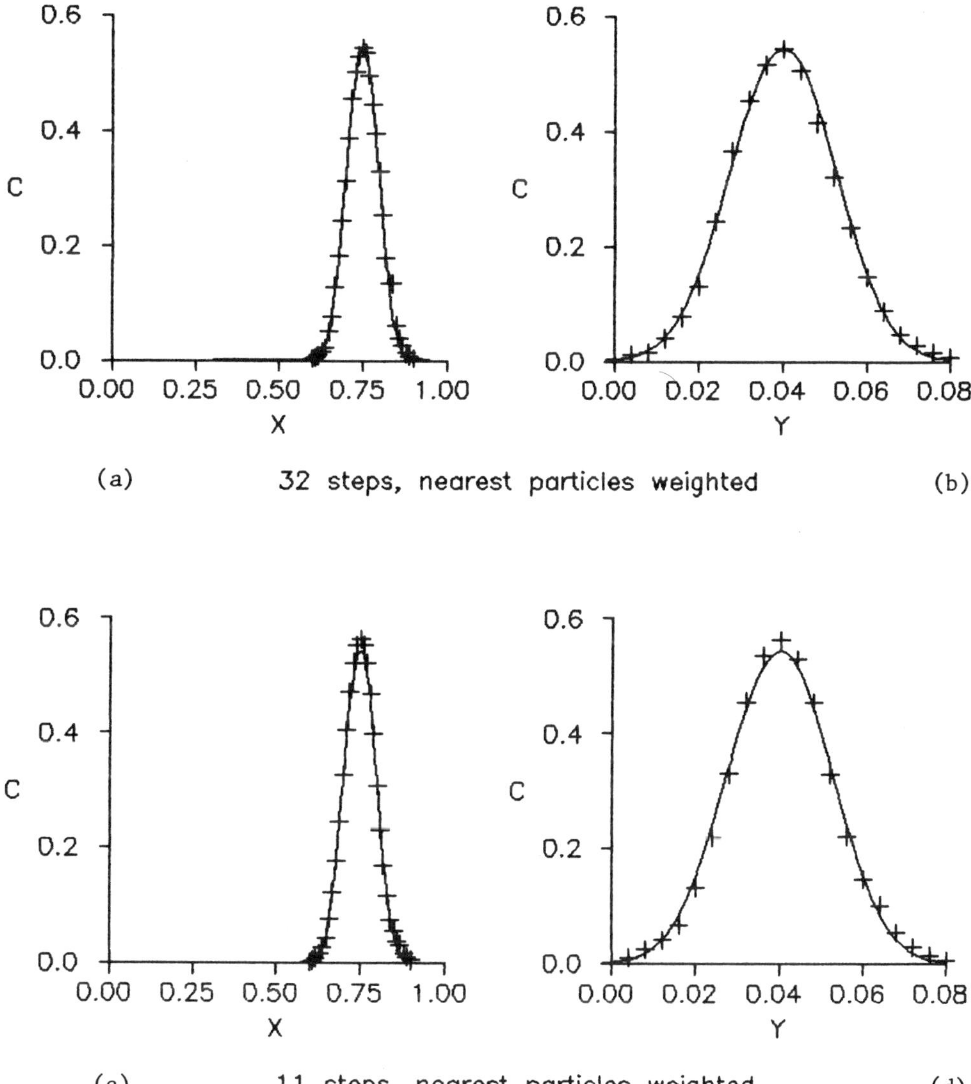

Figure 3. Simulation of problem 1. with only nearest adjacent particles included in weighting of nodal concentrations.

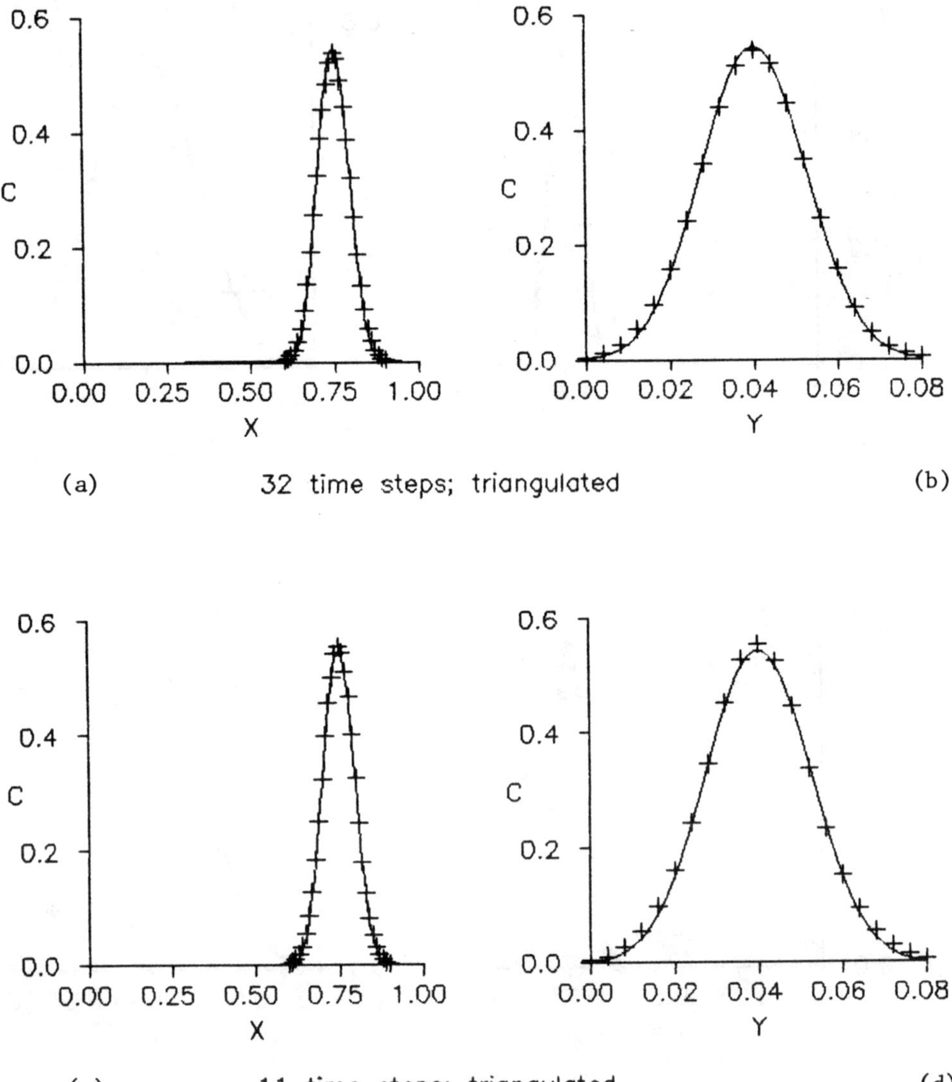

Figure 4. Simulation of problem 1. with triangulation of particles performed to determine advected nodal concentrations.

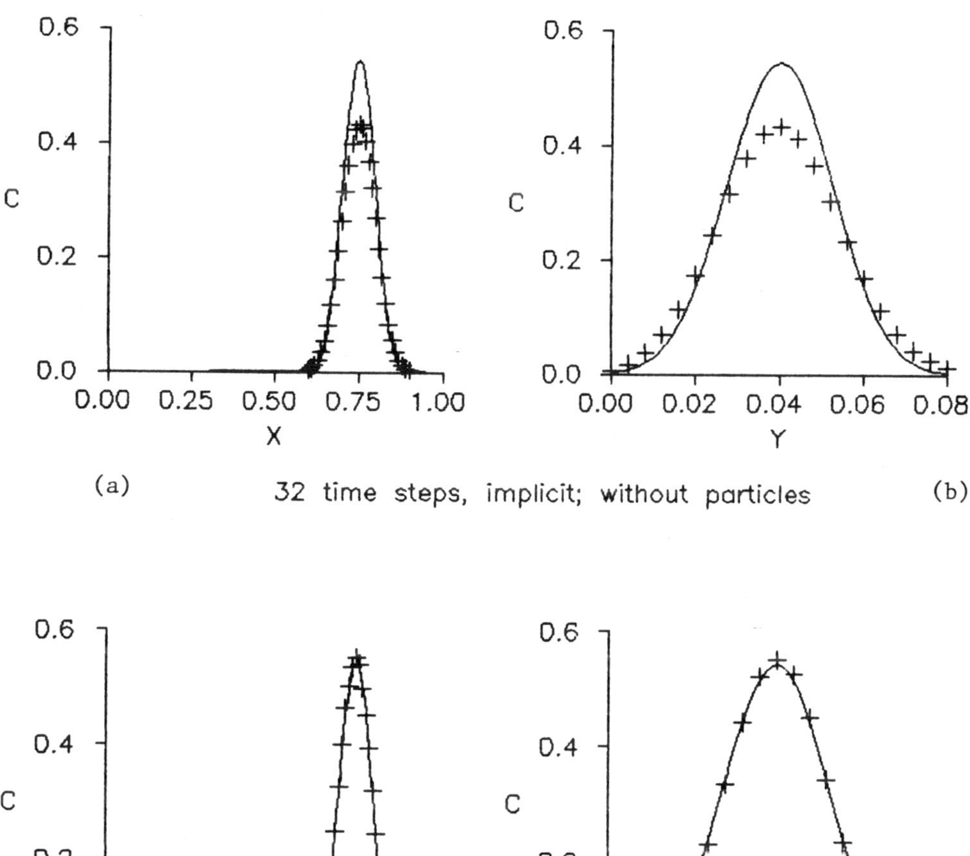

Figure 5. Simulation of problem 1. with implicit solution of the residual dispersion step.

Table 1. Departure of Numerical Solutions from Analytical Solution

Algorithm	32 time steps: deviation of peak (%)	32 time steps: deviation of mass (%)	11 time steps: deviation of peak (%)	11 time steps: deviation of mass (%)
No particles	-21	-1	-13	1
Weight all particles	1	3	3	3
Weight nearest particles	-.1	3	4	3
Triangulated	-1	1	4	2

The second example problem is similar to the first with the exception that dispersion is reduced by an order of magnitude (D_x = 0.1 and D_y = 0.01). Therefore, the Peclet numbers are increased by an order of magnitude; Pe_x is now 100 and Pe_y is now 13.33. Results from simulations with and without particles at time = 0.0006 are depicted in Figure 6. Once again, the simulation without particles suffers from numerical dispersion (the numerical peak of the concentration wave is about 18 percent below the concentration of the peak of the analytical solution). The mass of the simulated wave is virtually identical to the mass of the analytical solution after 11 time steps. Incorporating particles into the simulation of the problem results in a numerical concentration at the peak that is about 2 percent below the analytical solution while the mass of the numerical wave is about 1 percent above the analytical value.

The third example is similar to the first and second except that there is no dispersion of the wave; $D_x = D_y = 0$. For the case of pure advection, the Peclet number is infinite. Courant numbers for these simulations are the same as those for example 1. Figure 7 represents simulations with and without particles for 11 time steps. Numerical dispersion is evident in the results from the simulation without forward moving particles. When particles are incorporated into the simulation (458 for 11 steps), advection of the wave is virtually free of dispersion or distortion.

Often the results of a numerical-simulation method are sensitive to the orientation of the grid. Radial advection-dispersion from a circular source is presented as an example to examine the sensitivity of the results to the orientation of the grid. For this example, the analytical solution is radially symmetric about the circular source. The initial concentration is zero everywhere. The velocity from the source is steady and radial with a magnitude of 50/radius. The source is located at x = 75, y = 75 with a radius of 2.5 and begins emitting water with a concentration of 1.0 at time 0.0. The

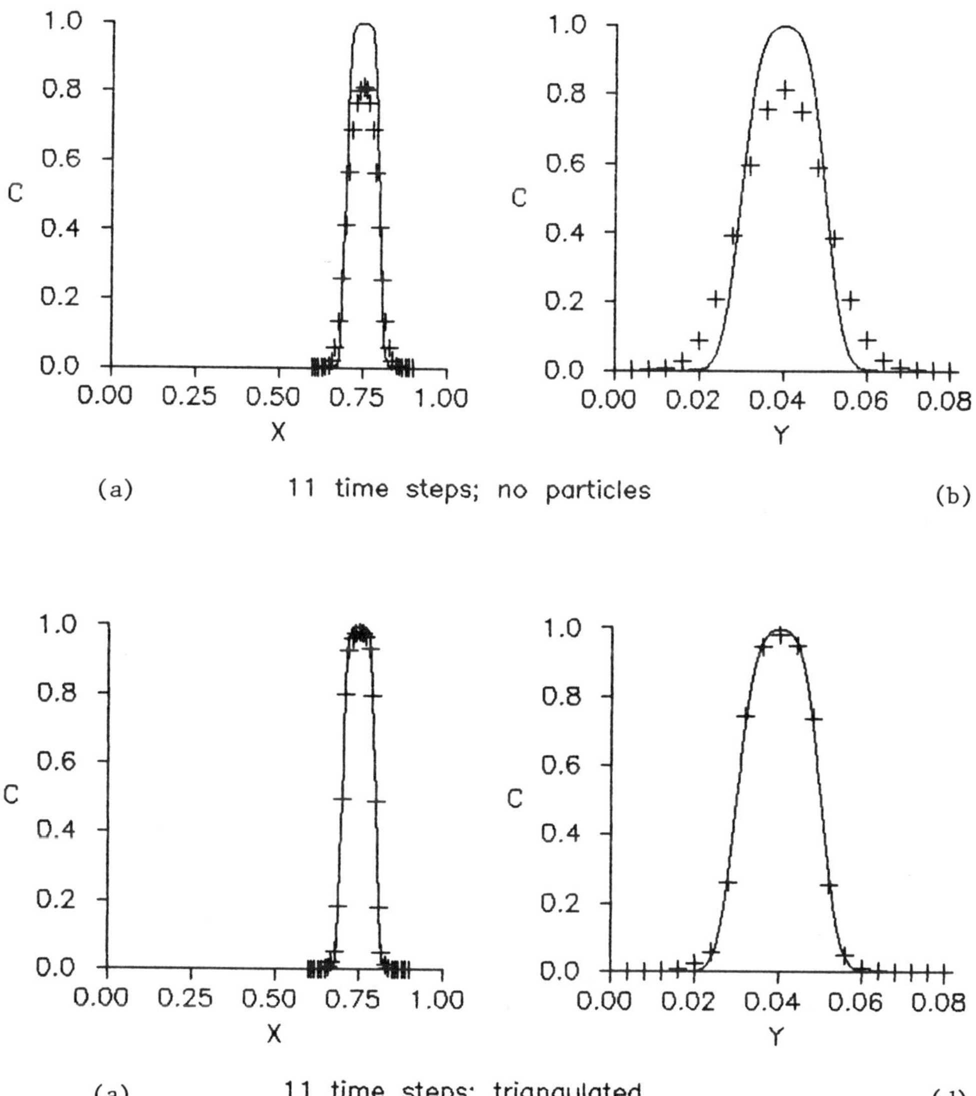

Figure 6. Simulation of problem 2.

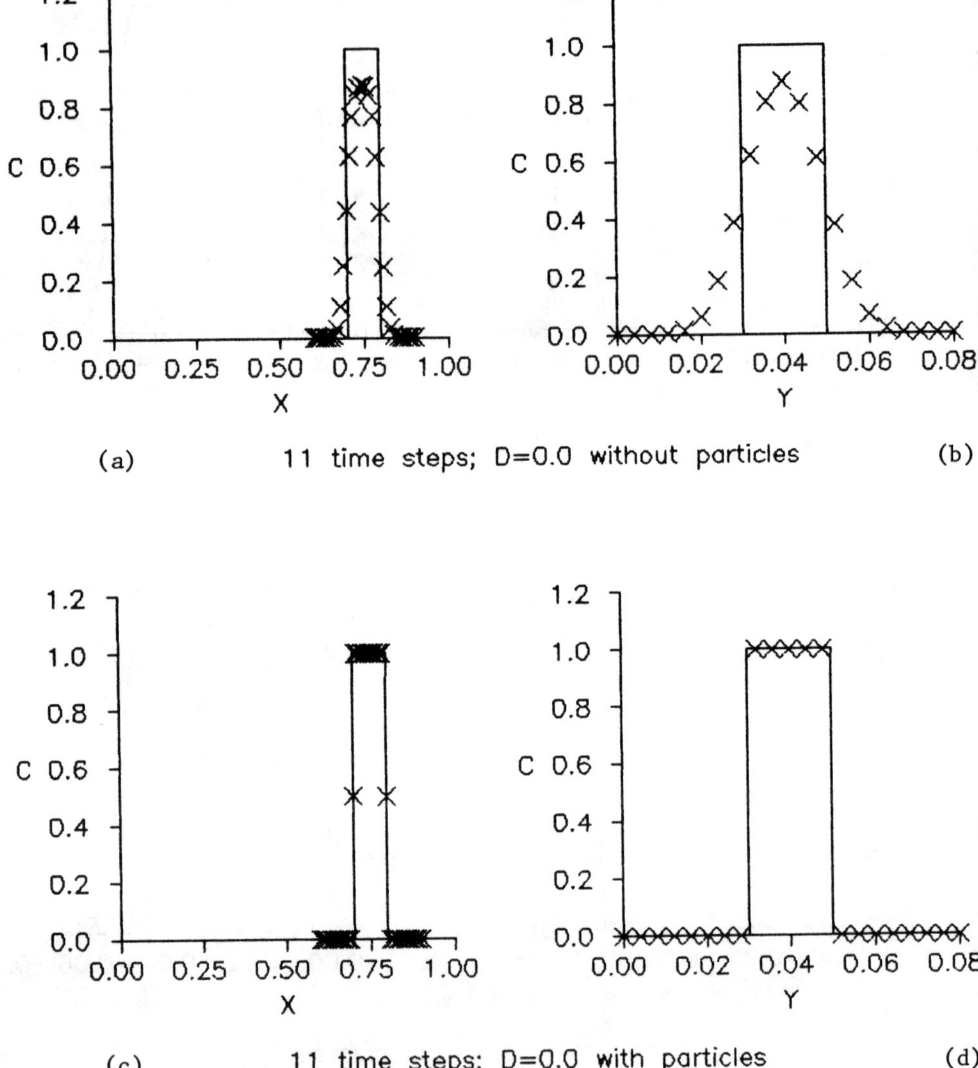

Figure 7. Simulation of problem 3.

longitudinal dispersivity is 5 and the transverse dispersivity is 0. Figure 8 depicts the results of simulations with and without particles at time = 5 over a rectangular grid with a distance between nodes of 5. The maximum Courant number is 2 and the maximum Peclet number is 20. The explicit-stability criterion given by Eq. (7.24) allows the residual dispersion problem to be solved explicitly at most nodes while requiring an implicit solution at only a few nodes. Figure 8 depicts the results of a numerical evaluation to the analytical solution given by Javandel, Doughty and Tsang (12) by crosses. The solid lines in Figure 8 connect nodal values of the numerical results along lines passing through the source; along y = 75, x = 75, y + x = 150, y = x, 2x - y = 75 and 2y - x = 75. Both simulations are relatively free of significant grid orientation effects.

9. CONCLUSIONS

Formal decomposition of the advection-dispersion equation results in two problems that can be addressed efficiently and accurately. The advection problem can be solved independent of the residual dispersion problem over each time step by an adaptive method that combines forward particle tracking with single-step reverse particle tracking. The results from the advection problem are used as input to a residual dispersion problem that primarily involves dispersion. The residual dispersion problem is solved by Galerkin finite elements on a fixed grid.

The adaptive scheme inserts forward-moving particles where they are needed to minimize numerical dispersion and prevent clipping of concentration peaks. Particles are automatically deleted when they are no longer needed. In the absence of forward-moving particles, the method relies on single-step reverse particle tracking to advect the concentration field.

The purely parabolic form of the residual dispersion problem may be approximated by a finite element formulation that consists of only diagonal or symmetric matrices. Mass-lumping of the capacity matrices allows the application of an economical mixed explicit-implicit scheme to solve the resulting finite element equations.

Application to three two-dimensional test problems with a uniform velocity field suggest that this Eulerian-Lagrangian formulation of the advection-dispersion equation can simulate over the entire range of Peclet numbers with Courant numbers well in excess of 1.

Results from a radial advection-dispersion problem suggest that the method is relatively free of the influence of grid orientation on the solution.

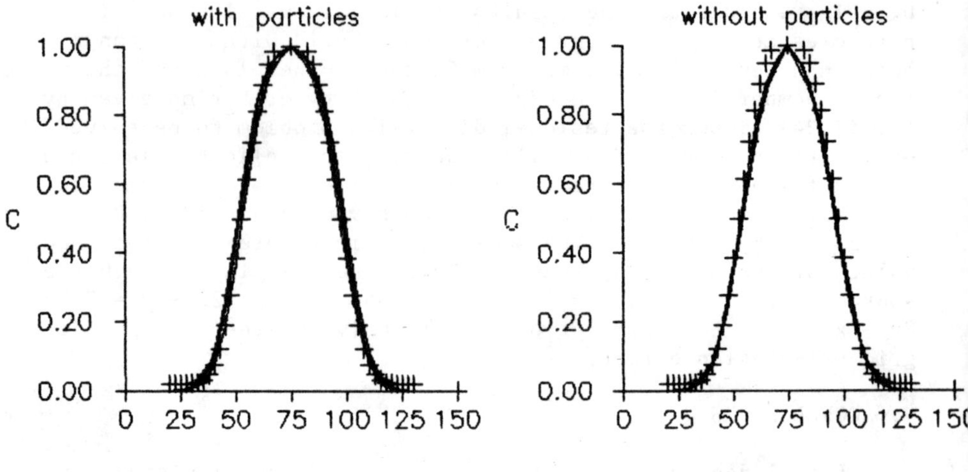

Figure 8. Simulation of problem 4.

Additional testing will be required to establish the sensitivity solutions to: the number of particles, the method for projecting particle concentrations to the nodes, the criterion for insertion or deletion of particles, and the influence of Peclet numbers on the duration of time steps.

10. REFERENCES

1. Baptista, A.M., Adams, E.E., and Stolzenbach, K.D., "The 2-D, Unsteady, Transport Equation Solved by the Combined Use of the Finite Element Method and the Method of Characteristics," *Proceedings of the 5th International Conference on Finite Elements in Water Resources*, June 1984, pp. 353-359.
2. Cheng, R.T., Casulli, V., and Milford, S.N., "Eulerian-Lagrangian Solution of the Convection-Dispersion Equation in Natural Coordinates," *Water Resources Research*, Vol. 20, No. 7, July 1984, pp. 944-952.
3. Daus, A.D., and Frind, E.O., "An Alternating Direction Galerkin Technique for Simulation of Contaminant Transport in Complex

Groundwater Systems," *Water Resources Research*, May 1985, pp. 653-664.

4. Doughty, C., Hellstrom, G., Tsang, C.F., and Claesson, J., "A Dimensionless Parameter Approach to the Thermal Behavior of an Aquifer Thermal Energy Storage System," *Water Resources Research*, Vol. 18, No. 3, June 1982, pp. 571-587.
5. Ewing, R.E., Russell, T.F., and Wheeler, M.F., "Convergence Analysis of an Approximation of Miscible Displacement in Porous Media by Mixed Finite Elements and a Modified Method of Characteristics," *Computer Methods in Applied Mechanics and Engineering*, Vol. 47, No. 1-2, Dec., 1984, pp. 73-92.
6. Garder, A.O., Peaceman, D.W., and Pozzi, A.L., Jr., "Numerical Calculation of Multidimensional Miscible Displacement by the Method of Characteristics," *Society of Petroleum Engineers Journal*, Vol. 4, No. 1, March 1964, pp. 26-36.
7. Gupta, M.M., Manohar, R.P., and Stephenson, J.W., "A Single Cell High Order Scheme for the Convection-Diffusion Equation with Variable Coefficients," *International Journal for Numerical Methods in Fluids*, Vol. 4, 1984, pp. 641-651.
8. Guven, O., Falta, R.W., Molz, F.J., and Melville, J.G., "Analysis and Interpretation of Single-Well Tracer Tests in Stratified Aquifers," *Water Resources Research*, May 1985, pp. 676-684.
9. Hasbani, Y., Livne, E., and Bercovier, M., "Finite Elements and Characteristics Applied to Advection-Diffusion Equations," *Computers and Fluids*, Vol. 11, No. 2, 1983, pp. 71-83.
10. Hinstrup, P., Kej, A., and Kroszynski, U., "A High Accuracy Two-Dimensional Transport-Dispersion Model for Environmental Applications," *Proceedings, XVIII International Association of Hydraulic Research Congress*, Vol. 13, 1977, pp. 129-137.
11. Holly, F.M., Jr., and Usseglio-Polatera,J., "Dispersion Simulation in Two-Dimensional Tidal Flow," *Journal of Hydraulic Engineering*, Vol. 110, No. 7, July 1984, pp. 905-926.
12. Javandel, I., Doughty, C., and Tsang, C.F., "Groundwater Transport: Handbook of Mathematical Models," *Water Resources Monograph 10*, American Geophysical Union, Washington, D.C., 1984, pp. 1-227.
13. Jensen, O.K., and Finlayson, B.A., "Solution of the Transport Equations Using a Moving Coordinate System," *Advances in Water Resources*, Vol. 3, No. 1, March 1980, pp. 9-18.
14. Konikow, L.F., and Bredehoeft, J.D., "Computer Model of Two-Dimensional Solute Transport and Dispersion in Ground Water," *Techniques of Water-Resources Investigations of the United States Geological Survey, Book 7*, Chapter C2, U.S. Geological Survey, Reston, Va., 1978, pp. 1-90.
15. Le Roux, A.Y., and Quesseveur, P., "Convergence of an Anti-diffusion Lagrange-Euler Scheme for Quasi-Linear Equations," *Society for Industrial and Applied Mathematics Journal for Numerical Analysis*, Vol. 21, No. 5, Oct., 1984, pp. 985-994.

16. McBride, G.B., and Rutherford, J.C., "Accurate Modeling of River Pollutant Transport," *Journal of Environmental Engineering*, American Society of Civil Engineers, Vol. 110, No. 4, Aug. 1984, pp. 808-827.
17. Neuman, S.P., "A Eulerian-Lagrangian Numerical Scheme for the Dispersion-Convection Equation Using Conjugate Space-Time Grids," *Journal of Computational Physics*, Vol. 41, No. 2, June 1981, pp. 270-294.
18. Neuman, S.P., "Adaptive Eulerian-Lagrangian Finite Element Method for Advection-Dispersion," *International Journal for Numerical Methods in Engineering*, Vol. 20, 1984, pp. 321-337.
19. Neuman, S.P., and Narasimhan, T.N., "Mixed Explicit-Implicit Iterative Finite Element Scheme for Diffusion-Type Problems: I. Theory," *International Journal for Numerical Methods in Engineering*, Vol. 11, 1977, pp. 309-323.
20. Neuman, S.P., and Sorek, S., "Eulerian-Lagrangian Methods for Advection-Dispersion," *Proceedings of the 4th International Conference on Finite Elements in Water Resources*, June 1982, pp. 14:41-14:68.
21. Patel, M.K., Markatos, N.C., and Cross, M., "A Critical Evaluation of Seven Schemes for Convection-Diffusion Equations," *International Journal for Numerical Methods in Fluids*, Vol. 5, 1985, pp. 225-244.
22. Prickett, T.A., Naymik, T.G., and Lonnquist, C.G., "A Random-Walk Solute Transport Model for Selected Groundwater Quality Evaluations," *Bulletin 65*, Illinois State Water Survey, Champaign, Illinois, 1981, pp. 1-103.
23. Rice, J.G., and Schnipke, R.J., "A Monotone Streamline Upwind Finite Element Method for Convection-Dominated Flows," *Computer Methods in Applied Mechanics and Engineering*, Vol. 48, 1985, pp. 313-327.
24. Sloan, S.W., and Houlsby, G.T., "An Implementation of Watson's Algorithm for Computing 2-Dimensional Delaunay Triangulations," *Advances in Engineering Software*, Vol. 6, No. 4, 1984, pp. 192-197.
25. Smith, R.M., and Hutton, A.G., "The Numerical Treatment of Advection: A Performance Comparison of Current Methods," *Numerical Heat Transfer*, Vol. 5, 1982, pp. 439-461.

11. LIST OF SYMBOLS

$\underset{\approx}{A}$	Finite-element 'dispersion matrix'
$\underset{\approx}{A}'$	Particle finite -element 'dispersion matrix'
a	Half-length of rectangular wave in example 1
$\underset{\approx}{B}$	Finite-element 'boundary matrix'
b	Half-width of rectangular wave in example 1
C	Prescribed concentration outside of boundary
CO	Initial concentration
c	Concentration
$\bar{c}$	Component of concentration due to advection (see equations 6.7 and 6.8)
$\dot{c}$	Component of concentration due to dispersion (see equations 6.7 and 6.8)
$\underset{\approx}{D}$	Dispersion tensor
$\underset{\approx}{F}$	Finite-element radioactive decay and velocity divergence matrix
L	Characteristic length
$\underset{\sim}{M}$	Particle finite-element 'mass-change vector'
$\underset{\sim}{n}$	Unit vector that is outward-normal to boundary Γ
Pe	Peclet number
Q	Boundary-condition source term
q	Source or sink strength
R	Retardation factor
Re	Flow region surrounded by Γ
$\underset{\sim}{V}$	Finite-element 'source vector'
$\underset{\approx}{W}$	Finite-element 'capacity matrix'
$\underset{\approx}{W}'$	Particle finite-element 'capacity matrix'
t	Time
t0	Initial time
$\underset{\sim}{v}$	Velocity vector
$\underset{\approx}{x}$	Location vector
α_c	Courant number
α	Scalar used in boundary condition specification
ε	Porosity
Γ	Boundary of region
λ	Radioactive decay
ξ	Finite element basis function

Subscripts

c	Associated with the initial center of the concentration peak

i	Associated with inflow or noflow portion of boundary Γ
k	Associated with time level k
m	Associated with node m
n	Associated with node n
o	Associated with outflow portion of boundary Γ
p	Associated with particle p
r	Associated with particle r emitted from a boundary condition
x	Associated with the x-direction
y	Associated with the y-direction

Superscripts:

k	Associated with time level k
N	Associated with the finite-element approximation of a continuous function

MOVING POINT TECHNIQUES

Christopher L. Farmer

MOVING POINT TECHNIQUES

Christopher L. Farmer

Oil Recovery Projects Division
United Kingdom Atomic Energy Authority
Atomic Energy Establishment
Winfrith, Dorchester, Dorset, DT2 8DH, United Kingdom

ABSTRACT

Moving point techniques provide accurate solutions to convection dominated convection-diffusion equations. These methods use a set of moving points for the simulation of convection and a fixed mesh for the simulation of diffusion and dispersion. The method introduces very low levels of numerical diffusion, overshoot or phase error.

Three classes of moving point techniques are described; (i) the modified method of characteristics (MMOC) (Hartree, Russell, Douglas, Wheeler, Huffenus, Khaletzky, Donea, Morton) (ii) the pure moving point method (MPM) (Garder, Peaceman, Pozzi) and (iii) the hybrid moving point method (HMPM) (Farmer, Norman). The HMPM uses the MMOC in smoothly varying situations and the MPM when fronts are spread over two or fewer mesh lengths. Numerical results are described which show that the finite element MMOC is a good method for smooth problems (fronts spread over three or more mesh lengths) and the HMPM a good method for problems with sharp fronts. Algorithms for the application of the MPM and the HMPM to the simulation of miscible and segregated displacement of fluid in a porous medium are given. The moving point method is shown to track fronts accurately even in time dependent fingering problems.

1. INTRODUCTION

The transport of a miscible contaminant through a water aquifer and the miscible displacement of oil by a solvent in an oil reservoir are among the problems which may be described by the *miscible displacement equations*. These equations consist of,

(i) an elliptic boundary value problem for the fluid pressure from which the fluid velocity may be obtained via Darcy's law and (ii) a parabolic initial-boundary value problem for the contaminant or solvent concentration.

The miscible displacement equations also describe the process of *immiscible segregated displacement* in which water in the presence of residual oil displaces oil in the presence of connate water. In this case the parabolic equation reduces to a hyperbolic equation.

The transport of contaminant or solvent is by the three processes of *convection, molecular diffusion* and *hydrodynamic dispersion*. The hydrodynamic dispersion is of a diffusive character but is dependent upon the velocity of the fluid.

A primary feature of the miscible displacement equations is that on the distance and time scales of interest in practical problems the convection transport mechanism dominates the diffusion and dispersion mechanisms. It is well known that numerical methods which work well for the diffusion equation produce unacceptable oscillatory errors when applied to convection dominated equations. The simplest 'cure' for this problem is to introduce a mesh dependent diffusion term which removes the spurious oscillations at the expense of introducing a diffusion or dispersion process which may, on practical meshes, dominate the physical diffusion and dispersion by orders of magnitude. More sophisticated cures can introduce significant phase errors.

The main purpose of the following review is to describe numerical methods which do not suffer from unacceptable levels of spurious oscillations, diffusion or phase error with a particular emphasis upon moving point techniques deriving from the method of Garder, Peaceman and Pozzi (13) which we refer to as the *GPP algorithm*.

The GPP algorithm uses a set of moving points which move with the fluid velocity and whose associated solution values are changed through time to allow for the effects of diffusion and dispersion. In addition to the set of moving points a fixed mesh is introduced upon which concentrations are assigned by interpolation from the moving points. The fixed mesh values are viewed as the desired approximation and are used to evaluate the changes in moving point concentrations due to diffusion and dispersion.

Garder, Peaceman and Pozzi showed their algorithm to be of high accuracy even on relatively coarse meshes. The algorithm has not, however, been widely used because numerical experiments indicate failure to converge under mesh refinement (Price, Cavendish and Varga (31)) and no analysis of the method was available (Neuman (26)) until recently (Farmer (10), Raviart (34)). The GPP algorithm is similar in character to the Particle-in-Cell method (Harlow (15)) and to particle methods in general (Hockney and Eastwood (18)) for which there was also no analysis available (Birkhoff (3)) until recently (Raviart (33)).

In Chapter 2 we state the miscible displacement equations and in Chapter 3 the segregated flow equations. Chapter 4 reviews the modified method of characteristics which is closely related to the moving point method and which is used as part of the hybrid moving point method. Chapter 5 describes the GPP and related algorithms which make minimal use of the fixed mesh and Chapter 6 the hybrid method which makes maximal use of the fixed mesh. Chapter 7 discusses the implementation of moving point algorithms and Chapter 8 describes numerical experiments. In Chapter 9 we review methods for the solution of the pressure and velocity equations.

2. THE MISCIBLE DISPLACEMENT EQUATIONS

2.1 The Convection-Diffusion Equation

Throughout the following we shall consider various functions of a spatial point $\underline{x}$ in 2-dimensional Euclidean space, E_2. The components of $\underline{x}$ will be denoted by x_i $(i = 1,2)$ or (x,y). The x_i notation for components is used in partial derivative symbols. All of the algorithms we shall describe generalise to 3-dimensional space. The symbol t denotes time.

Let Ω be a finite, simply connected, rectangular subset of E_2 with boundary $\partial\Omega = \partial\Omega_1 \cup \partial\Omega_2 \cup \partial\Omega_3$. We define Ω by

$$\Omega = \{\underline{x} : \underline{x} = (x,y)\ ,\ 0 \leqslant x \leqslant L_x,\ 0 \leqslant y \leqslant L_y\} \tag{2.1}$$

The assumption of a rectangular region is not essential but simplifies the exposition.

Let $\underline{u} = \underline{u}(\underline{x},t)$ be a continuous time-dependent solenoidal vector field in E_2 which will be interpreted as the Darcy velocity (volumetric flux) of fluid moving within a saturated porous medium. The components of $\underline{u}$ are denoted by u_i $(i = 1,2)$ or (u,v).

We wish to find a scalar field $c = c(\underline{x},t)$ satisfying the conditions

$$\phi\frac{\partial c}{\partial t} + \underline{u}\cdot\nabla c = \nabla\cdot D\nabla c \qquad \underline{x}\in\Omega,\ t > 0 \tag{2.2}$$

$$c(\underset{\sim}{x},o) = c_o(\underset{\sim}{x}) \qquad \underset{\sim}{x}\varepsilon\Omega \qquad (2.3)$$

$$c(\underset{\sim}{x},t) = 1 \qquad \underset{\sim}{x}\varepsilon\partial\Omega_1,\ t \geqslant 0 \qquad (2.4)$$

$$\frac{\partial c}{\partial n} = 0 \qquad x\varepsilon\partial\Omega_2 \cup \partial\Omega_3,\ t \geqslant 0 \qquad (2.5)$$

where D is a positive second rank tensor called the *dispersion tensor* but which includes the diffusion terms. We interpret c as the concentration of contaminant or solvent. ϕ is a scalar function of $\underset{\sim}{x}$ called the porosity. The symbol $\frac{\partial}{\partial n}$ denotes the outward pointing spatial normal derivative at a point on the boundary. The function c_o, which specifies the initial condition, is assumed to satisfy the boundary conditions. The dispersion tensor is, in general, a function of the Darcy velocity field. A particular example of a velocity dependent dispersion tensor may be found in Peaceman (29).

2.2 The Pressure Equation

We seek a scalar field $p = p(\underset{\sim}{x},t)$ and a solenoidal vector field $\underset{\sim}{u} = \underset{\sim}{u}(\underset{\sim}{x},t)$ satisfying the conditions

$$\nabla\cdot\underset{\sim}{u} = 0 \qquad \underset{\sim}{x}\varepsilon\Omega,\ t \geqslant 0 \qquad (2.6)$$

$$\underset{\sim}{u} = -\frac{k}{\mu}\cdot(\nabla p + \rho g \nabla H) \qquad \underset{\sim}{x}\varepsilon\Omega,\ t \geqslant 0 \qquad (2.7)$$

$$p = p_I(t) - \rho(1) g \frac{\partial H}{\partial x} y \qquad \underset{\sim}{x}\varepsilon\partial\Omega_1,\ t \geqslant 0 \qquad (2.8)$$

$$p = -\int_o^y \rho g \frac{\partial H}{\partial x}\, dy \qquad x\varepsilon\partial\Omega_2,\ t \geqslant 0 \qquad (2.9)$$

$$\underset{\sim}{u}\cdot\underset{\sim}{n} = 0 \qquad \underset{\sim}{x}\varepsilon\partial\Omega_3,\ t \geqslant 0 \qquad (2.10)$$

where k is an $\underset{\sim}{x}$-dependent positive second rank tensor called the permeability, p is an $\underset{\sim}{x}$ and t dependent scalar function called the pressure, ρ is a strictly positive function of c called the density, g is a real constant called the gravitational acceleration, H is an $\underset{\sim}{x}$-dependent scalar called the height above a datum, and μ is a strictly positive function of c called the viscosity. $\underset{\sim}{n}$ is a vector function of $\underset{\sim}{x}$ and is the outward pointing unit normal to $\partial\Omega$.

In some applications the function $p_I(t)$, called the inlet pressure, is controlled so that the total flow rate

$$Q(t) = - \int_{\partial\Omega_1} \underset{\sim}{u}\cdot\underset{\sim}{n}\, ds \tag{2.11}$$

is some prescribed function of t and where s parameterises $\partial\Omega_1$.

2.3 Remarks on the Derivation of the Miscible Displacement Equations

Eq. (2.6) is derived via a mass balance on the total fluid. Eq. (2.2) is derived via a mass balance on the contaminant or solvent. Strictly speaking these equations are only a rigorous consequence of mass conservation if ρ is independent of c or if D = 0. However, we assume that D and $\frac{d\rho}{dc}$ are small enough to justify the equations as an approximation.

The boundary and initial conditions are representative of those encountered in practice. Our use of these particular boundary conditions is not essential for the methods described in the following. The absence of source terms is not essential but in our applications of moving point techniques we have, so far, injected and produced fluids via the boundary.

2.4 The Convection-Diffusion Equation in a Moving Co-ordinate System

Let $\underset{\sim}{z} = \underset{\sim}{z}(\underset{\sim}{x},t)$ be a time dependent mapping of E_2 onto itself, which is differentiable and uniquely invertible with inverse $\underset{\sim}{x} = \underset{\sim}{x}(\underset{\sim}{z},t)$ satisfying the conditions

$$\phi(\underset{\sim}{x}) \frac{d}{dt} \underset{\sim}{x}(\underset{\sim}{z},t) = \underset{\sim}{u}^*(\underset{\sim}{x}(\underset{\sim}{z},t),t) \tag{2.12}$$

$$\underset{\sim}{x}(\underset{\sim}{z},o) = \underset{\sim}{x}_o(\underset{\sim}{z}) \tag{2.13}$$

where $\underset{\sim}{x}_o$ is a prescribed mapping of E_2 onto itself and $\underset{\sim}{u}^*$ is a continuous extension of $\underset{\sim}{u}$ from Ω to E_2.

Under this mapping the convection-diffusion Eq. (2.2) becomes

$$\phi(\underset{\sim}{x})\frac{d}{dt} a(\underset{\sim}{z},t) = \nabla\cdot(D\nabla)c(\underset{\sim}{x},t) \qquad \underset{\sim}{x}\varepsilon\Omega,\ t > 0 \tag{2.14}$$

$$a(\underset{\sim}{z},t) = c(\underset{\sim}{x},t) \qquad \underset{\sim}{x}\varepsilon\Omega,\ t \geqslant 0 \tag{2.15}$$

$$\underset{\sim}{x} = \underset{\sim}{x}(\underset{\sim}{z},t) \qquad \underset{\sim}{x}\varepsilon\Omega,\ t \geq 0 \tag{2.16}$$

with initial-boundary conditions on c as before.

We may interpret $a = a(\underset{\sim}{z},t)$ as the Lagrangian and $c = c(\underset{\sim}{x},t)$ as the Eulerian values of a field of concentration carried by a fluid particle $\underset{\sim}{z}$ located at $\underset{\sim}{x}$ at time t.

Moving point methods are obtained by discretising Eqs. (2.2) - (2.16).

3. THE SEGREGATED DISPLACEMENT EQUATIONS

3.1 Formulation as a Moving Boundary Problem

We consider the situation illustrated in Figure 1 where there are three regions of flow. In region I we have mobile oil and connate water, in region II mobile water and residual oil and in region III water which is present in a basal aquifer. A similar situation arises in the study of the time-dependent displacement of the water table in an aquifer where instead of oil we have air. Both of these situations are idealised in that we ignore the existence of a capillary fringe or, in other words, that the flow is *segregated*.

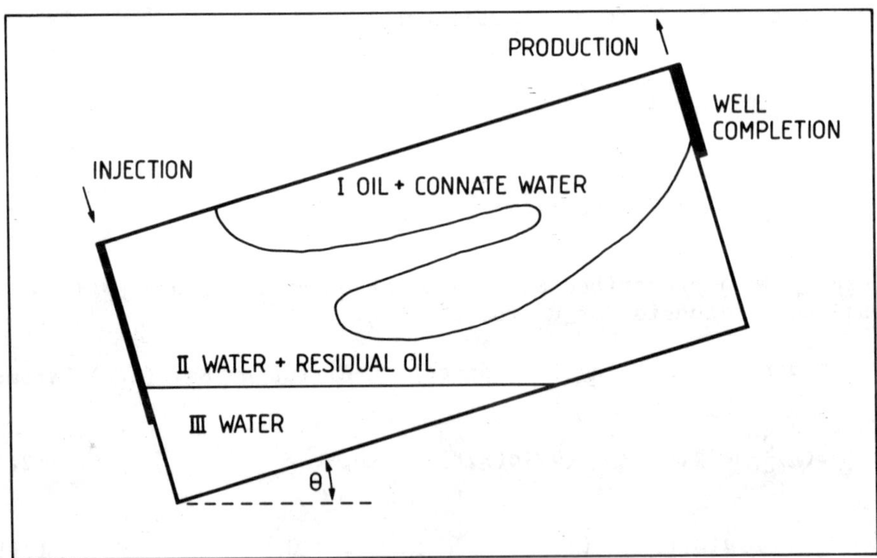

Figure 1. The segregated displacement of oil.

The assumption of segregated flow may in some circumstances be a good approximation and enables us to obtain a non-trivial exact solution for time-dependent two-dimensional problems. Such exact solutions are useful for checking the accuracy of numerical methods.

The interface between regions II and III is stationary but water can flow between these regions. The interface between regions I and II is time-dependent.

We parameterise the interface between zones I and II by a real parameter σ and introduce a vector function $\underset{\sim}{F}$ such that $\underset{\sim}{x} = \underset{\sim}{F}(\sigma,t)$ is the position at time t of the point σ in the front.

The equations of motion of the system illustrated in Figure 1 are as follows;

$$\nabla\cdot\underset{\sim}{u}_J = 0 \tag{3.1}$$

$$\underset{\sim}{u}_J = -\lambda_J \nabla\Psi_J \tag{3.2}$$

$$\lambda_J = \frac{k_J k}{\mu_J} \tag{3.3}$$

$$\Psi_J = p_J + \rho_J gH \tag{3.4}$$

where J = I, II, or III. λ_J is called the *mobility*, Ψ_J the *potential*, and k_J the *end-point relative permeability* of phase J. We note that $k_{III} = 1$.

At the interface we have the equations,

$$p_{II}(\underset{\sim}{x},t) = p_{III}(\underset{\sim}{x},t) \qquad \underset{\sim}{x}\varepsilon\partial II\cap\partial III,\ t \geqslant 0 \tag{3.5}$$

$$p_I(\underset{\sim}{x},t) = p_{II}(\underset{\sim}{x},t) \qquad \underset{\sim}{x} = \underset{\sim}{F}(\sigma,t),\ t \geqslant 0 \tag{3.6}$$

$$\phi_r \frac{d}{dt}\underset{\sim}{F}(\sigma,t) = \underset{\sim}{u}_{II}(\underset{\sim}{F},t) \qquad t \geqslant 0 \tag{3.7}$$

$$\phi_r = \phi(1 - S_{or} - S_{wc}) \tag{3.8}$$

where S_{or} and S_{wc} are the residual oil and connate water saturations respectively. k_J, S_{or} and S_{wc} are, in general, functions of $\underset{\sim}{x}$.

Eqs. (3.1) are the mass balances for the incompressible fluids. Eqs. (3.2) are Darcy's law generalised by the introduction of the relative permeability to the case when another phase is present. The relative permeabilities are the end point values of the relative permeabilities which appear in the formulation of the more realistic Buckley-Leverett equations with capillary pressure which we will refer to as the *full two-phase flow equations* (see, for example, Peaceman (10) for a derivation of these two-phase flow equations). Eqs. (3.5) and (3.6) are the statement of continuity of fluid pressure at the interfaces.

Eq. (3.7) is derived by considering the jump conditions for the motion of a discontinuity in a solution of the full two-phase flow equations. Alternatively mass balance considerations lead to Eq. (3.7) if one takes into account that on one side of the interface, water at saturation S_{wc} is immobile and on the other side, oil at saturation S_{or} is immobile. An equation alternative to Eq. (3.7) is obtained by replacing $\underset{\sim}{u}_{II}$ by $\underset{\sim}{u}_{III}$ in Eq. (3.7).

The segregated displacement equations generalise the model of Muskat (23) to the case where immobile fluids may be present.

3.2 Exact Solutions of the Segregated Flow Equations

3.2.1 The Dietz steady state solution

In the special case that the front does not break we may introduce a scalar function f = f(x,t) such that

$$\underset{\sim}{F} = (x, f(x,t)) \tag{3.9}$$

Equation (3.7) then implies that

$$\phi_r \frac{\partial f}{\partial t} + u \frac{\partial f}{\partial x} = v \tag{3.10}$$

where u and v are evaluated at the front.

We now look for solutions satisfying

$$\frac{\partial \Psi_J}{\partial y} = 0 \tag{3.11}$$

and find that the equation of the front is

$$f(x,t) = \tan\beta\left[\frac{u_\infty}{\Phi_r} t - x\right] + f_o \tag{3.12}$$

where $$\tan\beta = -\frac{\partial f}{\partial x} = \tan\Theta + \frac{(\alpha_I - \alpha_{II})}{g\cos\Theta\,(\rho_{II} - \rho_I)} \tag{3.13}$$

$$\alpha_J = -\frac{u_\infty}{\lambda_J} \tag{3.14}$$

and Θ is such that

$$H(\underset{\sim}{x}) = x\sin\Theta + y\cos\Theta \tag{3.15}$$

The parameter f_o defines the initial condition and u_∞ is the x-component of the Darcy flux at infinity.

This solution was derived by Dietz (4).

3.2.2 The Jacquard-Séguier time dependent solution

Jacquard and Séguier (20) have obtained analytical solutions for a class of segregated displacement problems in a channel of finite width. These solutions were obtained using conformal mapping and Green's function techniques. Jacquard and Séguier assume uniform permeability and porosity with constant mobility behind the front and a different constant mobility ahead of the front.

Jacquard and Séguier prove that if the equation of the front at time t = 0 is given by

$$\sinh\left[\frac{2\pi(x - x_o)}{L_y}\right] = \sinh(\gamma_o)\sin\left[\frac{2\pi|\sigma|}{L_y}\right] \tag{3.16}$$

then at later times

$$\sinh\left[\frac{2\pi}{L_y}(x - x_o - U_o t)\right] = \tag{3.17}$$

$$\sinh(\gamma_o)\sin\left[\frac{2\pi|\sigma|}{L_y}\right]\exp\left[\frac{2\pi\lambda U_0 t}{L_y}\right]$$

Where the parameter σ satisfies $-\frac{L_y}{2} \leq \sigma \leq \frac{L_y}{2}$, $U_0 = \frac{u_\infty}{\phi_r}$, γ_o and x_o are parameters controlling the initial size and position of a single viscous finger and $\lambda = (\lambda_I - \lambda_{II})/(\lambda_I + \lambda_{II})$.

3.3 Segregated Displacement as a Special Case of Miscible Displacement

Introduce a scalar function $c = c(\underset{\sim}{x},t)$ which takes the value 1 in the water and the value 0 in the oil. Introduce also the functions $\mu = \mu(c)$, $\rho = \rho(c)$ and $k_r = k_r(c,\underset{\sim}{x})$ which are strictly positive functions of c and satisfy

$$\begin{aligned} &\mu(1) = \mu_w, \quad \mu(0) = \mu_o \\ &\rho(1) = \rho_w, \quad \rho(0) = \rho_o \\ &k_r(c,\underset{\sim}{x}) = 1 \qquad \underset{\sim}{x}\varepsilon III \\ &k_r(1,\underset{\sim}{x}) = k_w, \quad k_r(0,\underset{\sim}{x}) = k_o \qquad \underset{\sim}{x}\varepsilon I \cup II \end{aligned} \tag{3.18}$$

where the suffices w and o refer to water and oil respectively. We then consider the equations

$$\nabla \cdot \underset{\sim}{u} = 0 \tag{3.19}$$

$$\underset{\sim}{u} = -\alpha \cdot (\nabla p + \rho g \nabla H) \tag{3.20}$$

$$\alpha = \frac{k_r(c,\underset{\sim}{x})k}{\mu(c)} \tag{3.21}$$

$$\phi_r \frac{\partial c}{\partial t} + \underset{\sim}{u} \cdot \nabla c = 0 \tag{3.22}$$

One may show, by an application of the transport theorem for a system with discontinuous fronts (Kotchine's Theorem; Truesdell and Toupin (37)), that when the function c possesses a discontinuity lying along a continuous curve, with c unity on one side and zero on the other, that Eqs. (3.18) - (3.22) imply equations (3.1) - (3.4).

We thus see that segregated displacement is a limiting case of miscible displacement when region III (water) is absent. In the general case when the basal aquifer, region III, is present one may think of segregated displacement as miscible displacement with zero diffusion and zero dispersion, with a sharp front and a non-separable mobility tensor $\alpha(c,\underset{\sim}{x})$. The mobility tensor is separable in the true miscible displacement because it is a product of a function of c with the $\underset{\sim}{x}$-dependent permeability tensor.

4. THE MODIFIED METHOD OF CHARACTERISTICS

4.1 Definitions

Introduce a set of points, $E^h{}_2$, called the *fixed mesh*, defined by

$$E^h{}_2 = \{\underset{\sim}{x}_m : \underset{\sim}{x}_m = (x_i, y_j) = (i,j)h - (\tfrac{3}{2},\tfrac{3}{2})h,\ m = (j-1)N_x + i,$$
$$1 \leqslant i \leqslant N_x,\ 1 \leqslant j \leqslant N_y\} \qquad (4.1)$$

where h is a positive quantity called the *mesh spacing* and i, j, m, N_x and N_y are integers. The integer m lies in the range $1 \leqslant m \leqslant M$ where $M = N_x N_y$. Points with $i = 1$ or N_x and points with $j = 1$ or N_y are called *fictitious points* and are used in imposing boundary conditions and implementing the moving point method.

The region, Ω_m, of E_2 defined by

$$\Omega_m = \{\underset{\sim}{x} : \underset{\sim}{x} = (x,y),\ x_i - \tfrac{h}{2} \leqslant x \leqslant x_i + \tfrac{h}{2},\ y_j - \tfrac{h}{2} \leqslant y \leqslant y_j + \tfrac{h}{2}\} \qquad (4.2)$$

is called the m-th or (i,j)-th *mesh cell*. Cells outside Ω are called *fictitious cells*.

The use of a constant mesh spacing is not essential but simplifies the description and implementation of algorithms.

Introduce the discrete time $t_n = n\tau$, $\tau \geqslant 0$ $(n = 0,1,2,\ldots)$ where τ is the *time step*.

4.2 Fixed Mesh Interpolation

Let F_m^n $(m = 1, 2, \ldots M)$ be the values of a scalar valued function defined on the fixed mesh at time t_n. We define a *fixed mesh interpolator* I_F^s which constructs a function G^n according to

$$G^n(\underset{\sim}{x}) = I_F^s[F_\ell^n, \underset{\sim}{x}] \qquad \underset{\sim}{x} \in \Omega \qquad (4.3)$$

where the index ℓ ranges over the indices of the fixed mesh points and s is the order of approximation, in some sense, of the interpolator.

The fixed mesh interpolator is assumed, in the following, to have the property that for each m,

$$G^n(\underset{\sim}{x}_m) = F^n_m \tag{4.4}$$

Examples of such interpolators are piecewise constant interpolation (s=1) and piecewise bilinear interpolation (s=2).

To perform bilinear interpolation we first introduce the functions $\alpha_i = \alpha_i(x)$ defined by

$$\alpha_i(x) = \begin{cases} 1 - (x - x_i)/h & x_i \leqslant x \leqslant x_{i+1} \\ 1 + (x - x_i)/h & x_{i-1} \leqslant x \leqslant x_i \\ 0 & x \notin [x_{i-1}, x_{i+1}] \end{cases} \tag{4.5}$$

From the functions α_i we construct the functions β_ℓ by

$$\beta_\ell(\underset{\sim}{x}) = \alpha_i(x)\alpha_j(y) \qquad \ell = (j-1)N_x + i \tag{4.6}$$

The bilinear interpolator I^2_F is then defined by

$$I^2_F[F^n_\ell, \underset{\sim}{x}] = \sum_\ell F^n_\ell \, \beta_\ell(\underset{\sim}{x}) \tag{4.7}$$

where the summation ranges over the indices $1 \leqslant \ell \leqslant M$.

4.3 Time Stepping Algorithm in the Modified Method of Characteristics

We suppose that at time t_n we have an approximation $c^n(\underset{\sim}{x})$ to the concentration and we wish to construct an approximation $c^{n+1}(\underset{\sim}{x})$ at time level n + 1. Focus attention at time t_{n+1} on a point $\underset{\sim}{x} \varepsilon \Omega$ and, by solving the equation (2.12) backward in time, find the position $\underset{\sim}{x}^*$ of this point at time t_n. Thus if a marker point is placed at $\underset{\sim}{x}^*$ at t_n it will arrive at $\underset{\sim}{x}$ at t_{n+1}. We call the displacement $\underset{\sim}{x} - \underset{\sim}{x}^*$ the *characteristic displacement* at time t_{n+1} and denote it by the symbol $\chi^{n+1} = \chi^{n+1}(\underset{\sim}{x})$.

The time stepping algorithm of the MMOC is then

$$c^{n+1}(\underset{\sim}{x}) = c^n(\underset{\sim}{x} - \underset{\sim}{\chi}^{n+1}(\underset{\sim}{x})) + \tau D^n_\omega \tag{4.8}$$

where $D^n_\omega = \frac{1}{\phi} \nabla\cdot(D^n\nabla)(\omega\, c^n(\underset{\sim}{x}) + (1-\omega)\, c^{n+1}(\underset{\sim}{x}))$

and $0 \leqslant \omega \leqslant 1$. The value of ω determines the extent to which the algorithm is explicit ($\omega = 1$) or implicit ($\omega = 0$). The dispersion tensor, D^n, depends on the time level via its dependence upon $\underset{\sim}{u}^n$.

4.4 Finite Difference Version of the Modified Method of Characteristics

Introduce a mesh function with values c^n_m which is interpreted as an approximation to the concentration at the point $\underset{\sim}{x}^m$ at time t_n.

At time $t = 0$ we assign c^o_m using the initial conditions. In the finite difference method we set $c^o_m = c_o(\underset{\sim}{x}_m)$. At time t_n suppose the c^n_m to be known. The c^{n+1}_m are constructed as follows.

Step 1. Solve the velocity-pressure equations to construct an approximate velocity field $\underset{\sim}{u}^n = \underset{\sim}{u}^n(\underset{\sim}{x})$. A method for doing this is described in chapter 9.

Step 2. Construct the set of *characteristic nodal displacements*, $\underset{\sim}{\chi}^{n+1}_m$, by solving the final value problem,

$$\phi(\underset{\sim}{x}_m - \underset{\sim}{\chi}_m)\,\frac{d\underset{\sim}{\chi}_m}{dt} = -\,\underset{\sim}{u}^n\,(\underset{\sim}{x}_m - \underset{\sim}{\chi}_m) \tag{4.9}$$

$$\underset{\sim}{\chi}_m(t_{n+1}) = 0 \tag{4.10}$$

to obtain $\underset{\sim}{\chi}^{n+1}_m = \underset{\sim}{\chi}_m(t_n)$. For example, if we use one step of an explicit Euler method we have

$$\underset{\sim}{\chi}^{n+1}_m(t_n) = \tau\underset{\sim}{u}^n(\underset{\sim}{x}_m)\,/\phi(\underset{\sim}{x}_m) \tag{4.11}$$

which is a suitable approximation if the velocity field is nearly constant. In general, if we wish to use an explicit Euler method, we must split the overall time step, τ, into sub-intervals. If this is inefficient we must use a higher-order method such as 4th order explicit Runge-Kutta and even then we may have to split the time step if the characteristics are very curved.

Step 3. Update the fixed mesh concentrations by,

$$c_m^{n+1} = I_F^2[c_\ell^n, x_m - \chi_m^{n+1}] + \frac{\tau}{\phi_m} \Delta\cdot(D^n\Delta)\ c_m^n \tag{4.12}$$

where $\Delta\cdot(D^n\Delta)$ is the central difference approximation to $\nabla\cdot(D^n\nabla)$ (see Peaceman (29)) and $\phi_m = \phi(x_m)$. We use bilinear interpolation for the fixed mesh interpolator. Since we are using an explicit version of the algorithm we must satisfy a stability condition on τ. This condition is the same as for the explicit finite difference algorithm for the diffusion equation (Peaceman (29)). An implicit version may be obtained by replacing c_m^n by c_m^{n+1} in the dispersion term of Eq. (4.12).

For fictitious points in fictitious cells adjacent to $\partial\Omega_1$ we use the Dirichlet boundary conditions. For fictitious points in fictitious cells adjacent to $\partial\Omega_2 \cup \partial\Omega_3$ we use block-centred reflection boundary conditions following Peaceman and Rachford (28). We must impose these boundary conditions at all times, including $t = 0$.

Algorithm (4.12) is a version of the modified method of characteristics (Douglas and Russell (6); Huffenus and Khaletzky (19); Hartree (16)). In one dimension the method reduces to first-order upstreaming and in two- or three-dimensions is equivalent to the tensor-viscosity method of Dukowicz and Ramshaw (7).

In one dimension with constant ϕ, u and D where D is for the rest of this subsection a constant diffusion-dispersion parameter and using Eq. (4.11) for the characteristic nodal displacement we find that,

$$\begin{aligned} \phi \frac{(c_m^{n+1} - c_m^n)}{\tau} &= -\frac{u}{h}(c_m^n - c_{m-1}^n) + D\frac{(c_{m-1}^n - 2c_m^n + c_{m+1}^n)}{h^2} \\ &= -\frac{u}{2h}(c_{m+1}^n - c_{m-1}^n) + (\frac{uh}{2} + D)\frac{(c_{m+1}^n - 2c_m^n + c_{m-1}^n)}{h^2} \end{aligned} \tag{4.13}$$

The coefficient $\frac{uh}{2}$ is the *coefficient of numerical diffusion* and in practical problems, because of computer memory limitations, may be orders of magnitude larger than D. The ratio,

$$Pe_h = uh/D \tag{4.14}$$

is called the *mesh Peclet number*.

Numerical diffusion causes excessive numerical smearing of solutions in regions of large spatial gradient. This is a serious difficulty in oil recovery simulation because the numerical smearing causes spurious stability leading to overestimates of recovery efficiency. In problems involving the pollution of water aquifers numerical diffusion may lead to overestimates of the spatial extent of contamination or to the suppression of fluctuation phenomena which can be of importance in risk assessment applications of numerical transport algorithms.

The use of a higher order, quadratic, fixed mesh interpolator does not greatly improve pointwise accuracy. This *two-point upstreaming* reduces numerical dispersion to a low level but introduces a phase error.

When ϕ and u are constant and D = 0 the two-point upstreaming scheme is

$$\phi \frac{(c_m^{n+1} - c_m^n)}{\tau} = - \frac{u}{2h} (3c_m^n - 4c_{m-1}^n + c_{m-2}^n) + \frac{\tau u^2}{2h^2} (c_{m+1}^n - 2c_m^n + c_{m-1}^n) \tag{4.15}$$

This algorithm is obtained by replacing the s = 2 interpolator in Eq. (4.12) by an s = 3 interpolator.

Assuming that the time truncation error can be reduced to negligible levels by choosing small enough τ, Eq. (4.15) is to order h^2 equivalent to

$$\phi\frac{\partial c}{\partial t} + u \frac{\partial c}{\partial x} = \frac{uh^2}{3} \frac{\partial^3 c}{\partial x^3} \tag{4.16}$$

Seeking solutions of the form

$$c(x,t) = \sin \kappa(x - \gamma t) \tag{4.17}$$

we find that

$$\gamma = \frac{u}{\phi}(1 + \frac{\kappa^2 h^2}{3}) \tag{4.18}$$

Thus waves travel with a velocity increased by a factor $\left(1 + \frac{\kappa^2 h^2}{3}\right)$. The wave number κ is large in sharp fronts so that such fronts propagate with a phase error. This analysis is verified by numerical experiments (Farmer (10)).

4.5 Finite Element Version of the Modified Method of Characteristics

As in the finite difference version we assign the c_m^o using the initial condition but in the sense of an L_2 projection. In other words we solve the equations

$$\sum_{\ell} c_{\ell}^{o}[\beta_{\ell}, \beta_m] = [c_o, \beta_m] \qquad (4.19)$$

where the index ℓ ranges over the indices of all the mesh points in Ω. The values of c_{ℓ}^{o} on fictitious points are assigned using the same discrete boundary conditions used in the finite difference version. The square bracket notation is defined by

$$[a,b] = \int_{\Omega} ab\, d\underset{\sim}{x} \qquad (4.20)$$

for arbitrary, integrable, scalar functions a and b.

We now suppose that at time t_n we know the c_m^n for each fixed mesh point. The c_m^{n+1} are constructed as follows.

Step 1 As for the finite difference version.

Step 2 As for the finite difference version.

Step 3 Using the Galerkin finite element method solve the elliptic boundary value problem

$$\phi(\underset{\sim}{x})c^{n+1}(\underset{\sim}{x}) = \phi(\underset{\sim}{x})I_F^2\,[c_{\ell}^{n}, \underset{\sim}{x} - \chi^{n+1}(\underset{\sim}{x})] + \tau\nabla\cdot(D^n\nabla)c^{n+1} \qquad (4.21)$$

to find $c^{n+1}(\underset{\sim}{x})$ subject to the boundary conditions (2.4) and (2.5) where $\chi^{n+1}(\underset{\sim}{x})$ is an interpolation of the characteristic nodal displacements. Since the finite element method will couple together grid points via the c^{n+1} term on the left hand side there is less advantage, compared to the finite difference version, in using an explicit approximation to the diffusion-dispersion term.

Essentially the same algorithm has been discussed by Douglas and Russell (6), Russell and Wheeler (35) and Bercovier et al (2) in relation to the convection-diffusion equation. Algorithms differ in their method of constructing χ^{n+1}. Morton and Parrott (24) and Morton (25) discuss a similar algorithm for pure convection and the generalisation to non-linear hyperbolic equations.

Implementation of the above algorithm requires evaluation of inner products of the form

$$[\beta_\ell(\underset{\sim}{x} - \chi^{n+1}(\underset{\sim}{x})),\ \beta_m(\underset{\sim}{x})] \tag{4.22}$$

This was accomplished using Lobatto quadrature by Russell and Wheeler (35).

The resulting algorithm is very much more accurate than the finite difference version, provided that no 'mass lumping' is performed. (Mass lumping is the process of constructing a diagonal approximation to the matrix arising from the discretisation of c^{n+1} in Eq. (4.21) by (i) summing the terms along rows off the main diagonal (ii) adding these sums to the main diagonal (iii) setting terms off the main diagonal to zero).

The full finite element modified method of characteristics (FEMMOC) shows excellent phase behaviour, very low levels of numerical diffusion and low levels of overshoot provided that any fronts are spread over three or more mesh cells (Russell and Wheeler (35)).

4.6 Approximate Versions of the Modified Method of Characteristics

4.6.1 2nd order FEMMOC

Expansion of algorithm (4.8) in a Taylor series in χ^{n+1} to 2nd order gives

$$c^{n+1}(\underset{\sim}{x}) = c^n(\underset{\sim}{x}) - \chi_i^{n+1} \frac{\partial c^n}{\partial x_i} + \frac{\chi_i^{n+1}\ \chi_j^{n+1}}{2} \frac{\partial^2 c^n}{\partial x_i \partial x_j} + \tau D_\omega^n \tag{4.23}$$

where we use the summation convention regarding repeated indices; χ^{n+1} and c^n are evaluated at $\underset{\sim}{x}$.

Löhner et al (22) describe an algorithm which differs from Eq. (4.23) only in the method for constructing χ^{n+1}. These authors have shown, for the special case D = 0, that spatial discretisation of Eq. (4.23) using the Galerkin finite element method with piecewise linear functions leads to an algorithm which has low levels of numerical diffusion, overshoot and phase error.

4.6.2 3rd order FEMMOC

Expansion of algorithm (4.8) with $D^n_\omega = 0$ to 3rd order in χ^{n+1} gives

$$c^{n+1}(\underset{\sim}{x}) = c^n(\underset{\sim}{x}) - \chi_i^{n+1} \frac{\partial c^n}{\partial x_i} + \frac{\chi_i^{n+1} \; \chi_j^{n+1}}{2} \frac{\partial^2 c^n}{\partial x_i \partial x_j} - \frac{\chi_i^{n+1} \; \chi_j^{n+1} \; \chi_k^{n+1}}{6} \frac{\partial^3 c^n}{\partial x_i \partial x_j \partial x_k} \tag{4.24}$$

The third order term in Eq. (4.24) may be approximated by substituting

$$-\chi_k^{n+1} \frac{\partial c^n}{\partial x_k} \approx c^{n+1}(\underset{\sim}{x}) - c^n(\underset{\sim}{x}) \tag{4.25}$$

which results in the scheme

$$\left(1 - \frac{\chi_i^{n+1} \; \chi_j^{n+1}}{6} \frac{\partial^2}{\partial x_i \partial x_j}\right) (c^{n+1}(\underset{\sim}{x}) - c^n(\underset{\sim}{x})) = -\chi_i^{n+1} \frac{\partial c^n}{\partial x_i} + \frac{\chi_i^{n+1} \; \chi_j^{n+1}}{2} \frac{\partial^2 c^n}{\partial x_i \partial x_j} \tag{4.26}$$

Donea (5) describes an algorithm which differs from Eq. (4.26) only in the method for constructing χ^{n+1}. Donea shows that spatial discretisation of Eq. (4.26) using the Galerkin finite element method with piecewise linear functions leads to an algorithm which again has very low levels of numerical diffusion, overshoot and phase error.

In Chapter 8 we will describe some numerical results comparing the Donea (5) method with first-order finite difference upstreaming and two moving point methods.

4.6.3 Remarks on semi-discrete finite element methods

As shown by Gresho et al (14) a standard semi-discrete Galerkin finite element method applied to the pure convection equation leads to excellent results provided (i) no mass lumping is performed (ii) fronts are smeared over three or more mesh cells (iii) the resulting initial value problem on the set of ordinary differential equations for the nodal concentrations is integrated with high accuracy. However, the high accuracy needed in the time integration leads to considerable computational expense.

The advantage of the various FEMMOC's over the semi-discrete Galerkin FEM is that very much longer time steps may be taken, without loss of accuracy, when updating the concentration field. Accurate construction of the characteristic nodal displacements may require short subsidiary time steps but this is only in solving a pair (triple in 3-D) of ordinary differential equations. Each mesh point has an associated pair of equations but these pairs of equations are not coupled together.

5. THE MOVING POINT METHOD

5.1 Definitions

Introduce a set of *moving points* $\{P_k : k = 1,2,\ldots,N\}$. N is the total number of moving points. Let $\underset{\sim}{X}_k^n = (X_k^n, Y_k^n)$ be the values of a two-component vector valued mesh function defined on the moving points P_k. $\underset{\sim}{X}_k^n$ is the *position vector* of *moving point* P_k *at time* t_n. Let A_k^n be the *concentration* on moving point P_k which is interpreted as an approximation to $c(\underset{\sim}{X}_k^n, t_n)$.

Let ν_m^n be the values of a scalar mesh function which is the total number of moving points with position vectors in the m-th mesh cell at time t_n. Cells in which $\nu_m^n = 0$ are referred to as *empty cells*.

If the k-th moving point, P_k, lies in the m-th mesh cell we shall write $k \varepsilon m$ and if point P_k is nearer to $\underset{\sim}{x}_m$ than any other moving point we write $k \bar{\varepsilon} m$.

5.2 Point Moving Algorithm

To construct $\underset{\sim}{X}_k^{n+1}$ from $\underset{\sim}{X}_k^n$ we solve the equation

$$\phi(\underset{\sim}{X}_k) \frac{d}{dt} \underset{\sim}{X}_k = \underset{\sim}{u}(\underset{\sim}{X}_k, t_n) \tag{5.1}$$

$$\underset{\sim}{X}_k(t_n) = \underset{\sim}{X}_k^n \tag{5.2}$$

to obtain $\underset{\sim}{X}_k^{n+1} = \underset{\sim}{X}_k(t_{n+1})$ (5.3)

The solution to the above initial value problem can be found using T steps of either (i) the first-order forward Euler algorithm or (ii) in cases where $\underset{\sim}{u}$ has strongly curved streamlines the classical 4-th order Runge-Kutta method with time step τ/T. T is an integer which may vary during the course of a calculation.

We denote the operation of moving the points by the symbol E and we write

$$\underset{\sim}{X}_k^{n+1} = E\left[\underset{\sim}{X}_k^n,\ \underset{\sim}{u}(\underset{\sim}{x},t_n),\ \tau/T\right] \tag{5.4}$$

5.3 Moving Point Interpolation

Let I_M^s denote the moving point interpolator which from moving point function values G_k^n and fixed mesh values F_ℓ^n constructs a function Q^n according to

$$Q^n(\underset{\sim}{x}) = I_M^s\left[G_k^n,\ F_\ell^n,\ \underset{\sim}{x}\right] \qquad \underset{\sim}{x}\varepsilon\Omega \tag{5.5}$$

where the indices k and ℓ range over the indices of the moving and fixed mesh points respectively and s is the order of the interpolator. We will relate G_k^n and F_ℓ^n to moving point and fixed mesh concentrations in later sections.

Some examples of moving point interpolation are;

5.3.1 NGP interpolation

*N*earest *g*rid *p*oint interpolation is a piecewise constant interpolation in which the concentration value on the moving point nearest to a mesh cell centre is assigned to the whole mesh cell. If the cell is empty we assign to the cell the fixed mesh value F_m^n. Thus,

$$Q^n(\underset{\sim}{x}) = \begin{cases} G_k^n & \underset{\sim}{x}\varepsilon\Omega_m,\ k\bar{\varepsilon}m,\ \nu_m^n > 0 \\ F_m^n & \underset{\sim}{x}\varepsilon\Omega_m,\ \nu_m^n = 0 \end{cases} \tag{5.6}$$

5.3.2 Averaging interpolation

Averaging interpolation assigns to mesh cells the average concentration of the moving points in that cell. If the cell is empty we assign the fixed mesh value F_m^n. Thus

$$Q^n(\underset{\sim}{x}) = \begin{cases} \sum_{j \varepsilon m} G_j^n & \underset{\sim}{x} \varepsilon \Omega_m, \ \nu_m^n > 0 \\ F_m^n & \underset{\sim}{x} \varepsilon \Omega_m, \ \nu_m^n = 0 \end{cases} \tag{5.7}$$

Averaging interpolation was introduced by Garder et al (13) although no prescription for the $\nu_m = 0$ case was stated.

NGP and averaging interpolation give rise to a piecewise constant fixed mesh function. When $\nu_m^n > 0$ these methods are locally first order in h. When $\nu_m^n = 0$ one cannot make a general statement about the order of accuracy.

Higher-order interpolation from moving points to the fixed mesh can, in principle, be performed by fitting an interpolation formula to moving point values. However, this is difficult from both numerical and implementation view points (see Franke (12) for a review of scattered data set interpolation). An s = 2 interpolator, which is easy to implement, has been described in Farmer (10).

In the context of the hybrid moving point method higher-order interpolation is much simpler and this is described in the next chapter.

5.4 The Pure Moving Point Method

At time t = 0 assign the moving point positions with, say, four points per fixed mesh cell. Moving point positions are most conveniently assigned using a uniform random number generator so that the initial number of points per cell is easy to adjust. Place moving points in all cells except fictitious cells adjacent to $\partial\Omega_2 \cup \partial\Omega_3$. The concentrations A_k^o are assigned using $A_k^o = c_o(\underset{\sim}{X}_k^o)$ and the fixed mesh concentrations using $c_m^o = c_o(\underset{\sim}{x}_m)$. Moving points in the fictitious cells adjacent to $\partial\Omega_1$ are given the appropriate Dirichlet boundary condition.

At time t_n we suppose c_m^n known for each fixed mesh cell, $\underset{\sim}{X}_k^n$ and A_k^n known for each moving point. The c_m^{n+1}, $\underset{\sim}{X}_k^{n+1}$ and A_k^{n+1} are constructed as follows;

Step 1. Compute the velocity mesh functions $u_{i+1/2,j}^n$ and $v_{i,j+1/2}^n$ where $u_{i+1/2,j}^n$ is an approximation to the x-component velocity at $x_i+h/2$ and $v^n_{i,j+1/2}$ an approximation to the y-component of velocity at $y_j+h/2$. Methods for constructing the velocity mesh function are described in Chapter 9.

The moving point method requires the provision of a velocity field everywhere in Ω and we construct a piecewise interpolation $\underset{\sim}{u}^n(\underset{\sim}{x}) = (u^n(\underset{\sim}{x}), v^n(\underset{\sim}{x}))$ by,

$$
\begin{aligned}
u^n(\underset{\sim}{x}) &= u^n_{i-1/2,j}\ (1 - \zeta) + u^n_{i+1/2,j}\ \zeta \\
v^n(\underset{\sim}{x}) &= v^n_{i,j-1/2}\ (1 - \eta) + v^n_{i,j+1/2}\ \eta
\end{aligned}
\tag{5.8}
$$

where
$$
\begin{aligned}
\zeta &= (x - x_i)/h + 1/2 \\
\eta &= (y - y_j)/h + 1/2
\end{aligned}
\tag{5.9}
$$

and $\underset{\sim}{x}\ \varepsilon\Omega_m$.

The interpolation (5.8) has the properties;

(i) it is solenoidal in cell m by virtue of the discrete zero divergence condition Eq. (9.1).

(ii) the velocity normal to the cell surface is continuous.

It is possible to construct a full bi-linear interpolation which has the above properties which is useful in test problems with non-constant velocity fields. However, practical problems involving oil reservoirs or water aquifers very often have permeability streaks in which the velocity does possess tangential discontinuities. Thus in practical problems we favour the interpolation (5.8).

Step 2 Update the moving point positions to obtain $\underset{\sim}{X}^{n+1}_k$ by

$$
\underset{\sim}{X}^{n+1}_k = E[\underset{\sim}{X}^n_k,\ \underset{\sim}{u}^n(\underset{\sim}{x}),\ \tau/T] \tag{5.10}
$$

Choose T so that over any time interval, τ/T, no moving point travels a distance of, say, more than h/3 in any direction.

Step 3 Calculate the changes, δc^n_m, in concentration due to hydrodynamic dispersion by,

$$\delta c_m^n = \tau\Delta\cdot(D^n\Delta)c_m^n \tag{5.11}$$

where $\Delta\cdot(D^n\Delta)$ is the central difference approximation to $\nabla\cdot(D^n\nabla)$. We must ensure that τ is small enough to satisfy the usual stability condition for an explicit central difference method for a diffusion equation.

Step 4 Update the moving point concentrations by setting, for each k,

$$A_k^{n+1} = A_k^n + I_F^2[\delta c_\ell^n, x_k^{n+1}] \tag{5.12}$$

Step 5 Construct the fixed mesh concentrations c_m^{n+1} by the moving point interpolation

$$c_m^{n+1} = I_M^s[A_k^{n+1}, c_\ell^n, x_m] \tag{5.13}$$

Thus for empty cells we use the fixed mesh concentration at the previous time step.

Step 6 Delete all moving points from the fictitious cells. Place, say, four moving points carrying the appropriate Dirichlet boundary condition in each fictitious cell adjacent to $\partial\Omega_1$.

5.5 Accuracy and Convergence of the Pure Moving Point Method

Averaging interpolation (s=1) was used in the original Garder et al (13) algorithm. We call such algorithms *pure* to distinguish them from the hybrid algorithms to be described later. Garder et al (13) showed that on problems with a high mesh Peclet number the pure moving point method was very accurate even when fronts were sharp. However Price, Cavendish and Varga (31) showed in numerical experiments that under mesh refinement the algorithm did not converge. Furthermore, as emphasised by Neuman (26), no analysis of the algorithm was available.

We have observed in numerical experiments that if an s=2 (Farmer (10)) moving point interpolator is used then the pure moving point method does converge. Unfortunately we can provide no numerical analysis to explain this.

However, in Farmer (10) we have shown that convergent pure moving point algorithms can be formulated by modifying the calculation of the change in moving point concentrations due to dispersion. This is accomplished by placing the centre of the difference stencil upon each moving point and using a mesh spacing $h^* \geq h$ for the central difference approximation. The mesh values on the stencil, other than the centre value on the moving point, are obtained by interpolation. Convergence is obtained by making $h \to 0$ faster than $h^* \to 0$. In this way we can construct a convergent algorithm even when using first order moving point interpolators such as NGP or averaging interpolation.

Our convergence proof relied on the assumptions of (i) periodic boundary conditions (ii) constant and diagonal dispersion tensor (iii) at least one moving point in each fixed mesh cell before performing the moving point interpolation. Although it may be possible to relax these assumptions we have developed the hybrid algorithm, described in the next chapter, as a way of improving the accuracy and efficiency of the moving point method.

5.6 Related Methods

The moving point method has some similarity with the random walk method of Prickett et al (32) in which diffusion/dispersion is approximated by imposing a suitable random walk upon the trajectories of the moving points. The fixed mesh concentrations are obtained from the expectation values of the ν_m^n.

Raviart (33), (34) has described a moving point method which does not use an intermediate fixed mesh. We are not, however, aware of any numerical tests or implementations of this algorithm. In our applications to oil reservoir simulation we require concentrations on a fixed mesh for the purpose of finding viscosities for the fixed mesh calculation of velocities. For this reason we require a moving point to fixed mesh interpolator.

5.7 Remarks on the Differences between the MMOC and the MPM

Let us suppose that we construct the moving point positions in the MPM and the characteristic nodal displacements in the MMOC without significant error. Further suppose the diffusion/dispersion tensor to be zero so that the problem is one of pure convection. In this case the concentrations carried by the moving points in the MPM are *exact at the moving points*. The error incurred in the MPM is then only that introduced by the moving point interpolator. Thus the error in the MPM does not grow in time. In the MMOC there is a gradual increase in error as a result of the repeated projection of the transported solution to the fixed mesh.

The essential difference between the MPM and the MMOC is that in the MPM, moving points retain their identity throughout a calculation whereas in the MMOC the moving points are defined anew at every time step.

6. THE HYBRID MOVING POINT METHOD

6.1 Introduction

The simpler fixed mesh methods, such as finite difference MMOC work well if the mesh Peclet number is small or if the solution is very smooth. The pure MPM works well if the mesh Peclet number is large and the solution is not very smooth. The hybrid moving point method is a combination of these two schemes in which at sufficiently low mesh Peclet number only the fixed mesh method is used. Since the fixed mesh method is convergent the hybrid scheme is, trivially, convergent.

6.2 Fixed Mesh Algorithm

Any convergent fixed mesh scheme is appropriate and a convenient choice is the first order finite difference MMOC. The moving point interpolator in the HMPM requires an estimate of the fixed mesh concentrations and this is obtained using the MMOC. Let $\tilde{c}_m^{n+1}$ denote this first estimate and we write

$$\tilde{c}_m^{n+1} = \Phi[c_\ell^n] \tag{6.1}$$

where the index ℓ ranges over the indices of the fixed mesh cells.

6.3 Moving Point Interpolation

Following the notation of section (5.3) we introduce an interpolation based upon a Taylor expansion about the nearest moving point to a cell centre using spatial derivatives evaluated on the fixed mesh,

$$Q^n(\underset{\sim}{x}) = \begin{cases} G_k^n - (\underset{\sim}{X}_k^n - \underset{\sim}{x})\cdot\nabla I_F^s[F_\ell^n,\underset{\sim}{x}] & \underset{\sim}{x}\varepsilon\Omega_m,\ k\bar{\varepsilon}m,\ \nu_m^n > 0 \\ I_F^2[F_\ell^n,\underset{\sim}{x}] & \underset{\sim}{x}\varepsilon\Omega_m,\ \nu_m^n = 0 \end{cases} \tag{6.2}$$

where $s \geqslant 3$.

In cases where F_m^n possesses large gradients the Taylor interpolator described above can overshoot whereas NGP or averaging interpolation will not overshoot. In our codes we combine a second order Taylor interpolator ($s = 3$ in Eq. (6.2)) with a lower order method according to the magnitude of the local concentration gradient. If the concentration gradient exceeds a user defined value we use averaging interpolation and otherwise the Taylor method.

In Farmer and Norman (11) we used NGP interpolation when the concentration gradient was large. However we have found in practical applications that averaging interpolation is preferable since it provides some local smoothing and provides information about the position of a discontinuity within a mesh cell.

6.4 Local Insertion-Deletion of Moving Points

In many problems of interest the concentration field may be such that it is rapidly changing at sharp fronts which only occur in a small fraction of Ω. In such cases the solution will be constant with values 0 or 1 over large subregions of Ω. When this happens it is only necessary for there to be moving points at or near to the fronts. We therefore implement an insertion-deletion strategy which controls the distribution of moving points.

Let,

$$S_m^n = (c_{i+1,j}^n - c_{i-1,j}^n)^2 + (c_{i,j+1}^n - c_{i,j-1}^n)^2 \tag{6.3}$$

where $m = (j-1)N_x + i$

and specify;

(i) the minimum, m_b, and maximum, M_b, allowable values of ν_m^n when $S_m^n < S^*$

(ii) the minimum, m_s, and maximum, M_s, allowable values of ν_m^n when $S_m^n \geqslant S^*$

where S^* is a slope parameter used to define the presence of sharp fronts.

The insertion-deletion procedure is then;

1. In each fixed mesh cell, if necessary, delete sufficient moving points to satisfy the condition

$$(\nu_m^n \leqslant M_b \text{ and } S_m^n < S^*) \text{ or } (\nu_m^n \leqslant M_s \text{ and } S_m^n \geqslant S^*).$$

It is convenient to delete moving points with the largest k-indices amongst the points in the m-th mesh cell.

2. If $(\nu_m^n < m_b$ and $S_m^n < S^*)$ or $(\nu_m^n < m_s$ and $S_m^n \geqslant S^*)$ then insert a single moving point with its position vector equal to $\underset{\sim}{x}_m$ and assign the concentration c_m^n to the moving point.

The operation of insertion-deletion will be denoted by the symbol

$$\underset{\sim}{X}_k^n = J[\underset{\sim}{X}_\ell^n,\ c_m^n] \tag{6.4}$$

6.5 The Hybrid Moving Point Method

The overall structure of the hybrid algorithm is of the implicit-pressure, explicit-saturation kind as used, for example by Peaceman and Rachford (28).

At time t = 0 we perform the steps;

(i) assign fixed mesh concentrations, c_m^o, for each m by setting

$$c_m^o = c_o(\underset{\sim}{x}_m) \tag{6.5}$$

(ii) introduce moving points uniformly at random within each cell thus assigning the $\underset{\sim}{X}_k^o$. The number of moving points in the m-th mesh cell will be

$$m_b \text{ if } S_m^o < S^*$$

$$\text{or} \quad m_s \text{ if } S_m^o \geqslant S^* \tag{6.6}$$

(iii) assign moving point concentrations, A_k^o, for each k by setting

$$A_k^o = c_o(\underset{\sim}{X}_k^o) \tag{6.7}$$

At time t_n we suppose c_m^n known for each fixed mesh cell, $\underset{\sim}{X}_k^n$ and A_k^n known for each moving point. The c_m^{n+1}, $\underset{\sim}{X}_k^{n+1}$ and A_k^{n+1} are constructed as follows;

Step 1 As Step 1 of the pure MPM in Section 5.4.

Step 2 As Step 1 of the pure MPM.

Step 3 Update the fixed mesh concentrations using the fixed mesh algorithm to obtain the first approximation

$$\tilde{c}_m^{n+1} = \Phi[c_\ell^n] \tag{6.8}$$

Step 4 Calculate the changes, δc_m^n, in concentration due to diffusion-dispersion by

$$\delta c_m^n = \tau\Delta\cdot(D^n\Delta)c_m^n \tag{6.9}$$

where $\Delta\cdot(D^n\Delta)$ is the central difference approximation to $\nabla\cdot(D^n\nabla)$. We must test that τ is small enough to satisfy the usual stability condition for an explicit central difference method for a diffusion equation.

δc_m^n can be obtained as part of Step 3.

Step 5 Update the moving point concentrations by setting, for each k,

$$A_k^{n+1} = A_k^n + I_F^2[\delta c_\ell^n, \underset{\sim}{x}_k^{n+1}] \tag{6.10}$$

Step 6 Construct the fixed mesh concentration c_m^{n+1} by the interpolation

$$c_m^{n+1} = \lambda I_M^s[A_k^{n+1}, c_\ell^{*n+1}, \underset{\sim}{x}^m] + (1 - \lambda)c_m^{*n+1} \tag{6.11}$$

where

$$c_\ell^{*n+1} = \begin{cases} \tilde{c}_\ell^{n+1} & \ell \geqslant m \\ c_\ell^{n+1} & \ell < m \end{cases} \tag{6.12}$$

and $\geqslant$ and $<$ in Eq. (6.12) denotes an ordering relation on the mesh cell indices. λ is a weighting factor, dependent upon the lowest upper bound, Pe_h, of the mesh Peclet number. The function λ is chosen via numerical experiments in such a way that the hybrid moving point method is trivially convergent. A suitable λ is

$$\lambda = \begin{cases} 1 & Pe_h \geqslant 10^{-3} \\ 0 & Pe_h < 10^{-3} \end{cases} \tag{6.13}$$

Step 7 Perform the insertion-deletion procedure

$$\underset{\sim}{X}_k^{n+1} = J[\underset{\sim}{X}_\ell^{n+1}, c_m^{n+1}] \tag{6.14}$$

7. THE IMPLEMENTATION OF MOVING POINT METHODS

7.1 Data Structures

For the fixed mesh algorithm we require arrays C, U and V for the fixed mesh concentrations and velocities. As workspace we require arrays CØ and DC to store the concentrations at the previous time step and the change in fixed mesh cell concentrations due to diffusion-dispersion.

For the moving point algorithm we require arrays A, X and Y for the concentrations and x-, y-coordinates of the moving points. As workspace we require arrays NN and NGP of dimension equal to the number of fixed mesh cells. NN and NGP are used to store the number of moving points in each cell and the index of the moving point nearest to a cell centre. We initialise NN and NGP to zero at each time step so that a zero in these arrays corresponds to an empty cell.

As explained in Section 4.1 we use fictitious cells surrounding the physical cells. We assign a zero velocity to the outward facing boundaries of these cells. Provided that the time step is not too long the moving points will remain within the physical and fictitious cells.

7.2 Scaling of Positions and Velocities

It is convenient to scale X, Y, U and V so that (i) the integer parts of X and Y give the i and j indices of the mesh cell occupied by the moving points and (ii) the non-integer part the coordinates in the local (ζ, η) system in the cell as defined by Eqs. (5.9). We scale U and V by the equivalent of dividing the velocities by the mesh lengths so that the inner loop executing the point moving algorithm contains no divisions.

The device of assigning zero velocity to the outward facing sides of the fictitious cells cannot be guaranteed to prevent points from leaving the defined region. Thus to prevent errors of addressing non-existent array elements it is wise to test the i and j indices of moving points and force moving points which enter fictitious cells to remain there during the execution of the point moving algorithm.

7.3 Point Insertion and Deletion

A convenient method of moving point deletion is to loop through the moving points as described in the following pseudo-code notation.

```
NP = 0
FOR K = 1 TO N
  X(K-NP) = X(K)
  Y(K-NP) = Y(K)
  A(K-NP) = A(K)
  L = (INT(Y(K)-1))*N_x + INT(X(K))
  NN(L) = NN(L) + 1
  If point K is in a fictitious cell or NN(L) exceeds the
  maximum allowable number of points then NP = NP+1
NEXT K
```

To insert moving points we first place the required number, m_b, in each fictitious cell from which fluid flows into Ω; that is, in the fictitious cells adjacent to $\partial\Omega_1$. If the number of points to be inserted is greater than one it is convenient to place these points using a uniform random number generator.

We then place a single moving point in each mesh cell in Ω where the number of moving points is less than the allowable minimum. The allowable minimum and maximum in general depends upon the local concentration gradient and so we are performing a type of adaptive mesh refinement which is very easy to implement.

8. NUMERICAL EXPERIMENTS

8.1 One Dimensional Convection

In this section we describe numerical investigations into the accuracy of algorithms for pure convection or, equivalently, infinite mesh Peclet number convection-diffusion/dispersion. In particular we examine accuracy on a fixed mesh as a function of the width of a front. This is essentially equivalent to examining accuracy under mesh refinement with a front of fixed width.

To investigate the effect of the smoothness of solutions upon the accuracy of numerical methods in simulating convection we will study the following test problem:

Find a function $c = c(x,t)$ satisfying the conditions,

$$\frac{\partial c}{\partial t} + u\frac{\partial c}{\partial x} = 0 \qquad 0 \leqslant x \leqslant L_x,\ t > 0 \tag{8.1}$$

$$u > 0 \qquad \frac{\partial u}{\partial x} = 0 \tag{8.2}$$

$$c(0,t) = 1 \qquad t \geqslant 0 \tag{8.3}$$

$$c(x,0) = e^{-(x/s)^2} \qquad 0 \leqslant x \leqslant L_x \tag{8.4}$$

where s is a parameter controlling the width of the front.

This problem has the exact solution

$$c(x,t) = \begin{cases} 1.0 & x \leqslant ut \\ e^{-(x-ut)^2/s^2} & x > ut \end{cases} \tag{8.5}$$

We will examine the accuracy of the finite difference, first-order upstreaming method, the Donea (5) Taylor-Galerkin MMOC, the HMPM with NGP interpolation, Eq. (5.6), and the HMPM with Taylor interpolation, Eq. (6.2).

Numerical experiments were performed using 40 mesh cells and a Courant number, $u\tau/h$, of 0.269. The moving point methods used 4 moving points per cell placed at $t = 0$ using a uniform random number generator. The results after 50 time steps are shown for various values of s in Figures 2-4.

We see that the first order upstreaming solution is very inaccurate, giving answers which are insensitive to the problem. The Taylor-Galerkin method is very good if the solution possesses some smoothness. However the HMPM with the Taylor interpolator (denoted by GNT, for Gregory-Newton-Taylor, on the figures) is the most accurate method on the smooth problems. As the solution becomes discontinuous the NGP moving point method is the most accurate of the algorithms tested. Nevertheless, the accuracy of the HMPM with the Taylor interpolator is good even when the solution is discontinuous. We note that in Figure 4 ($s = 0$) the exact solution is a step function; the figure shows a linear interpolation of the exact solution and so appears as a ramp function.

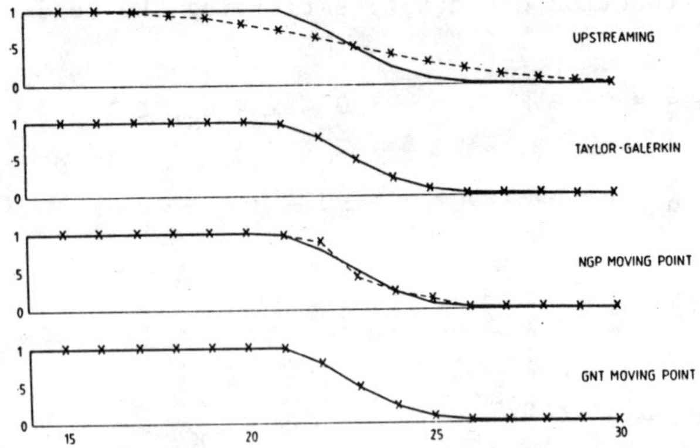

Figure 2. Linear interpolation of results of convection test with S = 25, ——— exact solution, ------ numerical solution.

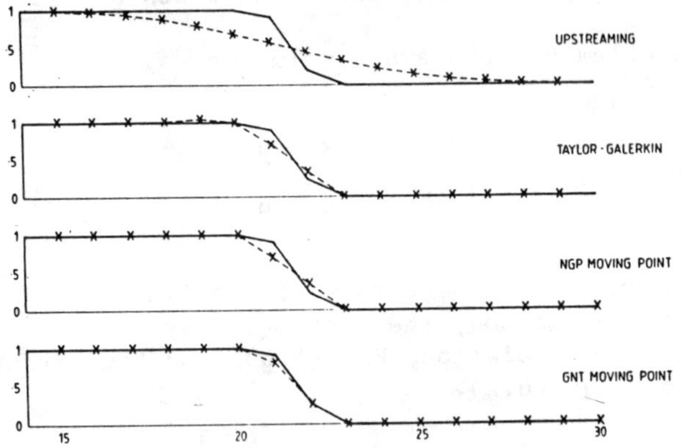

Figure 3. Linear interpolation of results of convection test with S = 10, ——— exact solution, ------ numerical solution.

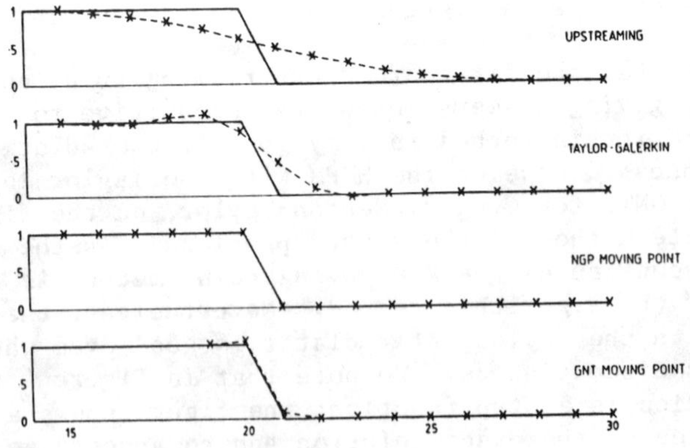

Figure 4. Linear interpolation of results of convection test with S = 0, ——— exact solution, ------ numerical solution.

8.2 One-Dimensional Convection-Diffusion

In this sub-section we assume constant ϕ, u and D where D is a constant diffusion-dispersion parameter.

A solution of the constant coefficient convection-dispersion equation in an infinite interval is

$$c_s(x,t) = \tfrac{1}{2} - \tfrac{1}{2}\,\mathrm{erf}\left[\frac{x - ut/\phi}{2\sqrt{Dt/\phi}}\right] \qquad (8.6)$$

If we choose constants u, D, x_o and t_o appropriately then the function

$$c(x,t) = c_s\,(x-x_o, t+t_o) \qquad (8.7)$$

is, for small enough t, a good approximation to the solution of the initial boundary value problem stated in Section 2.1. The solution (8.6) has the form of a smoothed step function with values between 0 and 1.

8.2.1 Effect of mesh refinement

We first investigate the effect of mesh refinement, holding all other parameters fixed. At the initial time we have placed 4 moving points in each mesh cell using a uniform random number generator and have performed no local refinement. We have set $M_b = 10$ and $m_b = 1$ to control the subsequent number of points per cell, although in the present example the number of points per cell cannot exceed 8. We have used Taylor interpolation (see Section 6.3) and in the y-direction there is one mesh cell.

The values of the fixed parameters are given in Table 1.

In Table 2 we show the maximum absolute errors

$$\varepsilon^n = \max_m \left| c(x_m, t_n) - c_m^n \right| \qquad (8.8)$$

after a time equal to 16 of the time steps used in the 40 mesh cell case. For the case of 40 cells we give the errors of two different realisations of the random number generator used to place the moving points at t = 0.

Table 1. Parameters used in numerical experiments on mesh refinement.

Parameter	Value	Dimensions
L_x	10^3	m
L_y	10^2	m
Front velocity	1	ft/day
Porosity	0.2	
D	$1.0\ 10^{-6}$	$m^2\ s^{-1}$
t_o	$9.0\ 10^7$	s
x_o	0.0	m

Table 2. Maximum absolute error as a function of mesh size.

No of Cells	h	τ	Pe_h	maximum absolute error
20	50	4.10^6	35.28	$1.274\ 10^{-1}$
40	25	2.10^6	17.64	$1.516\ 10^{-2}$
				$1.306\ 10^{-2}$
80	12.5	10^6	8.82	$3.156\ 10^{-3}$
160	6.25	$0.5\ 10^6$	4.41	$8.456\ 10^{-4}$

Price Cavendish and Varga (31) observed that the original algorithm of Garder et al (13) failed to converge under mesh refinement. The results reported in Table 2 indicate convergence of the hybrid moving point method even when $\lambda(h) = 1$, for all h. We have been unable to prove convergence in this special case but it is clear that the increase in the order of the moving point interpolator, from one to two, leads to an improved performance.

8.2.2 Error at low mesh Peclet numbers

To check the accuracy of the hybrid moving point algorithm at low mesh Peclet numbers we examined the maximum absolute errors using the fixed parameters in Table 3. We then varied the velocity and have displayed the error in Table 4.

Table 3. Parameters used for low mesh Peclet number experiments

Parameter	Value	Dimensions
No of cells	40	
h	25	m
τ	3.10^6	s
t_o	9.10^7	s
x_o	5.10^2	m
D	10^{-5}	$m^2 s^{-1}$

Table 4. Errors for moving point (λ=1) and upstreaming (λ=0) algorithms at low mesh Peclet numbers

Front Velocity (ft/day)	h	Pe_h	λ	ε^{16}
.01	25	.0176	1	$3.897\ 10^{-4}$
			0	$7.197\ 10^{-4}$
.001		.0018	1	$1.930\ 10^{-4}$
			0	$2.281\ 10^{-4}$

The results in Table 4 show the hybrid moving point method with Taylor interpolation to be more accurate than first order upstreaming even at low mesh Peclet numbers. On the basis of these results we have selected the $\lambda(h)$ weighting function given by Eq. (6.9).

Since the hybrid algorithm seems to converge, the weighting function $\lambda(h)$, Eq. (6.9), only serves the purpose of enabling us to assert that our algorithm is convergent; in practical problems the mesh Peclet number will often be such that the upstreaming algorithm is only used in empty cells.

8.2.3 Effect of varying the number of moving points per cell

Garder, Peaceman and Pozzi (13) found that the accuracy of a moving point method using averaging interpolation is not sensitive to the number of moving points per cell. The use of the higher order Taylor interpolator, Eq. (6.2), might be expected to increase this sensitivity so that as the number of moving points per cell is increased then accuracy improves. Unfortunately this proves not to be the case.

The results shown in Table 6 were obtained using the parameter values in Table 5. The use of S*=0 defines the whole flow region as a 'front' in the sense that at t=0 moving points are placed, uniformly at random, in all the mesh cells. The upper limits on the number of points per cell were chosen so that points were not deleted from the interior of the flow region.

In Table 6 we observe the interesting, and perhaps surprising, effect of a deterioration in accuracy as the number of moving points is increased above approximately 8 per cell. However, the method is inconsistent under this mode of refinement with a truncation error of order h^o; the observed convergence under overall mesh refinement, as observed in Section 8.2.1, is the surprising effect.

Table 5. Parameters used in testing accuracy when varying the number of moving points per cell.

Parameter	Value	Dimensions
No of cells	20	
h	50	m
τ	4.10^6	s
t_o	9.10^7	s
x_o	0	m
D	10^{-6}	$m^2 s^{-1}$
front velocity	1	ft/day
Pe_h	35.28	

Table 6. Effect on the maximum absolute error of varying the number of moving points per cell.

m_b	m_s	S*	ε^{54}	
4	4	0	5.220	10^{-2}
8	8	0	2.625	10^{-2}
16	16	0	3.962	10^{-2}
24	24	0	4.111	10^{-2}

8.2.4 Effect of local moving point refinement

Using the parameter values in Table 5 we investigated the effect of adjusting the parameters in the local moving point insertion-deletion algorithm. Results are shown in Table 7. The moving points are placed uniformly at random within each cell with a 'background' density of N_b/cell and a density of N_s/cell under the front at t=0. Fronts are defined by the slope parameter S* which was introduced in Section 6.4. M_b and m_b are upper and lower bounds on the densities of background points; M_S and m_s are upper and lower bounds on the numbers of points falling under a front.

The result in Table 7 for S* = 1.0 corresponds to a pure first-order upstreaming algorithm. The result with S* = 0.2 shows how considerable improvements in accuracy can be obtained with moving points placed only in the few cells occupied by a front. The result with S* = 0.0 shows that little accuracy is gained by using moving points throughout the flow region.

Table 7. Effect of local moving point refinement on maximum absolute errors

M_b	N_b	m_b	M_S	N_S	m_s	S*	No of cells with moving points at t=0	Total no of moving points used	ε^{54}
10	0	0	10	4	1	0	20	80	$5.665\ 10^{-2}$
						0.2	3	12	$6.084\ 10^{-2}$
						0.6	1	4	0.1544
						1.0	0	0	0.2268
10	0	1	10	4	1	0.6	20	23	$9.691\ 10^{-1}$

8.3 Two-Dimensional Solutions

In Section 3.2.2 we stated the exact solution of Jacquard and Séguier (20) for the growth of a viscous finger in a segregated displacement. These authors also give exact solutions for the fluid pressure in the oil and water phases. Numerical evaluation of these expressions showed that the pressure quickly becomes constant across the channel (i.e. $\frac{\partial p}{\partial y} \approx 0$) as one moves away from the front. It is therefore a good approximation to consider a long but finite channel in which to compare numerical and analytical solutions.

By comparing the exact position of the front with the numerically predicted position we are able to obtain a quantitative measure of the errors in our numerical procedures in a physically relevant case. In our numerical experiments we examine the areal case obtained by setting g=0.

Of particular interest are cases with an adverse mobility ratio, $\mu_o/\mu_w > 1$. However, as shown by Saffman and Taylor (36) such adverse mobility ratio problems have linear stability behaviour in which the growth parameters of modes are inversely proportional to their wavelength. Such behaviour is symptomatic of an ill-posed problem (Ockendon (27)). At the very best one expects numerical methods to be ill-conditioned on such problems.

Indeed if one attempts to construct numerical solutions corresponding to an adverse mobility ratio problem of the Jacquard-Séguier type one observes one of two things;

(i) On meshes with rectangular symmetry, with or without moving points, the solutions produce viscous 'ears' on the 'nose' of the viscous finger. (If moving point methods are used this result is only obtained using initial point positions with the same symmetry as the mesh). We have observed viscous ears in 1-point and 2-point upstreaming finite difference methods applied to this problem.

(ii) On regular meshes with rectangular symmetry but using moving points placed at random at t=0 the solutions produce multiply fingered fronts as observed in physical experiments.

We have investigated several procedures for regularising this problem. These have included the use of artifically high dispersion coefficients and the use of filters, neither of which works satisfactorily. However we have found that integration backwards in time will produce good results - even for finite dispersion cases in which the problem is ill-posed. The backward miscible displacement equations with non-zero D are ill-posed in the same sense that the backward heat problem is ill-posed. However in the case of the backward miscible displacement problem the numerics are well-conditioned if the mobility ratio is sufficiently adverse and the mesh is not too fine. The backward miscible displacement problem, with non-zero D, can, if necessary, be regularised by devices such as the quasi-reversibility method of Lattes and Lions (21). A detailed discussion of backward integration of miscible displacement problems can be found in Farmer (8).

In the case of zero diffusion-dispersion the solvent concentration satisfies a pure convection equation. For a given velocity field the Garder et al (13) algorithm is convergent. The results described in this section were obtained using the pure moving point method with averaging interpolation.

We have experimented on the Jacquard-Séguier problem with $\gamma_o = 30$ and $\mu_o/\mu_w = 2$. Four moving points per fixed mesh cell were placed on a 38 x 38 physical mesh at t = 0. Figures 5a - 5d show the development of a front backwards in time. The ruled region represents a viscous oil; the clear region lower viscosity water. A mesh cell, Ω_m^n, is defined as containing oil if $c_m^n < 0.5$ and water if $c_m^n \geqslant 0.5$. The bold line marks a linear interpolation of the exact Jacquard-Séguier solution. These pictures show that the front position is located by the numerical method to an accuracy of approximately $\pm h$ if the system is physically stable.

The moving point method will generally give this accuracy provided that the system to be simulated is physically stable on the mesh scale. For example the Dietz problem with gravity effects present, Eq. (3.12), may be solved by moving point methods also with an error of approximately $\pm h$ in the predicted front position.

Further numerical results, relating to the 2-D rotating cone test problem, may be found in Farmer and Norman (11).

9. NUMERICAL SOLUTION OF THE VELOCITY-PRESSURE EQUATIONS

9.1 Introduction

In this chapter we briefly review methods for solving the velocity-pressure equations, Eqs. (2.6)-(2.10). The quantity of greatest importance in miscible and segregated displacement is the velocity, rather than the pressure. For this reason methods which solve for the pressure and which then construct the velocity by numerical differentiation are, generally, not appropriate because the resulting interpolated velocity field is not solenoidal.

Thus we are interested in *mixed methods* which treat both the velocities and the pressures as unknowns. It is our intention in this chapter to merely indicate the structure of such methods since many reviews are already available; Aziz and Settari (1), Peaceman (30), Russell and Wheeler (35). The review of Russell and Wheeler (35) is of particular relevance since they discuss the solution of the velocity-pressure equations in the context of the modified method of characteristics for the concentration equation.

9.2 The Mixed Finite Difference Method

Integration of Eq. (2.6) over Ω_m and approximation of the velocity field by assuming that the normal component of velocity is constant along any face of Ω_m leads to the equations,

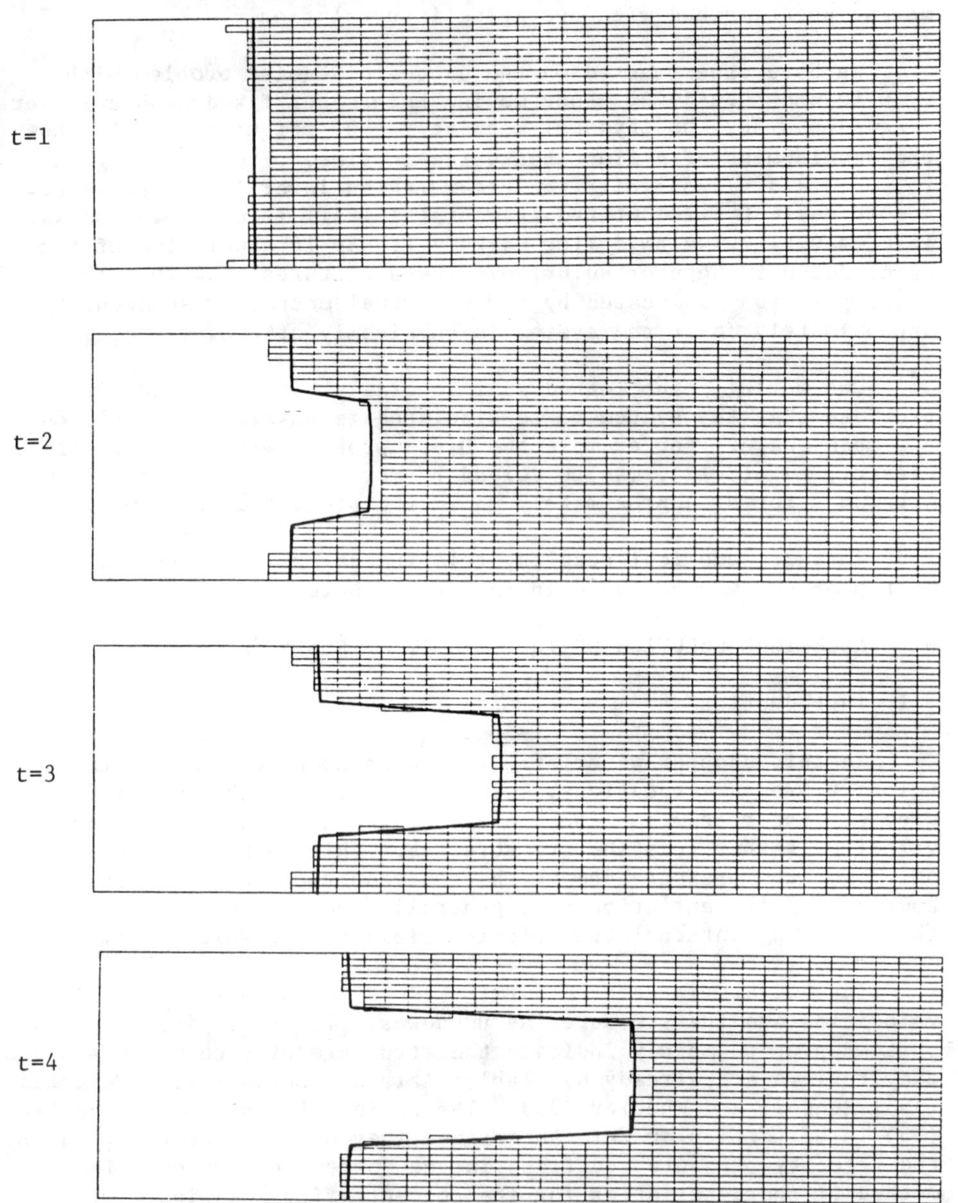

Figure 5. Jacquard-Séguier test problem. Ruled area, numerical prediction of oil-filled area; —— exact position of front.

$$u^n_{i+1/2,j} - u^n_{i-1/2,j} + v^n_{i,j+1/2} - v^n_{i,j-1/2} = 0 \tag{9.1}$$

where our notation is explained in Section 5.4.

To discretise the u component of $\underset{\sim}{u}$ in Darcy's law, Eq. (2.7), in zero gravity conditions, we consider two adjacent mesh cells in the x-direction. Let the mobility, α defined by Eq. (3.21) have x-components $\beta^n_{i+1,j}$ and $\beta^n_{i,j}$ at the centres of the two cells. Let the pressures at the centres of the two cells be $p^n_{i+1,j}$ and $p^n_{i,j}$. We then seek a mobility $\beta^n_{i+1/2,j}$ and pressure $p^n_{i+1/2,j}$ on the interface between the two cells satisfying the equations

$$\begin{aligned} u^n_{i+1/2,j} &= -\beta^n_{i+1/2,j}(p^n_{i+1,j} - p^n_{i,j})/h \\ &= -\beta^n_{i+1,j}(p^n_{i+1,j} - p^n_{i+1/2,j})/(h/2) \\ &= -\beta^n_{i,j}(p^n_{i+1/2,j} - p_{i,j})/(h/2) \end{aligned} \tag{9.2}$$

Solution of Eqs. (9.2) leads to the difference formula

$$u^n_{i+1/2,j} = -\beta^n_{i+1/2,j}(p^n_{i+1,j} - p_{i,j})/h \tag{9.3}$$

where $$\beta^n_{i+1/2,j} = \frac{2\beta^n{}_{i+1,j}\beta^n_{i,j}}{(\beta^n_{i+1,j} + \beta^n_{i,j})} \tag{9.4}$$

The coefficients $\beta^n_{i+1/2,j}$ are called *harmonic transmissibilities.*

The v-component of Darcy's law is discretised in an analogous way.

In the presence of gravity we again use harmonic transmissibilities with harmonic interpolation for $(\alpha\rho)^n{}_{i+1/2,j}$ and $(\alpha\rho)^n_{i,j+1/2}$ in the difference formula,

$$\begin{aligned} u^n_{i+1/2,j} = &- \beta^n_{i+1/2,j}(p^n_{i+1,j} - p^n_{i,j})/h \\ &- (\beta\rho)^n_{i+1/2,j}g(H_{i+1,j} - H_{i,j})/h \end{aligned} \tag{9.5}$$

Following Peaceman and Rachford (28) it is convenient to use reflection boundary conditions on H for the heights of fictitious points in mesh cells adjacent to $\partial\Omega_3$.

To obtain the velocity and pressure fields we substitute the discrete Darcy law into the discrete incompressibility condition to obtain a linear system for the $p^n_{i,j}$. The Dirichlet boundary condition, Eq. (2.9), is imposed weakly by assigning the surface pressure to the fictitious cells adjacent to $\partial\Omega_1$ and $\partial\Omega_2$. The Neumann boundary conditions, Eq. (2.10), are imposed using reflection boundary conditions by equating the pressures in the fictitious and physical cells either side of $\partial\Omega_3$.

The resulting linear equations are then solved by one of the many available methods described in, for example, Aziz and Settari (1), Peaceman (30), Hackbusch and Trottenberg (17).

The velocity field in the interiors of the mesh cells is given by Eqs. (5.8).

The proof of the convergence of the above method is discussed in Russell and Wheeler (35) where the interpretation of the method as a mixed finite element method is given. This interpretation leads to higher-order finite element based versions of the method which promise greater accuracy for an equivalent amount of numerical work.

10. CONCLUDING DISCUSSION

Fixed mesh finite difference methods, including the upstreaming methods,do not provide satisfactory ways of solving high Peclet number (convection dominated) convection-diffusion/ dispersion equations. Such methods suffer from one or all of the problems (i) oscillations (ii) phase errors (iii) numerical diffusion. Without modification of the algorithms these difficulties cannot be alleviated other than by excessive mesh refinement or an excessive increase in the order of the method.

In view of these difficulties moving point techniques and the finite element version of the modified method of characteristics have been developed. These new methods are very much more accurate and efficient, on a given mesh, than the corresponding finite difference algorithms. The new characteristics based methods have been shown to work well on multi-dimensional miscible displacement and segregated flow problems.

The numerical experiments described in the preceding chapters indicate that on a given fixed mesh the HMPM is more accurate than the Taylor-Galerkin FEMMOC. However, further work is required to clarify the relative merits of moving point techniques and the modified method of characteristics. The important feature of a method is, of course, the amount of computational work to achieve a given accuracy rather than the size of the mesh actually used. We look forward to research which will determine the relative efficiency of the competing methods.

Of considerable interest is the generalisation of the characteristics based methods to non-linear problems such as the full two-phase flow equations and to multi-component systems. We anticipate difficulties in this area in connection with the dependence of the vector field of characteristic displacements upon the spatial gradient of the permeability. However, we hope that our review will stimulate work to overcome these difficulties and lead to algorithms of improved accuracy and efficiency for the simulation of flow through porous media.

11. ACKNOWLEDGEMENTS

I would like to thank R A Norman for his assistance in performing the numerical experiments and Debbie Jones for typing the manuscript. This research was supported by the UK Department of Energy.

12. REFERENCES

1. Aziz, K., and Settari, A., *Petroleum Reservoir Simulation*, 1st ed., Applied Science Publishers, Ltd., London, 1979.
2. Bercovier, M., Pironneau, O., and Sasti, V., "Finite Elements and Characteristics for some Parabolic-Hyperbolic Problems," *Applied Mathematical Modelling*, Guildford, England, Vol. 7, April 1983, pp. 89-96.
3. Birkhoff, G., "Numerical Fluid Dynamics," *SIAM Review*, Vol. 25, 1983, pp. 1-34.
4. Dietz, D. N., "A Theoretical Approach to the Problem of Encroaching and By-Passing Edge Water," *Proceedings of the Akademie Van Wetenschappen, Amsterdam*, Vol. 56-B, 1953, pp. 83-92.
5. Donea, J., "A Taylor-Galerkin Method for Convective Transport Problems," *International Journal for Numerical Methods in Engineering*, Chichester, England, Vol. 20, 1984, pp. 101-119.
6. Douglas, Jr., J., and Russell, T. F., "Numerical methods for Convection-Dominated Diffusion Problems Based on Combining the Method of Characteristics with Finite Element or Finite Difference Procedures," *SIAM Journal on Numerical Analysis*, Vol 19, No 5, 1982, pp. 871-885.
7. Dukowicz, J. K., and Ramshaw, J. D., "Tensor Viscosity Method for Convection in Numerical Fluid Dynamics," *Journal of Computational Physics*, Vol. 32, 1979, pp. 71-79.
8. Farmer, C. L., "The Numerical Calculation of Unstable Miscible Displacement," *AEEW - R 1628*, AEE Winfrith, Dorchester, England, June 1983 (Reprinted April 1985).

9. Farmer, C. L., "A Moving Point Method for the Numerical Calculation of Miscible Displacement," *AEEW - R 1895*, AEE Winfrith, Dorchester, England, Dec. 1984.
10. Farmer, C. L., "A Moving Point Method for Arbitrary Peclet Number Multi-Dimensional Convection-Diffusion Equations," *IMA Journal of Numerical Analysis*, London, to appear.
11. Farmer, C. L., and Norman, R. A., "The Implementation of Moving Point Methods for Convection-Diffusion Equations," *Numerical Methods for Fluid Dynamics*, K. W. Morton and M. J. Baines, eds., Oxford University Press, Oxford, 1985.
12. Franke, R., "Scattered Data Interpolation: Tests of Some Methods," *Mathematics of Computation*, Vol. 38, No. 157, Jan. 1982, pp. 181-200.
13. Garder, Jr., A. O., Peaceman, D. W., and Pozzi, Jr., A. L., "Numerical Calculation of Multi-Dimensional Miscible Displacement by the Method of Characteristics," *Society of Petroleum Engineers Journal*, March 1964, pp. 26-36.
14. Gresho, P. M., Lee, R. L., and Sani, R. L., "Advection-Dominated Flows, with Emphasis on the Consequences of Mass Lumping," *Finite Elements in Fluids*, Vol 3, R. H. Gallagher, O. C. Zienkiewicz, J. T. Oden, M. M. Cecchi and C. Taylor, eds., John Wiley & Sons, Chichester, England, 1978, pp. 335-350.
15. Harlow, F. H., "The Particle-in-Cell Computing Method in Fluid Dynamics," *Methods in Computational Physics*, Vol 13, B. Alder, S. Fernbach, M. Rotenberg, eds., 1964, pp. 319-343.
16. Hartree, D. R. "Some Practical Methods of Using Characteristics in the Calculation of Non-Steady Compressible Flow," *Report LA-HU-1*, Department of Mathematics, Harvard University, Cambridge 38, Massachusetts, Sept. 1953.
17. Hackbusch, W., and Trottenberg, U., eds., *Multigrid Methods*, Springer-Verlag, Berlin, 1982.
18. Hockney, R. W., and Eastwood, J. W., *Computer Simulation Using Particles*, McGraw-Hill Inc, New York, 1981.
19. Huffenus, J. P., and Khaletzky, D., "The Lagrangian Approach to Advective Term Treatment and its Application to the Solution of the Navier-Stokes Equations," *International Journal for Numerical Methods in Fluids*, Chichester, England, 1981, pp. 365-387.
20. Jacquard, P., and Séguier, P., "Mouvement de Deux Fluides en Contact dans un Milieux Poreux," *Journal de Mecanique*, Vol. 1, No. 4, Dec. 1962, pp. 367-394.
21. Lattes, R., and Lions, J. L., *The Method of Quasi-Reversibility*, American-Elsevier, New York, 1969.
22. Löhner, R., Morgan, K., and Zienkiewicz, O. C., "The Solution of Non-Linear Hyperbolic Equation Systems by the Finite-Element Method," *International Journal for Numerical Methods in Fluids*, Vol. 4, Chichester, England, 1984, pp. 1043-1063.

23. Muskat, M., "Two Fluid Systems in Porous Media. The Encroachment of Water into an Oil Sand," *Physics*, Vol. 5, Sept. 1934, pp. 250-264.

24. Morton, K. W., and Parrott, A. K., "Generalised Galerkin Methods for First-Order Hyperbolic Equations," *Journal of Computational Physics*, Vol. 36, No. 2, July 1980, pp. 249-270.

25. Morton, K. W., "Generalised Galerkin Methods for Steady and Unsteady Problems," *Numerical Methods for Fluid Dynamics*, K. W. Morton and M. J. Baines, eds., Academic Press, London, 1982, pp. 1-32.

26. Neuman, S. P., "A Eulerian-Lagrangian Numerical Scheme for the Dispersion-Convection Equation Using Conjugate Space-Time Grids", *Journal of Computational Physics*, Vol. 41, 1981, pp. 270-294.

27. Ockendon, J. R., "Linear and Nonlinear Stability of a Class of Moving Boundary Problems," *Proceedings Pavia Bimestre*, (1979), 1st. Naz. Alta Matem. Fransesco Severi, E. Magenes, ed., Rome, Italy, Vol. 2, 1980, pp. 443-478.

28. Peaceman, D. W., and Rachford, H. H., "Numerical Calculation of Multi-Dimensional Miscible Displacement," *Society of Petroleum Engineers Journal*, Dec. 1962, pp. 327-339.

29. Peaceman, D. W., "Improved Treatment of Dispersion in Numerical Calculation of Multi-Dimensional Miscible Displacement," *Society of Petroleum Engineers Journal*, Sept. 1966, pp. 213-216.

30. Peaceman, D. W., *Fundamentals of Numerical Reservoir Simulation*, Elsevier Scientific Publishing Company, Amsterdam, The Netherlands, 1977.

31. Price, H. S., Cavendish, J. C., and Varga, R. S., "Numerical Methods of Higher-Order Accuracy for Diffusion-Convection," *Society of Petroleum Engineers Journal*, Sept. 1968, pp. 293-303.

32. Prickett, T. A., Naymik, T. G., and Lonnquist, C. G., "A 'Random-Walk' Solute Transport Model for Selected Groundwater Quality Evaluations," *ISWS/BUL-65/81*, Bulletin 65, State of Illinois, Illinois Department of Energy and Natural Resources, Champaign, 1981.

33. Raviart, P. A., "An Analysis of Particle Methods," *C.I.M.E. Course in Numerical Methods in Fluid Dynamics*, Como, July 1983, (to be published in Lecture Notes in Mathematics, Springer-Verlag).

34. Raviart, P. A., "Particle Numerical Models in Fluid Dynamics," *Numerical Methods for Fluid Dynamics*, K. W. Morton and M. J Baines, eds., Oxford University Press, Oxford, 1985.

35. Russell, T. F., and Wheeler, M. F., "Finite Element and Finite Difference Methods for Continuous Flows in Porous Media," *The Mathematics of Reservoir Simulation*, R. E. Ewing, ed., SIAM, Philadelphia, 1983, pp. 35-106.

36. Saffman, P.G., and Taylor, G.I., "The Penetration of a Fluid into a Porous Medium or Hele-Shaw Cell Containing a More Viscous Fluid," *Proceedings of the Royal Society of London*, Vol. A 245, 1958, pp. 312-329.
37. Truesdell, C., and Toupin, R., "The Classical Field Theories," *Handbuch der Physik*, S. Flugge, ed., Vol. III/I, pp. 226-693, Springer-Verlag, Berlin, 1960.

13. LIST OF SYMBOLS

A_k^n	Concentration on k-th moving point at time t_n
a	Lagrangian concentration field
c	Eulerian concentration field
c_o	Concentration field at t=0
c^n	Semi-discrete concentration field at time level n
$\underset{\sim}{c}_m^n$	Numerical concentration mesh function at $\underset{\sim}{x}_m$ at time t_n
c_m^n , c^{*n}_m	Intermediate numerical concentration mesh functions
c_s	Exact solution of linear convection-diffusion equation
$\underset{\approx}{D}$	Diffusion-dispersion tensor
D_w^n	Change in concentration per unit time in semi-discrete modified method of characteristics
E	Point moving operation
E_2^h	The fixed mesh
E_2	Two-dimensional Euclidean space
$\underset{\sim}{F}$	Vector front position in Dietz problem
F_m^n	General fixed mesh function evaluated at fixed mesh point m at time level n
f	y-component of front position in Dietz problem
f_o	Parameter in Dietz' exact solution
G^n	General function of $\underset{\sim}{x}$ at time level n
g	Gravitational acceleration
H	Height of $\underset{\sim}{x}$ above an horizontal datum
h	Mesh spacing
h^*	Large mesh spacing
I_F^s	Fixed mesh interpolator of order s
I_M^s	Moving point interpolator of order s
J	Moving point insertion-deletion operation
$\underset{\approx}{k}$	Permeability tensor
k_J	Relative permeability of J-th phase

k_r	Relative permeability in miscible displacement formulation of segregated displacement
L_x	Length of Ω
L_y	Width of Ω
M	Total number of fixed mesh points
M_b	Maximum allowable background moving point density (points/cell)
M_s	Maximum allowable moving point density under a front
m_b	Minimum allowable background moving point density
m_s	Minimum allowable moving point density in a front
N	Total number of moving points
N_b	Background density of moving points at t=0
N_s	Moving point density in fronts at t=0
N_x	Number of fixed mesh points in x-direction
N_y	Number of fixed mesh points in y-direction
$\underset{\sim}{n}$	Outward pointing unit normal on $\partial\Omega$
P_k	The k-th moving point
Pe_h	Least upper bound on the mesh Peclet number
p	Pressure field
p_I	Pressure on inlet face, $\partial\Omega_1$
p_J	Pressure in J-th phase
p_m	Numerical pressure mesh function at $\underset{\sim}{x}_m$ at time t_n
Q	Total flow rate
Q^n	General function of $\underset{\sim}{x}$ at time level n
S_m^n	Mesh function measuring the slope of c_m^n
S^*	Slope parameter defining presence of fronts
S_{or}	Residual oil saturation
S_{wc}	Connate water saturation
s	Width parameter in convection test problem
T	Number of sub-intervals within τ for point moving algorithms
t	Time
t_o	Parameter in exact solution of convection-diffusion equation
U_o	u_∞/ϕ_r
$\underset{\sim}{u}$	Darcy velocity vector field(volumetric flux)
$\underset{\sim}{u}^*$	Extension of $\underset{\sim}{u}$ from Ω to E_2

u	x-component of $\underset{\sim}{u}$
v	y-component of $\underset{\sim}{u}$
$\underset{\sim}{X}_k$	Exact position of k-th moving point
$\underset{\sim}{X}_k^n$	Numerical position of k-th moving point
X_k^n	x-component of $\underset{\sim}{X}_k^n$
x_o	Parameter in exact solution of convection-diffusion equation
$\underset{\sim}{x}^*$	Position vector of $\underset{\sim}{x}$ at previous time level
$\underset{\sim}{x}_m$	Position vector of the centre of the m-th fixed mesh cell, Ω_m
$\underset{\sim}{x}$	Position vector
$\underset{\sim}{x}_o$	Initial condition for Eq. (2.12)
x_i	x co-ordinate of i-th grid point in the x-direction or i-th component of $\underset{\sim}{x}$ when x_i occurs in a partial derivative symbol
x	x-component of $\underset{\sim}{x}$
Y_k^n	y-component of $\underset{\sim}{X}_k^n$
y	y-component of $\underset{\sim}{x}$
y_j	y-component of j-th grid point in the y-direction
$\underset{\sim}{z}$	Lagrangian co-ordinate vector
α	Mobility tensor
α_J	$-u_\infty/\lambda_J$
α_i	Function used in linear interpolation
$\beta^n_{i+\frac{1}{2},j}$	Harmonic transmissibility in x-direction
$\beta^n_{i,j+\frac{1}{2}}$	Harmonic transmissibility in y-direction
β_ℓ	Function used in bilinear interpolation
γ_o	Parameter in Jacquard-Séguier exact solution
γ	Parameter in exact solution of 2-nd order upstreaming model
δc_m^n	Change in concentration mesh function due to diffusion-dispersion
ε^n	Maximum absolute error at time level n
ζ	x-component of local co-ordinate in mesh cell m
η	y-component of local co-ordinate in mesh cell m
Θ	Dip angle
κ	Wave number in exact solution of 2-nd order upstreaming model
λ	Weighting function in hybrid moving point method
λ_J	Mobility tensor of the J-th phase
μ	Viscosity function
μ_J	Viscosity of the J-th phase

μ_o	Oil viscosity
μ_w	Water viscosity
ν_m^n	Number of moving points in m-th mesh cell at time level n
ρ	Fluid density function
ρ_J	Density of the J-th phase
ρ_o	Oil density
ρ_w	Water density
σ	Frontal parameter in segregated displacement
τ	The time step
Φ	Symbol representing the fixed mesh algorithm
ϕ	Porosity
ϕ_m	Porosity at $\underset{\sim}{x}_m$
ϕ_r	$\phi(1 - S_{or} - S_{wc})$
$\underset{\sim}{\chi}^n$	Exact characteristic displacement field at time level n
$\underset{\sim}{\chi}_m$	Exact characteristic nodal displacement of m-th fixed mesh point at arbitrary time
$\underset{\sim}{\chi}_m^n$	Numerical characteristic nodal displacement of m-th fixed mesh point at time level n
Ψ_J	Potential in the J-th phase
ω	Explicit-implicit weighting parameter in the modified method of characteristics
Ω	Physical flow region
Ω_m	m-th fixed mesh cell

Subscripts

J	J-th phase
i,j,k,ℓ,m	Fixed and moving point subscripts
o	Oil
w	Water
i+1/2	Mesh cell surface index in x-direction
j+1/2	Mesh cell surface index in y-direction

Superscripts

n	Time level
s	Order of interpolators

Special Symbols

ε	Set theoretic membership
$\cup$	Set theoretic union
$\cap$	Set theoretic intersection
$k\bar{\varepsilon}m$	Moving point k is the nearest moving point to the centre of cell m
$\partial\Omega$	Boundary of Ω

$\partial\Omega_1$	Part of $\partial\Omega$ with inward flux
$\partial\Omega_2$	Part of $\partial\Omega$ with outward flux
$\partial\Omega_3$	Part of $\partial\Omega$ with zero flux
∇	Spatial gradient operator
Δ	Numerical central difference gradient operator
[a,b]	Integral over Ω of the product of functions a and b
$\partial/\partial n$	Normal derivative at a point of $\partial\Omega$
$\partial/\partial t$	Time derivative
$\cdot$	Summation over a repeated index as in $k.\nabla p$
$\nabla\cdot(D\nabla)c$	Symbol for $\sum_{ij} \frac{\partial}{\partial x_i} D_{ij} \frac{\partial c}{\partial x_j}$

SUBJECT INDEX

A

B

C

D

E

F

G

H

I

J

K

L

M

N

O

P

R

S

T

U

V

W

Y

FUNDAMENTALS OF TRANSPORT PHENOMENA IN POROUS MEDIA

Edited by

Jacob Bear and M. Yavuz Corapcioglu

1984 by Martinus Nijhoff Publishers, Dordrecht, The Netherlands

TABLE OF CONTENTS